Does Algebra One Have to Be Taught
IN JUST ONE YEAR?

Consider this. The business world demands more math skills from us than ever before, and this is not going to change. Consider this. Mathematics test scores across the nation show that our students are not keeping pace with the demands for higher math skills in the workplace.

The way we teach is changing, too. Heterogeneous classrooms, students with different learning styles, and fewer remedial math courses challenge us to adopt alternative teaching strategies that will reach a wider range of students.

So does algebra one have to be taught in just one year? Absolutely not.

Enter *Algebra One Interactions,* a fun, motivational algebra one course paced over two years. This is an innovative program built for students with varying abilities: students who have difficulty mastering concepts, who need more practice and hands-on experience, or students who are ready for algebra at an earlier age.

With *Algebra One Interactions,* algebra is accessible to **all** students. Now all students can enjoy learning algebra and reach the summit of their classroom dreams. Now all students have the opportunity to succeed in a world that demands more mathematics skills of them than ever before.

*"**F**or the first time, I'm really getting into math. It's my subject now. I especially got interested by the **Critical Thinking** feature about extending statistics. Man, just doubling the length and width of a box really changed the volume."*

SUCCESS

HRW
ADVANCED
ALGEBRA

HRW GEOMETRY

HRW GEOMETRY

ALGEBRA ONE
INTERACTIONS

COURSE 2

HRW ALGEBRA

ALGEBRA ONE
INTERACTIONS

COURSE 1

ALGEBRA ONE
INTERACTIONS

COURSE 1

SETTING A PACE FOR SUCCESS

Establishing a Solid Base Camp

*"*In ***Algebra One Interactions,*** more than half of *Course 1* is devoted to exploring math concepts that help prepare students for algebra and geometry. By covering integers and data patterns, for example, students learn the basic building blocks of algebra. With paper-folding explorations, they learn geometry. I also like how Chapter 10 serves as a springboard out of linear functions into nonlinear functions in *Course 2*. The overlap really allows students to feel comfortable moving ahead—and believe me, that reassurance never happened in regular algebra.*"*

COURSE 1 TABLE OF CONTENTS

Climbing Higher and Having Fun at the Same Time

*"*This is definitely heads-up, real-world algebra. ***Algebra One Interactions*** lets my students literally get their hands around symbolic operations and helps them make the jump from the concrete to the abstract. In the *Exploring* lesson in Chapter 7, my class got to use algebra tiles to solve problems with money. You should see the lights go on in their eyes!*"*

COURSE 2 TABLE OF CONTENTS

*"*I never really enjoyed math because I never felt very comfortable with it. But hey, that's history. Now I see how math relates to me. For example, learning about parabolas with a graphics calculator was easy, but I also found out how parabolas can be used in real life.*"*

TEAM OF EXPERIENCED GUIDES

"With so many technology tools and teaching resources to support me in the classroom, I know exactly where I stand with my class and exactly where we're going next."

- **Solution Key**   Worked–out solutions for exercises in the Pupil's Edition.
- **Practice Workbook**   Additional practice for each lesson.
- **Reteaching Masters**   Alternative teaching strategies for re-presenting lesson concepts.
- **Technology Masters**   Computer and calculator activities that offer additional practice.
- **Technology Handbook**   Teacher's guide to using hand-held graphics calculators and computer software (One booklet for both courses).
- **Assessment Resources**   A useful collection that includes a Diagnostic Test, Chapter and Mid-Chapter Assessments, Quizzes, and Alternative Assessments that include performance-based assessments and games.
- **Lab Activities and Long-Term Projects**   Hands-on activities and projects that engage students inside and outside the classroom and complement regular classroom work.
- **Tech Prep Math Resources**   Technical career applications that review the content in each chapter and real-world group projects from various technical fields.

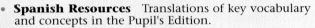

  - **Spanish Resources**   Translations of key vocabulary and concepts in the Pupil's Edition.
  - **Assessment Software**   Variety of assessment items and answers delivered on user-friendly software. (Versions available for both Macintosh® and Windows®.)
  - **Assessment Software Item Listing**   Printout of all the items and answers included on the Assessment Software.
  - **Teaching Transparencies**   Binder of transparencies that offers over 100 full-color visuals with suggested lesson plans for their use.
  - **Teaching Transparencies Directory**   Useful guide that makes it easier to review the quality and instructional value of color transparencies and lesson plans.
  - **HRW Algebra Tiles and Activities Booklet**   Helps students visualize mathematical expressions.

- **Algebra Explorations CD-ROM**   Multimedia CD-ROM software program integrated into *Algebra One Interactions*, allowing you to explore the world of functions in an alternative approach to the textbook.
- **Block-Scheduling Handbook**   Block-scheduling plans to help you organize indiviual work, group activities, and ongoing assessments.
- **Portfolio Writing Activities**   Creative writing activities related to the topics in each chapter.
- **Problem Solving/Critical Thinking Masters**   Stimulating problems that take the Pupil's Edition one step further and allow students to extend and enrich their knowledge.
- **Cooperative–Learning Activities**   Exploration activities, and games for cooperative learning groups.

*Two booklets, one for each course, except where otherwise noted.

*"The best part was getting to use the computer to explore absolute value functions, and using the **Algebra Explorations CD-ROM**, we got to help a rescue helicopter find a lost hiker. To me, this kind of math makes sense, this kind of math makes concepts come alive."*

TEACHER'S PLANNING GUIDE

The Trail Is Clearly Marked

" The *Teacher's Planning Guide* is just the thing for taking home and using to plan lessons. It gives me clear, no-nonsense suggestions for teaching in a heterogeneous classroom like mine.

The *Teacher's Answer Edition* is just what I needed, too—all the answers all in one place. It's the perfect reference tool for everyday use in the classroom—easy to use, quick, and I don't have to dig through a lot of teachers' shoptalk. "

" I wish I had two teachers' books like these for my other classes. By managing and organizing instruction this way, I save a lot of time. And the flexibility in planning really complements my teaching style. Finally, I can see my way into tomorrow's classes clearly and concisely. "

Teacher's Planning Guide—A Superb Compass for Every Lesson

" If other teachers are like me, they know how hard it is to create effective lesson plans—lesson plans that are relevant and also fun for the students. They take time. They take patience. And they also take a lot of experience.

For me, the **Algebra One Interactions** *Teacher's Planning Guide* points in exactly the right direction every time and in every situation. For each lesson, it provides two distinct approaches: *Using the Book* if I want my students to follow the material in the Pupil's Edition; and *Using Models* if I want to provide my class with more activity-based instruction. The *Teacher's Planning Guide* even has assessment and enrichment activities that make my job that much easier. "

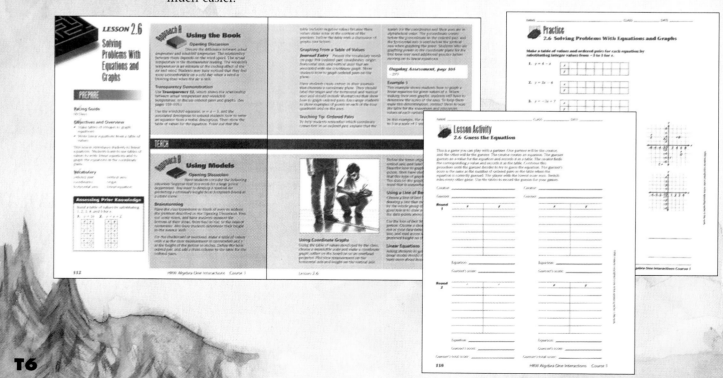

AND TEACHER'S ANSWER EDITION

PREPARE!

"When I opened up the *Teacher's Planning Guide*, I knew I had my hands on an invaluable reference tool I would turn to again and again. The *Prepare* feature took the butterflies out of my stomach because it put the lessons in context. A *Pacing Guide, Objectives and Overview, Vocabulary,* and *Assessing Prior Knowledge* provides crucial information that saves time and establishes goals for both me and my class."

TEACH AND ASSESS!

"Students are sharp—they know if you're prepared; they know if the lesson you're presenting is well thought out. If it's not, their eyes glaze over, their heads start nodding, and then you're doomed.

With *Teach* and *Assess*, that kind of class is not going to happen. You get sound pedagogy, quality teaching strategies that students can relate to."

"Each lesson provides two excellent approaches for teaching: *Using the Book*, if you want to stay close to the text for your instruction; and *Using Models*, for a more kinesthetic, hands-on approach to the content. *Opening Discussion* suggests questions to ask and explanations to give the class to assure that students are keyed into the material. Then, features such as *Journal Entry, Brainstorming, Defining, Cooperative Learning,* and *Explorations,* to name a few, establish definable, tangible goals as well as provide options for teachers to meet individual needs and class styles."

"The *Assignment Guide, Error Analysis,* and *Alternative Assessment* summarize sections and help you evaluate your students' progress through the lessons."

RETEACH AND ENRICH

"All students at some point need reinforcement of the skills they're learning. *Reteach* and *Enrich* provide superlative hands-on opportunities for practicing and applying skills. You'll find these features, in particular, help students bridge what to them can seem like an insurmountable gap between concrete and symbolic ways of thinking."

"*Reteach* and *Enrich* exemplify the ways **Algebra One Interactions** puts math back in the real world. *Reteach* in Chapter 1, for example, has students work in groups to resolve how many handshakes occur if ten people in a room shake hands with each person in the room. Talk about a real 'hands on' approach!"

"I like how at the end of each lesson you get three blackline masters right there: *Practice, Lesson Activity,* and *Basic Skills Practice*. These really provide a lot of extra help."

TEACHER'S ANSWER EDITION

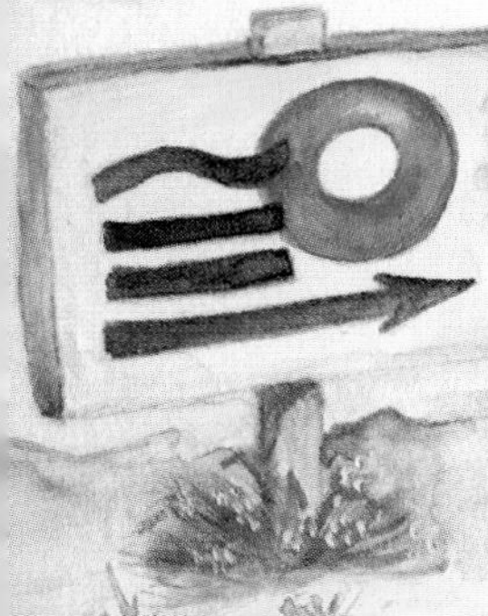

Teacher's Answer Edition—Follow the Signs of Progress

"Have you ever felt that teacher's editions were too big and bulky, unmanageable, because they tried to do too many things at once? Well, here's a teacher's edition that simply gives you the answers and that won't feel like you're lugging rocks back and forth to your classroom.

And it dovetails so well with the *Teacher's Planning Guide*. On the one hand, I get a treasure trove of great lesson plans, on the other, a crystal-clear teacher's edition. I couldn't ask for anything more."

"One word about the *Teacher's Answer Edition*: answers, answers, answers. Direct and to the point. No nonsense."

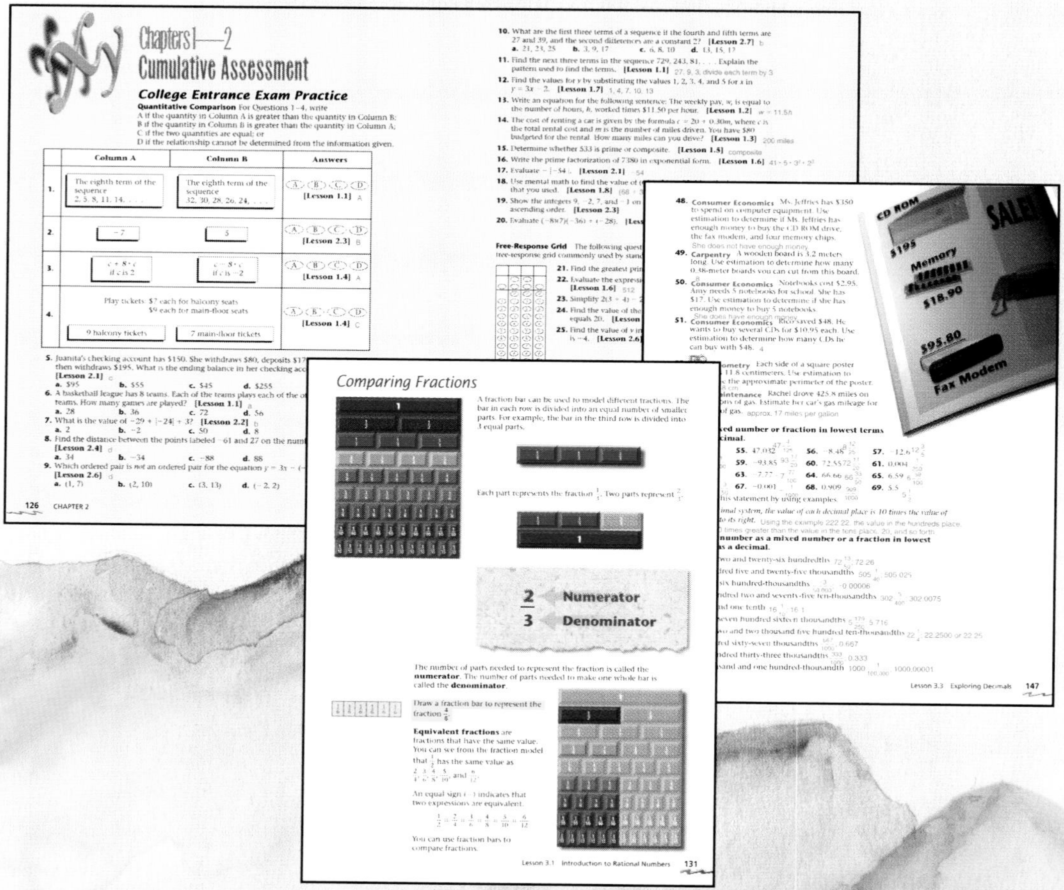

EFFECTIVE INTEGRATION AT EVERY CURVE IN THE TRAIL

Applications that put the fresh air back into algebra

"**W**hy do I have to learn this?' How many times have we all heard that question? With *Algebra One Interactions*, the answer is simple. Each lesson opens with *Why*—why, for example, matrices can be used to analyze payroll information for a small company or the results of a swimming competition."

"**A**lgebra doesn't happen out in space somewhere where students can't reach it and apply it to their own lives. *Algebra One Interactions* brings math down to earth with all the connections to science and social studies, business and economics, sports and leisure, language arts and life skills, cultures and technology. Then students can decide for themselves what interests them, and how or when they want to travel to the stars."

"**T**he groundspeed of the airplane was only 525.5 miles per hour. Flying into a head wind of 72 miles per hour slowed it down significantly, and by adding the groundspeed to the head wind I figured out the airspeed of the plane. To me, that's what learning math is all about. If I was the pilot, I would've taken the plane to a higher altitude for a faster, smoother ride."

Math connections that let students explore

"**O**pen your eyes and you start to see connections everywhere you look. This is especially true with *Algebra One Interactions*. Geometry, probability and statistics, transformations, and maximum and minimum are all effectively integrated into each bend and curve of the instruction, so students can see how algebra connects with other mathematical disciplines."

"**T**his summer some friends and I are planning a hiking trip up Guadalupe Peak. It won't be that tough though. I figured the slope of the trail with the information I received about the height and length of the trail at certain checkpoints. No problem, the statistical graph showed me the mountain is not as steep as I thought!"

STUDENTS ASCEND TO THE TOP AND COME ALIVE

Practice and assessment built into the instruction—now you're talking!

"Granted, practice and assessment aren't the most important things in life, but they matter in the world of mathematics. And it matters that you have a healthy variety of questions so you can track students progress before, during, and after instruction *and* so you can lead students on different pathways if they have diverse levels of interest and ability."

"*Algebra One Interactions* won me over with its method of placing ongoing questioning strategies at key points throughout the lessons. Look in the instruction and you'll find interactive, open-ended questions to challenge students understanding. You'll also find *Critical Thinking* and *Communicate* questions that extend concepts and *Try This* exercises that assure students understand what they're learning as they move through the lesson."

"The *Practice & Apply* section is great because it really provides a chance for students to sink their teeth into the concepts. It's not just drill and kill either. Interdisciplinary and cultural connections give the exercises and applications a healthy, real-world focus. When students finish, they move on to *Look Back* and *Look Beyond*. These practice features are superb for reviewing concepts and preparing for new ones. Then, go to *portfolio activities*, long-term *Chapter Projects*, and *Eyewitness Math* activities for alternative practice and assessment to stretch students imaginations."

"*Wow, check it out! I never knew math could be like this. I learned about the stock market, and by the time I got to the* **Practice & Apply** *homework, I knew all the answers.*"

The heart of algebra is active learning!

"For me, algebra is about getting students involved and having them experience success. I rely on *Algebra One Interactions* for this very reason. It gets students involved; it motivates them with kinesthetic learning models, technology, and ongoing questioning strategies that truly work."

"This book is great. For the first time, I understand how math is more than just equations on a page. I can add or subtract integers, for example. I can explore how integers are used in real life, and I can apply what I learn. I wish all my other classes were like this."

TECHNOLOGY—

The Summit is Yours

"The beauty of **Algebra One Interactions** is that it's 'technology ready.' When technology becomes available to a teacher, it can be used in the instruction since the option is built right in.

The benefit is obvious: students love using technology and having options. And my class becomes an interactive expedition so students can travel through and explore concepts, not just memorize them and repeat them back on tests."

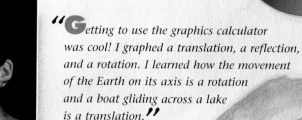

"Getting to use the graphics calculator was cool! I graphed a translation, a reflection, and a rotation. I learned how the movement of the Earth on its axis is a rotation and a boat gliding across a lake is a translation."

THE SIGN OF

PROGRESS

HRW MATHEMATICS

HRW
ALGEBRA ONE
INTERACTIONS

COURSE 1

TEACHER'S ANSWER EDITION

Integrating

MATHEMATICS
TECHNOLOGY
EXPLORATIONS
APPLICATIONS
ASSESSMENT

HOLT, RINEHART AND WINSTON
Harcourt Brace & Company

Austin • New York • Orlando • Atlanta • San Francisco • Boston • Dallas • Toronto • London

AUTHORS

Paul A. Kennedy

An associate professor in the Department of Mathematics at Southwest Texas State University, Dr. Kennedy is a leader in mathematics education reform. His research focuses on developing algebraic thinking using multiple representations and technology. He has numerous publications and is often invited to speak and conduct workshops on the teaching of secondary mathematics.

Diane McGowan

Respected throughout the state of Texas as an educational leader, author, and popular presenter, Ms. McGowan is a past president of the Texas Council of Teachers of Mathematics. She is a teacher at James Bowie High School in Austin, Texas, and has been the recipient of numerous awards for teaching excellence.

James E. Schultz

Dr. Schultz has over 30 years of successful experience teaching at the high school and college levels and is the Robert L. Morton Mathematics Education Professor at Ohio University. He helped establish standards for mathematics instruction as a co-author of the NCTM *Curriculum and Evaluation Standards for Mathematics* and *A Core Curriculum: Making Mathematics Count for Everyone.*

Kathy Hollowell

Dr. Hollowell is an experienced high school mathematics and computer science teacher who currently serves as Director of the Mathematics & Science Education Resource Center, University of Delaware. She is a former president of the Delaware Council of Teachers of Mathematics.

Irene "Sam" Jovell

An award-winning teacher at Niskayuna High School, Niskayuna, New York, Ms. Jovell served on the writing team for the New York State Mathematics, Science, and Technology Framework. A popular presenter at state and national conferences, Ms. Jovell's workshops focus on technology-based innovative education.

Cover Design: Katie Kwun

Portions of this work were published in previous editions.

(Acknowledgments appear on pages 759–760, which are extensions of the copyright page.)

HRW is a registered trademark licensed to Holt, Rinehart and Winston.

Printed in the United States of America
    5 6 7   041   00 99 98
ISBN: 0-03-051257-3

TEACHER'S ANSWER EDITION

The most common teacher's edition produced today is a Teacher's Wraparound Edition or a Teacher's Annotated Edition. Its intent is to provide the teacher with helpful suggestions and tips for classroom instruction along with the answers to all the exercises. Although useful in many situations, the standard teacher's edition does not provide the teacher with the assistance needed in the classroom for which *HRW Algebra One Interactions* is written.

Holt, Rinehart and Winston has developed a two-book system for *HRW Algebra One Interactions* that gives teachers the help they need to teach and manage in today's diverse classroom. One of the two teacher's books is this *Teacher's Answer Edition,* which is a duplicate of the *Pupil Edition* with nothing more than overprinted answers. When you need answers, that is exactly what you get.

The other book available to the teacher, is the *Teacher's Planning Guide.* This comprehensive teaching tool provides, in a spacious format, the level of detail that you need. The *Teacher's Planning Guide* provides an extensive presentation of teaching approaches that allows you to create the type of learning environment you want and need for your own classroom.

• •

REVIEWERS

Richard A. De Aguero
Miami Senior High School
Miami, Florida

Linda Bailey
Putnam City Schools
Oklahoma City, Oklahoma

Stephen G. Bolks
Saddleback High School
Santa Ana, California

Susan P. Brown
Robert A. Taft High School
Cincinnati, Ohio

Nancy S. Dockery
John Marshall High School
San Antonio, Texas

Vaughn W. Ekbom
Johnson Senior High School
St. Paul, Minnesota

John W. Ikerd
Yorba Middle School
Orange, California

James E. Kozman
Franklin Heights High School
Columbus, Ohio

Leon Lias, Sr.
Skyline High School
Dallas, Texas

Dorothy B. Link
Briggs High School
Columbus, Ohio

What Is the Philosophy and Purpose of HRW Algebra One Interactions?

The reform movement in school mathematics has caused a shift not only in the way mathematics is taught, but also in the particular mathematics courses offered. Many of the traditional mathematics courses such as general mathematics and pre-algebra are being eliminated from the high school curriculum. Algebra one, a course at one time reserved for average and above-average students, is becoming a required course for all high school students. However, publishers' algebra textbooks have catered only to the college bound student. Unable to cope with the abstract presentation of mathematical topics in these textbooks, the remainder of the student population was relegated to a computational skills-based program called general mathematics. Another popular remedial course, pre-algebra, veiled the computational skills practice with the introduction of algebraic symbols and a smattering of geometry topics. This structure in the mathematics curriculum provided students with the necessary number of credits for graduation and, in a time long past, provided students with the minimum mathematics skills needed for non-professional careers.

Prepare Students for the Future

General math and pre-algebra no longer provide a student with the math skills needed to enter into today's technological society. Today, an understanding of algebraic concepts is a minimal requirement for entry into most careers and is a gateway to all forms of higher mathematical study.

The goal is clear and noble and certainly in the best interest of the student. However, the algebra teacher is now faced with a broad base of students of divergent learning levels and learning styles.

Support the Teacher

How can the teacher manage a classroom setting that has suddenly undergone such change? It is obvious that teaching practices and teaching materials must undergo change to accommodate the new algebra classroom. *HRW Algebra One Interactions* is a program designed specifically to accommodate the divergent needs of the students found in this new setting.

Make Algebra Accessible to All

The instructional design of *HRW Algebra One Interactions* is to make algebra accessible to everyone. The lessons in this program are written in such a way that students are presented first with numbers and data, then with tables and graphs, and finally with abstract algebraic concepts. The instruction typically begins with the presentation of real-world data or with the use of manipulatives. In other words, instruction on important mathematical concepts begins with the concrete. As the instruction progresses, the students move to the tabular, then to the graphical,

GEOMETRY
Connection

Spreadsheet

Exploration

STATISTICS
Connection

CRITICAL
Thinking

Graphics
Calculator

EXAMPLE

and then to the symbolic. All levels of student abilities are accounted for in this approach. This is the essence of the instruction found in the student lessons. Labels and separate textbook features that describe students in terms of different ability labels will not be found in the student book. The provisions for different learning levels are embedded in the instructional design of the student lessons. For example, students are often asked to look for patterns by organizing observational data into tables, then they make conjectures from the data. As students generalize or synthesize their observations as conjectures, they progress from the concrete to the abstract. Students who do not possess acute memorization skills will be able to grasp algebraic thinking that was once reserved only for the college-bound student. Students historically labeled as advanced or accelerated will find the instruction more relevant and will gain a deeper understanding of the concepts. They will be more likely to show an interest in mathematically oriented subjects such as chemistry, physics, logic, economics, statistics, and computer science.

Extend Instructional Time

HRW Algebra One Interactions is a two-book algebra one program designed to be taken over a two-year period of time. Twice as much instructional time allows the teacher to take a hands-on, exploratory, and activity-oriented approach to instruction. Manipulatives and calculators will be used extensively, and students will be actively involved in using mathematics in real-world applications. Much group work and many projects will be incorporated into instruction. The program includes all the topics of an algebra one program. It contains a greater coverage of geometry topics and integrates the mathematics of transformations, statistics, probability, and maximum and minimum. *HRW Algebra One Interactions* also includes all the pre-algebra skills necessary for success in algebra.

Maximize Curriculum Options

Many districts will use this program as a paced or extended algebra one program for students who begin their algebra one study in grade nine and finish at the end of grade ten. Having completed *HRW Algebra One Interactions*, these students will have the prerequisite skills to move into geometry or possibly algebra two. They may or may not take an additional math course before graduation.

Other districts will use *HRW Algebra One Interactions* as an extended algebra one program for students who begin their algebra one study in grade eight. These students will complete the program at the end of grade nine. This allows students to take as many as three other mathematics courses before graduation, much as the traditional curriculum is organized.

And finally, students who are ready to begin a two-year algebra one program in the seventh grade will find *HRW Algebra One Interactions* the best way to complete an algebra one curriculum by the end of grade eight.

ASSIGNMENT GUIDE

Lesson	Exercises for Core Level	Exercises for Core-Plus Level
1.1	1–51	1–13 and 16–51
1.2	1–53	1–28 and 33–53
1.3	1–36	1–5, 7–12, and 14–36
1.4	1–39	1–15 and 18–39
1.5	1–45	1–13, 16–23, 26–35, and 37–47
1.6	1–48	1–6, 8–14, 17–19, and 23–48
1.7	1–43	1–8, 12–27, and 30–43
1.8	1–45	1–7, 10–19, 22–26, and 29–45
2.1	1–56	7–49, 24–34, and 37–57
2.2	1–47	1–10 and 16–49
2.3	1–90	1–12, 20–41, and 68–92
2.4	1–50	1–6, 9–25, 28–52
2.5	1–57	1–5 and 10–59
2.6	1–56	1–4, 9–22, and 27–58
2.7	1–38	1–6, 9–16, and 18–40
3.1	1–49	1–4, 6–13, 18–31, 36–47, and 49–51
3.2	1–43	1–4, 9–27, 30–43
3.3	1–93	1–20, 26–53, 57–78, and 81–87
3.4	1–67	1–10, 13–21, 23–40, and 54–69
3.5	1–81 and 83	1–11, 14–30, 33–37, 40–51, and 60–84
3.6	1–49	1–17 and 22–51
3.7	1–53	1–29 and 36–54
3.8	1–52	1–55 and 59–60
3.9	1–52	1–12, 15–37, and 40–53
4.1	1–64	1–7, 9–11, 13, 15–35, 40–43, and 45–64
4.2	1–56	1–2, 25–30, 32–40, and 42–58
4.3	1–83	1–10, 14–18, 20–21,23–24, 26–42, 46–57, and 64–83
4.4	1–42	1–6, 9–12, 14–22, 26–28, and 31–45
4.5	1–50	1–19, 24–31, and 34–50
4.6	1–66	1–9, 14–22, 27–32, 35–38, and 41–68
4.7	1–70	1–4, 5–46, and 49–75
4.8	1–52	1–52
5.1	1–47	1–13, 16–18, 20–25, 28–36, 39–48, and 50
5.2	1–60	1–7, 8–14, 21–26, 32–36, 40–51, and 61
5.3	1–61	1–6, 9–16, 21–39, and 43–62
5.4	1–56	1–5, 8–18, and 22–58
5.5	1–74	1–4, 7–21, 25–29, 32–37, 41–56, and 59–76
5.6	1–65	1–5, 8–21, 24–41, and 45–65
5.7	1–56 and 60	1–12, 16–35, 39–48, 50–55, and 60–61
6.1	1–71	1–7, 10–13, and 20–73
6.2	1–75	1–20 and 27–77
6.3	1–47	1–24 and 31–49

Lesson	Exercises for Core Level	Exercises for Core-Plus Level
6.4	1–48	1–19 and 24–49
6.5	1–65	1–11 and 16–69
6.6	1–76	1–78
7.1	1–56	1–19, 25–31, and 35–56
7.2	1–45	1–14, 17–30, and 34–46
7.3	1–41	1–5, 6–18 even, 19–25 odd, and 26–44
7.4	1–39	1–5, 8–13, and 16–39
7.5	1–58	1–16, 20–35, and 39–60
7.6	1–54	1–11, 16–38, and 40–54
8.1	1–44	1–6, 8–15, 17–22, and 24–44
8.2	1–65	1–15, 17–30, 33–46, and 49–67
8.3	1–46	1–7, 9–19, 21–35, and 38–47
8.4	1–56	1–23, 28–38, and 40–58
8.5	1–61	1–13, 16–29, and 31–64
8.6	1–50	1–7, 9–27, and 29–50
8.7	1–37	1–19 and 22–39
8.8	1–63	1–5 and 12–65
9.1	1–33	1–8 and 12–33
9.2	1–53	1–32 and 40–55
9.3	1–42	1–9, 11–16, and 21–44
9.4	1–56	1–11, 15–39, and 43–58
9.5	1–40	1–30 and 33–40
10.1	1–39	1–41
10.2	1–43	1–45
10.3	1–49	1–15, 17–33, and 36–50
10.4	1–48	1–13, 16–22, 25–35, and 37–55
10.5	1–36	1–17 and 19–37
10.6	1–63	1–5, 9–31, and 37–65
11.1	1–39	1–41
11.2	1–48	1–50
11.3	1–53	1–54
11.4	1–50	1–54
11.5	1–28	1–30
11.6	1–27	1–29
12.1	1–38	1–8, 12–17, and 21–39
12.2	1–52	1–15, 21–24, and 31–55
12.3	1–39	1–40
12.4	1–50	1–18, 11, 14–27, 30–34, and 37–51
12.5	1–41	1–11, 16–23, and 27–43
12.6	1–48	1–10, 14–26, and 30–49
12.7	1–65	1–6, 9–27, and 40–69

TABLE OF CONTENTS

Math Connections

Geometry 11, 13, 14, 15, 28, 41, 55, 84, 118   **Statistics** 35, 55, 100, 102

Applications

Science
Ecology 32
Physical Science 119
Physics 83, 114, 118
Science 49
Temperature 73, 96

Social Studies
Demographics 77
Geography 91

Language Arts
Communicate 12, 20,
  27, 34, 41, 48, 54, 60,
  76, 83, 90, 96, 101,
  108, 117

Business and Economics
Advertising 55
Banking 102, 124
Business 29, 84, 97
Fund-raising 35

Inventory 62
Rental 119
Sales 27, 61
Wages 61, 62, 69

Life Skills
Consumer Awareness 22
Consumer Economics 35
Home Improvement 54

Sports and Leisure
Entertainment 28, 35,
  62, 68
Recreation 109, 117
Scuba Diving 76
Sports 83, 84, 92, 124
Travel 29

Other
Photography 110

Math Connections

Geometry 133, 140, 141, 147, 155, 159, 160, 167, 185   **Maximum/Minimum** 233, 234, 238
Probability 168, 188   **Statistics** 144, 148, 174, 185

Applications

Science
Criminology 175
Genetics 188

Social Studies
Lifestyles 166
Political Science 171

Language Arts
Communicate 134, 139,
  146, 152, 159, 166, 174,
  183, 193, 208, 213, 220,
  230, 236, 244, 253, 260

Eyewitness Math 176
Business and Economics
Advertising 262
Agriculture 167, 238, 255
Architecture 268
Business 160, 195
Carpentry 147, 218
Construction 209, 237,
  238, 268
Engineering 231
Forestry 240
Investments 175

Manufacturing 144, 160
Metalworking 158
Sales 175
Stocks 130, 132, 135, 151
Wages 140, 175

Life Skills
Auto Maintenance 145,
  147
Baking 160
Consumer Economics 145,
  147
Cooking 167

Sports and Leisure
Crafts 143, 148
Games 178, 184
Recreation 139, 254
Sports 149, 167, 168,
  192, 231

Other
Academics 175
Cartography 262
Graphic Arts 258
Photography 153, 262

Math Connections

Geometry 277, 278, 279, 290, 295, 297, 298, 306, 325, 334, 339, 342, 344, 347, 349, 350, 361, 363, 372, 376   **Probability** 372   **Statistics** 284, 285, 294, 297, 363, 372, 376

Applications

Science
Physics 341
Temperature 318

Social Studies
Government 362, 372
Student Government 362

Language Arts
Communicate 277, 283, 289, 296, 304, 311, 317, 333, 342, 349, 355, 361, 370
Eyewitness Math 336

Business and Economics
Agriculture 362
Aviation 291
Banking 310, 311, 312, 351
Carpentry 338, 343
Discounts 350, 362, 368, 372
Fund-raising 284, 293, 356
Income Tax 371
Inventory 285

Investments 295
Sales 324
Sales Tax 297, 369
Savings 312, 355
Small Business 335
Stocks 352
Wages 334

Life Skills
Auto Maintenance 363
Consumer Economics 312, 356, 362, 367
Cooking 297, 364

Health 362

Sports and Leisure
Crafts 278
Hobbies 324, 350
Sports 297, 318, 340, 355, 356
Travel 297, 329, 343

Other
Photography 278

Math Connections

Geometry 388, 394, 400, 401, 402, 403, 404, 406, 413, 416, 425, 439, 441, 458, 471, 473, 474, 476, 477, 485 **Coordinate Geometry** 470 **Maximum/Minimum** 415 **Statistics** 386, 387, 394, 425

Applications

Science
Chemistry 397, 399, 424, 448, 485
Ecology 448
Life Science 385
Physics 455, 458
Temperature 472

Social Studies
Government 420, 424
Social Studies 465
Student Government 394

Language Arts
Communicate 387, 393, 398, 405, 411, 419, 432, 438, 447, 456, 463, 469, 476, 483
Eyewitness Math 442
Language Arts 470

Business and Economics
Carpentry 402
Engineering 441
Forestry 394
Fund-raising 394, 396, 398, 400, 441, 460, 464, 465, 485

Manufacturing 414
Rental 484
Sales Tax 388
Small Business 398, 399, 407
Wages, 394, 395, 398, 485

Life Skills
Consumer Economics 384, 388, 413, 457, 491
Health 419, 431

Sports and Leisure
Auto Racing 452
Ballet 411, 412

Crafts 417
Entertainment 406, 470, 485
Hobbies 394
Recreation 438, 466, 471, 484
Sports 399, 425, 452, 485
Theater 398
Travel 439, 490

Other
Academics 484
Time 391

Math Connections

Geometry 509, 523, 544, 548, 555   **Coordinate Geometry** 560, 575
Statistics 558

Applications

Science
Astronomy 569
Biology 506
Chemistry 503, 509, 518,
   522, 523, 530, 555
Life Science 563
Physics 543, 546, 549

Social Studies
Demographics 539, 565

Language Arts
Communicate 502, 508,
   515, 521, 528, 542, 547,
   554, 561, 567, 574
Eyewitness Math 556

Business and Economics
Agriculture 503
Banking 543
Fund-raising 499, 509,
   516, 535, 555

Investments 509, 514,
   516, 523, 540, 542,
   569, 580
Rental 569
Sales 503, 522, 529, 535
Small Business 496

Life Skills
Age 523
Consumer Economics 548,
   562, 569
Nutrition 503

Sports and Leisure
Crafts 530, 562
Music 554
Sports 523, 534, 568
Theater 516
Travel 555

Math Connections

Geometry 600, 617, 627, 642, 649, 657, 681   **Maximum/Minimum** 646, 648, 649
Probability 642   **Statistics** 618, 622, 663

Applications

Science
Biology 598, 606, 608
Chemistry 592

Social Studies
Demographics 591, 617
Geography 590, 622
Social Studies 588

Language Arts
Communicate 590, 597, 605, 613, 621, 626, 640,

647, 655, 661, 668, 673, 679
Business and Economics
Agriculture 620
Business 615, 688
Construction 642, 649, 655, 656, 663, 668, 672, 675
Economics 589, 614
Manufacturing 651, 659, 663, 669, 674, 688
Sales 615

Small Business 608
Stocks 592, 633
Wages 616

Life Skills
Consumer Economics 621, 627, 632, 641
Education 632
Fire Prevention 591
Health 591, 607

Sports and Leisure
Entertainment 598
Hobbies 613
Music 624
Recreation 677
Sports 598, 599, 600, 606, 607, 623, 642
Travel 640

Other
Academics 607
Communications 614

Appendix

CHAPTER 1

Data and Patterns in Algebra

Throughout history, people from all parts of the world have been fascinated with patterns. The early native tribes of the American Southwest were aware of the patterns that influenced their lives. They recorded these patterns by weaving them into rugs.

In the technological world of modern society, patterns provide a basis for discoveries in science and engineering. Scientists use patterns to study, understand, and predict nature. Mathematicians look for regular patterns when investigating systems of numbers. Patterns can provide a powerful tool for solving problems.

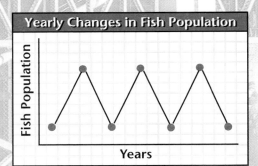

Yearly Changes in Fish Population

In biology, a mathematical pattern can describe the changes in a region's fish population. The graph has a striking resemblance to the pattern shown on the Native American blanket.

Patterns are used in the structure and design of modern cities and the transportation networks that connect them. This student is using a computer to create the design found in a trestle bridge.

PORTFOLIO ACTIVITY

1. Three dot patterns for a sequence of peaks are shown below. Construct the fourth and fifth dot patterns for this sequence of peaks. The fourth dot pattern should use 41 dots.

5   13   25

2. How many dots are in the pattern with 7 peaks? 85

3

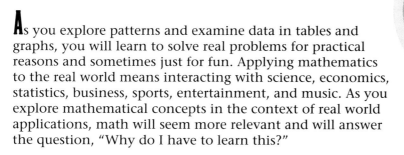

LESSON 1.0 Mathematical Power

why *As you explore and investigate ideas from algebra you will see how the power of mathematics is used to solve problems that occur in the everyday world. Put this power to work, and you will be successful in solving many problems.*

As you explore patterns and examine data in tables and graphs, you will learn to solve real problems for practical reasons and sometimes just for fun. Applying mathematics to the real world means interacting with science, economics, statistics, business, sports, entertainment, and music. As you explore mathematical concepts in the context of real world applications, math will seem more relevant and will answer the question, "Why do I have to learn this?"

As you explore, you will have the opportunity to see how technology can help you understand mathematics and process real-world data. This technology takes the form of scientific calculators, graphics calculators, graphics software, and spreadsheets.

Algebra in Marketing

Statistics and probability are used to determine buying patterns, project future sales, determine packaging size and style, and create new products.

Algebra in Science

Mathematics has been referred to as the language of science. Whether it is in the field of biology, chemistry, physics, earth science, or ecology, scientists cannot investigate or represent their findings without understanding and using mathematics.

Algebra in Medicine

A person's heart rate is a function of many physiological factors such as age, diet, and heredity. A medical professional must understand these relationships to perform his or her job effectively.

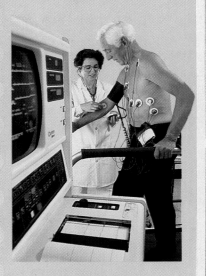

Advertising Costs

Category	Total ad costs (in billions)
Automotive	$5.9
Food	$3.5
Entertainment	$3.1
Travel & Hotels	$2.2
Snacks	$1.2

0 1 2 3 4 5 6
Total ad costs (in billions)

Algebra in Mass Media

Today's information comes to us in many forms—newspapers, television, and computer networks. Data is collected and stored within microseconds, making it possible to interpret the data and predict results in ways we never thought possible only a few years ago. As you explore mathematical concepts and relationships, you will learn how mathematics helps you make sense of real-world data.

Exploring With Algebra

As you explore, you may use various tools such as algebra tiles, calculators, and other manipulative devices. As you use tools, make tables, and draw graphs, you will be given instructions to construct models. These models define and represent mathematical concepts that you will use now and in the future. You will build your own mathematical power.

You can begin now . . .

Exploration *Graphing Changes*

1 The table shows the distances covered by a mountain-bike racer on a level but rough road. Assuming a constant rate of speed, compute the values missing from the table.

Time (hr)	0	0.5	1.0	1.5	2.0	2.5	3.0
Distance (mi)	0	10	20	30	?	?	?
					40	50	60

2 Write a sentence that tells how to find the distance if you know the time and rate of speed.
Distance is rate times time.

3 If the mountain-bike rider maintains the same speed, how far will the rider travel in 5 hours? 100 mi

4 A graph is another way to show the relationship between time and distance. It can also be used to make predictions. Use the graph to predict how long it will take to cycle 100 miles. 5 hr

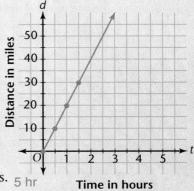

The graph in Step 4 describes the cyclist's movement at a constant rate. Examine the graphs below.

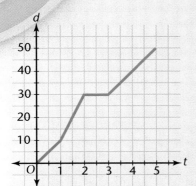

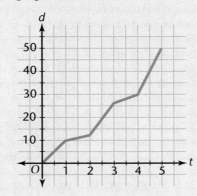

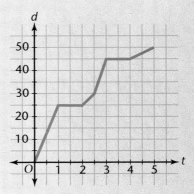

5 How does each graph describe the changes in the movement of a cyclist during each 5-hour trip?

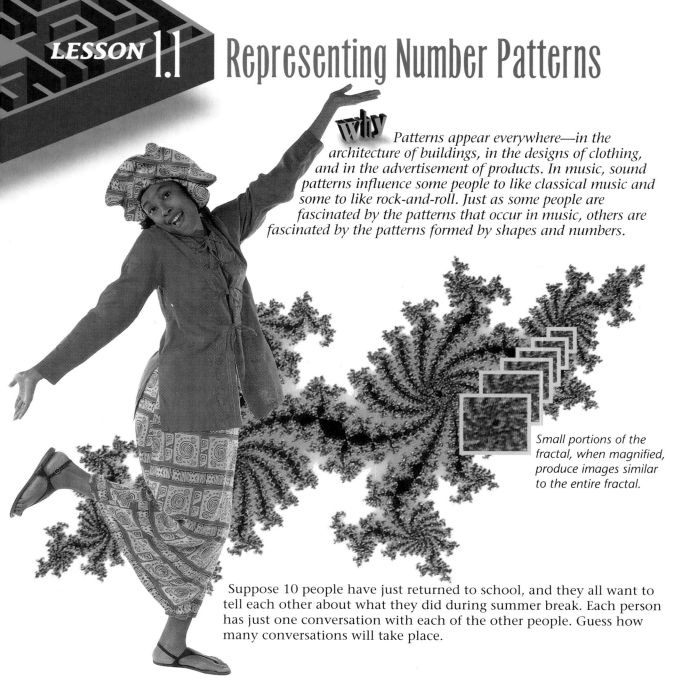

Representing Number Patterns

why *Patterns appear everywhere—in the architecture of buildings, in the designs of clothing, and in the advertisement of products. In music, sound patterns influence some people to like classical music and some to like rock-and-roll. Just as some people are fascinated by the patterns that occur in music, others are fascinated by the patterns formed by shapes and numbers.*

Small portions of the fractal, when magnified, produce images similar to the entire fractal.

Suppose 10 people have just returned to school, and they all want to tell each other about what they did during summer break. Each person has just one conversation with each of the other people. Guess how many conversations will take place.

Using Problem-Solving Strategies

If you are not sure how to solve a problem directly, experiment with various problem-solving strategies. For example, you can draw a picture, think of a simpler problem, make a table, or look for a pattern.

Think of a simpler problem. Begin with the most basic situation for the problem; then make regular changes to build a pattern. Start with just 1 person. How many conversations can 1 person have if talking to oneself doesn't count? Organize the results, and make a table.

2 people
1 conversation

3 people
3 conversations

When you add a fourth person, *D*, the same three conversations are possible, but there are also three new conversations possible—*A* with *D*, *B* with *D*, and *C* with D—for a total of 6.

4 people
6 conversations

Look for a pattern. When a regular number pattern emerges, organize the information in a table. Are there other patterns that might help solve this problem?

People	1	2	3	4	5
Conversations	0	1	3	6	?
Increase		+1	+2	+3	

Once you discover how the pattern in the number of conversations increases, continue the pattern to find the number of conversations possible among 5 people.

Reason logically from the pattern. When the fifth person joins, add 4 new conversations (one with each of the other four people) to the previous 6 conversations. Five people will have **6 + 4**, or **10**, conversations.

People	1	2	3	4	5
Conversations	0	1	3	6	10
Increase		+1	+2	+3	+4

When a sixth person joins, add 5 new conversations to the previous 10, and so on. Continue to build the pattern to 10 people.

People	1	2	3	4	5	6	7	8	9	10
Conversations	0	1	3	6	10	15	21	28	36	45
Increase		+1	+2	+3	+4	+5	+6	+7	+8	+9

When mathematicians study number sequences for patterns, they often make *conjectures*. A **conjecture** is a statement, based on observations, that they believe to be true. Mathematicians try to prove the conjecture or find a counterexample to show that the conjecture is not true.

The total number of conversations among any number of students is the sum of the numbers from 1 to one less than the number of students. Test this conjecture for 11, 12, 13, and 14 students.

Which figure completes the pattern? c

Pattern

    ?

| A | B | C | D |

Choices

Number Sequences

When a sequence shows a pattern, discover as much as you can about how the numbers relate to each other.

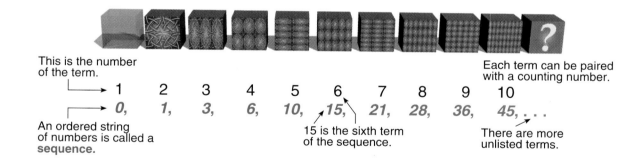

This is the number of the term.

1	2	3	4	5	6	7	8	9	10
0,	1,	3,	6,	10,	15,	21,	28,	36,	45, . . .

An ordered string of numbers is called a **sequence**.

15 is the sixth term of the sequence.

Each term can be paired with a counting number.

There are more unlisted terms.

> **EXAMPLE 1**

Predict the next three terms for each of the following sequences.

Ⓐ 15, 18, 21, 24, 27, . . .   Ⓑ 2, 6, 12, 20, 30, . . .

Solution ➤

Ⓐ The first sequence begins with 15. Each of the following terms is 3 more than the one before. A calculator can be helpful when you explore sequences. A calculator with a *constant feature* can easily provide more terms of the sequence by repeating an operation.

Calculator

For example, the first sequence can be generated by entering

15 [+] 3 [=] [=] [=] [=].

Continue pressing the *equals key* three more times. The next three terms of the sequence are **30**, **33**, and **36**.

On some *graphics* calculators, enter

15 [ENTER] [+] 3 [ENTER] [ENTER] [ENTER] [ENTER].

Continue pressing the *enter key* three more times to generate **30**, **33**, and **36**.

B
2   6   12   20   30
 └+4┘ └+6┘ └+8┘   └+10┘

The next three terms are found by adding **12, 14,** and **16.**

30 + 12 = 42   **42 + 14 = 56**   **56 + 16 = 72** ❖

The sequence in Example 1, Part B, appears when counting the number of dots in the pattern below.

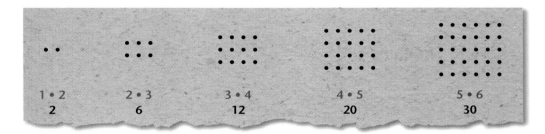

The next three terms are **6 · 7 = 42, 7 · 8 = 56,** and **8 · 9 = 72.**

Why do you think the numbers in the sequence 2, 6, 12, 20, 30, . . . are called **rectangular numbers?** Because each can be visualized with a rectangle one unit longer than it is wide

GEOMETRY
Connection

EXAMPLE 2

Find the sum $1 + 2 + 3 + 4 + 5$ using a geometric pattern.

Solution ➤

There are many ways to find this sum. This method shows how geometric patterns can be used to find the sum of $1 + 2 + 3 + 4 + 5$.

Represent the sum of 1 to 5 with a triangle of dots. Form a rectangle of dots by copying the triangle and rotating it.

The total number of dots is $5 \cdot 6 = 30$. There are half as many dots in the triangle. $\frac{5 \cdot 6}{2} = 15$

The sum of the numbers from 1 to 5 is 15. ❖

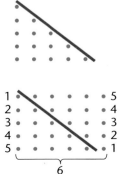

Spreadsheet

Technology can simplify the exploration of sequences and can provide tools to calculate sums of sequences. Try using a computer *spreadsheet* to add the numbers 1 to 9. A spreadsheet allows you to control lists or tables of data. Familiarize yourself with the layout of this spreadsheet.

Spreadsheet title

Cell reference
(Shows where the results appear) — C2

Sum of numbers

=SUM (A2:A10) — **Formula bar**

Column identifiers

	A	B	C
1	Numbers		Sum
2	1		45
3	2		
4	3		
5	4		
6	5		
7	6		
8	7		
9	8		
10	9		

Active cell

Row

Cell

Row identifiers   **Column**   **(Spreadsheet)**

Try This   Find the sum $1 + 2 + 3 + 4 + 5 + 6 + 7 + 8 + 9$ using a geometric pattern. Then use your understanding of the number pattern to find a simple way to calculate the sum $1 + 2 + 3 + 4 + 5 + \cdots + 100$.

Answers: 45, 5050. The pattern leads to the formula $\frac{n(n + 1)}{2}$ for the sum of consecutive whole numbers from 1 to n.

EXERCISES & PROBLEMS

Communicate

1. Explain how you would set up a table to solve the following problem. If each of the teams A, B, C, and D play each other, one of the games will be A versus B. Show how you would list all the games.

2. Describe how to find the sum of the numbers 1 through 20.

Explain how you would find the pattern for the following sequences.

3. 3, 7, 11, 15, 19, 23, . . .   4. 1, 1, 2, 3, 5, 8, 13, . . .

5. Give four problem-solving strategies that are used in Lesson 1. Choose one strategy from your list, and explain how it helps in solving problems.

Practice & Apply

6. If each of the teams *A*, *B*, *C*, *D*, and *E* play each other, one of the games will be *A* versus *B*. List all of the games, and tell how many there are. 10

7. In 1993, after Penn State joined the Big Ten athletic conference, the Big Ten had 11 teams. If each of the 11 teams played each of the other teams, how many games would there be? Make a table similar to the conversation problem. 55

8. If an athletic conference has 12 teams and each of the teams plays each of the other teams, how many games will there be? 66

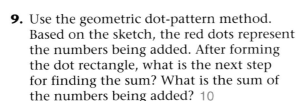

9. Use the geometric dot-pattern method. Based on the sketch, the red dots represent the numbers being added. After forming the dot rectangle, what is the next step for finding the sum? What is the sum of the numbers being added? 10

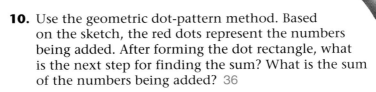

10. Use the geometric dot-pattern method. Based on the sketch, the red dots represent the numbers being added. After forming the dot rectangle, what is the next step for finding the sum? What is the sum of the numbers being added? 36

11. Use the geometric dot-pattern method to find the sum $1 + 2 + 3 + 4 + 5 + 6 + 7$. Include the sketch. 28

12. Find the sum $1 + 2 + 3 + 4 + 5 + \cdots + 40$. Think of the sketch without actually drawing it. $\frac{40 \cdot 41}{2} = 820$

13. How can you find the sum $1 + 2 + 3 + 4 + 5 + \cdots + 50$? $\frac{50 \cdot 51}{2} = 1275$

Find the next three terms in each sequence.
Then explain the pattern used to find the terms for each.

14. 4, 9, 14, 19, 24, . . .
29, 34, 39
15. 7, 16, 25, 34, 43, . . .
52, 61, 70
16. 9, 19, 29, 39, 49, . . .
59, 69, 79
17. 2, 4, 8, 16, 32, . . .
64, 128, 256
18. 5, 7, 9, 11, 13, . . .
15, 17, 19
19. 3, 9, 27, 81, 243, . . .
729, 2187, 6561
20. 8, 10, 12, 14, 16 . . .
18, 20, 22
21. 16, 8, 4, 2, 1, . . .
$\frac{1}{2}, \frac{1}{4}, \frac{1}{8}$
22. 5, 12, 19, 26, 33, . . .
40, 47, 54

23. Solve the conversation problem for 10 people by drawing a sketch. Draw 10 dots around a circle, and connect them with segments to represent conversations. How many segments did you draw? 45

24. Geometry Diagonals are segments other than sides that connect the vertices of a polygon. Draw a hexagon and its diagonals. How many diagonals does a regular hexagon have? How does this problem relate to the conversation problem? 9

Karl Friedrich Gauss (1777–1855) was one of the greatest of all mathematicians.

Cultural Connection: Europe When Gauss was very young, a teacher told him to find the sum of all the whole numbers from 1 to 100, thinking this would keep him busy for a long time. But Gauss very cleverly paired the numbers and observed that $1 + 99 = 100$, $2 + 98 = 100$, $3 + 97 = 100$, and so on.

25. How many pairs could be formed in this way without repeating any numbers? 49

26. Which whole numbers from 1 to 100 would not appear in any of the pairs? 50 and 100

27. Based on Exercises 25 and 26, what is the sum of the numbers from 1 to 100? 5050

Geometry The numbers in each array are square numbers. The first four square numbers are 1, 4, 9, and 16.

28. Find the 10th square number. 100

29. Find the 20th square number. 400

1   4   9   16

30. Suppose 8 cities are to be linked by phone lines, with 1 phone line between each pair of cities. How many phone lines will there be? 28

31. Use the geometric dot-pattern method to find the sum $1 + 3 + 5 + 7 + 9$. 25

Geometry The numbers in each array are triangular numbers. The first four triangular numbers are 1, 3, 6, and 10.

32. Find the 10th triangular number. 55

33. Find the 20th triangular number. 210

1   3   6   10

34. Show visually that the square number 16 is the sum of two triangular numbers. Which two? 10 and 6

35. Show visually that 100 is the sum of two triangular numbers. Which two? 55 and 45

36. Copy and complete the following table.

Number	1	3	5	7	9	11	13	15	17
Sum	1	4	9	16	? 25	? 36	? 49	? 64	? 81

37. Examine the pattern in the table in Exercise 36. Guess the sum of the first 100 odd numbers. Explain your method. 10,000

38. Copy and complete the following table.

Number	1	2	3	4	5	6	7
Cube of the number	1	$2^3 = 8$	$3^3 = 27$	$4^3 = 64$	? 125	? 216	? 343
Sum of the cubes	1	9	36	100	? 225	? 441	? 784

39. Examine the pattern in the table in Exercise 38. Guess the sum of the first 10 cubes. Explain your method. 3025

Look Back

David and three friends collected 196 aluminum cans.

40. They divided the cans among themselves equally into 4 bags. How many cans are in each bag? 49

41. The 4 friends take the cans to the recycling center, where they can get $0.20 a pound for the cans. The total weight of the cans is 9 pounds. How much money will they get all together? $1.80

42. How much will each of the 4 friends get from recycling the cans if they split the money equally? $0.45

43. If a teacher collects $2.75 from each of 28 students for a field trip, how much does the teacher collect? $77

Recycled aluminum cans are compressed before shipping.

44. The citywide concert sold 6702 tickets to students and 3749 tickets to parents. How many tickets were sold all together? 10,451

45. 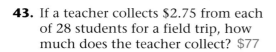 **Geometry** What is the area of a square with sides of 23 centimeters? 529 sq cm

46. What are some fractions that are equivalent to $\frac{3}{5}$? $\frac{6}{10}, \frac{30}{50}, \frac{-3}{-5}$
Answers may vary.

Find the value of the following expressions.

47. $6(9 + 4)$ **48.** $\sqrt{225}$
    78   15

49. Kim had $4.75 and spent $3.12. How much money does she have left? $1.63

Look Beyond

50. The following pattern was first explored in China and Iran and was later called Pascal's Triangle. Extend the pattern by finding rows 6, 7, and 8.
row 6: 1, 6, 15, 20, 15, 6, 1   row 7: 1, 7, 21, 35, 35, 21, 7, 1

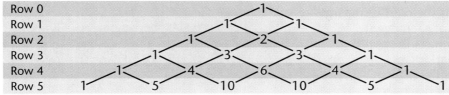

row 8: 1, 8, 28, 56, 70, 56, 28, 8, 1   Sums: 1, 2, 4, 8, 16, 32, 64, 128, 256

51. Find the sum of the numbers in rows 0–8 of Pascal's Triangle.

Exploring Problems With Tables and Equations

why *Tables can be used to organize data and information. Equations are often written to describe the data presented in tables. Tables and equations can be used to study data in more depth.*

2 flags
1 Section

3 flags
2 Sections

4 flags
3 Sections

Suppose your school is sponsoring the local Special Olympics. Your principal decides to form several committees to plan the events. To help in planning, the committees collect data and organize their data in a table.

Raul and Andrea are on the committee for planning the location of each event. To find out how many flags they will need, they begin by marking off 10-meter sections for different events. Raul and Andrea discover that they need **2** flags for **1** section, **3** flags for **2** sections, **4** flags for **3** sections, and so on.

Raul and Andrea notice that the number of flags needed is 1 more than the number of sections. They use this information to make a table.

How can you find the number of flags needed for 30 sections?

30 + 1 = 31 flags

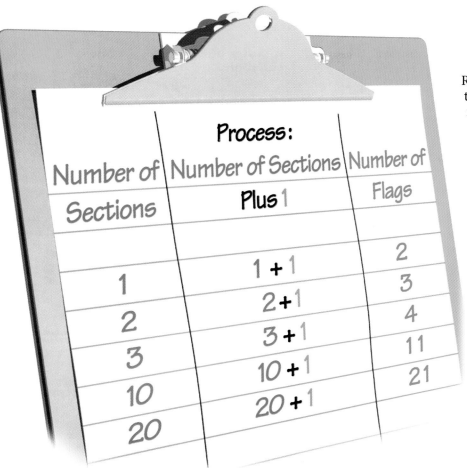

Number of Sections	Process: Number of Sections Plus 1	Number of Flags
1	1 + 1	2
2	2 + 1	3
3	3 + 1	4
10	10 + 1	11
20	20 + 1	21

CRITICAL Thinking

Find the number of sections formed with 25 flags. Explain the process you used. 25 flags will form 24 sections. The number of flags is always one more than the number of sections.

The verbal expression *number of sections plus 1* can be written as an *algebraic expression*. To write an **algebraic expression,** replace words with letters or symbols.

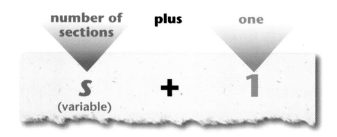

number of sections   plus   one

s
(variable)   **+**   **1**

The letter s represents the number of sections. A letter that represents a number is called a **variable.**

You can use the algebraic expression **s** + **1** to write an *equation* for the following sentence.

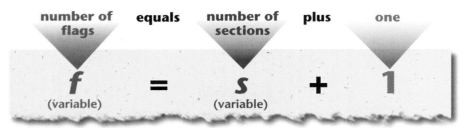

number of flags	equals	number of sections	plus	one
f (variable)	=	**s** (variable)	+	**1**

This equation can be written $f = s + 1$ or $s + 1 = f$.

To find the number of flags, **f**, substitute the *actual number* of sections, **s**, in the equation $f = s + 1$. For example, to find the number of flags needed for **6** sections, substitute **6** for **s**.

$$f = s + 1$$
$$6$$

Since $6 + 1 = 7$, the number of flags is 7.

•Exploration 1 *Solving Problems With One Operation*

George is on the committee that is in charge of reserved seating for the Special Olympics. To determine the number of people they will be able to seat in the stands, George starts a table.

George knows that each row in the stands can seat 12 people.

1 Copy and complete the table. Use parentheses to show multiplication.

Number of rows	Process: 12 times the number of rows	Number of people
1	12(1)	12
2	12(2)	24
3	? 12(3)	? 36
4	? 12(4)	? 48
5	? 12(5)	? 60
10	? 12(10)	? 120

2 How many people can be seated in 15 rows? How many rows are needed to seat 240 people? 180 people; 20 rows

3 Examine the pattern in the table. Let the variable *r* represent the number of rows. Now write an algebraic expression for the Process column. 12(*r*) or 12*r*

4 Complete the sentence with a verbal expression to describe the number of people that can be seated if the number of rows is known.

The number of people is ___?___.
12 times the number of rows

5 Write an equation for your sentence using *r* for the number of rows and *p* for the number of people.

$$p = \frac{?}{12r} \; ❖$$

The expression *12 times r* can be written algebraically as

$$12(r), \qquad 12 \cdot r, \qquad \text{or} \qquad 12r.$$

Give three ways to write the equation for the number of people that can be seated. $p = 12r; p = 12 \cdot r; p = 12(r)$

•Exploration 2 *Solving Problems With Two Operations*

Julie and Andy are participating in a walk-a-thon to raise money for the Special Olympics. A sponsor will give them $10 for the entry fee, plus $3 for each mile walked. Julie and Andy start a table to determine the amount they will raise from this sponsor for the number of miles they walk.

1 Copy and complete the table.

2 How much money will Julie raise if she walks 8 miles? If Andy raised $28, how many miles did he walk? $34; 6 miles

3 Let *m* represent the number of miles walked. Write an expression for the Process column. $10 + 3m$

4 Complete the sentence with a verbal expression to describe the amount of money raised when you know the number of miles walked.

$10 plus 3 times the number of miles walked
The amount of money raised is ___?___.

Number of miles walked	Process: $10 for entry, plus $3 for each mile	Amount of money raised
1	10 + 3(1)	$13
2	10 + 3(2)	? $16
3	? 10 + 3(3)	? $19
4	? 10 + 3(4)	? $22
5	? 10 + 3(5)	? $25
10	? 10 + 3(10)	? $40

5 Write an equation for your sentence using *a* for the amount of money raised and *m* for the number of miles walked.

$$a = \frac{?}{10 + 3m} \; ❖$$

Erin and Logan are on the committee in charge of the awards. They purchase medium-sized medals from a trophy shop.

Refer to the costs for trophies listed in the photo below. To write an equation for the cost of the awards, first identify the process used to find the cost.

The verbal expression describing the process is *$25 plus $0.75 times the number of medals.* Let *m* represent the number of medals, and let *c* represent the cost.

cost	is	$25	plus	$0.75	times	number of medals
c	=	25	+	0.75	•	m

The equation for the cost of the awards, $c = 25 + 0.75m$, contains two operations, **addition** and **multiplication**. ❖

EXERCISES & PROBLEMS

Erin and Logan are on the committee in charge of the awards. They purchase medium-sized medals from a trophy shop.

Communicate

1. Explain what variables are and how they are used.

2. Explain how the algebraic expression $5 + 2h$ can represent a verbal expression such as *$5 plus $2 per hour.*

3. Describe what it means to substitute values for *h* in the expression $5 + 2h$.

4. Write the equation $C = 5 + 2h$ in sentence form when *C* represents the cost of renting a go-cart, and *h* represents the number of hours the go-cart is used.

5. Explain how a table can be used to solve a problem.

6. Give an algebraic expression for the cost of large-sized medals. Then explain how to find the cost for 100 large-sized medals.

Practice & Apply

7. Write an algebraic expression for the following verbal expression.

Eight dollars entry fee per person 8p

8. Write an equation for the following sentence.

The total wages equal the number of hours times $9 per hour. w = 9h

9. In the Application on page 20, how many small-sized medals can Erin and Logan purchase if the total cost is $50.50? 85

Thermometer Readings

Reading r	Process	Temperature t
10°F	10 − 5	5°F
20°F	20 − 5	? 15
30°F	30? − 5	? 25
40°F	40? − 5	? 35
50°F	50? − 5	? 45
60°F	60? − 5	? 55
100°F	100? − 5	? 95

10. Copy and complete the table for each thermometer reading.

11. The temperature is 5 degrees lower than the reading on the thermometer. Write an expression for the process column. r − 5

12. Write an equation for the relationship between the thermometer reading and the temperature, using the variables r and t. t = r − 5

Cross-Country Biking

Number of hours h	Process	Number of miles m
1	12.5(1)	12.5
2	12.5(?)	?
3	12.5(?)	?
4	12.5(?)	?
5	12.5(?)	?
10	12.5(?)	?

The speedometer shows the steady speed at which Mary rides her bike cross-country.

13. Using the variables h and m, write an equation for the number of miles that Mary rides her bike.

$$m = \underline{\ ?\ } \quad 12.5h$$

Surfboard Rental

Number of hours h	Process	Cost c
1	5 + 1.5(1)	?
2	5 + 1.5(?)	?
3	5 + 1.5(?)	?
4	5 + 1.5(?)	?
5	5 + 1.5(?)	?
10	5 + 1.5(?)	?

14. Julia rents surfboards at the beach for $5 plus $1.50 per hour.

Write an equation for the cost, using the variables h and c.

$$c = \underline{\ ?\ } \quad 5 + 1.5h$$

Suzette is setting up pens for the show animals at the county fair. She connects sections of fence to build a row of pens.

Use the diagram of pens to complete Exercises 15–18.

Number of pens	Number of sections	Process
1	4	1 + 3(1)
2	? 7	1 + 3(2)
3	? 10	1 + 3(3)
4	? 13	1 + 3(4)
5	? 16	1 + 3(5)
10	? 31	1 + 3(10)

4 sections   7 sections

1 pen   2 pens

15. Copy and complete the table.

16. Add a process column to your table, and fill in the new column.

17. Let *p* represent the number of pens. Write an expression for the process. $1 + 3p$

18. Use the expression you wrote for Exercise 17 to write an equation for the number of sections, *s*.
$s = 1 + 3p$

Write an algebraic expression that represents each verbal expression.

19. The number of pounds of aluminum cans to recycle, *p*, multiplied by the amount per pound, 50¢ $0.50p$

20. The cost of a sports car, $85,000, decreased by the trade-in value of another car, *t* $85{,}000 - t$

21. The number of videos, 12, times the price per video, *p* $12p$

22. The number of hours in *d* days $24d$

23. The amount, *a*, of tax withheld, subtracted from a total wage of $1245 $1245 - a$

Consumer Awareness The cost to park at the airport lot is $1.30 per hour, or $0.65 per half hour. Parking in the private lot next to the airport is $2.00 plus $0.60 per hour, or $2.00 plus $0.30 per half-hour.

24. Let *c* represent the cost for parking. Write an equation for the cost to park at the airport lot for *h* hours. $c = 1.30h$

25. Let *c* represent the cost for parking. Write an equation for the cost to park at the private lot for *h* hours. $c = 2 + 0.60h$

26. Technology Using a graphics calculator, build a table of airport parking-lot fees based on 0.5 to 5 hours in increments of half-hours. Record the table.

27. Technology Using a graphics calculator, build a table of private parking-lot fees based on 0.5 to 5 hours in increments of half-hours. Record the table.

28. In terms of half-hours, when is the airport lot cheaper and when is the private lot cheaper?

Build a table of values by substituting 1, 2, 3, 4, 5, and 10 for x.

29. $y = 2x$ **30.** $y = x - 1$ **31.** $y = x + 3$ **32.** $y = 2x - 2$

33. $y = 5x$ **34.** $y = x + 2$ **35.** $y = 3x - 3$ **36.** $y = 2x + 1$

37. $y = 4x$ **38.** $y = 2x - 1$ **39.** $y = 3x + 6$ **40.** $y = 4x - 2$

41. $y = 3x$ **42.** $y = 3x + 2$ **43.** $y = 4x + 3$ **44.** $y = -0.2x - 7$

45.  **Portfolio Activity** Complete the portfolio activity given on page 3.

Look Back

Find the next three terms in each sequence. [Lesson 1.1]

46. 1, 3, 5, 7, 9, . . . 11, 13, 15 **47.** 1, 1, 2, 3, 5, 8, 13, . . . 21, 34, 55

48. How many 2-foot pieces can be cut from a 12-foot board? 6

49. A liter contains approximately 34 ounces. How many 12-ounce cans of soda would it take to fill a 3-liter bottle? 8.5

50. A school bus seats 60 students. How many buses are needed to take 268 students to a football game? 5 buses

51. The distance around a track is 400 meters. How many laps must be run for a 6-kilometer race? 15

52. Five tennis players play each other once. How many games are played? 10

53. Find the sum $1 + 2 + 3 + 4 + 5 + 6 + 7 + 8 + 9 + 10$. 55

Look Beyond

54. Square numbers are formed by multiplying a number by itself. For example, since $1 \cdot 1$ is 1 and $2 \cdot 2$ is 4, the numbers 1 and 4 are called square numbers.

Complete the table of square numbers.

Number n	Process: $n \cdot n$	Square number n^2
1	$1 \cdot 1$	1
2	$2 \cdot 2$	4 ?
3	$3 \cdot 3$	9 ?
4	$4 \cdot 4$?	16 ?
5	$5 \cdot 5$?	25 ?
10	$10 \cdot 10$?	100 ?

LESSON 1.3 Solving Problems Using Tables and Charts

why *Information that is organized in a table can often be shown on a chart. Tables and charts can help you visualize how the information is related.*

Spanish Club
Coupon Books
$25.00 each
10% Goes to the Spanish Club

The Spanish Club is planning a trip. To raise money for this trip, each member will sell discount coupon books. To determine how many coupon books each member needs to sell to earn enough money for the trip, you can write an equation and make a table.

EXAMPLE 1

For each coupon book that Sarah sells, the club will earn $2.50 profit. To earn a total of $125, how many books does Sarah need to sell?

Solution ➤

The club's profit is $2.50 times the number of coupon books she sells. Let p represent the **profit**, and let b represent the number of **books**. Then write an equation using p and b.

profit	is	$2.50	times	number of books sold
p	=	2.50	•	b

Recall that this equation can also be written $p = 2.50b$.

Make a table to organize the information. Substitute values of b in the equation $p = 2.50b$ to find the profit, p.

Number of books b	Process: $2.50b$	Profit p
10	2.50(10)	25.00
20	2.50(20)	50.00
30	2.50(30)	75.00
40	2.50(40)	100.00
50	2.50(50)	125.00
60	2.50(60)	150.00

By examining the table, you can see that the club will earn $125 profit if Sarah sells 50 coupon books. ❖

Try This The drama club at Derrik's school is selling greeting cards to earn money for a trip. For each box of greeting cards that Derrik sells, the club will earn $3.00 profit. To earn a total of $165, how many boxes of cards does Derrik need to sell? 55 boxes

A bar chart is a convenient way to display information from a table.

Exploration *Using Bar Charts*

The following table shows the data from the first and third columns of the table from Example 1.

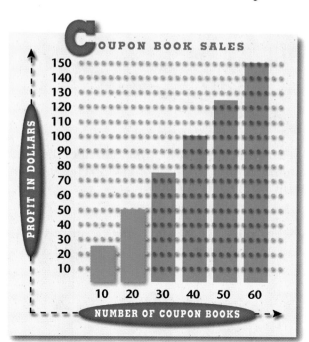

Books, b	10	20	30	40	50	60
Profit, p	25	50	75	100	125	150

Another way to represent data in a table is to make a bar chart. The numbers of coupon books sold are listed on the horizontal axis. The amounts of profit are listed on the vertical axis.

The height of a vertical bar represents the amount of profit made from selling that number of coupon books. For example, the vertical bar over the number 10 represents the profit from selling 10 coupon books. Since the height of that bar is 25, the profit from selling 10 coupon books is $25.

 Copy and complete the bar chart for 30, 40, 50, and 60 coupon books, using the data in the table.

4. From a bar chart, we can estimate the height of a bar at a given location on the horizontal axis. We can also, given the height of a bar, estimate the location of a bar on the horizontal axis.

2 Use your bar chart to estimate how much profit is made by selling 35 coupon books. $87.50

3 Use the equation $p = 2.50b$ to find the number of coupon books that must be sold to earn a profit of $75. Find the number of coupon books that must be sold to earn a profit of $100. Use your bar chart to estimate the number of coupon books that must be sold to earn a profit of $87.50. 30 books; 40 books; 35 books

4 Explain how you can use a bar chart to estimate or find unknown information. ❖

You can use a graphics calculator to build a table of values.

EXAMPLE 2

Graphics Calculator

Cheryl uses her cellular phone for 120 minutes in July. What is Cheryl's bill for July?

Cellular Phone Service
JULY SPECIAL
Only $25.00 per month
plus
$.40 per minute

Solution ➤

First write an equation that describes the cost in terms of the number of minutes. Let y represent the monthly cost. Let x represent the number of minutes.

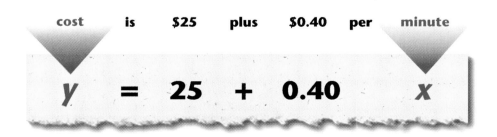

cost	is	$25	plus	$0.40	per	minute

$$y = 25 + 0.40x$$

Enter the right-hand side of the equation,

$$25 + 0.40x,$$

into your calculator.

Create a table of values for the equation entered.

For X, use a minimum of 60 and a difference of 10.

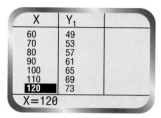

The table shows that Y_1 is 73 when X is 120. So Cheryl's bill for July is $73.00. ❖

Try This Rick uses a competing cellular phone company. His cellular phone service costs $15 per month plus $0.50 per minute. How much does it cost Rick to use his phone for 120 minutes in one month? $75.00

How can you find the maximum number of minutes that Cheryl can use her cellular phone in July without exceeding $100 in charges?
To find the maximum number of minutes, look further down the table described in Example 2. Cheryl can use her phone a maximum of 187 minutes.

x	y
185	$99
186	$99.40
187	$99.80
188	$100.20

EXERCISES & PROBLEMS

Communicate

Sales The sale price, s, of an item is determined by multiplying the original price, p, by 0.60. Use this information for Exercises 1–4.

1. Explain how to write an equation that represents the relationship between the original price, p, and the sale price, s.

2. How can you use a table to find the original price, p, of an item that has a sale price of $72.00?

3. Describe how to represent the relationship between the original price, p, and the sale price, s, with a bar chart.

4. How can you use a bar chart to find or estimate the original price, p, of an item that has a sale price of $90.00?

5. Describe a real-world situation that could be represented by the equation $2h + 4 = 32$.

Practice & Apply

Make a table of values using 10, 20, 30, and 40 for the variable in the expression. Include a process column.

6. $2.7x$ **7.** $3x - 2$ **8.** $4.25h + 10$ **9.** $12c - 4$

Technology Use a graphics calculator to make a table of values for each equation. Use the table to determine the value of *d* when *c* has the indicated value.

10. $c = 7d - 4$, *c* is 31. 5   **11.** $c = 5.6d$, *c* is 67.2. 12

12. $c = 3.6d + 24$, *c* is 88.8. 18   **13.** $c = 0.6d + 18$, *c* is 78. 100

Entertainment The cost of renting a pair of in-line skates at the boardwalk is given by the formula $c = 5 + 2.25h$, where *c* is the total cost and *h* is the number of hours the skates are rented.

14. Make a table to show the total cost of renting the skates for 1, 2, 3, 4, and 5 hours.

15. Suppose you can afford to spend $14 renting $14 = 5 + 2.25h$
the in-line skates. What equation describes this situation? For how many hours can you rent the skates? 4

16. Make a bar chart using the data from your table.

17. Use the bar chart from Exercise 16 to determine how many hours you can rent the in-line skates if you have $23.
8

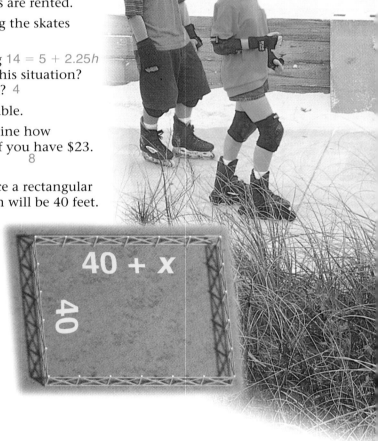

Geometry Suppose you are going to fence a rectangular region of your yard. One dimension of the region will be 40 feet. The other dimension will be 40 feet or more.

18. Write an equation that describes the total length of fencing around this region.
$f = 2(40) + 2(40 + x)$

19. Copy and complete the table to show possible total amounts of fencing.

40 + x

40

x	Process	Amount of fencing
0	?	?
5	?	?
10	?	?
15	?	?
20	?	?
25	?	?

20. Suppose the amount of fencing needed is 240 feet. What are the dimensions of the rectangular region? What is the value of *x*? $x = 40$; 40 ft by 80 ft

21. Suppose one dimension of the region will be 50 feet, and the other dimension will be 50 feet or more. Write an equation that describes the total length of fencing around the region. $f = 2(50) + 2(50 + x)$

Travel The bar chart shows that the distance traveled by a car at a constant rate depends on the time traveled.

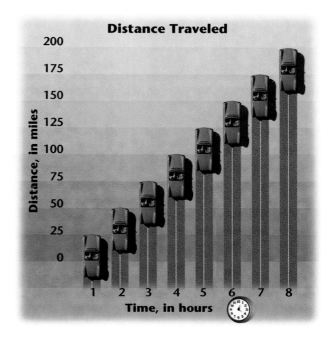

Distance Traveled

Use the bar chart for Exercises 22–26.

22. Make a table for the data in the bar chart.

23. At what constant rate (speed) is the car traveling? 25 mph

24. If the car travels for 3 hours, how many miles does it travel? 75 miles

25. If the car travels 240 miles, how many hours does it travel? 9.6 hr

26. Write an expression for the distance, d, in terms of the number of hours. $25h$

Business The amount, a, that a mechanic charges for his services is given by the formula $a = 24 + 12.5h$, where a is the amount in dollars and h is the number of hours the mechanic works.

27. Make a table to show the amount charged for 1, 2, 3, 4, and 5 hours.

27.

h	1	2	3	4	5
a	$36.50	$49	$61.50	$74	$86.50

28. Suppose the mechanic charges $124. How many hours will the mechanic work for $124? 8

29. Make a bar chart using the data from your table.

30. Use your bar chart to find how many hours the mechanic will work for $100. Assume that the mechanic charges only by whole hours. 6 hr

Look Back

31. There are six high schools in the school district. If each school's football team plays each of the other teams only once, how many games will be played? **[Lesson 1.1]** 15

Write an algebraic expression for each phrase.   [Lesson 1.2]

32. A five-dollar entry fee per person $5p$

33. Two dollars for each ride plus a three-dollar entry fee $2r + 3$

Write an algebraic equation for each situation.   [Lesson 1.2]

34. The cost, c, is $4.50 per pound, p. $c = 4.5p$

35. The profit, p, is $5 per person, x. $p = 5x$

Look Beyond

36. Is the expression $16 - 8 \cdot 2$ equal to 16 or 0? What does a scientific calculator give for an answer? 0; 0

Exploring Variables and Equations

why *You want to rent a bicycle, but you don't know how long you can rent it with the money you have. In mathematics, one technique used to solve this kind of problem begins when you represent the unknown number in the problem with a symbol. You then work with the symbol until the algebra reveals the symbol's value.*

•Exploration 1 *Using a Variable*

For how many hours can you rent a bike if you have $15 to spend?

This relatively common problem provides an example that can be used to explore the basic tools of algebra. Examine the pattern for the bike rental.

Hours	1	2	3	4	5	...	h
Cost in $	2	4	6	?	?	...	?

1 What is the cost for 1 hour of bike rental? $2.00

2 What is the cost for 2 hours of bike rental? $4.00

3 What is the cost for 4 hours of bike rental? $8.00

4 When you rent a bike for 5 hours, what number do you multiply by 5 to get the value for the *cost*? 2

5 Let *h* stand for the number of hours. Follow the same pattern that you use to generate the values for *cost*. What number would you multiply times *h*? 2

6 What should the *cost* expression be in this table for *h* hours? 2*h*

7 When *h* is 5, what is the value of 2*h*? ❖ 10

You can find the value of 2*h* when *h* is 6 by substitution. Replace *h* with 6, and simplify.

$$2h = 2 \cdot 6 = 12$$

Extend the table.

Hours (*h*)	1	2	3	4	5	6	7	8
Cost (*c*)	2	4	6	8	10	12	?	?

14   16

What is the cost for bike rental if the hours rented are halfway between 7 and 8 hours and you can pay by the half-hour? For how many hours can you rent a bike at the park if you have $15? $15; $7\frac{1}{2}$

If the cost for renting a bike is $3 an hour, how many hours can you rent a bicycle for $15? Try other values to see how they affect the hour and cost table. 5 hours

Another way to represent the cost for renting the bicycle is to use the letter *c*. Since both *c* and 2*h* represent *cost,* they are equivalent. This is written as *c* = 2*h* or 2*h* = *c*.

Recall that a variable is a letter or other symbol that can be replaced by any number or other expression. Variables combine with numbers and operations (addition, subtraction, multiplication, and division, for example) to form expressions. Two equivalent expressions, when separated by an equal sign, form an **equation.**

Examples of variables	*Examples of expressions*	*Examples of equations*
a, x, y	$3x + 4, 2h,$	$3x + 4 = 8,$
c, h, d, t	$-16t^2 + 4t$	$c = 2h,$ or $2h = c$

When you solve an equation, you find the value or values of the variable that make the equation a true statement. The solution for 2*h* = 15 is $7\frac{1}{2}$ because $2 \cdot 7\frac{1}{2} = 15$.

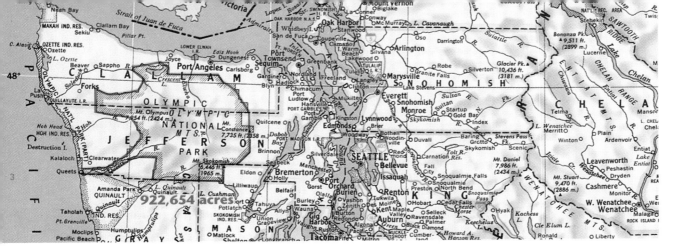

Ecology The land area of Olympic National Park is 922,654 acres. How long would it take to lose forest land equal to the area of Olympic National Park if 57,600 acres were destroyed each day?

•Exploration 2 A Diminishing Sequence

Calculator

On a calculator with a constant feature, enter the area of Olympic National Park. Subtract 57,600 repeatedly, counting the number of times you subtract.

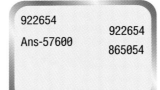

```
922654
              922654
Ans-57600
              865054
```

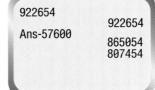

```
922654
              922654
Ans-57600
              865054
              807454
```

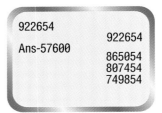

```
922654
              922654
Ans-57600
              865054
              807454
              749854
```

As you continue to subtract, you can see how quickly the forest area is reduced. Continue repeating the subtraction until the entire forest area is depleted.

To express this problem as an equation, you need to interpret and relate the information from the problem.

1 How many acres of forest land are equal to the size of Olympic National Park? 922,654 acres

2 What does repeatedly subtracting 57,600 represent?
The number of acres depleted each day

3 How would you express *the number of acres depleted each day,* using the variable *d* to represent days? 57,600*d*

4 If you let *f* represent the number of acres remaining, how much forest is left when *d* is 10? 346,654 acres

5 Explain what it means when *f* is 0. Zero acres of forest remain.

6 How many days would it take to completely destroy an area the size of Olympic National Park at the given rate? ❖ 17 days

In one region, there are 15,000 acres of forest and an average of 16 acres are destroyed every day. The projected amount of forest remaining after 30 days is 14,520 acres. However, a survey shows that there are actually 13,923 acres of forest remaining after 30 days. What could account for the difference between the projected amount and the actual amount?
Even though only 16 acres of forest are actually destroyed daily, this damage to the overall health of the forest accelerates the destruction of other acreage.

APPLICATION

Danica places 3 photos on the cover of her picture album. On all of the other pages she places 4 photos. She has 178 photos in all. How many pages will she need if she uses all of her photos? This problem can be solved by writing an equation and using a guess-and-check strategy.

Let p equal the number of pages in Danica's picture album. To solve this problem, solve for p in the equation $4p + 3 = 178$. Guess the number of pages, substitute the number you guess for p, and check your guess. Keep track of your attempts in a table. Recall that in algebra, parentheses are used to show multiplication.

Guess 1: Try $p = 10$.
$$4p + 3 = 178$$
$$4(10) + 3 = 43$$

43 is too small. Try a larger number for p.

Guess 2: Try $p = 50$.
$$4p + 3 = 178$$
$$4(50) + 3 = 203$$

203 is too large. Try a smaller number for p.

Guess 3: Try $p = 40$.
$$4p + 3 = 178$$
$$4(40) + 3 = 163$$

163 is too small. You will find that the number for p is between 43 and 44.

For 178 photos, 43 pages is not enough. The problem asks for the number of pages Danica will use, so the answer is 44 pages. One page more than 43 is needed for the extra photos even though the last page will not be filled. ❖

Always be careful to interpret your answer in a way that makes sense. How many concert tickets can you buy for $15? The answer is 1, not $1\frac{7}{8}$. You cannot buy $\frac{7}{8}$ of a concert ticket, and you do not have enough money to buy 2 tickets.

EXERCISES & PROBLEMS

Communicate

1. Define *variable, expression,* and *equation.*

2. Explain how to write an equation to solve the following problem: Jeff has 3 more apples than Phil, and together they have 9 apples. How many apples does Phil have?

3. Discuss how you would set up the equation for the following problem: If 12 pencils cost $1.92, find the cost of 1 pencil.

4. Tell how you would set up the equation for the following problem: If tickets for a concert cost $10 each, how many can you buy with $35?

5. Explain how you would use the guess-and-check strategy to solve the equation $10x + 3 = 513$.

Practice & Apply

6. Guess and check to solve the equation $3x + 4 = 49$. 15

7. Guess and check to solve the equation $4x + 3 = 49$. 11.5

8. Is $14y$ an expression or a variable? expression

If pencils cost 20 cents each, find the cost of

9. 0 pencils.
   0¢

10. 4 pencils.
    80¢

11. 10 pencils.
    200¢ or $2

12. p pencils.
    $20p$¢

Technology Tell how to generate each sequence on a scientific calculator.

13. 6, 8, 10, 12, 14, . . .

14. 15, 25, 35, 45, 55, . . .

15. 100, 90, 80, 70, 60, . . .

16. 52, 48, 44, 40, 36, . . .

Make a table to show the substitutions of 1, 2, 3, 4, and 5 for the variables in the expressions below.

17. $4x$

18. $5y$

19. $7s + 4$

20. $3n$

17.

x	1	2	3	4	5
$4x$	4	8	12	16	20

18.

y	1	2	3	4	5
$5y$	5	10	15	20	25

19.

s	1	2	3	4	5
$7s + 4$	11	18	25	32	39

20.

n	1	2	3	4	5
$3n$	3	6	9	12	15

Beatrice solved the equation $5x + 7 = 102$ and got 19 for x.

21. Explain what Beatrice did to get this answer.

22. How can you determine if the answer is correct?
    Substitute 19 for x in $5x + 7 = 102$.

23. If tickets for a concert cost $11 each, how many tickets can you buy with $126? 11

24. If tickets for a concert cost $9 each, how many can you buy with $135? 15

25. Consumer Economics Write an equation that models the following situation. How many apples can you buy with 99 cents if apples cost 30 cents each? $30a = 99$

26. Solve the equation you wrote in Exercise 25. Does the exact solution to the equation answer the problem posed about the apples? Explain.

27. Entertainment How many $5 movie tickets can you buy with $32? 6

28. Fund-raising How many $5 raffle tickets must you sell to raise $32? 7

29. If 5 people split the cost of a $32 birthday gift equally, how much should each person pay? $6.40

 Statistics The average number of people in a family in 1995 was 3.18.

30. How many families had exactly 3.18 people? zero

31. How large would a typical family be? 3

 **Look Back**

32. How much of each ingredient should be used if the recipe is tripled?

Date and Nut Bars
1 Cup pecans or walnuts
1/2 Cup dates
3/4 Cup sifted all-purpose flour
3 eggs
1 1/2 Cups brown sugar, firmly packed
3/4 Teaspoon baking powder
1/4 Teaspoon salt

33. A can of a soft drink contains 355 milliliters of liquid. If 5 people split 2 cans evenly, how many milliliters will each get? 142

34. If each pizza is cut into 12 pieces, how many pieces will there be in 4 pizzas? 48

Cultural Connection: Africa Thousands of years ago Diophantus, a great mathematician and Hypatia, the first known woman mathematician, investigated the relationship between square numbers and triangular numbers.

35. Find at least three more examples of how a square number can be represented as the sum of two triangular numbers. **[Lesson 1.1]** $10 + 15 = 25; 15 + 21 = 36; 21 + 28 = 49$

36. Are there any square numbers that cannot be represented as the sum of two triangular numbers? **[Lesson 1.1]** 1

37. In Lesson 1.1 you found that the sum $1 + 2 + 3 + 4 + 5$ is $\frac{5 \cdot 6}{2}$.

Examine the next two cases, the sum of the first 6 numbers and the sum of the first 7 numbers. Make a conjecture about the sum of the first n numbers. **[Lesson 1.1]** Conjecture: $1 + 2 + 3 + \ldots + n = \frac{n(n+1)}{2}$

38. Build a table of values for the equation $y = 4x + 1$ by substituting 1, 2, 3, 4, and 5 for x. **[Lesson 1.2]**

x	1	2	3	4	5
$y = 4x + 1$	5	9	13	17	21

Look Beyond

39. If x^2 means x times x, then guess and check to solve for x in the equation $x^2 = 256$. 16

Exploring Factors and Divisibility Patterns

WHY *Factors give you valuable information about a number. For instance, the factors of 36 determine the possible dimensions of a booth that is 36 square feet.*

Eric rents a booth at a crafts fair. He is allowed to set up 36 square feet in any way that he wants. Eric decides to make his booth 9 feet long and 4 feet wide.

Natural numbers are the counting numbers, such as 1, 2, 3, 4, 5, 6, and so on.

The numbers 4 and 9 are *factors* of 36.

$36 \div 4 = 9$ and $36 \div 9 = 4$

A number is a **factor** of a second number if it divides into the second number evenly, without a remainder.

Exploration 1 Finding Factors

Area = length · width;
9 ft · 4*ft* = 36 square feet

Rectangle dimensions:
  1 ft × 36 ft
  2 ft × 18 ft
  3 ft × 12 ft
  4 ft × 9 ft
  6 ft × 6 ft

1 How do you know from the blueprint that Eric's booth will be 36 square feet?

2 Using only natural numbers, draw pictures showing all the possible plans for a booth that Eric could have. Label the length and the width of each room.

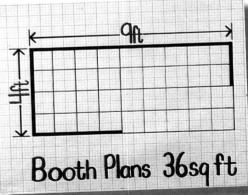

Booth Plans 36sq ft

The blueprint shows Eric's plan for the dimensions of his booth.

3 Copy the table below. Use your pictures from Step 2 to complete the table.

Width, w	Length, l	Area, A
1 ft	36 ft	36 sq ft
2 ft	18 ft	36 sq ft
3 ft	? 12	? 36
4 ft	? 9	? 36
6 ft	? 6	? 36

4 Explain how you can use the dimensions of a rectangle to find its area. Multiply the length times the width.

5 Let *l* represent the **length**, and let **w** represent the **width**. Fill in the blank to complete the equation for the area of a rectangle.

$$A = \underline{?}\ l \cdot w$$

6 Is Eric likely to set up a booth that is 1 foot by 36 feet? Is Eric likely to set up a booth that is 2 feet by 18 feet? Explain. ❖ no   no
Such booth dimensions will not provide enough room for Eric and his crafts.

In Exploration 1 you found the possible dimensions of a rectangular booth with an area of 36 square feet by looking at the factors of 36.

To find all of the factors of any natural number, you can use division. Continue to divide until the factors begin to repeat.

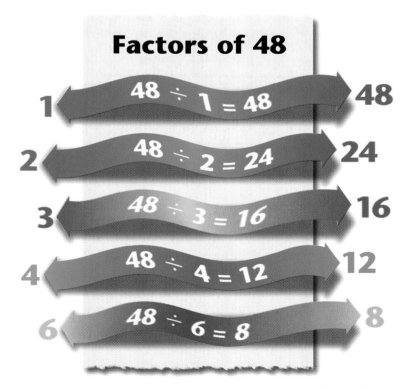

Factors of 48

1	$48 \div 1 = 48$	48
2	$48 \div 2 = 24$	24
3	$48 \div 3 = 16$	16
4	$48 \div 4 = 12$	12
6	$48 \div 6 = 8$	8

·Exploration 2 Primes, Composites, and Squares

There are many number properties that can be discovered by investigating a table of numbers and their factors.

1 Copy and extend the table of natural numbers and their factors. Include all factors for natural numbers from 1 to 20.

Natural number	Factors of that number
1:	1
2:	1, 2
3:	1, 3
4:	1, 2, 4
5:	1, 5
6:	1, 2, 3, 6

2 List all of the numbers with *exactly* 2 factors. These numbers are called *prime numbers*. 2, 3, 5, 7, 11, 13, 17, 19

3 List all of the numbers with *more than* 2 factors. These numbers are called *composite numbers*. 4, 6, 8, 9, 10, 12, 14, 15, 16, 18, 20

4 Explain why the natural number 1 is *neither* a prime nor a composite number. 1 has only a single factor, so it is neither prime (2 factors) nor composite (more than 2 factors).

5 List all of the numbers with an odd number of factors.

For example, 4 has *three* factors: 1, 2, and 4. 1, 4, 9, 16

6 The numbers you found in Step 5 are called *square numbers*. **Explain why 1, 4, 9, 16, . . . are called *square numbers*.** A number of dots equal to a square number can form a square grid.

7 Copy and extend the table below to include 1^2 through 10^2.

Sum of consecutive odd numbers, S	Number of addends, n	Square n^2
1 = 1	1	$1^2 = 1 \cdot 1 = 1$
1 + 3 = _?_ 4	2	$2^2 = 2 \cdot 2 = 4$
1 + 3 + 5 = _?_ 9	3	$3^2 = 3 \cdot 3 = 9$
1 + 3 + 5 + 7 = _?_ 16	4	$4^2 = 4 \cdot 4 = $ _?_ 16
1 + 3 + 5 + 7 + 9 = _?_ 25	5	$5^2 = 5 \cdot 5 = $ _?_ 25

9. The area of a square with side length n is n^2, so $S = n^2$ produces a square number.

8 Describe what the table indicates about adding odd numbers to get square numbers? The sum of the first n odd numbers equals the square number $n \cdot n$, or n^2.

9 Show geometrically why $S = n^2$ describes square numbers. ❖

In Exploration 2, you found three important sets of numbers: *prime* numbers, *composite* numbers, and *square* numbers.

PRIME, COMPOSITE, AND SQUARE NUMBERS

A **prime number** is a natural number with exactly two factors, itself and 1.

A **composite number** is a natural number that is greater than 2 and that is not prime. The natural number 1 is neither prime nor composite.

A **square number** has an odd number of factors and can be represented as a square with an area equal to the square number.

The square numbers 1, 4, 9, and 16 are illustrated above.

•Exploration 3 *Divisibility Patterns*

1 Natural numbers that are *even* are divisible by 2. Explain why 2796 is divisible by 2. The number 2796 is divisible by what other natural numbers less than 10? 2796 is divisible by 2 because it ends in an even digit, 6. It is also divisible by 3, 4, and 6.

2 Copy and complete the table below.

Number	Is the number divisible by 3?	Sum of digits	Is the sum divisible by 3?
24	yes	2 + 4 = 6	yes
36	? yes	3 + 6 = ? 9	? yes
49	no	4 + 9 = ? 13	no
105	? yes	1 + 0 + 5 = ? 6	? yes
156	? yes	1 + 5 + 6 = ? 12	? yes
313	? no	3 + 1 + 3 = ? 7	? no
675	? yes	6 + 7 + 5 = ? 18	? yes
11,100	? yes	1 + 1 + 1 + 0 + 0 = ? 3	? yes

3 How do you think you can use the sum of a number's digits to test for divisibility by 3? If the sum of a number's digits is divisible by 3 then so is the number.

4 If a number is divisible by 6, is it divisible by 2? If a number is divisible by 6, is it divisible by 3? Do you think that if a number is divisible by 2 and by 3, then it is divisible by 6? Explain.

5. If a number ends in 0 or 5, it is divisible by 5. If a number ends in 0, it is divisible by 10.

5 Compare the last digit of each number in the two lists below.

Multiples of 5: 5, 10, 15, 20, 25, . . .
Multiples of 10: 10, 20, 30, 40, 50, . . .

How do you think you can tell if a number is divisible by 5? by 10?

6 Examine the list of multiples of 9. How do you think you can use the sum of a number's digits to test for divisibility by 9?

Multiples of 9: 9, 18, 27, 36, 45, . . .
If the sum of a number's digits are divisible by 9, then so is the number.

CRITICAL *Thinking*

How can you use the rules of divisibility to find all of the factors of 540?

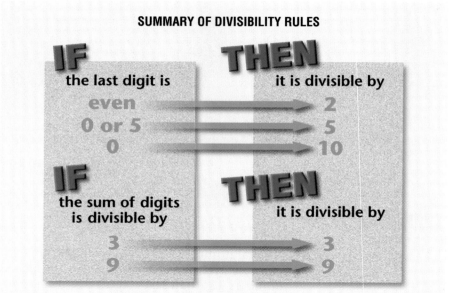

SUMMARY OF DIVISIBILITY RULES

IF the last digit is
even
0 or 5
0

THEN it is divisible by
2
5
10

IF the sum of digits is divisible by

THEN it is divisible by
3
9
3
9

EXTENSION

Determine which numbers of 2, 3, 5, 6, 9, and 10 are factors of 150.

Is 2 a factor? Since 150 is an even number, it is divisible by 2.

Is 3 a factor? Since the sum of the digits in 150 is 6, and 6 is divisible by 3, 150 is divisible by 3.

Is 5 a factor? Since the last digit of 150 is 0, it is divisible by 5.

Is 6 a factor? Since 150 is divisible by 2 *and* 3, it is divisible by 6.

Is 9 a factor? Since the sum of the digits in 150 is 6, and 6 is *not* divisible by 9, 150 is not divisible by 9.

Is 10 a factor? Since the last digit of 150 is 0, it is divisible by 10.

So the numbers 2, 3, 5, 6, and 10 are factors of 150. The number 9 is *not* a factor of 150. ❖

EXERCISES & PROBLEMS

Communicate

1. What are the natural numbers?

2. Explain the difference between a factor and a product.

3. Describe how you can use the areas of rectangles to find the factors of 72.

4. Define *prime number* and *composite number*.

5. How can you tell whether the number 1001 is prime or composite?

6. Describe how you can tell whether 24,570 is divisible by 2, 3, 5, 6, 9, or 10.

Practice & Apply

7. **Geometry** There are 360° in a circle. Make a list of the factors of 360. Why do you think we use 360 for the number of degrees in a circle? 360 is used, most likely, because it has so many factors.

Factors of 360: 1, 2, 3, 4, 5, 6, 8, 9, 10, 12, 15, 18, 20, 24, 30, 36, 40, 45, 60, 72, 90, 120, 180, and 360.

8. Explain why every even natural number greater than 2 has at least three factors. Each even, natural number greater than 2 has a factor of 2, in addition to itself and 1.

9. **Geometry** A family wants to build a house with an area of 2500 square feet. Draw five different rectangular floor plans. Floor plans: 20 ft × 125 ft, 25 ft × 100 ft, 50 ft × 50 ft, 500 ft × 5 ft, and 250 ft × 10 ft

Write a list of factors for each number. Circle the prime factors.

10. 48   **11.** 56   **12.** 72   **13.** 84   **14.** 51

If the number is divisible by 2, 5, and 10, write *yes*. If the number is not divisible by all three of these numbers, write *no*.

15. 235   **16.** 522   **17.** 730   **18.** 2895   **19.** 6234
no   no   yes   no   no

If the number is divisible by 2, 3, and 6, write *yes*. If the number is not divisible by all three of these numbers, write *no*.

20. 1728   **21.** 2500   **22.** 7318   **23.** 27,912   **24.** 60,992
yes   no   no   yes   no

If the number is divisible by 3, 6, and 9, write *yes*. If the number is not divisible by all three of these numbers, write *no*.

25. 108 yes   **26.** 3507 no   **27.** 51,726 no   **28.** 3345 no   **29.** 4266 yes

Determine whether each number is divisible by 15.

30. 205 no   **31.** 195 yes   **32.** 375 yes   **33.** 140 no   **34.** 360 yes

35. Without dividing, determine whether 3 people can share a $1132 prize evenly. Explain your answer. no; 1132 is not divisible by 3.

Determine whether each number is prime, composite or neither.

36. 61
prime

37. 57
composite

38. 185
composite

39. 1
neither

40. 372
composite

41. Find two prime numbers with a sum that is a prime number. $2 + 3 = 5$

42. *Twin primes* are consecutive primes, such as 3 and 5, that have a difference of 2. List three more pairs of *twin primes*. 5 and 7, 11 and 13, 17 and 19

Look Back

43. Find the next four terms in the sequence 1, 5, 9, 13, . . .
[Lesson 1.1] 17, 21, 25, 29

44. Build a table of values for the equation $y = 2x + 1$ for natural-number x-values from 1 to 10. What do you notice about the y-values? **[Lesson 1.2]**

45. If you go the speed limit shown on Route 66, how many miles can you drive in 2.5 hours? **[Lesson 1.3]**
150 mi

Look Beyond

46. **Technology** In Exercise 9, you found five different rectangular floor plans with an area of 2500 square feet. Use the square root key on your calculator, $\boxed{\sqrt{}}$, to find the length of the sides of a square with an area of 2500 square feet. 50 ft

47. Compare the perimeters of the five rectangles from Exercise 9 with the perimeter of the square from Exercise 46. Which perimeter is smallest?
square

why *Exponents are used to write large numbers in science. For example, the distance from the Earth to Saturn is usually written using exponents.*

Scientists use the Hubble Space Telescope to view the planet Saturn, which is 762,700,000 miles away from Earth when it is closest to Earth.

An **exponent** indicates how many times a number, called the base, is used as a factor. In Lesson 1.5, you used 2 as an exponent to write square numbers. For example, $3 \cdot 3$ is 3^2, or *three squared*. That is, the area of a square with sides that are 3 units long is 3^2, or 9, square units.

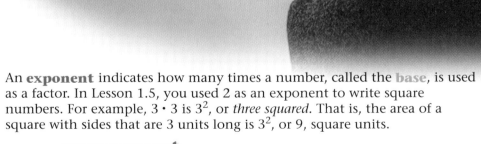

exponent
$$3^2 = 3 \cdot 3 = 9$$
base

Using Exponents

Units of volume are called **cubic units.** A cube with a 1-centimeter edge has a volume of 1 cubic centimeter.

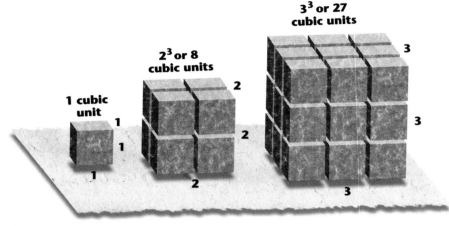

1 cubic unit

2³ or 8 cubic units

3³ or 27 cubic units

You can use exponents to show the volume of cubes. The volume of a cube with an edge of 2 is $2 \cdot 2 \cdot 2$, or 3 factors of 2, and can be written as 2^3. The expression 2^3 is a **power** and is read *two to the third power* or *two cubed.* Explain why the expression *two cubed* is used to mean *two to the third power.*   The expression two cubed comes from finding the volume of a cube with edges of length 2: $2 \cdot 2 \cdot 2 = 2^3$.

Calculator

You can use a calculator to compute powers quickly. Some examples of keys that can be used to compute powers on a calculator include the following:

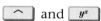

To evaluate 2^3 on some graphics calculators, enter the following keystrokes.

On some scientific calculators, you must use the following keystrokes.

Use exponents to show the volumes of the cubes below. $5^3, 6^3, 7^3$

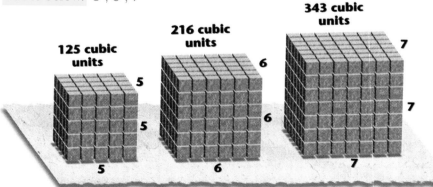

125 cubic units

216 cubic units

343 cubic units

EXAMPLE 1

Calculator

Use the formula $V = e^3$ to find the volume of a cube with edges of **4.5** centimeters.

Solution ➤

Use the formula $V = e^3$ to find the volume of a cube. Substitute **4.5** for e in the formula $V = e^3$.

$$V = e^3$$
$$V = (4.5)^3$$
$$V = \underbrace{(4.5)(4.5)(4.5)}_{}$$

3 factors

Then use your calculator to evaluate $(4.5)^3$ or $(4.5)(4.5)(4.5)$.

$$V = 91.125$$

The volume of the cube with edges of 4.5 centimeters is 91.125 cubic centimeters. ❖

Try This Find the volume of a cube with edges of 5.75 centimeters.
$5.75^3 \approx 190.109$ cubic cm

EXAMPLE 2

Evaluate each power.

Ⓐ 3^4 Ⓑ 2^5

Solution ➤

Ⓐ $$3^4 = \underbrace{(3)(3)(3)(3)}_{} = 81$$

4 factors

Ⓑ $$2^5 = \underbrace{(2)(2)(2)(2)(2)}_{} = 32$$

5 factors   ❖

$10^4 = 10,000$
Try This   Evaluate each power: 10^4 and 2^6. $2^6 = 64$

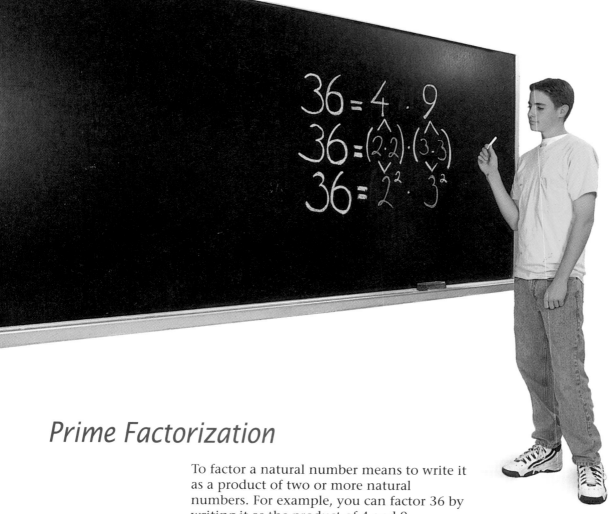

Prime Factorization

To factor a natural number means to write it as a product of two or more natural numbers. For example, you can factor 36 by writing it as the product of 4 and 9.

$$36 = 4 \cdot 9$$

Prime factorization is a process of writing a natural number as a product of only prime numbers. One way to find the prime factorization of 36 is to factor 4 and 9.

$$36 = 4 \cdot 9$$
$$36 = (2 \cdot 2)(3 \cdot 3)$$

Since $2 \cdot 2 \cdot 3 \cdot 3$ contains only prime numbers, it is the prime factorization of 36.

You can write the prime factorization of 36 using exponents. This is called **exponential form.**

$$36 = (2 \cdot 2) \cdot (3 \cdot 3) \longleftarrow \text{prime factorization}$$
$$36 = 2^2 \cdot 3^2 \longleftarrow \text{exponential form}$$

Another way to find the prime factorization of a number is shown in Example 3.

EXAMPLE 3

Write the prime factorization of 120 in exponential form.

Solution ➤

Begin by dividing 120 by the smallest prime number possible. Continue until the quotient is a prime number.

You can write your division process from left to right.

$$120 \div 2 = 60$$

$$60 \div 2 = 30$$

$$30 \div 2 = 15$$

$$15 \div 3 = 5$$

You can also write your division process from top to bottom.

$$2 \overline{)120}$$

$$2 \overline{)60}$$

$$2 \overline{)30}$$

$$3 \overline{)15}$$

$$5$$

The prime factors of 120 are 2, 2, 2, 3, and 5.

$$120 = (2 \cdot 2 \cdot 2) \cdot 3 \cdot 5 \qquad\qquad 120 = 2^3 \cdot 3 \cdot 5$$

The prime factorization of 120 in exponential form is $120 = 2^3 \cdot 3 \cdot 5$. ❖

Try This Write the prime factorization of 378 in exponential form. $2 \cdot 3^3 \cdot 7$

For large numbers you can use a calculator and the divisibility rules to find the prime factorization.

Calculator

EXAMPLE 4

Write the prime factorization of 12,306.

Solution ➤

Since the number is even, first divide by 2. Then continue to divide by the smallest prime number possible until the quotient is a prime number.

12,306 ÷ 2 = 6153 (6153 is divisible by 3.)

6153 ÷ 3 = 2051 (2051 is divisible by 7.)

2051 ÷ 7 = 293 (293 is a prime number.)

So the prime factorization of 12,306 is $2 \cdot 3 \cdot 7 \cdot 293$. ❖

Try This Write the prime factorization of 43,650. $2 \cdot 3^2 \cdot 5^2 \cdot 97$

CRITICAL
Thinking

How can you use the prime factorization of a number to find all the factors of a number? Calculate all possible products of that number's prime factors.

EXERCISES & PROBLEMS

Communicate

1. Explain what exponents are and how they are used.

2. Why is 2^3 not equal to 6? *2^3 equals $2 \cdot 2 \cdot 2$, or 8, not $2 \cdot 3$, or 6.*

3. Describe two ways to find the prime factorization of a number. Give numerical examples.

4. Why is the expression *5 squared* used to mean *5 to the second power?*

5. Explain how you can use your calculator to find the prime factors of a number such as 43,650.

Practice & Apply

Use exponents to rewrite the expression.

6. $5 \times 5 \times 5 \times 5$ 5^4

7. $2 \cdot 2 \cdot 2 \cdot 2 \cdot 3 \cdot 3 \cdot 5$ $2^4 \cdot 3^2 \cdot 5$

8. $x \cdot x \cdot x \cdot y \cdot y$ $x^3 y^2$

9. $x \cdot x + 3$ $x^2 + 3$

Evaluate each power.

10. 2^4 16

11. 2^5 32

12. 2^6 64

13. 10^3 1000

14. 10^6 1,000,000

15. 8^4 4096

Use the formula $V = e^3$ to find the volume of a cube with the following edges.

16. 6 inches 216 cu in.

17. 4 centimeters 64 cu cm

18. 3.2 meters 32.768 cu m

19. 1.8 yards 5.832 cu yd

Write the prime factorization for each number.

20. 54 $2 \cdot 3^3$

21. 100 $2^2 \cdot 5^2$

22. 39 $3 \cdot 13$

23. 240 $2^4 \cdot 3 \cdot 5$

24. 399 $3 \cdot 7 \cdot 19$

25. 63 $3^2 \cdot 7$

26. 121 11^2

27. 200 $2^3 \cdot 5^2$

28. 91 $7 \cdot 13$

29. 101 101

30. 136 $2^3 \cdot 17$

31. 82 $2 \cdot 41$

32. The diagram below the photograph of Apollo 13 indicates the average distance from Earth to the Moon. Write the prime factorization of this number in exponential form. $2927 \cdot 3^4$

237,087 miles

Earth

Evaluate each expression if *a* is 2, *b* is 3, and *c* is 5.

33. abc 30   **34.** a^2bc 60   **35.** a^2b^3c 540   **36.** ab^3c^2 1350

Technology **Use your calculator to find the prime factorization of each number. Write your answers in exponential form.**

37. 32,292 $2^2 \cdot 3^3 \cdot 13 \cdot 23$ **38.** 45,220 $2^2 \cdot 5 \cdot 7 \cdot 17 \cdot 19$

39. 83,571 $3 \cdot 89 \cdot 313$ **40.** 48,364 $2^2 \cdot 107 \cdot 113$

41. Science The approximate age of the Earth is 4,500,000,000 years. Write the prime factorization of this number in exponential form. $2^8 \cdot 3^2 \cdot 5^9$

42. Science The Sun is about 150,000,000 kilometers from the Earth. Write the prime factorization of this number in exponential form. $2^7 \cdot 3 \cdot 5^8$

43. Science The temperature of the Sun is about 6000 degrees Celsius. Write the prime factorization of this number in exponential form. $2^4 \cdot 3 \cdot 5^3$

44. What is the smallest number whose prime factorization consists of seven different prime numbers?
510,510 = $2 \cdot 3 \cdot 5 \cdot 7 \cdot 11 \cdot 13 \cdot 17$

Look Back

45. Find the next four terms in the following sequence. **[Lesson 1.1]**

2, 5, 4, 7, 6, 9, 8, 11, 10, 13, 12

Use the following table for Exercises 46–48.

x	1	2	3	4	5	10	100
y	2	7	12	17	?	?	?
					22	47	497

46. Copy and complete the table. **[Lesson 1.2]**

47. Make a bar chart to represent the data in the table. **[Lesson 1.3]**

Look Beyond

48. Is the expression 67 + 53 equal to 53 + 67? yes
Is the expression 6 · 12 equal to 12 · 6? yes
Why or why not? For both addition and multiplication, the order does not matter. We say that they are both commutative.

Moon

LESSON 1.7

Order of Operations

why *Calculators and computers have dramatically changed the way people work with numbers. New technology has made computation and mathematical exploration easier and more exciting. Even without technology, you need to understand the order for performing operations to get accurate results.*

Mary plans to buy carpet for two rooms. One room is 12 feet by 15 feet, and the other room is 20 feet by 14 feet. She uses her calculator to find the number of square feet.

12 [×] 15 [+] 20 [×] 14 [=] | 2800 |

Mary knows that 2800 square feet is way too much. Her entire house is less than 2000 square feet. But calculators don't make mistakes. What is wrong?

CRITICAL *Thinking*

CT– The correct answer is 460. Mary's calculator added 20 to 12 · 15, and then multiplied that result by 14.

What is the correct answer for Mary's computation? What steps did her calculator use to get 2800?

To avoid misunderstandings and errors, mathematicians have agreed on certain rules for computation called the *order of operations*. One rule requires that multiplication be performed before addition. By following this rule, Mary can make the correct calculation.

$$12 \times 15 + 20 \times 14 =$$
$$180 + 280 =$$
$$460 \text{ square feet}$$

The Order of Operations

1. Perform all operations enclosed in symbols of inclusion (parentheses, brackets, braces, and bars) from innermost outward.
2. Perform all operations with exponents.
3. Perform all multiplications and divisions in order from left to right.
4. Perform all additions and subtractions in order from left to right.

Calculators that follow the order of operations use **algebraic logic.** You can use a computation like $2 + 3 \cdot 4$ to test for algebraic logic.

Key in 2 [+] 3 [×] 4 [=]. If the answer is 14, the calculator uses algebraic logic. On some calculators you might get 20. These calculators do not use algebraic logic, but you can use parentheses to get the correct answer.

2 [+] [( ] 3 [×] 4 [ )] [=]

You can also enter the multiplication first.

3 [×] 4 [+] 2 [=]

The examples you will see in this book assume that you have a calculator with algebraic logic.

EXAMPLE 1

How do you use a calculator with algebraic logic to work Mary's problem?

Graphics Calculator

Solution ➤
Key in the numbers and operations just as they appear. 12 [×] 15 [+] 20 [×] 14 [ENTER].
The answer is 460. ❖

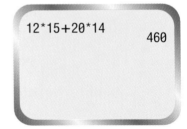

Using Inclusion Symbols

Symbols like parentheses, ( ), brackets, [ ], braces, { }, and the fraction bar are called **symbols of inclusion.** These symbols group numbers and variables. Treat any grouped numbers and variables as a single quantity. Operations should always be done within the innermost symbols of inclusion first. Then work outward.

EXAMPLE 2

Insert inclusion symbols to make $30 + 4 \div 2 - 1 = 16$ true.

Solution ➤

Use parentheses to group 30 and 4 before dividing by 2. Begin with the operations in the innermost symbols of inclusion.

$$[(30 + 4) \div 2] - 1 \overset{?}{=} 16$$
$$[34 \div 2] - 1 \overset{?}{=} 16$$
$$17 - 1 \overset{?}{=} 16$$
$$16 = 16 \qquad \text{True} ❖$$

Try This Insert inclusion symbols to make $5 + 30 \div 7 - 4 = 15$ true.
$5 + [30 \div (7 - 4)] = 15$

In algebra you are often asked to *evaluate* an expression. To do this, you replace the variables with the numbers that are assigned to those variables. Then proceed with the computation using the order of operations.

EXAMPLE 3

Evaluate $5x^2 + 7y$ when x is 3 and y is 2.

Solution ➤

Replace x with 3 and y with 2. $5x^2 + 7y$
First, square the 3. $5 \cdot 3^2 + 7 \cdot 2$
Perform all of the multiplications. $5 \cdot 9 + 7 \cdot 2$
Finally, add the results. $45 + 14$
$45 + 14 = 59$ ❖

EXAMPLE 4

Show the keystrokes and the answer for $\dfrac{57 + 95}{16} + \dfrac{220}{88 + 104}$.

Solution ➤

Since the entire quantity $57 + 95$ is divided by 16, place parentheses around $57 + 95$. Do the same for $88 + 104$.

$$(57 + 95) \div 16 + 220 \div (88 + 104)$$

The keystrokes are shown.

The answer is 10.646 to 3 decimal places. Is this reasonable? The first fraction is about $\frac{160}{16}$, or 10. The second fraction is about $\frac{200}{200}$, or 1. Thus, the estimated sum is about $10 + 1$, or 11. The answer 10.646 is reasonable. ❖

Scientific Calculator

EXAMPLE 5

On Monday, John's father borrowed $3 and said that he would repay John double the amount plus $1 on Friday. The next Monday, John's father borrowed the amount he gave John the previous Friday and said that he would repay him double that amount plus $1 on Friday. The next week he did the same. On the third Friday, John figured his father owed him 2{2[2(3) + 1] + 1} + 1 dollars. How much money should John receive that Friday?

Solution ➤

When you simplify an expression that contains several pairs of inclusion symbols, begin with the innermost pair and work outward.

$$2\{2[2(3) + 1] + 1\} + 1$$
$$2\{2[7] + 1\} + 1$$
$$2\{15\} + 1$$
$$31$$

If you use a calculator or a computer, the expression will usually contain only parentheses. The example would appear as 2(2(2(3) + 1) + 1) + 1. This does not affect the way you simplify the expression. ❖

If this pattern continues for one more week, how much money will John get on the fourth Friday? $63

Pattern Exploration and Technology

Calculator

Complete the pattern using a calculator.

1111 · 1111 = 1234321   1 · 1 = 1
11111 · 11111 = 123454321   11 · 11 = ? 121
111111 · 111111 = 12345654321 111 · 111 = ? 12321

Predict the next three numbers in the pattern. Check your prediction with your calculator. What happens when you try to compute 111111 · 111111? Continue developing the pattern. What happens as the number of ones increases? What are the limitations of using a calculator to explore this pattern?

EXERCISES & PROBLEMS

Communicate

1. Explain how $3 + 2 \cdot 4$ can give two different answers.
2. What two possible answers might you get for $20 \div 2 \cdot 5$?
3. Which answer to $20 \div 2 \cdot 5$ is correct? Why?
4. Describe the order of the steps for simplifying $\{[3(8 - 4)]^2 - 6\} \div (4 - 2)$.
5. Explain why rules called *the order of operations* are necessary for computation.
6. Explain why 3 is a reasonable estimate for $\frac{173 + 223}{151 - 21}$.

Practice & Apply

Place inclusion symbols according to the correct order of operations to make each equation true. Tell how you would use a calculator to check your answers.

7. $(28 - 2) \cdot 0 = 0$　　　　**8.** $59 - 4 \cdot (6 - 4) = 51$

Simplify using the correct order of operations. If answers are not exact, round them to three decimal places.

9. $57 \cdot 29 + 89$　1742　　**10.** $72(98) + 12$　7068　　**11.** $89 + 57 \cdot 29$　1742

12. $3(15) + 9$　54　　**13.** $43 \cdot 32 + 91 \cdot 67$　7473　　**14.** $45(75) + 9(24)$　3591

15. $\frac{28 + 59}{97 - 17}$　1.0875　　**16.** $\frac{97 - 17}{72 + 7}$　1.013　　**17.** $\frac{43 \cdot 91}{8 \cdot 25}$　19.565

18. $157 - 29 + 23 \cdot 9$　335　　**19.** $91 \div 7 + 6$　19　　**20.** $187 - 34 \div 17$　185

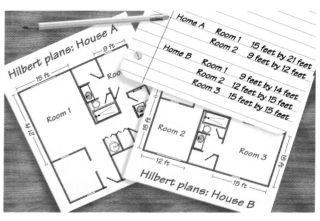

Home Improvement Mr. Hilbert owns two rental properties, House A and House B. He wants to carpet two rooms in House A and three rooms in House B. He finds a carpet store selling carpet for $12.99 per square yard. The dimensions of the rooms in his homes are shown on the blueprint. HINT: $1 \text{yd}^2 = 9 \text{ft}^2$

21. Find the total number of square feet of carpet needed for House A.　423 sq ft

22. Find the total number of square feet of carpet needed for House B.　531 sq ft

23. Find the cost of carpeting the rooms in House A.　$610.53

24. Find the cost of carpeting the rooms in House B.　$766.41

Evaluate each expression.

25. $2(5 + 4) \div 9$ 2

26. $12 - 7 \cdot 3 + 9^2$ 72

27. $3 - 1 + 24 \div 6$ 6

28. $7 + 6 \div 2 \cdot 10$ 37

Given that a is 5, b is 3, and c is 4, evaluate each expression.

29. $a + b - c$ 4

30. $a^2 + b^2$ 34

31. $a^2 - b^2$ 16

32. $(a + b) \cdot c$ 32

33. $a^2 - b - c$ 18

34. $a^2 - (b + c)$ 18

35. Technology Using a scientific calculator, what answer do you get if you

enter 32 ⟦ × ⟧ 38 ⟦ ÷ ⟧ 11 ⟦ × ⟧ 19 ⟦ = ⟧ ? 2100.3636

Explain how the calculator gets this answer. Perform operations from left to right.

36. **Statistics**
The teacher's grade
book shows the results
of an algebra quiz. Find
the average for the class.
Show your method.

87.5

$$\frac{100 \cdot 4 + 90 \cdot 12 + 80 \cdot 7 + 70 \cdot 0 + 60 \cdot 1}{24}$$

Look Back

37. Advertising A photographer arranges cans of soup in a large triangle for
a supermarket ad. The top row contains 1 can, and each of the rows contains
1 can more than the row above it. If there are 10 rows of cans, how many cans
will the photographer need to form the triangle display? **[Lesson 1.1]** 55

Find the next two terms of each sequence. [Lesson 1.1]

38. 2, 5, 8, 11, 14, . . . 17, 20 **39.** 59, 54, 49, 44, 39, . . . 34, 29 **40.** 3, 6, 12, 24, 48, . . . 96, 192

Look Beyond

41. **Geometry** The perimeter of a rectangle is the distance around the border,
or the sum of twice the length and twice the width. The area is the product of
the length and width. For what length and width will the perimeter and the area
be the same number? 3 by 6 or 4 by 4

Recall that an exponent indicates repeated multiplication.
Indicate whether each equation is true or false.

42. $(3 + 4)^2 = 3 + 4^2$ F

43. $(3 + 4)^2 = (3 + 4)(3 + 4)$ T

44. $(3 + 4)^2 = 3^2 + 2(3)(4) + 4^2$ T

45. $(3 + 4)^2 = 3^2 + 4^2$ F

LESSON 1.8
Exploring Properties and Mental Computation

why
Mental computation is an important skill. Applying number properties makes it easier to compute and estimate quickly without a calculator.

Felicia works in the summer at the Excellent Adventures river-tube rental. Each morning she checks the inventory of tubes stored in three sheds. Felicia counts 44, 27, and 56 rafts. She quickly computes the sum 127 mentally. How do you think Felicia used mental math to compute the sum 44 + 27 + 56?

The Commutative and Associative Properties

When you are using mental computations, it is helpful to know about patterns that occur in number operations or in number properties.

•Exploration 1 *The Commutative Property*

1 Complete the tables by continuing the pattern.

Sums	
Column A	**Column B**
56 + 11	11 + 56
28 + 32	32 + 28
13 + 87	87 + 13
29 + 21 ?	21 + 29
52 + 18	? 18 + 52
11 + 49 ?	49 + 11

Products	
Column C	**Column D**
6 · 11	11 · 6
8 · 9	9 · 8
10 · 54	54 · 10
9 · 20	? 20·9
12·4 ?	4 · 12
16 · 7	? 7·16

2 Compare the sums in Columns A and B. How are they alike and how are they different? The sums are written in the opposite order but have the same value.

3 Compare the products in Columns C and D. How are they alike and how are they different? The products are written in the opposite order but have the same value.

4 Fill in the blank for the generalization about the pattern in the Sums table.

added
Any two numbers can be ? in either order.

5 Fill in the blank for the generalization about the pattern in the Products table.

multiplied
Any two numbers can be ? in either order. ❖

CRITICAL Thinking Do you think that any two numbers can be *subtracted* in either order? Do you think that any two numbers can be *divided* in either order? Explain. No; no. For example $7 - 3 = 4$, but $3 - 7 = -4$. Similarly, $12 \div 6 = 2$, but $6 \div 12 = \frac{1}{2}$.

Exploration 2 *The Associative Property*

1 Complete the following tables by continuing the pattern.

Sums		Products	
Column A	**Column B**	**Column C**	**Column D**
$(12 + 8) + 15$	$12 + (8 + 15)$	$(2 \cdot 5) \cdot 16$	$2 \cdot (5 \cdot 16)$
$(39 + 4) + 16$	$39 + (4 + 16)$	$(20 \cdot 5) \cdot 13$	$20 \cdot (5 \cdot 13)$
$(24 + 36) + 7$	$24 + (36 + 7)$	$(3 \cdot 11) \cdot 10$	$3 \cdot (11 \cdot 10)$
$(11 + 28) + 12$?	$11 + (28 + 12)$	$(5 \cdot 4) \cdot 12$	$5 \cdot (4 \cdot 12)$
$(15 + 5) + 18$	$15 + (5 + 18)$?	$(8 \cdot 6) \cdot 5$?	$8 \cdot (6 \cdot 5)$
$(33 + 7) + 25$?	$33 + (7 + 25)$	$(8 \cdot 11) \cdot 10$	? $8 \cdot (11 \cdot 10)$

2 Find and compare the sums in Columns A and B. How are they alike and how are they different? The sums are grouped differently but evaluate to the same numbers.

3 Find and compare the products in Columns C and D. How are they alike and how are they different? The products are grouped differently but evaluate to the same numbers.

4 Fill in the blank for the generalization about the pattern in the Sums table.

addition
The way numbers are grouped for ? does not affect the sum.

5 Fill in the blank for the generalization about the pattern in the Products table.

multiplication
The way numbers are grouped for ? does not affect the product. ❖

CT– Yes; yes. For example $(15 - 6) - 2 = 7$, but $15 - (6 - 2) = 11$. Similarly, $(16 \div 4) \div 2 = 2$, but $16 \div (4 \div 2) = 8$.

CRITICAL Thinking Do you think that the way numbers are grouped for *subtraction* affects the difference? Do you think that the way numbers are grouped for *division* affects the quotient? Explain.

You can use the commutative and associative properties to regroup expressions for quick mental computation.

APPLICATION

At the beginning of the lesson, Felicia used mental computation to check the inventory of tubes stored in three sheds. You can use the commutative and associative properties to compute the total number of rafts mentally.

$$(44 + 27) + 56$$
$$= (27 + 44) + 56 \quad \text{Commutative Property of Addition}$$
$$= 27 + (44 + 56) \quad \text{Associative Property of Addition}$$
$$= 27 + 100$$
$$= 127$$

You can also use the commutative and associative properties to compute products mentally.

$$4 \cdot (27 \cdot 25)$$
$$= 4 \cdot (25 \cdot 27) \quad \text{Commutative Property of Multiplication}$$
$$= (4 \cdot 25) \cdot 27 \quad \text{Associative Property of Multiplication}$$
$$= 100 \cdot 27$$
$$= 2700 \quad ❖$$

The Distributive Property and Mental Computation

The *Distributive Property* is another common number property.

•Exploration 3 *The Distributive Property*

 Complete the table by continuing the pattern.

Column A	Column B
18(5 + 4)	18(5) + 18(4)
9(15 − 7)	9(15) − 9(7)
14(6 + 12)	14(6) + 14(12)
5(12 − 8) ?	5(12) − 5(8)
8(15 + 2)	8(15) +?8(2)
? 7(13 − 11)	7(13) − 7(11)

2. The two columns evaluate the same number, but in Column A, add or subtract first and then multiply. In Column B, the products are found first and the sum or difference second.

3. The product of a sum is equal to the sum of products. The product of a difference is equal to the difference of products.

2 Examine the table. Compare the expressions in Columns A and B. How are they alike and how are they different?

3 Make a generalization about this pattern. ❖

Exploration 4 *Mental Computation*

Examine the table below, reading each row from left to right. Copy and complete the table.

	Column A	Column B	Column C	Column D
1	4(15)	4(10 + 5)	4 · 10 + 4 · 5	40 + 20 = 60
2	8(17)	8(10 + 7)	8 · 10 + 8 · 7	80 + 56 $\underset{?}{=}$ 136
3	5(22)	5(10 + 10 + 2)	5·10 + 5·10 + 5·2 $\overset{?}{}$	50 + 50 + 10 = 110 $\overset{?}{}$
4	12(25)	(10 + 2)25	10 · 25 + 2 · 25	250 + 50 = 300 $\overset{?}{}$
5	21(7)	(20 + 1)7	20·7 + 1·7 $\overset{?}{}$	140 + 7 = 147 $\overset{?}{}$
6	6(128)	6 (100 + 20 + 8) $\overset{?}{}$	6·100 + 6·20 + 6·8 $\overset{?}{}$	600 + 120 + 48 = 768 $\overset{?}{}$

7. In the table, "large" numbers are written as sums of smaller, easier numbers such as 10 or 7. The multiplication is then distributed across the sum, and the products are added together.

7 Describe the process shown in each row from left to right.

8 Explain how the Distributive Property can be used in mental computation of products. ❖ It allows the multiplication of easy numbers, such as 20 and 8, rather than more difficult numbers, such as 28.

APPLICATION

Bill and Tracy rent tubes to go down the river on the weekend. On Saturday they rent the tubes for 6 hours, and on Sunday they rent the tubes for 5 hours. To find the total rental fee that each pays for the entire weekend, multiply the sum of hours that a tube is rented by the hourly rate.

$$2(6 + 5) = 2 \cdot 11$$
$$= 22$$

So the tube rental will cost Bill and Tracy $22.00 each for the weekend.

You can also find the total rental fees for the weekend by first multiplying the hourly rate for tube rental by 6 hours and by 5 hours. Then add the two days' rental fee for the weekend's total.

$$2(6 + 5) = 2 \cdot 6 + 2 \cdot 5$$
$$= 12 + 10$$
$$= 22$$

The second way of finding the total rental fees illustrates the *Distributive Property of Multiplication Over Addition.* ❖

SUMMARY

The **Number Properties** you explored in this lesson are given below.

For any numbers *a* and *b*, the **Commutative Property** states that
$$a + b = b + a \text{ and } a \cdot b = b \cdot a.$$

For any numbers *a*, *b*, and *c*, the **Associative Property** states that
$$(a + b) + c = a + (b + c) \text{ and } (a \cdot b) \cdot c = a \cdot (b \cdot c).$$

For any numbers *a*, *b*, and *c*, the **Distributive Property of Multiplication Over Addition** states that
$$a(b + c) = ab + ac \text{ and } (b + c)a = ba + ca.$$

For any numbers *a*, *b*, and *c*, the **Distributive Property of Multiplication Over Subtraction** states that
$$a(b - c) = ab - ac \text{ and } (b - c)a = ba - ca.$$

EXERCISES & PROBLEMS

Communicate

1. Give examples of how you can use the commutative and associative properties to help you compute mentally.

2. Explain how to use mental math and the distributive property to find each product.
   a. 8(21)   **b.** 11(35)   **c.** 14(22)

3. Refer to the Application on page 59. The next weekend Janis and Kayla rent tubes for 5 hours on Saturday and 3 hours on Sunday. Explain how to use the distributive property to compute the total weekend rental fee for each.

4. Explain how the Distributive Property of Multiplication Over Subtraction can be used in the real world.

Practice & Apply

Complete each step, and name the property used.

5. (24 + 27) + 56
= (27 + _?_) + 56   24   Commutative Property
= 27 + (24 + _?_)   56   _?_ Property   Assoc
= 27 + _?_   80
= _?_   107

6. 25 · (27 · 4)
= 25 · (_?_ · 27)   4   _?_ Property   Comm
= (_?_ · 4) · 27   25   _?_ Property   Assoc
= _?_ · _?_   100·27
= _?_   2700

7. 25(2 + 4)   2, 25
= 25 · _?_ + _?_ · 4   _?_ Property   Dist
= _?_ + _?_   50 + 100
= _?_   150

Use mental math to find each sum or product. Show your work and explain each step.

8. (27 + 98) + 73   **9.** (45 · 32) · 0

10. (87 · 5) · 2   **11.** 50 · (118 · 20)

12. (688 + 915) + 312   **13.** (25 · 78) · 4

14. 2 · (129 · 5)   **15.** (133 + 52) + 67

Name the property illustrated.

16. 32 + 17 = 17 + 32   Comm

17. 13 · 21 + 13 · 9 = 13(21 + 9)   Dist

18. 6(4.7 − 2) = 6(4.7) − 6(2)   Dist

19. 4(5x) = (4 · 5)x   Assoc

20. 8.2(2 + 5.3) = (2 + 5.3)8.2   Comm

21. (6 − 3)5 = 6 · 5 − 3 · 5   Dist

22. Wages Phil makes $9.50 per hour. He works 7 hours on Monday and 8 hours on Tuesday. Show two methods to compute his total wages.

23. Wages Marcia earns $15 an hour tutoring math students. She tutored 4 hours on Friday and 6 hours on Saturday. Using two different methods, find the total Marcia earned for tutoring on Friday and Saturday.

24. Sales At a music store, 876 CDs are sold during the holiday sale. Explain how you can use the Distributive Property to compute the total mentally.

Entertainment A skating rink charges $8 per person. For a group of 30 or more people, the rink charges $6 per person.

25. On Saturday, 25 members of the French Club go skating. Find the total cost for skating, using two different methods. $200

26. The next weekend, a group of 49 students go skating. Find the total cost for skating, using two different methods. $294

Use the Distributive Property and mental math to complete each product.

27. $4 \cdot 28 = 4(20 + 8)$
$ = 4 \cdot 20 + 4 \cdot 8$
$ = 80 + 32$
$ = \underline{\ ?\ }$ 112

28. $40 \cdot 28 = 40(20 + 8)$
20 $ = 40 \cdot \underline{\ ?\ } + 40 \cdot \underline{\ ?\ }$ 8
$ = \underline{\ ?\ } + \underline{\ ?\ }$ 800 + 320
$ = \underline{\ ?\ }$ 1120

29. $9 \cdot 680 = 9(600 + 80)$
9 $ = \underline{\ ?\ } \cdot 600 + \underline{\ ?\ } \cdot 80$ 9
$ = \underline{\ ?\ } + \underline{\ ?\ }$ 5400 + 720
$ = \underline{\ ?\ }$ 6120

30. $90 \cdot 680 = 90(600 + 80)$ $90 \cdot 600 + 90 \cdot 80$
$ = \underline{\ ?\ } \cdot \underline{\ ?\ } + \underline{\ ?\ } \cdot \underline{\ ?\ }$
$ = \underline{\ ?\ } + \underline{\ ?\ }$ 54,000 + 7200
$ = \underline{\ ?\ }$ 61,200

31. $95 \cdot 99 = 95(100 - 1)$
$ = \underline{\ ?\ } \cdot \underline{\ ?\ } - \underline{\ ?\ } \cdot \underline{\ ?\ }$ $95 \cdot 100 - 95 \cdot 1$
$ = \underline{\ ?\ } - \underline{\ ?\ }$ 9500 - 95
$ = \underline{\ ?\ }$ 9405

32. $68 \cdot 70 = (70 - 2) \cdot 70$
$ = \underline{\ ?\ } \cdot \underline{\ ?\ } - \underline{\ ?\ } \cdot \underline{\ ?\ }$
$ = \underline{\ ?\ } - \underline{\ ?\ }$
$ = \underline{\ ?\ }$

32.
$70 \cdot 70 - 2 \cdot 70$
$4900 - 140$
4760

33. Wages Janie works 5 hours on Saturday and 4.5 hours on Sunday. She earns $12 an hour. Use the Distributive Property to compute Janie's total wages for the weekend. Use the order of operations to check the answer. $114

34. Inventory Sally works at a bakery. Each morning she has to total all of the baked goods that are delivered. How can Sally use mental math to total the baked goods shown below?
$188 = (53 + 21 + 36) + (26 + 24) + 28$

35. Give an example of the Distributive Property of Multiplication Over Subtraction. $7(100 - 2) = 7 \cdot 100 - 7 \cdot 2 = 686$
Answers may vary.

Use the Distributive Property to rewrite each expression.

36. $xy + wy$
$(x + w)y$

37. $rs + rq$
$r(s + q)$

38. $9xy + 21xy$
$(3 + 7)\,3xy$

 *Look Back*

Make a table of values by substituting 1, 2, 3, 4, 5, and 10 for x. Make a bar chart from the data in your table. [Lesson 1.3]

39. $y = 3x$

40. $y = 2x - 1$

41. Charlie's recording studio charges bands $100, plus $45 per hour of recording. Write an equation for the total charge, c, based on the number of hours, h, of recording. Make a table of charges for hours 1 through 8. **[Lesson 1.4]**

42. Write the prime factorization of 360. **[Lesson 1.6]** $2^3 \cdot 3^2 \cdot 5$

43. Find the volume of a cube with a 6-centimeter edge. **[Lesson 1.6]** $6^3 = 216$ cu cm

41.

h	$c = 100 + 45h$
1	$145
2	$190
3	$235
4	$280
5	$325
6	$370
7	$415
8	$460

Look Beyond

Use the Distributive Property to complete each product.

44. $25 \cdot 76 = (20 + 5) \cdot (70 + 6)$
$= 20 \cdot (70 + 6) + 5 \cdot (70 + 6)$
$= 20 \cdot 70 + 20 \cdot 6 + 5 \cdot 70 + 5 \cdot 6$
$= \underline{?} + \underline{?} + \underline{?} + \underline{?}$ $1400 + 120 + 350 + 30$
$= \underline{?}$ 1900

45. $26 \cdot 34 = (30 - 4) \cdot (30 + 4)$
$= (30 - 4) \cdot 30 + (30 - 4) \cdot 4$
$= 30 \cdot 30 - 4 \cdot 30 + 30 \cdot 4 - 4 \cdot 4$
$= \underline{?} - \underline{?} + \underline{?} - \underline{?}$ $900 - 120 + 120 - 16$
$= \underline{?}$ 884

Patterns in numbers are sometimes closely related to patterns in art. The work of a quilt maker, for example, is filled with mathematically intricate designs. This project will give you a method for creating artistic patterns from numbers.

Many patterns can be created by using the remainders from dividing numbers. Divide the whole numbers from 0 to 9 by 4, and examine the remainders.

Repeating Patterns

Number	0	1	2	3	4	5	6	7	8	9
Remainder	0	1	2	3	0	1	2	3	0	1

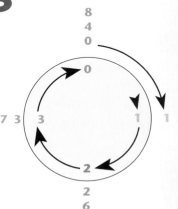

Notice how the remainders repeat. A pattern that repeats is sometimes called a cycle. This is like a clock that cycles through the hours from 1 to 12 and back to 1. In this case, the cycle goes from 0 to 3 before repeating. For example, 4 brings you back to 0; 5 brings you to 1; and so on.

Building a Design Table

In the left and top margins on a piece of graph paper, place the whole numbers from 0 to 9. Make a standard multiplication table. On another piece of graph paper, make a table of the remainders when each product is divided by 4.

Standard Multiplication Table

X	0	1	2	3	4	5	6	7	8	9
0	0	0	0	0	0	0	0	0	0	0
1	0	1	2	3	4	5	6	7	8	9
2	0	2	4	6	8	10	12	14	16	18
3	0	3	6	9	12	15	18	21	24	27
4	0	4	8	12	16	20	24	28	32	36
5	0	5	10	15	20	25	30	35	40	45
6	0	6	12	18	24	30	36	42	48	54
7	0	7	14	21	28	35	42	49	56	63
8	0	8	16	24	32	40	48	56	64	72
9	0	9	18	27	36	45	54	63	72	81

Remainder Table

X	0	1	2	3	4	5	6	7	8	9
0	0	0	0	0	0	0	0	0	0	0
1	0	1	2	3	0	1	2	3	0	1
2	0	2	0	2	0	2	0	2	0	2
3	0	3	2	1	0	3	2	1	0	3
4	0	0	0	0	0	0	0	0	0	0
5	0	1	2	3	0	1	2	3	0	1
6	0	2	0	2	0	2	0	2	0	2
7	0	3	2	1	0	3	2	1	0	3
8	0	0	0	0	0	0	0	0	0	0
9	0	1	2	3	0	1	2	3	0	1

Creating the Designs

1. Choose a color for each number in the remainder table.

2. Create an artistic design for each number in the remainder table.

3. On one grid, replace the numbers with the colored squares. On the second grid replace the numbers with the artistic designs.

0 1 2 3

0 1 2 3

Activity 1

Create a pattern graph using the remainders from division by 5. Be creative in your choice of designs to replace the remainders. Extend your table beyond nine numbers for the left and top margins to enlarge the pattern.

Activity 2

On another piece of paper, create a different pattern. Use an operation other than multiplication. Try addition or some combination of operations. Use remainders from division by a number greater than 5. Vary the colors. Be creative.

Activity 3

Compare the patterns. See what conclusions you can draw from comparing the patterns for addition with the patterns for multiplication. How do the patterns compare when you use more numbers? What other interesting information can you find from the patterns?

Chapter 1 Review

Vocabulary

algebraic expression	17	Distributive Property	60	prime factorization	46
algebraic logic	51	equation	31	prime number	39
Associative Property	60	exponent	43	rectangular numbers	11
base of an exponent	43	factor	36	square number	39
Commutative Property	60	natural numbers	36	symbols of inclusion	51
composite number	39	order of operations	51	variable	17
conjecture	10				

Key Skills & Exercises

Lesson 1.1

➤ **Key Skills**

Generate the terms of a sequence.

Examine the sequence 6, 15, 24, 33, 42, . . . The first term is 6. Each term is 9 more than the term before. Add 9 to find the next three terms.

$$42 + 9 = 51 \qquad 51 + 9 = 60 \qquad 60 + 9 = 69$$

➤ **Exercises**

Find the next three terms of the sequence. Then explain the pattern used to find the terms.

1. 1, 4, 7, 10, 13, . . .
16, 19, 22
$t_n = t_{n-1} + 3$

2. 1, 4, 16, 64, 256, . . .
1024; 4096; 16,384
$t_n = 4t_{n-1}$

3. 27, 9, 3, 1, $\frac{1}{3}$, . . . $\frac{1}{9}, \frac{1}{27}, \frac{1}{81}$
$t_n = \frac{1}{3}t_{n-1}$

Lesson 1.2

➤ **Key Skills**

Solve problems with two operations.

The total charge for in-line skate rental is *a base charge of $6, plus $3 for each hour*. To find the total charge for 1, 2, and 3 hours, make a table.

In-Line Skate Rentals

Number of hours	Process: $6 base charge plus $3 for each hour	Charge
1	6 + 3(1)	$9
2	6 + 3(2)	$12
3	6 + 3(3)	$15

Write an equation for a linear relationship.

Write an equation for the charge of in-line skate rentals using the variables *c* and *h*.

charge equals $6 base charge plus $3 for each hour

$$c \quad = \quad 6 \qquad + \qquad 3h$$

► **Exercises**

The charge for bicycle rental is $8, plus $2 per hour.

4. Copy and complete the table. Write an equation for the rental charge. $c = 8 + 2h$

5. What is the charge to rent a bicycle for 6 hours? $20

6. How long can you rent a bicycle for $26? 9 hr

Bicycle Rental

Number of hours, h	Process	Charge, c
1	8 + 2(1)	? 10
2	8 + 2(2)	? 12
3	3 8 + 2(?)	? 14
5	8 + 2(5)?	? 18
10	8 + 2(10)?	? 28

Lesson 1.3

► **Key Skills**

Use tables and bar charts to solve a problem.

The charge for renting a camcorder is $10, plus $15 per day.

Camcorder Rental

Number of days, d	Process: 10 + 15d	Charge, c
1	10 + 15(1)	25
2	10 + 15(2)	40
3	10 + 15(3)	55
4	10 + 15(4)	70

► **Exercises**

An electronics store pays Felix a base salary of $200 per week, plus $40 for each VCR that he sells. If x represents the number of VCRs that he sells in a week and y represents his total pay in one week, then an equation for his weekly pay is $y = 200 + 40x$.

7. Make a table to show Felix's weekly pay for 1, 2, 3, 4, and 5 VCRs sold.

8. Suppose that he earns $520 in one week. How many VCRs did he sell that week? 8

9. Make a bar chart using the data from your table.

10. Use your bar chart to find how many VCRs he needs to sell in order to earn at least $350 in one week. 4

7.

x	y
1	$240
2	$280
3	$320
4	$360
5	$400

Lesson 1.4

► **Key Skills**

Evaluate an expression using substitution, and use the results to solve an equation.

Make a table by substituting the values 1, 2, 3, 4, and 5 for x in the expression $2x + 3$. Then use the table of values to solve the equation $2x + 3 = 9$.

x	1	2	3	4	5
2x + 3	5	7	9	11	13

$2x + 3$ is 9 when x is 3.

11.

x	1	2	3	4	5
y	6	14	22	30	38

► **Exercises**

11. Make a table by substituting the values 1, 2, 3, 4, and 5 for x in the expression $8x - 2$.

12. Use the table of values to solve the equation $8x - 2 = 14$. 2

Lesson 1.5

► **Key Skills**

Find all of the factors of a given number.

To find all of the factors of 45, divide by 1, 2, 3, and so on until the factors begin to repeat.

► **Exercises**

Write a list of factors for each number. Circle the prime factors.

13. 18   14. 32   15. 75   16. 98

13. 1, ②, ③, 6, 9, 18
14. 1, ②, 4 , 8, 16, 32
15. 1, ③, ⑤, 15, 25, 75
16. 1, ②, ⑦, 14, 49, 98

Lesson 1.6

➤ Key Skills

Write the prime factorization of a number using exponents.
The prime factorization of 6363 in exponential form is $3^2 \cdot 7 \cdot 101$.

➤ Exercises

Write the prime factorization of each number in exponential form.

17. 56
$2^3 \cdot 7$

18. 102
$2 \cdot 3 \cdot 17$

19. 29
29

20. 720
$2^4 \cdot 3^2 \cdot 5$

Lesson 1.7

➤ Key Skills

Perform calculations in the proper order.

The order of operations are: parentheses, exponents, multiplication and division from left to right, and finally addition and subtraction from left to right.

$$5(7-4) - 6^2 \div 3 + 1$$
$$5(3) - 6^2 \div 3 + 1$$
$$5(3) - 36 \div 3 + 1$$
$$15 - 12 + 1$$
$$4$$

➤ Exercises

Simplify.

21. $17 - 4 \cdot 3$
5

22. $32 - 24 \div 6 - 4$
24

23. $3 \cdot 4^2 - [24 \div (6-4)]$
36

Lesson 1.8

➤ Key Skills

Identify number properties.

Commutative Property

$$18 + 6 = 6 + 18$$
$$17 \cdot 5 = 5 \cdot 17$$

Associative Property

$$(18 + 6) + 4 = 18 + (6 + 4)$$
$$(17 \cdot 5) \cdot 2 = 17 \cdot (5 \cdot 2)$$

Distributive Property

$$6(5 + 20) = 6(5) + 6(20)$$

Find a sum or product using metal math.

To simplify $(18 + 6) + 4$, think

$$(18 + 6) + 4 = 18 + (6 + 4)$$
$$= 18 + 10, \text{ or } 28$$

To simplify $(17 \cdot 5) \cdot 2$, think

$$(17 \cdot 5) \cdot 2 = 17 \cdot (5 \cdot 2)$$
$$= 17 \cdot 10, \text{ or } 170$$

To simplify $7(31)$, think

$$7(31) = 7(30 + 1)$$
$$= 7(30) + 7(1)$$
$$= 210 + 7, \text{ or } 217$$

➤ Exercises

Use mental math to find each sum or product. Identify the property or properties that you use.

24. $(27 + 8) + 12$
47, Assoc

25. $(25 \cdot 87) \cdot 4$
8700, Comm, Assoc

26. $(6.2 + 7.1) + 3.8$
17.1, Comm, Assoc

27. $(63 \cdot 20) \cdot 5$
6300, Assoc

Applications

28. Entertainment Tickets to an amusement park cost $13 per person. How many tickets can be bought for $98? 7

Suppose that the cost of a ballet ticket is $22 per person.

29. What is the cost of of 4 tickets? of 8 tickets? $88, $176

30. Represent the cost of tickets by c and the number of tickets by t. Then write an equation for the cost of an order of tickets. $c = 22t$

Chapter 1 Assessment

Find the next three terms of each sequence.

1. 4, 8, 16, 32, 64, . . .   **2.** 49, 40, 32, 25, 19, . . .
128, 256, 512   14, 10, 7

If notebooks cost 59¢, find the cost of the following. $1.18 $2.95 $7.08
3. 2 notebooks   **4.** 5 notebooks   **5.** 12 notebooks

6. Use the variables c to represent cost and n to represent the number of notebooks purchased. Write an equation to model the cost of notebooks. $c = 59n$

7. How many notebooks can you buy with $14.75?   25

Wages Alice sells memberships to an exercise club. Her weekly salary is $250, plus $20 for each membership sold. If x represents the number of memberships that she sells in a week, and y represents her total pay in that week, then an equation for her weekly pay is $y = 250 + 20x$.

8. Make a table to show Alice's weekly pay for 1, 2, 3, 4, and 5 memberships sold that week.

x	1	2	3	4	5
y	270	290	310	330	350

9. Suppose that she earns $330 in one week. How many memberships did she sell that week?   4

10. Make a bar chart using the data from your table.

11. Use your bar chart to find how many memberships Alice needs to sell in one week to earn at least $340 in that week.   5

12. Make a table of values by substituting the values 1, 2, 3, 4, and 5 for the variable in the expression $5x + 7$.

x	1	2	3	4	5
$5x + 7$	12	17	22	27	32

13. Use your table of values from item 12 to solve the equation $5x + 7 = 22$.   3

Write a list of factors for each number. Circle the prime factors.

14. 20   **15.** 42   **16.** 76

14. 1, ②, 4, ⑤, 10, 20
15. 1, ②, ③, 6, ⑦, 14, 21, 42
16. 1, ②, 4, ⑲, 38, 76

Determine whether each number is prime or composite.

17. 19   **18.** 30   **19.** 31
prime   composite   prime

Write the prime factorization of each number in exponential form.

20. 8 2^3 **21.** 81 3^4 **22.** 105 $3 \cdot 5 \cdot 7$ **23.** 252 $2^2 \cdot 3^2 \cdot 7$

24. Simplify $3 + 27 \div 3^2 - (7 + 5)$. -6

Use mental math to find each sum or product.

25. $(9 + 37) + 11$ 57   **26.** $20 \cdot (5 \cdot 19)$ 1900

CHAPTER 2

Patterns With Integers

The highest known land elevation is the top of Mount Everest which stands at 29,028 feet above sea level in south-central Asia. The lowest land elevation is the Mariana Trench, which is 36,198 feet below sea level in the Pacific Ocean east of the Philippine Islands.

Elevation can be represented with integers. The height of Mount Everest can be written as 29,028, and the depth of the Mariana Trench can be written as −36,198.

A scuba diver explores Fairy Basslets and the coral reefs from about 30 feet below sea level in Maldives.

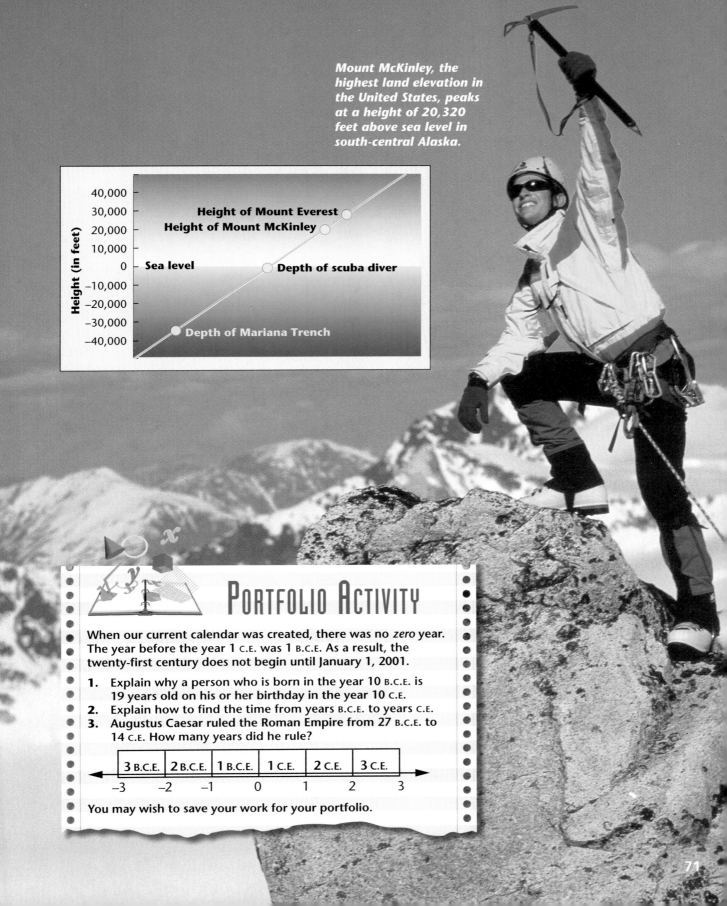

Mount McKinley, the highest land elevation in the United States, peaks at a height of 20,320 feet above sea level in south-central Alaska.

Height (in feet)

40,000
30,000 — Height of Mount Everest
20,000 — Height of Mount McKinley
10,000
0 — Sea level   Depth of scuba diver
−10,000
−20,000
−30,000
−40,000   Depth of Mariana Trench

PORTFOLIO ACTIVITY

When our current calendar was created, there was no *zero* year. The year before the year 1 C.E. was 1 B.C.E. As a result, the twenty-first century does not begin until January 1, 2001.

1. Explain why a person who is born in the year 10 B.C.E. is 19 years old on his or her birthday in the year 10 C.E.
2. Explain how to find the time from years B.C.E. to years C.E.
3. Augustus Caesar ruled the Roman Empire from 27 B.C.E. to 14 C.E. How many years did he rule?

3 B.C.E.	2 B.C.E.	1 B.C.E.	1 C.E.	2 C.E.	3 C.E.
−3	−2	−1	0	1	2	3

You may wish to save your work for your portfolio.

Integers and the Number Line

14,491 ft Mount Whitney

WHY Integers are used to describe situations involving temperature, profit or loss, time before or after the liftoff of a space shuttle, and elevation of land. In all of these examples, integers are used to count positive and negative units.

0 ft Sea level

−282 ft Death Valley

The highest point in California is Mount Whitney, which has an elevation of 14,491 feet. Just over 100 miles away is Death Valley, the lowest point in California. Death Valley has an elevation of 282 feet *below sea level*, or −282 feet. The elevation of Mount Whitney can be represented by the *positive* integer 14,491. The elevation of Death Valley can be represented by the *negative* integer −282. Sea level is represented by 0, which is *neither* a positive integer nor a negative integer.

INTEGERS

The set of integers consists of all positive whole numbers, all negative whole numbers, and zero.

EXAMPLE 1

Write each of the following verbal expressions as an integer.

A 25 below 0°C   **B** A profit of $2500

Solution ➤

A 25 below 0°C can be modeled by a negative integer. It can be written as −25.

B A profit of $2500 can be modeled by a positive integer. It can be written as 2500. ❖

Try This Write each of the following verbal expressions as an integer.

a. 35 above 0°F   35
b. A loss of $4000   −4000

EXAMPLE 2

Temperature Find the change in temperature from 7:00 A.M. to 5:00 P.M. at the top of Mount McKinley.

The thermometers show the temperatures in degrees Fahrenheit at the top of Mount McKinley at two different times in one day.

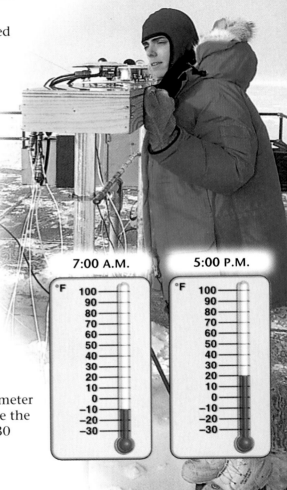

Solution ➤

Look at the distance on the thermometer between the two temperatures. Since the distance is 30 degrees, the number 30 describes the distance.

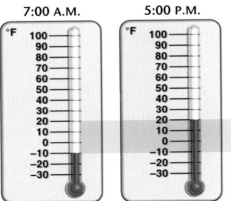

Thus, the temperature at 5:00 P.M. is 30 degrees greater than the temperature at 7:00 A.M. ❖

The thermometer in Example 2 is a vertical number-line model for integers. However, integers are usually represented on a horizontal number line.

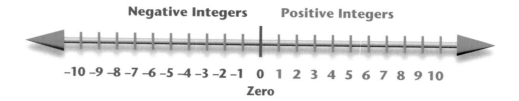

Negative Integers   **Positive Integers**

–10 –9 –8 –7 –6 –5 –4 –3 –2 –1 0 1 2 3 4 5 6 7 8 9 10
Zero

OPPOSITES
Two integers are **opposites** if they
*   are on *opposite* sides of zero,
*   and are the same distance from zero.

Opposites are always the same distance from 0.
For example, 5 and −5 are both 5 units from zero on the number line.

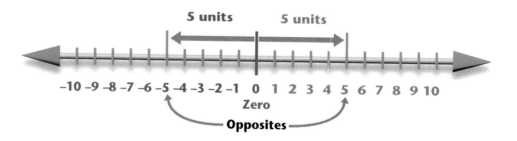

5 units   **5 units**

–10 –9 –8 –7 –6 –5 –4 –3 –2 –1 0 1 2 3 4 5 6 7 8 9 10
Zero
Opposites

The opposite of 5 is −(5), or −5.
The opposite of −5 is −(−5), or 5.
The opposite of zero is zero, so −(0) = 0.

The distance between an integer and zero is called its *absolute value*. In mathematics, the symbol | | indicates the absolute value of a number.

$|5| = 5$ *The absolute value of 5 is 5.*
$|-5| = 5$ *The absolute value of −5 is 5.*

ABSOLUTE VALUE

For any number x,

if x is a positive integer or zero, then the absolute value of x is x, or $|x| = x$; and

if x is a negative number, then the absolute value of x is the opposite of x, or $|x| = -x$.

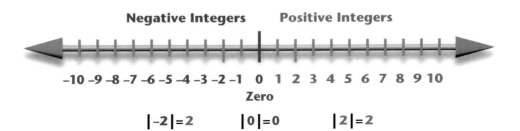

Negative Integers   **Positive Integers**

$$-10\ -9\ -8\ -7\ -6\ -5\ -4\ -3\ -2\ -1\quad 0\quad 1\ 2\ 3\ 4\ 5\ 6\ 7\ 8\ 9\ 10$$

Zero

$$|-2| = 2 \qquad |0| = 0 \qquad |2| = 2$$

Graphics Calculator

The expression **ABS** is used to tell computers and graphics calculators to compute absolute value.

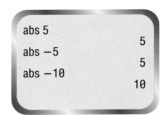

```
abs 5
                  5
abs -5
                  5
abs -10
                 10
```

EXAMPLE 3

Find the absolute value.

A $|-8|$ **B** $|0|$ **C** $|14|$

Solution ➤

A The distance from 0 to -8 on the number line is 8 units. So $|-8| = 8$.

B The distance from 0 to 0 on the number line is 0 units. So $|0| = 0$.

C The distance from 0 to 14 on the number line is 14 units. So $|14| = 14$. ❖

Try This   Find the absolute value of each integer.

a. $|-1|$ 1   **b.** $|10|$ 10   **c.** $|-6|$ 6

CRITICAL *Thinking*

Name two different integers that have the same absolute value. How are these two integers related? Explain why any two integers related in this way must have the same absolute value.

Exercises & Problems

Communicate

1. Describe the set of integers.
2. Give three examples of how integers are used.
3. Explain and give an example of what is meant by the opposite of a number.
4. Discuss and give examples of what is meant by the absolute value of a number.
5. Explain how |−10| and |10| are related.

   |−10| and |10| are both equal to 10. Since −10 and 10 are opposites, they must be the same distance from zero.

Practice & Apply

Determine whether each number is an integer.

Integer | | | Not an int. | | Integer

6. 5   **7.** −3   **8.** 0   **9.** −3.6   **10.** $\frac{1}{2}$ **11.** 7

Integer   Integer   Not an int.

Write an integer to represent each amount.

12. 20 degrees above zero   20
13. a withdrawal of $45   −45
14. a depth of 500 meters   −500
15. a price increase of $3   3
16. a weight loss of 9 pounds   −9
17. a gain of 15 yards   15
18. a growth of 2 centimeters   2
19. a loss of $20   −20
20. a temperature drop of 6 degrees   −6
21. 10 years ago   −10

Write the opposite of each number.

22. 17   −17   **23.** −17   17   **24.** 0   0   **25.** −12   12   **26.** 6   −6

Find the absolute value.

27. |−4| 4   **28.** |0| 0   **29.** |−3| 3   **30.** |25| 25   **31.** |−8| 8

32. |14| **33.** |−2| **34.** |−30| **35.** |−25| **36.** |18|
    14   2   30   25   18

Scuba Diving   Write the indicated change in a diver's position as a positive or negative integer.

37. Starts at −10 feet and ends at −40 feet   −30
38. Starts at −80 feet and ends at −60 feet   20
39. Starts at −90 feet and ends at −10 feet   80
40. Starts at −20 feet and ends at −70 feet   −50
41. Starts at −62 feet and ends at −17 feet   45

SEA LEVEL 0

−10 ft

−20 ft

−30 ft

−40 ft

−50 ft

−60 ft

−70 ft

−80 ft

−90 ft

−100 ft

Demographics The bar graph below shows the populations of four cities in 1980 and 1990.

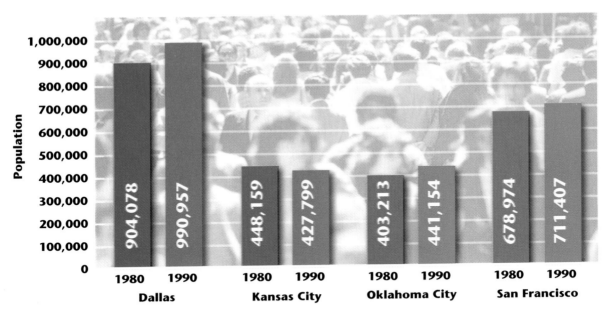

Write the change in population from 1980 to 1990 for each city using a positive or negative integer.

42. San Francisco 32,433

43. Kansas City −20,360

44. Dallas 86,879

45. Oklahoma City 37,941

46. Suppose the temperature is −20°F at 6:00 A.M. and 30°F at 2:00 P.M. What is the change in temperature from 6:00 A.M. to 2:00 P.M.? 50°F

Look Back

Make a table of values for each equation by substituting 1, 2, 3, 4, 5, and 10 for x. [Lesson 1.3]

47. $y = 2x + 1$

x	1	2	3	4	5	10
y	3	5	7	9	11	21

48. $y = 5x - 1$

x	1	2	3	4	5	10
y	4	9	14	19	24	49

Evaluate each expression if a is 3, b is 4, and c is 2. [Lesson 1.4]

49. ab^2c 96

50. a^2bc 72

51. abc^2 48

52. a^4c^3 648

53. Find the volume of a cube with edges of 7 centimeters. [Lesson 1.6]
343 cu cm

Use mental math to find each sum or product. [Lesson 1.8]

54. $5 \cdot 13 \cdot 6$ 390

55. $78 + (23 + 22)$ 123

56. $8 \cdot 59$ 472

Look Beyond

57. **Technology** Graph the equation $y = |x|$, using the ABS feature on a graphics calculator. Describe the graph. The graph looks like a "V" with the vertex at the origin.

Exploring Integer Addition

why *Several species of jellyfish stay at or below sea level for long periods of time. As a jellyfish rises, its depth changes. The changing depth of the rising jellyfish is just one example of a value that can be found by adding numbers.*

A jellyfish swims upward by expanding its body like an opening umbrella and then pulling it together rapidly. This squeezes water out from beneath the body and moves the jellyfish upward.

Using Models to Add Integers

represents 1

represents −1

Tiles can be used to represent integers.

By placing tiles on a working area, called a **mat**, you can model integers.

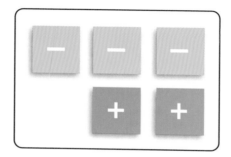

A pair that consists of one positive tile and one negative tile is called a **neutral pair**. The value of a neutral pair is zero.

neutral pair

You can find the value of a tile model by removing neutral pairs.

When you remove the neutral pairs from the model shown below, the value is −1. Why?

Create three other tile models that have a value of −1.

When the neutral pairs are removed, one −1 tile is left. Other tile models with a value of −1 may vary.

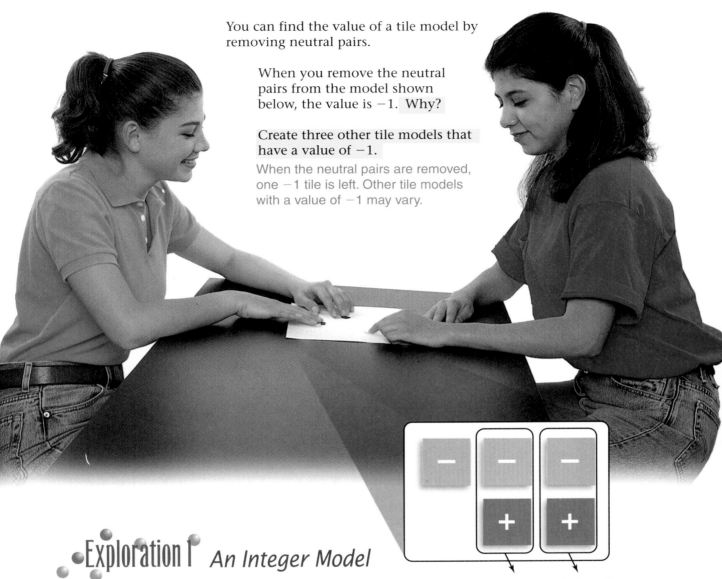

•Exploration 1 *An Integer Model*

For Steps 1 and 2, place tiles in the arrangements shown below. Remove neutral pairs, and find the value of each tile model.

1   −2

2   2

3️⃣ Create four tile models that each have a value of zero. Compare these four tile models. How are they alike, and how are they different?

4️⃣ Explain why the integers 3 and −3 are called opposites. Name three other pairs of opposites.

5️⃣ Explain why more than one tile arrangement can represent the same integer. ❖

You can use a tile model to add integers.

To model the sum 2 + (−3),
start with 2 positive tiles.

Add 3 negative tiles.

Remove any neutral pairs formed.

What is 2 + (−3)? −1

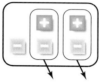

Again, use a tile model to add integers.

To model the sum −5 + (3), start with
5 negative tiles.

Add 3 positive tiles.

Remove any neutral pairs formed.

What is −5 + (3)? −2

Write each addition expression that is modeled
below. Then find the sum for each model.

2 + (−5) = −3   4 + (−3) = 1

Exploration 2 Adding Integers

1 Use tiles to model $(-4) + (-2)$. Record the sum.

2 Use tiles to model $2 + 4$. Record the sum.

3 How are the processes you used to find the sums in Steps 1 and 2 alike? How are they different?

4 Explain how to find $12 + 40$ and $(-25) + (-40)$ without tiles. Find each sum, and check your answer with a calculator.

5 Describe how to add two positive integers. Describe how to add two negative integers. Add the absolute values. Use the common sign.

6 Use tiles to model $-4 + 2$ and $-4 + 7$. Record each sum.

7 Use tiles to model $6 + (-3)$ and $6 + (-10)$. Record each sum.

8 How are the processes you used to find the sums in Steps 6 and 7 alike? How are they different?

9 Explain how to find $-12 + 40$ and $25 + (-40)$ without tiles. Find each sum, and check your answer with a calculator.

10 Explain how to add a positive integer and a negative integer. ❖

Find the difference of the absolute values. Use the sign of the integer with the greater absolute value.

CRITICAL Thinking

Find the sum $3 + (-3)$. What happens when you add a number and its opposite?

$3 + (-3) = 0$. When a number is added to its opposite, the sum is zero.

The illustrations below help to explain how to add integers with like signs and how to add integers with unlike signs.

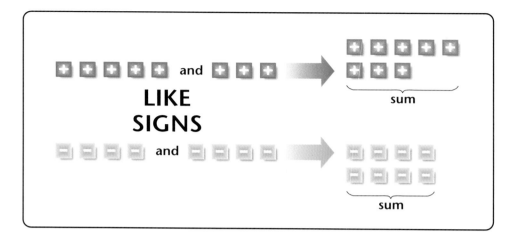

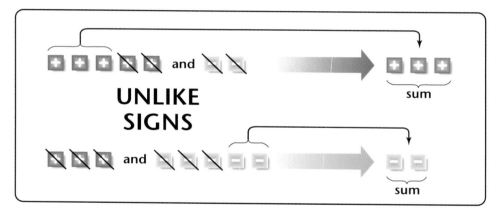

You can use the idea of absolute value to write rules for adding integers.

SUMMARY OF THE RULES FOR ADDING INTEGERS

Like Signs: (+ and +) or (− and −)
Find the sum of the absolute values, and use the sign common to both integers.

Unlike Signs: (+ and −) or (− and +)
Find the difference of the absolute values, and use the sign of the integer with the greater absolute value.

SUMMARY OF THE PROPERTY OF OPPOSITES

For any number a,

$$a + (-a) = 0.$$

Exercises & Problems

Communicate

1. Explain how neutral pairs can be used to add integers.
2. Describe how to add two integers that have the same sign.
3. Describe how to add two integers that have unlike signs.
4. Explain how the absolute value of an integer is used to add integers.

Practice & Apply

Use algebra tiles to find the following sums.

5. $-5 + (-2)$ −7   **6.** $-3 + 3$ 0   **7.** $2 + (-6)$ −4   **8.** $-2 + (-6)$ −8

9. $8 + (-3)$ 5   **10.** $-4 + (-5)$ −9 **11.** $-1 + 5$ 4   **12.** $-1 + (-2)$ −3

Find each sum.

13. $-28 + 50$ 22   **14.** $17 + (-34)$ −17   **15.** $38 + (-72)$ −34

16. $54 + (-16)$ 38   **17.** $-13 + (-18)$ −31   **18.** $31 + (-69)$ −38

19. $(-33) + (-5)$ −38   **20.** $43 + (-51) + 8$ 0   **21.** $-61 + (-15) + 9$ −67

22. $14 + (-29) + (-12)$ **23.** $-43 + 82 + |-19|$ **24.** $-308 + |-24| + (-29)$
   −27   58   −313

Substitute 2 for a, −3 for b, and 5 for c. Evaluate each expression.

25. $(a + b) + c$ 4   **26.** $|a + b| + c$ 6   **27.** $a + |b + c|$ 4

28. $a + (b + c)$ 4   **29.** $a + (c + b)$ 4   **30.** $c + |a + b|$ 6

Physics Electrons have a charge of −1, and protons have a charge of +1. The total charge of an atom is the sum of its electron charges and its proton charges. Find the total charge of each atom.

31. 16 protons, 18 electrons −2

32. 10 protons, 10 electrons 0

33. Sports The Lincoln High football team completes two series of downs with the indicated gains and losses, measured in yards. Find the total gain or loss.
Gain of 6 yards

Down	1	2	3	4
Gain/Loss	8	−4	−3	15

Down	1	2	3	4
Gain/Loss	−3	−5	4	−6

34. Business In business, a loss is recorded by placing the amount inside the parentheses. What is the yearly profit or loss for the Family Shoe Store? ($197.03)

1st Quarter	2nd Quarter	3rd Quarter	4th Quarter
($389.75)	($794.28)	$1796.50	($809.50)

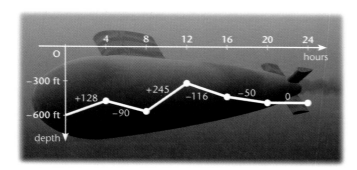

35. The graph starts at -600 feet. It shows the change in depth of a submarine over six 4-hour intervals. What is the depth of the submarine after 24 hours?

-483 ft

Look Back

36. Find the sum of $1 + 2 + 3 + \cdots + 40$.   **[Lesson 1.1]**
820

Find the next two terms in each number sequence.   [Lesson 1.1]

37. 40, 37, 34, 31, _?_, _?_
28, 25

38. $-5, -1, 3, 7,$ _?_, _?_
11, 15

39. 1, 3, 6, 10, _?_, _?_
15, 21

40. **Geometry** What is the perimeter of a square with a side of 3.5 centimeters? 14 cm

Calculate.   [Lesson 1.7]

41. $15 - 21 + 3 + 4$ 1   **42.** $[3(4 - 2)^2] + 7$ 19   **43.** $12 + 3^2 + (9 - 6)$ 24

44. Sports A football running back's carries in yards for the first quarter are $-2, 8, 12, -6,$ and -1. Find three different ways to total the yardage for the first quarter.   **[Lesson 1.8]**

Determine whether each number is an integer.   [Lesson 2.1]

45. 5
Integer

46. $-16,000$
Integer

47. -3.2
Not an integer

48. 0
Integer

Look Beyond

49. Recall that adding the same number repeatedly can be represented by multiplication.

$$2 + 2 + 2 + 2 + 2 = 5(2) = 10$$

With this in mind, represent $7(-2)$ as an addition problem. Then find the value of $7(-2)$. -14

50. Recall that -14 is the opposite of 14 and 14 is the opposite of -14. Use this information and the method in Exercise 49 to find the product $-[7(-2)]$. 14

LESSON 2.3 Solving Equations and Comparing Integers

why *Solving integer equations using the guess-and-check method leads to a powerful understanding of the way integers are ordered. You must understand the order of integers to compare them.*

7:00 A.M.

3:00 P.M.

$$-10 + x = 40$$

Using Guess-and-Check

Recall that temperature can be shown on a number line. The thermometers represent vertical number lines.

You can see that the temperature is $-10°F$ at 7:00 A.M. Throughout the day, the temperature rises. By 3:00 P.M., the temperature is $40°F$.

Explain how the equation $-10 + x = 40$ describes this situation.

You can use a problem-solving method called guess-and-check to solve the equation $-10 + x = 40$ and find the increase in temperature.

EXAMPLE 1

Use guess-and-check to solve the equation $-10 + x = 40$ and find the increase in temperature.

Solution ➤

Choose a possible value for x. Substitute your guess into the equation, and check the result.

Guess 1 Try substituting **30** for x.

$$-10 + x = 40$$
$$-10 + \mathbf{30} \stackrel{?}{=} 40$$
$$20 \stackrel{?}{=} 40 \qquad \text{False}$$

Since 20 does not equal 40, the solution is *not* 30. Try a larger number.

Guess 2 Try substituting **50** for x.

$$-10 + x = 40$$
$$-10 + \mathbf{50} \stackrel{?}{=} 40$$
$$40 \stackrel{?}{=} 40 \qquad \text{True}$$

Since 40 equals 40, the solution is 50. ❖

Try This Use guess-and-check to solve the equation $-1 + x = -6$. $x = -5$

The National Weather Service uses a chart to determine windchill. Windchill describes the chill a person actually feels on exposed skin as a result of the combination of wind and temperature. The table below shows windchill temperatures associated with an actual temperature of 32°F.

Wind speed (mph)	0	5	10	15	20	25	30	35
Windchill temperature (°F)	32	29	18	11	6	3	0	−2

With a 35-mile-per-hour wind, the actual temperature of 32°F feels like −2°F. How much is the decrease from the actual temperature to this windchill temperature?

Explain how the equation $32 - x = -2$ models this situation.

EXAMPLE 2

Use guess-and-check to find the decrease from the actual temperature to the windchill temperature.

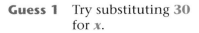

35 mph
Actual wind speed

32°F   -2°F
Actual   Windchill
temperature   temperature

Solution ➤

Use guess-and-check to solve $32 - x = -2$. Choose a possible value for x. Substitute your guess into the equation, and check.

Guess 1 Try substituting **30** for x.

$$32 - x = -2$$
$$32 - 30 \overset{?}{=} -2$$
$$2 \overset{?}{=} -2 \quad \text{False}$$

Since 2 does *not* equal -2, the solution is *not* 30. When you subtract 30 from 32, the result, 2, is too big. Try *subtracting a larger number* so that the result will be smaller.

Guess 2 Try substituting **40** for x.

$$32 - x = -2$$
$$32 - 40 \overset{?}{=} -2$$
$$32 + (-40) \overset{?}{=} -2$$
$$-8 \overset{?}{=} -2 \quad \text{False}$$

Since -8 does *not* equal -2, the solution is *not* 40. When you subtract 40 from 32, the result, -8, is too small. Now try subtracting a number between 30 and 40.

Guess 3 Try substituting **34** for x.

$$32 - x = -2$$
$$32 - 34 \overset{?}{=} -2$$
$$32 + (-34) \overset{?}{=} -2$$
$$-2 = -2 \quad \text{True}$$

Since $-2 = -2$, the solution is 34.

Thus, with a 35-mile-per-hour wind, a 32°F actual temperature is *decreased* by 34°F, or changed by -34°F, to a windchill temperature of -2°F. ❖

Try This Use guess-and-check to find the change from an actual temperature of 40°F to a windchill temperature of 16°F. $x = 24$

Lesson 2.3   Solving Equations and Comparing Integers   **87**

Ordering Integers

Temperature can help you understand how integers are ordered. For example, another way to say $-10°F$ *is colder than* $20°F$ is to say -10 *is less than* 20. The phrase *is less than* can be represented by the symbol <.

-10 is less than 20

$-10 \quad < \quad 20$

You can write the same relationship using the *greater than* symbol, >.

20 is greater than -10

$20 \quad > \quad -10$

The statements $-10 < 20$ and $20 > -10$ are called *inequalities*.

INEQUALITY

An inequality is a mathematical statement that contains the symbol
$>, \geq, <, \leq,$ or $\neq$.

A number-line model is a useful tool for demonstrating inequalities.

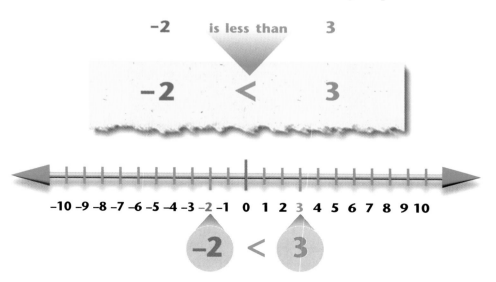

Since -2 is *to the left of* 3 on the number line, -2 is *less than* 3.

Also, since 3 is *to the right of* -2 on the number line, 3 is *greater than* -2.

EXAMPLE 3

Write two inequalities for each pair of integers.

A −6, 5

B −6, −8

Solution ➤

A Since −6 is *less than* 5, write −6 < 5.
Since 5 is *greater than* −6, write 5 > −6.

B Since −6 is *greater than* −8, write −6 > −8.
Since −8 is *less than* −6, write −8 < −6. ❖

If you can find a positive integer, x, to add to an integer, a, so that the sum
is equal to another integer, b, then $a < b$ and $b > a$.

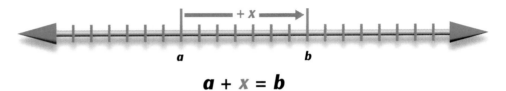

$$a + x = b$$

Note that this idea applies to all numbers, not just integers.

EXAMPLE 4

Solve each equation to determine whether each inequality is true.

A Use guess-and-check to solve $-2 + x = 4$. Use your solution to
determine whether $-2 < 4$.

B Use guess-and-check to solve $-3 + x = 0$. Use your solution to
determine whether $-3 < 0$.

Solution ➤

A The solution to $-2 + x = 4$ is **6**. Since $-2 + 6 = 4$ and **6** is a positive
integer, the inequality $-2 < 4$ is true.

B The solution to $-3 + x = 0$ is **3**. Since $-3 + 3 = 0$ and **3** is a positive
integer, the inequality $-3 < 0$ is true. ❖

Try This Solve $-7 + x = -1$ to determine whether $-7 < -1$. *$x = 6$; Since*
$-7 + 6 = -1$ and 6 is positive, $-7 < -1$

CRITICAL
Thinking

Explain how a number line can be used to show that if the solution to
$2 + x = -5$ is -7, then $-5 < 2$ and $2 > -5$.

EXERCISES & PROBLEMS

Communicate

1. Describe how you can use temperature or elevation to illustrate integer inequality. Give an example of each.

2. Point A is to the left of point B on the number line. Explain why the number represented by point A is less than the number represented by point B.

3. How can you use the solution to the equation $-4 + x = -5$ to determine whether $-4 > -5$ is true?

4. Explain how understanding the order of integers helps you make guesses when solving an equation by guess-and-check.

Practice & Apply

Use guess-and-check to solve each equation.

5. $x + 5 = -1$ -6 6. $-1 + x = -4$ -3

7. $4 + x = 2$ -2 8. $x + 1 = -1$ -2

9. $-3 + x = -4$ -1 10. $-7 + x = -2$ 5

11. $x + (-4) = 0$ 4 12. $-4 + x = -4$ 0

13. $6 = 2 + x$ 4 14. $-6 = x + (-3)$ -3

Write two inequalities for each pair of integers. Use both the < and > symbols.

15. $5, -5$ 16. $-3, -1$ 17. $0, -2$ 18. $-3, -7$ 19. $4, -8$

20. $-1, -2$ 21. $4, -1$ 22. $-4, -2$ 23. $7, -9$ 24. $-8, 10$

25. $-7, 8$ 26. $-15, 15$ 27. $8, -21$ 28. $-16, -22$ 29. $14, -7$

30. $101, -236$ 31. $16, -17$ 32. $-18, -9$ 33. $18, 9$ 34. $-200, -1$

Show each set of integers on a number line. Then list the integers in ascending order (from least to greatest).

35. $5, -5, 1, -3$ 36. $-2, -1, 0, -4, -10, 2$ 37. $0, -2, 2, -3, 3$

38. $4, 6, 3, 2$ 39. $-4, -6, -3, -2, 1$ 40. $0, -5, 5, -4, 6$

41. $30, 10, 40, 20$ 42. $1, -2, 3, -4, 5$ 43. $-10, -20, 30, 40, -50$

Solve each equation to show that each inequality is true.

44. Solve $-5 + x = -2$ to show that $-5 < -2$. $x = 3$

45. Solve $-3 + x = -1$ to show that $-3 < -1$. $x = 2$

46. Solve $-1 + x = 4$ to show that $-1 < 4$. $x = 5$

47. Solve $2 + x = 5$ to show that $2 < 5$. $x = 3$

48. Solve $-8 + x = 7$ to show that $-8 < 7$. $x = 15$

49. Solve $5 + x = 12$ to show that $5 < 12$. $x = 7$

50. Solve $-3 + x = 4$ to show that $-3 < 4$. $x = 7$

51. Solve $-6 + x = -3$ to show that $-6 < -3$. $x = 3$

Geography   Use the map to complete Exercises 52–55.

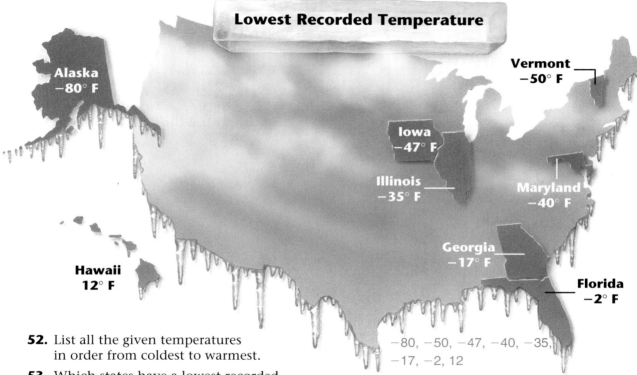

Lowest Recorded Temperature

Alaska −80° F
Vermont −50° F
Iowa −47° F
Illinois −35° F
Maryland −40° F
Georgia −17° F
Hawaii 12° F
Florida −2° F

52. List all the given temperatures in order from coldest to warmest. $-80, -50, -47, -40, -35, -17, -2, 12$

53. Which states have a lowest recorded temperature that is colder than $-40°F$?   Alaska, Vermont, and Iowa

54. How many states have a lowest recorded temperature that is warmer than $-25°F$?   Georgia, Hawaii, and Florida

55. Write an inequality to compare the lowest recorded temperatures in Alaska and in Hawaii. $-80 < 12$

Use guess-and-check to solve each equation.

56. $-4 + x = -2$ 2

57. $y + 12 = 10$ -2

58. $5 + x = 2$ -3

59. $32 - x = -2$ 34

60. $32 + y = -21$ -53

61. $19 - x = -9$ 28

62. $2x = -8$ -4

63. $32 - y = -2$ 34

64. $2z - 1 = -7$ -3

65. $3x = -12$ -4

66. $-12 + 2y = -6$ 3

67. $2z + 1 = -7$ -4

In flag football, the offense has four downs to gain at least 10 yards by running or passing. Once 10 yards or more have been gained, the offense gets another four downs.

Sports While playing offense in flag football, Melinda lost 3 yards on the first down. Tim lost 5 yards on the second down.

68. Write an inequality to compare the yardage lost by Tim and Melinda. $-5 < -3$

69. Write an integer to represent Tim and Melinda's total loss in yardage. -8

70. Write an equation to represent that, after the second down, the offense needs to gain x number of yards in order to earn another first down. $-8 + x = 10$

In each case, determine whether $a < b$, $a = b$, or $a > b$. Assume that a is positive.

71. $a + a = b$ $a < b$

72. $a + (2a) = b$ $a < b$

73. $a + 0 = b$ $a = b$

74. $a - a = b$ $a > b$

Look Back

Use mental math to find each sum or product.   [Lesson 1.8]

75. $(23 + 48) + 77$ 148   **76.** $(4 \cdot 32) \cdot 25$ 3200   **77.** $2 \cdot (95 \cdot 5)$ 950   **78.** $54 + (75 + 46)$ 175

Use the Distributive Property and mental math to find each product.   [Lesson 1.8]

79. $24 \cdot 90$ 2160   **80.** $26 \cdot 30$ 780   **81.** $40 \cdot 49$ 1960   **82.** $20 \cdot 99$ 1980

Write the absolute value of each integer.   [Lesson 2.1]

83. -4 4   **84.** 0 0   **85.** -3 3   **86.** 25 25   **87.** -8 8   **88.** 16 16

Look Beyond

Use guess-and-check to find all possible solutions to each equation.

2 or -2

89. $2x = -6$ -3 **90.** $\frac{x}{2} = -14$ -28 **91.** $x^2 = 4$ **92.** $2x = -4$ -2

LESSON 2.4

Exploring Integer Subtraction

Why Each time Mark writes a check, the bank subtracts money from his account. When he deposits his paycheck, the bank adds money to his account. Many bank transactions can be modeled by positive or negative numbers.

Suppose Mark starts with $50 in his account and writes a check for $20. This transaction can be represented in two ways.

  subtraction: $50 - 20$ or   addition: $50 + (-20)$

In either case, Mark will have $30 left.

  $50 - 20 = 30$ or $50 + (-20) = 30$

Using Tiles to Subtract Integers

You can model subtraction of integers by taking away tiles.

6 − 4 = 2   **−4 − (−2) = −2**

In Lesson 2.2 you learned that adding the same number of positive and negative tiles, or neutral pairs, does not change the *value* of the tiles you have. The result is the same as adding 0. Why do you think the number 0 is called the **additive identity**? Adding zero to a number does not change its value.

ADDITION PROPERTY OF ZERO

For any number a,

$a + 0 = a = 0 + a.$

The next activity involves using the Addition Property of Zero. The tile model shows how it is possible to subtract, or *take away,* −4 from 2.

$$2 - (-4) = \underline{\;?\;}$$

1. Start with 2 positive tiles.

2. Since you want to subtract −4, add 4 neutral pairs. The total value of the tiles is still 2.

3. Now you can subtract, or *take away,* the 4 negative tiles and rewrite the original problem as 2 + 4.

Thus, 2 − (− 4) = 2 + 4 = 6.

CRITICAL Thinking

Describe how to use algebra tiles to calculate −2 − 4. What is the addition expression that is equivalent to −2 − 4? Start with two negative tiles. Add 4 neutral pairs; then remove 4 positive tiles. Six negative tiles remain, so the solution is −6. This expression is equivalent to −2 + (−4).

•Exploration *Subtraction With Tiles*

Use tiles to subtract 5 from 3.

1 Begin with 3 positive tiles.

2 The number you want to subtract is 5, so add 5 neutral pairs. What value did you add to the expression when you added these neutral pairs? zero

3 How many and what kind of tiles should you remove? 5 positive tiles

4 Remove the tiles. How many of each kind are left? 3 positive; 5 negative

5 What addition expression is equivalent to 3 − 5? What is the sum? 3 + (−5) = −2

6 Use this procedure to calculate − 3 − (−4). 1

7 Explain how to subtract an amount with a greater absolute value from an amount with a lesser absolute value, such as −5 − (−8). ❖

7. Add the opposite of the number being subtracted. −5 − (−8) becomes −5 + 8.

Relating Addition and Subtraction

When working with integers, you can use addition to calculate a difference. To subtract an integer, add its opposite. The same is true for any number.

THE DEFINITION OF SUBTRACTION

For all numbers a and b,

$$a - b = a + (-b).$$

EXTENSION

You will sometimes need to determine the distance between points on the number line. One method uses absolute value. Find the distance between points A and B.

Subtract the values and then find the absolute value of the difference.

The distance between -4 and 2 is $|2 - (-4)| = 6$, or $|(-4) - 2| = 6$.

In general, if you let a represent the value for the point A and let b represent the value for the point B, the distance between points on the number line is $|b - a|$ or $|a - b|$. ❖

What is the distance between -3 and 5 on the number line?   8

Calculator

On all calculators $\boxed{-}$ is used for the subtraction operation. Some calculators use $\boxed{+/-}$ or $\boxed{\pm}$ to change the sign of a number. Other calculators use $\boxed{(-)}$ to find the opposite of a number.

How does your calculator treat the difference between subtraction and the opposite of a number? What keystrokes on your calculator let you calculate $-35 - 27$?

EXERCISES & PROBLEMS

Communicate

1. What is the effect of adding 0 to a number?
   leaves the value unchanged
2. How does adding neutral pairs help to model
   a subtraction problem? ensures there are enough
   (+) or (−) tiles available to remove
3. Mark's account balance is $5. Explain what
   happens if Mark writes a check for $25 dollars.
   His balance will be − 20 dollars.
4. For the problem −4 − (−7), how many neutral pairs
   do you add to solve the problem? Why?

5. What is meant by adding the opposite?

6. Explain how addition can replace subtraction when you use integers.
   When subtracting an integer, we can add its opposite instead.

Practice & Apply

Evaluate each expression.

7. $67 − 3$ 64   8. $42 − (−9)$ 51   9. $−10 − (−21)$ 11   10. $−35 − 17$ − 52

11. $33 − (−33)$ 66   12. $−78 + (−45)$ 13. $990 − (−155)$ 14. $−97 − 88$ − 185
                                          − 123   1145
15. $−43 + 23 + (−43)$ 16. $−77 − 77 + 5$ 17. $108 + (−18) − 8$ 18. $85 − (−12) − (−9)$
        − 63   − 149   82   106

Substitute 5 for x, −3 for y, and −10 for z. Evaluate each expression.

19. $x − y$ 8   20. $x + y − z$ 12   21. $(x − z) − y$ 18   22. $x − (z − y)$ 12

23. $y − x$ − 8   24. $y − x + z$ − 18 25. $(x + y) − (x − y)$ 26. $y − y − y − y$
                                                      − 6   6

Find the distance between each pair of points on the number line.

27. 4, 9 5 units   28. −6, 15 21 units 29. −47, −23 24 units 30. −12, 74 86 units

*Mike has a balance of $145
in his savings account.*

31. If Mike withdraws $37, how much money
    does he have left in his account? $108

DATE	DEPOSIT	WITHDRAWAL	BALANCE
3/16	62.50		62.50
4/19	55.00		117.50
5/15	80.00		197.50
6/20		24.50	173.00
7/20	12.00		185.00
8/16		40.00	145.00

32. **Temperature** Theresa noticed that the
    temperature in her freezer was 5°F. She
    lowered the thermostat by 7°F. Later, she
    set the thermostat another 2°F lower.
    What temperature would you expect in
    the freezer after Theresa finished changing
    the thermostat? − 4° F

This table shows that the windchill temperature depends on the air temperature and the speed of the wind.

Wind Speed in Miles per Hour	10	20	30
Air Temperature 20°F	3	−10	−18
Air Temperature 10°F	−9	−24	−33
Air Temperature 0°F	−22	−39	−49
Air Temperature −10°F	−34	−53	−64

33. The air temperature is 20°F. The wind speed is 20 miles per hour. What is the windchill temperature? − 10° F

34. The air temperature is − 10°F. How much colder does it feel when the wind speed is 20 miles per hour than when there is no wind? 43° F colder

35. The air temperature is − 10°F. The wind speed increases from 10 to 30 miles per hour. How many degrees does the temperature seem to drop? 30° F

36.  **Portfolio Activity** Complete the activity given on page 71.

Look Back

38.

m	50	100	150	200
c	$52.50	$65	$77.50	$90

Business A car rental firm charges $40 plus 25 cents per mile to rent a car. [Lesson 1.3]

37. Using m for the number of miles, write an equation for the cost, c, of renting a car. $c = 40 + 0.25m$

38. Use a table to find the cost of renting a car for 50, 100, 150, and 200 miles.

39. How many miles can you drive the car if you can spend $72 on the rental? 128 miles

If the number is divisible by 3, 6, and 9, write *yes*. If the number is *not* divisible by all three numbers, write *no*. [Lesson 1.5]

40. 123 no   **41.** 3009 no   **42.** 4224 no   **43.** 6318 yes

Simplify. [Lesson 1.7]

44. $36 − 12 ÷ 3 − 20$ **45.** $3 · 5 + 7 ÷ 2$ **46.** $28 ÷ 2 · 7 + 4$ 102
12   18.5

Write the opposite and the absolute value of each integer. [Lesson 2.1]

47. −8 8, 8   **48.** 0 0, 0   **49.** 6 − 6, 6   **50.** −25 25, 25

Look Beyond

51. One point on the number line is at 4. The location of the other point is not known. Let the unknown point be at x. If the distance between the points is 7, what values could x be? (HINT: Draw the number line, and try different values of x.) − 3 or 11

− 6 or 14

52. What could x be if the distance between the points in Exercise 51 is 10?

LESSON 2.5

Exploring Integer Multiplication and Division

why *Measures of gain and loss in financial affairs, time before and after blastoff, and distance above and below sea level are reported using positive and negative numbers. Calculations using this data often involve multiplication and division. When calculating with positive and negative numbers, it is important to know what happens to the sign of the result.*

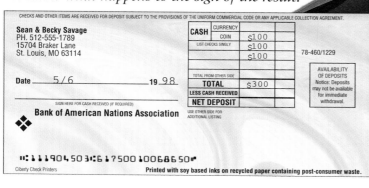

CHECKS AND OTHER ITEMS ARE RECEIVED FOR DEPOSIT SUBJECT TO THE PROVISIONS OF THE UNIFORM COMMERCIAL CODE OR ANY APPLICABLE COLLECTION AGREEMENT.

Sean & Becky Savage
PH. 512-555-1789
15704 Braker Lane
St. Louis, MO 63114

Date ___5/6___ 19 _98_

SIGN HERE FOR CASH RECEIVED (IF REQUIRED)

Bank of American Nations Association

CASH	CURRENCY		
	COIN	$100	
	LIST CHECKS SINGLY	$100	
		$100	
TOTAL FROM OTHER SIDE			
TOTAL		$300	
LESS CASH RECEIVED			
NET DEPOSIT			

78-460/1229

AVAILABILITY
OF DEPOSITS
Notice: Deposits
may not be available
for immediate
withdrawal.

USE OTHER SIDE FOR
ADDITIONAL LISTING

⑆111904503⑆617500100686 50⑈

Ciberty Check Printers   Printed with soy based inks on recycled paper containing post-consumer waste.

Examine what happens to an account when transactions are made. Explain how the signs of the integers relate to the transactions. A positive integer is like a deposit, and a negative integer is like a withdrawal.

Transaction	Representation	Result
Add 3 *deposits* of $100	$(3)(100) = 300$	Increase $300
Add 4 *withdrawals* of $20	$(4)(-20) = -80$	Decrease $80
Remove 5 *deposits* of $50	$(-5)(50) = -250$	Decrease $250
Remove 2 *withdrawals* of $10	$(-2)(-10) = 20$	Increase $20

•Exploration 1 *Multiplying Integers*

 Complete each pattern.

a.
$(2)(3) = 6$
$(2)(2) = 4$
$(2)(1) = 2$
$(2)(0) = 0$
$(2)(-1) = \underline{?}\ -2$
$(2)(-2) = \underline{?}\ -4$
$(2)(-3) = \underline{?}\ -6$

b.
$(3)(3) = 9$
$(2)(3) = 6$
$(1)(3) = 3$
$(0)(3) = 0$
$(-1)(3) = \underline{?}\ -3$
$(-2)(3) = \underline{?}\ -6$
$(-3)(3) = \underline{?}\ -9$

c.
$(3)(-3) = -9$
$(2)(-3) = -6$
$(1)(-3) = -3$
$(0)(-3) = 0$
$(-1)(-3) = \underline{?}\ 3$
$(-2)(-3) = \underline{?}\ 6$
$(-3)(-3) = \underline{?}\ 9$

2 Identify the sign for each statement.
   a. (positive) • (positive) = __?__ positive
   b. (positive) • (negative) = __?__ negative
   c. (negative) • (positive) = __?__ negative
   d. (negative) • (negative) = __?__ positive

3 Find the products.
   a. (12)(4) = __?__ 48   **b.** (13)(−3) = __?__ −39
   c. (−81)(6) = __?__ −486   **d.** (−16)(−5) = __?__ 80

4 To multiply two integers, first multiply their absolute values. positive
   a. What is the sign of the product when two integers have *like signs*?
   b. What is the sign of the product when two integers have *unlike signs*? ❖ negative

CRITICAL Thinking   Is the product positive or negative when an even number of negative numbers are multiplied? Is the product positive or negative when an odd number of negative numbers are multiplied? Explain your answers.
Positive; negative. Multiplying any two negative integers gives a positive result.
For an odd number of negative integers, there is one number that cannot be
Multiplication and division are related.   paired with another negative integer.

Multiplication Fact	Related Division Facts
6 • 5 = 30	a. 30 ÷ 5 = 6
	b. 30 ÷ 6 = 5

•Exploration 2 Dividing Integers

1 Find the products. Examine all the signs.
   a. (8)(7) = __?__ 56   **b.** (5)(−3) = __?__ −15
   c. (−4)(2) = __?__ −8   **d.** (−8)(−1) = __?__ 8

2. a. 56 ÷ 7 = 8
      56 ÷ 8 = 7

   b. (−15) ÷ (−3) = 5
      (−15) ÷ 5 = −3

   c. (−8) ÷ 2 = −4
      (−8) ÷ (−4) = 2

   d. 8 ÷ (−1) = −8
      8 ÷ (−8) = −1

2 Write the related division facts for **a**, **b**, **c**, and **d** in Step 1. Compare the signs in each multiplication problem with the signs of the related division facts.

3 Identify the sign for each statement.
   a. (positive) ÷ (positive) = __?__ positive
   b. (positive) ÷ (negative) = __?__ negative
   c. (negative) ÷ (positive) = __?__ negative
   d. (negative) ÷ (negative) = __?__ positive

4 Find the quotients.
   a. (48) ÷ (2) = __?__ 24   **b.** (18) ÷ (−2) = __?__ −9
   c. (−24) ÷ (−3) = __?__ 8   **d.** (−49) ÷ (7) = __?__ −7

5 To divide two integers, first divide their absolute values.
   a. What is the sign of the quotient when the two integers have *like signs*? positive
   b. What is the sign of the quotient when the two integers have *unlike signs*? ❖ negative

Zero is also an integer. As you know, zero has several special properties.

PROPERTIES OF ZERO

Let a represent any number.

1. **The product of any number and zero is zero.**
$$a \cdot 0 = 0 \text{ and } 0 \cdot a = 0$$

2. **Zero divided by any nonzero number is zero.**
$$0 \div a = \frac{0}{a} = 0, a \neq 0$$

3. **A number divided by zero is undefined.**
That is, IT IS IMPOSSIBLE TO DIVIDE BY ZERO.

STATISTICS
Connection

APPLICATION

What is the average score for the bowling team?
Is a guess of 133 reasonable?

To see how accurate the guess is, find the differences between each actual score and the guess of 133. Add the differences, and divide the total by 5 because there are 5 scores.

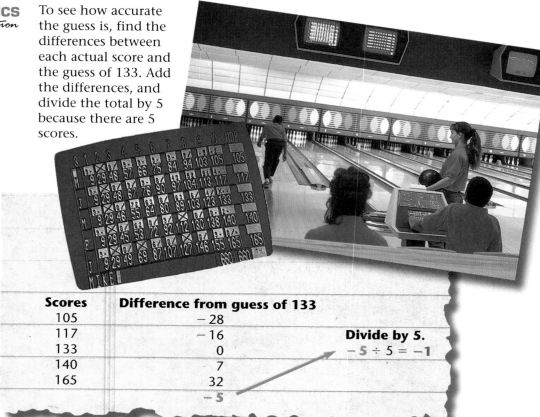

Scores	Difference from guess of 133	
105	− 28	
117	− 16	**Divide by 5.**
133	0	$-5 \div 5 = -1$
140	7	
165	32	
	− 5	

The average difference is **−1**. Add **−1** to the guess of 133. The average bowling score is 133 + **(−1)**, or 132. The guess is reasonable.

You can check the answer by adding all of the scores and dividing by 5. ❖

Exercises & Problems

Communicate

For Exercises 1 and 2, discuss what happens to an account balance when the indicated transactions are made.

1. Add 4 withdrawals of $25. Decrease by $100

2. Subtract 3 withdrawals of $10. Increase by $30

3. Explain how to write a bank transaction involving deposits or withdrawals that can be modeled by the expression $(3)(-5)$. Adding 3 withdrawals of $5

4. Discuss how to write two division problems and their answers based on the multiplication problem $(3)(-5) = -15$. $(-15) \div (-5) = 3$ or $(-15) \div 3 = -5$

5. Explain how to find the average of 95, 119, 110, 130, 141, and 155 using differences. Is a guess of 120 reasonable?

Practice & Apply

Evaluate.

6. $(-12)(-6)$ 72

7. $(-12) - (-6)$ -6

8. $(-8) - (-2)$ -6

9. $(-6.6) \div (3)$ -2.2

10. $(-0.8)(-2)$ 1.6

11. $(-8) \div (-2)$ 4

12. $(-8) + (-2)$ -10

13. $(-22) \div (-1)$ 22

14. $(-12) + (-6)$ -18

15. $(-1.2) \div (-6)$ 0.2

16. $(-5)[6 + (-6)]$ 0

17. $(7)(4)(-6)$ -168

18. $(-9) - [8 + (-3)]$ -14

19. $(-54)(-115)$ 6210

20. $(-8)(-2)(-3)$ -48

21. $(-225) \div (-5)$ 45

22. $(-47)(23)$ -1081

23. $(-2108) \div (124)$ -17

24. $(-6942) \div (-78)$ 89

25. $(-90) \div (-15) \times (-3)$ -18

26. $\frac{(7)(-1)}{-7}$ 1

27. $\frac{(-6)(-12)}{3}$ 24

28. $\frac{(-100)(-5)}{-25}$ -20

29. $\frac{(-1)(10)(-80)}{-4}$ -200

Tell whether each statement is true or false.

30. The sum of two negatives is negative. True

31. The difference of two negatives is always negative. False

32. The product of two negatives is negative. False

33. The quotient of two negatives is negative. False

34. The average of a set of negative numbers is negative. True

Banking Juan opened a savings account with $20. He made 4 deposits of $20 each and made withdrawals of $10, $15, and $25.

35. What is the total increase in Juan's account? $80

36. What is the total decrease in Juan's account? $50

37. What is the balance of Juan's account after these four deposits and three withdrawals? $50

Statistics A running back in football ran 12 times for the following yardage: $-2, -2, -1, 0, 0, 2, 2, 3, 3, 3, 6, 16$.

38. What is his total *net* yardage? HINT: To get the net yardage, subtract the total losses from the total gains. 30 yards

39. What is the *median* (middle value) number of yards in a run? HINT: Order the yardage from smallest to largest to find the middle value. 2 yards

40. What is the *mode* (most frequent value) for the number of yards in a run? 3 yards

41. What is the *mean* (average) number of yards in a run? 2.5 yards

42. What is his longest run? 16 yards

43. Find the average of 95, 104, 87, 120, 102, 100, and 99 using differences. Is a guess of 100 reasonable? 101; 100 is a reasonable guess.

44. **Statistics** Five people were asked to estimate the number of beans in a jar. Their errors were $-135, -43, -22, 38,$ and 111. What is the average error? Make a guess, and then use differences to find the average error. -10.2

Look Back

Evaluate. [Lessons 2.2, 2.4]

45. $-2 + (-7)$ -9 **46.** $-8 - (-3)$ -5 **47.** $|-7|$ 7 **48.** $|13|$ 13

Write two inequalities for each pair of integers. Use both the < and > symbols. [Lesson 2.3]

49. $6, -2$ $\begin{array}{l}-2 < 6 \\ 6 > -2\end{array}$ **50.** $-7, 7$ $\begin{array}{l}-7 < 7 \\ 7 > -7\end{array}$ **51.** $0, -9$ $\begin{array}{l}-9 < 0 \\ 0 > -9\end{array}$ **52.** $1, -3$ $\begin{array}{l}-3 < 1 \\ 1 > -3\end{array}$

Use guess-and-check to solve each equation. [Lesson 2.3]

53. $x + 2 = -2$ -4 **54.** $-5 + x = -1$ 4 **55.** $x + (-7) = 0$ 7

Look Beyond

Simplify.

56. $\dfrac{6x + 12}{2}$ $3x + 6$ **57.** $\dfrac{-5y - 25}{-5}$ $y + 5$ **58.** $\dfrac{3w + 24}{-3}$ $-w - 8$ **59.** $\dfrac{5y - 45}{-5}$ $-y + 9$

LESSON 2.6 Solving Problems With Equations and Graphs

Why *Many real-world relationships, such as that between actual temperature and windchill temperature, can be represented with equations and graphs.*

An anemometer is a gauge for determining the force or speed of the wind. The relationship between the actual temperature and the windchill temperature depends on the wind speed.

Using Integers to Graph Equations

Given a constant wind speed of 7 miles per hour, a windchill table shows that an actual temperature of 32°F is reduced to a windchill temperature of 27°F. This relationship between the actual temperature and the windchill temperature (given a 7-mile-per-hour wind speed) can be approximated by a linear equation.

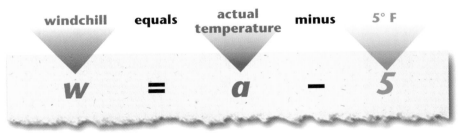

windchill	equals	actual temperature	minus	5° F
w	=	*a*	−	5

Recall from Lesson 1.3 that you can use an equation to make a table of values. Substitute values of *a*, the actual temperature, to find the corresponding values of *w*, the windchill temperature. These two values, *a* and *w*, can be combined to form an *ordered pair, (a, w)*.

Actual temperature, *a*	Windchill temperature, *w* = *a* − 5	Ordered pair, (*a*, *w*)
15	*w* = 15 − 5 *w* = 10	(15, 10)
10	*w* = 10 − 5 *w* = 5	(10, 5)
5	*w* = 5 − 5 *w* = 0	(5, 0)
0	*w* = 0 − 5 *w* = −5	(0, −5)
−5	*w* = −5 − 5 *w* = −10	(−5, −10)
−10	*w* = −10 − 5 *w* = −15	(−10, −15)

Graphing From a Table of Values

You can make a graph to illustrate this relationship between windchill temperature and actual temperature.

An **ordered pair,** such as (**15**, **10**), (**0**, **−5**), or (**−10**, **−15**) from the table, is a pair of *coordinates* written with parentheses. In an ordered pair, **coordinates** give the address of a point on a graph.

The **origin** is the point where the *horizontal axis* and the *vertical axis* intersect.

When the **first number** in an ordered pair is a positive number, move to the *right* of the origin, and when it is a negative number, move to the *left* of the origin.

When the **second number** in the ordered pair is a positive number, move *up* from the origin, and when it is a negative number, move *down* from the origin.

The points (**15**, **10**) and (**− 10**, **−15**) are graphed here.

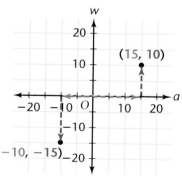

When you graph all of the points from the windchill table and connect them, you form a line. You can use the graph of this line to find the corresponding actual temperatures, *a*, and windchill temperatures, *w*, for a wind speed of 7 miles per hour.

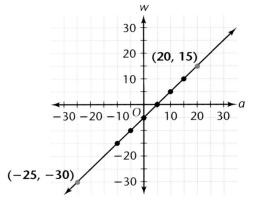

For example, the point (20, 15) is the ordered pair (*a*, *w*) representing an actual temperature of 20°F and a windchill temperature of 15°F.

Also, the point (−25, −30) represents an actual temperature of −25°F and a windchill temperature of −30°F.

Use the graph to find the actual temperature, *a*, for a windchill temperature, *w*, of −25°F.   − 20° F

The line graphed above is represented by the equation $w = a - 5$. This equation is a *linear equation* because its graph is a line.

LINEAR EQUATION
A **linear equation** is an equation whose graph is a line.

EXAMPLE 1

Graph the linear equation $y = 2x - 3$, based on integer values of *x* from −3 to 3.

Solution ➤

First build a table of values. Substitute integer values of *x* from −3 to 3.

x	y = 2x − 3	(x, y)
−3	y = 2(−3) − 3 = −9	(−3, −9)
−2	y = 2(−2) − 3 = −7	(−2, −7)
−1	y = 2(−1) − 3 = −5	(−1, −5)
0	y = 2(0) − 3 = −3	(0, −3)
1	y = 2(1) − 3 = −1	(1, −1)
2	y = 2(2) − 3 = 1	(2, 1)
3	y = 2(3) − 3 = 3	(3, 3)

Graph the ordered pairs from the table.

Connect the points to show the graph of the linear equation $y = 2x - 3$.

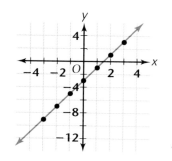

Try This   Graph the equation $y = -2x + 3$, based on integer values of x from -3 to 3.

Graphics Calculator

You can use a graphics calculator to make a table and a graph for the equation $y = 2x - 3$.

First enter the equation.

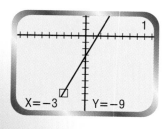

Then create a table using integer values of x beginning with -3.

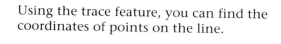

Then graph the equation $y = 2x - 3$.

Using the trace feature, you can find the coordinates of points on the line.

Writing Linear Equations

Linear equations can be used to describe many real-world situations, such as the distance traveled in a given amount of time.

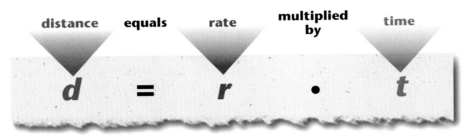

distance	equals	rate	multiplied by	time
d	$=$	r	$\bullet$	t

The data in the following table show the distance in miles that a plane travels in 0 to 5 hours. The linear equation $d = 250t$ describes this situation.

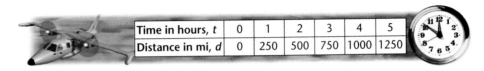

Time in hours, t	0	1	2	3	4	5
Distance in mi, d	0	250	500	750	1000	1250

How fast is the plane flying?   250 miles per hour

Patterns in data can often be described using linear equations.

EXAMPLE 2

Write a linear equation to describe the cost of raft rental.

Time (in hours), x	0	1	2	3
Raft rental (in dollars), y	10	15	20	25

Solution ➤

This data set can be described by this sentence:

The cost of raft rental is $10, plus $5 per hour.

So the equation $y = 10 + 5x$ describes the data.

Graphics Calculator

You can use a graphics calculator to see that each equation is correct. First enter the equation. Then use the table feature.

The **x-values** and the **y-values** in the table above can be found in the table on the graphics calculator. ❖

X	Y₁
0	10
1	15
2	20
3	25
4	30
5	35
6	40

X=0

Try This   The following table gives the cost per day and deposit, for renting a steam carpet cleaner. Write a linear equation for the total amount needed, y.

$y = 40 + 16x$

Time (in days), x	0	1	2	3
Cost (in dollars), y	40	56	72	88

In Example 2, the equation for the cost of raft rental is $y = 10 + 5x$. Suppose the equation $y = 12 + 4x$ describes the cost of raft rental at another shop. If you plan to use the raft for 2 hours, which rental shop is more expensive? Explain.   They are the same price. At the first shop, the cost of a rental for two hours is $10 + 5(2)$ or $20. At the second shop, the cost is $12 + 4(2)$ or $20. The first shop has a lower initial charge but a higher hourly rate, while the second shop has a higher initial charge but a lower hourly rate.

EXERCISES & PROBLEMS

Communicate

1. Explain how to form ordered pairs for an equation from a table of values.   Pick a value for your starting variable. Using that number, compute the value of the other variable from the equation. List the values as ordered pairs.

2. Explain how to graph ordered pairs.

3. Describe how you can use a linear equation to describe a pattern in a table. Give two examples.

4. Use examples to explain how to use a graph of a linear equation to find more data in a given situation.

Practice & Apply

Make a table of values and ordered pairs for each equation by substituting integer values from −3 to 3 for x.

5. $y = 10 - x$
6. $y = 10 + x - 2$
7. $y = -1 - x$
8. $y = 5 - x$

9. $y = x - (-1)$
10. $y = x - 4$
11. $y = 2x - 12$
12. $y = -3x + 12$

13. $y = 5x - 30$
14. $y = -12x$
15. $y = x \div (-1)$
16. $y = x \div (-2)$

17. $y = 2x + 4$
18. $y = -3x + 1$
19. $y = 4x - 3$
20. $y = 2x - (-5)$

Write the ordered pair that represents each point.

21. A $(-5, 2)$

22. B $(2, 3)$

23. C $(4, -4)$

24. D $(-3, -6)$

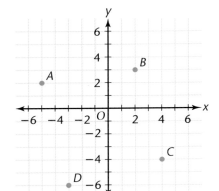

Graph each equation using integer values of x from −2 to 3. Use your graph to find the value of y when x is −3.

25. $y = x - 2$
26. $y = 4 - x$
27. $y = x + 2$
28. $y = x - 5$

29. $y = 2x - 3$
30. $y = 1 - 2x$
31. $y = -2x - 1$
32. $y = 2x - 8$

33. Recreation Using the information in the table below, write a linear equation that describes the cost of bicycle rental.
$$y = 4x + 20$$

BICYCLE RENTAL COST				
Time (in hours), x	0	1	2	3
Cost (in dollars), y	20	24	28	32

34. Photography Write a linear equation that represents the total amount needed for renting an underwater camera, including the deposit.

$y = 45x + 650$

Camera Rental Costs

Time (in days), x	0	1	2	3
Cost (in dollars), y	650	695	740	785

Write a linear equation to describe each set of data.

35.

Time (in hours), x	0	1	2	3
Cost (in dollars), y	12	16	20	24

$y = 12 + 4x$

36.

Time (in days), x	0	1	2	3
Cost (in dollars), y	0	18	36	54

$y = 18x$

37.

Time (in hours), x	0	1	2	3
Distance (in miles), y	0	60	120	180

$y = 60x$

38.

Time (in years), x	0	1	2	3
Weight (in pounds), y	0	24	48	72

Complete each table. Then write a linear equation to describe the relationship between x and y. Use a graphics calculator to check your work.

$y = x + 1$

39.

x	-3	-2	-1	0	1	2	3
y	-2	-1	0	1	2	?	?

3 4

40.

x	-3	-2	-1	0	1	2	3
y	9	6	3	0	-3	?	?

$y = -3x$

$-6, -9$

41.

x	-3	-2	-1	0	1	2	3
y	-5	-2	1	4	?	?	?

$y = 3x + 4$

7, 10, 13

Look Back

Use the bar chart below for Exercises 42–44. [Lesson 1.3]

42. Make a table for the data in the bar chart.

43. What was the difference in CD and tape sales in 1991? 27 more tapes

44. What was the difference in CD and tape sales in 1994? 317 more CD's

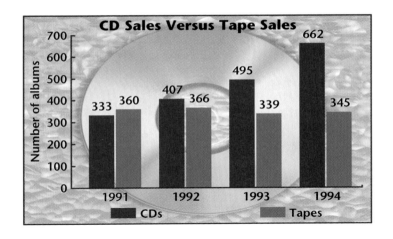

42.

Year	CD sales	Tape sales
1991	333	360
1992	407	366
1993	495	339
1994	662	345

Write the prime factorization of each number. [Lesson 1.6]
45. 49 7^2 **46.** 72 $2^3 \cdot 3^2$ **47.** 17 17   **48.** 81 3^4

Write two inequalities for each pair of integers. Use both the $<$ and $>$ symbols. [Lesson 2.3]

6 > −7

49. −2, −5 −2 > −5   **50.** −3, 1 −3 < 1   **51.** 0, −3 0 > −3   **52.** 6, −7 −7 < 6
    −5 < −2   1 > −3   −3 < 0
53. −4, −7   **54.** 4, −1   **55.** −2, −3   **56.** −5, −2
  −7 < −4   −1 < 4   −3 < −2   −5 < −2
  −4 > −7   4 > −1   −2 > −3   −2 > −5

Look Beyond

57. Using the equations $y = -x + 1$ and $y = 2x - 8$, make a table of values and ordered pairs. Then graph the equations on the same set of axes.

58. Locate the point of intersection of the lines. Write the coordinates of the point of intersection. (3, −2)

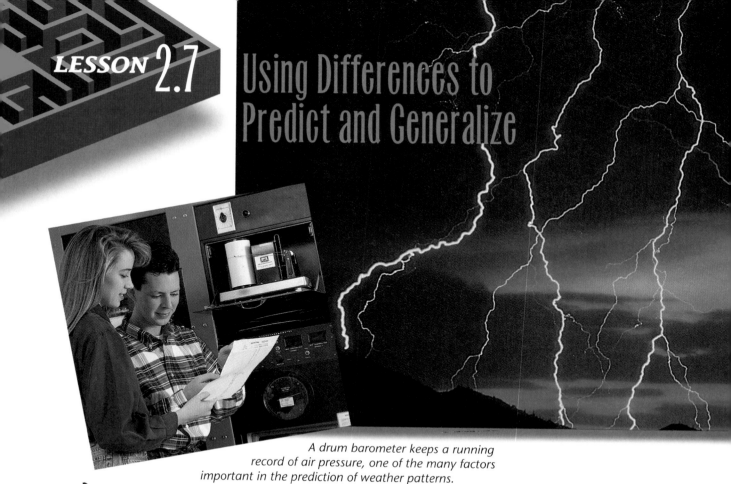

Using Differences to Predict and Generalize

A drum barometer keeps a running record of air pressure, one of the many factors important in the prediction of weather patterns.

Using Differences to Identify Patterns

why

People often make predictions to answer questions they have about the future. We use data from the past and regular patterns in the present to make predictions about the future.

Recall from Lesson 1.1 the problem of finding the number of possible conversations among 5 people. Examine the table again. Look at the first differences between consecutive terms in the sequence 0, 1, 3, 6, 10, . . .

People   1   2   3   4   5
Conversations   0   1   3   6   10

           1   2   3   4   ◄—— First differences

Use the first differences to find the second differences.

People   1   2   3   4   5
Conversations   0   1   3   6   10

           1   2   3   4   ◄—— First differences
              1   1   1   ◄—— Second differences

You can see from the table that the second differences are *constant*. When the first differences are greater than 0, the sequence is increasing. When the first differences are less than 0, the sequence is decreasing.

If you know the differences, you can predict other terms of the sequence.

EXAMPLE 1

Find the next two terms of each sequence.

A 80, 73, 66, 59, 52, . . .   **B** 1, 4, 9, 16, 25, . . .

Solution ➤

A Find the first differences.

80　　73　　66　　59　　52

　　−7　　−7　　−7　　−7 ◀—— First differences are constant.

Notice that each term decreases by 7. The first difference is represented by − 7. Subtract 7 from the previous term to find each new term.

52 − 7 = 45   45 − 7 = 38

The next two terms are 45 and 38.

B Find the first differences.

1　　4　　9　　16　　25

　3　　5　　7　　9 ◀—— First differences are not constant.

Since the first differences are not constant, find the second differences.

1　　4　　9　　16　　25

　3　　5　　7　　9

　　2　　2　　2 ◀—— Second differences are constant.

Now work backward to find the first differences. To find each of the next first differences, add 2 to the previous first difference.

9 + 2 = 11   11 + 2 = 13

1　　4　　9　　16　　25

　3　　5　　7　　9　　11　　13 ◀—— First differences

　　2　　2　　2　　2　　2 ◀—— Second differences are constant.

Continue working backward to find the next two terms of the sequence. To find each of the next terms, add the first difference to the previous term.

25 + 11 = 36   36 + 13 = 49

1　　4　　9　　16　　25　　36　　49

　3　　5　　7　　9　　11　　13 ◀—— First differences

　　2　　2　　2　　2　　2 ◀—— Second differences are constant.

The next two terms are 36 and 49. ❖

CRITICAL
Thinking

The numbers in the sequence 1, 4, 9, 16, . . . are square numbers. The first term is $1 \cdot 1 = 1^2$, or 1. The second term is $2 \cdot 2 = 2^2$, or 4. What is the relationship between the term number and the value of that term?
The value of the term is the term number squared. For example, the fourth term is 4^2.

EXAMPLE 2

Physics   The set of data shown in the table is from the flight of a small rocket during the first 4 seconds of its flight. The flight of the rocket ends when it hits the ground 14 seconds after takeoff. Use the method of finding differences to find the maximum height during the rocket's flight.

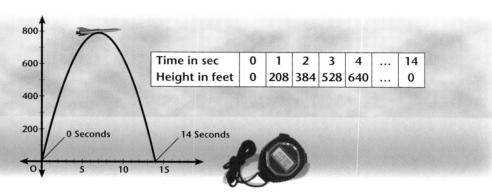

Time in sec	0	1	2	3	4	...	14
Height in feet	0	208	384	528	640	...	0

Solution ➤

Look at the differences in the table to discover the pattern.

Time in seconds	0	1	2	3	4
Height in feet	0	208	384	528	640

208   176   144   112 ◄— First differences
   −32 −32 −32 ◄— Second differences

The second differences are each a constant − 32 feet. This means the first differences *decrease* by 32 feet each time. Use the strategy of **working backward** from the second differences to extend the table.

Time in seconds	0	1	2	3	4	5	6	7	8	9	10
Height in feet	0	208	384	528	640	720	768	784	768	720	640

208 176 144 112 80 48 16 −16 −48 −80
  −32 −32 −32 −32 −32 −32 −32 −32 −32

After 7 seconds, the numbers for the height begin to repeat in reverse order. By continuing the table to 14 seconds, the pattern shows points along the complete path of the rocket.

The rocket's highest altitude is reached halfway through the flight.

Time in seconds	11	12	13	14
Height in feet	528	384	208	0

The table shows that the highest point reached is 784 feet. This occurs at 7 seconds. Why does this information seem reasonable? ❖

Try This Suppose a rocket flight takes 15 seconds from launch to return to the Earth. After how many seconds do you think it will reach its maximum height?
7.5 seconds

Using a spreadsheet can extend your ability to use differences to discover sequence patterns, especially when the differences and calculations become more complicated.

Spreadsheet Compute the first difference of 208 in cell C2 by subtracting the value in cell B2 from the value in cell B3. What instructions do you think are needed to compute the values in cells C3 and D2 of the

	A	B	C	D
	seconds	height	1st diff	2nd diff
2	0	0	208	−32
3	1	208	176	−32
4	2	384	144	−32
5	3	528	112	−32
6	4	640	80	
7	5	720		

spreadsheet? Compute the value in C3 by subtracting the value in B3 from the value in B4. Compute the value in D2 by subtracting the value in C2 from the value in C3.

Using Differences to Identify Linear Equations

The method of differences can be used to write equations that describe linear patterns.

The table below shows the distance in knots that an ocean liner can travel at a constant speed in a given number of hours.

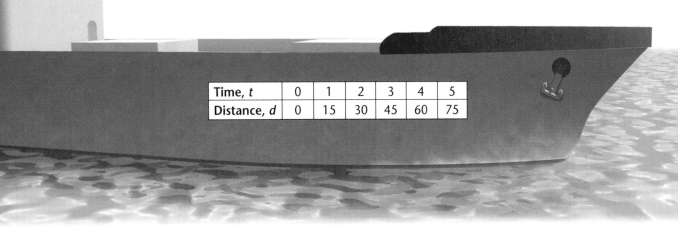

Time, t	0	1	2	3	4	5
Distance, d	0	15	30	45	60	75

CT– Since the distance at time 0 is 0 knots, and since the first differences show that each additional hour adds 15 knots, the table obeys the equation $d = 15t$.

Find the first differences, and examine the table.

Time (in hours), t	0	1	2	3	4	5
Distance (in knots), d	0	15	30	45	60	75

15 15 15 15 15 ◄─── First differences

CRITICAL *Thinking* Explain how to use the differences above to write a linear equation for the relationship between the time in hours, t, and the distance in miles, d.

•Exploration•

Using Linear Equations to Generalize

Suppose you want to rent a rototiller to prepare your garden for planting.

Walt and his grandson rent a rototiller for $20, plus $7 per hour.

1 Explain why the linear equation $c = 20 + 7h$ describes this situation.
The cost is $20 plus $7 for each hour.

2 Copy and complete the table. Find the first differences for the cost of renting this rototiller.

Time (in hours), h	0	1	2	3	4	5
Cost (in dollars), c	20	27	?	?	?	?

? 7   ? 34 7   ? 41 7   ? 48 7   ? 55 ← **First differences**

3 Explain how you can use the first differences from the table to write the equation for the cost.

4 The data in the next table show a rototiller rental fee at *another* store.

Time (in hours), h	0	1	2	3	4	5
Cost (in dollars), c	25	31	?	?	?	?

? 6   ? 37 6   ? 43 6   ? 49 6   ? 55 ← **First differences**

Copy and complete the table. Find the first differences. Use the differences to write a linear equation for the cost. $C = 25 + 6h$

5. For a 3 hour rental, the first store is less expensive. For a 5 hour rental the cost is the same. The second store has a higher initial cost but a lower hourly rate.

5 Compare the two rental plans. Which one would you use if you were going to rent a rototiller for 3 hours? for 5 hours? Explain. ❖

If the first differences in a sequence of terms are constant, then there is a linear equation that describes the sequence of terms.

Recreation The data in the table show the fee to rent a canoe. To write a linear equation for the cost of canoe rental, first find the differences.

Time (in hours), t	0	1	2	3	4	5
Cost (in dollars), c	10	18	26	34	42	50
First Differences →		8	8	8	8	8

Roc-n-Row Canoes

Since the first differences are constant at 8, the hourly rate is $8.

Notice that the cost of canoe rental begins at $10 for 0 hours. Then $10 is the initial charge.

So the cost, c, of renting a canoe is an initial charge of $10 plus $8 per hour, t.

$$c = 10 + 8t \; ❖$$

CRITICAL *Thinking*

The table shows the cost to rent a canoe from another company.

Time (in hours), t	0	1	2	3	4	5
Cost (in dollars), c	0	12.50	25.00	37.50	50.00	62.50

Which company is less expensive? Explain. The second company is less expensive for rentals of 2 hours or less, but becomes more expensive as the rental time lengthens. This occurs because the second company has no initial charge with a higher hourly rate.

EXERCISES & PROBLEMS

Communicate

1. Describe the method for predicting the next two terms in the following sequence: 88, 76, 64, 52, 40, . . .

2. Suppose a rocket takes 17 seconds to return to the Earth after launching. Explain how to find the time it takes the rocket to reach its maximum height.

3. Explain why you can stop calculating additional differences when the differences become constant.

4. Explain how to work backward from second differences to extend a sequence.

5. Describe how to use first differences to write linear equations. Give two examples.

1. The first differences are all −12, so the next two terms are 28 and 16.

2. Since a rocket takes as long to fall as it does to rise, it takes half of the 17 seconds, or 8.5 seconds, to reach maximum height.

3. Once the differences are constant, any higher order differences will be 0. This does not help to predict the pattern for the original sequence.

Practice & Apply

6. Find the first and second differences for the sequence 20, 27, 36, 47, 60.

Find the next two terms of each sequence.

7. 18, 32, 46, 60, 74, .88, 102.

8. 33, 49, 65, 81, 97, . .113, 129

9. 20, 21, 26, 35, 48, .65, 86.

10. 30, 31, 35, 42, 52, .65, 81

11. 100, 94, 88, 82, 76, .70, 64.

12. 44, 41, 38, 35, 32, .29, 26

13. 12, 12, 18, 31, 53, 87, . . .
137, 208

14. 1, 7, 23, 50, 89, . . .
141, 207

Apply the method of finding differences until you get a constant. Which differences are constant? What is it?

15. 1, 2, 3, 4, 5, . . .   The first differences are a constant 1.

16. $1^2, 2^2, 3^2, 4^2, 5^2$, . . . HINT: This is the same as 1, 4, 9, 16, 25, . . . Second differences; 2

17. $1^3, 2^3, 3^3, 4^3, 5^3$, . . . HINT: This is the same as 1, 8, 27, 64, 125, . . .
Third differences; 6

18. Tell how many differences you will have to compute before you get a constant for $1^4, 2^4, 3^4, 4^4, 5^4, 6^4$, . . . Fourth differences

19. The first three terms of a sequence are 7, 11, and 16. Find the first and second differences. Assuming the second differences are constant, what are the next three terms?   22, 29, 37

20. The first three terms of a sequence are 2, 6, and 12. The second differences are a constant 2. What are the next three terms of the sequence?   20, 30, 42

21. The third and fourth terms of a sequence are 15 and 23. If the second differences are a constant 2, what are the first 5 terms of the sequence?   5, 9, 15, 23, 33

22. If the second differences of a sequence are a constant 3, the first of the first differences is 7, and the first term is 2, find the first 5 terms of the sequence.
2, 9, 19, 32, 48

23. Physics If a rocket lands after 20 seconds of flight, at what time do you think it reaches its maximum height? How can you tell?   10 seconds; a rocket takes as long to rise as it does to fall.

24. Suppose a rocket takes 24 seconds to reach its maximum height. How many seconds after launching do you think it hits the ground?   48 seconds

Technology Complete the spreadsheet that follows Example 2. What spreadsheet instruction formula should be in the following cells?

25. C6   C6 = B7 − B6

26. D5   D5 = C6 − C5

27. ▱▽   **Geometry** Complete this table for the perimeter of a square with the length of a side given in centimeters.

Side length	1	2	3	4	5	6	7	8
Perimeter	4	8	12	16	?	?	?	?

20, 24, 28, 32

28. Apply the method of differences to find the next four terms of the sequence 4, 8, 12, 16, . . . Compare the results with the previous exercise. Do you get the same answer? 20, 24, 28, 32; Yes

29. Physical Science As the altitude increases, the boiling point of water decreases according to the table below. For example, in Las Vegas, Nevada, at an altitude of 2180 feet, the boiling point of water is about 98° Celsius. Estimate the boiling point of water in Colorado Springs, Colorado, which is at an altitude of 6170 feet. Approx. 94° C

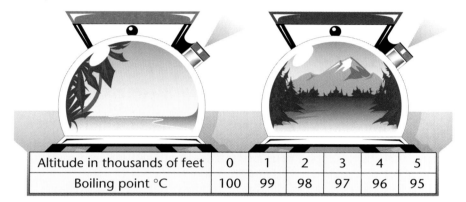

Altitude in thousands of feet	0	1	2	3	4	5
Boiling point °C	100	99	98	97	96	95

Look Back

30. Find the sum of 1 + 2 + 3 + 4 + . . . + 18. **[Lesson 1.1]**   171

31. Find the next three terms in the following pattern. **[Lesson 1.1]**

$$43, 49, 55, 61, 67, . . . \quad 73, 79, 85$$

Rental A truck rental firm charges $25 plus 30 cents per mile to rent a moving truck. **[Lesson 1.3]**

32. Write an equation that shows the cost, c, of renting the truck in terms of the number of miles, m. $c = 0.3m + 25$

33. Use a table to evaluate your expression for 25, 50, 75, 100, and 125 miles.

34. How many miles can you drive the truck if you have budgeted $58 for the rental?   110 miles

Evaluate. [Lesson 2.5]

35. $(-11)(-6)$ **36.** $(-12) \div (-4)$ **37.** $369 \div (-3)$ **38.** $(25)(-4)$
66   3   −123   −100

Look Beyond

The method of finding differences works only for certain kinds of patterns. Try it on the following patterns and explain what happens. Predict the next three terms of each pattern without using differences. Explain your method.

39. 1, 2, 4, 8, 16, 32, . . .   **40.** 1, 10, 100, 1000, 10,000, . . .

Northern Extremes

A region's climate can be described by many factors. Average high and low temperatures are part of a region's climate. The table below shows the average high and low temperatures in Calgary, Alberta. Temperatures in Canada are recorded in degrees Celsius.

Month	Jan	Feb	Mar	Apr	May	June	July	Aug	Sept	Oct	Nov	Dec
Average high	−7	−2	3	11	17	20	24	23	18	12	3	−2
Average low	−15	−14	−9	−3	3	7	9	8	4	−1	−8	−13
Average range	8	12	12	14	14	13	15	15	14	13	11	11

The average range is the difference between the average high and the average low. This can also be described as the average increase from low to high.

The range for June is

$$20 - 7 = 13°C.$$

The range for January is

$$-7 - (-15) = 8°C.$$

Why do you think the range is less in January?

A double-line graph displays the data.

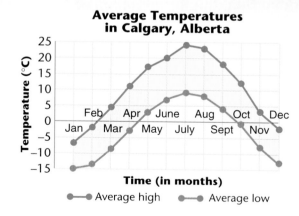

Average Temperatures in Calgary, Alberta

Activity 1

Finding the Range and Building Line Graphs

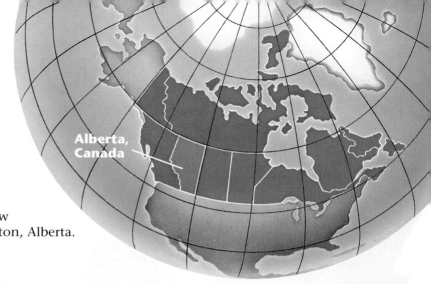

Alberta, Canada

The table below shows the average high and low temperatures in Edmonton, Alberta.

Month	Jan	Feb	Mar	Apr	May	June	July	Aug	Sept	Oct	Nov	Dec
Average high	−8	−6	1	11	18	21	24	22	17	11	1	−7
Average low	−18	−17	−10	−2	4	8	11	8	4	−1	−9	−16
Average range	?	?	?	?	?	?	?	?	?	?	?	?
	10	11	11	13	14	13	13	14	13	12	10	9

1. Copy and complete the table with the average range for each month.

2. Construct a double-line graph. Show the average high and low temperatures in Edmonton, Alberta for each month.

3. Find the average *annual* high, the average *annual* low, and the average *annual* range for Calgary and for Edmonton. Compare the average *annual* ranges. Calgary: high 10°C, low $-2\frac{2}{3}$°C, range $12\frac{2}{3}$°C; Edmonton: high $8\frac{3}{4}$°C, low $-3\frac{1}{6}$°C, range $11\frac{11}{12}$°C

4. Without looking at a map, explain how you can determine whether Edmonton or Calgary is farther north? Explain your reasoning. Edmonton appears to be farther north because its average annual temperatures are lower.

Activity 2

Research

Find the average monthly high and low temperatures for another Canadian city. Build a table and double-line graph for the new data. Compare your data and graph with the information about Alberta. Answers vary.

Chapter 2 Review

Vocabulary

absolute value	75	inequality	88	opposites	74
Addition Propery of Zero	93	integers	72	Properties of Zero	100
additive identity	93	linear equation	105	Property of Opposites	82
first and second differences	112	neutral pair	78	subtraction	95

Key Skills & Exercises

Lesson 2.1

➤ **Key Skills**

Represent integers on a number line.

To find the change in position of a person walking from point *A* to point *B*, look at the distance between the two points. The number 9 describes the distance.

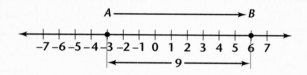

➤ **Exercises**

Examine the diagram of an elevator shaft at the right. Write the indicated change in an elevator's position as a positive or negative integer.

Elevator shaft

15
14
13
12
11
10
9
8
7
6
LOBBY 5
4
3
2
1

1. Starts at the first floor below the lobby, and ends at the sixth floor above the lobby 7
2. Starts at the third floor below the lobby, and ends at the fifth floor above the lobby 8
3. Starts at the second floor below the lobby, and ends at the eighth floor above the lobby 10

Lesson 2.2

➤ **Key Skills**

Add integers.

To find the sum of -3 and -7, first write the absolute values, 3 and 7. Since the signs are *like*, find the *sum* of the absolute values, and use the common sign, negative.

$$-3 + (-7) = -10$$

To find the sum of -3 and 7, first write the absolute values, 3 and 7. Since the signs are *unlike*, find the *difference* of the absolute values, and use the sign of 7, the integer with the greater absolute value.

$$-3 + 7 = 4$$

➤ **Exercises**

Find the sum.

4. $8 + (-9)$ -1 **5.** $-29 + (-37)$ -66 **6.** $-17 + 23$ 6   **7.** $-41 + (-5) + 19$ -27

Lesson 2.3

➤ Key Skills

Solve an equation using guess-and-check.

To solve $-8 + x = -6$, try substituting 4 for x.

$$-8 + 4 \overset{?}{=} -6$$
$$-4 \neq -6 \qquad \text{Try a smaller number.}$$
$$-8 + 2 \overset{?}{=} -6$$
$$-6 = -6$$

The solution is 2.

Order integers using inequality symbols and a number line.

To determine which number, -4 or -1, is the greater number, use a number line.

Since -4 is to the left of -1, write $-4 < -1$.

➤ Exercises

Use guess-and-check to solve each equation.

8. $x + 3 = -3$ -6 **9.** $-7 + x = 5$ 12   **10.** $9 + x = 2$ -7 **11.** $x + (-2) = -5$ -3

Show each set of integers on a number line. Then list the integers in ascending order (from least to greatest).

12. $8, -6, -3, 0$ **13.** $4, -4, 5, -5, 1$

Lesson 2.4

➤ Key Skills

Subtract integers.

To subtract -8, add the opposite.   To subtract 19, add the opposite.

$$-16 - (-8) = -16 + 8 = -8 \qquad -38 - 19 = -38 + (-19) = -57$$

➤ Exercises

Find each difference.

14. $9 - (-15)$ 24   **15.** $48 - (-48)$ 96   **16.** $-13 - 28$ -41
17. $39 - (-18)$ 57   **18.** $-67 - (-42)$ -25 **19.** $-23 - (-72)$ 49
20. $-43 - (-42) - 53$ -54 **21.** $8 - 14 - 27$ -33 **22.** $54 - (-42)$ 96
23. $-44 - 63$ -107 **24.** $25 - 53$ -28 **25.** $94 - (-33)$ 127

Lesson 2.5

➤ Key Skills

Multiply and divide integers.

Use the rules for multiplying and dividing integers to evaluate the following expressions.

$$(-4)(-3) = 12 \qquad\qquad (-72) \div (-9) = 8$$
$$(4)(-3) = -12 \qquad\qquad (72) \div (-9) = -8$$
$$(-4)(3) = -12 \qquad\qquad (-72) \div (9) = -8$$
$$(4)(3) = 12 \qquad\qquad (72) \div (9) = 8$$

➤ Exercises

Evaluate.

26. $(-12)(-5)$ 60   **27.** $54 \div (-9)$ -6 **28.** $(-6)(8)$ -48
29. $-64 \div (-4)$ 16   **30.** $(-121) \div (-11)$ 11   **31.** $(-6)(-3) \div (-9)$ -2
32. $45[8 + (-8)]$ 0   **33.** $(-5)(-5)(-1)(1)$ -25 **34.** $(-45) \div [(-3)(-3)]$ -5
35. $12(-2) \div [(-2)(3)]$ 4 **36.** $-5[(-4) -5]$ 45   **37.** $[(-4)(-7) \div (2)](-7)$ -98

Lesson 2.6

➤ Key Skills

Graph a linear equation based on a table of integer values.

To graph the linear equation
$y = -2x + 3$, build a table of values.

Then graph the ordered pairs
from the table.

x	$y = -2x + 3$	(x, y)
-2	$y = -2(-2) + 3 = 7$	$(-2, 7)$
-1	$y = -2(-1) + 3 = 5$	$(-1, 5)$
0	$y = -2(0) + 3 = 3$	$(0, 3)$
1	$y = -2(1) + 3 = 1$	$(1, 1)$

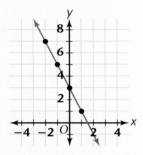

➤ Exercises

Graph each equation using integer values of x from -2 to 3. Begin by making a table of values. Use your graph to find the value of y when x is -3.

38. $y = x + 3$ **39.** $y = 3 - x$ **40.** $y = 2x - 4$

Lesson 2.7

➤ Key Skills

Use differences to identify linear equations.

In the table, determine whether the first differences are constant.

The first differences are constant, so there is a linear equation for the table.

Bicycle Rental

Time (in hours)	0	1	2	3	4
Cost (in dollars)	5	11	17	23	29

6 6 6 6

The hourly rate is $6. Since the cost for 0 hours is $5, a deposit of $5 is charged. So, the cost, c, is $5 plus $6 per hour, t.

$$c = 5 + 6t$$

➤ Exercises

41. Explain how you know that there is a linear equation to describe the cost of ski rental. Then write a linear equation for the cost of ski rental.   Equation is linear because first differences are constant; $c = 9 + 7t$.

Ski Rental

Time (in hours), t	0	1	2	3	4
Cost (in dollars), c	9	16	23	30	37

Applications

42. Sports The Madison High football team completed 4 downs with the gains and losses shown at the right. Find the total gain or loss.   4 yard gain

Yards Gained/Lost

Down	1	2	3	4
Gain/Loss	3	-7	-1	9

Banking Alice opened a checking account with a deposit of $600, followed by 3 deposits of $80 each and then by 5 withdrawals of $110 each.

43. What is Alice's account balance immediately after the 3 deposits? $840

44. What is Alice's account balance immediately after the 5 withdrawals? $290

Chapter 2 Assessment

Write the indicated change in position on the number line as a positive or negative number.

1. Starts at −5 and ends at 3. 8

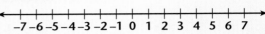

$$-7\ -6\ -5\ -4\ -3\ -2\ -1\ \ 0\ \ 1\ \ 2\ \ 3\ \ 4\ \ 5\ \ 6\ \ 7$$

2. Starts at 4 and ends at −2. −6
3. Starts at −1 and ends at −6. −5

Find each sum.

4. $6 + (-5)$ 1
5. $-13 + (-7)$ −20
6. $-11 + 31$ 20
7. $4 + (-11) + 6$ −1
8. $-1 + (-7) + (-8)$ −16
9. $31 + (-56) + (-7)$ −32
10. $14 + |14| + (-6)$ 22
11. $-114 + |14| + |-15|$ −85

Use guess-and-check to solve each equation.

12. $x + 5 = -8$ −13
13. $-4 + x = 12$ 16
14. $x + (-41) = 42$ 83
15. $3 + x = 1$ −2
16. $-13 + x = -6$ 7
17. $18 + x = -18$ −36
18. $-47 = x + (-54)$ 7
19. $6 + x = 17$ 11

20. Show each integer on a number line. Then list the integers in ascending order.

$$9, -8, 5, -6, -4, 7$$

Find each difference.

21. $13 - (-13)$ 26
22. $17 - (-23)$ 40
23. $-51 - (-49)$ −2
24. $1 - 2 - 3$ −4
25. $-9 - (-4)$ −5
26. $12 - 15 - (-2)$ −1
27. $45 - (-55) - 21$ 79
28. $4 - (-4)$ 8

Find the distance between the following pairs of points on the number line.

29. 3, 11 8
30. −5, 10 15
31. −61, −97 36
32. 58, −34 92
33. −9, 12 21
34. −35, −45 10
35. −61, 85 146
36. 43, −92 135

Evaluate.

37. $99(-2)$ −198
38. $-102 - 17$ −119
39. $-20[-16 + (-14)]$ 600
40. $(-3)(-3)(-3)$ −27
41. $-7(27) \div (-9)$ 21
42. $-80 \div [(-20)(-2)]$ −2
43. $\dfrac{(-5)(-15)}{5}$ 15
44. $\dfrac{(-2)(12)(-28)}{21}$ 32

45. Make a table of values for $y = -2x + 5$ using integer values of x from −3 to 3. Then graph the equation.

46. The second and third terms of a sequence are 9 and 20. The second differences are a constant 4. What are the first five terms of the sequence? 2, 9, 20, 35, 54

47. The table shows the distance that a turtle can travel at a constant speed in a given number of hours. Explain how you know that there is a linear equation for the data in the table. Then write a linear equation for the data in the table.

Time (in hours), t	0	1	2	3	4
Distance (in yards), d	0	26	52	78	104

Chapters 1-2
Cumulative Assessment

College Entrance Exam Practice

Quantitative Comparison For Questions 1–4, write

A if the quantity in Column A is greater than the quantity in Column B;

B if the quantity in Column B is greater than the quantity in Column A;

C if the two quantities are equal; or

D if the relationship cannot be determined from the information given.

	Column A	Column B	Answers
1.	The eighth term of the sequence 2, 5, 8, 11, 14, . . .	The eighth term of the sequence 32, 30, 28, 26, 24, . . .	(A) (B) (C) (D) **[Lesson 1.1]** A
2.	−7	5	(A) (B) (C) (D) **[Lesson 2.3]** B
3.	$c + 8 \cdot c$ if c is 2	$c - 8 \cdot c$ if c is −2	(A) (B) (C) (D) **[Lesson 1.4]** A
4.	Play tickets: $7 each for balcony seats $9 each for main-floor seats 9 balcony tickets	7 main-floor tickets	(A) (B) (C) (D) **[Lesson 1.4]** C

5. Juanita's checking account has $150. She withdraws $80, deposits $170, and then withdraws $195. What is the ending balance in her checking account? **[Lesson 2.1]** c

   a. $95   **b.** $55   **c.** $45   **d.** $255

6. A basketball league has 8 teams. Each of the teams plays each of the other teams. How many games are played?   **[Lesson 1.1]** a

   a. 28   **b.** 36   **c.** 72   **d.** 56

7. What is the value of $-29 + |-24| + 3$?   **[Lesson 2.2]** b

   a. 2   **b.** −2   **c.** 50   **d.** 8

8. Find the distance between the points labeled −61 and 27 on the number line. **[Lesson 2.4]** d

   a. 34   **b.** −34   **c.** −88   **d.** 88

9. Which ordered pair is *not* an ordered pair for the equation $y = 3x - (-4)$? **[Lesson 2.6]** d

   a. (1, 7)   **b.** (2, 10)   **c.** (3, 13)   **d.** (−2, 2)

10. What are the first three terms of a sequence if the fourth and fifth terms are 27 and 39, and the second differences are a constant 2? **[Lesson 2.7]** b
 a. 21, 23, 25   **b.** 3, 9, 17   **c.** 6, 8, 10   **d.** 13, 15, 17

11. Find the next three terms in the sequence 729, 243, 81, . . . Explain the pattern used to find the terms. **[Lesson 1.1]** 27, 9, 3; divide each term by 3.

12. Find the values for y by substituting the values 1, 2, 3, 4, and 5 for x in $y = 3x - 2$. **[Lesson 1.7]** 1, 4, 7, 10, 13

13. Write an equation for the following sentence: The weekly pay, w, is equal to the number of hours, h, worked times $11.50 per hour. **[Lesson 1.2]** $w = 11.5h$

14. The cost of renting a car is given by the formula $c = 20 + 0.30m$, where c is the total rental cost and m is the number of miles driven. You have $80 budgeted for the rental. How many miles can you drive? **[Lesson 1.3]** 200 miles

15. Determine whether 533 is prime or composite. **[Lesson 1.5]** composite

16. Write the prime factorization of 7380 in exponential form. **[Lesson 1.6]** $41 \cdot 5 \cdot 3^2 \cdot 2^2$

17. Evaluate $- |-54|$. **[Lesson 2.1]** -54

18. Use mental math to find the value of $(68 + 26) + 32$. Identify the property that you used. **[Lesson 1.8]** $(68 + 32) + 26 = 100 + 26 = 126$; Associative property

19. Show the integers 9, -2, 7, and -1 on a number line. Then list the integers in ascending order. **[Lesson 2.3]**

20. Evaluate $(-8)(7)(-36) + (-28)$. **[Lesson 2.5]** 1988

Free-Response Grid   The following questions may be answered using a free-response grid commonly used by standardized test services.

21. Find the greatest prime factor of 341. **[Lesson 1.5]** 31

22. Evaluate the expression $x^3 y^z$ if x is 2, y is 4, and z is 3. **[Lesson 1.6]** 512

23. Simplify $2(3 + 4) - 2 + 3^2(4) - 6 + 1$. **[Lesson 1.7]** 43

24. Find the value of the expression $15h + 20$ when h equals 20. **[Lesson 1.3]** 320

25. Find the value of y in the equation $y = 3x - (-25)$ when x is -4. **[Lesson 2.6]** 13

CHAPTER 3

LESSONS

Rational Numbers and Probability

Rational numbers can be written as fractions, decimals, and percents. You can find these numbers in a variety of real-world applications. For example, the results of surveys and opinion polls are often given in rational numbers.

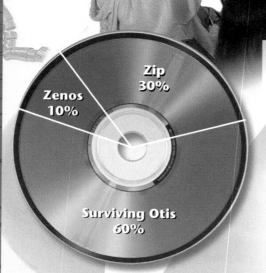

Zip
30%

Zenos
10%

Surviving Otis
60%

Three out of five voters chose Surviving Otis to be the band that plays at the spring dance.

Students at a high school take a survey to decide which of the top three bands will play at the spring dance. After the votes are counted, the results are given in percents and ratios.

Out of 540 voters, 60%, or 324, chose
Surviving Otis to play at the spring dance.

PORTFOLIO ACTIVITY

Leonardo Fibonacci, a famous mathematician who lived in the Middle Ages, discovered a special sequence of numbers that became known as the *Fibonacci sequence.* Each term in this sequence, beginning with the third term, is the sum of the two previous terms.

$$1, 1, 2, 3, 5, 8, 13, 21, 34, 55, 89, \ldots$$

Consecutive Fibonacci numbers can be used to approximate the dimensions of special rectangles called **golden rectangles**. Golden rectangles are special because they are very pleasing to the eye. The ratio of the length to width of a golden rectangle is approximately 1.618034. This rational number can represent the **golden ratio**.

Use consecutive pairs of Fibonacci numbers to approximate the dimensions of golden rectangles. Compare the ratio of each pair of dimensions with the golden ratio, ≈1.618034.

Find five rectangles in the real world with sides whose lengths are in ratios close to the golden ratio.

Length	Width	Ratio of dimensions
5	3	$\frac{5}{3} \approx 1.667$
8	5	8/5 = ? 1.6
13	8	13/8 = ? 1.625

Introduction to Rational Numbers

Software Today +$\frac{1}{2}$

Music Mart +$\frac{1}{4}$

A-1 Video +$\frac{3}{4}$

Almundy Security +$\frac{3}{8}$

adbak ustral

Mi Se −$\frac{1}{2}$

Why *Fractions are used to describe many everyday situations. Systems of measurement, such as the English system of inches and feet, rely on fractions. Stock analysts and stock owners use fractions to measure market changes.*

Stocks As a graduation present, Michael received *stock shares* in four different corporations. A **stock share** represents a small piece of ownership of a corporation. Michael reads the business section of the newspaper frequently to determine the changes in each of his stocks. A **stock change** is an increase or decrease in the price of a stock share. The changes are listed as fractions of a dollar. The changes in his stocks one Monday are listed above.

To find out which stock share shows the most growth, you need to know how to compare the fractions.

Comparing Fractions

A fraction bar can be used to model different fractions. The bar in each row is divided into an equal number of smaller parts. For example, the bar in the third row is divided into 3 equal parts.

Each part represents the fraction $\frac{1}{3}$. Two parts represent $\frac{2}{3}$.

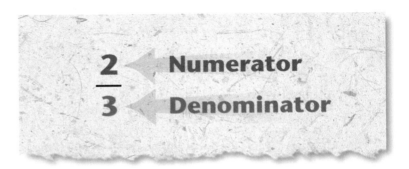

$$\frac{2}{3} \quad \text{Numerator / Denominator}$$

The number of parts needed to represent the fraction is called the **numerator**. The number of parts needed to make one whole bar is called the **denominator**.

Draw a fraction bar to represent the fraction $\frac{4}{6}$.

Equivalent fractions are fractions that have the same value. You can see from the fraction model that $\frac{1}{2}$ has the same value as $\frac{2}{4}, \frac{3}{6}, \frac{4}{8}, \frac{5}{10}$, and $\frac{6}{12}$.

An equal sign (=) indicates that two expressions are equivalent.

$$\frac{1}{2} = \frac{2}{4} = \frac{3}{6} = \frac{4}{8} = \frac{5}{10} = \frac{6}{12}$$

You can use fraction bars to compare fractions.

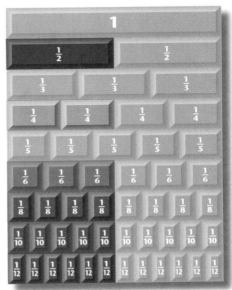

EXAMPLE 1

Economics Use fraction bars to write, in order from least to greatest, the increases in Michael's stocks: $\frac{1}{2}$, $\frac{1}{4}$, $\frac{3}{4}$, and $\frac{3}{8}$. Which stock increased in value the most?

Solution ➤

From the model, you can see that from least to greatest the fractions are

$$\frac{1}{4}, \frac{3}{8}, \frac{1}{2}, \text{ and } \frac{3}{4}.$$

Since the greatest number is $\frac{3}{4}$, the stock share for A-1 Video increased in value the most. ❖

Try This Use fraction bars to write, in order from least to greatest, the fractions $\frac{2}{3}$, $\frac{9}{12}$, $\frac{1}{2}$, and $\frac{8}{12}$. $\frac{1}{2}, \frac{2}{3}$ $= \frac{8}{12}, \frac{9}{12}$

Fraction bars can be used to estimate fraction values.

EXAMPLE 2

Use fraction bars to tell whether each fraction is closest to $\frac{1}{3}$, $\frac{1}{2}$, or 1.

Ⓐ $\frac{3}{10}$ **Ⓑ** $\frac{5}{6}$

Solution ➤

Ⓐ Compare $\frac{3}{10}$ with $\frac{1}{3}$, $\frac{1}{2}$, and 1.

Since 3 of the $\frac{1}{10}$ bars are closest in size to the $\frac{1}{3}$ bar, $\frac{3}{10}$ is closest to $\frac{1}{3}$.

Ⓑ Compare $\frac{5}{6}$ with $\frac{1}{3}$, $\frac{1}{2}$, and 1.

Since 5 of the $\frac{1}{6}$ bars are closest in length to the 1 bar, $\frac{5}{6}$ is closest to 1. ❖

Using Equivalent Fractions

GEOMETRY
Connection

Area models can be used to model equivalent fractions. Notice that in the watermelon models shown, it takes three of the $\frac{1}{12}$ pieces to equal one $\frac{1}{4}$ piece.

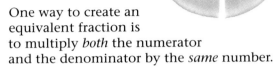

$\frac{1}{4}$ $\frac{1}{12}$ $\frac{1}{12}$ $\frac{1}{12}$ $\frac{1}{12}$

Use this information to complete the following statement.

$$\frac{1}{4} = \frac{?}{12} \quad 3$$

One way to create an equivalent fraction is to multiply *both* the numerator and the denominator by the *same* number.

$$\frac{1}{4} = \frac{1\cdot 3}{4\cdot 3} = \frac{3}{12}$$

The multiplication $\frac{1\cdot 3}{4\cdot 3} = \frac{3}{12}$ is the same as $\frac{1}{4}\cdot\frac{3}{3} = \frac{3}{12}$ and $\frac{1}{4}\cdot 1 = \frac{3}{12}$. So $\frac{3}{12}$ has the same value as $\frac{1}{4}$.

Notice that $\frac{3}{3} = 1$. Explain why multiplying $\frac{1}{4}$ by $\frac{3}{3}$ results in the same value as $\frac{1}{4}$. Since $\frac{3}{3} = 1$, multipying by $\frac{3}{3}$ does not change the value of $\frac{1}{4}$.

EXAMPLE 3

Show that $\frac{2}{5}$ and $\frac{8}{20}$ are equivalent fractions.

CT— You can use division to show that fractions are equivalent, since dividing any number by one does not change its value.

Solution ➤

Multiply $\frac{2}{5}$ by $\frac{4}{4}$ to get $\frac{8}{20}$.

$$\frac{2}{5} = \frac{2}{5}\cdot\frac{4}{4} = \frac{8}{20} \; \diamondsuit$$

Try This Show that $\frac{5}{7}$ and $\frac{10}{14}$ are equivalent fractions. $\frac{5}{7}\cdot\frac{2}{2} = \frac{10}{14}$

CRITICAL
Thinking

Explain why you can also use division to show that the fractions are equivalent.

$$\frac{10\div 2}{14\div 2} = \frac{5}{7} \quad \text{and} \quad \frac{10}{14}\div\frac{2}{2} = \frac{5}{7}$$

Use division to find two equivalent fractions for $\frac{18}{30}$. $\frac{18\div 2}{30\div 2} = \frac{9}{15}, \frac{18\div 6}{30\div 6} = \frac{3}{5}$

EXERCISES & PROBLEMS

Communicate

Trish wants to buy $\frac{1}{3}$ of a pound of Parmesan cheese, and Bud is buying $\frac{2}{6}$ of a pound of Parmesan cheese.

1. Explain what it means for two fractions to be equivalent. Give an example of two equivalent fractions. Include a model in your example.

2. Explain why multiplying $\frac{3}{4}$ by $\frac{2}{2}$ produces a fraction equivalent to $\frac{3}{4}$.

3. Show how you can use fraction bars to find equivalent fractions.

4. Compare the amounts of cheese that Trish and Bud are buying. Is one amount larger? Explain.

Practice & Apply

Use fraction bars to tell whether each fraction is closest to 0, $\frac{1}{2}$, or 1.

5. $\frac{7}{10}$ $\frac{1}{2}$ **6.** $\frac{2}{3}$ $\frac{1}{2}$ **7.** $\frac{5}{12}$ $\frac{1}{2}$

Compare the fractions, using the symbols $<$, $>$, or $=$.

8. $\frac{1}{2}$ and $\frac{4}{8}$ $=$ **9.** $\frac{2}{3}$ and $\frac{6}{12}$ $>$ **10.** $\frac{1}{3}$ and $\frac{1}{6}$ $>$ **11.** $\frac{4}{5}$ and $\frac{30}{40}$ $>$

12. $\frac{1}{2}$ and $\frac{5}{6}$ $<$ **13.** $\frac{2}{3}$ and $\frac{5}{6}$ $<$ **14.** $\frac{1}{3}$ and $\frac{1}{21}$ $>$ **15.** $\frac{3}{12}$ and $\frac{1}{4}$ $=$

Tell whether each fraction is closest to $\frac{1}{4}$, $\frac{1}{2}$, or $\frac{3}{4}$.

16. $\frac{7}{10}$ $\frac{3}{4}$ **17.** $\frac{2}{3}$ $\frac{3}{4}$ **18.** $\frac{2}{5}$ $\frac{1}{2}$ **19.** $\frac{7}{8}$ $\frac{3}{4}$ **20.** $\frac{6}{8}$ $\frac{3}{4}$

21. $\frac{7}{12}$ $\frac{1}{2}$ **22.** $\frac{5}{10}$ $\frac{1}{2}$ **23.** $\frac{8}{12}$ $\frac{3}{4}$ **24.** $\frac{10}{11}$ $\frac{3}{4}$ **25.** $\frac{8}{10}$ $\frac{3}{4}$

Compare each fraction with $\frac{1}{2}$, using the symbols $<$, $>$, or $=$.

26. $\frac{3}{10}$ $<$ **27.** $\frac{2}{3}$ $>$ **28.** $\frac{3}{5}$ $>$ **29.** $\frac{5}{8}$ $>$

30. $\frac{3}{7}$ $<$ **31.** $\frac{9}{11}$ $>$ **32.** $\frac{7}{12}$ $>$ **33.** $\frac{4}{12}$ $<$

Economics As with integers, fractions can be positive or negative. Such fractions are used to list stock-market changes. Increases in stocks are listed as positive fractions, while decreases are listed as negative fractions. Stock market analysts often rank stocks from the lowest performing stock to the highest performing stock.

	Stock	Change		Stock	Change
💾	Software Today	$+\frac{1}{2}$	🎓	Edu-Corp.	$+\frac{3}{4}$
🎵	Music Mart	$+\frac{1}{4}$	📹	A-1 Video	$-\frac{3}{4}$
🎬	Cine Video	$-\frac{1}{4}$	💻	Compu Ware	$-\frac{1}{2}$
🔒	Almundy Security	$-\frac{3}{8}$	R	DNB Rentals	$+\frac{7}{8}$

Today's Most Active Stocks

34. Rank the stock shares for A-1 Video, Almundy Security, and Cine Video from lowest performing to highest performing. A-1 Video, Almundy Security, Cine Video

35. Rank the stock shares for Music Mart, CompuWare, and Software Today from highest performing to lowest performing. Software Today, Music Mart, CompuWare

36. Rank all eight of the stock shares from lowest performing to highest performing. A-1 Video, CompuWare, Almundy Security, Cine Video, Music Mart, Software Today, Edu Corp, DNB Rentals

37. Write five fractions that are equivalent to $\frac{2}{3}$. Answers may vary. Examples: $\frac{4}{6}, \frac{6}{9}, \frac{8}{12}, \frac{10}{15}, \frac{12}{18}$

Look Back

Compare the numbers, using the symbols <, >, or =. [Lesson 2.3]

38. -1 and 3 39. -3 and -5 40. -45 and -50 41. -100 and -98
    <   >   >   <

Use guess-and-check to solve each equation. [Lesson 2.3]

42. $x + 5 = -1$ 43. $3 - x = -2$ 44. $2x = -6$ 45. $2x + 1 = 5$
    $x = -6$ $x = 5$ $x = -3$ $x = 2$

The data in the table below show the rental cost for a surfboard. [Lesson 2.6]

Time (in hours), x	0	1	2	3	4	5
Cost (in dollars), y	5.00	7.50	10.00	12.50	15.00	17.50

46. Find the first differences. 2.50

47. Use the first differences to write a linear equation that describes the cost. $y = 5 + 2.50x$

Look Beyond

Cultural Connection: Africa Ancient Egyptians expressed fractional amounts using sums of different unit fractions. For example, they would represent $\frac{2}{3}$ as $\frac{1}{2} + \frac{1}{6}$. Express each of the following amounts as the sum of two or more unit fractions. HINT: You may wish to use fraction bars. Answers may vary.

48. $\frac{3}{4}$ $\frac{1}{4} + \frac{1}{2}$ 49. $\frac{5}{6}$ $\frac{1}{2} + \frac{1}{3}$ 50. $\frac{11}{24}$ $\frac{1}{3} + \frac{1}{8}$ 51. $\frac{7}{9}$ $\frac{1}{3} + \frac{4}{9}$

LESSON 3.2 Using Equivalent Fractions

WHY *When you know how to write equivalent fractions with different denominators, you can compare fractions.*

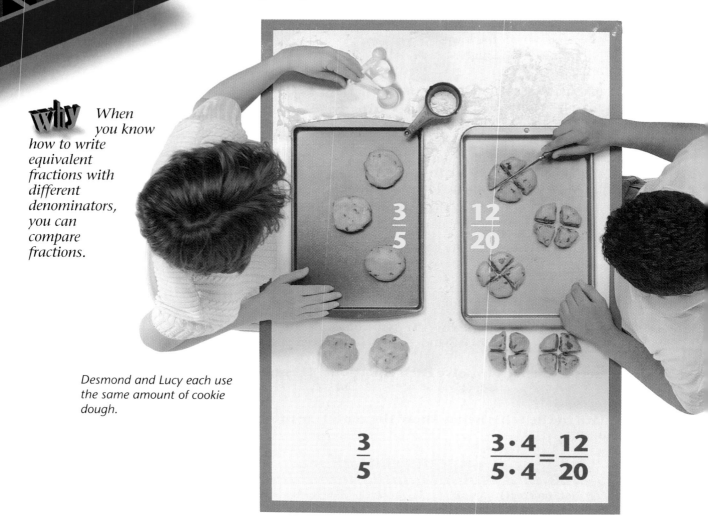

$$\frac{3}{5}$$

$$\frac{12}{20}$$

Desmond and Lucy each use the same amount of cookie dough.

$$\frac{3}{5} \qquad \frac{3 \cdot 4}{5 \cdot 4} = \frac{12}{20}$$

Notice from the cookies that Desmond and Lucy are making that $\frac{3}{5}$ and $\frac{12}{20}$ are equivalent fractions. There are many fractions that are equivalent to $\frac{3}{5}$.

Fractions in Lowest Terms

All of the fractions listed below are equivalent.

$$\frac{3}{5} \qquad \frac{6}{10} \qquad \frac{9}{15} \qquad \frac{12}{20} \qquad \frac{15}{25} \qquad \frac{60}{100}$$

Only one of the fractions, $\frac{3}{5}$, is in *lowest terms*.

A fraction is in **lowest terms** when the only factor common to both the numerator and denominator is **1**.

$$\frac{3}{5} \rightarrow \begin{array}{l} \text{factors of 3: } \textbf{1} \text{ and 3} \\ \text{factors of 5: } \textbf{1} \text{ and 5} \end{array}$$
The only common factor is **1**. $\Bigg\} \quad \frac{3}{5}$ is in lowest terms.

Follow these steps to express a fraction, such as $\frac{12}{20}$, in lowest terms.

1. Find the largest integer that will divide evenly into both the numerator and the denominator. This integer is called the **greatest common factor (GCF)**.

Factors of 12: 1, 2, 3, **4**, 6, and 12
Factors of 20: 1, 2, **4**, 5, 10, and 20

You can see that **4** is the greatest common factor (GCF) of 12 and 20.

2. Divide both the numerator, 12, and the denominator, 20, by the GCF, **4**, to form an equivalent fraction in lowest terms.

$$\frac{12 \div 4}{20 \div 4} = \frac{3}{5}$$

CRITICAL Thinking

Is it possible for a fraction to be in lowest terms if both the numerator and the denominator are even numbers? Explain. No. If both terms are even then they can be divided by 2.

EXAMPLE 1

Express each fraction in lowest terms.

A $\frac{8}{16}$ **B** $\frac{9}{15}$ **C** $\frac{60}{100}$

Solution ➤

A The GCF of 8 and 16 is 8. Divide both numerator and denominator by 8.
$$\frac{8}{16} = \frac{8 \div 8}{16 \div 8} = \frac{1}{2}$$
$\frac{8}{16}$ is written in lowest terms as $\frac{1}{2}$.

B The GCF of 9 and 15 is 3. Divide both numerator and denominator by 3.
$$\frac{9}{15} = \frac{9 \div 3}{15 \div 3} = \frac{3}{5}$$
$\frac{9}{15}$ is written in lowest terms as $\frac{3}{5}$.

C The GCF of 60 and 100 is 20. Divide both numerator and denominator by 20.
$$\frac{60}{100} = \frac{60 \div 20}{100 \div 20} = \frac{3}{5}$$
$\frac{60}{100}$ is written in lowest terms as $\frac{3}{5}$. ❖

Try This   Express each fraction in lowest terms.

a. $\frac{5}{15}$ $\frac{1}{3}$ **b.** $\frac{14}{21}$ $\frac{2}{3}$ **c.** $\frac{75}{100}$ $\frac{3}{4}$

The Least Common Denominator

Equivalent fractions can be used to compare one fraction with another. To compare two fractions, first express each fraction with the same denominator, the *least common denominator*. The **least common denominator (LCD)** of two fractions is the smallest number that is a multiple of both denominators.

EXAMPLE 2

Use the least common denominator (LCD) to compare $\frac{3}{4}$ and $\frac{5}{6}$, using the symbols <, >, or =.

Solution ➤

Step 1
Find the least common multiple of the denominators 4 and 6.

Multiples of 4: 4, 8, 12, 16, 20, 24,...

Common Multiple 12

Common Multiple 24

Multiples of 6: 6, 12, 18, 24, 30, 36,...

The least common multiple of 4 and 6 is 12.
Thus, the least common denominator (LCD) of $\frac{3}{4}$ and $\frac{5}{6}$ is 12.

Step 2
Express each fraction with the LCD, 12. Find the number that you can multiply the numerator and denominator of each fraction by to get a denominator of 12.

$$\frac{3}{4} \cdot \frac{3}{3} = \frac{9}{12} \qquad \frac{5}{6} \cdot \frac{2}{2} = \frac{10}{12}$$

Least Common Denominator
12

Step 3
Once the denominators of the fractions are the same, compare the numerators. Since $9 < 10$, $\frac{9}{12} < \frac{10}{12}$. So, $\frac{3}{4} < \frac{5}{6}$. ❖

$\frac{25}{60} < \frac{36}{60}$; therefore $\frac{5}{12} < \frac{3}{5}$.

Try This Use the least common denominator to compare $\frac{5}{12}$ and $\frac{3}{5}$ using the symbols <, >, or =.

EXERCISES & PROBLEMS

Communicate

1. Explain how to find the greatest common factor (GCF) of two numbers. List all factors of both numbers and pick the largest factor that appears in both lists.
2. How can you use the GCF to write a fraction in lowest terms? Find the GCF of both the numerator and denominator. Then divide the numerator and denominator by that number.
3. Explain how to find the least common denominator (LCD) of two fractions. List several multiples of the denominators of the two fractions. The LCD is the smallest number that appears in both lists.
4. How can you use the LCD to compare two fractions using the symbols <, >, or =? Write the fractions with the least common denominator. Then compare the numerators of both fractions.

Practice & Apply

Write each fraction in lowest terms.

5. $\frac{8}{10}$ $\frac{4}{5}$ 6. $\frac{12}{18}$ $\frac{2}{3}$ 7. $\frac{5}{45}$ $\frac{1}{9}$ 8. $\frac{108}{126}$ $\frac{6}{7}$ 9. $\frac{17}{34}$ $\frac{1}{2}$ 10. $\frac{13}{23}$ $\frac{13}{23}$

11. $\frac{48}{64}$ $\frac{3}{4}$ 12. $\frac{150}{600}$ $\frac{1}{4}$ 13. $\frac{32}{88}$ $\frac{4}{11}$ 14. $\frac{36}{45}$ $\frac{4}{5}$ 15. $\frac{56}{72}$ $\frac{7}{9}$ 16. $\frac{48}{72}$ $\frac{2}{3}$

Use the LCD to compare each pair of fractions, using the symbols <, >, or =.

17. $\frac{1}{2}$ and $\frac{4}{5}$ < 18. $\frac{2}{3}$ and $\frac{4}{5}$ < 19. $\frac{1}{3}$ and $\frac{3}{16}$ > 20. $\frac{1}{2}$ and $\frac{6}{14}$ >

21. $\frac{2}{3}$ and $\frac{4}{15}$ > 22. $\frac{25}{32}$ and $\frac{5}{8}$ > 23. $\frac{2}{3}$ and $\frac{5}{16}$ > 24. $\frac{7}{10}$ and $\frac{2}{3}$ >

25. **Recreation** The lengths of different ski trails are often given in tenths of a mile. However, these distances are sometimes given in halves, fifths, or fourths. Suppose Joliet sees the sign shown here. Explain how to use equivalent fractions to compare the lengths of the different trails. Write fractions equivalent to those shown that have denominators common to the other fractions. (Fox Chase is the smallest trail; Moose Alley is the longest.)

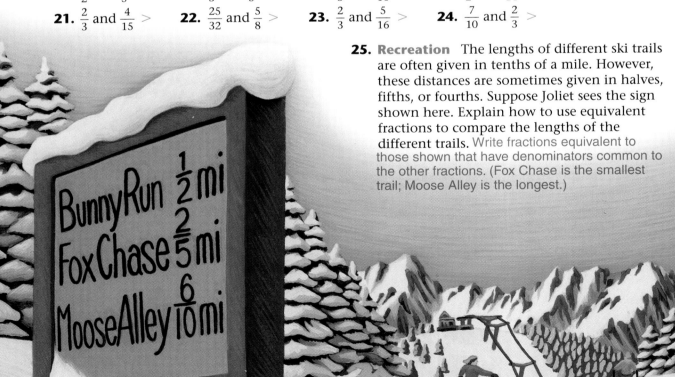

Write the list of fractions in order from least to greatest.

26. $\frac{3}{8}, \frac{2}{3}, \frac{1}{4}, \frac{1}{8},$ and $\frac{1}{2}$

27. $-\frac{5}{8}, -\frac{2}{3}, \frac{1}{4}, -\frac{3}{4}, \frac{1}{8},$ and $-\frac{1}{2}$

28. $\frac{1}{4}, \frac{5}{8}, \frac{1}{2}, \frac{5}{16}, \frac{3}{8},$ and $\frac{7}{16}$

29. $\frac{1}{3}, \frac{2}{5}, \frac{4}{5}, \frac{1}{2}, \frac{7}{10},$ and $\frac{2}{3}$

26. $\frac{1}{8}, \frac{1}{4}, \frac{3}{8}, \frac{1}{2}, \frac{2}{3}$

27. $-\frac{3}{4}, -\frac{2}{3}, -\frac{5}{8}, -\frac{1}{2}, \frac{1}{8}, \frac{1}{4}$

28. $\frac{1}{4}, \frac{5}{16}, \frac{3}{8}, \frac{7}{16}, \frac{1}{2}, \frac{5}{8}$

29. $\frac{1}{3}, \frac{2}{5}, \frac{1}{2}, \frac{2}{3}, \frac{7}{10}, \frac{4}{5}$

Unit fractions are fractions that have a numerator of 1. Use the following list of fractions for Exercises 30–32.

$$\frac{1}{8} \quad \frac{1}{3} \quad \frac{1}{4} \quad \frac{1}{5} \quad \frac{1}{6} \quad \frac{1}{2} \quad \frac{1}{100}$$

30. Write the unit fractions in order from least to greatest. $\frac{1}{100}, \frac{1}{8}, \frac{1}{6}, \frac{1}{5}, \frac{1}{4}, \frac{1}{3}, \frac{1}{2}$

31. What happens to the size of a unit fraction as the denominator increases? The fraction gets smaller.

32. Explain how you can order unit fractions by comparing denominators. The larger the denominator, the smaller the fractions. You can order unit fractions by listing the denominators from greatest to least. Ordering unit fractions in this way lists them from least to greatest.

Look Back

33–36. Only whole-number dimensions given

□⊘ **Geometry** Give all of the possible dimensions for a rectangle with the given area. [Lesson 1.5]

33. 546 square feet

34. 82 square centimeters
$1 \times 82; 2 \times 41$

35. 113 square inches
1×113

36. 91 square meters
$1 \times 91; 7 \times 13$

33. 1×546
2×273
3×182
6×91
7×78
13×42
14×39
21×26

Substitute 4 for _a_, −2 for _b_, and −3 for _c_. Evaluate each expression. [Lesson 2.2]

37. $|a + b| + c$ -1

38. $a + (c + b)$ -1

39. $a + |b + c|$ 9

40. $b + |b + c| + (-a)$ -1

41. $a + (-c) + (-b)$ 9

42. $a + (-b) + (-a)$ 2

Look Beyond

43. Wages Lorel works part time in a factory. His gross earnings are $225.00 per week. The deductions from his gross earnings (earnings before witholdings) are shown below on the pay stub from his paycheck. What is his net pay (earnings after witholdings)? $177.55

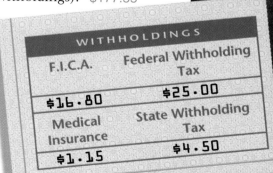

WITHHOLDINGS

F.I.C.A.	Federal Withholding Tax
$16.80	$25.00
Medical Insurance	State Withholding Tax
$1.15	$4.50

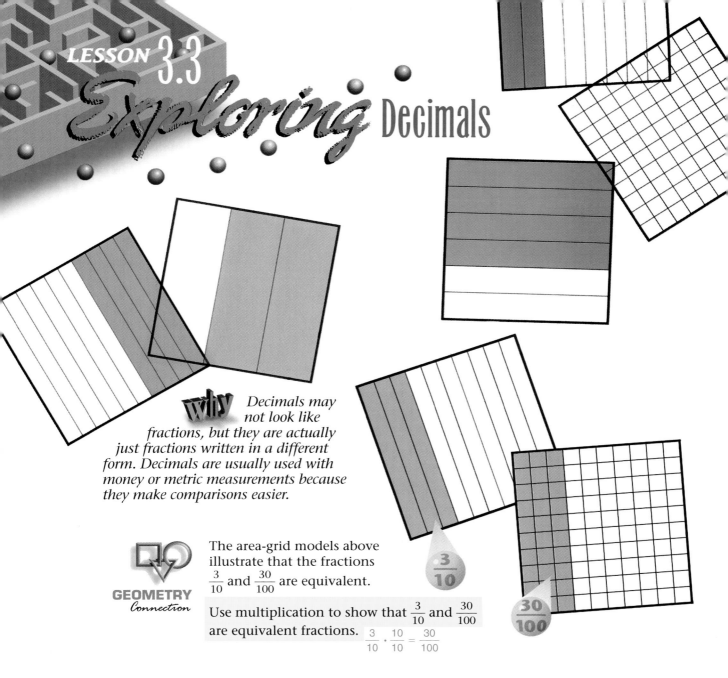

Exploring Decimals

why *Decimals may not look like fractions, but they are actually just fractions written in a different form. Decimals are usually used with money or metric measurements because they make comparisons easier.*

GEOMETRY
Connection

The area-grid models above illustrate that the fractions $\frac{3}{10}$ and $\frac{30}{100}$ are equivalent.

.3 / 10

30 / 100

Use multiplication to show that $\frac{3}{10}$ and $\frac{30}{100}$ are equivalent fractions. $\frac{3}{10} \cdot \frac{10}{10} = \frac{30}{100}$

Place Value

In decimal notation, $\frac{3}{10}$ is written 0.3, and $\frac{30}{100}$ is written 0.30.

The value of a digit in a decimal is determined by its position, or *place*, in the decimal. For example, in 0.30 the digit 3 is in the *tenths* place. It indicates the number of tenths. The digit 0 is in the *hundredths* place. It indicates the number of hundredths.

0.3 — tenths

0.30 — hundredths

This place-value chart shows the *place value* for some of the digits in a decimal.

Place Value

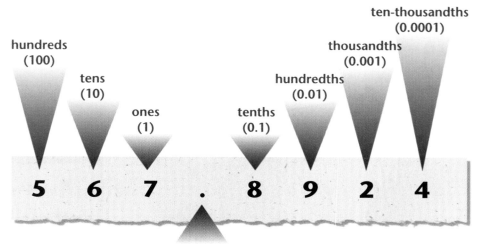

hundreds (100)

tens (10)

ones (1)

tenths (0.1)

hundredths (0.01)

thousandths (0.001)

ten-thousandths (0.0001)

5 6 7 . 8 9 2 4

Decimal Point

You use place value to read decimals. For example, the decimal 567.8924 from the place-value chart above is read as follows.

five hundred sixty-seven and eight thousand nine hundred twenty-four ten-thousandths

Notice that the word *and* is only used to indicate the location of the decimal point. This decimal number can also be written as a *mixed number.*

$$567.8924 = 567 \frac{8924}{10,000}$$

A **mixed number** is a number written with both a whole number and a fraction.

Cultural Connection: Asia, Europe The idea of place value was developed by a Hindu mathematician around 500 B.C.E. It was made popular in a book about the Hindu number system written by the Arabic mathematician Al-Khowarizmi in 825 B.C.E. However, at that time there was no efficient way to represent fractions. So European scholars were slow to adopt the Hindu number system. It was not until the seventeenth century that John Napier introduced the decimal point into the European system.

A fraction can be changed to a decimal by dividing the numerator of the fraction by its denominator. For example, $\frac{1}{2}$ is the same as $1 \div 2$. When 1 is divided by 2, the result is 0.5.

The decimal 0.5 is called a **terminating decimal** because the decimal number *terminates,* or ends, after the 5.

A **repeating decimal** is a nonterminating decimal that repeats a pattern. For example, the number 0.454545 . . . is a repeating decimal because the digits 4 and 5 form a pattern that repeats without ending.

Exploration 1 *Terminating and Repeating Decimals*

1 Use division to write each fraction as a decimal rounded to the thousandths place.

a. $\frac{4}{5}$ **b.** $\frac{4}{11}$ **c.** $\frac{5}{8}$ **d.** $\frac{7}{9}$ **e.** $\frac{5}{12}$ **f.** $\frac{3}{4}$

0.8   0.364   0.625   0.778   0.417   0.750

2 Which fractions in Step 1 are equivalent to terminating decimals? Change $\frac{5}{8}$ to a fraction by dividing without a calculator. What part of the division process causes the decimals to terminate?

3 Which fractions in Step 1 are equivalent to repeating decimals? Change $\frac{4}{11}$ and $\frac{5}{12}$ to decimals by dividing without a calculator. What part of the division process causes the decimals to repeat? ❖

There are two ways to write repeating decimals. One way is to place a bar above the digit or digits that repeat: $0.\overline{18}$. A second way is to write the repeating digit or digits several times followed by three dots: 0.1818 . . .

 **CRITICAL** *Thinking*

All fractions can be changed to decimals that either repeat or terminate. However, some fractions, such as $\frac{5}{17}$, $\frac{7}{17}$, and $\frac{1}{97}$ *appear* to be equivalent to nonterminating decimals that do not repeat. What do these fractions have in common? They have repeating patterns that are too large to be shown on a calculator.

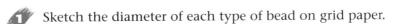 **Exploration 2** *Comparing Decimals*

Crafts   The diameters of ceramic and glass jewelry beads are measured in centimeters. The glass beads measure 0.35 centimeters in diameter, and the ceramic beads measure 0.4 centimeters in diameter. Which type of bead is larger? Compare the decimals 0.35 and 0.4 to find out.

1 Sketch the diameter of each type of bead on grid paper.

2 Complete the following statement. $0.4 = \frac{4}{10} = \frac{?}{100} = ?$ $\frac{40}{100} = 0.40$

3 Which decimal is larger, 0.35 or 0.4? Explain. 0.4 is larger since 0.40 is larger than 0.35.

4 Write an equivalent fraction for each of the three decimals 0.35, 0.4, and 0.358. Use the equivalent fractions to write the decimals in order from least to greatest. Describe how to compare the two decimals 0.35 and 0.4 with the decimal 0.358.

5. Write each decimal as a fraction with a common denominator or as decimals with the same number of decimal places.

5 What method can you use to compare any group of decimals? ❖

APPLICATION

STATISTICS
Connection

Mrs. Maylan, the high school counselor, needs to rank the grade-point averages of five seniors in order from greatest to least. The averages are as follows: 3.3, 3.003, 3.03, 3.04, and 3.042.

First write each decimal with the same number of decimal places.

Since the greatest number of decimal places is three, write each decimal with three decimal places.

3.3	**3.003**	**3.03**	**3.04**	**3.042**
⬇	⬇	⬇	⬇	⬇
3.300	**3.003**	**3.030**	**3.040**	**3.042**

Then write the numbers in order, and change them back to the original form.

3.300	**3.042**	**3.040**	**3.030**	**3.003**
⬇	⬇	⬇	⬇	⬇
3.3	**3.042**	**3.04**	**3.03**	**3.003**

Thus, the decimals in order from greatest to least are as follows:

3.3, 3.042, 3.04, 3.03, 3.003 ❖

Exploration 3 *Estimating Decimals*

Manufacturing

1 Refer to the caption about the manufactured bolts below. What is the manufacturer's cost to make one bolt? The cost of these bolts can be determined by division. Divide $123 by 1000. Record your answer in both dollars and cents. $0.123 or $0.12; 12.3¢ or 12¢

A manufacturer estimates that the cost to make 1000 bolts is $123.

2. 12¢ is closer to the actual cost because 0.120 is closer to 0.123 than 0.130.

2 Complete the following statement.

The cost to make one bolt is between $0. 12 ? , or 12 ? cents, and $0. 13 ?, or 13 ? cents. Which of these costs do you think is closer to the actual cost? Explain.

3 Why is the approximate cost per bolt to the nearest cent a useful estimate? $0.12; this estimate is useful for finding the price of an individual bolt or for amounts that are not multiples of 1000.

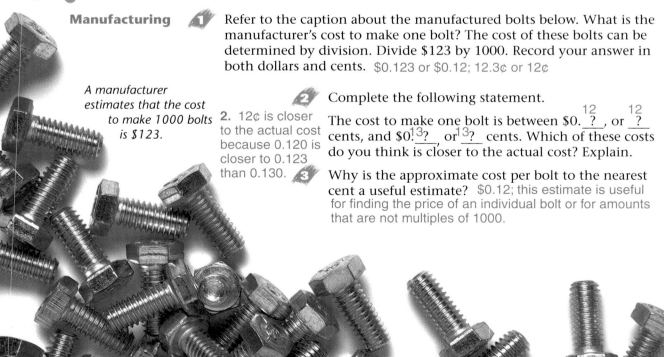

On the number line 0.123 is closer to 0.120 than to 0.130. Therefore, the price of a bolt is closer to 12¢ each.

Use the number line below to justify your answers in Step 2. ❖

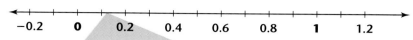

-0.2   0   0.2   0.4   0.6   0.8   1   1.2

0.120   0.123   0.130

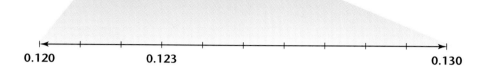

APPLICATION

Consumer Economics

Marcus wants to buy 8 discount tapes. He has $30 to spend. According to the sale price shown, does he have enough money to buy the tapes?

To estimate, round up the cost of the tapes to $4. Then the total is 8 × 4 = $32.

But the tapes actually cost $0.11 less than $4.00. So the actual amount can be found as shown.

$$32.00 - 8 \times 0.11 = 32.00 - 0.88$$
$$= 31.12$$

The actual cost for the tapes is $31.12, so Marcus does not have enough money to buy 8 discount tapes. ❖

TAPE SALE
$3.89 each

APPLICATION

Auto Maintenance

Jimmy drove 346.9 miles on 11.8 gallons of gas. He wants to *estimate* the gas mileage, which is the number of miles he can drive on 1 gallon of gas.

The gas mileage is $\frac{346.9}{11.8}$ miles per gallon. You can find the gas mileage by dividing. However, to *estimate*, first round 11.8 to 12 and round 346.9 to a number that is divisible by 12.

$$\frac{346.9}{11.8} \begin{array}{c} \rightarrow \\ \rightarrow \end{array} \frac{348}{12} = 29$$

The estimated gas mileage of Jimmy's car is 29 miles per gallon. ❖

EXERCISES & PROBLEMS

Communicate

1. Describe how to use an area grid to compare the decimals 0.4 and 0.04.

2. Explain how you can estimate a solution to the following problem. Matthew bought 11.6 gallons of gas at a cost of $1.199 per gallon. What is an approximate total cost for the gasoline?

3. How is the number 1097.683 said? Explain what each digit in this number represents.

4. How are the fractions $\frac{2}{3}$ and $\frac{3}{5}$ different in terms of their decimal equivalents?

5. Describe a method for arranging the following decimals from least to greatest:

$$-5.05 \quad 5.5 \quad -5.2 \quad 5.499$$

Practice & Apply

Write each decimal as a fraction or mixed number in lowest terms.

6. 0.23 $\frac{23}{100}$ 7. 0.8 $\frac{4}{5}$ 8. 2.1 $2\frac{1}{10}$ 9. 5.7 $5\frac{7}{10}$ 10. 0.955 $\frac{191}{200}$ 11. 0.864 $\frac{108}{125}$

12. 0.572 $\frac{143}{250}$ 13. 3.24 $3\frac{6}{25}$ 14. 5.08 $5\frac{2}{25}$ 15. 71.3 $71\frac{3}{10}$ 16. 125.4 $125\frac{2}{5}$ 17. 32.531 $32\frac{531}{1000}$

18. 81.44 $81\frac{11}{25}$ 19. 894.32 $894\frac{8}{25}$ 20. 62.14 $62\frac{7}{50}$ 21. 0.005 $\frac{1}{200}$ 22. 4.015 $4\frac{3}{200}$ 23. 45.45 $45\frac{9}{20}$

Write each fraction as a terminating or repeating decimal.

24. $\frac{3}{8}$ 0.375 25. $\frac{4}{5}$ 0.8 26. $-\frac{7}{10}$ −0.7 27. $\frac{5}{12}$ 0.41$\overline{6}$ 28. $-\frac{2}{3}$ −0.$\overline{6}$ 29. $-\frac{1}{9}$ −0.$\overline{1}$

30. $5\frac{3}{4}$ 5.75 31. $-7\frac{1}{2}$ −7.5 32. $3\frac{2}{9}$ 3.$\overline{2}$ 33. $9\frac{7}{12}$ 9.58$\overline{3}$ 34. $-4\frac{1}{11}$ −4.$\overline{09}$ 35. $-\frac{1}{7}$ −0.$\overline{142857}$

Order the decimals from the least to the greatest.

36. 0.4 0.38 0.49 0.472 0.425

37. −0.035 −0.35 −0.5 −0.05 −0.53   −0.53, −0.5, −0.35, −0.05, −0.035

38. 5.32 5.2 4.97 5.037 5.3

39. 6.091 6.01 6.009 6.9 6.19   6.009, 6.01, 6.091, 6.19, 6.9

40. −7.11 −7.011 −7.105 −7.01 −7.1   −7.11, −7.105, −7.1, −7.011, −7.01

41. −18.9 −19 −18.78 −19.25 −18.03   −19.25, −19, −18.9, −18.78, −18.03

Order the numbers from the greatest to the least.

42. 7.3 $7\frac{3}{8}$ 7.045 $7\frac{5}{12}$ $7\frac{1}{3}$

43. 0.15 $\frac{3}{8}$ −0.35 $\frac{3}{4}$ $\frac{679}{1000}$ $\frac{3}{4}, \frac{679}{1000}, \frac{3}{8}, 0.15, -0.35$

44. −5.6 $-5\frac{5}{8}$ −5.55 $-5\frac{2}{3}$ $-5\frac{2}{9}$

45. −0.6 $-\frac{2}{3}$ −0.75 $-\frac{3}{5}$ −0.5 $-\frac{1}{3}$

46. $8\frac{1}{9}$ 8.72 $8\frac{6}{11}$ 8.56 $8\frac{4}{9}$

47. $\frac{14}{11}$ 1.303 $1\frac{1}{3}$ 0.133 $1\frac{1}{5}$

48. Consumer Economics Ms. Jeffries has $350 to spend on computer equipment. Use estimation to determine if Ms. Jeffries has enough money to buy the CD ROM drive, the fax modem, and four memory chips. *She does not have enough money.*

49. Carpentry A wooden board is 3.2 meters long. Use estimation to determine how many 0.38-meter boards you can cut from this board. *8*

50. Consumer Economics Notebooks cost $2.95. Amy needs 5 notebooks for school. She has $17. Use estimation to determine if she has enough money to buy 5 notebooks. *She does have enough money.*

51. Consumer Economics Rico saved $48. He wants to buy several CDs for $10.95 each. Use estimation to determine how many CDs he can buy with $48. *4*

52. Geometry Each side of a square poster measures 11.8 centimeters. Use estimation to determine the approximate perimeter of the poster. *approx. 48 cm*

53. Auto Maintenance Rachel drove 425.8 miles on 25.3 gallons of gas. Estimate her car's gas mileage for 1 gallon of gas. *approx. 17 miles per gallon*

Write a mixed number or fraction in lowest terms for each decimal.

54. 3.2 $3\frac{1}{5}$

55. 47.032 $47\frac{4}{125}$

56. -8.48 $-8\frac{12}{25}$

57. -12.6 $-12\frac{3}{5}$

58. 5.002 $5\frac{1}{500}$

59. -93.85 $-93\frac{17}{20}$

60. 72.55 $72\frac{11}{20}$

61. 0.004 $\frac{1}{250}$

62. 0.56 $\frac{14}{25}$

63. -7.77 $-7\frac{77}{100}$

64. 66.66 $66\frac{33}{50}$

65. 6.59 $6\frac{59}{100}$

66. 72.06 $72\frac{3}{50}$

67. -0.001 $-\frac{1}{1000}$

68. 0.909 $\frac{909}{1000}$

69. 5.5 $5\frac{1}{2}$

70. Explain this statement by using examples.

In our decimal system, the value of each decimal place is 10 times the value of the place to its right. Using the example 222.22, the value in the hundreds place, 200, is 10 times greater than the value in the tens place, 20, and so forth.

Write each number as a mixed number or a fraction in lowest terms and as a decimal.

71. seventy-two and twenty-six hundredths $72\frac{13}{50}$; 72.26

72. five hundred five and twenty-five thousandths $505\frac{1}{40}$; 505.025

73. negative six hundred-thousandths $\frac{-3}{50,000}$; -0.00006

74. three hundred two and seventy-five ten-thousandths $302\frac{3}{400}$; 302.0075

75. sixteen and one tenth $16\frac{1}{10}$; 16.1

76. five and seven hundred sixteen thousandths $5\frac{179}{250}$; 5.716

77. twenty-two and two thousand five hundred ten-thousandths $22\frac{1}{4}$; 22.2500 or 22.25

78. six hundred sixty-seven thousandths $\frac{667}{1000}$; 0.667

79. three hundred thirty-three thousandths $\frac{333}{1000}$; 0.333

80. one thousand and one hundred-thousandth $1000\frac{1}{100,000}$; 1000.00001

Statistics A batting average is the ratio of the number of hits to the number of times at bat. Use the baseball cards for Exercises 81–84.

81. Compute the batting average for Rodriguez. Express the answer in decimal form with decimals rounded to the nearest thousandth.
0.372

82. Compute the batting average for Czarnecki. Express the answer in decimal form with decimals rounded to the nearest thousandth.
0.331

83. Compute the batting average for Washington. Express the answer in decimal form with decimals rounded to the nearest thousandth.
0.374

84. List the batting averages for all three players in order from the least to the greatest.
Czarneki (0.331), Rodriguez (0.372), Washington (0.374)

Ricky Rodriguez
Wt. 148 Ht. 5'10"

GAMES	34
TIMES AT BAT	121
RUNS	28
HITS	45

Bill Washington
Wt. 155 Ht. 6'0"

GAMES	31
TIMES AT BAT	115
RUNS	9
HITS	43

Paul Czarnecki
Wt. 150 Ht. 5'11"

GAMES	32
TIMES AT BAT	127
RUNS	19
HITS	42

Look Back

85. Evaluate $a^2 - b^2$ when a is -4 and b is 2.   **[Lesson 1.7]** 12

Determine whether the given number is a solution to the equation. Write *true* or *false*.   **[Lesson 2.3]**

86. $x - 7 = 4$ false
$x = 3$

87. $y - 8 = 9$ false
$y = 1$

88. $2y + 6 = 4$ true
$y = -1$

89. $3y - 4 = 6$ false
$y = 3$

90. $5n - 2n = 9$ true
$n = 3$

91. $4y + 2y = 36$ false
$y = 3$

92. Use guess-and-check to solve the equation $3x + 2 = 11$.   **[Lesson 2.3]** $x = 3$

93. Crafts Helen has a piece of material that is $3\frac{1}{3}$ yards long. She needs a piece of material that is $3\frac{3}{8}$ yards long for a project. Can Helen use the $3\frac{1}{3}$-yard piece of material she has for her project?   **[Lesson 3.1]** no

Look Beyond

Find a value for *n* to make each statement a true proportion.

94. $\frac{1}{2} = \frac{n}{10}$ 5

95. $\frac{2}{9} = \frac{n}{18}$ 4

96. $\frac{n}{4} = \frac{3}{6}$ 2

97. $\frac{8}{12} = \frac{6}{n}$ 9

LESSON 3.4

Exploring Addition and Subtraction of Rational Numbers

Why *Sometimes you need an exact answer to a problem. At other times, an estimate is all that is necessary. This is especially true when solving problems that involve rational numbers.*

A jogging trail near Andrea's house makes a $\frac{3}{4}$-mile loop. Every afternoon Andrea jogs around the loop 3 times. What is the total distance Andrea jogs every afternoon?

Exploration 1 *Fractions With Like Denominators*

Sports

$\frac{3}{4} + \frac{3}{4} + \frac{3}{4}$

1 Write an addition expression that you can use to find the total distance that Andrea jogs each afternoon.

2 Copy the number line below. Show the fraction $\frac{3}{4}$ on the number line. Then use the number-line model and your expression from Step 1 to compute the total distance.

| $\frac{1}{4}$ | $\frac{1}{4}$ | $\frac{1}{4}$ | $\frac{1}{4}$ | $\frac{1}{4}$ | $\frac{1}{4}$ | $\frac{1}{4}$ | $\frac{1}{4}$ | $\frac{1}{4}$ | $\frac{1}{4}$ | $\frac{1}{4}$ | $\frac{1}{4}$ |

0   $\frac{3}{4}$ 1 $\frac{3}{4}$ $\frac{3}{4}$ 2   3

$\frac{3}{4} + \frac{3}{4} + \frac{3}{4} = \frac{?}{4} \; \frac{9}{4}$

3 Copy and complete the equations below to simplify your answer from Step 2.

Total distance in miles:
$$\frac{?}{4} = \frac{8}{4} + \frac{?}{4} \; \frac{9}{4}; \frac{1}{4}$$
$$= 2 + \frac{?}{4} \; \frac{1}{4}$$
$$= ? \; 2\frac{1}{4}$$

Use a number-line model to find $\frac{5}{8} + \frac{5}{8}$. Write the answer in lowest terms.

 Find $\frac{4}{5} - \frac{2}{5}$ using a number-line model.

 Explain how to add or subtract fractions that have the same denominator. ❖ If denominators are the same, add or subtract the numerators.

EXTENSION

A fraction such as $\frac{9}{4}$, which has a numerator that is greater than the denominator, is called an **improper fraction**. Improper fractions can be written as mixed numbers.

Since $\frac{9}{4} = \frac{8}{4} + \frac{1}{4}$, and $\frac{8}{4}$ is 2, the improper fraction $\frac{9}{4}$ can be written as the mixed number $2\frac{1}{4}$.

Calculator

You can also write an improper fraction as a mixed number by dividing the numerator by the denominator.

$$\frac{9}{4} = 9 \div 4 \qquad\qquad 9 \boxed{\div} 4 \boxed{=} 2.25$$

$$= 2 \text{ remainder } 1$$

$$= 2\frac{1}{4}$$

❖

You can use a fraction model to discover the method for adding fractions with unlike denominators.

Exploration 2 *Fractions With Unlike Denominators*

 Explain why the model below shows that $\frac{1}{2} + \frac{1}{3} = \frac{5}{6}$.

Since $\frac{1}{2} = \frac{3}{6}$ and $\frac{1}{3} = \frac{2}{6}$, $\frac{3}{6} + \frac{2}{6} = \frac{5}{6}$.

 Explain how Steps 1–3 at the right show that $\frac{1}{2} + \frac{1}{3} = \frac{5}{6}$.

Multiply by 1 to get a common denominator.

Step 1 $\quad \frac{1}{2} + \frac{1}{3} = \frac{1 \cdot 3}{2 \cdot 3} + \frac{1 \cdot 2}{3 \cdot 2}$

Step 2 $\qquad\qquad = \frac{3}{6} + \frac{2}{6}$ Add.

Step 3 $\qquad\qquad = \frac{5}{6}$ Simplify.

3 Explain how to use the fraction bars to find the difference $\frac{1}{2} - \frac{1}{3}$.

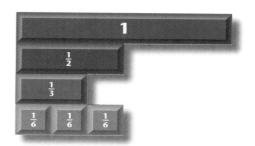

The bars show that $\frac{1}{2} = \frac{3}{6}$ and $\frac{1}{3} = \frac{2}{6}$.

Since $\frac{3}{6} - \frac{2}{6} = \frac{1}{6}$,

then $\frac{1}{2} - \frac{1}{3} = \frac{1}{6}$.

 4 Use equivalent fractions to write $\frac{1}{2} - \frac{1}{3}$ with common denominators. Explain each step of the process.

 5 Explain how to add and subtract fractions with unlike denominators. ❖ First find a common denominator for the fractions. Then add or subtract the fractions with like denominators.

APPLICATION

Stocks   Rachel is looking at the weekly change in one of her stocks. She is interested in adding the daily changes to compute the net change in her stock for the week. What is the total net change after Friday?

The daily changes in Rachel's stocks are $+\frac{3}{4}, -\frac{1}{4}, -\frac{1}{2}, -\frac{3}{8}$, and $+\frac{7}{8}$.

First add the fractions with like denominators.

$$\frac{3}{4} + \left(-\frac{1}{4}\right) = \frac{2}{4} \qquad \left(-\frac{3}{8}\right) + \frac{7}{8} = \frac{4}{8}$$

The fractions $\frac{2}{4}$ and $\frac{4}{8}$ both simplify to $\frac{1}{2}$. Add $\frac{1}{2} + \frac{1}{2} + \left(-\frac{1}{2}\right)$ to get the total change after Friday, which is $\frac{1}{2}$. ❖

You can use a variety of strategies to add several fractions. In the application above, the fractions with like denominators are added first. Another method involves combining the positive fractions first, combining the negative fractions second, and then finding a final sum.

Positive Fractions	Negative Fractions
$\frac{3}{4} + \frac{7}{8}$	$-\frac{1}{4} + \left(-\frac{1}{2}\right) + \left(-\frac{3}{8}\right)$
$\downarrow \qquad \downarrow$	$\downarrow \qquad \downarrow \qquad \downarrow$
$\frac{6}{8} + \frac{7}{8} = \frac{13}{8}$	$-\frac{2}{8} + \left(-\frac{4}{8}\right) + \left(-\frac{3}{8}\right) = -\frac{9}{8}$

Final Sum: $\frac{13}{8} + \left(-\frac{9}{8}\right) = \frac{4}{8}$, or $\frac{1}{2}$ ❖

SUMMARY

To add or subtract fractions with unlike denominators:

- Find the Least Common Denominator, LCD, of the denominators of all of the fractions.

- Use the LCD to write equivalent fractions so that all of the fractions have the same denominator.

- Add or subtract.

EXERCISES & PROBLEMS

Communicate

1. Explain why $\frac{1}{3} + \frac{1}{4}$ is not equal to $\frac{2}{7}$.

2. Use a model to show why $\frac{1}{3} + \frac{1}{4} = \frac{7}{12}$.

3. Explain how to use equivalent fractions to add three or more fractions. Give an example.

4. Give an example of a problem containing fractions that requires an exact answer.

5. Give an example of a problem containing fractions that can be solved by estimating.

Practice & Apply

The payroll chart shows the hours worked by 6 employees. Find the total hours each person worked Monday through Friday.

6. Andrews $27\frac{3}{4}$ **7.** Bolla $22\frac{1}{4}$

8. Garza $22\frac{3}{4}$ **9.** Holland $23\frac{1}{4}$

10. Tate $32\frac{1}{4}$ **11.** Wuest $20\frac{1}{4}$

Find the total wages earned by each person for working Monday through Friday.

12. Andrews $201.19

13. Bolla $150.19

14. Garza $176.31

15. Holland $176.24

16. Tate $256.39

17. Wuest $134.66

	Payroll Chart					
Name	**Wage (per hr)**	**Mon hours**	**Tue hours**	**Wed hours**	**Thur hours**	**Fri hours**
Andrews	$7.25	5	$6\frac{1}{4}$	$8\frac{3}{4}$	$4\frac{1}{2}$	$3\frac{1}{4}$
Bolla	$6.75	$1\frac{1}{4}$	$3\frac{1}{2}$	$7\frac{1}{4}$	$5\frac{3}{4}$	$4\frac{1}{2}$
Garza	$7.75	$7\frac{1}{4}$	$3\frac{3}{4}$	$1\frac{1}{2}$	$2\frac{1}{2}$	$7\frac{3}{4}$
Holland	$7.58	$3\frac{1}{2}$	$2\frac{1}{2}$	$4\frac{1}{4}$	6	7
Tate	$7.95	$7\frac{1}{2}$	$7\frac{1}{4}$	$6\frac{1}{4}$	$8\frac{3}{4}$	$2\frac{1}{2}$
Wuest	$6.65	6	$4\frac{1}{2}$	$6\frac{1}{2}$	$3\frac{1}{4}$	0

Find the hours worked by all employees for each day.

18. Mon $30\frac{1}{2}$ **19.** Tue $27\frac{3}{4}$ **20.** Wed $34\frac{1}{2}$ **21.** Thur $30\frac{3}{4}$ **22.** Fri 25

23. Technology Create a spreadsheet for the payroll chart shown on page 152. Compute each employee's daily wages and weekly wages by using the hourly wages listed in the table.

24. Photography Estimate the perimeter of Wilson's picture frame, at the right. Then compute the exact perimeter. 56; $55\frac{5}{8}$

Estimate the perimeter of each picture frame. Then compute the exact perimeter.

25. $8\frac{3}{8}$ by $10\frac{5}{16}$ 38; $37\frac{3}{8}$ **26.** $5\frac{3}{16}$ by $7\frac{3}{4}$ 26; $25\frac{7}{8}$

27. $10\frac{5}{8}$ by $13\frac{1}{2}$ 48; $48\frac{1}{4}$ **28.** $60\frac{1}{2}$ by $36\frac{3}{4}$ 194; $194\frac{1}{2}$

Find two mixed numbers that have the given sum.

29. $4\frac{1}{3}$ **30.** $6\frac{3}{8}$ **31.** $-\frac{1}{2}$

32. $11\frac{4}{7}$ **33.** $9\frac{7}{8}$ **34.** $-1\frac{1}{2}$

Find two mixed numbers that have the given difference.

35. $4\frac{1}{3}$ **36.** $6\frac{3}{8}$ **37.** $-\frac{1}{2}$

38. $11\frac{4}{7}$ **39.** $9\frac{7}{8}$ **40.** $-1\frac{1}{2}$

Add or subtract. Check each of your answers using a calculator.

41. $\frac{1}{2} + \frac{4}{5}$ $1\frac{3}{10}$ **42.** $\frac{2}{3} - \frac{7}{12}$ $\frac{1}{12}$ **43.** $\frac{1}{3} + \frac{3}{16}$ $\frac{25}{48}$ **44.** $-\frac{1}{2} + \frac{2}{3} + 1\frac{2}{3}$ $1\frac{5}{6}$

45. $\frac{3}{4} - \frac{1}{4} + 1\frac{3}{8}$ $1\frac{7}{8}$ **46.** $3\frac{3}{4} + 2\frac{3}{16}$ $5\frac{15}{16}$ **47.** $4\frac{1}{2} - 2\frac{3}{4}$ $1\frac{3}{4}$ **48.** $8\frac{7}{10} - \frac{7}{8}$ $7\frac{33}{40}$

49. $5 - \frac{3}{16}$ $4\frac{13}{16}$ **50.** $-\frac{2}{3} - \frac{7}{12}$ $-1\frac{1}{4}$ **51.** $-\frac{1}{3} + 1\frac{3}{4}$ $1\frac{5}{12}$ **52.** $-\frac{1}{2} + \left(-1\frac{7}{8}\right) + \left(-\frac{2}{3}\right)$ $-3\frac{1}{24}$

53. $9\frac{5}{6} - \frac{5}{6}$ 9 **54.** $6\frac{4}{5} - 9\frac{7}{8}$ $-3\frac{3}{40}$ **55.** $-9\frac{5}{6} + 3\frac{11}{12}$ $-5\frac{11}{12}$ **56.** $-\frac{2}{3} + \left(1\frac{7}{8}\right) + \left(-8\frac{1}{2}\right)$ $-7\frac{7}{24}$

Look Back

Write a list of all of the factors for each number. Circle the prime factors.   [Lesson 1.5]

57. 82 1, ②, ㊶, 82 **58.** 54 1, ②, ③, 6, 9, **59.** 75 1, ③, ⑤, 15, **60.** 63 1, ③, ⑦, 9, 21, 63
                              18, 27, 54   25, 75

Find each sum.   [Lesson 2.2]

61. $-15 + (-12)$ -27 **62.** $-61 + (-8) + 7$ -62 **63.** $24 + (-41) + (-9)$ -26

64. $|-3| + (-7)$ -4 **65.** $6 + (-12) + |7|$ 1 **66.** $201 + (-32) + (-198)$ -29

Look Beyond

Use guess-and-check to solve each equation.

67. $x - \frac{1}{2} = 1\frac{2}{3}$ $2\frac{1}{6}$ **68.** $x - \frac{3}{4} = 10\frac{1}{4}$ 11 **69.** $\frac{x}{4} = 6\frac{1}{4}$ 25

Multiplying and Dividing Rational Numbers

1 mi

1 mi

$\frac{1}{2}$ mi

$\frac{1}{2}$ mi

Physical models can be used to show multiplication and division with rational numbers. For example, the area of a rectangle provides an excellent model for multiplication.

The streets of Colorado Springs, Colorado, are laid out in square blocks.

Using Area Models

2 units

3 units

The figure at the left shows an *area model* of a 2-by-3 rectangle. To find the area of a rectangle, multiply the length by the width.

What is the area, in square units, of this 2-by-3 rectangle? 6 square units

You can also use an area model to illustrate the method used to multiply fractions. The map of Colorado Springs shows a region that has dimensions of $\frac{1}{2}$ mile by $\frac{1}{2}$ mile. This map can be used as an area model. In this area model, the $\frac{1}{2}$-mile-by-$\frac{1}{2}$-mile region of Colorado Springs is $\frac{1}{4}$ of the total area. This model shows that $\frac{1}{2} \cdot \frac{1}{2} = \frac{1}{4}$.

EXAMPLE 1

GEOMETRY
Connection

Use transparent area-grid models to find $\frac{1}{2} \cdot \frac{3}{4}$.

Solution ➤

Overlap a $\frac{1}{2}$-area grid and a $\frac{3}{4}$-area grid.

The overlapping region is $\frac{3}{8}$. The model shows that

$$\frac{1}{2} \cdot \frac{3}{4} = \frac{3}{8}.$$

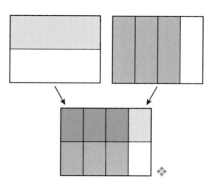

Try This Use transparent area-grid models to find $\frac{1}{2} \cdot \frac{2}{3}$.

Examine the product $\frac{1}{2} \cdot \frac{3}{4} = \frac{3}{8}$ from Example 1. It appears that the product of two fractions can be found by multiplying their numerators and denominators.

$$\frac{1}{2} \bullet \frac{3}{4} = \frac{3}{8}$$

Calculator

You can check your results from area grids using a calculator. Change each fraction to a decimal, and check your results.

```
(1/2)*(3/4)
                    .375
1/2*3/4
                    .375
```

$$\frac{1}{2} \quad \text{x} \quad \frac{3}{4} \quad = \quad \frac{3}{8}$$

1 ÷ 2 =   3 ÷ 4 =   3 ÷ 8 =

$$.5 \quad \text{x} \quad .75 \quad = \quad .375$$

CRITICAL
Thinking

When multiplying $\frac{1}{2} \cdot \frac{1}{3}$, why does changing the fractions to decimals before multiplying result in an approximate answer rather than an exact answer?

Multiplication With Mixed Numerals

A neighborhood park has dimensions of $2\frac{1}{2}$ blocks by $2\frac{1}{2}$ blocks. You can use mixed numbers to compute the area of this park. First change the mixed number to an improper fraction.

$2\frac{1}{2}$ $\frac{5}{2}$

$$2\frac{1}{2} \cdot 2\frac{1}{2} = \frac{5}{2} \cdot \frac{5}{2}$$
$$= \frac{25}{4}$$
$$= 6\frac{1}{4}$$

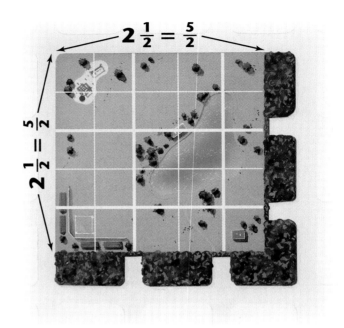

$2\frac{1}{2} = \frac{5}{2}$

$2\frac{1}{2} = \frac{5}{2}$

EXAMPLE 2

Find the product $4\frac{1}{2} \cdot 2\frac{1}{3}$. Write your answer as a mixed number in lowest terms. Use mental math and estimation to check the product.

Solution ➤

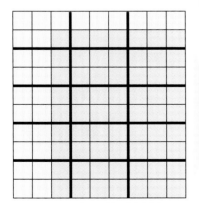

$4\frac{1}{2} = \frac{9}{2}$

$2\frac{1}{3} = \frac{7}{3}$

Write each mixed number as an improper fraction.

Then multiply.

Write your answer as a mixed number.

$$4\frac{1}{2} \cdot 2\frac{1}{3} = \frac{9}{2} \cdot \frac{7}{3}$$
$$= \frac{63}{6}$$
$$= 10\frac{3}{6}, \text{ or } 10\frac{1}{2}$$

To estimate, replace $2\frac{1}{3}$ with **2** and with **3** in the expression $4\frac{1}{2} \cdot 2\frac{1}{3}$. The product is between $4\frac{1}{2} \cdot 2 = 9$ and $4\frac{1}{2} \cdot 3 = 13\frac{1}{2}$. Since $2\frac{1}{3}$ is closer to 2 than to 3, the answer should be closer to 9 than to $13\frac{1}{2}$. Therefore, the product $10\frac{1}{2}$ is reasonable. ❖

Try This   Find the product $3\frac{1}{3} \cdot 1\frac{4}{5}$. Write your answer as a mixed number. Use mental math and estimation to check the product.

Graphics
Calculator

The multiplication in Example 2 can be done in one step on a graphics calculator.

Explain how the keystrokes for performing this multiplication on a graphics calculator compare with the keystrokes used on a scientific calculator.

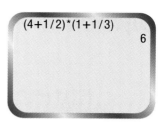

(4+1/2)*(1+1/3)

6

CRITICAL
Thinking

Examine the calculations shown below. Then use the same method to write an alternate solution to Example 2.

$$\frac{9}{5} \cdot \frac{10}{3} = \frac{3 \cdot 3}{5} \cdot \frac{2 \cdot 5}{3}$$

$$= \frac{3 \cdot 3 \cdot 2 \cdot 5}{5 \cdot 3}$$

$$= \frac{2 \cdot 3 \cdot 3 \cdot 5}{3 \cdot 5}$$

$$= 2 \cdot 3 \cdot \frac{3}{3} \cdot \frac{5}{5}$$

$$= 2 \cdot 3 \cdot 1 \cdot 1$$

$$= 6$$

Reciprocals and Division

Reciprocals are pairs of numbers whose product is 1. For example, $\frac{4}{7}$ and $\frac{7}{4}$ are reciprocals because $\frac{4}{7} \cdot \frac{7}{4} = \frac{28}{28} = 1$.

EXAMPLE 3

Name the reciprocal of each number. Show that the product of each number and its reciprocal is 1.

A $\frac{1}{2}$ **B** $\frac{2}{3}$ **C** $4\frac{4}{5}$

Solution ➤

A The reciprocal of $\frac{1}{2}$ is $\frac{2}{1}$, or 2. $\frac{1}{2} \cdot \frac{2}{1} = \frac{2}{2}$, or 1

B The reciprocal of $\frac{2}{3}$ is $\frac{3}{2}$. $\frac{2}{3} \cdot \frac{3}{2} = \frac{6}{6}$, or 1

C First write the mixed number $4\frac{4}{5}$ as an improper fraction, $\frac{24}{5}$.
The reciprocal of $\frac{24}{5}$ is $\frac{5}{24}$. $\frac{24}{5} \cdot \frac{5}{24} = \frac{120}{120}$, or 1 ❖

Division With Fractions

How many half dollars make $6? Although you can probably answer 12 very quickly, finding the answer actually involves dividing by a fraction.

You can see from the illustration above that $6 \div \frac{1}{2} = 6 \cdot 2 = 12$. Notice that 2 is the reciprocal of $\frac{1}{2}$. To divide a number by a fraction, multiply the number by the reciprocal of the fraction.

EXAMPLE 4

Metalworking

One iron scroll is made from $2\frac{1}{2}$ feet of metal. If a metal worker buys 10 feet of metal, how many iron scrolls can he make from it?

Solution ➤

The model below shows that $2\frac{1}{2}$ is equal to $\frac{5}{2}$ and shows the division of 10 by $2\frac{1}{2}$.

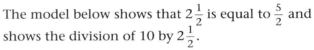

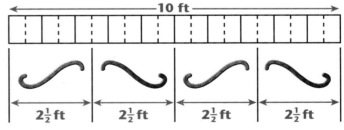

Use the reciprocal of $2\frac{1}{2}$ to write the product. Then find the quotient.

$$10 \div 2\frac{1}{2} = 10 \div \frac{5}{2}$$
Change $2\frac{1}{2}$ to an improper fraction.

$$= \frac{10}{1} \cdot \frac{2}{5}$$
Rewrite the expression with multiplication.

$$= \frac{20}{5}$$
Multiply.

$$= 4$$
Simplify.

With 10 feet of metal, the metal worker can make 4 iron scrolls. ❖

EXERCISES  PROBLEMS

Communicate

1. Explain how you can use an area model to show multiplication with fractions. Give two examples.
2. What are reciprocals? Explain how reciprocals are used in division. Give two examples.
3. Is it true that the quotient $4\frac{1}{2} \div \frac{2}{3}$ should be greater than $4\frac{1}{2}$ because the divisor, $\frac{2}{3}$, is less than 1? Explain.
4. Explain why the reciprocal of -1 is -1.

Practice & Apply

Use an area model to illustrate and find each product. Use a calculator to check your answers.

5. $4 \cdot \frac{1}{3}$ **6.** $\frac{2}{3} \cdot \frac{1}{2}$ **7.** $1\frac{1}{2} \cdot 1\frac{1}{3}$ **8.** $2\frac{1}{2} \cdot 1\frac{1}{2}$

Geometry Estimate the area of each rectangle. Then find the actual area. Use a calculator to check your answers.

9. $2\frac{3}{4}$ in. $5\frac{1}{3}$ in.
15 sq. in.;
$14\frac{2}{3}$ sq. in.

10. $12\frac{5}{11}$ ft $16\frac{1}{2}$ ft
204 sq. ft.;
$205\frac{1}{2}$ sq. ft.

11. $\frac{1}{2}$ mi $2\frac{3}{4}$ mi
$1\frac{1}{2}$ sq. mi.;
$1\frac{3}{8}$ sq. mi.

Estimate each product. Then find the actual product.

12. $9 \cdot 2\frac{2}{3}$ 27; 24   **13.** $2\frac{1}{4} \cdot 8$ 16; 18   **14.** $3\frac{1}{3} \cdot 12$ 36; 40   **15.** $10 \cdot 1\frac{3}{5}$ 20; 16

16. $1\frac{1}{2} \cdot 2\frac{1}{3}$ 4; $3\frac{1}{2}$ **17.** $3\frac{3}{4} \cdot 1\frac{2}{3}$ 8; $6\frac{1}{4}$ **18.** $4\frac{1}{2} \cdot 2\frac{2}{3}$ 15; 12   **19.** $6\frac{1}{4} \cdot 1\frac{3}{5}$ 12; 10

20. $8\frac{1}{5} \cdot 7\frac{1}{2}$ 64; $61\frac{1}{2}$ **21.** $2\frac{2}{5} \cdot 4\frac{1}{6}$ 8; 10   **22.** $2\frac{5}{8} \cdot 5\frac{1}{4}$ 15; $13\frac{25}{32}$ **23.** $1\frac{5}{6} \cdot 3\frac{1}{9}$ 6; $5\frac{19}{27}$

24. $8\frac{1}{4} \cdot 15\frac{3}{5}$ 120; $128\frac{7}{10}$ **25.** $7\frac{2}{3} \cdot 12\frac{1}{8}$ 96; $92\frac{23}{24}$ **26.** $2\frac{1}{6} \cdot 11\frac{1}{3}$ 22; $24\frac{5}{9}$ **27.** $3\frac{1}{5} \cdot 8\frac{2}{3}$ 27; $27\frac{11}{15}$

Find the reciprocal of each number. Show that the product of each number and its reciprocal is 1. (See Additional Answers)

28. $\frac{2}{3}$ $\frac{3}{2}$ **29.** $2\frac{1}{2}$ $\frac{2}{5}$ **30.** $1\frac{2}{3}$ $\frac{3}{5}$ **31.** $4\frac{4}{7}$ $\frac{7}{32}$ **32.** $5\frac{2}{5}$ $\frac{5}{27}$

Business At the pet store shown below, large sacks of birdseed are divided into smaller quantities.

33. How many $2\frac{1}{2}$-pound bags are there in one large sack? 20

34. How many $3\frac{1}{3}$-pound bags are there in one large sack? 15

35. How many $1\frac{2}{3}$-pound bags are there in one large sack? 30

36. How many pounds of birdseed are left in the large sack after filling as many $1\frac{1}{2}$-pound sacks as possible? $\frac{1}{2}$ pound

37. Baking A baker has $17\frac{1}{2}$ cups of flour on hand. If it takes $2\frac{1}{3}$ cups to make a loaf of bread, how many loaves can the baker make? 7

Charlie is dividing a 50-pound sack of birdseed into smaller sacks to sell individually.

Manufacturing A factory produces 1200 units each hour. Copy and complete the table to determine the number of units produced in each interval. For example, 5 minutes is $\frac{5}{60}$, or $\frac{1}{12}$, of an hour. So find the number of units produced in $\frac{1}{12}$ of an hour.

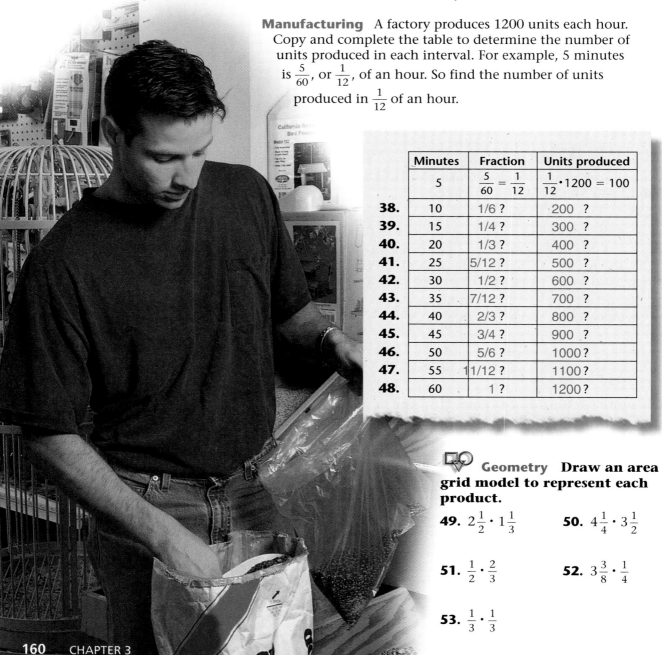

	Minutes	Fraction	Units produced
	5	$\frac{5}{60} = \frac{1}{12}$	$\frac{1}{12} \cdot 1200 = 100$
38.	10	1/6 ?	200 ?
39.	15	1/4 ?	300 ?
40.	20	1/3 ?	400 ?
41.	25	5/12 ?	500 ?
42.	30	1/2 ?	600 ?
43.	35	7/12 ?	700 ?
44.	40	2/3 ?	800 ?
45.	45	3/4 ?	900 ?
46.	50	5/6 ?	1000 ?
47.	55	11/12 ?	1100 ?
48.	60	1 ?	1200 ?

Geometry **Draw an area grid model to represent each product.**

49. $2\frac{1}{2} \cdot 1\frac{1}{3}$ **50.** $4\frac{1}{4} \cdot 3\frac{1}{2}$

51. $\frac{1}{2} \cdot \frac{2}{3}$ **52.** $3\frac{3}{8} \cdot \frac{1}{4}$

53. $\frac{1}{3} \cdot \frac{1}{3}$

Multiply by the reciprocal to find each quotient. Use a calculator to check the quotient.

54. $\frac{1}{2} \div \frac{2}{3}$ $\frac{3}{4}$ **55.** $7 \div 2\frac{1}{2}$ $2\frac{4}{5}$ **56.** $5 \div 1\frac{2}{3}$ 3 **57.** $12 \div \frac{3}{4}$ 16

58. $9 \div 2\frac{1}{4}$ 4 **59.** $2\frac{2}{3} \div 8$ $\frac{1}{3}$ **60.** $3\frac{1}{3} \div 2$ $1\frac{2}{3}$ **61.** $10 \div 1\frac{2}{3}$ 6

62. $4\frac{1}{2} \div 2\frac{1}{4}$ 2 **63.** $3\frac{3}{4} \div 1\frac{2}{3}$ $2\frac{1}{4}$ **64.** $1\frac{1}{2} \div 3$ $\frac{1}{2}$ **65.** $6\frac{1}{4} \div \frac{2}{3}$ $9\frac{3}{8}$

66. $3\frac{1}{5} \div \frac{4}{15}$ 12 **67.** $2\frac{1}{3} \div 1\frac{5}{6}$ $1\frac{3}{11}$ **68.** $4\frac{1}{2} \div 2\frac{7}{10}$ $1\frac{2}{3}$ **69.** $\frac{9}{10} \div 1\frac{1}{5}$ $\frac{3}{4}$

Look Back

70. Find the next three terms in the sequence 4, 9, 16, 25, . . . Explain the pattern used to find the terms. **[Lesson 1.1]** 36, 49, 64; The sequence is 2^2, 3^2, 4^2, 5^2 . . .

71. Find the greatest prime factor of 76. **[Lesson 1.5]** 19

72. Evaluate the expression $x + y \cdot z$ when x is -3, y is 14, and z is 2. **[Lesson 1.7]** 25

73. Find the values for y by substituting the values 1, 2, 3, 4, and 5 for x in $y = 5x - 1$. **[Lesson 1.7]**

73.

x	1	2	3	4	5
y	4	9	14	19	24

74. Find the absolute value $|-24|$. **[Lesson 2.1]** 24

75. Evaluate $(-4)(3)(-16) + (-8)$. **[Lesson 2.5]** 184

Build a table of values by substituting -3, -2, -1, 0, 1, 2, and 3 for x in each equation. Sketch a graph on grid paper. [Lesson 2.6]

76. $y = 2x - 1$ **77.** $y = 4 - x$ **78.** $y = -4x + 12$

Add or subtract. [Lesson 3.4]

79. $\frac{1}{2} - \frac{4}{5}$ $-\frac{3}{10}$ **80.** $2\frac{2}{3} + 1\frac{7}{12}$ $4\frac{1}{4}$ **81.** $6 - \frac{3}{4}$ $5\frac{1}{4}$

Look Beyond

Technology   Use your graphics calculator to create a table and graph for the reciprocal function, $Y = 1/X$. Use intervals of $\frac{1}{2}$ (or 0.5) for X in the table.

82. Explain why you can use the equation $Y = 1/X$ to create a table and graph of reciprocals.

83. Explain why there is an error message in the table when $X = 0$.

84. Explain why the point (0.5, 2) is on the graph.

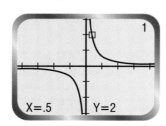

Exploring Ratios

Why Ratios are used to compare many different types of data. For example, ratios can be used to express the number of servings per recipe or the number of goals made per the number of goals attempted. Graphs can be used to represent ratios and to make predictions based on ratios.

Rita keeps the statistics for the hockey team at her high school. She wants to show the number of goals made compared with the number of goals attempted by each player for the season. A **ratio** is a comparison of two quantities using division. Rita decides to use ratios to record this information.

Rita noticed that Reggie has a record of 2 goals for every 7 attempts. She wrote this as the ratio 2 to 7, or $\frac{2}{7}$. Rita decides to investigate what will happen if Reggie maintains the same record for the entire season. Notice that a ratio can be written as a fraction. A **rational number** is any number that can be written as a ratio of two integers in which the denominator is not zero.

Exploration 1 Ratios

1 Use yellow algebra tiles to represent the number of successes and blue algebra tiles to represent the number of attempts. Use the tiles to represent the ratio 2 to 7.

2 Use tiles to represent an equivalent ratio that uses 4 yellow tiles. How many blue tiles should you use? Explain. 14 blue tiles. In the equivalent ratio, the number of tiles used of each color is doubled.

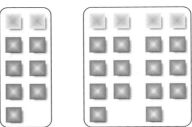

3 Describe what might be expected in a game in which Reggie has 21 attempts. Expect 6 goals.

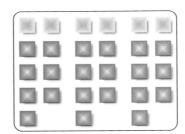

4 Copy the table below. Use the algebra tiles to complete the table for some expected numbers.

				28		63
Number of attempts	7	14	21	?	35	?
Number of goals	2	4	?	8	?	18
			6		10	

5 Describe how you used tiles and the ratio $\frac{2}{7}$ to complete the table in Step 4.

6 How many successes would you expect if Reggie's total number of attempts in a season is 70? if his total number of attempts is 84? 20; 24

7 Describe the pattern you found when completing the table in Step 4. What common factors do the ratios share? Every ratio simplifies to $\frac{2}{7}$.

8 Given the ratio $\frac{2}{7}$ in the table, how do you know what number to multiply in the numerator to get the equivalent ratio $\frac{?}{21}$? ❖
Multiply by 3, since $3 \times 7 = 21$.

CT– About 6 goals. Estimate using the proportion $\frac{1}{6} = \frac{x}{36}$.

CRITICAL *Thinking*

Suppose another player on the hockey team has a 1-to-6 ratio of goals made to goals attempted. How many goals would you expect him to make out of 35 attempts? Explain your reasoning.

A graph is a useful tool for displaying information described by a ratio.

Exploration 2 Graphing Data

1 Sean has a record of 3 goals for every 10 attempts. Copy and complete the table for this ratio.

Number of attempts	10	20	30	40	50	60	70	80	90	100
Number of goals	3	6	9	12	15	18	21	24	27	30

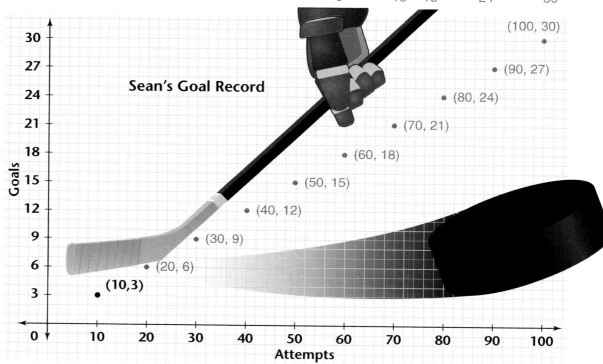

Sean's Goal Record

(10,3) (20, 6) (30, 9) (40, 12) (50, 15) (60, 18) (70, 21) (80, 24) (90, 27) (100, 30)

Goals

Attempts

2 On a piece of graph paper, label a grid so that the horizontal axis represents the number of attempts, and the vertical axis represents the number of goals. Plot the points represented in the table. For example, the point (10, 3) represents 10 attempts and 3 goals made.

3 Use your graph to complete this sentence. Sean made _?_ goals for every 90 attempts. 27

4 Connect the points on your graph. Use your graph to determine the expected number of goals made for 24 attempts, 36 attempts, and 58 attempts. 7, 11, 17

5 Use your graph to determine the expected number of attempts necessary to make 16 goals, 24 goals, and 28 goals. 53, 80, 93

6 Explain how to use your graph to determine the expected number of goals made if you know the number of attempts. Explain how to use your graph to determine the expected number of attempts if you know the number of goals made. ❖

Proportions

Notice that in the completed table from Exploration 1, the ratio for 7 attempts, $\frac{2}{7}$, is a fraction that is equivalent to the ratio for 21 attempts.

$$\frac{2}{7} = \frac{6}{21}$$

This equation is called a *proportion*. A **proportion** is an equation containing two or more equivalent ratios. The process for finding equivalent ratios is the same as the process for finding equivalent fractions.

$$\frac{2}{7} \cdot \frac{3}{3} = \frac{6}{21}$$

You used equivalent fractions to solve proportions when filling in the table. For instance, in solving $\frac{2}{7} = \frac{8}{?}$, you used the multiplier $\frac{4}{4}$ to get $\frac{2}{7} \cdot \frac{4}{4} = \frac{8}{28}$.

Ty runs at a rate of 2 miles in 11 minutes. Describe how to draw a graph to represent this rate. Describe how you would use your graph to determine how long it takes Ty to run 5 miles.

APPLICATION

Examine the table below.

	Ratio	Woods High
Number of freshmen in band	4	n
Number of freshmen	15	240

The proportion that can be solved to find the number of Woods High freshmen in band is shown.

$$\frac{4}{15} = \frac{n}{240}$$

You can use equivalent fractions to solve the proportion for n, the number of Woods High freshmen in band.

$$\frac{4}{15} \cdot \frac{16}{16} = \frac{n}{240}$$

Since 15 times 16 equals 240, 4 times 16 must equal n.

$$\frac{4}{15} \cdot \frac{16}{16} = \frac{64}{240}$$

So n is 64.

Therefore, 64 Woods High freshmen are in band. ❖

EXERCISES & PROBLEMS

Communicate

1. Define the term *ratio*.
2. Describe how a table could be used to solve this problem: The ratio of men to women in the state legislature is 7 to 2. If there are 56 men in the legislature, how many women are there?
3. Explain how a multiplier can be used to solve the problem in Exercise 2.
4. Explain how the property of equivalent ratios can be used to solve a ratio problem.
5. Describe how to determine if two ratios are equivalent.

Practice & Apply

Complete each proportion using the property of equivalent ratios.

6. $\frac{3}{7} = \frac{9}{?}$ 21

7. $\frac{6}{8} = \frac{?}{40}$ 30

8. $\frac{11}{12} = \frac{?}{144}$ 132

9. $\frac{?}{54} = \frac{7}{9}$ 42

10. $\frac{24}{?} = \frac{8}{9}$ 27

11. $\frac{36}{42} = \frac{9}{?}$ 10.5

12. $\frac{?}{72} = \frac{6}{9}$ 48

13. $\frac{49}{?} = \frac{7}{15}$ 105

14. $\frac{3}{4} = \frac{30}{?}$ 40

15. $\frac{49}{56} = \frac{?}{8}$ 7

16. $\frac{63}{49} = \frac{9}{?}$ 7

17. $\frac{20}{30} = \frac{?}{3}$ 2

18. $\frac{?}{66} = \frac{12}{11}$ 72

19. $\frac{3}{4} = \frac{123}{?}$ 164

20. $\frac{75}{?} = \frac{15}{16}$ 80

21. $\frac{?}{156} = \frac{5}{12}$ 65

Out of 55,600 households surveyed:

3200 own at least two cars.

1220 have children under two years of age.

23,350 have at least one dog.

Lifestyles Write a ratio to describe each situation. Then write an equivalent ratio in lowest terms.

22. The ratio of the number of households with at least two cars to the number of households surveyed $\frac{3200}{55,600} = \frac{8}{139}$

23. The ratio of the number of households with children under two years of age to the number of households surveyed $\frac{1220}{55,600} = \frac{61}{2780}$

24. The ratio of the number of households with at least one dog to the number of households surveyed $\frac{23,350}{55,600} = \frac{467}{1112}$

Determine each unknown quantity.

25. Cooking A recipe for 48 meatballs calls for 2 pounds of beef. How many pounds of beef do you need to make 36 meatballs? $1\frac{1}{2}$ lbs

26.  **Geometry** The ratio of a rectangle's length to its width is 3 to 2. If the length is 24 centimeters, what is the width? 16 cm

27. Sports A baseball player had 3 hits in 9 times at bat. Suppose the player continues to hit at this rate. How many hits would you expect him to have in 45 times at bat? 15 hits

Geometry The Nichols are planning to build a rectangular greenhouse. They decide to use a specific ratio for the length and the width. The graph shows the relationship between the possible lengths and widths.

28 If the length is 8 feet, what is the width? 6 ft

29. If the length is 12 feet, what is the width? 9 ft

30. What is the ratio of the length to the width in lowest terms?
$\frac{4}{3}$

Agriculture A citrus grower uses a market analysis to determine that he should plant 7 orange trees for every 4 grapefruit trees.

31. What is the ratio of the number of orange trees to the number of grapefruit trees? $\frac{7}{4}$

32. How many grapefruit trees should he plant if he plants 70 orange trees? 40

33. How many grapefruit trees should he plant if he plants 98 orange trees? 56

34. How many orange trees should he plant if he plants 160 grapefruit trees? 280

35. If the total number of trees he plants is 264, how many of each type should he plant? $o = 168$ $g = 96$

36. Make a graph to show the ratio of the number of orange trees to the number of grapefruit trees.

37. Use your graph to predict the number of grapefruit trees that the citrus grower will plant if he plants 28 orange trees. 16

167

Sports A volleyball player's statistics indicate that she misses 1 serve for every 4 successful serves.

38. What is the ratio of missed serves to successful serves? $\frac{1}{4}$

39. What is the ratio of the number of successful serves to the total number of attempts? $\frac{4}{5}$

40. How many successful serves would you expect if there were a total of 30 attempts? 24

41. Make a graph to show the ratio of the number of successful serves to the total number of attempts.

42. Use your graph to predict the number of successful serves expected for 15 attempts. 12

43.  **Portfolio Activity** Complete the problem in the portfolio activity given on page 129.

Look Back

44. The data in the following table show a plumber's fee, y, for working x hours. Use the first differences to write a linear equation. **[Lesson 1.3]** $y = 12x + 20$

x	1	2	3	4	5	6
y	32	44	56	68	80	92

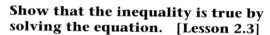**Show that the inequality is true by solving the equation.** **[Lesson 2.3]**

45. $-5 < 4$; $-5 + x = 4$ $x = 9$
Since x is positive $-5 < 4$.

46. $-7 < -3$; $-7 + x = -3$ $x = 4$
Since x is positive $-7 < -3$.

Build a table of values for each equation. Use the values −2, −1, 0, 1, 2, 3, and 4 for x. Graph the points on a coordinate plane, and draw a line through the points. **[Lesson 2.6]**

47. $y = 2x + 3$

48. $y = 3x - 1$

49. A board is $13\frac{5}{8}$ feet long. Another board is $13\frac{3}{4}$ feet long. Which board is longer? **[Lesson 3.1]** $13\frac{3}{4}$

Look Beyond

Probability A number cube has six faces numbered from 1 to 6. If the number cube is rolled six times, how many times would you expect each of the following events to occur?

50. The number rolled is 5.
1 time

51. The number rolled is an even number. 3 times

LESSON 3.7 Percent

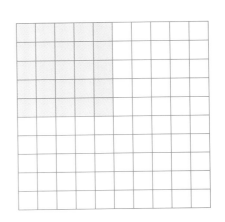

why *The next time you read a newspaper, notice how many times a percent is used. Percents are found in tables, charts, and graphs. They are also in ads, on display signs, and in financial statements.*

The expression *twenty-five percent* is written 25%. A **percent** is a ratio that compares a number with 100. A percent is another way to represent a fraction or a decimal.

Twenty-Five Percent of Pet Owners Say They Prefer to Have Their Pets Groomed Professionally.

Writing Percents as Decimals and Fractions

In the model, 25 out of the 100 squares are shaded. The ratio of shaded squares to total squares is $\frac{25}{100}$ which can be written as a fraction, decimal, or percent.

$$\frac{25}{100} = 0.25 = 25\%$$

EXAMPLE 1

Write each percent as a decimal.

Ⓐ 30%   **Ⓑ** 82%

Solution ➤

Ⓐ 30% means 30 out of 100. So $30\% = \frac{30}{100} = 0.30$, or 0.3.

Ⓑ 82% means 82 out of 100. So $82\% = \frac{82}{100} = 0.82$. ❖

CRITICAL *Thinking*

Describe a method you can use to change a percent to a decimal in one step. Will your method work with all percents? Why or why not? Divide the percent by 100. The method will not work in one step if the percent contains a fraction.

A fraction-bar model can also be used to show percents. The percent bar below shows 25%. Notice that 1 out of 4 parts is colored. So 25% is $\frac{25}{100} = \frac{1}{4}$.

$$\frac{1}{4} = 25\%$$

$$\frac{4}{4} = 100\%$$

Percents can be greater than 100%. The percent bar below shows 125%.

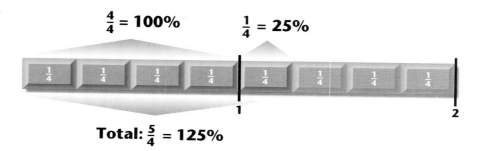

$$\frac{4}{4} = 100\% \qquad \frac{1}{4} = 25\%$$

Total: $\frac{5}{4}$ **= 125%**

All of the first bar is shaded. In the second bar, 1 out of 4 parts is shaded. The fraction shown is $1\frac{1}{4}$ or $\frac{5}{4}$. So 125% is $\frac{125}{100} = \frac{5}{4}$, or $1\frac{1}{4}$.

EXAMPLE 2

Use percent bars to write each percent as a fraction in lowest terms.

Ⓐ 40%   **Ⓑ** 120%

Solution ➤

Ⓐ 40% is $\frac{40}{100} = \frac{2}{5}$, which means 2 out of 5.

$$\frac{2}{5} = 40\%$$

Ⓑ 120% is $\frac{120}{100} = \frac{6}{5} = 1\frac{1}{5}$.

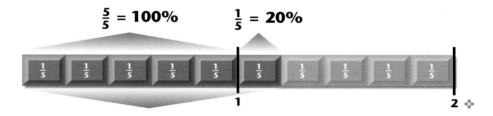

$$\frac{5}{5} = 100\% \qquad \frac{1}{5} = 20\%$$

Try This Use percent bars to write 60% and 180% as fractions in lowest terms.

$\frac{3}{5}, 1\frac{4}{5}$

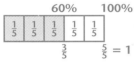

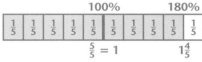

EXAMPLE 3

Political Science An opinion poll indicates that 3 out of 5 registered voters will vote for Bessy Smith in the next city election. Based on this poll, what *percent* of voters can be expected to vote for Bessy Smith in the next city election?

Solution ➤

Draw a percent bar to represent 3 out of 5.

$$\frac{3}{5} = ?$$

$$\frac{5}{5} = 100\%$$

Three out of five is $\frac{3}{5}$. To write the fraction as a decimal, write $\frac{3}{5}$ as an equivalent fraction with a denominator of 100.

$$\frac{3}{5} = \frac{3 \cdot 20}{5 \cdot 20} = \frac{60}{100} = 60\%$$

Therefore, 60% of the voters can be expected to vote for Bessy Smith in the next city election. ❖

Writing Decimals and Fractions as Percents

Recall that a percent can be written as a ratio that compares a number with 100. You can use this definition to write decimals and fractions as percents.

EXAMPLE 4

Write each decimal as a percent.

A 0.7   **B** 0.823   **C** 1.65

Solution ➤

Any decimal can be changed to a percent by multiplying by $\frac{100}{100}$.

A $0.7 \cdot \frac{100}{100}$

$= \frac{0.7}{1} \cdot \frac{100}{100}$

$= \frac{70}{100}$

$= 70\%$

B $0.823 \cdot \frac{100}{100}$

$= \frac{0.823}{1} \cdot \frac{100}{100}$

$= \frac{82.3}{100}$

$= 82.3\%$

C $1.65 \cdot \frac{100}{100}$

$= \frac{1.65}{1} \cdot \frac{100}{100}$

$= \frac{165}{100}$

$= 165\%$ ❖

Try This   Write the decimals 0.48, 2.05, and 0.9 as percents.
48%, 205%, 90%

CRITICAL *Thinking*

A shortcut to changing a decimal, such as 0.823, to a percent is to multiply it by 100, and write the percent sign.

$$0.823 \cdot 100 = 82.3, \text{ so } 0.823 = 82.3\%$$

Describe a shortcut for changing a decimal to a percent by using mental math. Explain why this shortcut works. **CT–** Since multiplying a decimal by 100 moves the decimal point to the right two places, simply move the decimal point without multiplying. Then add a percent sign.

EXAMPLE 5

Write each fraction as a percent.

A $\frac{5}{8}$ **B** $\frac{1}{3}$

Solution ➤

Scientific Calculator

A To change $\frac{5}{8}$ to a percent, first divide 5 by 8. Multiply the quotient, 0.625, by 100. Add a percent sign to the product.

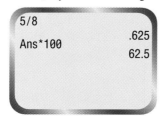

```
5/8
              .625
Ans*100
              62.5
```

or   5 ÷ 8 × 100 =   62.5

The percent is 62.5%.

B To change $\frac{1}{3}$ to a percent, first divide 1 by 3. Multiply the quotient, 0.333 . . . , by 100.

1/3	
	.3333333333
Ans*100	
	33.33333333

or   1 ÷ 3 × 100 =   33.33333333

Since the repeating decimal 33.333 . . . is equal to $33\frac{1}{3}$, the percent is $33\frac{1}{3}$%. ❖

Try This   Write each fraction as a percent.

a. $\frac{3}{8}$ **b.** $\frac{2}{3}$

37.5% $66\frac{2}{3}$%

Estimating With Percents

Certain percents are considered benchmarks. These are percents that can be used to estimate answers to problems. You should become very familiar with these percents.

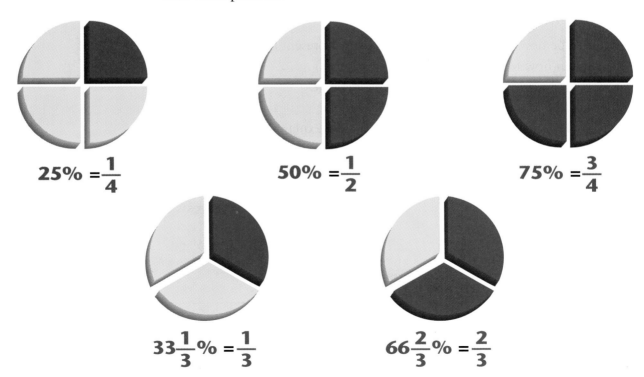

$25\% = \dfrac{1}{4}$ $50\% = \dfrac{1}{2}$ $75\% = \dfrac{3}{4}$

$33\dfrac{1}{3}\% = \dfrac{1}{3}$ $66\dfrac{2}{3}\% = \dfrac{2}{3}$

EXAMPLE 6

Of the 191 members in the freshman class, 72% are involved in at least one extracurricular activity. Use estimation to find approximately how many freshman students are involved in an extracurricular activity.

STATISTICS *Connection*

Solution ➤

72% **of** 191 means 72% **times** 191.

72% is close to 75%, and 75% is equivalent to $\frac{3}{4}$.

191 is close to 200.

So multiply $\frac{3}{4}$ and 200 to estimate 72% of 191.

$$\frac{3}{4} \cdot 200 = 150$$

According to these statistics, *approximately* 150 freshman students are involved in an extracurricular activity. ❖

EXERCISES & PROBLEMS

Communicate

1. Describe how a percent bar can be used to represent 80%.

2. Explain how to change a percent to a fraction.

3. Describe how to change a fraction to a percent.

4. Explain how to estimate 24% of 397.

5. Which is greater, 40% of 100 or 10% of 400? Explain your reasoning.

Practice & Apply

Use a percent bar to write each percent as a fraction in lowest terms.

6. 20%   **7.** 60%   **8.** 90%   **9.** 75%   **10.** 125%   **11.** 140%

What percent does each model represent?

12.

0   1
80%

13.
0      $66\frac{2}{3}\%$ 1

14.
0   1

75%

15.
0      $66\frac{2}{3}\%$ 1

Change each decimal to a percent.

16. 0.8 80%   **17.** 0.9 90%   **18.** 0.82 82%   **19.** 0.16 16%   **20.** 3.5 350%   **21.** 7.2 720%

22. 0.815 81.5% **23.** 0.128 12.8% **24.** 0.359 35.9% **25.** 2.89 289% **26.** 8.14 814% **27.** 7.035 703.5%

28. 1.133113.3% **29.** 1.055105.5% **30.** 0.1 10%   **31.** 0.01 1%   **32.** 0.546 54.6%   **33.** 8.15 815%

Change each fraction to a percent.

34. $\frac{7}{8}$ 87.5%   **35.** $\frac{3}{4}$ 75%   **36.** $\frac{7}{12}$ 58.3%   **37.** $\frac{5}{6}$ 83.3%   **38.** $3\frac{9}{20}$ 345%   **39.** $2\frac{7}{10}$270%

40. $\frac{9}{12}$ 75%   **41.** $\frac{2}{3}$ 66.67%   **42.** $\frac{125}{250}$ 50%   **43.** $\frac{4}{5}$ 80%   **44.** $\frac{1}{8}$ 12.5%   **45.** $\frac{1}{5}$ 20%

46. Academics Last year, 79% of the students at a certain school passed the state mathematics exam. If there are currently 1988 students at this school, about how many students can be expected to pass the state mathematics exam this year? 1570

47. Criminology Last year, 48% of the crimes committed in a certain state were robberies or burglaries. If there are 263 crimes this year, how many can be expected to be robberies or burglaries? 126

48. Wages The employees of Humboldt Creamery are offered a 9% raise next year. If an employee's salary is $1498.78 per month this year, what will it be after the increase next year? $1633.67

49. Sales The retail price of an appliance at Merrell's Appliances is determined by increasing the wholesale price by 35%. If the wholesale price is $239, by how much is the price increased? $83.65

50. Investments An investment is expected to yield a profit of 9.8% per year. If the investment is $4050, what is the approximate yield for the year? $396.90

Look Back

51. Write an equation for the following sentence: The weekly pay, *w*, is equal to the number of hours worked, *h*, times $9.50 per hour.   **[Lesson 1.2]**
$w = 9.50\,h$

Cultural Connection: Africa The earliest woman mathematician whose name we know is Hypatia. She was a professor at the University of Alexandria, Egypt, about 400 C.E. In her commentary on Diophantus, an earlier Egyptian mathematician, Hypatia spoke of pentagonal numbers.

52. The sequence for pentagonal numbers is 1, 5, 12, 22, 35, . . . Find the number of differences it takes to reach a constant. What is the constant?   **[Lesson 2.7]** 2nd differences; 3

53. What kind of function will generate the terms of the pentagonal sequence?   **[Lesson 2.7]** quadratic

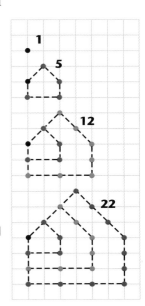

Look Beyond

54. 5 times; the coin will land heads up $\frac{1}{2}$ of the time.

54. If you toss a coin 10 times, how many times do you expect it to land heads up? Why? Explain how your reasoning involves rational numbers.

Hot Hand OR Hoop-la?

'Hot Hands' Phenomenon: A Myth?

The gulf between science and sports may never loom wider than in the case of the hot hands.

Those who play, coach or otherwise follow basketball believe almost universally that a player who has successfully made his last shot or last few shots—a player with hot hands—is more likely to make his next shot. An exhaustively statistical analysis led by a Stanford University psychologist, examining thousands of shots in actual games found otherwise: the probability of a successful shot depends not at all on the shots that came before.

To the psychologist, Amos Tversky, the discrepancy between reality and belief highlights the extraordinary differences between events that are random and events that people perceive as random. When events come in clusters and streaks, people look for explanations; they refuse to believe they are random, even though clusters and streaks do occur in random data.

To test the theory, researchers got the records of every shot taken from the field by the Philadelphia 76ers over a full season and a half. When they looked at every sequence of two shots by the same player—hit–hit, hit–miss, miss–hit or miss–miss, they found that a hit followed by a miss was actually a tiny bit likelier than a hit followed by a hit.

They also looked at sequences of more than two shots. Again, the number of long streaks was no greater than would have been expected in a random set of data, with every event independent of its predecessor.

Researchers Survey of 100 Basketball Fans

No 9%

Yes 91%

Does a player have a better chance of making a shot after having just made his last two or three shots than he does after having just missed his last two or three shots?

No 16%

Yes 84%

Is it important to pass the ball to someone who has just made several (2, 3, or 4) shots in a row?

Discuss Do you agree with the majority on the two survey questions? Why or why not?

Cooperative Learning

To form your own opinion about the hot-hand study, it helps to know how the researchers tabulated their data. For simplicity, you will use a portion of their data, which shows field-goal attempts by two players for several games. Both players usually make about half of their shots from the field.

A—11100010 01110111010010 000011101 0100000101 010101001110 1001111 01001010001 011110110

B—10101000100 11100000110011 0101010111001111 111001001101 0001101010 10010001111

(1 = hit   0 = miss   space = new game)

1. First look at what happens after a shot is made within each game.

   a. Copy and complete Table 1, using the data for either Player A or Player B.

   b. What percentage of the time was a hit followed by another hit? What percentage of the time was a hit followed by a miss?

   c. What would you expect your results in **b** to be if there were no such thing as a *hot hand*, that is, if the shots were hit or missed just by chance? Explain. (Assume that the player makes 50% of his shots on average.)

2. Now look at what happens after a shot is missed within each game.

   a. Copy and complete Table 2. Use the same player you used for Table 1.

   b. What percentage of the time was a miss followed by a hit? What percentage of the time was a miss followed by another miss?

   c. What would you expect your results in **b** to be if the shots were hit or missed just by chance?

TABLE 1	
After a hit	
Number of times next shot is made	Number of times next shot is missed

TABLE 2	
After a miss	
Number of times next shot is missed	Number of times next shot is made

3. Do you think there is such a thing as a hot hand in basketball? Why or why not? How does your belief affect the strategy you would use if you were coaching a basketball team?

LESSON 3.8 Experimental Probability

why *People in all cultures seem to enjoy games. Many games involve an element of luck or chance. Chance is often introduced by number cubes or spinners. You can use experimental probability to determine a number which represents the chance that an event will happen.*

Discovering Experimental Probability

Cultural Connection: Americas One American Indian game, *shaymahkewuybinegunug,* involves tossing sticks. It is played by members of the Ojibwa, or Chippewa, tribe. This game uses five flat sticks that have carved pictures of snakes on one side and that are plain on the other. Players take turns tossing the sticks to earn points.

Exploration 1 *Probability and the Coin Game*

Games To play an adaptation of the Ojibwa game, use three coins in place of the sticks. Each person tosses the three coins 10 times. Each time the three coins are the same, that person gets a point. After each person completes 10 tosses, the person with the most points is the winner.

or

1 Play the game. How many times did all 3 of your coins come up alike?

2 If you make 20 tosses, how many times do you think all 3 coins will come up alike?

3 Perform an experiment to check your guess. Toss 3 coins 20 times, and record how often the 3 coins come up alike.

Experimental probability is usually represented as a fraction written in lowest terms. The numerator is the number of times all 3 coins come up alike. The denominator is the total number of tosses, 20. Probabilities are sometimes expressed as decimals or percents.

4 Give the experimental probability of your result in Step 3. How does your result compare with results from other groups? Notice that there can be many different experimental probabilities.

5 Combine the data for the entire class. What is the experimental probability for the entire class? Is your result close to the result for the entire class?

6 In this *experiment,* each toss of 3 coins is a *trial.* Let *t* represent the number of trials. When all 3 coins come up alike, it is considered a *successful event.* Let *f* represent the number of times a successful event occurs. Express the experimental probability, *P*, in a formula using *f* and *t.* ❖ $P = \frac{f}{t}$

•Exploration 2 *Probability From a Number Cube*

1 The faces on an ordinary number cube are numbered 1 to 6. If the cube is rolled 10 times, guess how many times a 5 will appear on the top of the cube.

2 Roll one cube 10 times. Count how many times you get a 5.

3 Define an *event* and a *trial* in this experiment.

4 What is the experimental probability of getting a 5 in 10 trials in this experiment? ❖

Tell how to find the experimental probability for each of the following events.

• 3 on one roll of a number cube
• heads on a toss of a coin

CT– Have each student in class roll one cube. Use a ratio to compare the number of 3s to the number of students. The coin toss can be done in like manner. The *experimental probability* is the given ratio.

Experimental probability varies when an experiment is conducted several times.

EXPERIMENTAL PROBABILITY

Let t be the number of trials in the experiment. Let f be the number of times a successful event occurs.

The **experimental probability**, P, of the event is given by

$$P = \frac{f}{t}.$$

•Exploration 3 *A Fair Game With Two Number Cubes*

1 A two-player game involves rolling two number cubes and finding the sum. If the sum is 5, 6, 7, 8, or 9, player A gets a point. If the sum is 2, 3, 4, 10, 11, or 12, player B gets a point. Guess which player has a better chance of winning. Explain your answer.

2 When a game is fair, each player has the same chance of winning. Design and conduct a fair game using two number cubes. ❖
Answers may vary. For example, player A gets a point if the sum is even. Player B gets a point if the sum is odd.

1. Player A can get a point in 24 ways.
Player B can get a point in only 12 ways.

Cultural Connection: Asia
Even in prehistoric times people played games of chance and strategy. They must have wondered what the probability of winning would be if they followed a certain strategy. But one of the earliest records of the collection of statistical data had geographical implications.

About 2100 B.C.E., a calculation of the surface area of a plot of land at Umma, Mesopotamia (Iraq), was engraved on a clay tablet.

Random Numbers

A key ingredient in experimental probability is the use of random numbers. In games, you may have used simple random-number generators such as spinners. For making selections, you may have tossed a coin or drawn a number from a hat. There are also books that contain tables of random numbers. Some calculators and computers are useful because they can generate random numbers that can be adapted to fit specific conditions.

Spreadsheet

Spreadsheet software and calculators typically use a command like RAND() or RAND to generate random numbers. The basic random numbers are decimals from 0 to 1, including 0 but not including 1. On some graphics calculators, as you continue to press ENTER you get more random numbers.

A6	=RAND()	
	A	**B**
1	0.30261695	
2	0.08310659	
3	0.77245826	
4	0.70603577	
5	0.34317643	
6	**0.66527653**	

Spreadsheet

Rand	
	.0078387869
	.9351587791
	.1080114624
	.0062633066
	.5489861799
	.8555803143

Graphics calculator

A6	=RAND()*5	
	A	**B**
1	1.40817388	
2	2.17844473	
3	3.12062745	
4	3.38549538	
5	0.02076914	
6	**4.06690849**	

Rand*5	
	4.885421232
	1.391544132
	1.376071474
	.6089506775
	.2629490302
	3.611896579

If each random number is multiplied by 5, the result is six new random numbers. The new random numbers range in value from 0 to 5. Can these new random numbers ever include 5? Explain. No: since the numbers generated by the RAND function do not include 0 or 1, multiplying by 5 will generate numbers between 0 and 5, but not including 0 or 5.

When the integer value function INT is applied to the random-number generator, random integers from 0 through 4 are produced.

A6	=INT(RAND()*5)	
	A	**B**
1	1	
2	0	
3	2	
4	3	
5	4	
6	**0**	

int (rand*5)	
	3
	1
	4
	2
	1
	2

Thus the spreadsheet command INT(RAND()*5) gives random integers from 0 through 4. Most programs use a symbol like = or @ to signal a formula.

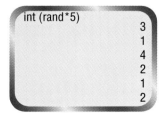

You can adapt the output of a random-number generator to your needs by changing the command.

> ### GENERATING RANDOM INTEGERS
> The command INT(RAND * K) + A generates random integers from a list of K consecutive integers beginning with A.

EXAMPLE 1

Determine the possible numbers that can appear at random when you use the command below.

INT(RAND*2) + 3

Graphics Calculator

Solution ➤

Examine the function in steps from the inside out.

(A) **RAND** results in a decimal value from 0 to 1 excluding 1.

(B) **RAND*2** results in a decimal value from 0 to 2 excluding 2.

(C) **INT(RAND*2)** results in the integer 0 or 1.

(D) **INT(RAND*2) + 3** results in the integer 3 or 4. ❖

Try This Generate sets of random numbers with 3 different commands. Describe the numbers that can possibly appear for each. Answers may vary.

EXAMPLE 2

Suppose you are given two random numbers from 1 to 100.
Find the experimental probability that at least one of them is less than or equal to 40.

Solution ➤

Generate 10 pairs of random numbers from 1 to 100. Count how many times at least one of each pair is less than or equal to 40.

Spreadsheet

This spreadsheet shows 10 pairs of random integers from 1 to 100 generated by the command INT(RAND()*100)+1.

In 7 of the 10 trials, at least one of the two numbers is less than or equal to 40. So the experimental probability is $\frac{7}{10}$. ❖

C11		=INT(RAND()*100)+1	
	A	**B**	**C**
1	Trial	1st Number	2nd Number
2	1	98	68
3	2	33	82
4	3	21	17
5	4	94	14
6	5	87	36
7	6	73	83
8	7	56	73
9	8	65	12
10	9	18	58
11	10	99	**18**

CT– Increasing the number of trials will usually cause the experimental probability to be closer to the actual (theoretical) probability.

CRITICAL *Thinking*

Will you get the same experimental probability if you use a greater number of trials? How does the number of trials affect the probability?

EXERCISES & PROBLEMS

Communicate

1. Explain what is meant by experimental probability.

2. Describe an experiment to find the experimental probability of getting at least 3 heads on a toss of 4 coins.

3. Is it possible for two groups that conduct the same probability experiment to get different results? Explain or give an example.

4. Is it possible for someone to conduct the same probability experiment twice and get different results? Explain or give an example.

5. If RAND generates a number from 0 to 1 (including 0, but not 1), describe the numbers you get from INT(RAND*7) + 1.

6. Look at the spreadsheet data for Example 2. Tell how to find the experimental probability that both numbers are less than 50.

Practice & Apply

7. Describe an experiment to find the experimental probability that if 4 coins are flipped, there will be either 4 heads or 4 tails.

8. Describe an experiment to find the experimental probability that if 2 number cubes are rolled, at least one of them will show a 6.

9. To determine an experimental probability, Fred conducted 15 trials and Ted conducted 16 trials. Is it possible that they arrived at the same experimental probability? Explain.

Two coins were flipped 20 times with the following results.

Trial	1	2	3	4	5	6	7	8	9	10	11	12	13	14	15	16	17	18	19	20
Coin 1	H	H	H	H	T	H	T	T	T	H	H	H	T	T	H	T	T	T	H	H	H
Coin 2	T	H	H	T	T	T	T	T	H	T	T	T	H	H	H	H	H	H	T	H

According to the data, find each experimental probability.

10. Both coins are alike. $\frac{7}{20}$

11. Both coins are heads. $\frac{1}{4}$

12. At least one coin is heads. $\frac{9}{10}$

13. Neither coin is heads. $\frac{1}{10}$

Look at the spreadsheet data for Example 2. Find the experimental probability that

14. both numbers are less than 80. $\frac{2}{5}$

15. the first number is greater than the second number. $\frac{3}{5}$

Games Design and conduct experiments to determine the following experimental probabilities. Describe each experiment carefully, and give your results for

16. getting tails when a coin is flipped.

17. getting 2 tails when a coin is flipped twice.

18. getting at least 3 when a number cube is rolled.

19. getting a multiple of 3 when a number cube is rolled.

20. getting an odd sum when two number cubes are rolled.

21. getting "doubles" (the same number on both cubes) when two number cubes are rolled.

22. getting the same result at least 3 times in a row when a coin is tossed 5 times.

23. not getting the same result twice in a row when a coin is tossed 5 times.

Technology The command RAND generates a decimal value from 0 to 1, not inclusive. Describe the output of each command.

24. RAND*2 $0 < n < 2$

25. INT(RAND*2) Integers 0 and 1

26. INT(RAND*2)+1
Integers 1 and 2

27. 100*(INT(RAND*2)+1) Integers 100 and 200

Before graphics calculators were used, engineers and scientists used books of tables to find data such as random numbers.

For Exercises 28–30, use the command INT(RAND*5)+10, where RAND generates a number from 0 to 1, not inclusive.

28. How many different numbers are possible? 5

29. What is the least number? 10

30. What is the greatest number?
14

Technology Write commands to generate random numbers from each of the following lists, where RAND generates a number from 0 to 1, including 0 but not 1. If possible, check your results using technology. Adapt the command to suit the computer or calculator you are using.

33. INT(RAND*6)+1

31. 0, 1, 2 INT(RAND*3)

32. 0, 1, 2, 3, 4, 5, 6, 7, 8, 9 INT(RAND*10)

33. 1, 2, 3, 4, 5, 6

34. 1, 2, 3, 4, 5, 6, 7, 8, 9, 10 INT(RAND*10) +1

Look Back

x	0	1	2	3	4	5	6
y	3	5	7	9	11	13	15

35. Build a table of values for the equation $y = 2x + 3$ by substituting 0, 1, 2, 3, 4, 5, and 6 for x.   **[Lesson 1.2]**

36. The cost of renting a car is given by the formula $c = 27 + 0.25m$, where c is the total rental cost and m is the number of miles driven. You have $95 budgeted for the rental. How many miles can you drive? **[Lesson 1.3]** 272 miles

37. ⬙ **Geometry** The dimensions of a rectangular playing field are 128 yards by 67 yards. Estimate the area.   **[Lesson 1.5]** 9100 sq yds

38. Write the prime factorization of 368 in exponential form.   **[Lesson 1.6]** $2^4 \cdot 23$

39. Use mental math to find the sum $19 + (21 + 22)$. Identify the property that you used.   **[Lesson 1.8]** $40 + 22 = 62$; Assoc Prop

40. Show the integers 5, -7, 4, and -2 on a number line. Then list the integers from least to greatest.   **[Lesson 2.3]** $-7, -2, 4, 5$

Use guess-and-check to solve each equation.   [Lesson 2.6]

41. $-1 = 4 - x$
$x = 5$

42. $x + 5 = -2$
$x = -7$

43. $2x + 1 = 1$
$x = 0$

44. Use differences to predict the next three terms in the sequence 1, 5, 14, 30, 55, . . . **[Lesson 2.7]** 91, 140, 204

45. What are the first three terms of a sequence if the fourth and fifth terms are 15 and 27 and the second differences are a constant 2?   **[Lesson 2.7]**
$-9, -3, 5$

Compare. Use $<$, $>$, or $=$.   [Lessons 3.1, 3.3]

46. $0.25 \overset{<}{} \frac{3}{8}$

47. $\frac{7}{100} \overset{<}{} \frac{3}{15}$

48. $121.01 \overset{<}{} 121.2$

49. A stock posted the following changes in one week: $+\frac{3}{4}$, $-\frac{1}{8}$, $-\frac{3}{4}$, $-\frac{3}{8}$, and $+\frac{1}{2}$. What was the net change in the stock for that week? **[Lesson 3.4]** 0

50. ⬙ **Statistics** In the school election, 237 out of 415 students voted for Sam. Estimate the percent of students who voted for Sam. **[Lesson 3.7]** approx. 60% since $\frac{240}{400} = 60\%$

Look Beyond

51. How many pairs of numbers are possible from one red cube and one blue cube if each cube has faces numbered 1 through 6? 36

52. How many of the pairs in Exercise 51 form a sum of 3? 2

Find the value of each expression.

53. $1^2 - 0^2$ 1

54. $2^2 - 1^2$ 3

55. $3^2 - 2^2$ 5

56. $4^2 - 3^2$ 7

57. $5^2 - 4^2$ 9

58. $6^2 - 5^2$ 11

Use the pattern from Exercises 53–58 to find the value of each expression.

59. $100^2 - 99^2$ 199

60. $a^2 - (a - 1)^2$ $2a - 1$

LESSON 3.9 Theoretical Probability

why *Many situations have theoretical probabilities: tossing a coin, rolling a die, or choosing a person at random. Theoretical probabilities allow you to predict what is most likely to happen in an actual situation or experiment. Theoretical probability has valuable applications in many fields, including genetics.*

You can use theoretical probability to determine the chances of a family having male or female children.

If a coin is tossed, what is the probability that the result will be heads? Recall from Lesson 3.8 that experimental probability is the number of times an *event* occurs divided by the number of *trials*.

For example, suppose Gail tosses a penny 10 times. The result is 4 heads and 6 tails. Then the experimental probability of a coin toss resulting in heads is the number of times the event occurred, 4, divided by the number of trials, 10.

$$P(\text{heads}) = \frac{\text{event}}{\text{trials}} = \frac{4}{10}, \text{ or } 40\%$$

5 or $\frac{1}{2}$ of the times

Each time an experiment such as one toss of a coin, is performed, the result is called the **outcome**. In the experiment of tossing a coin 10 times, how many times would you *expect* heads to occur?

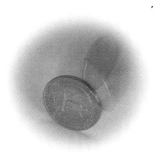

Theoretical probability is a measure of what you *expect* to occur. There are two possible outcomes for the experiment of tossing a coin one time: heads or tails. A **sample space** for an experiment is the set of possible outcomes for that experiment.

So the sample space, *S*, for the experiment of tossing a coin one time is:

$$S = \{\text{heads, tails}\}.$$

THEORETICAL PROBABILITY

When each outcome listed in a sample space has an equal chance of occurring, the **theoretical probability**, *P*, that a particular outcome, or event, *E*, will occur is given by

$$P(E) = \frac{\text{number of elements in the desired event}}{\text{number of elements in the sample space}}.$$

The event heads and the event tails are equally likely. To find the theoretical probability of one coin toss resulting in heads, count the elements involved.

$$P(\text{heads}) = \frac{\text{number of elements in the desired event}}{\text{number of elements in the sample space}} = \frac{1}{2}, \text{ or } 50\%$$

How can you use this theoretical probability to predict the number of heads that will result from tossing a coin 500 times? Since heads should occur $\frac{1}{2}$ of the time, mult. 500 by $\frac{1}{2}$. Toss a coin 10 times, and record the results. Toss a coin 20 times, and record the results. Toss a coin 30 times, and record the results. (You can also use a random number generator to simulate the experiments.)

Compare the experimental probabilities of tossing the coin 10, 20, and 30 times with the theoretical probabilities of each experiment. What do you notice about the experimental probability as the number of trials increases? As the number of trials increases, the closer the experimental probability is to the theoretical probability.

CRITICAL Thinking

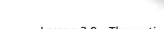

EXAMPLE 1

PROBABILITY
Connection

A game requires a player to roll an ordinary 6-sided number cube. What is the theoretical probability that the player will roll 1 *or* 5?

$$P(1 \text{ } or \text{ } 5) = ?$$

Solution ➤

There are 6 sides of the number cube, so there are 6 elements in the sample space that are equally likely.

$$S = \{1, 2, 3, 4, 5, 6\}$$

The desired event is rolling 1 *or* 5, so there are 2 elements in the desired event.

1	2	3
4	5	6

To find the theoretical probability of rolling 1 *or* 5 with one roll of a 6-sided number cube, count the elements.

$$P(1 \text{ } or \text{ } 5) = \frac{\text{number of elements in the desired event}}{\text{number of elements in the sample space}} = \frac{2}{6}, \text{ or } \frac{1}{3}$$

The theoretical probability of rolling 1 *or* 5 with one roll is $\frac{1}{3}$. ❖

If you roll an ordinary 6-sided number cube 210 times, how many times would you predict the roll to be 1 *or* 5? 70 times

CRITICAL
Thinking

In Example 1, the theoretical probability of rolling a 1 is $\frac{1}{6}$, and the theoretical probability of rolling a 5 is $\frac{1}{6}$. Explain why *adding* these two probabilities, $\frac{1}{6} + \frac{1}{6}$, also gives the theoretical probability of rolling 1 *or* 5. Since the probabilities $P(1)$ and $P(5)$ do not affect each other, the probability of $P(1 \text{ or } 5)$ is the same as $P(1) + P(5)$.

Independent Events

Suppose you conduct an experiment involving two events. If the outcome of one event does not affect the probability of the other event occurring, the events are called **independent events**.

EXAMPLE 2

Genetics Mr. and Mrs. Sanderson are expecting their first child. Assume there is an equally likely chance that the Sandersons will have a boy or girl. If the couple has 2 children, what is the probability of having 2 girls?

Solution ➤

Whether the *first* baby is a boy or girl has no effect on whether the *second* baby is a boy or a girl. The event boy (B) and the event girl (G) are independent events.

To determine the sample space, draw a diagram.

	Second child	
	Boy (B)	Girl (G)
First child Boy (B)	BB	BG
Girl (G)	GB	GG

You can see that there are 4 elements in the sample space, S.

$$S = \{BB, BG, GB, GG\}$$

To find the theoretical probability that both children will be girls, count the elements.

$$P(GG) = \frac{\text{number of elements in the desired event}}{\text{number of elements in the sample space}} = \frac{1}{4}, \text{ or } 25\%.$$

The theoretical probability that the Sandersons will have two girls is $\frac{1}{4}$. ❖

EXAMPLE 3

Julian and Gilberto perform an experiment involving two independent events: tossing a coin and spinning a spinner. The spinner has 3 equal regions. Find the theoretical probability of tossing heads and spinning a 1.

$$P(H1) = ?$$

Number 1 (1)

Heads (H)

Solution ➤

	Second event		
	1	2	3
First event H	H1	H2	H3
T	T1	T2	T3

You can see that there are 6 elements in the sample space, S.

$$S = \{H1, H2, H3, T1, T2, T3\}$$

To find the theoretical probability of tossing heads and spinning a 1, count the elements.

$$P(H1) = \frac{\text{number of elements in the desired event}}{\text{number of elements in the sample space}} = \frac{1}{6}$$

So the theoretical probability of tossing heads and spinning a 1 is $\frac{1}{6}$, or approximately 17%. ❖

EXAMPLE 4

Ten cards numbered 1 through 10 are placed in a container. One card is drawn at random, then replaced. A second card is drawn. What is the theoretical probability that the first number selected is greater than 5 *and* that the second number selected is a prime number?

Leonard replaces the card into the container before the second card is drawn.

Solution ➤

Count the elements in the sample space.

The first draw has 10 possible outcomes, so the sample space for the probability of the first event has 10 elements.

Since the first card is *replaced* before the second card is selected, the events are independent and the sample space for the probability of the second event also has 10 elements.

Therefore, the sample space for this experiment has 100 elements.

| | | **Second event** | | | | | | | | |
	1	**2**	**3**	**4**	**5**	**6**	**7**	**8**	**9**	**10**
1	1, 1	1, 2	1, 3	1, 4	1, 5	1, 6	1, 7	1, 8	1, 9	1, 10
2	2, 1	2, 2	2, 3	2, 4	2, 5	2, 6	2, 7	2, 8	2, 9	2, 10
3	3, 1	3, 2	3, 3	3, 4	3, 5	3, 6	3, 7	3, 8	3, 9	3, 10
4	4, 1	4, 2	4, 3	4, 4	4, 5	4, 6	4, 7	4, 8	4, 9	4, 10
5	5, 1	5, 2	5, 3	5, 4	5, 5	5, 6	5, 7	5, 8	5, 9	5, 10
6	6, 1	6, 2	6, 3	6, 4	6, 5	6, 6	6, 7	6, 8	6, 9	6, 10
7	7, 1	7, 2	7, 3	7, 4	7, 5	7, 6	7, 7	7, 8	7, 9	7, 10
8	8, 1	8, 2	8, 3	8, 4	8, 5	8, 6	8, 7	8, 8	8, 9	8, 10
9	9, 1	9, 2	9, 3	9, 4	9, 5	9, 6	9, 7	9, 8	9, 9	9, 10
10	10, 1	10, 2	10, 3	10, 4	10, 5	10, 6	10, 7	10, 8	10, 9	10, 10

(**First event** labels rows 1–10)

Count the number of desired outcomes.

The model has rows 6 through 10 shaded to illustrate all of the possible ways that the *first* card drawn can be greater than 5.

The model also has columns 2, 3, 5, and 7 shaded to illustrate all of the possible ways that the *second* card drawn can be a prime number.

The 20 *overlapping* shaded areas illustrate all of the possible ways that *both* events can occur.

P(1st number > 5 *and* 2nd number is prime)

$$= \frac{\text{number of elements in the desired event}}{\text{number of elements in the sample space}}$$

$$= \frac{20}{100}, \text{ or } \frac{1}{5} ❖$$

The theoretical probability of drawing a number that is greater than 5 on the first draw is $\frac{5}{10}$, or $\frac{1}{2}$. The theoretical probability of drawing a prime number on the second draw is $\frac{4}{10}$, or $\frac{2}{5}$. Explain why *multiplying* these two probabilities, $\frac{1}{2} \cdot \frac{2}{5}$, also gives the theoretical probability of the first number being greater than 5 *and* the second number being a prime number.

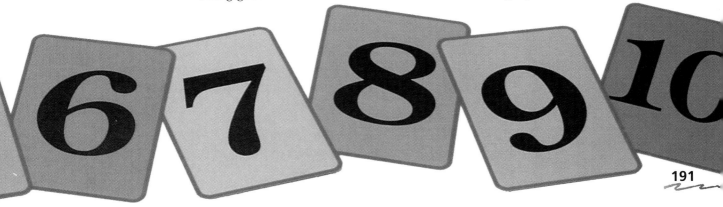

Dependent Events

When the outcome of one event *does* affect the possible outcomes of another event, the events are called **dependent events**.

In Example 4, suppose the first card selected is *not replaced* before the second card is selected. The second event would then have only 9 possible outcomes. Why does this make the two events dependent? Because what happens in the 2nd event depends on what happens in the 1st event.

EXAMPLE 5

Sports Alan, Debbie, Lucia, and Phil want to play doubles in tennis. To form tennis teams, they randomly draw from 2 yellow and 2 orange tennis balls in a bucket. The two people with yellow tennis balls will be partners, and the two with orange tennis balls will be partners. What is the probability that the first 2 tennis balls drawn are orange?

Solution ➤

	O_1	O_2	Y_1	Y_2
O_1	O_1O_1	O_1O_2	O_1Y_1	O_1Y_2
O_2	O_2O_1	O_2O_2	O_2Y_1	O_2Y_2
Y_1	Y_1O_1	Y_1O_2	Y_1Y_1	Y_1Y_2
Y_2	Y_2O_1	Y_2O_2	Y_2Y_1	Y_2Y_2

The sample space for these two dependent events can also be represented with a grid. Let O_1 and O_2 represent the two orange balls, and let Y_1 and Y_2 represent the two yellow balls.

Once a ball is removed from the sample space, it cannot be drawn again. Therefore, the outcomes O_1O_1, O_2O_2, Y_1Y_1, and Y_2Y_2 are impossible and cannot be elements in the sample space.

There are 12 elements in the sample space.

There are 2 elements in the desired event.

P(1st ball is orange *and* 2nd ball is orange)

$$= \frac{\text{number of elements in the desired event}}{\text{number of elements in the sample space}}$$

$$= \frac{2}{12}, \text{ or } \frac{1}{6}$$

Therefore, the probability that the first 2 tennis balls drawn are orange is $\frac{1}{6}$. ❖

20 times

If this experiment were performed 120 times, how many times would you predict that the 2 orange balls would be chosen in the first 2 draws?

EXERCISES & PROBLEMS

Communicate

1. Explain the difference between experimental and theoretical probability.
2. Give an example of independent events.
3. Give an example of dependent events.
4. Carlos tosses 2 coins. Describe how to find the sample space for this experiment.
5. Mary rolls 2 6-sided number cubes. Explain how to find the theoretical probability that she will roll a 3 on each cube.

Practice & Apply

A bag contains 12 marbles: 6 red (R), 4 green (G), and 2 yellow(Y). One marble is randomly drawn from the bag. Use this information for Exercises 6–15.

6. $P(G) = \underline{\ ?\ } \frac{1}{3}$
7. $P(R) = \underline{\ ?\ } \frac{1}{2}$
8. $P(Y) = \underline{\ ?\ } \frac{1}{6}$
9. $P(not\ G) = \underline{\ ?\ } \frac{2}{3}$
10. $P(not\ R) = \underline{\ ?\ } \frac{1}{2}$
11. $P(not\ Y) = \underline{\ ?\ } \frac{5}{6}$
12. $P(G\ or\ R) = \underline{\ ?\ } \frac{5}{6}$
13. $P(G\ or\ Y) = \underline{\ ?\ } \frac{1}{2}$
14. $P(R\ or\ Y) = \underline{\ ?\ } \frac{2}{3}$
15. Show that $P(G) + P(R) + P(Y) = 1$.

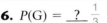

$$\frac{4}{12} + \frac{6}{12} + \frac{2}{12} = 1$$

16. Work with a group to find experimental probabilities for drawing one marble in 100 trials. Compare the experimental probabilities with the theoretical probability. *Answers may vary.*
17. What is the theoretical probability that you will roll a number that is greater than 3 in one roll of a 6-sided number cube? $\frac{1}{2}$
18. What is the theoretical probability that you will roll a number that is greater than 3 in two rolls of a 6-sided number cube? $\frac{3}{4}$
19. What is the theoretical probability that you will roll a number that is greater than 3 in only one of two rolls of a 6-sided number cube? $\frac{1}{2}$
20. What is the theoretical probability that you will roll a number that is less than 3 in two rolls of a 6-sided number cube? $\frac{5}{9}$

21. Calvin performs an experiment in which he tosses a coin and rolls a 6-sided number cube. What is the theoretical probability that he will get tails on the coin and roll a 3 on the number cube? $\frac{1}{12}$

22. Lara performs an experiment in which she flips a coin and rolls a 6-sided number cube. What is the theoretical probability that she will get heads on the coin and roll a 4 on the number cube? $\frac{1}{12}$

23. Sean performs an experiment in which he flips two coins and rolls an ordinary 6-sided number cube. What is the theoretical probability that he will get tails on both coins and roll a 7 on the number cube? 0

Consider the spinner with equal regions A, B, C, and D. Find the following probabilities for one spin.

24. $P(A) = \underline{\quad?\quad}$ $\frac{1}{4}$ **25.** $P(B \text{ or } C) = \underline{\quad?\quad}$ $\frac{1}{2}$

Find the following probabilities for two spins.

26. $P(D \text{ and } A) = \underline{\quad?\quad}$ $\frac{1}{16}$ **27.** $P(B \text{ or } C) = \underline{\quad?\quad}$ $\frac{1}{2}$

28. What is the probability that the spinner will land on A in either of the two spins? $\frac{7}{16}$

29. What is the probability that the spinner will land on B and then on C? $\frac{1}{16}$

A bag contains 12 marbles: 6 red (R), 4 green (G), and 2 yellow (Y). Two marbles are randomly drawn. Use a grid to find the following probabilities.

		First marble returned (independent events)	First marble *not* returned (dependent events)
30.	$P(R, \text{ then } R)$	? 1/4	? 5/22
31.	$P(R, \text{ then } G)$	? 1/6	? 2/11
32.	$P(R, \text{ then } Y)$	? 1/12	? 1/11
33.	$P(G, \text{ then } R)$	? 1/6	? 2/11
34.	$P(G, \text{ then } G)$	? 1/9	? 1/11
35.	$P(G, \text{ then } Y)$	? 1/18	? 2/33
36.	$P(Y, \text{ then } R)$	? 1/12	? 1/11
37.	$P(Y, \text{ then } G)$	? 1/18	? 2/33
38.	$P(Y, \text{ then } Y)$	? 1/36	? 1/66
39.	SUM	? 1	? 1

40. Shelley performs an experiment in which she rolls two 6-sided number cubes. What is the theoretical probability that she will roll a 4 on both of the number cubes? $\frac{1}{36}$

41. There are 4 marbles in a bag: 2 blue and 2 green. Tonia picks 2 marbles from the bag and does *not* replace them. What is the probability that she picked a blue marble and a green marble in that order? $\frac{1}{3}$

42. There are 5 marbles in a bag: 2 blue and 3 green. Gabriel picks 2 marbles from the bag and does *not* replace them. What is the probability that he picked a blue marble and then picked another blue marble? $\frac{1}{10}$

43. There are 5 marbles in a bag: 2 red, 2 yellow, and 1 black. Suzanne picks 2 marbles from the bag and does *not* replace them. What is the probability that she picked a black marble and a red marble in that order? $\frac{1}{10}$

44. From the experiment in Exercise 43, what is the theoretical probability that Suzanne picked a red marble and a yellow marble in that order? $\frac{1}{5}$

45. Hollie and Becky perform an experiment of rolling two 6-sided number cubes. What is the probability of rolling a 6 on the first number cube and a 3 on the second number cube? $\frac{1}{36}$

Look Back

46. Business   The amount a car rental agency charges for the use of a car is given by the formula $C = 0.12m + 50$, where m represents the number of miles the car is driven. How much would you expect to pay if you drove the car 240 miles? **[Lesson 1.3]** $78.80

Simplify each expression. [Lessons 2.2, 2.4]

47. $67 - (-92) + (-23)$ 136   **48.** $(-155) + 1245 - (-145)$ 1235

Find the next six fractions in the pattern. Then write each fraction as a decimal, and order the fractions from greatest to least. [Lesson 3.2]

49. $\frac{1}{2}$ $\frac{1}{3}$ $\frac{1}{4}$ $\frac{1}{5}$ $\frac{1}{6}$ **50.** $\frac{1}{2}$ $\frac{2}{3}$ $\frac{3}{5}$ $\frac{5}{8}$ $\frac{8}{13}$

51. Out of 55,700 people, 5600 were unemployed. What is the ratio of the total number of people to the number of people *employed*? **[Lesson 1.3]** $\frac{557}{501}$

Look Beyond

On March 12, 1996, a newspaper reported that a satellite was out of control and was expected to hit the Earth.

52. The Earth is 30% land. What is the probability that the satellite would hit water? $\frac{7}{10}$

53. The paper reported that 11 other similarly sized objects had fallen to Earth in 1996 and that all of them had hit water. What is the probability that 12 objects in a row would hit water? $\left(\frac{7}{10}\right)^{12}$

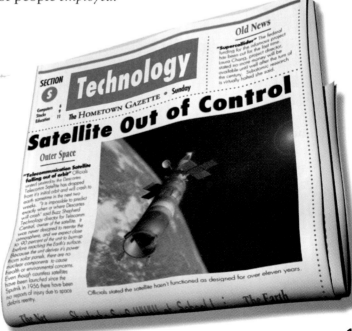

Fractions of a Dollar

$\frac{1}{8} = 0.125 \rightarrow 12\frac{1}{2}¢$

$\frac{1}{4} = 0.25 \rightarrow 25¢$

$\frac{3}{8} = 0.375 \rightarrow 37\frac{1}{2}¢$

$\frac{1}{2} = 0.5 \rightarrow 50¢$

$\frac{5}{8} = 0.625 \rightarrow 62\frac{1}{2}¢$

$\frac{3}{4} = 0.75 \rightarrow 75¢$

$\frac{7}{8} = 0.875 \rightarrow 87\frac{1}{2}¢$

In stock reports, dollar amounts are listed in fractions and can be converted to decimals and to cents.

Buying shares of stock is a way for you to own part of a company. If the company profits, the value of your stock increases and you profit.

The *closing values* of individual shares are printed each day in the newspaper, along with the *net change* from the previous day.

STOCK Corporation	Closing value	Net change
A-1 Video	$47\frac{1}{8}$	$-1\frac{1}{4}$
Alumundy Security	$30\frac{1}{2}$	$-\frac{1}{4}$
DNB Rentals	$17\frac{3}{4}$	$-\frac{1}{2}$
Software Today	$27\frac{1}{4}$	$+1\frac{1}{8}$

In the listings shown at the left, Software Today closed at $27\frac{1}{4}$, or \$27.25. The stock rose $1\frac{1}{8}$, or \$1.125 from the previous day. To find the value of a share of Software Today stock on the previous day, subtract $1\frac{1}{8}$ from $27\frac{1}{4}$.

$$27\frac{1}{4} - 1\frac{1}{8} = 27\frac{2}{8} - 1\frac{1}{8}$$
$$= 26\frac{1}{8}$$

On the previous day, one share of stock of Software Today was worth $26\frac{1}{8}$, or \$26.125.

The value rose by $1\frac{1}{8}$, so if you owned 100 shares of Software Today stock, you would profit by $100\left(1\frac{1}{8}\right) = 100(1.125) = \$112.50$ on that day.

Changes in stock are easy to see when represented with a line graph. The table below shows the performance of Software Today stock for one week, and the corresponding line graph follows. The percent change for the week shows the net change in the value of the stock for the entire week as a percent of the opening value.

Opening value $26\frac{1}{8}$	Mon	Tues	Wed	Thurs	Fri	Net change for the week	Percent change for the week $= \dfrac{\text{Net change}}{\text{Opening value}}$
	$27\frac{1}{4}$	$26\frac{7}{8}$	$26\frac{3}{8}$	$27\frac{5}{8}$	$27\frac{3}{8}$		
Net change	$+1\frac{1}{8}$	$-\frac{3}{8}$	$-\frac{1}{2}$	$+1\frac{1}{4}$	$-\frac{1}{4}$	$+1\frac{1}{4}$	$\dfrac{1\frac{1}{4}}{26\frac{1}{8}} = \dfrac{1.25}{26.125} = 0.0478$, or 4.78%

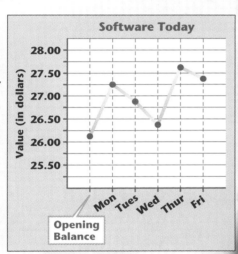

One hundred shares of Software Today stock would have profited $100\left(1\frac{1}{4}\right)$, or $125, for the week.

Building a Table and Graph

Make a table and a line graph for Almundy Security. Make them similar to those shown for Software Today. Compute the weekly profit for 100 shares. The opening value for the week is $30\frac{1}{4}$. The net changes for Monday through Friday are $+\frac{1}{4}$, $-\frac{3}{4}$, $+1\frac{1}{2}$, $+1\frac{1}{8}$, and $-\frac{3}{8}$.

Stock Simulation   Answers may vary.

Select three real companies from newspaper stock listings, and simulate the purchase of 100 shares at an opening value. Follow the stocks for one month. Create a table and a line graph of the daily values and net changes. Compute weekly profit and loss and weekly percent changes. Finally, compute a monthly profit and loss and monthly percent change in value for each of the three stocks. You may want to use a spreadsheet to organize your data, create your graphs, and complete your computations.

Chapter 3 Review

Vocabulary

Key Skills & Exercises

Lesson 3.1

➤ **Key Skills**

Compare two fractions.

To determine whether $\frac{2}{3}$ is closer to $\frac{1}{2}$ or to $\frac{5}{8}$, compare the fraction blocks for $\frac{2}{3}$, $\frac{1}{2}$, and $\frac{5}{8}$.

$\frac{2}{3}$ is closer to $\frac{5}{8}$.

$\frac{1}{8}$	$\frac{1}{8}$	$\frac{1}{8}$	$\frac{1}{8}$	$\frac{1}{8}$	$\frac{1}{8}$	$\frac{1}{8}$	$\frac{1}{8}$

$\frac{1}{3}$	$\frac{1}{3}$	$\frac{1}{3}$

$\frac{1}{2}$	$\frac{1}{2}$

➤ **Exercises**

Tell whether each fraction is closest to $\frac{1}{3}$, $\frac{2}{3}$, or 1.

1. $\frac{3}{8}$ $\frac{1}{3}$

2. $\frac{1}{4}$ $\frac{1}{3}$

3. $\frac{7}{10}$ $\frac{2}{3}$

4. $\frac{5}{6}$ 1

Lesson 3.2

➤ **Key Skills**

Write a fraction in lowest terms.

To express $\frac{15}{21}$ in lowest terms, divide the numerator and denominator by the GCF of 15 and 21, which is 3.

$$\frac{15 \div 3}{21 \div 3} = \frac{5}{7}$$

Use the LCD to compare two fractions.

To compare $\frac{2}{3}$ and $\frac{3}{5}$, express each fraction with a denominator of 15, the LCD.

$$\frac{2}{3} \underline{\ ?\ } \frac{3}{5}$$

$$\frac{10}{15} > \frac{9}{15} \text{ so } \frac{2}{3} > \frac{3}{5}$$

➤ **Exercises**

Write each fraction in lowest terms.

5. $\frac{9}{18}$ $\frac{1}{2}$

6. $\frac{21}{27}$ $\frac{7}{9}$

7. $\frac{26}{52}$ $\frac{1}{2}$

8. $\frac{42}{54}$ $\frac{7}{9}$

Use the LCD to compare each pair of fractions using the symbol <, >, or =.

9. $\frac{1}{2}$ and $\frac{2}{5}$ $>$

10. $\frac{2}{3}$ and $\frac{5}{12}$ $>$

11. $\frac{3}{8}$ and $\frac{1}{4}$ $>$

12. $\frac{4}{5}$ and $\frac{9}{10}$ $<$

Lesson 3.3

➤ Key Skills

Write a decimal as a fraction or mixed number, and write a fraction as a decimal.

The decimal 0.25 is written as $\frac{25}{100}$. To change the fraction to a decimal, divide the numerator by the denominator.

$$\frac{3}{4} = 3 \div 4 = 0.75$$

Order numbers from least to greatest or from greatest to least.

To order decimals and mixed numbers from least to greatest, first rewrite all of them with the same number of decimal places. Then write the decimals in order.

➤ Exercises

Write each decimal as a fraction or mixed number in lowest terms.

13. 0.49 $\frac{49}{100}$ **14.** 6.4 $6\frac{2}{5}$ **15.** 2.94 $2\frac{47}{50}$ **16.** 0.64 $\frac{16}{25}$

Write each fraction as a terminating or repeating decimal.

17. $\frac{7}{8}$ 0.875 **18.** $\frac{1}{12}$ $0.08\overline{3}$ **19.** $-6\frac{1}{3}$ $-6.\overline{3}$ **20.** $\frac{2}{7}$ $0.\overline{285714}$

Order the numbers from least to greatest.

21. 0.06, 0.008, 0.4, 0.0203
0.008, 0.0203, 0.06, 0.4

22. 2.14, $2\frac{1}{7}$, 2.4, 2.1
2.1, 2.14, $2\frac{1}{7}$, 2.4

Lesson 3.4

➤ Key Skills

Add and subtract fractions.

To add fractions, first write them with a common denominator using the LCD. Then add the numerators.

$$\frac{1}{3} + \frac{1}{6} = \frac{1 \cdot 2}{3 \cdot 2} + \frac{1}{6}$$
$$= \frac{2}{6} + \frac{1}{6} = \frac{3}{6}, \text{ or } \frac{1}{2}$$

➤ Exercises

Add or subtract. Write your answer in lowest terms.

23. $\frac{1}{4} + \frac{1}{2}$ $\frac{3}{4}$ **24.** $\frac{5}{6} - \frac{2}{3}$ $\frac{1}{6}$ **25.** $\frac{1}{5} + \frac{3}{8}$ $\frac{23}{40}$ **26.** $9\frac{1}{3} - \frac{7}{9}$ $8\frac{5}{9}$

Lesson 3.5

➤ Key Skills

Multiply and divide rational numbers. Write your answer in lowest terms.

To multiply or divide fractions, first change mixed numbers to improper fractions. To divide fractions, multiply the first fraction by the reciprocal of the second fraction.

$$5\frac{2}{3} \div \frac{3}{4} = \frac{17}{3} \cdot \frac{4}{3}$$
$$= \frac{68}{9}, \text{ or } 7\frac{5}{9}$$

➤ Exercises

Find each product or quotient.

27. $5 \div 1\frac{1}{4}$ 4 **28.** $4\frac{1}{3} \cdot 6$ 26 **29.** $8\frac{1}{2} \div 9$ $\frac{17}{18}$ **30.** $2\frac{3}{4} \div 3\frac{5}{8}$ $\frac{22}{29}$

Lesson 3.6

> ### ➤ Key Skills

Solve proportions by using equivalent fractions.

Solve $\frac{3}{20} = \frac{?}{80}$.

Since $20 \cdot 4 = 80$, multiply the numerator by 4.

$$\frac{3}{20} = \frac{?}{80}$$

$$\frac{3 \cdot 4}{20 \cdot 4} = \frac{12}{80} \quad \text{so} \quad \frac{3}{20} = \frac{12}{80}.$$

> ### ➤ Exercises

Solve the proportions.

31. $\frac{2}{5} = \frac{10}{?}$ 25   **32.** $\frac{5}{9} = \frac{?}{45}$ 25   **33.** $\frac{?}{82} = \frac{1}{2}$ 41   **34.** $\frac{54}{?} = \frac{6}{17}$ 153

Lesson 3.7

> ### ➤ Key Skills

Write decimals and fractions as percents.

To change $\frac{2}{3}$ to a percent, first change it to a decimal. $\frac{2}{3} = 0.\overline{6666}$

Multiply the decimal by 100. $= 66.\overline{66}\%$

Simplify by writing the repeating decimal as a fraction. $= 66\frac{2}{3}\%$

> ### ➤ Exercises

Change each fraction or mixed number to a percent.

35. $\frac{5}{8}$ 62.5%   **36.** $\frac{3}{20}$ 15%   **37.** $\frac{7}{10}$ 70%   **38.** $5\frac{1}{3}$ $533\frac{1}{3}\%$

Lesson 3.8

> ### ➤ Key Skills

Calculate experimental probability.

Two coins are flipped 5 times with the following results.

Trial	1	2	3	4	5
Coin 1	H	T	T	T	H
Coin 2	T	T	H	H	H

The experimental probability that coin 1 will land heads on any toss is $\frac{2}{5}$.

The experimental probability that coin 2 will land heads on any toss is $\frac{3}{5}$.

> ### ➤ Exercises

Use the data above to find each experimental probability.

39. At least one coin is heads. $\frac{4}{5}$ **40.** Both coins are tails. $\frac{1}{5}$

Lesson 3.9

> ### ➤ Key Skills

To determine theoretical probabilities.

When each possible outcome is equally likely, the theoretical probability, P, of event E of occurring is $P(E) = \dfrac{\text{number of elements in the desired event}}{\text{number of elements in the sample space}}$.

> ### ➤ Exercises

There are 2 blue marbles and 3 green marbles in a bag. Tony picks one marble, and then picks another marble. What is the probability that he picked 2 green marbles:

41. if he *did* replace the first marble picked? $\frac{9}{25}$ **42.** if he *did not* replace the first marble picked? $\frac{3}{10}$

Chapter 3 Assessment

1. Tell whether the fraction $\frac{9}{10}$ is closest to $\frac{2}{3}$, $\frac{7}{8}$, or 1. $\frac{7}{8}$

2. Use the LCD to compare the fractions $\frac{2}{3}$ and $\frac{5}{8}$. $\frac{2}{3} > \frac{5}{8}$

Order the numbers from least to greatest.

3. 0.16, 0.06, 0.016, 0.106   0.016, 0.06, 0.106, 0.16

4. 1.8, 1.08, $1\frac{7}{8}$, 1.78   1.08, 1.78, 1.8, $1\frac{7}{8}$

Write each fraction in lowest terms.

5. $\frac{7}{14}$ $\frac{1}{2}$

6. $\frac{35}{40}$ $\frac{7}{8}$

7. $\frac{29}{92}$ $\frac{29}{92}$

8. Write the fraction $\frac{3}{7}$ as a terminating or repeating decimal. $0.\overline{428571}$

Add or subtract.

9. $\frac{2}{3} + \frac{5}{6}$ $1\frac{1}{2}$

10. $5\frac{5}{6} - 4\frac{1}{4}$ $1\frac{7}{12}$

11. $\frac{1}{8} + \frac{1}{9}$ $\frac{17}{72}$

12. $\frac{1}{4} - \frac{1}{3}$ $-\frac{1}{12}$

13. $\frac{2}{3} + 2\frac{1}{6}$ $2\frac{5}{6}$

14. $\frac{5}{7} + \frac{4}{8}$ $1\frac{3}{14}$

Find each product.

15. $5 \cdot 4\frac{1}{3}$ $21\frac{2}{3}$

16. $8\frac{5}{8} \cdot 3$ $25\frac{7}{8}$

17. $1\frac{3}{4} \cdot 2\frac{1}{5}$ $3\frac{17}{20}$

18. $-4\frac{1}{3} \cdot 6$ -26

19. $5\frac{1}{4} \cdot -\frac{3}{5}$ $-3\frac{3}{20}$

20. $\frac{1}{2} \cdot \frac{1}{2} \cdot \frac{1}{3}$ $\frac{1}{12}$

Complete the proportions using the property of equivalent fractions.

21. $\frac{2}{7} = \frac{?}{21}$ 6

22. $\frac{5}{15} = \frac{?}{3}$ 1

23. $\frac{3}{7} = \frac{12}{?}$ 28

24. $\frac{?}{8} = \frac{18}{24}$ 6

Write the decimals as fractions or mixed numbers in lowest terms.

25. 1.96 $1\frac{24}{25}$

26. 0.6 $\frac{3}{5}$

27. 2.25 $2\frac{1}{4}$

28. 8.55 $8\frac{11}{20}$

29. 0.75 $\frac{3}{4}$

Change each number to a percent.

30. 0.43 43%

31. 2.03 203%

32. $\frac{7}{20}$ 35%

33. $1\frac{3}{8}$ 137.5%

34. If you roll two 6-sided number cubes, what is the probability that you will roll a 5 on one number cube and less than a 4 on the other number cube? $\frac{1}{12}$

35. If you roll one 6-sided number cube two times, what is the probability that you will roll a 3 on one number cube and greater than a 4 on the other number cube? $\frac{1}{18}$

36. There are 3 red marbles and 2 green marbles in a bag. If you pick two marbles *without* replacing them, what is the probability that you will draw a green marble first and then a red marble? $\frac{3}{10}$

CHAPTER 4

LESSONS

Geometry Connections

The word *geometry* comes from two Greek words: *geo,* which means "Earth," and *meter,* which means "measure." The ancient Greeks studied geometry to help them survey land, or measure the Earth.

Today, surveyors, architects, builders, graphic artists, construction workers, engineers, and many people use geometry every day.

PORTFOLIO ACTIVITY

Graphic artists use scale factors to enlarge and reduce artwork. Scale factors in computer graphics programs are often given as percents. On page 263, you will be asked to investigate the method of using scale factors that graphic artists use.

Enlarge 150%

Scale

Width:	150%	Percent	Options
Height:	150%	Percent	OK
☒ Constrain			Cancel

Lines and Angles

Ideas from geometry are found throughout algebra and in the real world. For example, the crosspieces on a kite form special angles called right angles.

A **point** shows an exact location.

A•
Point *A*

A **line** consists of all the points that extend infinitely in two opposite directions.

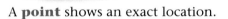

Line *MN*, or $\overleftrightarrow{NM}$
Line *l*, or *l*

A **ray** is part of a line that contains one endpoint and all the points extending in one direction from the endpoint. A ray is named in one direction only.

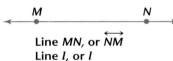

Ray *MN*, or $\overrightarrow{MN}$

A **segment** consists of two endpoints on a line and all the points between the endpoints.

M ——————————————— N

Segment *MN*, or $\overline{NM}$

A **plane** consists of all the points on a flat surface that extend infinitely in all directions.

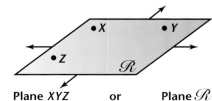

Plane *XYZ*   or   Plane $\mathcal{R}$

•Exploration 1 Intersecting Lines and Angles

1 Fold a sheet of paper in half. Label the crease line *l*. You may want to draw over the crease so that it can be seen easily.

2 Fold the paper again in a different direction so that the new crease line crosses line *l*. Label this new crease line *m*. The point where line *m* crosses line *l* is the **point of intersection**. Label the point of intersection point *A*. Because lines *l* and *m* cross, they are called **intersecting lines**.

3 Intersecting lines form *angles*. An **angle** is formed by two rays that intersect at a common point, called the **vertex** of the angle.

4 angles

4 How many angles on your paper have a vertex at *A*? ❖ Answers may vary; students may also count straight angles.

•Exploration 2 Perpendicular and Parallel Lines

1 Fold a sheet of paper, and label the crease line *l*. Draw point *B* anywhere on line *l*. Fold your paper through *B* so that line *l* matches with itself on both sides of the crease. Label this new crease line *n*.

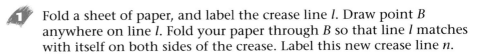

2 Lines *l* and *n* intersect in a special way. They are *perpendicular* to each other. You can write this as *l* ⊥ *n*. **Perpendicular lines** intersect to form square corners (also called *right angles*). Describe the angles on your paper formed at vertex *B*. All angles appear to be the same size, with square corners.

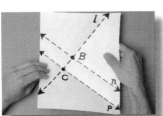

3 Fold your paper through another point, *C*, on line *l* so that line *l* again matches up with itself. Label this crease line *p*. Lines *p* and *n* are called **parallel lines.** You can write this as *p* ∥ *n*. Describe the relationship between lines *p* and *n*. They are side-by-side and always the same distance apart; they never meet.

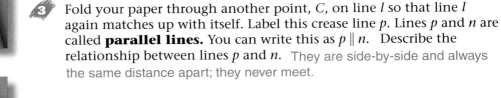

4 Fold line *l* through two more points so that it matches up with itself each time, forming two more lines parallel to *n*. Are these new lines also parallel to *p*? Based on your observations, write a definition of parallel lines. ❖ Yes; parallel lines never meet and are always the same distance apart.

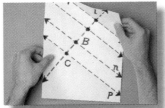

Angles

Angles are usually named using three letters, with the vertex named in the middle. They can also be named by just the vertex or by a number.

∠FDE, ∠EDF, or∠D

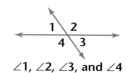

∠1, ∠2, ∠3, and ∠4

A *protractor* can be used to measure an angle in degrees. Angles with the same measure are said to be congruent.
Are ∠F and ∠G congruent?   yes

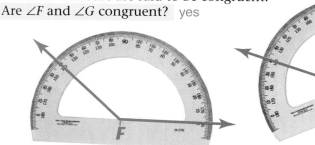

CRITICAL Thinking

There are two sets of numbers on a protractor. How do you know which numbers to use when measuring an angle?

A *straight angle* consists of two opposite rays that form a straight line. A **straight angle** has a measure of 180°.

Straight angle ∠ABC

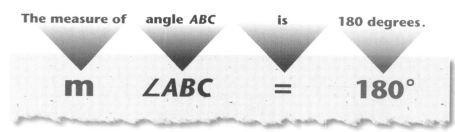

The measure of   angle *ABC*   is   180 degrees.

$$m \quad \angle ABC \quad = \quad 180°$$

A **right angle** has a measure of 90°.

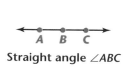

right angle symbol

Right angle ∠XYZ

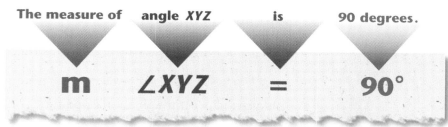

The measure of   angle *XYZ*   is   90 degrees.

$$m \quad \angle XYZ \quad = \quad 90°$$

An **acute angle** has a measure of less than 90°.

Acute angle ∠MNP

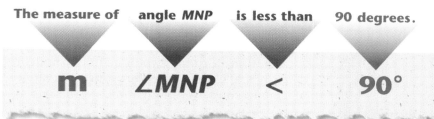

The measure of   angle *MNP*   is less than   90 degrees.

$$m \quad \angle MNP \quad < \quad 90°$$

An **obtuse angle** has a measure of greater than 90°.

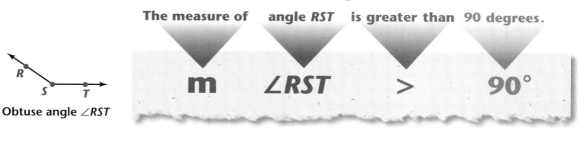

The measure of   angle *RST*   is greater than 90 degrees.

$$\text{m} \quad \angle RST \quad > \quad 90°$$

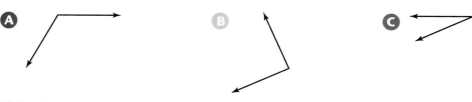

Obtuse angle ∠*RST*

EXAMPLE

Classify each angle as right, acute, or obtuse.

A

B

C

Solution ➤

Compare the measure of each angle with 90°.

A The measure of this angle is greater than 90°, so it is an obtuse angle.

B This angle, shown below, appears to have a measure of 90°. Copy the angle and extend the rays of the triangle. After measuring with a protractor, the angle can be classified as a right angle.

C The measure of this angle is less than 90°, so it is an acute angle. ❖

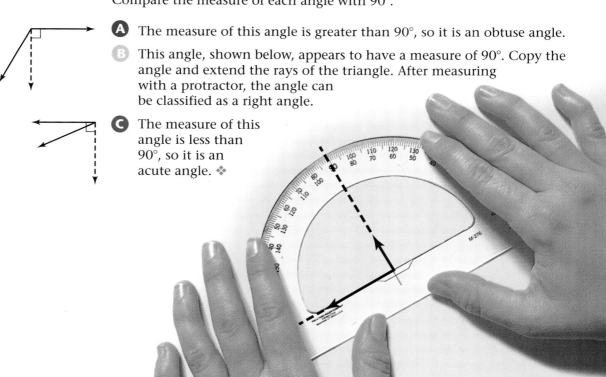

EXERCISES & PROBLEMS

Communicate

1. Describe the terms *point, line, ray, segment,* and *plane* in your own words.

2. What real-world objects can you use to model angles? Which of these objects can be used to model acute, right, and obtuse angles?

3. What is the difference between perpendicular and parallel lines?

4. Suppose two lines intersect to form four angles that each have the same measure. What type of angles are formed? Explain.

5. How can you construct right angles by paper folding?

Practice & Apply

Use a protractor to measure each angle in Exercises 6–14.

6. 45°

7. 120°

8. 40°

9. 135°

10. 90°

11. 170°

12. 90°

13. 115°

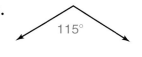

14. 75°

For Exercises 15–20, name the geometric figure from this lesson that best describes each real-world object.

15. a football field
parallel lines

16. a stream of water from a sprinkler ray

17. a speck of dust
point

18. the horizon over the open ocean line

19. a side of a box
part of a plane

20. an edge of a box line segment

For Exercises 21–26, name each figure in all possible ways.

21.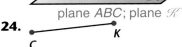
plane *ABC*; plane *K*

22.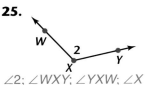
ray *PD*
ray *PF*
$\overrightarrow{PD}$
$\overrightarrow{PF}$

23.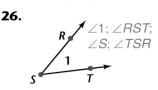
line *n*; line *ML*;
line *LM*; $\overleftrightarrow{ML}$, $\overleftrightarrow{LM}$

24.
C ●———————● K
segment *CK*; segment *KC*;
$\overline{CK}$; $\overline{KC}$

25.
∠2; ∠*WXY*; ∠*YXW*; ∠*X*

26.
∠1; ∠*RST*;
∠*S*; ∠*TSR*

Use the figure at the right for Exercises 27–31.

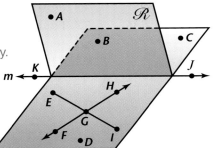

27. Name the intersection point of $\overline{EI}$ and $\overrightarrow{FH}$. G

28. Is point *D* contained in plane *R*? no

29. Segment *EI* is contained in plane __?__. Answers may vary.
plane *CBD*

30. Name a line that is the intersection of
two planes. $\overleftrightarrow{KJ}$, $\overleftrightarrow{JK}$, line *KJ*, line *JK*

31. Point *C* is in plane __?__. Answers may vary.
plane *CBD*

**Use a protractor to draw angles with the following measures.
Label each angle as either acute, obtuse, right, or straight.**

32. 35°   **33.** 67°   **34.** 103°   **35.** 170°   **36.** 93°   **37.** 72°

38. 45°   **39.** 90°   **40.** 100°   **41.** 180°   **42.** 75°   **43.** 160°

44. Construction Why are roofs with acute angles found in climates with
lots of snow? Why is a flat roof unsuitable for a snowy climate?

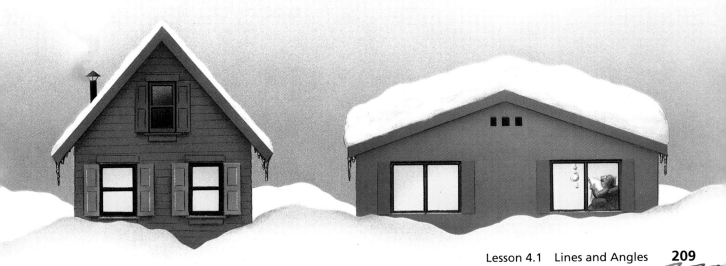

Use the figure at the right for Exercises 45–50.

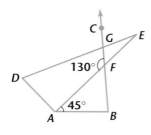

45. ∠W has sides _?_ and _?_. $\overrightarrow{WT}$; $\overrightarrow{WV}$

46. The vertex of ∠XYZ is _?_. Y

47. ∠W can also be called _?_ or _?_. ∠TWV; ∠VWT

48. To measure an angle, use a _?_. protractor

49. Two angles are congruent if _?_. Their measures are equal.

50. To draw an angle congruent to ∠W, use a _?_. protractor or a compass and straight edge

Use the figure at the left for Exercises 51–56.

51. Name three straight angles. ∠DGE; ∠AFE; ∠GFB

52. Name an acute angle with its measure given in degrees. ∠FAB

53. Name an obtuse angle with its measure given in degrees. ∠GFA

54. Find the measure of ∠GFE. 50°

55. Find the measure of ∠AFB. 50°

56. Is ∠ABF a right angle? Explain.

Look Back

Evaluate each expression. [Lessons 3.4, 3.5]

57. $\frac{2}{3} - 2.25$ $-1\frac{7}{12}$ **58.** $\frac{2}{3} \cdot 2.25$ $1\frac{1}{2}$ **59.** $\frac{2}{3} \div 2.25$ $\frac{8}{27}$

Write each ratio as a percent. [Lesson 3.7]

60. 2:3 $66\frac{2}{3}$% **61.** 8:12 $66\frac{2}{3}$% **62.** 4:5 80%

63. At Northlake Middle School, 162 of the 645 students walk to school. Estimate the percent of students who walk to school. [Lesson 3.7]

Approx 2.5%

Look Beyond

64. Measure the sides and angles of each triangle. Give each triangle a name based on your measurements. Side and angle measures may vary.

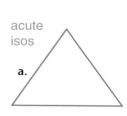

acute isos

a.

acute right

b.

scalene right

c.

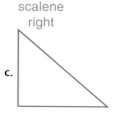

obtuse isos

d.

Exploring Angles

Why *There are many useful applications for angles and parallel lines. For example, a hologram uses properties of parallel lines to reflect light in different directions. As a laser beam bounces off parallel mirrors, it changes direction.*

A boy admires circular holograms at the Museum of La Villette in Paris.

•Exploration Special Kinds of Angles

Geometry Graphics

In this exploration, you will need geometry graphics technology or folding paper, a pencil, and a protractor.

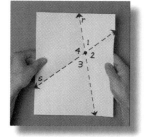

1 Fold two intersecting lines, *r* and *s*, that are not perpendicular. Number the angles as shown below at the left.

2 Angles 2 and 4 are *vertical angles*. Angles 1 and 3 are also *vertical angles*. Describe these angles. **Write a definition of *vertical angles* from your description.** Vertical angles are those opposite each other, formed by intersecting lines.

3 Measure angles 1–4. Compare m∠1 with m∠3. Compare m∠2 with m∠4. **Based on your observations, write a property of vertical angles.** Do you think your property is true for all vertical angles? Vertical angles have equal measures; yes.

4 Angles 1 and 2 are *adjacent angles*. Angles 2 and 3, angles 3 and 4, and angles 1 and 4 are also pairs of adjacent angles. Describe these angles. **Write a definition of *adjacent angles* from your description.** Adj angles have a common side, common vertex, and no common interior points.

5 Measure and compare the following pairs of angles:
∠1 and ∠2   ∠2 and ∠3   ∠3 and ∠4   ∠1 and ∠4
The angles in each pair are *supplementary angles*. **From your observations, write a definition of *supplementary angles*.** ❖

CRITICAL Thinking If two intersecting lines are perpendicular, what can you say about the adjacent angles and vertical angles that are formed?
They are congruent right angles.

Supplementary and Complementary Angles

In Exploration 1, you observed supplementary angles that are also adjacent angles. However, supplementary angles are not always adjacent. For example, ∠A and ∠B are supplementary. Why? Because 135 + 45 = 180°

Two angles with a combined measure of 90 degrees are called *complementary angles*. ∠C and ∠D are complementary since 30 + 60 = 90°.

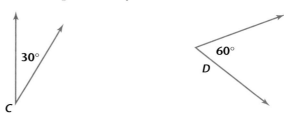

Draw a picture of complementary angles that are also adjacent.

EXTENSION

Suppose you know that the measure of ∠1 in the diagram is 50°. You can use this information to find the measures of ∠2, ∠3, and ∠4.

Since ∠1 and ∠2 form right angles, they are complementary.

Substitute 50 for m∠1.

$$m\angle1 + m\angle2 = 90°$$
$$50 + m\angle2 = 90°$$

Since 50 + 40 = 90:

$$m\angle2 = 40°$$

Since ∠1 and ∠4 are vertical angles, they are congruent.
$$m\angle1 = m\angle4 = 50°$$

Since ∠3 and ∠4 are complementary angles, the sum of their measures is 90°.

Substitute 50 for m∠4.

$$m\angle3 + m\angle4 = 90°$$
$$m\angle3 + 50 = 90°$$

Since 40 + 50 = 90:

$$m\angle3 = 40° ❖$$

SUMMARY OF ANGLE TYPES

Vertical angles are the opposite angles formed whenever two lines intersect. All vertical angles have equal measure.

Adjacent angles are two angles in the same plane with a common vertex and a common side, but with no interior points in common.

Supplementary angles are two angles with a combined measure of 180°. One angle is said to *supplement* the other. Supplementary angles may or may not be adjacent.

Complementary angles are two angles with a combined measure of 90°. One angle is said to be the *complement* of the other. Complementary angles may or may not be adjacent.

 CRITICAL *Thinking*

How can you use supplementary angles 1 and 2 and supplementary angles 2 and 3 to explain why vertical angles 1 and 3 must be congruent?
Since m∠1 + m∠2 = m∠2 + m∠3, subtract m∠2 from each side of the equation to find m∠1 = m∠3.

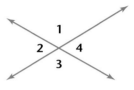

EXERCISES & PROBLEMS

Communicate

1. Describe supplementary angles.

2. Describe complementary angles.

3. Explain why vertical angles are congruent. Use the fact that supplementary angles have a combined measure of 180° in your explanation.

4. Explain why perpendicular lines form both congruent adjacent angles and congruent vertical angles.

5. Give a real-world example of complementary angles.

6. Give a real-world example of supplementary angles.

Practice & Apply

Use Figure 1 for Exercises 7–13.

7. Find m∠2. 74.5°   **8.** Find m∠4. 15.5°   **9.** Find m∠3. 74.5°

10. Name an acute angle. Answers may vary; ∠2, ∠3, ∠4

11. Name an obtuse angle. ∠1

12. Name a pair of vertical angles. ∠2 and ∠3 or ∠4 and 15.5°

13. Name a pair of complementary angles.
∠3 and ∠4 or ∠2 and 15.5° or ∠2 and ∠4 or ∠3 and 15.5°

Use Figure 2 for Exercises 14–23.

14. Find m∠BFC. 65°   **15.** Find m∠BFD. 155°

16. Find m∠DFE. 25°   **17.** Find m∠CFD. 90°

18. Find m∠BFE. 180°

19. Name two straight angles. ∠DFA; ∠EFB

20. Name two right angles. ∠DFC; ∠CFA

21. Name two pairs of vertical angles. ∠DFB and ∠EFA; ∠DFE and ∠BFA

22. Name two pairs of complementary angles.

23. Name four pairs of supplementary angles.

Figure 1

Figure 2

Use the information below for Exercises 24–27.

If two angles are complementary, one angle is said to be the *complement* of the other. If two angles are supplementary, one angle is said to be the *supplement* of the other.

24. The complement of an angle is 15°. What is the measure of the angle? 75°

25. The complement of an angle is 82°. What is the measure of the angle? 8°

26. The supplement of an angle is 35°. What is the measure of the angle? 145°

27. The supplement of an angle is 125°. What is the measure of the angle? 55°

28. Two congruent angles are complementary. What is the measure of each angle? 45°

29. Two congruent angles are supplementary. What is the measure of each angle? 90°

30. Three congruent angles form a straight angle. What is the measure of each angle? 60°

31. One of two complementary angles is twice the measure of the other. What is the measure of each angle? 60° and 30°

32. What is the measure of the angle between two faces of the quartz crystal shown? 120°

33. How does this method of crystal measurement use supplementary angles?

34. Suppose that the measure of the angle between the faces of another crystal is 140°. Draw a diagram showing the position of the goniometer measuring the angle between two faces of this crystal.

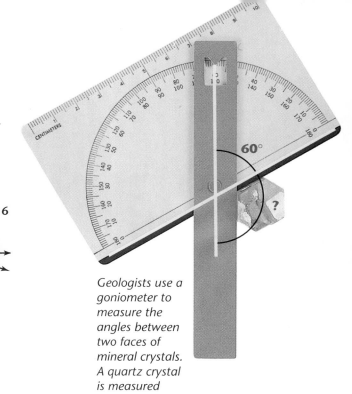

60°

In the figure at the right, m∠1 = 12.5°. Find the measure of each angle.

35. m∠7 90°   **36.** m∠2 90°

37. m∠3 77.5° **38.** m∠4 12.5°

39. m∠5 77.5° **40.** m∠6 102.5°

Geologists use a goniometer to measure the angles between two faces of mineral crystals. A quartz crystal is measured here.

41. Is it true that if two angles are supplementary, then one of them must be obtuse? Explain.
No, both could be 90°

Look Back

Find the absolute value. [Lesson 2.1]

42. $|10|$ 10   **43.** $|-7|$ 7   **44.** $|-0.8|$ 0.8   **45.** $\left|-\dfrac{2}{3}\right|$ $\dfrac{2}{3}$ **46.** $|0.01|$ 0.01

Change each fraction to a decimal. [Lesson 3.3]

47. $\dfrac{2}{9}$ $0.\overline{2}$ **48.** $\dfrac{2}{5}$ 0.4   **49.** $\dfrac{1}{3}$ $0.\overline{3}$ **50.** $\dfrac{3}{8}$ 0.375   **51.** $\dfrac{5}{2}$ 2.5

Change each fraction to a percent. [Lesson 3.7]

52. $\dfrac{1}{8}$ 12.5%   **53.** $\dfrac{3}{4}$ 75%   **54.** $\dfrac{2}{5}$ 40%   **55.** $\dfrac{9}{3}$ 300%   **56.** $\dfrac{8}{9}$ $88.\overline{8}$%

Look Beyond

Write an equation for each figure. Then solve for *x*.

57.

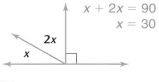

2x
x

$x + 2x = 90$
$x = 30$

58.

2x 3x

$2x + 3x = 180$
$x = 36$

Exploring Parallel Lines and Triangles

why *Many useful geometric properties can be discovered through measurement and observation. Properties of parallel lines and triangles are often used in art and architecture.*

On this bridge, you can find triangles, parallel lines, perpendicular lines, and transversals.

Transversals

A line that crosses two other lines in the same plane is called a **transversal**.

Exploration 1 Angles Formed by Transversals

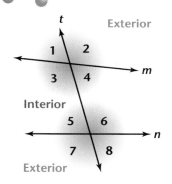

1 Which line, *m*, *n*, or *t*, is a transversal? *t*

2 Angles 1, 2, 7, and 8 are called **exterior angles**. Angles 3, 4, 5, and 6 are called **interior angles**. Write a definition for *exterior angles* and a definition for *interior angles*.

3 Angles 1 and 5 are **corresponding angles**. Angles 4 and 8 are also corresponding angles. Name two more pairs of corresponding angles. ∠2 and ∠6
∠3 and ∠7

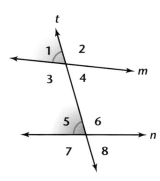

Corresponding angles

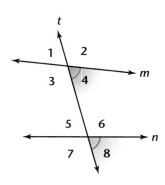

Corresponding angles

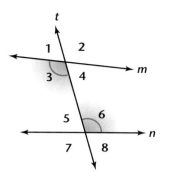

Alternate interior angles

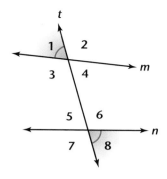

Alternate exterior angles

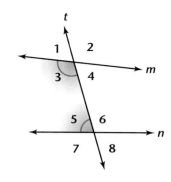

Consecutive interior angles

Angles 3 and 6 are **alternate interior angles.** Angles 1 and 8 are **alternate exterior angles.** Angles 3 and 5 are **consecutive interior angles.** Name another pair of alternate interior angles, another pair of alternate exterior angles, and another pair of consecutive interior angles. ❖ Alt int angles: ∠4 and ∠5; alt ext angles: ∠2 and ∠7; consecutive int angles: ∠4 and ∠6

Exploration 2 *Parallel Lines and Transversals*

For this exploration, you will need geometry graphics technology or a protractor.

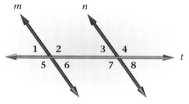

Figure 1

In Figure 1, *t* is a transversal, and lines *m* and *n* are parallel.

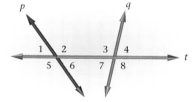

Figure 2

In Figure 2, *t* is a transversal, and lines *p* and *q* are **not** parallel.

Geometry Graphics

1 Measure two pairs of alternate interior angles in Figure 1. Measure two pairs of alternate interior angles in Figure 2. What conclusion can you make about the measures of alternate interior angles that are formed by parallel lines? Alt int angles are congruent.

2 Measure four pairs of corresponding angles in Figure 1. Measure four pairs of corresponding angles in Figure 2. What conclusion can you make about the measures of corresponding angles that are formed by parallel lines? Corr angles are congruent when lines are parallel.

3 Measure two pairs of alternate exterior angles in Figure 1. Measure two pairs of alternate exterior angles in Figure 2. What conclusion can you make about the measures of alternate exterior angles that are formed by parallel lines? They are congruent.

4 Measure two pairs of consecutive interior angles in Figure 1. Measure two pairs of consecutive interior angles in Figure 2. Find the sum of the measures for each pair. What conclusion can you make about the sum of the measures of consecutive interior angles that are formed by parallel lines? ❖ Their sum is 180°.

Carpenters use a *miter box* to cut lumber at different angles. The photo below shows a board being cut at a 45° angle. Using the properties of parallel lines and transversals, the carpenter can find the measures of the other angles.

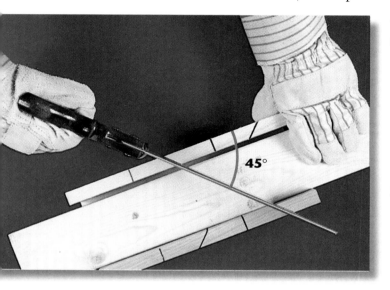

The 45° angle and its adjacent angle are supplementary.

$$45 + x = 180°$$
$$x = 135°$$

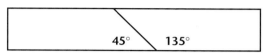

Each of these angles forms a pair of alternate interior angles with another angle formed by the transversal.

Triangles

A **triangle** is a closed figure consisting of three line segments that intersect only at their endpoints. Triangles can be classified as **right**, **acute**, or **obtuse** according to the measure of the largest angle.

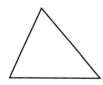
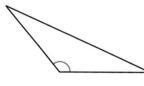

Right triangle   **Acute triangle**   **Obtuse triangle**

A triangle can also be classified according to the number of sides that are congruent to each other.

Segments that are the same length are **congruent** segments. Slash marks are used to indicate congruent sides of a triangle.

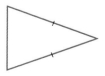

An **isosceles triangle** has *two* congruent sides.   An **equilateral triangle** has *three* congruent sides.   A **scalene triangle** has *no* congruent sides.

•Exploration 3 Angles of a Triangle

You will need a pencil, paper, a protractor, and a pair of scissors.

1 Draw and cut out three triangles: one right, one acute, and one obtuse. Measure the angles in each triangle with a protractor. Find the sum of the measures.   180°, 180°, 180°

2 Cut or tear off the bottom two angles of each triangle and place them adjacent to the third angle as shown. What type of angle is formed by placing all three angles side by side? What do you think is the sum of measures for these three angles?
Straight angle, 180°

3 What is the sum of all the angle measures of a triangle? ❖ 180°

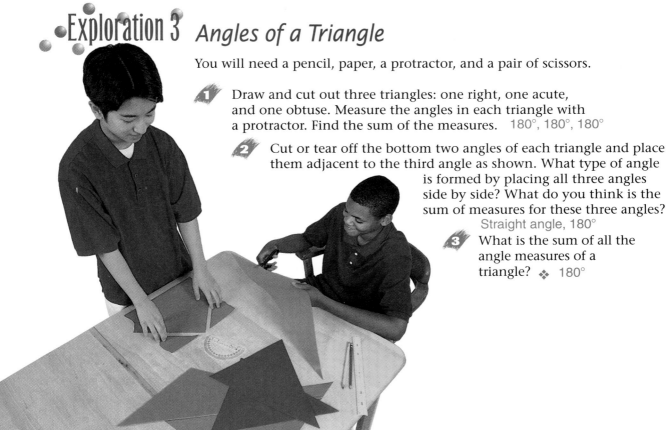

•Exploration 4 Sides of a Triangle

For this exploration, you will need a ruler and uncooked spaghetti or paper strips.

1 Draw three triangles, as in Step 1 of Exploration 3, but do not cut them out.

2 Measure the sides of each triangle. Compare the length of each side with the measure of the angle opposite each side. What can you conclude about the measure of the angle opposite the longest side? the shortest side? Longest side is opposite largest angle; shortest side is opposite smallest angle.

3 The base angles of an isosceles triangle are the angles that are opposite the congruent sides. What can you conclude about the measures of the base angles of an isosceles triangle? Draw several isosceles triangles, and measure the base angles to check your conclusion.
They are congruent.

Base angles

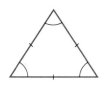

4 What can you conclude about the measures of the angles of an equilateral triangle, based on the lengths of the sides? Draw several equilateral triangles, and measure the angles to check your conclusion. All angles are congruent and 60°.

5 *In a triangle, the sum of two side lengths is always larger than the length of the third side.* Using uncooked spaghetti or paper strips, try to form triangles using the measures given below (in either centimeters or inches). Which of the measures below cannot be the measures of the sides of a triangle? Why?

    a. 2, 2, 2 yes   **b.** 3, 4, 5 yes
    c. 1, 2, 3 no   **d.** 2, 2, 5 no
    e. 4, 4, 5 yes   **f.** 1, 2, 4 no

6 Consider the relationship between the angle measures and side measures of a triangle. Explain why the sum of the measures of any two sides will always be greater than the measure of the third side. ❖

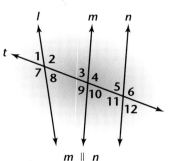

EXERCISES & PROBLEMS

Communicate

1. Explain how carpenters use parallel lines and transversals in determining angle measures when cutting lumber.

2. Using the figure at the right, name two corresponding angles that are congruent and two corresponding angles that are not congruent. Explain why.

3. Using the figure at the right, explain why angles 10 and 11 are supplementary but angles 8 and 9 are *not* supplementary.

4. How can you determine if you can draw a triangle with side lengths of 33, 33, and 75?

5. Explain why the side opposite the right angle in a right triangle must be the longest side.

6. Why are the base angles of an isosceles triangle congruent?

m ∥ *n*

Practice & Apply

In the figure at the right, _m_ ∥ _n_, and m∠7 = 40°.

7. Find m∠2. 140°   **8.** Find m∠3. 40°
9. Find m∠4. 40°   **10.** Find m∠5. 140°
11. Find m∠6. 40°   **12.** Find m∠1. 140°
13. Find m∠8. 140°
14. List all pairs of alternate interior angles. ∠2 and ∠5; ∠6 and ∠3
15. List all pairs of alternate exterior angles. ∠1 and ∠8; ∠4 and ∠7
16. List all pairs of corresponding angles.
17. List all pairs of consecutive interior angles. ∠2 and ∠3; ∠5 and ∠6
18. Explain why ∠5 and ∠6 are supplementary.
19. Name 10 pairs of supplementary angles.

Classify each triangle according to its angles and according to its sides.

20.

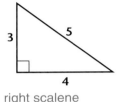

right scalene

21.

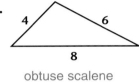

obtuse scalene

22.

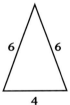

acute isosceles

Find the missing angle measures in each triangle.

23.

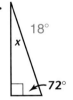

24.

25.

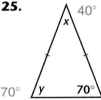

For Exercises 26–31, answer _yes_ or _no_ to each question. Explain your reasoning.

26. Can a right triangle be obtuse? no
27. Can a right triangle be isosceles? yes
28. Can a triangle have more than one right angle? no
29. Can an obtuse triangle be equilateral? no
30. Can an isosceles triangle be obtuse? yes
31. Can an isosceles triangle be equilateral? yes

Use right triangle _ABC_ for Exercises 32 and 33.

32. What is the measure of ∠C? 58°
33. How is ∠A related to ∠C? complementary
34. If all of the angles in a triangle are congruent, what is the measure of each angle? 60°

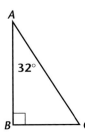

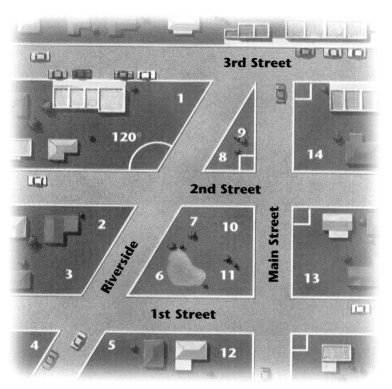

3rd Street

120°

2nd Street

Main Street

Riverside

1st Street

For Exercises 35–39, use the map at the left.

35. Name the streets that are parallel. Explain how you know that they are parallel.

36. How many angles have the same measure as angle 11? Which of these angles are numbered?
8; ∠10, ∠12, ∠13, ∠14

37. How many angles are supplementary to angle 10? Which of these angles are numbered?
8; ∠11, ∠12, ∠13, ∠14

38. How many angles have a measure of 120°? Which of these angles are numbered? 4; ∠3, ∠5, ∠7

39. How many angles are supplements of 120°? Which of these angles are numbered? 5; ∠1, ∠2, ∠4, ∠6, ∠8

Lines p and q are parallel. If m∠2 = x and m∠4 = $2x$, find each of the following angle measures.

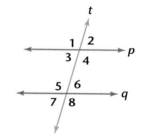

40. m∠1 120°   **41.** m∠3 60°   **42.** m∠6 60°

43. m∠5 120°   **44.** m∠7 60°   **45.** m∠8 120°

For Exercises 46–51, segment DE is parallel to segment CB, and m∠ADE = m∠AED. Find the indicated angle measures.

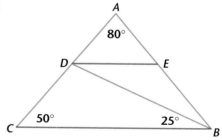

46. m∠ADE 50°   **47.** m∠AED 50°   **48.** m∠DEB 130°

49. m∠BDE 25°   **50.** m∠CDB 105°   **51.** m∠ABD 25°

Identify each pair of angles as alternate interior, alternate exterior, corresponding, or none of these.

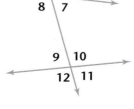

52. ∠6 and ∠10

53. ∠7 and ∠9

54. ∠6 and ∠12

55. ∠5 and ∠10

56. ∠8 and ∠12

57. ∠5 and ∠7

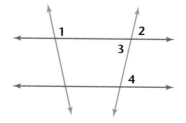

58. ∠1 and ∠2

59. ∠2 and ∠3

60. ∠1 and ∠3

61. ∠4 and ∠1

62. ∠2 and ∠4

63. ∠3 and ∠4

64. In Figure 1, ∠4 is called an **exterior angle** of triangle *ABC*. Angles 1 and 2 are called the **remote interior angles** of ∠4. Measure the four angles in the figure. How does the sum of the measures of ∠1 and ∠2 compare with the measure of ∠4?

65. Figure 2 shows two parallel lines and a triangle. How can you use the alternate interior angles in this figure to show that the following equation is true?

$$m\angle 1 + m\angle 2 + m\angle 3 = 180°$$

66. Is it true that if two angles are supplementary, then one of them must be obtuse? Explain. No, both could be right.

67. Explain why the hypotenuse of a right triangle is the longest side. Draw and label three right triangles to support your explanation.

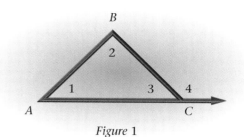

Figure 1

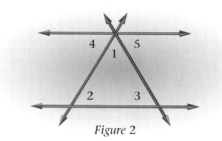

Figure 2

Look Back

Evaluate each expression for *n*-values of 6, 8, and 10.   [Lesson 1.7]

68. $n - 2$
4, 6, 8

69. $(n - 2)180$
720, 1080, 1440

70. $\frac{(n - 2)180}{n}$
120, 135, 144

Compare. Use <, >, or =.   [Lesson 2.3]

71. -2 and 5
<

72. 3 and -3
>

73. -15 and 2
<

Complete the proportions by using the property of equivalent ratios. [Lesson 3.6]

74. $\frac{2}{3} = \frac{6}{n}$ 9

75. $\frac{n}{8} = \frac{3}{12}$ 2

76. $\frac{14}{x} = \frac{35}{45}$ 18

Write each percent as a decimal. [Lesson 3.7]

77. 25%
0.25

78. 5%
0.05

79. 4.5%
0.045

80. 112%
1.12

81. 66.6%
0.666

Look Beyond

Write and solve an equation to find the angle measures of each triangle.

82.

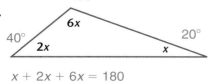

120°
6x
40°
2x
20°
x

$x + 2x + 6x = 180$

83.

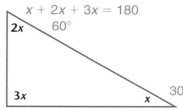

$x + 2x + 3x = 180$
2x 60°
90° 3x 30° x

LESSON 4.4
Exploring Polygons

why
Studying geometry in nature reveals a variety of geometric shapes called polygons. Throughout history, humans have adapted what they learn from nature for use in their lives.

A closed figure consisting of segments that intersect only at their endpoints is called a **polygon**. These endpoints are called **vertices**.

A **convex polygon** is a polygon in which any line segment connecting any two interior points of the polygon is in the interior of the polygon. If a polygon is not a convex polygon, then it is a **concave polygon**.

Convex polygon   **Concave polygon**

In this book, the word *polygon* will be used to mean a convex polygon unless otherwise stated. Polygons are classified according to the number of sides. The word *polygon* means "many sides." A polygon with three sides is a triangle. Some other polygons are shown below.

4 sides   **5 sides**   **6 sides**

    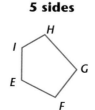   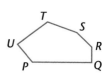

Quadrilateral *ABCD*   **Pentagon *EFGHI***   **Hexagon *PQRSTU***

Regular polygons are both equilateral and equiangular. That is, all of their sides are congruent, and all of their angles are congruent. Some regular polygons are shown below.

Regular quadrilateral (square)   **Regular pentagon**   **Regular hexagon**

Describe a regular triangle. What is it usually called?   All three sides are congruent, and all three angles are also congruent; an equilateral triangle. The names of some other polygons are listed below.

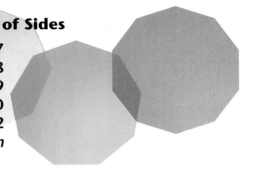

Name of Polygon	Number of Sides
heptagon	7
octagon	8
nonagon	9
decagon	10
dodecagon	12
n-gon	n

CRITICAL Thinking

Is it possible to have a polygon with congruent sides but *non*congruent angles? Is it possible to have a polygon with congruent angles but *non*congruent sides? Explain.   Yes; yes; a rectangle has noncongruent sides, but congruent angles; a rhombus (diamond) has congruent sides but noncongruent angles.

•Exploration 1 *Properties of Parallelograms*

Geometry Graphics

For this exploration you will need geometry graphics technology or a ruler and a protractor.

A quadrilateral with both pairs of opposite sides parallel is called a **parallelogram**.

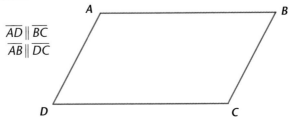

$\overline{AD} \parallel \overline{BC}$
$\overline{AB} \parallel \overline{DC}$

   Draw or trace parallelogram *ABCD*.

2 Measure the sides of parallelogram *ABCD*. Make a conjecture about the opposite sides of a parallelogram. They are congruent.

3 Measure the angles of parallelogram *ABCD*. Make a conjecture about the opposite angles of a parallelogram. They are congruent.

4 Draw another parallelogram with different side and angle measures. If you are using geometry graphics software, drag quadrilateral *ABCD* to change the measurements. Repeat Steps 2 and 3 for your new parallelogram. Do the results justify the conjectures you made in Steps 2 and 3? yes

5 The **diagonals** of a parallelogram are the segments that connect the vertices of the opposite angles. The diagonals of parallelogram *ABCD* are shown below. Draw the diagonals on your copy of parallelogram *ABCD*, and measure them. Are the diagonals of a parallelogram congruent? no

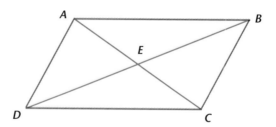

6 Measure the lengths of $\overline{AE}$, $\overline{EC}$, $\overline{DE}$, and $\overline{EB}$. What can you conclude about the lengths of $\overline{AE}$ and $\overline{EC}$? What can you conclude about the lengths of $\overline{DE}$ and $\overline{EB}$? Test your conjecture by drawing and measuring the diagonals of several other parallelograms. ❖
$\overline{AE}$ and $\overline{EC}$ are equal in length; $\overline{DE}$ and $\overline{EB}$ are equal in length.

•Exploration 2 *Special Quadrilaterals*

Geometry Graphics

For this exploration you will need geometry graphics technology or a ruler and a protractor.

1 A **rectangle** is a special parallelogram with four right angles. Draw a rectangle on grid paper. Are all of the properties you discovered about the measures of the sides and angles of parallelograms true for rectangles? Explain. Yes; opposite sides are congruent.

2 Draw and measure the diagonals of the rectangle. What can you conjecture about the diagonals of a rectangle? Draw several other rectangles to test your conjecture. Diags of a rectangle are congruent.

3 A *rhombus* and a *square* are two special parallelograms. A **rhombus** is a parallelogram with four congruent sides. A **square** is a rectangle with four congruent sides. Measure the angles of the square shown below. Explain why a square is both a rectangle and a rhombus.

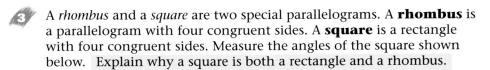

Rhombus   **Square**

4 Draw or trace the rhombus, and draw the two diagonals. Measure the angles formed by the diagonals. Make a conjecture about the angles formed by the diagonals of a rhombus. Test your conjecture for several other rhombuses. Is your conjecture true for squares also?
Diagonals form 90° angles (perpendicular); yes

5 A **trapezoid** is a quadrilateral with exactly one pair of parallel sides. An **isosceles trapezoid** is a trapezoid with congruent nonparallel sides. Can a trapezoid be classified as a parallelogram? Why or why not? No; a paralellogram has 2 pairs of parallel sides, but a trapezoid only has 1 pair.

Trapezoid   **Isosceles trapezoid**

Which property belongs to each quadrilateral? Copy and complete the following table.

Properties of Quadrilaterals					
Property	Parallelogram	Rectangle	Rhombus	Square	Trapezoid
Opposite sides are always congruent.	yes	yes	yes	yes	no
6 Opposite angles are always congruent.	? yes	? yes	? yes	? yes	? no
7 Diagonals always bisect each other.	yes ?	yes ?	yes ?	? yes	no ?
8 Diagonals are always congruent.	? no	yes ?	no ?	? yes	no ?
9 Diagonals are always perpendicular.	? no	? no	? yes	? yes	? no
10 All four sides are always congruent.	? no	no ?	yes ?	? yes	no ?
11 All angles are always 90°.	? no	? yes	? no	? yes	? no

Exploration 3 *Angles of Polygons*

In Lesson 4.3, you discovered that the sum of the measures of the angles of a triangle is 180°.

 Recall that a rectangle has four right angles. What is the sum of the measures of the four angles in a rectangle? 360°

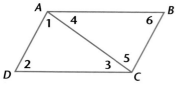

You can use the triangles formed by diagonal $\overline{AC}$ of quadrilateral *ABCD* to find the sum of the measures of the angles of the quadrilateral.

$$m\angle 1 + m\angle 2 + m\angle 3 = \underline{\ ?\ } \quad 180°$$
$$m\angle 4 + m\angle 5 + m\angle 6 = \underline{\ ?\ } \quad 180°$$

Therefore, the sum of the angle measures of quadrilateral *ABCD* is

$$m\angle 1 + m\angle 2 + m\angle 3 + m\angle 4 + m\angle 5 + m\angle 6 = \underline{\ ?\ }. \quad 360°$$

Quadrilateral	**Pentagon**	**Hexagon**	**Heptagon**	**Octagon**

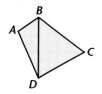 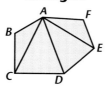 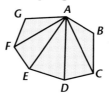 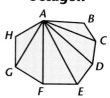

Copy the following table. Using the polygons shown above, complete the table.

Number of sides	Number of triangles formed	Sum of angle measures
4	? 2	360° ?
5	? 3	540° ?
6	? 4	720° ?
7	? 5	900° ?
8	? 6	1080° ?
n	? *n* − 2	? $(n-2)180$

9. number of triangles = *n* − 2, where *n* is the number of sides

 Compare the first two columns in the table. What is the relationship between the number of sides of a polygon and the number of triangles formed by its diagonals? How many triangles can be formed by the diagonals of a polygon with *n* sides?

10 Compare the last two columns in the table. What is the relationship between the number of triangles formed by the diagonals of a polygon and the sum of the angles in the polygon? How would you find the sum of the angle measures in an *n*-gon?

11 Write a formula for the sum of the angle measures in an *n*-gon. Use your formula to find the sum of the angle measures in a 20-gon and in a 100-gon. $S = 180(n - 2)$; 3240; 17,640

12 How can you find the measure of one angle in a regular quadrilateral (square)? in a regular pentagon?

13 Write a formula to find the measure of one angle in a regular polygon. Use your formula to find the measure of the following:
a. one angle in a regular hexagon 120°
b. one angle in and a regular octagon ❖ 135°

Angle measure in reg polygon $= \dfrac{(n - 2)180}{n}$

EXTENSION

Spreadsheet

You can use a graphics calculator or spreadsheet and the relationships you explored in Exploration 2 to build a table.

On a graphics calculator, use the expression $(X - 2)180$ for Y_1 to find the sum of the angle measures in a regular polygon, and use $((X - 2)180)/X$ for Y_2 to find the measure of each angle in a regular polygon.

X	Y_1	Y_2
3	180	60
4	360	90
5	540	108
6	720	120
7	900	128.57
8	1080	135
9	1260	140

X=3

	A	B	C
	X	(X−2)180	((X−2)180)/X
1	X	(X−2)180	((X−2)180)/X
2	3	180	60
3	4	360	90
4	5	540	108
5	6	720	120
6	7	900	128.5714286
7	8	1080	135
8	9	1260	140

❖

CRITICAL Thinking

As the number of sides of a regular polygon increases, what happens to the measure of each angle in the polygon? Does the measure have a limit? Explain. As the number of sides increases, the size of the angle increases. The angle in a regular polygon cannot reach 180° because the sides would not form a polygon.

EXERCISES & PROBLEMS

Communicate

1. Explain how quadrilaterals are classified.

2. Is a rectangle a parallelogram? Explain.

3. Is a rectangle a square? Is a square a rectangle? Explain.

4. Describe how to find the sum of the measures of the angles in a polygon by using the diagonals from one vertex.

Practice & Apply

Copy each parallelogram. Then label all of the side lengths and all of the angle measures for each parallelogram.

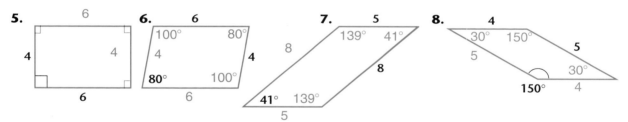

5.

6.

7.

8.

Classify each polygon as concave or convex, and classify the polygon according to the number of sides.

Convex hexagon   Convex heptagon   Concave hexagon

9.   10.   11.   12.   13.

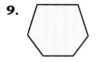      

Convex octagon   Concave dodecagon

Decide whether each statement is true or false. Use a figure to justify your answer.

14. Every rectangle is a parallelogram. True

15. Every square is a rectangle. True

16. Every parallelogram is a rectangle. False

17. The diagonals of all parallelograms are congruent. False

18. A quadrilateral with four congruent sides is a parallelogram. True

19. The adjacent interior angles of any parallelogram are supplementary. True

20. The diagonals of any rectangle are perpendicular. False

Find the sum of the angle measures in a polygon with the given number of sides.

21. 7 sides 900°   **22.** 5 sides 540°   **23.** 10 sides 1440°

24. 8 sides 1080°   **25.** 9 sides 1260°

Find the measure of one angle in each regular polygon.

26. triangle 60°   **27.** hexagon 120°   **28.** pentagon 108°

29. octagon 135°   **30.** heptagon $128\frac{4}{7}°$

31. Sports On a baseball field, home plate is shaped like a pentagon. It has three right angles, and the other two angles, ∠1 and ∠2, are congruent. Find m∠2. 135°

32. Engineering Gloria is designing a hot tub in the shape of a regular hexagon enclosed in a rectangle. A sketch of her design is shown below. Copy the figure and label each angle with its measure.

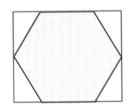

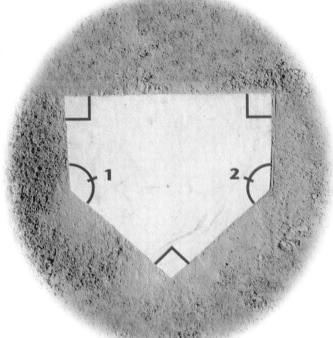

Look Back

Use guess-and-check to solve for x. [Lesson 2.3]

33. $x + 2 = -1$ −3   **34.** $5 - x = -1$ 6   **35.** $2x = -7$ $-3\frac{1}{2}$ **36.** $2x + 1 = 8$ $3\frac{1}{2}$

Evaluate each expression when l is $2\frac{1}{2}$ and w is 4. [Lessons 3.4, 3.5]

37. $l \cdot w$ 10   **38.** $l + w + l + w$ 13   **39.** $2l + 2w$ 13

Use mental math to compute each percentage. [Lesson 3.7]

40. 25% of 400 100   **41.** 10% of 50 5   **42.** 75% of 60 45

Look Beyond

Build a table of values for each equation by substituting −3, −2, −1, 0, 1, 2, and 3 for x. Then use the values in the table to sketch a graph of each equation.

43. $y = x^2 + 1$ **44.** $y = x^2 - 2$ **45.** $y = x^2 + 2x + 1$

LESSON 4.5

Exploring Perimeter and Area

Why The ability to find the perimeter and area of objects is an important part of the work of many professionals. Builders and landscapers are just two examples of professionals who rely on these measurements to complete their work.

In order to know how many tiles it will take to tile the edge of the rectangular swimming pool, Julio finds the perimeter.

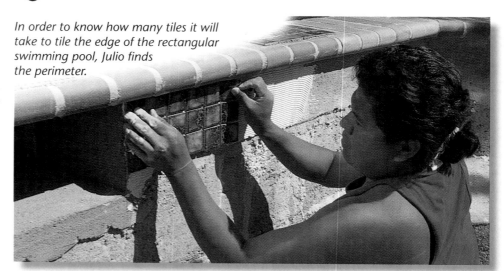

In Lesson 1.5 you explored areas and perimeters of squares and rectangles.

4 ft

| 4 ft | 16 sq ft |

Area = side · side
$$A = s^2$$
$$= 4^2$$
$$= 16$$

12 ft

| 3 ft | 36 sq ft |

Area = length · width
$$A = l \cdot w$$
$$= 12 \cdot 3$$
$$= 36$$

Exploration 1 Perimeter Formulas

1 What is the perimeter of the 4-by-4 square shown above? 16 feet

2 On grid paper, draw a square with sides 5 units long. What is the perimeter of the square? 20 units

3 What is the perimeter of a square with a side length of 25 feet? 100 feet

4 Use the results of Steps 1–3 to help you complete a formula for the perimeter of a square with sides of length *s*.

$$P = \underline{\ ?\ } \quad 4s$$

5 What is the perimeter of the 3-by-12 rectangle shown above? 30 feet

232 CHAPTER 4

6 Copy and complete the table below to find a formula for the perimeter of a rectangle.

Length	Width	Perimeter
12	3	$P = 2(12) + 2(3) = \underline{?}$ 30
5	4	$P = 2(5) + 2(4) = \underline{?}$ 18
2	7	$P = 2(\underline{?}) + 2(\underline{?}) = \underline{?}$ 2, 7, 18
ℓ	w	$P = \underline{?}$ $2\ell + 2w$

•Exploration 2 *Area and Fixed Perimeter*

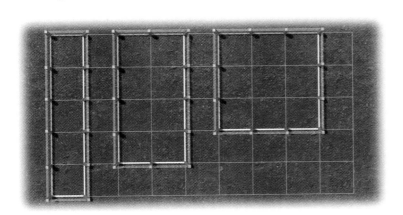

Ed has 12 yards of wire fence to build a pen for his potbellied pig.

MAXIMUM MINIMUM *Connection*

1 Consider all possible designs for a rectangular pen. Which pen do you think would work best? Why? Square; it has larger area.

2 Copy and complete the table to find the dimensions and area of each pen.

Fixed perimeter	Length	Width	Area
12	1	5	5 square yards
12	2	? 4	8 sq yd ?
12	3	? 3	9 sq yd ?

3 Which shape has the greatest area? 3-by-3 square

4 Suppose you have 16 yards of wire fence. What dimensions would you choose to create a pen with the greatest area? Why? 4-by-4; square

5 Make a generalization about the shape with a fixed perimeter and greatest possible area. Use two other examples to test your generalization. ❖ Given a fixed perimeter, a square has the largest area.

Lesson 4.5   Perimeter and Area   **233**

Exploration 3 *Perimeter and Fixed Area*

MAXIMUM MINIMUM *Connection*

The Bienivides City Council wants to build a recreation building at City Park with an area of 14,400 square feet. Cynthia works in the design department of a building contractor who is bidding on the project. She considers the possible dimensions of a basic rectangular building with an area of 14,400 square feet. In order to estimate the building cost, she must consider both the area, 14,400 square feet, and the perimeter, or distance around the building.

1 What is the perimeter of the building shown below? 520 feet

2 Explain why you can use the formula $P = 2w + 2l$ for the perimeter of a rectangle with width w and length l.

180 ft

80 ft

Copy and complete the table of dimensions and perimeters.

	Width, w	Length, l	Area = wl	Perimeter = $2w + 2l$
3	90	? 160 ft	14,400 square feet	? 500 ft
4	100	? 144 ft	14,400 square feet	? 488 ft
5	110	?130.91 ft	14,400 square feet	? 481.82 ft
6	120	? 120 ft	14,400 square feet	? 480 ft
7	130	?110.77 ft	14,400 square feet	? 481.54 ft
8	140	?102.86 ft	14,400 square feet	? 485.72 ft
9	150	? 96 ft	14,400 square feet	? 492 ft

10 Which building described in the table has the smallest perimeter?
What are the width and the length of this building?
120 ft-by-120 ft building

11 Suppose Cynthia designs a room that has an area of 400 square feet.
What dimensions would you choose to build the room with the
smallest perimeter? Why?

12 Make a generalization about the shape with a fixed area and the
smallest possible perimeter. Use two other examples to test your
generalization. ❖ Given a fixed area, the smallest perimeter is the
square with that area.

APPLICATION

Spreadsheet

In order to be cost effective, Cynthia realizes that she must keep the
perimeter of the recreation building around 480 feet. You can make a
spreadsheet or use the table feature of a graphics calculator to find the
dimensions to use.

The spreadsheet formulas are shown here.

The formula in this
column gives the
length by dividing
the area by the width.

$l = \frac{A}{w}$

Perimeter			
	A	B	C
1	width	length	Perimeter = 2w + 2l
2	20	= 14,400/A2	= 2*A2 + 2*B2
3	21	= 14,400/A3	= 2*A3 + 2*B3
4	22	= 14,400/A4	= 2*A4 + 2*B4

The formula in this
column gives the
perimeter.

$P = 2w + 2l$

Using a spreadsheet or
the table feature of a
graphics calculator, you
find that a perimeter of
480 feet can be obtained
with the dimensions
120 feet by 120 feet. ❖

X	Y$_1$	Y$_2$
80	180	520
90	160	500
100	144	488
110	130.91	481.82
120	120	480
130	110.77	481.54
140	102.86	485.71

X=120

EXERCISES & PROBLEMS

Communicate

1. Give two everyday examples of perimeter.
2. Give two everyday examples of area.
3. Explain the difference between perimeter and area.
4. Can two rectangles have the same perimeter but different areas? Give examples to justify your answer.
5. Can two squares have the same perimeter but different areas? Explain.
6. Can two rectangles have the same area but different perimeters? Give examples to justify your answer.
7. Can two squares have the same area but different perimeters? Explain. No; since the lengths of all the sides are equal, squares with the same area have the same perimeter.

Practice & Apply

Estimate the perimeter and area of each figure. Answers may vary.

8.
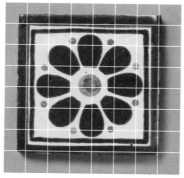
≈26 units; ≈ 42 sq units

≈21 units; ≈28 sq units

9.

≈15 units; ≈ 10 sq units

≈16 units; ≈14 sq units

10.
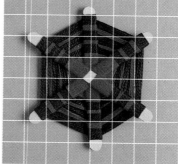
≈20 units; ≈25 sq units

≈20 units; ≈24 sq units

11.
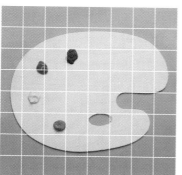

12.

13.
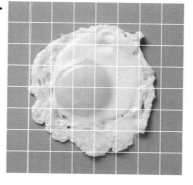

**Use the formulas for perimeter and area to compute the
perimeter and area of each rectangle described.**

14. $l = 95$ meters, $w = 48$ meters

15. $l = 2.5$ yards, $w = 5.75$ yards

16. $l = 15\frac{3}{16}$ inches, $w = 15\frac{3}{16}$ inches

17. $l = 8.5$ centimeters, $w = 4.4$ centimeters

18. $l = 2.5$ meters, $w = 5.75$ meters

19. $l = 18$ inches, $w = 2\frac{1}{2}$ inches

20. $l = 110$ feet, $w = 110$ feet

21. $l = 2.5$ miles, $w = 5.2$ miles

22. $l = 5\frac{2}{3}$ yards, $w = 1\frac{1}{2}$ yards

23. $l = 1\frac{3}{4}$ inches, $w = \frac{7}{8}$ inch

*In this new neighborhood, the
builders must build according
to the restrictions shown on
the blueprint.*

Construction   Using the restrictions shown on the blueprint,
find the dimensions, perimeter, and area of the largest possible
floor plan for each above lot size.

24. $a = 80$ feet, $b = 100$ feet

25. $a = 100$ feet, $b = 90$ feet

26. $a = 100$ feet, $b = 500$ feet

27. $a = 100$ feet, $b = 120$ feet

**Copy and complete the table. Then draw a picture of each
rectangle described (in centimeters) in the table.**

	Length	Width	Perimeter	Area
28.	4 cm	? 5 cm	18 cm	? 20 sq cm
29.	4 cm	? 4 cm	? 16 cm	16 sq cm
30.	? 1 cm	1.5 cm	5 cm	? 1.5 sq cm
31.	? 4 cm	2.5 cm	? 13 cm	10 sq cm
32.	? 2 cm	20 cm	? 44 cm	40 sq cm
33.	16 cm	? 2.5 cm	37 cm	? 40 sq cm

Agriculture Copy and complete the table of values to show the width, in feet, and area, in square feet, of each pen that Howie can make.

	Fixed perimeter	Length	Width	Area
34.	250	20	?	?
35.	250	40	?	?
36.	250	60	?	?
37.	250	80	?	?
38.	250	100	?	?
39.	250	120	?	?

40. Maximum/Minimum Which pen described in the table has the largest area? 60 × 65

41. Maximum/Minimum Which pen described in the table has the smallest area? 120 × 5

Howie has 250 yards of fencing. He decides to use it to build a rectangular pen in his field.

42. Technology Use a graphics calculator and the equation Y = X(125 − X) to build a table of values, where Y is the area, X is the width, and (125 − X) is the length. Use this table to determine if there is a pen with an area larger than any of those described in the table above. Yes; width = 62.5 yd

43. Construction Gloria and Phil design and build portable storage buildings. Design five rectangular plans that have an area of 600 square feet. Which floor plan has the smallest perimeter? Explain.

Look Back

Use guess-and-check to solve. [Lesson 2.3]

44. $60 = \frac{600}{x}$ 10   **45.** $8 = \frac{x}{100}$ 800   **46.** $50 = 20 + 2h$ 15

47. Find the measure of one angle in a regular pentagon. [Lesson 4.4]
108°

Look Beyond

Evaluate each expression if b is $2\frac{1}{2}$ and h is 4.

48. $\frac{1}{2}(b \cdot h)$ 5   **49.** $\frac{bh}{2}$ 5   **50.** $\frac{b}{h}$ $\frac{5}{8}$

Exploring Area Formulas

Why **Suppose you need to estimate the area destroyed by a forest fire. Knowing how to compute the areas of parallelograms, triangles, trapezoids, and other common shapes can help you estimate and compute the areas of irregular shapes.**

Area of a Parallelogram

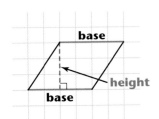

The **bases of a parallelogram** can be either pair of the parallel sides. The **height of a parallelogram** is the perpendicular length between two bases. This parallelogram has a base of 4 units and a height of 3 units.

Exploration 1 *Finding the Area of a Parallelogram*

For this exploration you will need grid paper and scissors (optional).

1 Copy the parallelogram on grid paper. Estimate the area by counting the number of squares inside the parallelogram. 12 sq units

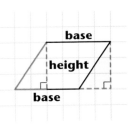

2 If you cut out the right triangle from the left side of the parallelogram, and move it to the right side of the parallelogram, what new figure is formed? rectangle

3 What is the relationship between the area of the parallelogram and the area of the rectangle? *The areas are equal.*

4 How can you use the base and height of a parallelogram to compute its area? Write a formula for the area of a parallelogram. *$A = bh$*

5 Use grid paper to draw and label four different parallelograms with an area of 12 square units. ❖

APPLICATION

Forestry Refer to the area of forest on page 239. The U.S.D.A. Forest Service plans to replant trees in portions of the burned area. The local 4-H Club is responsible for replanting an area in the shape of a parallelogram.

1050 ft

950 ft

To find the area of a parallelogram with a base of 950 feet and a height of 1050 feet, substitute 950 for b and 1050 for h in the formula.

$$A = b \cdot h$$
$$= 950 \cdot 1050$$
$$= 997,500$$

The area of the region that the 4-H Club is responsible for is 997,500 square feet. ❖

Area of a Triangle

A diagonal separates a parallelogram into two triangles. You can use this fact to discover the formula for the area of a triangle.

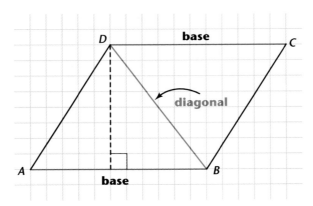

Exploration 2 *Finding the Area of a Triangle*

For this exploration you will need grid paper and scissors.

1 Sketch parallelogram *ABCD* below on grid paper. Cut out the parallelogram.

2 Cut along the diagonal to form the two triangles. Compare the size and shape of each. What can you conclude about the areas *ABD* and *CDB* of both triangles? They are the same shape and size.

3 What is the relationship between the area of the parallelogram and the area of the triangle? The area of each triangle is $\frac{1}{2}$ the area of the parallelogram.

4 Describe how you can use the base and height of a parallelogram to compute the area of a triangle. Write a formula for the area of a triangle. $A = \frac{1}{2}bh$.

5 Notice that an obtuse triangle can also be half of a parallelogram.

Draw another parallelogram that is made of two obtuse triangles.

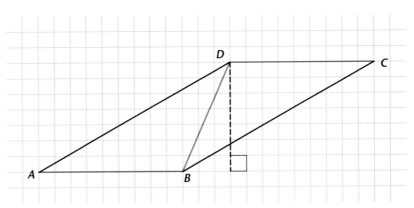

6 Draw a picture on grid paper to show that the right triangle shown is half of a parallelogram. What kind of parallelogram is formed? Notice that the height is one of the sides of the right triangle. What kind of right triangle forms half of a square? What kind of triangle forms half of a rhombus?

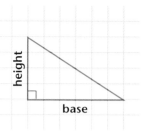

7 In this figure, two triangles have a base of 2 units and a height of 4 units. What is the area of each triangle? Draw and label four different triangles, each with the same area as the triangles shown here.

8 On grid paper, draw and label two different right triangles, two different obtuse triangles, and two different acute triangles, each with an area of 8 square centimeters. ❖

EXTENSION

Each triangle pictured below has a base of 2 units and a height of 4 units. Find the area of each of these triangles.

Substitute 2 for b and 4 for h in the formula for the area of a triangle.

$$A = \frac{1}{2}bh$$

$$= \frac{1}{2}(2)(4)$$

$$= 4$$

The area of each triangle is 4 square units. ❖

Area of a Trapezoid

A **trapezoid** is a shape with exactly two parallel sides, called **bases**. The lengths of these bases are indicated by b_1 and b_2.

Notice that the bases have different lengths. The **height of a trapezoid** is the perpendicular length between the bases.

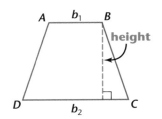

Exploration 3 — Finding the Area of a Trapezoid

For this exploration you will need grid paper.

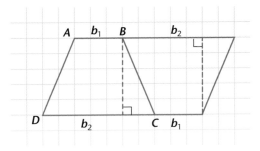

1 Sketch trapezoid $ABCD$ at the left on grid paper. Estimate the area by counting the number of squares inside the trapezoid. 25 sq units

2 Draw an upside-down copy of the trapezoid next to the original as shown. What shape is formed by both trapezoids together? What are the base and height of this new shape? Parallelogram; 10 units, 5 units

3 How can you find the area of the new shape formed by the two trapezoids? base times height

4 Explain how you can use this new shape to find the area of the original trapezoid. Write an equation for the area of a trapezoid.

$$A = \frac{1}{2}(b_1 + b_2)h$$

5 The two trapezoids in the picture below each have a height of 3 units and bases that total 5 units. What is the area of each trapezoid? On grid paper, draw three more trapezoids with the same area as those shown.

6 On grid paper, draw and label four different trapezoids, each with an area of 28 square units. ❖

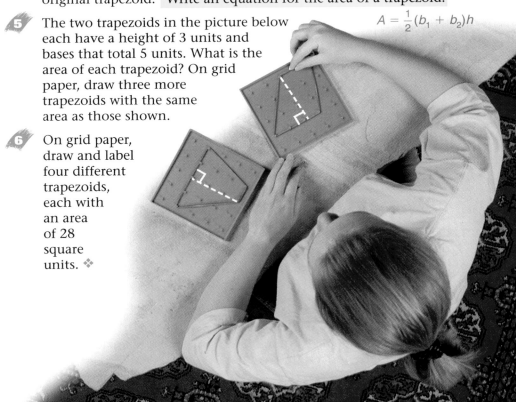

APPLICATION

The city council is planning to designate an area along the coastline as a public park. Estimate the area of this park.

Find the area of a trapezoid that closely resembles the shape of the park.

To find the area of the trapezoid, substitute 40 for b_1, 70 for b_2, and 90 for h in the formula for the area of a trapezoid.

$$A = \frac{(b_1 + b_2)h}{2}$$

$$= \frac{(40 + 70)90}{2}$$

$$= \frac{110 \cdot 90}{2}$$

$$= \frac{9900}{2}$$

$$= 4950$$

The area of the park is approximately 4950 square feet. ❖

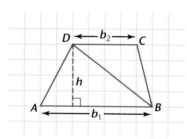

CRITICAL
Thinking

Use the areas of triangles ABD and BCD to show that the area of trapezoid $ABCD$ is $\frac{(b_1 + b_2)h}{2}$.

EXERCISES & PROBLEMS

Communicate

1. Draw a picture to show how you can use the area of a rectangle to find the area of a parallelogram.

2. Explain why the area of a triangle is the base times the height divided by 2.

3. Draw a picture to show how you can use the area of a parallelogram to find the area of a trapezoid.

4. Describe the differences among finding the areas of an acute triangle, an obtuse triangle, and a right triangle.

Practice & Apply

Use the grid to find the area of each figure. Estimate the perimeter.

5.

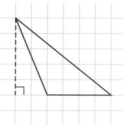

A = 10 sq units
P ≈ 16 units

6.

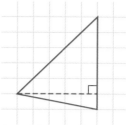

A = 15 sq units
P ≈ 16 units

7.

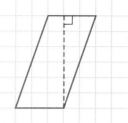

A = 18 sq units
P ≈ 18 units

8.

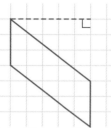

A = 15 sq units
P ≈ 16 units

9.

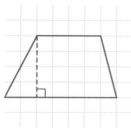

A = 22 sq units
P ≈ 19 units

10.

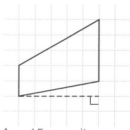

A = 15 sq units
P ≈ 16 units

11.

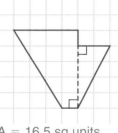

A = 16.5 sq units
P ≈ 17 units

12.

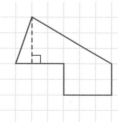

A = 15 sq units
P ≈ 18 units

13.

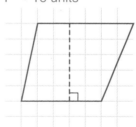

A = $27\frac{1}{2}$ sq units
P ≈ 21 units

14. On grid paper, draw and label four different parallelograms, each with an area of 8 square units.

15. On grid paper, draw and label four different triangles, each with an area of 8 square units.

16. On grid paper, draw and label four different trapezoids, each with an area of 8 square units.

22. 10 sq mm   **25.** 44 sq cm

Find the area of each triangle whose base and height are given.

4.25 sq mm

17. b = 9 centimeters, h = 4.8 centimeters
18. b = 1.7 millimeters, h = 5 millimeters

21.6 sq cm

19. b = $5\frac{1}{2}$ inches, h = $5\frac{1}{2}$ inches $15\frac{1}{8}$ sq in.
20. b = $8\frac{3}{4}$ feet, h = 5 feet $21\frac{7}{8}$ sq ft

21. b = 6 inches, h = 4 inches 12 sq in.
22. b = 2.5 millimeters, h = 8 millimeters

23. b = $2\frac{2}{3}$ inches, h = $5\frac{1}{2}$ inches $7\frac{1}{3}$ sq in.
24. b = $7\frac{1}{4}$ feet, h = 5 feet $18\frac{1}{8}$ sq ft

25. b = 8 centimeters, h = 11 centimeters
26. b = 8.4 meters, h = 12.2 meters 51.24 sq m

Find the area of each parallelogram whose base and height are given.

27. b = 8 centimeters, h = 14.4 centimeters 115.2 sq cm

28. b = 12$\frac{1}{4}$ feet, h = 5$\frac{3}{4}$ feet 70$\frac{7}{16}$ sq ft

29. b = 13 millimeters, h = 7.1 millimeters 92.3 sq mm

30. b = 11 centimeters, h = 4.25 centimeters 46.75 sq cm

31. b = 18 meters, h = 9.4 meters 169.2 sq m

32. b = 12.5 feet, h = 5.75 feet 71.875 sq ft

33. b = 18 inches, h = 10$\frac{1}{8}$ inches 182$\frac{1}{4}$ sq in.

34. b = 4.9 centimeters, h = 9.2 centimeters 45.08 sq cm

Find the area of each trapezoid whose base and height are given.

35. b_1 = 4 feet, b_2 = 2.5 feet, h = 11 feet 35.75 sq ft

36. b_1 = 24 centimeters, b_2 = 12 centimeters, h = 10 centimeters 180 sq cm

37. b_1 = 12.7 centimeters, b_2 = 11 centimeters, h = 18.3 centimeters 216.855 sq cm

38. b_1 = 6 feet, b_2 = 9$\frac{3}{4}$ feet, h = 15 feet 118$\frac{1}{8}$ sq ft

39. b_1 = 3.4 millimeters, b_2 = 6.7 millimeters, h = 8.9 millimeters 44.945 sq mm

40. b_1 = 18 meters, b_2 = 12 meters, h = 5.2 meters 78 sq m

Use the grid to find the area of each figure. Describe the method that you used for each. Methods may vary.

41.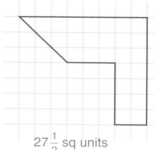

27$\frac{1}{2}$ sq units

42.

18 sq units

43.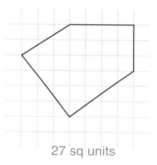

27 sq units

Copy and complete the table. The table below describes four different triangles.

	Base	Height	Area
44.	8	? 3	12
45.	? 5	10	25
46.	3	? 4	6
47.	? 20	1.5	15

Copy and complete the table. The table below describes four different trapezoids.

	Base$_1$	Base$_2$	Height	Area
48.	10	5	? 2	15
49.	5	? 3	4	16
50.	? 24	12	2	36
51.	8	12	? 5	50

←40 ft→

90 ft

←————— 150 ft —————→

All four faces of the pyramid are congruent.

For Exercises 52–56, use the trapezoid to approximate the shape of this pyramid.

52. Find the approximate area of one face of this pyramid. 8550 sq ft

53. Write a formula for finding the total area of all four sides of the pyramid.

54. What is the area of the bottom of the pyramid? Explain your method.

55. What is the area of the top of the pyramid? Explain your method.

56. A **shear** is a transformation in which you slide one vertex of a triangle in a direction parallel to its base in order to create a new triangle.

Use grid paper to explain why the new triangle formed by a shear has the same area as the original triangle.

Look Back

Evaluate the expression $a^2 + b^2$ for each of the following values. [Lesson 1.7]

57. $a = 3, b = 4$ 25

58. $a = 4.5, b = 6$ 56.25

59. $a = 3, b = 3$ 18

Use guess-and-check to solve. [Lesson 2.3]

60. $x + (-3) = -5$ −2

61. $x - 4 = -1$ 3

62. $x + 3 = 5$ 2

Use second differences to determine the next three terms in each sequence. [Lesson 2.7]

63. 5, 0, −3, −4, −3 0, 5, 12

64. 17, 7, 1, −1, 1 7, 17, 31

65. 21, 11, 5, 3, 5 11, 21, 35

66. 3, 6, 10, 15, 21 28, 36, 45

Look Beyond

67. What happens to the area of a triangle when you double its base and height? when you triple its base and height?

68. What happens to the area of a parallelogram when you double its base and height? when you triple its base and height?

Square Roots and the "Pythagorean" Right-Triangle Theorem

why *Architects and builders use a carpenter's square to construct and test right angles. This tool is used to construct right angles that are a part of triangles, squares, and other shapes. Architects and builders also use the special relationship that exists among the three sides of a right triangle.*

Gene hires a carpenter to build a square patio with an area of 12 square yards.

Square Roots

The area of Gene's new patio will be *greater than* a 3-yard-by-3-yard square with an area of 9 square yards and will be *less than* a 4-yard-by-4-yard square with an area of 16 square yards. The carpenter estimates that the side of the patio will be 3.5 yards.

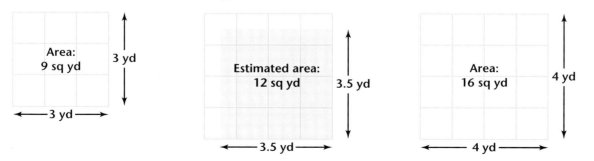

Area:
9 sq yd

3 yd

3 yd

Estimated area:
12 sq yd

3.5 yd

3.5 yd

Area:
16 sq yd

4 yd

4 yd

Is 3.5 yards a reasonable estimate for the side of the square with an area of 12 square yards? Why or why not?

There are two numbers, 4 and −4, that when squared equal 16.

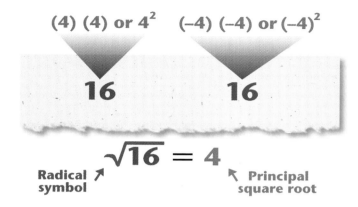

(4) (4) or 4²   **(−4) (−4) or (−4)²**

16   **16**

$$\sqrt{16} = 4$$

Radical ↗   **↖ Principal**
symbol   **square root**

The **principal square root** is a number greater than or equal to zero. In this chapter, *square root* will always refer to the principal square root.

The principal square root of 9 is 3.

$$\sqrt{9} = 3$$

The principal square root of 25 is 5.

$$\sqrt{25} = 5$$

What is the principal square root of 81?   9

Calculator

In Lesson 1.5, you studied square numbers, called **perfect squares**. The square root of a *perfect square* is always a rational number. The square root of a number that is *not* a perfect square is *not* a rational number.

For example, 12 is *not* a perfect square, and $\sqrt{12}$ is *not* a rational number. This type of square root can be estimated with a calculator.

12 √x̄ =   3.464101615

You can also use perfect squares (1, 4, 9, 16, 25, and so on) to estimate square roots. For example, you know that $\sqrt{12}$ is between $\sqrt{9}$ and $\sqrt{16}$. So an estimate for $\sqrt{12}$ is between 3 and 4.

What are the keystrokes you use to find $\sqrt{12}$ on your calculator?

2nd √   1   2 ENTER

> **EXAMPLE 1**

Use perfect squares to estimate $\sqrt{46}$ to the nearest tenth. Then use a calculator to find $\sqrt{46}$ to the nearest hundredth.

Solution ➤

Since $\sqrt{46}$ is between $\sqrt{36}$ and $\sqrt{49}$, a reasonable estimate for $\sqrt{46}$ is between 6 and 7. Since 46 is closer to 49 than to 36, estimate $\sqrt{46}$ to be closer to 7 than to 6. A reasonable estimate is about 6.8.

On a calculator, $\sqrt{46} \approx 6.782329983$.

So to the nearest hundredth, $\sqrt{46}$ is 6.78. ❖

> **Note: The symbol $\approx$ means *approximately equal to*.**

Try This   Use perfect squares to estimate $\sqrt{76}$ to the nearest tenth. Then use a calculator to find $\sqrt{76}$ to the nearest hundredth.

$\sqrt{64} < \sqrt{76} < \sqrt{81}$, so $8 < \sqrt{76} < 9$; $\sqrt{76} \approx 8.72$

The "Pythagorean" Right-Triangle Theorem

People in ancient Babylon, India, and China discovered a remarkable fact about right triangles. This fact is named after the Greek mathematician Pythagoras (6th century B.C.E.), who, among others, proved that the fact was true. The discovery about right triangles is illustrated in the picture below.

What are the lengths of sides *a*, *b*, and *c* of the right triangle below? What are the areas of squares on each side of the triangle? How do the areas relate to each other?

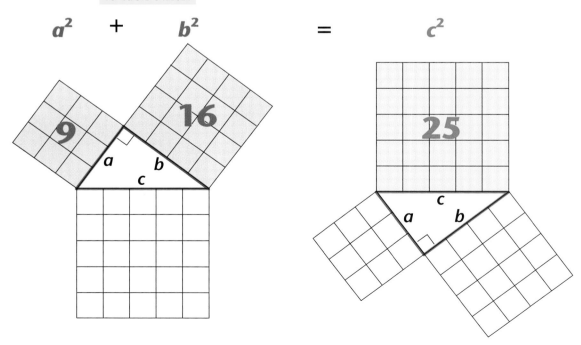

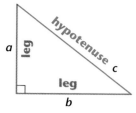

In a right triangle, the sides can be labeled a, b, and c, as shown above. The sides a and b are called **legs**. The longest side, c, is called the **hypotenuse**.

The relationship among the lengths of the three sides of a right triangle is shown below.

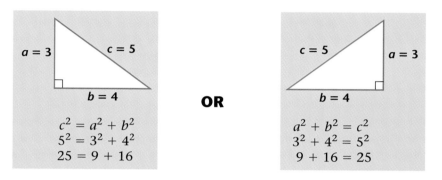

$$c^2 = a^2 + b^2$$
$$5^2 = 3^2 + 4^2$$
$$25 = 9 + 16$$

OR

$$a^2 + b^2 = c^2$$
$$3^2 + 4^2 = 5^2$$
$$9 + 16 = 25$$

The equation $a^2 + b^2 = c^2$ describes a relationship that is called the **"Pythagorean" Right-Triangle Theorem**. This theorem states that for any right triangle, the square of the length of the hypotenuse is equal to the sum of the squares of the lengths of the legs.

The **converse of the "Pythagorean" Right-Triangle Theorem** states that if the square of the length of one side of a triangle equals the sum of the squares of the lengths of the other two sides, then the triangle is a right triangle.

EXAMPLE 2

The lengths of the sides of two triangles are given. Use the converse of the "Pythagorean" Right-Triangle Theorem to decide if each triangle is a right triangle.

A 5, 12, 13   **B** 2, 2, 3

Solution ➤

A The longest side is the hypotenuse. So let 5, 12, and 13 be a, b, and c, respectively. Using substitution, see if $a^2 + b^2 = c^2$ is true.

$$a^2 + b^2 = c^2$$
$$5^2 + 12^2 \stackrel{?}{=} 13^2$$
$$25 + 144 = 169 \quad \text{True}$$

Since this is a true statement, a triangle with sides 5, 12, and 13 is a right triangle.

B The longest side is the hypotenuse. So let 2, 2, and 3 be *a*, *b*, and *c*, respectively. Using substitution, see if $a^2 + b^2 = c^2$ is true.

$$a^2 + b^2 = c^2$$
$$2^2 + 2^2 \stackrel{?}{=} 3^2$$
$$4 + 4 = 9 \quad \text{False}$$

Since this is *not* a true statement, a triangle with sides 2, 2, and 3 is *not* a right triangle. ❖

CRITICAL *Thinking*

The side lengths 2, 2, and 3 from Example 2 do not form the sides of a right triangle. Are they the side lengths of an acute triangle or an obtuse triangle? Explain your reasoning. Since $2^2 + 2^2 < 3^2$, the triangle is obtuse.

The "Pythagorean" Right-Triangle Theorem can be used to find the length of one side of a right triangle if the lengths of the other two sides are given.

EXAMPLE 3

Use the "Pythagorean" Right-Triangle Theorem to find the missing length in each right triangle.

A
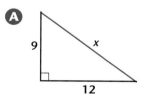

B
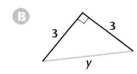

C
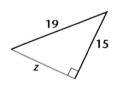

Solution ➤

A
$$9^2 + 12^2 = x^2$$
$$81 + 144 = x^2$$
$$225 = x^2$$
$$\sqrt{225} = x$$
$$15 = x$$

B
$$3^2 + 3^2 = y^2$$
$$9 + 9 = y^2$$
$$18 = y^2$$
$$\sqrt{18} = y$$
$$4.24 \approx y$$

C
$$z^2 + 15^2 = 19^2$$
$$z^2 + 225 = 361$$
$$z^2 = 361 - 225$$
$$z^2 = 136$$
$$z = \sqrt{136}$$
$$z \approx 11.66$$

Since *x* is the longest side of the triangle, it must be longer than 12. Since 15 is greater than 12, the answer is reasonable.

Since *y* is the longest side of the triangle, it must be longer than 3. Since 4.24 is greater than 3, the answer is reasonable.

Since *z* is the shortest side of the triangle, it must be less than 15. Since 11.66 is less than 15, the answer is reasonable. ❖

EXERCISES & PROBLEMS

Communicate

1. Explain what square roots are.
2. State the "Pythagorean" Right-Triangle Theorem in words.
3. How can you use the "Pythagorean" Right-Triangle Theorem to determine distances without measuring directly?
4. The "Pythagorean" Right-Triangle Theorem and its converse are two of the most important mathematical ideas used in the everyday world. Give three real-world examples of how they can be used.

Practice & Apply

Estimate each square root to the nearest whole number and to the nearest tenth. Then use a calculator to estimate to the nearest hundredth. Estimations may vary.

5. $\sqrt{5}$ 2, 2.2, 2.24
6. $\sqrt{10}$ 3, 3.1, 3.16
7. $\sqrt{12}$ 3, 3.4, 3.46
8. $\sqrt{20}$ 4, 4.4, 4.47
9. $\sqrt{27}$ 5, 5.2, 5.20
10. $\sqrt{39}$ 6, 6.3, 6.24
11. $\sqrt{42}$ 6, 6.5, 6.48
12. $\sqrt{50}$ 7, 7.1, 7.07
13. $\sqrt{60}$ 8, 7.7, 7.75
14. $\sqrt{72}$ 8, 8.5, 8.49
15. $\sqrt{95}$ 10, 9.8, 9.75
16. $\sqrt{110}$ 10, 10.5, 10.49
17. $\sqrt{143}$ 12, 11.9, 11.96
18. $\sqrt{170}$ 13, 13.1, 13.04
19. $\sqrt{180}$
20. $\sqrt{200}$ 14, 14.2, 14.14
21. $\sqrt{84}$ 9, 9.2, 9.17
22. $\sqrt{149}$
23. $\sqrt{99}$ 10, 9.9, 9.95
24. $\sqrt{74}$ 9, 8.6, 8.60
25. $\sqrt{212}$ 15, 14.5, 14.56
26. $\sqrt{23}$ 5, 4.8, 4.80
27. $\sqrt{130}$ 11, 11.5, 11.40
28. $\sqrt{255}$ 16, 15.9, 15.97

19. 13, 13.5, 13.42
22. 12, 12.2, 12.21

The lengths of the sides of a triangle are given. Use the converse of the "Pythagorean" Right-Triangle Theorem to determine if each triangle is a right triangle.

29. 1, 1.5, 2 no
30. 5, 12, 13 yes
31. 6, 8, 10 yes
32. 30, 40, 50 yes
33. 5, 6, 8 no
34. 10, 24, 26 yes
35. 12, 16, 20 yes
36. 7, 8, 10 no
37. 18, 24, 30 yes
38. 4, 6, 7 no
39. 27, 36, 45 yes
40. 7, 7, 9 no
41. 12, 13, 18 no
42. 40, 96, 104 yes
43. 1.5, 2, 2.5 yes
44. 15, 20, 25 yes
45. 8, 10, 13 no
46. 10, 24, 26 yes
47. 0.3, 0.4, 0.5 yes
48. 9, 9, 13 no

For Exercises 49–54, use the "Pythagorean" Right-Triangle Theorem to find the unknown side of each right triangle.

49.
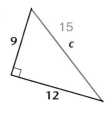
15
9
c
12

50.
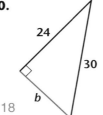
24
30
b
18

51.
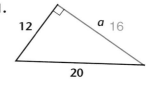
12
a 16
20

52.
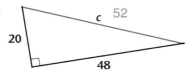
c 52
20
48

53.
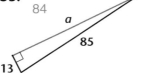
84
a
85
13

54.
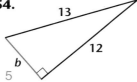
13
12
b
5

Find the unknown side of each right triangle in the table. Round your answers to the nearest tenth.

	Leg	Leg	Hypotenuse
55.	8	12	? 14.4
56.	16	20.5?	26
57.	10	20	? 22.4
58.	15	30	? 33.5
59.	4	4	? 5.7

60. A ladder is placed 10 feet from a building. How long is the ladder if it reaches 30 feet up the building? About 31.6 ft

61. What is the diagonal length of a rectangular pool with dimensions of 30 feet by 40 feet? 50 ft

62. Recreation   A hiker leaves camp and follows the route shown below. How far is she from camp? About 5.7 mi

camp
4 mi due south
?
4 mi due east

63. Agriculture A farmer plants pecan trees in rows so that each tree is 30 feet from the next. He measures the length of the diagonal of a square formed by 4 trees to be sure that it is square. What is the length of the diagonal if the trees form a square? About 42.4 feet

64. Suppose the farmer plants the trees 40 feet apart. What should the length of the diagonal be if 4 trees form a square? About 56.6 feet

65. The cable company buries a line diagonally across a lot. The lot measures 105 feet by 60 feet. How long is the line? About 120.9 feet

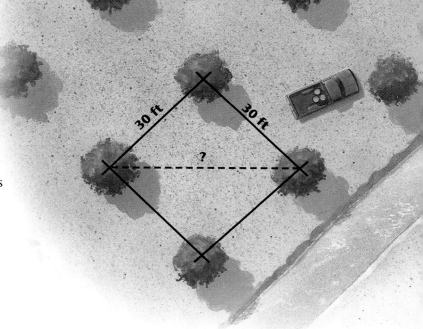

Look Back

Copy and complete each table. Then write an equation for each relationship. Use a calculator to check your equation. [Lesson 1.2]

66.

x	−2	−1	0	1	2	3	4
y = ?	−5	−4	−3	−2	−1	? 0	? 1

$y = x - 3$

67.

x	−2	−1	0	1	2	3	4
y = ?	−3	−1	1	? 3	? 5	? 7	? 9

$y = 2x + 1$

Find the area of each figure. [Lesson 4.5]

68. a right triangle with legs of 3 and 4 6 sq units

69. a parallelogram with a base of 6 and a height of 4.5 27 sq units

70. a trapezoid with bases of 4 and 6, and a height of 3 15 sq units

Look Beyond

Estimate the length from point *A* to each of the indicated points, using right triangles in your work when possible. Then use a calculator to compute distances to the nearest hundredth.

71. *AB* 5; 5   **72.** *AC* 5.6; 5.66   **73.** *AD* 5; 5   Answers may vary.

74. *AE* 9.4; 9.43   **75.** *AF* 9.2; 9.22

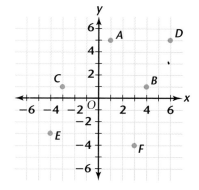

20 ft

5 ft

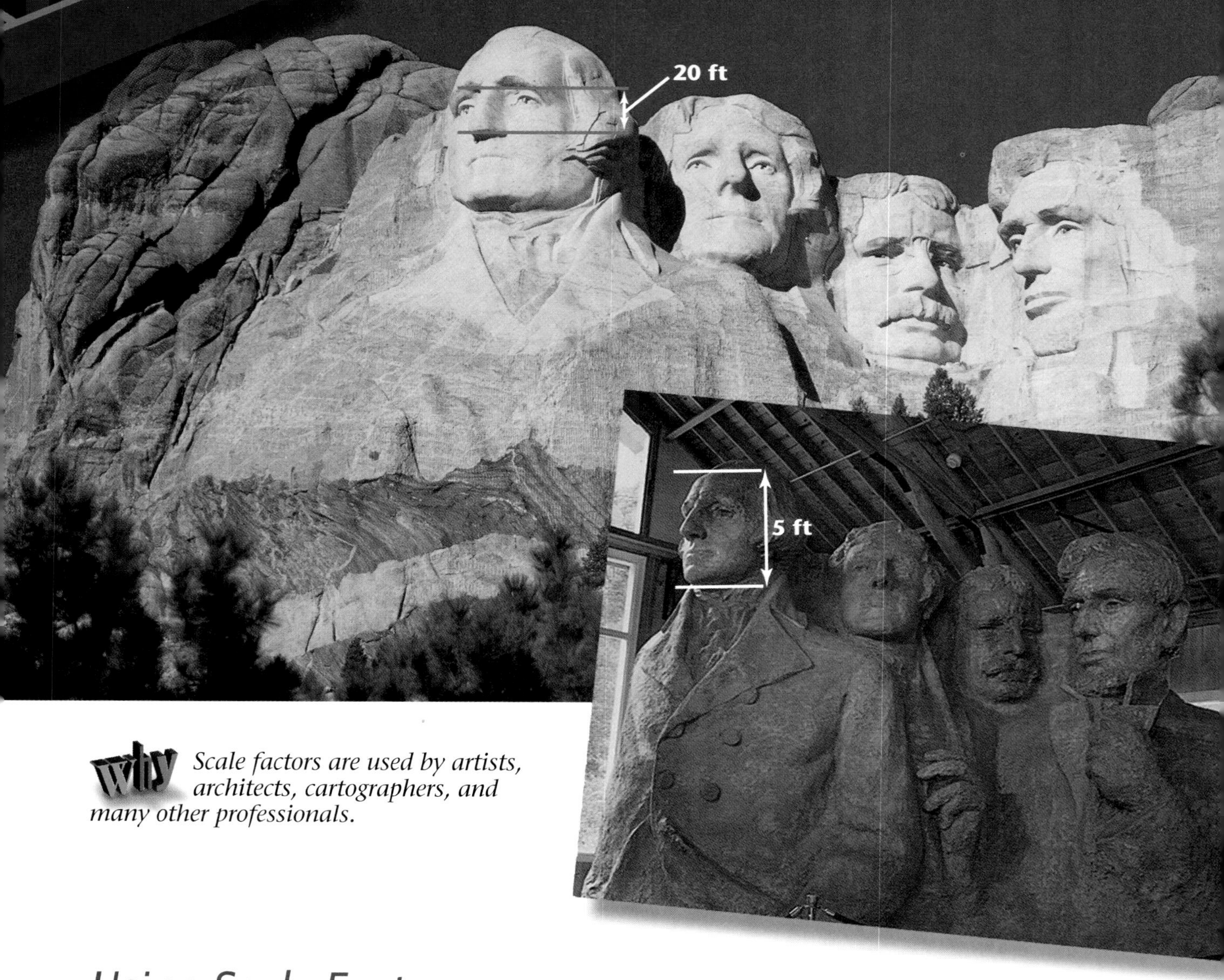

WHY *Scale factors are used by artists, architects, cartographers, and many other professionals.*

Using Scale Factors

The famous sculptor Gutzon Borglum designed the scale model for Mount Rushmore and supervised the mammoth task of carving the images of Washington, Jefferson, Roosevelt, and Lincoln on the face of the steep mountain. Borglum used the ratio 1 to 12 to convert his plaster model to the measurements that were used for the mountain sculpture. The number 12 is called a **scale factor** because it can be used to convert the measurements of the model to the measurements of the mountain. Conversely, the number 12 can also be used to convert measurements of the mountain to measurements of the model.

EXAMPLE 1

Use the scale factor 12 to complete the table.

		Plaster model	Mountain sculpture
A	Forehead to chin	5 feet	?
B	Length of nose	?	20 feet

Solution ➤

A The distance from the forehead to the chin on the plaster model is 5 feet. Multiply 5 by the scale factor to change the plaster-model measurement of 5 feet to the measurement of the mountain sculpture.

$$5 \cdot 12 = 60$$

The length from the forehead to the chin on the mountain sculpture is 60 feet.

B The length of the nose on the mountain sculpture is 20 feet. Divide 20 by the scale factor 12 to change the mountain-sculpture measurement of 20 feet to the measurement of the plaster model.

$$\frac{20}{12} = 1\frac{2}{3}$$

The length of the nose on the plaster model is $1\frac{2}{3}$ feet. ❖

Try This Use the scale factor of 12 to complete the table below.

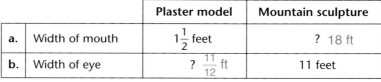

		Plaster model	Mountain sculpture
a.	Width of mouth	$1\frac{1}{2}$ feet	? 18 ft
b.	Width of eye	? $\frac{11}{12}$ ft	11 feet

Assume that a scale factor of 10 was used to create the larger pencil from the smaller pencil shown. How many feet long is the larger pencil?

60 in., or 5 ft

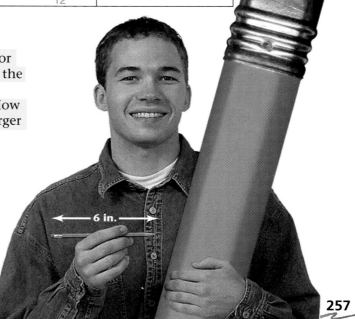

6 in.

Constructing Similar Figures

Gutzon Borglum's plaster model and mountain sculpture are called **similar figures** because they have the same shape. Graphic artists use scale factors to enlarge and reduce computer graphics. They create similar figures using this process. The graphics window below displays similar rectangles. To form the larger window, the small window is enlarged by a scale factor of 2, or 200%.

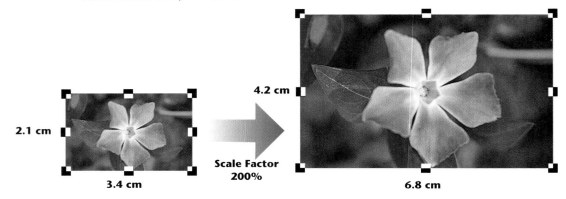

2.1 cm

4.2 cm

3.4 cm

Scale Factor
200%

6.8 cm

What is the ratio of the width of the larger window to the width of the smaller window? What is the ratio of the length of the larger window to the length of the smaller window? $\frac{2}{1}, \frac{2}{1}$

The widths of the two similar rectangles are called **corresponding sides**. The lengths of the two similar rectangles are also corresponding sides.

Notice that the ratio $\frac{\text{width of larger window}}{\text{width of smaller window}}$ is equal to $\frac{6.8}{3.4}$, or 2.

Also, the ratio $\frac{\text{length of larger window}}{\text{length of smaller window}}$ is equal to $\frac{4.2}{2.1}$, or 2.

Recall from Lesson 3.6 that if two ratios are the same, the corresponding sides are said to be proportional. The proportion $\frac{6.8}{3.4} = \frac{4.2}{2.1}$ is used to show this relationship. All similar figures have proportional sides and congruent angles.

EXAMPLE 2

Graphic Arts   In order to fit a triangle graphic onto a page, a graphic artist reduces the figure by a scale factor of 75%. Find the dimensions of the reduced figure.

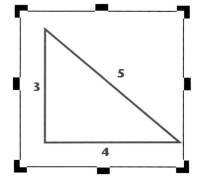

3

5

4

x 75%

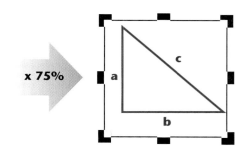

a

c

b

Solution ➤

Find *a*, *b*, and *c* by multiplying the corresponding side by the scale factor of 75% written as both a decimal and a fraction.

The scale factor as a decimal is 0.75.

Find *a*.	Find *b*.	Find *c*.
$a = 3 \cdot 0.75$	$b = 4 \cdot 0.75$	$c = 5 \cdot 0.75$
$a = 2.25$	$b = 3$	$c = 3.75$

The scale factor as a fraction is $\frac{3}{4}$.

Find *a*.	Find *b*.	Find *c*.
$a = 3 \cdot \frac{3}{4}$	$b = 4 \cdot \frac{3}{4}$	$c = 5 \cdot \frac{3}{4}$
$a = \frac{9}{4}$, or $2\frac{1}{4}$	$b = \frac{12}{4}$, or 3	$c = \frac{15}{4}$, or $3\frac{3}{4}$ ❖

EXAMPLE 3

Determine the scale factor of the similar figures. Then find the unknown lengths *x*, *y*, and *z*.

Solution ➤

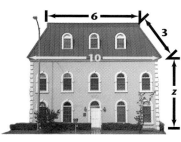

To find the scale factor, find the ratio of the corresponding sides when both lengths are known.

$$\frac{9}{6} = 1.5$$

Multiply the scale factor by the corresponding side to find *x* and *y*.

To find *x*: $x = 10 \cdot 1.5$
$x = 15$

To find *y*: $y = 3 \cdot 1.5$
$y = 4.5$

Divide the length of the side of the enlarged figure by the scale factor to find *z*.

$$z = \frac{8}{1.5}$$

$$z = 5.333 \ldots, \text{ or } 5\frac{1}{3}$$

To make sure that corresponding sides are proportional, write down ratios and divide with a calculator.

$$\frac{15}{10} = 1.5 \qquad \frac{4.5}{3} = 1.5 \qquad \frac{8}{5\frac{1}{3}} = 1.5$$

Calculator

8 ÷ (5 + 1 ÷ 3) = 1.5 ❖

EXERCISES & PROBLEMS

Communicate

1. Explain scale factors. Give two examples of how they are used in the real world.
2. Describe similar figures.
3. Explain how you can use scale factors to create similar figures.
4. Draw two similar triangles. Show that the sides are proportional by computing the ratios of the corresponding sides.

Practice & Apply

Use the scale factor to find the length and width of each rectangle.

	Scale factor	Length	Width
	original	8 cm	10 cm
5.	2	16 cm	?
6.	3	?	30 cm
7.	4	?	?
8.	$\frac{1}{2}$	?	?
9.	0.8	?	?

	Scale factor	Length	Width
	original	12 cm	15 cm
10.	25%	?	?
11.	175%	?	?
12.	0.6	?	?
13.	$\frac{3}{4}$	?	?
14.	2.5	?	?

15. **Technology** Use a calculator or spreadsheet to compute the perimeter of each rectangle in Exercises 5–14. How does the ratio $\frac{\text{perimeter of similar rectangle}}{\text{perimeter of original rectangle}}$ compare with the scale factor?

16. **Technology** Use a calculator or spreadsheet to compute the area of each rectangle in Exercises 5–14. How does the ratio $\frac{\text{area of similar rectangle}}{\text{area of original rectangle}}$ compare with the scale factor?

The right triangles described in the tables below are similar right triangles. Find the scale factor and the length of the unknown leg.

	Leg	Leg	Scale factor
	4 cm	8 cm	original
17.	40 cm	? 80	? 10
18.	? 20	40 cm	? 5
19.	4 cm	? 8	? 1

	Leg	Leg	Scale factor
	3 cm	4 cm	original
20.	? 9	12 cm	? 3
21.	27 cm	? 36	? 9
22.	? 33	44 cm	? 11

260 CHAPTER 4

23. Technology Use a calculator or spreadsheet to compute the area of each triangle in Exercises 17–22. How does the ratio $\dfrac{\text{area of similar triangle}}{\text{area of original triangle}}$ compare with the scale factor?

24. Technology Use a calculator or spreadsheet to compute the hypotenuse and perimeter of each triangle in Exercises 17–22. How does the ratio

$\dfrac{\text{perimeter of similar triangle}}{\text{perimeter of original triangle}}$

compare with the scale factor?

Use similar right triangles and the lengths given in the photo for Exercises 25–28 below.

13 m

1.1 m

3.4 m

25. Katrina is 1.4 meters tall. How long is the shadow of the tree behind her? About 10 m

26. Suppose the shadow of the building is measured at the same time that Katrina's shadow is measured. Find the height of the building. About 4.3 m

27. About 8.9 m

27. Suppose Katrina's height is 1.6 meters. Find the length of the tree's shadow.

28. Suppose Katrina's height is 1.5 meters. Find the height of the building. 4.64 m

29. The triple 3-4-5 is called a Pythagorean triple because 3, 4, and 5 are the lengths of the sides of a right triangle. Use scale factors and a 3-4-5 right triangle to create four other Pythagorean triples: two enlargements and two reductions.

A negative on a roll of 35-millimeter film actually has the dimensions 36 millimeters long by 24 millimeters wide.

30. Photography Sam is enlarging a negative from a roll of 35 millimeter film into an 12-centimeter-by-8-centimeter print. What is the scale factor? (1 cm = 10 mm) About 3.33

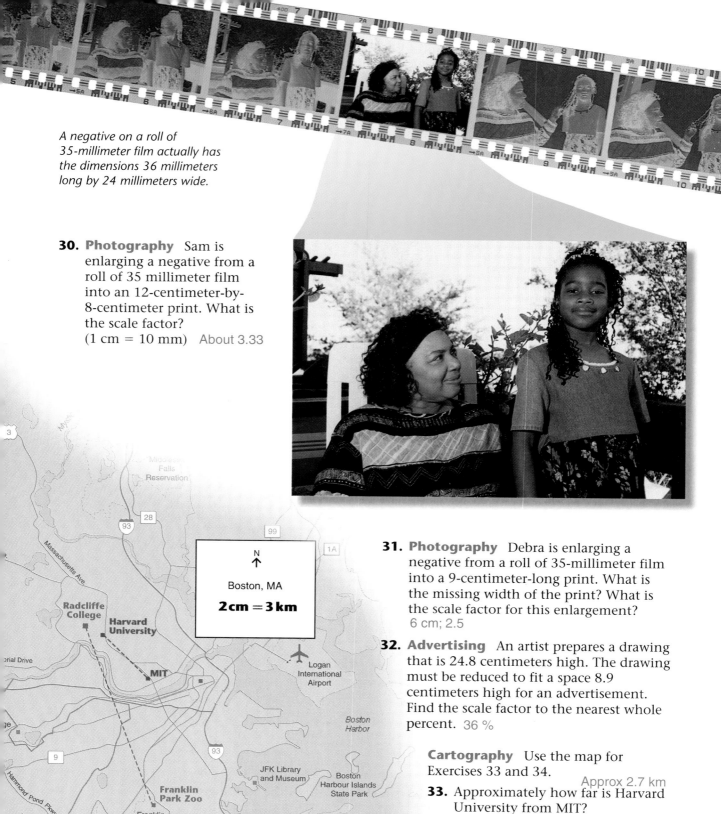

31. Photography Debra is enlarging a negative from a roll of 35-millimeter film into a 9-centimeter-long print. What is the missing width of the print? What is the scale factor for this enlargement?
6 cm; 2.5

32. Advertising An artist prepares a drawing that is 24.8 centimeters high. The drawing must be reduced to fit a space 8.9 centimeters high for an advertisement. Find the scale factor to the nearest whole percent. 36 %

Cartography Use the map for Exercises 33 and 34.
Approx 2.7 km

33. Approximately how far is Harvard University from MIT?

34. Approximately how far from the Franklin Park Zoo is Radcliff College? Approx 7.8 km

N↑

Boston, MA

2 cm = 3 km

Portfolio Activity Suppose an artist uses a scale factor of 150% to enlarge a 4-centimeter-by-8-centimeter rectangular graphic.

35. What are the dimensions of the enlarged graphic? Compute the perimeter and area of the graphic and the enlargement.

36. How do the perimeter and area of the enlargment compare with the original graphic?

37. Copy and complete the table.

38. Explain how you can use the scale factor to find the perimeter and area of an enlarged or reduced figure.

Scale factor (%)	Scale factor (decimal)	Width	Length	Perimeter	Area
100%	1	4 cm	8 cm	?	?
150%	1.5	?	?	?	?
200%	?	?	?	?	?
300%	?	?	?	?	?
75%	?	?	?	?	?
50%	?	?	?	?	?

Look Back

For Exercises 39–44, use mental math to estimate each of the following. Then use a calculator to compute. Estimations may vary. [Lessons 3.3, 3.5, 3.7]

39. 26% of 348 = _?_ 90.48

40. $\frac{458}{1020} = $ _?_ % 44.9

41. 34 feet · 93 feet = _?_ square feet 3162

42. $\frac{8.12}{96.03} = $ _?_ % 8.5

43. $7\frac{7}{16}$ inches · $9\frac{3}{8}$ inches = _?_ square inches 69.7265625

44. $\frac{\$8.48}{\$0.25} = $ _?_ $33.92

45. The complement of an angle is 18°. What is the angle measure? **[Lesson 4.2]** 72°

46. The complement of an angle is 15°. What is the angle measure? **[Lesson 4.2]** 75°

47. The supplement of an angle is 45°. What is the angle measure? **[Lesson 4.2]** 135°

48. One of two supplementary angles is twice the measure of the other. What is the measure of each angle? **[Lesson 4.2]** 60°; 120°

49. What is the sum of the angle measures of a decagon? **[Lesson 4.4]** 1440°

Look Beyond

Multiply the coordinates of the original triangle ABC by the given scale factor to construct triangle $A'B'C'$. Graph each triangle. Measure the lengths of sides $\overline{AB}$, $\overline{BC}$, $\overline{A'B'}$, and $\overline{B'C'}$.

	A	B	C	AB	BC
Original	(−2, 1)	(−2, −3)	(1, −3)	4	3

	Scale factor	A'	B'	C'	A'B'	B'C'
50.	2	?	?	?	?	?
51.	3	?	?	?	?	?
52.	4	?	?	?	?	?

Designing a Park

*S*uppose your
community has set aside a piece of land with an area
of 1 square mile on which to build a park. You are on the
committee to design the park. Your committee has decided that the
basic design will include a one-way road, three parking lots, and a
jogging trail. The road will run diagonally across the park. The parking
lots will be at each end of this road and in the center of the park. The
jogging trail will be constructed around the perimeter of the park.

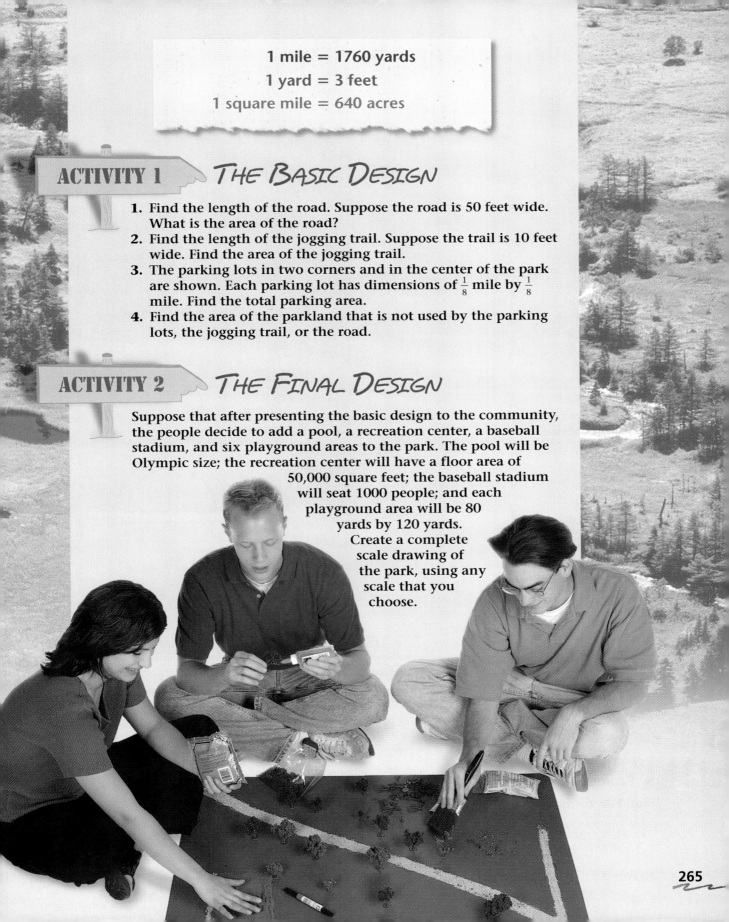

1 mile = 1760 yards
1 yard = 3 feet
1 square mile = 640 acres

ACTIVITY 1 ▷ THE BASIC DESIGN

1. Find the length of the road. Suppose the road is 50 feet wide. What is the area of the road?
2. Find the length of the jogging trail. Suppose the trail is 10 feet wide. Find the area of the jogging trail.
3. The parking lots in two corners and in the center of the park are shown. Each parking lot has dimensions of $\frac{1}{8}$ mile by $\frac{1}{8}$ mile. Find the total parking area.
4. Find the area of the parkland that is not used by the parking lots, the jogging trail, or the road.

ACTIVITY 2 ▷ THE FINAL DESIGN

Suppose that after presenting the basic design to the community, the people decide to add a pool, a recreation center, a baseball stadium, and six playground areas to the park. The pool will be Olympic size; the recreation center will have a floor area of 50,000 square feet; the baseball stadium will seat 1000 people; and each playground area will be 80 yards by 120 yards. Create a complete scale drawing of the park, using any scale that you choose.

265

Chapter 4 Review

Vocabulary

acute angle	206	parallel	205	right angle	206
acute triangle	218	perfect square	249	right triangle	218
complementary angles	213	perpendicular	205	scale factor	256
concave	224	polygon	224	scalene triangle	218
convex	224	principal square root	249	similar figures	260
equilateral triangle	218	"Pythagorean" Right-		supplementary angles	213
hypotenuse	251	Triangle Theorem	250	transversal	216
isosceles triangle	218	quadrilateral	224	trapezoid	243
obtuse triangle	218				

Key Skills & Exercises

Lesson 4.1

➤ **Key Skills**

Use a protractor to measure angles, and classify angles as acute, right, or obtuse.

To measure $\angle ABC$, place a protractor so that $\overrightarrow{BC}$ runs from the center of the protractor to the 0° mark. Read the angle measure at the point where $\overrightarrow{BA}$ crosses the outer edge of the protractor.

The angle measure is 120°. Since 120 > 90, the angle is obtuse.

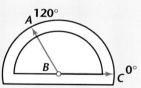

➤ **Exercises**

Use a protractor to draw each angle. Identify the angle as *acute*, *obtuse*, or *right*.

1. 90°   **2.** 138°   **3.** 67°

Lesson 4.2

➤ **Key Skills**

Find the complement or supplement of an angle.

Angles 1 and 2 are complementary, angles 1 and 3 are supplementary, and m∠1 is 13.5°. To find m∠2 and m∠3, solve the following equations.

$$13.5 + m\angle 2 = 90°$$
$$m\angle 2 = 76.5°$$

and

$$13.5 + m\angle 3 = 180°$$
$$m\angle 2 = 166.5°$$

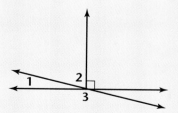

➤ **Exercises**

Find the complement and supplement of each angle.

4. 65° 25°, 115° **5.** 89° 1°, 91°   **6.** 36.4° 53.6°, 143.6°

7. The supplement of an angle is 135°. What is the measure of the angle? 45°

Lesson 4.3

➤ Key Skills

Identify pairs of angles that are alternate interior, alternate exterior, consecutive interior, or corresponding.

In the figure at the right, the transversal t crosses lines m and n, forming angles 1 through 8. Angles 3 and 6 are alternate interior angles; angles 1 and 8 are alternate exterior angles; angles 3 and 5 are consecutive interior angles; and angles 4 and 8 are corresponding angles.

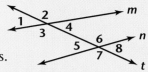

➤ Exercises

Using the figure above, classify each pair of angles as alternate interior, corresponding, alternate exterior, or consecutive interior.

8. $\angle 1$ and $\angle 5$
corresponding

9. $\angle 2$ and $\angle 7$
alt ext

10. $\angle 4$ and $\angle 6$
consec int

11. $\angle 4$ and $\angle 5$
alt int

Lesson 4.4

➤ Key Skills

Recognize properties of special quadrilaterals.

A square can be classified as a parallelogram (because opposite sides are parallel), a rectangle (because all angles are 90°), and as a rhombus (because all sides are congruent).

➤ Exercises

12. Which of the following are properties of rectangles?
   a. The diagonals are congruent. yes
   b. The opposite sides are parallel and congruent. yes
   c. The diagonals are perpendicular. no

Lesson 4.5

➤ Key Skills

Use a formula to find the perimeter or area of a rectangle. To find the perimeter of a rectangle with a width of 2.4 feet and a length of 3.6 feet, use the formula $P = 2w + 2l$.

To find the area of a rectangle with a width of 2.4 feet and length of 3.6 feet, use the formula $A = wl$.

$P = 2w + 2l$
$P = 2(2.4) + 2(3.6)$
$P = 4.8 + 7.2 = 12$ feet

$A = wl$
$A = (2.4)(3.6)$
$\quad = 8.64$ square feet

➤ Exercises

Find the perimeter and the area of each rectangle with width w and length l.

13. $l = 100$ yd
$w = 42$ yd
284 yd; 4200 sq yd

14. $l = 6.1$ m
$w = 7.9$ m
28 m; 48.19 sq m

15. $l = 3\frac{3}{8}$ in.
$w = 5\frac{1}{4}$ in.
$17\frac{1}{4}$ in.; $17\frac{23}{32}$ sq in.

Lesson 4.6

➤ Key Skills

Use a formula to find the area of a parallelogram, triangle, or trapezoid.

Parallelogram	Triangle	Trapezoid
$A = bh$	$A = \frac{1}{2}bh$	$A = \frac{1}{2}(b_1 + b_2)h$

➤ Exercises

Use a formula to compute the area of the indicated polygons.

16. Triangle: $b = 10$ mm, $h = 24$ mm 120 sq mm

17. Trapezoid: $b_1 = 7$ in., $b_2 = 9$ in., $h = 5\frac{1}{2}$ in. 44 sq in.

18. Parallelogram: $b = 14.5$ yd, $h = 6.3$ yd 91.35 sq yd

Lesson 4.7

➤ Key Skills

Use the "Pythagorean" Right-Triangle Theorem to find the unknown side of a triangle.

If a right triangle has legs a and b and hypotenuse c, then by the "Pythagorean" Right-Triangle Theorem, $a^2 + b^2 = c^2$.

$$a^2 + b^2 = c^2$$
$$3^2 + 4^2 = 5^2$$
$$9 + 16 = 25$$
$$25 = 25$$

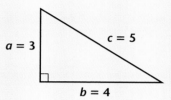

➤ Exercises

19. The lengths of the legs of a right triangle are 4 and 7.5. Find the length of the hypotenuse. 8.5

20. The length of the hypotenuse and one leg of a right triangle are 30 and 12, respectively. Find the length of the other leg. Approx 27.5

Lesson 4.8

➤ Key Skills

Use scale factors to find the missing side lengths of similar figures.

To find the missing side lengths of two similar triangles, first find the scale factor. Then multiply the scale factor by the smaller length to find the larger corresponding length.

The scale factor is $\frac{5}{2}$, or 2.5.

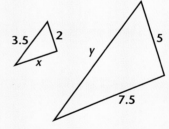

To find y: $y = (3.5)(2.5)$
$y = 8.75$

To find x: $2.5x = 7.5$
$x = 3$

➤ Exercises

The lengths of the two legs of a right triangle are 6 and 9. For each similar right triangle, find the scale factor and the length of the missing leg.

21. shorter leg: 18
3; 27

22. longer leg: 18
2; 12

23. shorter leg: 21
3.5; 31.5

Applications

24. Construction A pipe is buried diagonally across a perpendicular intersection of two streets. The intersection is a square with a side length of 52 feet. How long is the pipe?
73.5 ft

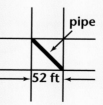

25. Architecture To estimate the height of a building, Janice finds the length of its shadow and the length of her own shadow. Her height is 5 feet, the length of her shadow is 8 feet, and the length of the building's shadow is 36 feet. How tall is the building? $22\frac{1}{2}$ ft

Chapter 4 Assessment

Draw each angle by using a protractor. Then identify each angle as *acute*, *obtuse* or *right*.

1. 45°   **2.** 155°   **3.** 89°   **4.** 91°

Find each of the following angle measures.

5. the complement of 23° 67°

6. the supplement of 23° 157°

7. the supplement of 178° 2°

Using the figure at the right, list all pairs of the indicated types of angles.   **8.** ∠6 and ∠3, ∠5 and ∠8   **9.** ∠1 and ∠2, ∠4 and ∠7

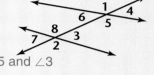

8. alternate interior   **9.** alternate exterior

10. corresponding   **11.** consecutive interior

10. ∠1 and ∠8, ∠6 and ∠7, ∠4 and ∠3, ∠5 and ∠2   **11.** ∠6 and ∠8, ∠5 and ∠3

Draw each figure described.

12. right scalene triangle   **13.** obtuse isosceles triangle

14. acute isosceles triangle   **15.** a convex quadrilateral

16. concave hexagon   **17.** equilateral triangle

Use formulas to find the perimeter and area of each rectangle with width *w* and length *l*.

18. $l = 250$ cm, $w = 15$ cm   **19.** $l = 14.5$ ft, $w = 6.8$ ft

$P = 530$ cm, $A = 3750$ sq cm $P = 42.6$ ft, $A = 98.6$ sq ft

Use a formula to find the area of each polygon.

20. triangle: $b = 4\frac{1}{8}$ m, $h = 3\frac{3}{4}$ m $7\frac{47}{64}$ sq m

21. trapezoid: $b_1 = 3$ ft, $b_2 = 8$ ft, $h = 5.5$ ft 30.25 sq ft

22. parallelogram: $b = 5.7$ cm, $h = 2.1$ cm 11.97 sq cm

Two sides of a right triangle are given. Use the "Pythagorean" Right-Triangle Theorem to determine the third side. Side *c* is the hypotenuse.

23. $a = 3$ cm, $c = 5$ cm 4 cm

24. $a = 7.5$ ft, $b = 6$ ft 9.6 ft

25. $c = 15$ in., $b = 10$ in. 11.2 in.

26. The two triangles shown below are similar. Find the scale factor and the values $\frac{3}{2}$, $a = 6\frac{2}{3}$, $b = 10\frac{1}{2}$ of *a* and *b*.

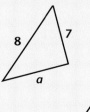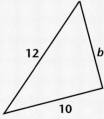

Chapters 1-4 Cumulative Assessment

College Entrance Exam Practice

Quantitative Comparison For Questions 1–4, write
A if the quantity in Column A is greater than the quantity in Column B;
B if the quantity in Column B is greater than the quantity in Column A;
C if the two quantities are equal; or
D if the relationship cannot be determined from the information given.

	Column A	Column B	Answers
1.	$23 \cdot 43 + 23 \cdot 59$	$23 \cdot (43 + 59)$	C ⟨A⟩ ⟨B⟩ ⟨C⟩ ⟨D⟩ **[Lesson 1.7]**
2.	$-17 - (-46)$	$-17 - 46$	A ⟨A⟩ ⟨B⟩ ⟨C⟩ ⟨D⟩ **[Lesson 2.4]**
3.	$(-144) \div 48$	$144 \div (-48)$	C ⟨A⟩ ⟨B⟩ ⟨C⟩ ⟨D⟩ **[Lesson 2.5]**
4.	$\dfrac{2}{3}$	$\dfrac{5}{6}$	B ⟨A⟩ ⟨B⟩ ⟨C⟩ ⟨D⟩ **[Lesson 3.1]**

5. What is an equation for the sentence, charge, c, equals a \$10 base charge plus \$5 per hour, h? **[Lesson 1.3]** b
  a. $c = 10 + 5 + h$ **b.** $c = 10 + 5h$ **c.** $h = 10 + 5c$ **d.** $c = 5 + 10h$

6. When you apply the method of finding differences to the sequence 4, 19, 44, 79, 124, what is the value of the constant that you get?
  [Lesson 2.7] b
  a. 5 **b.** 10 **c.** 15 **d.** 25

7. Which of the following has a value of $8\frac{5}{8}$? **[Lesson 3.4]** d
  a. $5\frac{1}{4} + 2\frac{7}{8}$ **b.** $12\frac{1}{8} - 3\frac{3}{8}$ **c.** $6\frac{3}{4} + 1\frac{5}{6}$ **d.** none of these

8. What is 0.509 represented as a percent? **[Lesson 3.7]** b
  a. 5.09% **b.** 50.9% **c.** 0.00509% **d.** 509%

9. Classify the two marked angles in the diagram at right. **[Lesson 4.3]** b
  a. alternate interior angles
  b. alternate exterior angles
  c. corresponding angles
  d. consecutive interior angles

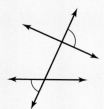

10. Write the fraction $\frac{301}{473}$ in lowest terms.   **[Lesson 3.2]** $\frac{7}{11}$

11. Order the following numbers from greatest to least.   **[Lesson 3.3]**

$$4.14, 4.014, 4\frac{1}{7}, 4.139, 4.4 \quad 4.4, 4\frac{1}{7}, 4.14, 4.139, 4.014$$

12. Walter is filling bags with $3\frac{1}{2}$ pounds of pecans. He has 56 pounds of pecans. How many bags can he fill?   **[Lesson 3.5]** 16

13. At last year's cookout, 18 pounds of beef were needed to prepare 72 hamburgers. This year, 120 hamburgers are expected to be consumed. How many pounds of beef will be required?   **[Lesson 3.6]** 30

14. Find the theoretical probability that you will roll a number greater than 2 if you roll an ordinary 6-sided number cube 1 time.   **[Lesson 3.9]** $\frac{2}{3}$

15. Identify the angle at the right as acute, obtuse, right, or straight.   **[Lesson 4.1]**
obtuse

16. The complement of an angle measures 45°. What is the measure of the angle?   **[Lesson 4.2]** 45°

17. Which special quadrilateral is both a rectangle and a rhombus?
[Lesson 4.4] square

18. Find the perimeter and area of a rectangle with a length of 6.8 feet and a width of 2.7 feet.   **[Lesson 4.5]** 19 ft; 18.36 sq ft

19. Find the area of the triangle at the right.
[Lesson 4.6] $7\frac{1}{2}$ sq units

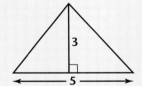

Free-Response Grid   The following questions may be answered using a free-response grid commonly used by standardized test services.

20. Use guess-and-check to solve the equation
$-8 + x = -3$.   **[Lesson 2.3]** 5

21. Write the decimal 0.36 as a fraction in lowest terms.
[Lesson 3.3] $\frac{9}{25}$

22. One of two supplementary angles is twice the measure of the other. What is the measure of the greater angle?
[Lesson 4.2] 120°

23. The lengths of the two legs of a right triangle are 7 centimeters and 24 centimeters. Find the length, in centimeters, of the hypotenuse.   **[Lesson 4.7]** 25

24. The lengths of two legs of a right triangle are 4 feet and 15 feet. The shorter leg of a triangle that is similar to the first triangle is 8 feet long. Find the length, in feet, of the longer leg of the second triangle.
[Lesson 4.8] 30 ft

CHAPTER 5

Addition and Subtraction in Algebra

An equation is a statement that two mathematical expressions are equal. If each of these expressions were placed on a mathematical "scale," they would balance.

In this chapter you will study a number of mathematical properties. These properties maintain the balance in an equation as you add and subtract to solve the equation.

PORTFOLIO ACTIVITY

A squirrel and a dog weigh the same as 2 cats. A dog weighs as much as 2 squirrels and a cat. How many squirrels will it take to equal the dog's weight?

Find the solution to the problem any way you can; then use algebra to find the solution. Remember, if one quantity equals another, they can replace each other.

Adding Expressions

why *In algebra you work with expressions that contain numbers and variables. When you add expressions, you can write the sum in a simplified form.*

Juan is collecting comic books. He has 3 cartons of comics and 4 loose comics in his room. He has 2 more cartons of comics and 2 loose comics in the trunk of his car.

How can you write an expression that describes the total number of cartons and comics? Let x represent the number of comic books in a carton.

In his room	3 cartons of comics and 4 comics	$3x + 4$
In the car	2 cartons of comics and 2 comics	$2x + 2$
All together	5 cartons of comics and 6 comics	$5x + 6$

The quantity $5x + 6$ is an example of an **algebraic expression.** Both $5x$ and 6 are **terms** of the expression $5x + 6$. The terms $3x$ and $2x$ are **like terms** because each term contains the same form of the variable, x. The number 5 is a factor of $5x$. The 5 is also called the **coefficient** of x. The number 6, which represents a fixed amount in the expression $5x + 6$, is often referred to as a **constant.**

Tiles can be used to model an expression containing a variable, such as x.

Represents positive x Represents negative x

EXAMPLE 1

Simplify $(3x + 4) + (2x - 1)$ using tiles.

Solution ➤

Represent each of the expressions using the appropriate tiles. Then combine like tiles.

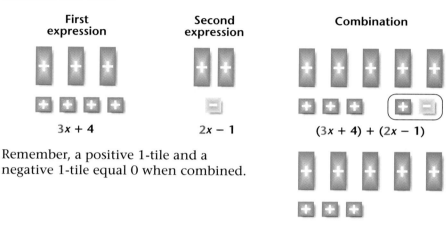

First expression	Second expression	Combination
$3x + 4$	$2x - 1$	$(3x + 4) + (2x - 1)$

Remember, a positive 1-tile and a negative 1-tile equal 0 when combined.

$5x + 3$

When you simplify the expression, the result is $5x + 3$. ❖

How would you model $(2x - 3) + (4x + 5)$? What is the result?

The Distributive Property

Combining like terms, such as $3x$ and $2x$, involves an important property of numbers called the Distributive Property.

The Distributive Property works in two directions. To multiply $3 \cdot (5 + 2)$, the multiplication is distributed over the addition.

$$3 \cdot (5 + 2) = 3 \cdot 5 + 3 \cdot 2 = 21$$

When used in reverse, a common factor can be removed from the terms.

$$3 \cdot 5 + 3 \cdot 2 = 3 \cdot (5 + 2) = 21$$

This can be shown on a number line.

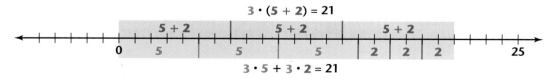

DISTRIBUTIVE PROPERTY
For all numbers a, b, and c,
$a(b + c) = ab + ac$ and $(b + c)a = ba + ca$.

The Distributive Property can be used to simplify an expression that contains like terms.

$$3x + 2x = (3 + 2)x$$
$$= 5x ❖$$

Explain how to add like terms, such as $3x$ and $2x$, using the following number line:

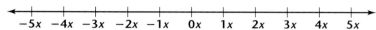

Each interval on the number line represents x. Begin at the point on the number line marked $3x$. To add $2x$, move two intervals to the right. The final point is marked $5x$. Thus, $3x + 2x = 5x$.

Rearranging Terms

Changing the order of two terms is called *commuting* the terms.

> **COMMUTATIVE PROPERTY FOR ADDITION**
> For any numbers a and b,
>
> $$a + b = b + a.$$

Now you can add the expressions in Example 1 without using tiles for models.

$(3x + 4) + (2x - 1)$	Given
$(3x + 4) + (2x + (-1))$	Definition of Subtraction
$3x + (4 + 2x) + (-1)$	**Associative Property**
$3x + (2x + 4) + (-1)$	**Commutative Property**
$(3x + 2x) + (4 + (-1))$	**Associative Property**
$5x + 3$	Combine like terms.

You can use the Associative and Commutative Properties to rearrange terms. For example, $(3x + 4) + (2x + (-1))$ can be rearranged as $(3x + 2x) + (4 + (-1))$. The middle three steps of the process shown above are combined.

$(3x + 4) + (2x - 1)$	Given
$(3x + 2x) + (4 - 1)$	**Rearrange terms.**
$5x + 3$	Combine like terms.

To check the addition, replace the variable x with a number that makes the expression possible to evaluate. For example, replace x with 10, and evaluate each expression.

$$(3x + 4) + (2x - 1) = 5x + 3$$
$$[3(10) + 4] + [2(10) - 1] \overset{?}{=} 5(10) + 3$$
$$53 = 53 \qquad \text{True}$$

a. $7b - 2$ b. $3b + 4$ c. $b + 7$

Try This   Add each expression to $5b + 1$.
a. $2b - 3$ **b.** $-2b + 3$ **c.** $6 - 4b$

EXAMPLE 2

GEOMETRY
Connection

The lengths of the sides of triangle *ABC* are shown in the diagram. What is the perimeter of triangle *ABC*?

Solution ➤

To find the perimeter, add the lengths of the sides.

$5x + y$

$4x + 8y + 8z$

A

B

$9x - 2z$

C

$$P = (5x + y) + (4x + 8y + 8z) + (9x - 2z) \qquad \text{Definition of Perimeter}$$
$$= (5x + 4x + 9x) + (y + 8y) + (8z - 2z) \qquad \text{Rearrange terms.}$$
$$= 18x + 9y + 6z \qquad \text{Combine like terms.}$$

The perimeter is $18x + 9y + 6z$. ❖

EXERCISES & PROBLEMS

Communicate

1. Define an *algebraic expression*. Give two examples.

2. Discuss what is meant by the term *coefficient*. Give two examples.

3. What is the Commutative Property for Addition? Give an example.

4. Describe the steps for using the Distributive Property to add $3x + 8x$.

5. Identify which of the following are like terms. Explain why.

$$7x, 2x, 3z, 5, 7y, 3x, -z, 3y, \text{ and } 23$$

6. Discuss how you would use algebra tiles to model $(5x + 2) + (3x - 4)$.

Practice & Apply

Add.

7. $(5a - 2) + (3a - 6)$ $8a - 8$

8. $(2x + 3) + (7 - x)$ $x + 10$

9. $(4x + 5) + (x + 9y)$ $5x + 9y + 5$

10. $(1.1a + 1.2b) + (2a - 0.8b)$ $3.1a + 0.4b$

11. $\left(\frac{x}{2} + 1\right) + \left(\frac{x}{3} - 1\right)$ $\frac{5x}{6}$

12. $\left(\frac{2m}{5} + \frac{1}{2}\right) + \left(\frac{m}{10} + \frac{5}{2}\right)$ $\frac{m}{2} + 3$

13. $(2a + 3b + 5c) + (7a - 3b + 5c)$ $9a + 10c$

14. $(x + y + z) + (2w + 3y + 5)$ $x + 4y + z + 2w + 5$

15. If 6 *x*-tiles and 3 1-tiles are combined with 2 *x*-tiles and 1 1-tile, describe the result algebraically.

Photography A picture is surrounded by a frame that is 3 inches wide. Find the total height of the frame and the picture if the height of the picture is

16. 12 inches. 18 in.

17. 20 inches. 26 in.

18. *h* inches. (6 + *h*) in.

A lid that is 1 inch thick is placed on top of a box. Find the total height of the box and lid if the height of the box is

19. 10 inches. 11 in.

20. 27.5 inches. 28.5 in.

21. *x* inches. (*x* + 1) in.

22. On Tuesday, John bought 2 large boxes of cassette tapes and 3 additional tapes. The next week he bought 1 large box and 7 additional tapes. Represent the boxes and tapes with algebraic expressions, and determine the sum of the expressions. (2*b* + 3) + (*b* + 7) = 3*b* + 10

23. Each month for 7 months Don buys 2 boxes of cookies. In all, he let his friend have a total of 14 cookies. Let *b* represent the number of cookies in a box. Represent in an algebraic expression the total number of cookies that Don has for himself during the 7 months. 14*b* − 14

24. Crafts A piece of fabric is cut into 12 equal lengths with 3 inches left over. Identify the variable, and write the length of the original piece as an expression. ℓ = length of pieces; 12ℓ + 3

25. If *m* is an unknown integer, represent the next two consecutive integers in terms of *m*. *m* + 1; *m* + 2

26. If *n* is an unknown odd integer, represent the next two odd integers in terms of *n*. *n* + 2; *n* + 4

27. If apples cost 30 cents each, bananas cost 25 cents each, and plums cost 20 cents each, write an expression to find the total cost of *a* apples, *b* bananas, and *p* plums. 30*a* + 25*b* + 20*p* is the cost in cents.

Geometry Write an algebraic expression for the perimeter of each figure. Then simplify.

28.

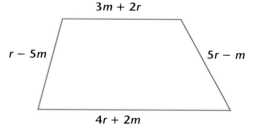

3*m* + 2*r*

r − 5*m*

5*r* − *m*

4*r* + 2*m*

29.

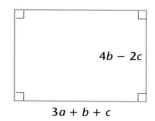

4*b* − 2*c*

3*a* + *b* + *c*

30.

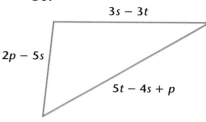

3*s* − 3*t*

2*p* − 5*s*

5*t* − 4*s* + *p*

31. 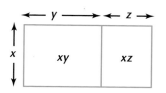 **Geometry** Write an expression for the area of the whole rectangle. Explain how the Distributive Property can be used to find this area.

Use the Distributive Property to write an equivalent expression for each of the following:

32. $3(5 + 7)$ $15 + 21$ **33.** $6(x - 3)$ $6x - 18$ **34.** $(x + y)n$ $xn + yn$

35. $(4 \cdot 3 + 4 \cdot 5)$ **36.** $4x - 4 \cdot 7$ **37.** $az + bz$ $(a + b)z$
     $4(3 + 5)$ $4(x - 7)$

38. Explain how you can use the Distributive Property to find $25 \cdot 12$.
Write $25 \cdot 12$ as $25(10 + 2)$. Then distribute 25 to get $250 + 50 = 300$.

Look Back

39. Find the next three terms in the sequence 10, 20, 40, 80, 160, $\underline{\ ?\ }$, $\underline{\ ?\ }$, $\underline{\ ?\ }$.
[Lesson 1.1]   320, 640, 1280

40. Technology Use a graphics calculator to make a table of values for the equation $w = 5h + 12$. Use the table to determine the value of h when w is 202.   **[Lesson 1.3]** 38

Simplify.

41. $12 + 3 \cdot 5$ **[Lesson 1.7]** 27   **42.** $|-12| + |-3|$ **[Lesson 2.2]** 15

43. $(-10)(-2) \div -5$ **[Lesson 2.5]** **44.** $16 \div 5\frac{1}{3}$ **[Lesson 3.5]** 3
                    -4

45. Suppose you use the method of differences on a sequence, and a constant appears in the second difference. What type of function would you expect this to be?   **[Lesson 2.7]** Quadratic

46. Find the area of a trapezoid with bases of 8 inches and 12 inches and a height of 16 inches.   **[Lesson 4.6]** 160 sq in.

47. Use the "Pythagorean" Right-Triangle Theorem to find side c if a is 7 and b is 9.   **[Lesson 4.7]** Approximately 11.4

Look Beyond

48. Find $(3a + 2b + 4p) - (5a + 3p)$. $-2a + 2b + p$

49. Draw a diagram to model an area represented by $(x + y)(x + y)$.

50. In the center of the city, the blocks are in the shape of a grid. You are at the bank at point A, and you want to walk to the post office at point B. If all the different ways you take are exactly 7 blocks long, how many different ways are there to get from A to B? 35

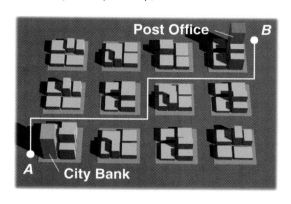

LESSON 5.2 Subtracting Expressions

why *You can subtract one integer from another by adding its opposite. The same procedure can also be used to subtract expressions.*

Bonnie bought 5 cartons of soda and 7 extra cans of soda for her party. Her guests drank 3 cartons and 4 cans of soda.

How do you write an algebraic expression that represents the total number of cans left after the party? Let x represent the number of cans in a carton.

Before the party, there were 5 cartons and 7 cans.

$$5x + 7$$

During the party, the guests drank 3 cartons and 4 cans of soda.

$$3x + 4$$

After the party, the amount left is 2 cartons and 3 cans.

$$2x + 3$$

$$(5x + 7) - (3x + 4) = 2x + 3$$

The algebraic expression $2x + 3$ represents the total number of cans left after the party.

Modeling Subtraction With Tiles

Subtraction can be modeled with tiles. Recall that a tile pair having opposite signs is sometimes referred to as a neutral pair.

Neutral pairs

EXAMPLE 1

Use algebra tiles to simplify $(2x + 3) - (4x - 1)$.

Solution ➤

Start with 2 positive x-tiles and 3 positive 1-tiles to represent $2x + 3$.

$2x + 3$

Now subtract $(4x - 1)$.

You need to subtract 4 positive x-tiles and 1 negative 1-tile. Add 2 neutral pairs of x-tiles and 1 neutral pair of 1-tiles. Now you can take away 4 positive x-tiles and 1 negative 1-tile.

$(2x + 3) - (4x - 1)$

This leaves 2 negative x-tiles and 4 positive 1-tiles.

$(2x + 3) - (4x - 1) = -2x + 4$ ❖

$-2x + 4$

Explain how to use neutral pairs to simplify $(3x - 5) - (6x + 2)$.

Adding the Opposite of an Expression

You can subtract a number from another by adding its opposite. This is also true for expressions.

DEFINITION OF SUBTRACTION FOR EXPRESSIONS
To subtract an expression, add its opposite.

When you subtract a quantity by adding the opposite, first change the sign of *each term* in the quantity that you are subtracting. Then rearrange terms to group the like terms, and add.

$(2x + 3) - (4x - 1)$	Given
$(2x + 3) + (-4x + 1)$	Definition of Subtraction
$(2x - 4x) + (3 + 1)$	Rearrange terms.
$-2x + 4$	Combine like terms.

EXAMPLE 2

Simplify.

A $(7m + 2) - (3m + 5)$ **B** $(10d - 3) - (4d + 1)$ **C** $(4x - 2) - (5x - 3)$

Solution ➤

A $(7m + 2) - (3m + 5)$ Given
$(7m + 2) + (-3m - 5)$ Definition of Subtraction
$(7m - 3m) + (2 - 5)$ Rearrange terms.
$4m - 3$ Combine like terms.

B $(10d - 3) - (4d + 1)$ Given
$(10d - 3) + (-4d - 1)$ Definition of Subtraction
$(10d - 4d) + (-3 - 1)$ Rearrange terms.
$6d - 4$ Combine like terms.

C $(4x - 2) - (5x - 3)$ Given
$(4x - 2) + (-5x + 3)$ Definition of Subtraction
$(4x - 5x) + (-2 + 3)$ Rearrange terms.
$-x + 1$ Combine like terms. ❖

Try This   Simplify $(3x - 4) - (-2x + 3)$. $5x - 7$

A negative sign in front of a variable does not necessarily mean that its *value* is negative.

EXAMPLE 3

Decide whether the expression $-W$ is positive or negative when

A W is 5.   **B** W is -7.   **C** W is 0.   **D** W is x.

Solution ➤

Substitute each value for W, and examine the value of $-W$.

A $-W = -(5) = -5$ The opposite of a positive 5 is negative 5. The value is negative.

B $-W = -(-7) = 7$ The opposite of a negative 7 is positive 7. The value is positive.

C $-W = -(0) = 0$ The value of 0 is neither positive nor negative.

D $-W = -(x)$ It is impossible to tell whether the expression is positive or negative. Since x is a variable, the value will depend on the value that x represents. ❖

CRITICAL *Thinking*

Examine the expression $-(5x + 2y)$. The negative sign means to find the opposite of *each term*. Explain why this might be considered a special case of the Distributive Property.

Try This If a is -4 and b is 3, find $-(a + b)$. 1

EXAMPLE 4

Simplify $(8x + 4y - z) - (6y + 3z - 5x)$.

Solution ➤

$(8x + 4y - z) - (6y + 3z - 5x)$	Given
$(8x + 4y - z) + (-6y - 3z + 5x)$	Definition of Subtraction
$(8x + 5x) + (4y - 6y) + (-z - 3z)$	Rearrange terms.
$13x - 2y - 4z$	Combine like terms. ❖

EXERCISES & PROBLEMS

Communicate

Tell how to represent each subtraction with algebra tiles.

1. $4x - 3x$ **2.** $(3x + 2) - (2x + 1)$ **3.** $(5x + 3) - (2x + 4)$

4. For the following problem, assign a variable to represent a ream of paper. Express the information by using algebraic expressions. Explain how to subtract the expressions. Then answer the question.

> Ms. Green had 5 reams of paper and 100 loose sheets.
> She gave Mr. Black 2 reams of paper and 50 loose sheets.
> How much did she have left?

5. Explain what is meant by "adding the opposite" when subtracting expressions.

6. Explain how to perform the following subtraction:
$$(7y + 9x + 3) - (3y + 4x - 1)$$

7. Describe the set of values for m that make $-6m$ positive.
Any negative value of m

Practice & Apply

Find the opposite of each expression.

8. 17 -17 **9.** -13 13   **10.** $2x$ $-2x$ **11.** -6 6   **12.** $9y + 2w$ $-9y - 2w$ **13.** $5a + 3b$ $-5a - 3b$

14. $2n - 3m$ $-2n + 3m$ **15.** $9c - 5d$ $-9c + 5d$ **16.** $-7x + 9$ $7x - 9$ **17.** $-7r + 4s$ $7r - 4s$ **18.** $-3p - q$ $3p + q$ **19.** $-j - k$ $j + k$

Simplify.

20. $9x - 3x$ 6x   **21.** $8y - 2y$ 6y   **22.** $5c - (3 - 2c)$ 7c $- 3$ **23.** $7d - (1 - d)$ 8d $- 1$

24. $(7r + 2s) + (9r + 3s)$ 16r $+ 5s$ **25.** $(9k + 2k) + (11j - 2j)$ 11k $+ 9j$

26. $(2a - 1) - (5a - 5)$ $-3a + 4$ **27.** $(9v - 8w) - (8v - 9w)$ v $+ w$

28. $(2x + 3) - (4x - 5) + (6x - 7)$ 4x $+ 1$ **29.** $(4y + 9) - (8y - 1) + (7 - y)$ $-5y + 17$

Tell whether each statement is true or false. Then explain why.

30. $x - (y + z) = x - y + z$

31. $x - (y - z) = x - y - z$

32. $-x + (y - z) = -x - y + z$

33. $-x$ is a negative number

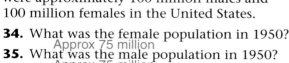 **Statistics** The graph at the right is called a stacked bar graph. It shows the population of the United States by gender. For example, it shows that in 1970 there were approximately 100 million males and 100 million females in the United States.

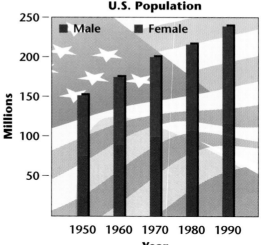

U.S. Population

34. What was the female population in 1950?
    Approx 75 million

35. What was the male population in 1950?
    Approx 75 million

36. What was the total population in 1950?
    Approx 150 million

37. What was the female population in 1980?
    Approx 110 million

38. What was the male population in 1980?
    Approx 110 million

39. What was the total population in 1980?
    Approx 220 million

Tell whether the given statement about the stacked bar graph is true or false.

40. The population rose during each 10-year period. True

41. The number of males was approximately equal to the number of females during each 10-year period. True

42. The number of females rose much faster than the number of males. False

Fund-raising Students at Valley View High School are selling fruitcakes for a fund-raiser. They earn $3 for each deluxe fruitcake and $2 for each regular fruitcake. How much would the students earn for selling

43. 89 deluxe fruitcakes and 234 regular fruitcakes? $735

44. d deluxe fruitcakes and r regular fruitcakes? $(3d + 2r)$ dollars

Suppose students earn $3.75 for each deluxe fruitcake and $2.50 for each regular fruitcake. How much would they earn for selling

45. 89 deluxe fruitcakes and 234 regular fruitcakes? $918.75

46. d deluxe fruitcakes and r regular fruitcakes? $(3.75d + 2.50r)$ dollars

47. How much did Joni and Wang earn together from selling fruitcakes? $185

48. Who earned more, Joni or Wang? How much more? Wang; $2.50 more

Fund Raiser Records

Fruitcake	Deluxe	Regular
	$3.75	$2.50
Joni	9	23
Wang	7	27

49. Inventory The school kitchen had 11 cases of canned juice plus 3 extra cans of juice. After lunch they had 6 cases of canned juice and no extra cans. How much juice was distributed during lunch? 5 cases and 3 cans

Cultural Connection: Africa The idea of adding and subtracting numbers has been around for thousands of years. Early Egyptians represented addition by showing feet walking toward the number, and they represented subtraction by showing feet walking away from the number.

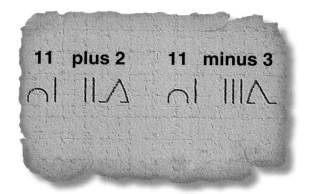

50. How would you represent 11 + 5 using the early Egyptian notation if 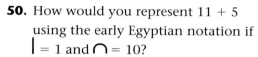 = 1 and ∩ = 10?

51. Represent 23 − 12 using the early Egyptian notation.

Look Back

52. Write an algebraic equation for the following verbal expression: The total charge was 16 times the number of hours. **[Lesson 1.2]** $c = 16h$

53. Write a list of factors for 72. **[Lesson 1.5]** 1, 2, 3, 4, 6, 8, 9, 12, 18, 24, 36, 72

54. Find the prime factorization of 150. Write the answer in exponential form. **[Lesson 1.6]** $2 \cdot 3 \cdot 5^2$

Place parentheses and brackets to make each equation true. [Lesson 1.7]

55. $28 \div 2 - 4 \cdot 1 = 10$ **56.** $16 \div (5 + 3) \div 2 = 1$ **57.** $40 \cdot (2 + 10 \cdot 4) = 1680$

58. The first reading on a gauge is 100. The changes recorded at 1-hour intervals are $-4, +51, 0, +7, -12, -78, +2, -13, -1$. What is the final reading on the gauge? **[Lesson 2.2]** 52

59. Use guess-and-check to solve the equation $x - 7 = -3$. **[Lessons 2.6]** 4

60. Find the area of a triangle with a base of 15 centimeters and a height of 21 centimeters. **[Lesson 4.6]** 157.5 sq cm

Look Beyond

61. **Statistics** Fred has test scores of 87, 74, and 90. How many points does he need on the next test to have an average of 85? 89

62. Paul and Dan leave from one end of a lake, and sail around the lake in the same direction. Paul sails at 12 miles per hour, while Dan sails at 10 miles per hour. The lake is 2 miles around, and they leave at noon. When will the two first meet again at the starting point? 1:00 P.M.

LESSON 5.3

Exploring Addition and Subtraction Equations

why *You have solved equations by the guess-and-check method and by graphing. It is important to develop a technique for solving equations more easily. Algebra tiles can be used to model this technique.*

Morton Bastrop
2   5

How many goals does Morton need to tie Bastrop? An equation can be used to model this problem. If x represents the number of goals needed, the problem can be modeled by the equation below.

$$x + 2 = 5$$

Algebra tiles can be used to solve this equation.

•Exploration 1 *Solving Addition Equations Using Neutral Pairs*

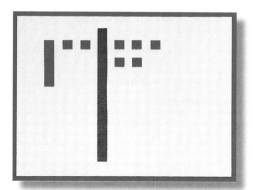

1 Use the diagram at the left as a guide. To solve the equation $x + 2 = 5$, you need to isolate the x-tile. This can be done if the two positive 1-tiles on the left side are eliminated. What tiles should you add to each side of the strip to form neutral pairs on the left side? **2 negative 1-tiles**

2 Sketch a picture of tiles to model $x + 2 + (-2) = 5 + (-2)$.

3 Remove the neutral pairs. Sketch a picture of the result.

4 Simplify the equation $x + 2 + (-2) = 5 + (-2)$. $x = 3$

5 What is the solution to the equation $x + 2 = 5$? ❖ 3

You can check your solution to the equation in Exploration 1, $x + 2 = 5$, by substituting tiles or by substituting numbers.

Substitute tiles.	**Substitute numbers.**
Replace the x-tile with your solution, 3 positive 1-tiles.	Rewrite the equation by substituting the solution, 3, for x.

$$(\boldsymbol{x}) + 2 = 5$$
$$(\boldsymbol{3}) + 2 = 5$$
$$5 = 5$$

Each side represents the number 5.

You solved the equation $x + 2 = 5$ by forming neutral pairs of tiles. You can also solve the same equation by *taking away* tiles.

•Exploration 2 Solving Addition Equations Using Subtraction

1 Use algebra tiles to set up the equation $x + 2 = 5$.

2 Isolate the x-tile by taking away all of the positive 1-tiles on the left side and the same number of 1-tiles on the right side. What is the result? 1 positive x-tile on the left side and 3 positive 1-tiles on the right side.

3 Isolate the x-variable in the equation $x + 2 = 5$ by subtracting 2 from each side of the equation. What is the result? $x = 3$

4 Explain why $x + 2 + (-2) = 5 + (-2)$ is the same as $x + 2 - 2 = 5 - 2$. ❖ Definition of Subtraction

CRITICAL Thinking

A table of values for the equation $y = x + 2$ is shown on a graphics calculator. How can the table be used to check your solution to the equation in Explorations 1 and 2? Find the y-value of 5. The corresponding x-value of 3 is the solution.

X	Y₁
−1	1
0	2
1	3
2	4
3	**5**
4	6
5	7
Y₁=5	

Tiles can be used to solve the equation $x + 4 = 2$. Set up the tiles to model the equation. In the steps that follow, describe how the tiles are used and why.

Use tiles to set up the equation $x + 4 = 2$.

Form neutral pairs by adding 4 negative 1-tiles to the left side and to the right side.

Remove neutral pairs.

The result is 1 x-tile on the left side and 2 negative 1-tiles on the right side.

Therefore, the solution to the equation $x + 4 = 2$ is -2. ❖

•Exploration 3 Solving Subtraction Equations Using Tiles

 Use algebra tiles to set up the equation $x - 3 = 4$.

How many positive 1-tiles must be added to each side to isolate the x-tile? 3

 Simplify the model by removing neutral pairs. What is the solution?
1 positive x-tile on the left and 7 positive 1-tiles on the right, or $x = 7$

 Write the changed equation. $x - 3 + \underline{?} = 4 + \underline{?}$
$x - 3 + 3 = 4 + 3$

 Check your solution. ❖ $x - 3 = 4$
$7 - 3 = 4$
$4 = 4$ True

To solve $x - 2 = -3$, use tiles to model the equation.

Form neutral pairs by adding 2 positive 1-tiles to each side.

Remove neutral pairs.

The result is 1 x-tile on the left side and
1 negative 1-tile on the right side.

Therefore, the solution to the equation $x - 2 = -3$ is -1. ❖

Describe two ways to check this solution.

EXERCISES & PROBLEMS

Communicate

1. Explain how to isolate the x-tile when solving an equation, and give examples.

2. Discuss how neutral pairs are used to solve equations.

3. Describe how to solve $x - 3 = 7$ in two different ways using tiles.

4. Describe how to solve the equation $x + 4 = -2$ using tiles.

5. Describe how to check the solution to the equation $x - 3 = 5$ using tiles.

6. Explain why the equation $x - 4 = 7$ can be rewritten as $x + (-4) = 7$.
   Definition of Subtraction

Practice & Apply

Write the equation modeled by each set of tiles.

7.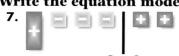
   $x - 3 = 2$

8.
   $x + 3 = -5$

9.
   $-6 = 2 + x$

10.
    $2 + z = -3$

Solve each equation using algebra tiles. Sketch a picture of each step, and write the solution.

11. $x - 4 = -5$

$x = -1$

12. $x + 2 = -3$

$x = -5$

13. $3 = -4 + x$

$7 = x$

14. $5 = x + 7$

$-2 = x$

15. $4 + x = -4$

$x = -8$

16. $-8 = 5 + x$

$-13 = x$

17. $5 + x = -3$

$x = -8$

18. $x - 3 = -8$

$x = -5$

Solve each equation.

19. $x + 3 = 4$ 1

20. $t - 5 = -2$ 3

21. $x + 5 = -2$ −7

22. $3 = x + 2$ 1

23. $-3 = y - 6$ 3

24. $-4 = x - 2$ −2

25. $x - 3 = -3$ 0

26. $x - 2 = -2$ 0

27. $m + 6 = -1$ −7

28. $a + 4 = -3$ −7

29. $-5 = y + 2$ −7

30. $-3 = w + 3$ −6

31. $x + 7 = 2$ −5

32. $-4 = n + 1$ −5

33. $y - 8 = 10$ 18

34. $h - 65 = 65$ 130

35. $b + 10 = -6$ −16

36. $5 + t = -7$ −12

37. $9 = v - 3$ 12

38. $d - 5 = 11$ 16

39. $5 = -8 - x$ −13

40. $y - 4 = -5$ −1

41. $3 = p + (-2)$ 5

42. $(-5) + t = 9$ 14

43. **Geometry** Write an equation to show that the perimeter, P, of this rectangle is 52. $21 + w + 21 + w = 52$

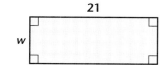

The airspeed, a, of this biplane is indicated by the speedometer below.

Aviation The groundspeed, g, of an airplane can be determined by subtracting the head-wind speed, h, from the plane's airspeed, a. The equation is $g = a - h$. Use this equation for Exercises 44–46.

44. Find the groundspeed, g, of this biplane if it is flying into a head wind, h, of 30 miles per hour. 150 mph

45. Flight 461 is flying into a head wind, h, of 94.7 miles per hour. The groundspeed, g, is 475.2 miles per hour. Find the airspeed of Flight 461. 569.9 mph

46. Flight 386 has a groundspeed, g, of 525.5 miles per hour. The plane is flying into a head wind, h, of 72 miles per hour. What is the airspeed of Flight 386? 597.5 mph

Look Back

Write two inequalities for each pair of integers. Use both the < and > symbols.   [Lessons 2.1, 2.3]

47. 12, 18 $12 < 18; 18 > 12$ **48.** 16, 10 $16 > 10; 10 < 16$ **49.** 9, 0 $9 > 0; 0 < 9$ **50.** -3, 0 $-3 < 0; 0 > -3$

51. $-8, -12$ $-8 > -12; -12 < -8$ **52.** $-17, -8$ $-17 < -8; -8 > -17$ **53.** $-15, 7$ $-15 < 7; 7 > -15$ **54.** 18, -4 $18 > -4; -4 < 18$

Simplify each expression.   [Lessons 2.2, 2.4, 2.5]

55. $(-5)(3) + (-2)(-3)$ -9 **56.** $-24 \div -12$ 2   **57.** $(-8)^2 + (-4)^2$ 80   **58.** $[(-8) + (-4)]^2$ 144

59. The triangles below are similar. Find the missing side lengths.   **[Lesson 4.8]**

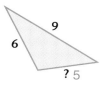
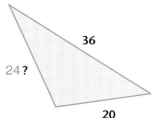

Add or subtract. [Lessons 5.1, 5.2]

60. $(2x - 7) + (9x - 6)$ $11x - 13$ **61.** $(4x - 3) - (5 - x)$ $5x - 8$

Look Beyond

62. Describe how you could use tiles to solve the equation $4x - 2 = 3x + 1$.

Addition and Subtraction Equations

why *Many problems can be modeled algebraically by writing an equation. A sequence of steps that isolates the unknown on one side of the equal sign provides the solution.*

How many degrees must the temperature shown above rise to reach the freezing point, 32°F?

If Riva has $20, how much more money does she need to buy the boots?

Although the two situations are different, each can be modeled by the same equation, $x + 20 = 32$.

Recall that tiles can be used to demonstrate the steps in solving the equation. Suppose the x-tile represents an unknown number of 1-tiles. The x-tile and the 1-tiles represent the equation $x + 20 = 32$.

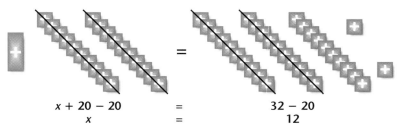

$$x + 20 - 20 = 32 - 20$$
$$x = 12$$

After you take away 20 tiles from each side, the amounts on each side of the equal sign remain equal. The x-tile is now alone on one side of the equal sign. Count the remaining tiles on the other side of the equal sign to find the number of 1-tiles that the x-tile represents.

Subtracting the same number of tiles from each side of the equation models the Subtraction Property of Equality.

CRITICAL *Thinking*

How can you use a model to represent the equation $3x + 4 = 2x + 10$?

SUBTRACTION PROPERTY OF EQUALITY

If equal amounts are subtracted from the expressions on each side of an equation, the expressions remain equal.

EXAMPLE 1

Fund-raising

The school band needs new uniforms. Write and solve an equation to find how much more money the band must raise.

Solution ➤

Organize the information.
Examine the problem. Identify the known and unknown information. Then write a sentence using words and operation symbols.

(the amount still to be raised) + (the amount they already have) = (the goal)

From the banner, you can see that the amount they have already raised is $2344.10 and the goal is $5000.00.

Let x represent the amount that is still to be raised.

Write the equation.
Replace the words you wrote in the first step with the variable and the known information from the problem.

$$x + 2344.10 = 5000.00$$

Solve the equation.
Use the Subtraction Property of Equality.

$$x + 2344.10 - 2344.10 = 5000.00 - 2344.10$$
$$x = 2655.90$$

The equation is solved when the variable is isolated on one side of the equation.

Check the answer.
Replace the variable x in the original equation with the value you found, and see if it makes a true statement. Be sure that the solution is reasonable and that it answers the question in the problem.

$$2655.90 + 2344.10 \stackrel{?}{=} 5000.00$$
$$5000.00 = 5000.00 \quad \text{True}$$

The band must raise $2655.90 more to reach the goal. ❖

Solving Subtraction Equations

Some equations contain subtraction. To solve $x - 2 = 8$, use a procedure similar to that used for equations containing addition.

Represent the equation with tiles. To solve the equation, add 2 positive tiles to each side of the equation, and remove neutral pairs. The unknown value is 10.

Equation	Solution

$$x - 2 = 8 \qquad\qquad x - 2 + 2 = 8 + 2$$
$$x = 10$$

$$x = 9$$

Solve with tiles. **a.** $x - 5 = 4$ **b.** $x - (-3) = 5$ $x = 2$

ADDITION PROPERTY OF EQUALITY
If equal amounts are added to the expressions on each side of an equation, the expressions remain equal.

In this dart game, the range of scores was 47 points, and the lowest score was 52.

EXAMPLE 2

Write and solve an equation to find the highest score in this dart game.

Solution ➤

Organize.
In statistics, the *range* is the difference between the highest and lowest scores.

range = highest score − lowest score

Let H represent the highest score.

Write.
Replace the words in the range formula with the variable, H, and the known values from the problem.

range = highest score − lowest score
$$47 = H - 52$$

STATISTICS
Connection

Solve.

$47 = H - 52$	Given
$47 + 52 = H - 52 + 52$	Addition Property of Equality
$99 = H$	Simplify.

Check.
Substitute 99 for H.

Since $47 = 99 - 52$, the highest score is 99. ❖

Literal Equations

Scientists often express numerical relationships as formulas. For example, $C = \frac{5}{9}(F - 32)$ relates Fahrenheit and Celsius temperatures. Because formulas often contain different letters that represent variables, they are called **literal equations**.

EXAMPLE 3

Investments The formula $A = P + I$ shows that the total amount, A, of money you receive from an investment equals the principal, P, (the money you started with) plus the interest, I. Write a formula for the interest, I, based on the principal, P, and the amount, A.

Solution ➤

To isolate I, subtract P from *both sides* of the equation.

$A = P + I$	Given
$A - P = P + I - P$	Subtraction Property of Equality
$A - P = I$	Group like terms and simplify.
$I = A - P$	Rewrite with I on the left. ❖

Literal equations that contain *subtraction* are treated in much the same way as equations that contain *addition*. In business, the equation $P = s - c$ is a formula for profit, where s represents the selling price and c represents the cost.

EXAMPLE 4

Solve the equation $P = s - c$ for s.

Solution ➤

To solve the equation for s, add c to both sides of the equation.

$P = s - c$	Given
$P + c = s - c + c$	Addition Property of Equality
$P + c = s$	Simplify.
$s = P + c$	Rewrite with s on the left. ❖

Explain how you would solve $a = x - b$ for x.

Try This   The figure shows the sector of a circle with center O and radius R. The formula for the length h is $h = R - k$. Solve this equation for k.

GEOMETRY
Connection

$k = R - h$

Programs to solve equations on a computer or a calculator already exist. For example, to solve the equation $23 = C + 14$ using the Maple™ software program, type in solve($23 = C + 14$);. The computer will display 9.

Computer

To solve the equation $5 - 2r = 7 - (3r - 8)$, type in solve($5 - 2 * r = 7 - (3 * r - 8)$);. The computer will display 10. Note that $2 * r$ is entered for $2r$. What do you think will result if you type in solve($4 + 3 * t = 6 + 4 * t$);? The computer will display -2.

EXERCISES & PROBLEMS

Communicate

1. Explain when to use the Addition Property of Equality and when to use the Subtraction Property of Equality.

2. Give a situation that can be modeled by
   a. $x + 20 = 50$.   **b.** $x - 20 = 50$.

3. Explain how to solve each equation using algebra tiles.
   a. $x + 6 = 10$ **b.** $x - 6 = 10$

4. Explain how to solve $s + x = r$ for s.

5. Explain how to solve $m + b = n$ for m.

Practice & Apply

Solve each equation. You may use algebra tiles.

6. $x + 9 = 6$ -3 **7.** $x - 7 = 3$ 10   **8.** $x - 10 = -4$ 6   **9.** $x + 6 = -4$ -10

State which property you would use to solve each equation. Then solve.

10. $a - 16 = 15$ Add; 31   **11.** $t + 29 = 11$ Subt; -18 **12.** $m + 54 = 36$ Subt; -18 **13.** $r - 10 = -80$ Add; -70

14. $l - 27 = 148$ Add; 175   **15.** $b - 109 = 58$ Add; 167   **16.** $y + 37 = -110$ Subt; -147 **17.** $396 = z + 256$ Subt; 140

18. $x + \frac{3}{4} = \frac{5}{4}$ Subt; $\frac{1}{2}$ **19.** $7.4 + t = 5.2$ Subt; -2.2 **20.** $r + 5.78 = 7$ Subt; 1.22   **21.** $\frac{3}{8} - x = \frac{3}{4}$ Add; Subt; $-\frac{3}{8}$

Solve each equation for a.

22. $a + b = c$ $a = c - b$ **23.** $a - b = c$ $a = c + b$ **24.** $a + b = -c$ $a = -c - b$ **25.** $a - b = -c$ $a = -c + b$

26. If $a - \frac{2}{3} = 4$, what is the value of $3a$?   14

27. If $2.5 + s = 5.3$, what is the value of $2s - 3$?   2.6

Exercises 28–30 refer to the chart.

28. Solve equation a. $x = 61$

29. Solve equation b. $x = 131$

30. Compare the steps in solving the two equations.

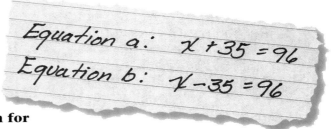

Equation a: $x + 35 = 96$

Equation b: $x - 35 = 96$

For Exercises 31–41, write an equation for each situation, and solve the equation.

31. Sports The first- and second-string running backs on a football team ran for a total of 94 yards. If the first-string back gained 89 yards, how many yards did the second-string back gain or lose? $89 + y = 94; y = 5; 5$ yd gain

32. The first- and second-string running backs on a football team ran for a total of 89 yards. If the first-string running back gained 94 yards, how many yards did the second-string running back gain or lose? $94 + y = 89; y = -5; 5$ yd loss

33. The first-string running back on a football team ran for a total of 94 yards. If the second-string running back on a football team ran for a total of 89 yards, what was their combined yardage? $94 + 89 = y; y = 183; 183$ combined

34. Carl sees on the calendar that his birthday, December 21, is the 355th day of the year and that the current day, October 15, is the 288th day of the year. In how many days is his birthday? $288 + d = 355; d = 67; 67$ more days

35. Cooking A recipe for turkey gravy says you should add water to the drippings to get $1\frac{1}{2}$ cups of liquid. If there is $\frac{7}{8}$ of a cup of drippings, how much water should be added? $\frac{7}{8} + c = \frac{3}{2}; c = \frac{5}{8}; \frac{5}{8}$ cup water

$m + 149 = 23{,}580; m = 23{,}431; 23{,}431$ miles

36. Travel If the odometer on Sarah's car at registered 23,580 when she finished   start a vacation trip of 149 miles, what did it read when she started the trip?

37. If the odometer on Elizabeth's car registered 23,580 when she started a trip of 149 miles, what did it read when she finished her trip?
$23{,}580 + 149 = m; m = 23{,}729; 23{,}729$ mi at finish

38. Sales Tax Sam bought two items costing $12.98 and $14.95 plus tax. He got 39¢ change from $30. How much was the tax?
$12.98 + 14.95 + t = 30 - 0.39; t = 1.68; \$1.68$ tax

39. **Statistics** The range of a set of scores was 28. If the highest score was 47, what was the lowest score?
$28 = 47 - \ell; \ell = 19;$ lowest score was 19.

40. **Geometry** Supplementary angles are pairs of angles with measures that total 180°. Find the measure of an angle whose supplement is an angle with a measure of 92 degrees. $92 + a = 180; a = 88; 88°$

41. Two of the angles of a triangle each measure 50 degrees. Find the measure of the third angle. (HINT: The sum of the measures of all the angles of a triangle is 180 degrees.) $50 + 50 + a = 180; a = 80; 80°$

42. 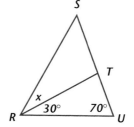 **Geometry** The angles *SRU* and *SUR* of triangle *RSU* each measure 70°. Write and solve an equation to find the measure of angle *SRT*. $30 + x = 70; x = 40; 40°$

43. What would the measure of angle *SRT* be in triangle *SRU* if the measure of angle *TRU* were 18°?
$18 + x = 70; x = 52; 52°$

 **Geometry**

44. Find *x* if *S* lies on line *RT*.
$148°$

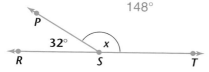

45. Find *y* if angle *UVW* is a right angle.
$62°$

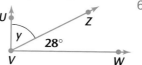

Technology Some computer programs can solve equations, as shown in the following examples. The line after > is what you type in. The next line is the answer that the computer displays when you press ENTER.

> solve(x + 98765432123456789 = 444444444444444444);

345679012320987655

> solve(A = P + I,I);

A − P

Give commands to solve the following equations.

46. $x + 12345.6789 = 55555555$
>solve(x + 12345.6789 = 55555555)

47. $A + B = C + D$ (Solve for *D*.)
>solve(A + B = C + D,D)

48. **Portfolio Activity** Complete the problem in the portfolio activity on page 273.

Look Back

49. State the pattern and find the next three numbers in the sequence 6, 0, 12, 6, 18, 12, _?_, _?_, _?_. **[Lesson 1.1]**

50. If poster board costs 39¢ for each piece, how many pieces of poster board can Shannon buy with $2.50? **[Lesson 1.4]** 6

51. Order the rational numbers from least to greatest. **[Lesson 3.1]**
$\frac{1}{8}, \frac{1}{4}, \frac{1}{2}, \frac{5}{8}, \frac{3}{4}$ $\frac{5}{8} \quad \frac{1}{4} \quad \frac{1}{8} \quad \frac{3}{4} \quad \frac{1}{2}$

52. Find the perimeter of a square with sides 2.5 inches long. **[Lesson 4.5]**
10 in.

Simplify. [Lessons 5.1, 5.2]

53. $(3x − 7) − (x − 4)$
$2x − 3$

54. $2(8 − 7y) + 5(2y)$
$−4y + 16$

55. $−4(n − 2) − (−3n + 4)$
$−n + 4$

Look Beyond

Solve each equation using algebra tiles.

56. $x − 3 = 8$ $x = 11$ **57.** $2x = 12$ $x = 6$ **58.** $2x + 5 = 11$ $x = 3$

Exploring Polynomials

Why *You have seen how to solve real-world problems by solving algebraic equations. Algebraic equations contain algebraic expressions that are also called polynomials. Solving equations usually involves operations with polynomials.*

monomials

constants

variables

-8 $\frac{3}{4}a^2bc$

$5x$ y $2c^2$

4 -2 $\frac{1}{3}$

y a b c

Recall that in mathematics, a numeral, such as 4, -2, or $-\frac{1}{3}$, is called a constant. A **monomial** is an algebraic expression that is either a constant, a variable, or a product of a constant and one or more variables.

Like Terms

Monomials are also called terms.

Like terms are two or more terms that contain the same variable(s). If the terms differ by at least one variable they are called **unlike terms**.

Like terms	Unlike terms
$3x$ and $2x$	$3y$ and $2x$
-8 and 1	-8 and $5s$
$5a^2c$ and $-4a^2c$	$5ac^2$ and $-4a^2c$

A **polynomial** is a term or a sum or difference of terms. A polynomial with two terms is called a **binomial**. A polynomial with three terms is called a **trinomial**.

Monomials	Binomials	Trinomials
$2x$	$3x + 2$	$2x^2 + 4x - 1$
$\frac{2}{3}x^2$	$4x - \frac{1}{3}$	$2a - 4b + 5$
$0.8a^2b^3$	$-5z + \frac{2}{7}x$	$5 - 4x^2y + 3xy^2$

You can use algebra tiles to represent polynomials. In Lessons 5.1 and 5.2, you used algebra tiles to add and subtract expressions, or polynomials.

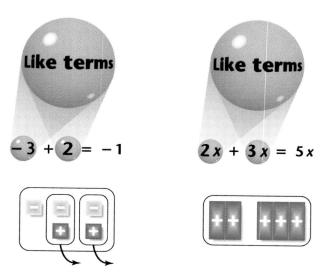

•Exploration 1 *Checking by Substitution*

Part I

1. Substitute a value for x to show that the statement $2x + 3x = 5x$ is true. Answers may vary. $2(2) + 3(2) = 5(2); 10 = 10$ True

2. Explain why this tile model shows the difference $2x - 3x$.

3. Use tiles to simplify the polynomial $2x - 3x$. Complete the following algebraic equation:

$$2x - 3x = \underline{\ ?\ }$$

Check your answer by substituting -3, 4, and 10 for x to show that the statement is true.

4. Describe how to simplify the polynomial $-2x + 5x - x$. How can you check your answer?

5. Use tiles to show that $x + 3$ does not equal $3x$. Check by substituting three different numbers for x in the inequality $x + 3 \neq 3x$.

6. Substitute values for a and b to show that $a + b \neq ab$.

7. Substitute values for a and b to show that $2a + 5b \neq 7ab$.

Part II

  You can also use an x^2-tile to model terms with x^2. Use tiles to simplify the polynomial $-3x^2 + 2x^2$, and complete the following algebraic equation. $-3x^2 + 2x^2 = -x^2$

$$-3x^2 + 2x^2 = -x^2$$
$$-3(10)^2 + 2(10)^2 = -(10)^2$$
$$-300 + 200 = -100$$
$$-100 = -100 \ \text{True}$$

$$-3x^2 + 2x^2 = -x^2$$
$$-3(-10)^2 + 2(-10)^2 = -(-10)^2$$
$$-300 + 200 = -100$$
$$-100 = -100 \ \text{True}$$

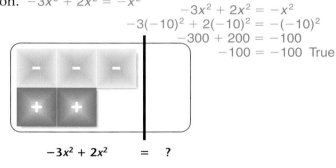

$$-3x^2 + 2x^2 \quad = \quad ?$$

Check your answer by substituting 10 and -10 for x to show that the statement is true.

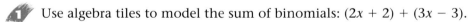 Describe how to simplify the polynomial $-2x^2 + 5x^2 - x^2$. How can you check your answer?

 Substitute values for a and b to show that $(a + b)^2 \neq a^2 + b^2$. ❖

Answers may vary. Let $a = 1$ and $b = 2$.
$$(a + b)^2 \neq a^2 + b^2$$
$$(1 + 2)^2 \neq 1^2 + 2^2$$
$$9 \neq 5 \ \text{True}$$

Exploration 2 *Binomial Addition*

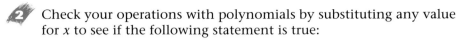 Use algebra tiles to model the sum of binomials: $(2x + 2) + (3x - 3)$.

$$(2x + 2) \quad + \quad (3x - 3) \qquad\qquad 5x - 1$$

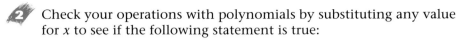 Check your operations with polynomials by substituting any value for x to see if the following statement is true:

$$(2x + 2) + (3x - 3) = 5x - 1$$

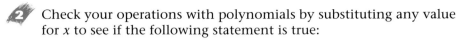 Substitute another value for x to show that the following statement is true:

$$(2x + 2) + (3x - 3) = 5x - 1$$

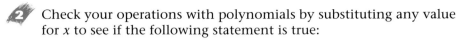 Use tiles to simplify the polynomial expression $(-3x - 1) + (2x + 5)$, and complete the following algebraic equation:

$$(-3x - 1) + (2x + 5) = \underline{\ ?\ }$$

Check your answer by substituting 10 for x to show that the statement is true.

5 Recall from Lesson 5.2 the Definition of Subtraction for Expressions, which states that to subtract expressions, you add the opposite. Use this definition and algebra tiles to simplify the polynomial expression $2x - (3x - 1)$, and complete the following equation:

$$2x - (3x - 1) = \underline{\ ?\ }$$

Check your answer by substituting -3, 4, and 10 for x to show that the statement is true.

6 Write a binomial addition expression that can be represented by these tiles. $(2x^2 - 3x) + (-3x^2 + 2x)$

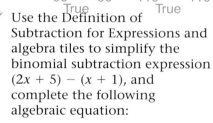

Simplify the expression and write an algebraic equation.

$$\underline{\ ?\ } = \underline{\ ?\ }$$
$$(2x^2 - 3x) + (-3x^2 + 2x) = -x^2 - x$$

Check your answer by substituting -10 and 10 for x to show that your statement is true.
$$-90 = -90 \qquad -110 = -110$$
$$\text{True} \qquad\qquad \text{True}$$

7 Use the Definition of Subtraction for Expressions and algebra tiles to simplify the binomial subtraction expression $(2x + 5) - (x + 1)$, and complete the following algebraic equation:

$$(2x + 5) - (x + 1) = \underline{\ ?\ }$$
$$x + 4$$

Check your answer by substituting 10 and 15 for x to show that the statement is true.
$$14 = 14 \quad \text{True} \qquad 19 = 19 \quad \text{True}$$

8 Explain how to simplify the polynomial addition expression $(2x^2 - x) + (-5x^2 - x)$. Explain how to simplify the polynomial subtraction expression $(2x^2 + x) - (5x^2 - x)$. Explain how to check your answers. ❖

In Exploration 2 you developed a method for adding and subtracting binomials.

EXTENSION

Use the Definition of Subtraction for Expressions to simplify the expression $(2x^2 + x) - (5x^2 + x)$.

$$(2x^2 + x) - (5x^2 + x) = (2x^2 + x) + (-5x^2 - x)$$
$$= 2x^2 + (-5x^2) + x - x$$
$$= -3x^2$$

So $(2x^2 + x) - (5x^2 + x) = -3x^2$. Substitute 10 for x to check your answer.

$$(2x^2 + x) - (5x^2 + x) = -3x^2$$
$$[2(10)^2 + 10] - [5(10)^2 + 10] \stackrel{?}{=} -3(10)^2$$
$$(200 + 10) - (500 + 10) \stackrel{?}{=} -300$$
$$210 - 510 \stackrel{?}{=} -300$$
$$-300 = -300$$

Graphics Calculator

You can use a calculator to help with the calculations when you check your answer. On a graphics calculator, you can see the numbers that you entered into the calculator after you get the result.

```
((2(10)²+10)−(5(10)²+10)
                     −300
−3(10)²
                     −300
```

You can also use a graphics calculator to check the equation $(2x^2 + x) - (5x^2 + x) = -3x^2$ by building a table of values. Enter the left side of the equation for Y_1 and the right side of the equation for Y_2. The Y_1 and Y_2 columns in the table should have the same corresponding values. ❖

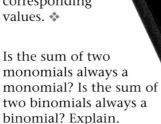

CRITICAL Thinking

Is the sum of two monomials always a monomial? Is the sum of two binomials always a binomial? Explain.

No, $3x + 2y = 3x + 2y$, a binomial
No, $(4x + 2y) + (3z + 5y)$
$= 4x + 7y + 3z$, a trinomial

EXERCISES & PROBLEMS

Communicate

1. What is the difference between a monomial, a binomial, and a trinomial?

2. Give two examples of combining like terms to simplify polynomials.

3. Explain how you can use the Definition of Subtraction for Expressions to subtract binomials. Give two examples.

4. Explain how to check your addition and subtraction by substitution. Substitute different values of x (the variable) back into the equation to see that it is true.

Practice & Apply

Simplify each polynomial by combining like terms. Check by substituting two different values for x. If the polynomial is already simplified, write *simplified*.

5. $2x - 3x - 4$ $-x - 4$

6. $1 + 3x - 1$ $3x$

7. $4x + 3 - 2x$ $2x + 3$

8. $2x^2 - 5 - x^2$ $x^2 - 5$

9. $-3b + b + 1$ $-2b + 1$

10. $5a + 6 - 2a$ $3a + 6$

11. $x^2 + 2x^2 - 3x$ $3x^2 - 3x$

12. $-x^2 - 3x^2 + 2x$ $-4x^2 + 2x$

13. $2x^2 + 3x - x^2$ $x^2 + 3x$

14. $2x^2 - x - 4x$ $2x^2 - 5x$

15. $-4x + 3 + 2x^2 - x$

16. $-3x^2 + 1 + 3x - 1$ $-3x^2 + 3x$

17. $4x + 3 - 2x^2 - 5$

18. $2x^2 - 5 - x^2 + 4$ $x^2 - 1$

19. $-3a + 8 - 2b$ Simplified

20. $5a^2 + 6 - 2a$ Simplified

21. $x^2 + 3x - 2x^2 + 4x$

22. $2x^2 - (-4x + 3)$ $2x^2 + 4x - 3$

23. $2x^2 + (-x^2 - 2)$ $x^2 - 2$

24. $-x^2 - x - (-3x^2 - x)$ $2x^2$

15. $2x^2 - 5x + 3$

17. $-2x^2 + 4x - 2$

21. $-x^2 + 7x$

For each tile model, write the sum of two binomials that it represents. Then find the sum.

25.

$(-2x - 3) + (2x + 2) = -1$

26.

$(x^2 - 2) + (3x^2 + 1) = 4x^2 - 1$

27.

$(-x^2 + 1) + (x^2 - 1) = 0$

28.

$(x^2 + 2x) + (2x^2 + 3x) = 3x^2 + 5x$

29.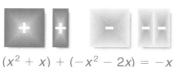

$(x^2 + x) + (-x^2 - 2x) = -x$

30.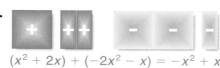

$(x^2 + 2x) + (-2x^2 - x) = -x^2 + x$

31. Check your answer for Exercise 25 by substituting 3 for x. $-6 - 3 + 6 + 2 = -1; -1 = -1$ True

32. Check your answer for Exercise 26 by substituting -3 for x.
$9 - 2 + 27 + 1 = 36 - 1; 35 = 35$ True

33. Check your answer for Exercise 27 by substituting 10 for x. $-100 + 1 + 100 - 1 = 0; 0 = 0$ True

34. Check your answer for Exercise 28 by substituting -5 for x.
$25 - 10 + 50 - 15 = 75 - 25; 50 = 50$ True

35. Check your answer for Exercise 29 by substituting 5 for x.
$25 + 5 + (-25) - 10 = -5; -5 = -5$ True

36. Check your answer for Exercise 30 by substituting 7 for x.
$49 + 14 + (-98) - 7 = -49 + 7; -42 = -42$ True

37. **Technology** Check the sum in Exercise 29 by using a graphics calculator to build a table of values. Which row of your table corresponds to your answer for Exercise 35?

38. **Technology** Check the sum in Exercise 30 by using a graphics calculator to build a table of values. Which row of your table corresponds to your answer for Exercise 36?

Simplify each binomial addition expression. Check by substituting two different values for x.

39. $(2x + 5) + (3x + 4)$ $5x + 9$ **40.** $(-3x + 1) + (-3x + 1)$ $-6x + 2$

41. $(4x + 3) + (2x - 5)$ $6x - 2$ **42.** $(2x^2 - 5) + (x^2 + 4)$ $3x^2 - 1$

43. $(-3b + 8) + (2b + 1)$ $-b + 9$ **44.** $(5a + 6) + (-2a - 4)$ $3a + 2$

45. $(x^2 + 3x) + (-2x^2 + 4x)$ $-x^2 + 7x$ **46.** $(-x^2 - x) + (-3x^2 - 2x)$ $-4x^2 - 3x$

47. $(2x^2 + 3x) + (-x^2 - 2)$ $x^2 + 3x - 2$ **48.** $(2x^2 - x) + (-4x + 3)$ $2x^2 - 5x + 3$

Simplify each binomial subtraction expression. Check by substituting two different values for x.

49. $(2x + 5) - (3x + 4)$ $-x + 1$ **50.** $(-3x + 1) - (-3x + 1)$ 0

51. $(4x + 3) - (2x - 5)$ $2x + 8$ **52.** $(2x^2 - 5) - (x^2 + 4)$ $x^2 - 9$

53. $(-3b + 8) - (2b + 1)$ $-5b + 7$ **54.** $(5a + 6) - (-2a - 4)$ $7a + 10$

55. $(x^2 + 3x) - (-2x^2 + 4x)$ $3x^2 - x$ **56.** $(-x^2 - x) - (-3x^2 - 2x)$ $2x^2 + x$

57. $(2x^2 + 3x) - (-x^2 - 2)$ $3x^2 + 3x + 2$ **58.** $(2x^2 - x) - (-4x + 3)$ $2x^2 + 3x - 3$

59. 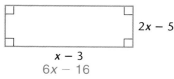 **Geometry** If x is 4 units, which of the following is greater: the perimeter of a square whose sides are each $(3x - 1)$ units or the perimeter of a rectangle whose length is $(x + 9)$ units and whose width is $(2x - 3)$ units? perimeter of the square

For Exercises 60–63, the lengths of sides are given. Write a simplified expression for the perimeter of each figure.

60.

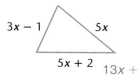

2x − 5

x − 3

6x − 16

61.

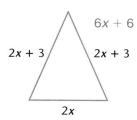

6x + 6

2x + 3   2x + 3

2x

62.

3x − 1   5x

5x + 2

13x + 1

63.

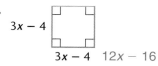

3x − 4

3x − 4   12x − 16

Look Back

Write the prime factorization of each number.   [Lesson 1.6]

64. 450 $2 \cdot 3^2 \cdot 5^2$ **65.** 484 $2^2 \cdot 11^2$ **66.** 318 $2 \cdot 3 \cdot 53$ **67.** 18,900 $2^2 \cdot 3^3 \cdot 5^2 \cdot 7$

For Exercises 68–70, use the following list of fractions, each of which has a numerator that is 1 less than the denominator.
[Lesson 3.1]

$$\frac{7}{8} \qquad \frac{2}{3} \qquad \frac{3}{4} \qquad \frac{4}{5} \qquad \frac{5}{6} \qquad \frac{1}{2} \qquad \frac{99}{100}$$

68. Write the fractions in order from least to greatest. $\frac{1}{2}, \frac{2}{3}, \frac{3}{4}, \frac{4}{5}, \frac{5}{6}, \frac{7}{8}, \frac{99}{100}$

69. What happens to the size of this type of fraction as the denominator increases? The value of the fraction increases.

70. Explain how you can order these types of fractions by comparing denominators. Order the denominators from least to greatest.

Name the reciprocal of each number. Show that the product of each number and its reciprocal is 1.   [Lesson 3.5]

71. $\frac{1}{3}$ **72.** $\frac{3}{4}$ **73.** $2\frac{1}{2}$

Look Beyond

Simplify each trinomial addition or subtraction expression. You may use tiles. Check your result by substituting 10 for x.

74. $(2x^2 - 3x + 5) + (-3x^2 + x - 4)$ **75.** $(2x^2 - 3x + 5) - (-3x^2 + x - 4)$

76. Technology Use a graphics calculator to check your results to Exercises 74 and 75 by building a table of values. Copy the table of values.

LESSON 5.6
Exploring Inequalities

WHY *In an equation, two mathematical expressions represent equal quantities. To solve real-world problems, you may have to compare two mathematical expressions that are not equal. This type of statement is called an inequality.*

The school science club has an account in which students deposit money they have earned from fund-raising. Each student has a separate account for his or her contributions to the club. Mark deposits $3 and Kendra deposits $7.

•Exploration 1 *Ordering Numbers*

1 Mark and Kendra each began with an account balance of $0. Whose account has more money after the first deposit? How much more?
Kendra has $4 more.

2 Copy the number line below. Kendra's deposit is labeled with the letter ***k***. Circle and label Mark's deposit with the letter ***m***. Circle 3 on the number line, and label it *m*.

$$-4 \ -3 \ -2 \ -1 \ \ 0 \ \ 1 \ \ 2 \ \ 3 \ \ 4 \ \ 5 \ \ 6 \ \ ⑦ \ \ 8 \ \ 9 \ \ 10 \ 11 \ 12$$
                                                                                        k

3 Randall is another student in the club. He had to use more money than he put into the account, so he owes the club $2. Represent Randall's balance using an integer. Label Randall's balance on the number line with the letter *r*.
Circle −2 on the number line, and label it *r*.

4 How much more than Randall does Kendra have in her account?
Kendra has $9 more.

5 Jessica deposits $8 into her account. Label this number on the number line with the letter *j*. Circle 8 on the number line, and label it *j*.

The amount of Jessica's deposit is greater than Kendra's. This can be written with symbols as

$$j > k, \text{ read "} j \text{ is greater than } k.\text{"}$$
$$\text{or}$$
$$k < j, \text{ read "} k \text{ is less than } j.\text{"} \quad k > m; m < k$$

6 Use symbols to compare Kendra's deposit with Mark's deposit. ❖

Lesson 5.6 Exploring Inequalities **307**

Kendra's account for the science club starts with a $7 deposit and grows as she makes each deposit. A number line can be used to represent this relationship. The statement "k is greater than or equal to 7" is written as $k \geq 7$, and represented on a number line below.

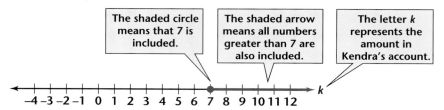

The shaded circle means that 7 is included.

The shaded arrow means all numbers greater than 7 are also included.

The letter k represents the amount in Kendra's account.

How is $A \leq 7$ represented on a number line?

EXTENSION

A mathematical statement containing the symbols $<$, $\leq$, $>$, or $\geq$ is called an **inequality**. You can graph on a number line to represent an inequality.

To represent the inequality $m < 7$, shade the numbers to the left of 7. Leave the circle at 7 unshaded to indicate that 7 is not included.

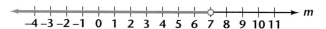

To represent the inequality $m \geq 4$, shade the numbers to the right of 4. Shade the circle at 4 to indicate that 4 is included.

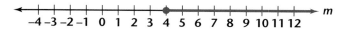

To represent the inequality $m \leq -1$, shade the numbers to the left of -1. Shade the circle at -1 to indicate that -1 is included.

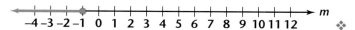

•Exploration 2 *Properties of Inequalities*

Copy and complete the table.

	Inequality	Number	Add number to each side of inequality.	Simplify.	True or false?
	$5 < 12$	2	$5 + 2 < 12 + 2$	$7 < 14$	true
1	$2 \leq 7$	3	$2 + 3 \leq ? 7 + 3$	$5 \leq ? 10$	? true
2	$8 > 3$	-4	$8 + (-4) > ? 3 + (-4)$	$4 > ? -1$	? true
3	$5 \geq -3$	4	$5 + 4 \geq ? -3 + 4$	$9 \geq ? 1$	? true
4	$-7 \leq -1$	8	$-7 + 8 \leq ? -1 + 8$	$1 \leq ? 7$	? true
5	$-12 < 7$	-5	$-12 + (-5) ? < 7 + (-5)$	$-17 ? < 2$	? true

6 What happens to an inequality when you add an equal amount to each side? The inequality remains true.

Copy and complete the table.   **10.** $-7 - 8 \le -1 - 8$; $-15 \le -9$; true

Inequality	Number	Subtract number from each side of inequality.	Simplify.	True or false?
5 < 12	2	5 − 2 < 12 − 2	3 < 10	true
2 ≤ 7	3	2 − 3 ≤?7 − 3	−1 ?≤ 4	? true
8 > 3	−4	8 − (−4) >?3 − (−4)	12> ?7	? true
5 ≥ −3	4	5 − 4 ≥?−3 − 4	1 ≥?−7	? true
−7 ≤ −1	8	?	?	?
−12 < 7	−5	−12 − (−5)?<7 − (−5)	−7 ?<12	? true

7
8
9
10
11

12 What happens to an inequality when you subtract an equal amount from each side? ❖ The inequality remains true.

The properties you discovered in Exploration 2 are called the *Addition and Subtraction Properties of Inequality*.

Randall's debt of $2 is represented on the number line by −2. For Exploration 3, suppose Randall must add enough to his account so that he has *at least $20* in it. The expression *at least $20* means he will have *$20 or more* in the account.

Science Club Accounts

Student	Account Balance
Kendra	$7.00
Mark	$3.00
Randall	−$2.00
Jessica	$8.00

Exploration 3 *Variable Inequalities*

Copy and complete the table.

Amount to be added to account	Addition process	Inequality to determine if this amount is at least $20	Simplify.	Is this amount at least $20?
5	5 + (−2)	5 + (−2) ≥ 20	3 ≥ 20	no
10	?	?	?	?
15	?	?	?	?
20	?	?	?	?
25	?	?	?	?
30	?	?	?	?

1
2
3
4
5

6 What is the least amount that Randall may add to his account to have *at least $20* in his account? $22

7 How many different possible amounts can Randall add to have *at least $20* in his account? Infinite; any amount ≥ $22

8 Use x to represent the amount of money that Randall will add to his account. Write an expression for the total amount of money in Randall's account. $x - 2$

9 Write an inequality to represent the relationship between the expression from Step 8 and $20. ❖ $x - 2 \ge 20$

In Exploration 3, you solved the inequality $x + (-2) \geq 20$ by guess-and-check. The inequality can also be solved using the Addition Property of Inequality that you discovered in Exploration 2.

EXTENSION

Start with the inequality. $x + (-2) \geq 20$
Add 2 to each side of the inequality. $x + (-2) + 2 \geq 20 + 2$
Write the result. $x \geq 22$

Randall must deposit *at least* $22.

Graphics Calculator

The solution to the inequality can also be found using a graphics calculator. Enter the equation $Y_1 = X + -2$, and create a table. Examine the table to find the values of x that make y equal to or greater than 20.

X	Y₁
20	18
21	19
22	**20**
23	21
24	22
25	23
26	24

$Y_1=20$

When x is greater than or equal to 22, the value of y is greater than or equal to 20. So the solution is all x-values greater than or equal to 22, or $x \geq 22$. ❖

APPLICATION

Banking   The total amount of money in the science-club account is $780. To meet the school regulations for accounts, the amount may be no more than $1200. How much more can be added to the account?

If a represents the amount added, the total amount allowed in the account can be represented by $a + 780$. The fact that the total amount in the account may be no more than $1200 means that the total amount is less than or equal to $1200. This problem can be modeled by the following inequality:

$$a + 780 \leq 1200$$

To solve the inequality for a, subtract 780 from each side.

$$a + 780 \leq 1200$$
$$a + 780 - 780 \leq 1200 - 780$$
$$a \leq 420$$

The amount added to the account must be less than or equal to $420. ❖

Graphics Calculator

You can also solve this inequality by using a graphics calculator.

Enter the equation $Y_1 = X + 780$, and create a table. Examine the table to find the values of x that make y less than or equal to 1200.

X	Y₁
400	1180
410	1190
420	**1200**
430	1210
440	1220
450	1230
460	1240

$Y_1=1200$

The solution is all x-values less than or equal to 420, or $x \leq 420$.

The graphics calculator shows the intersection of the lines $Y_1 = X - 3$ and $Y_2 = 4$.

What inequalities can be solved by examining this graph? Write each inequality using symbols.
$x - 3 > 4; x - 3 < 4; x - 3 \geq 4; x - 3 \leq 4;$

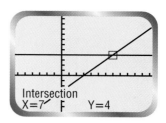

Intersection
X=7 F   Y=4

EXERCISES & PROBLEMS

Communicate

1. If two numbers are labeled on a number line, how can you tell which number is greater?
The number on the right is greater.
2. Explain how to graph the inequality $x > 3$ on a number line.
Open circle at 3, shade to the right.
3. How are the inequalities $x > -2$ and $x \geq -2$ different? $x > -2$ does not include -2; $x \geq -2$ does include -2.
4. Write a description of the inequality graphed on the number line below.
$x \geq 2$

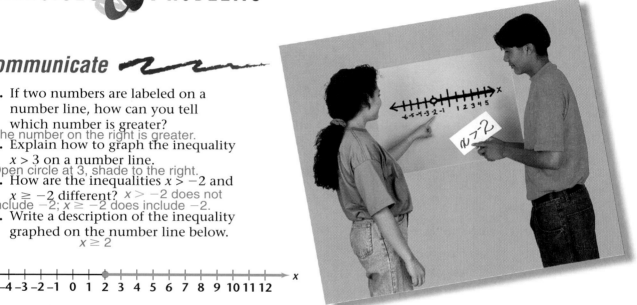

5. Describe the steps you would take to solve the inequality $x - 5 < 3$. Add 5 to each side, and simplify. The result is $x < 8$.

Practice & Apply

Graph each inequality on a number line.

6. $x < 3$ 7. $x \leq -1$ 8. $x > 4$ 9. $x \geq 5$ 10. $x \leq 5$ 11. $x \geq -12$

12. $x > -5$ 13. $x \geq 0$ 14. $x \geq -3$ 15. $x \leq 2$ 16. $4 \leq x$ 17. $-3 > x$

18. $x < -2$ 19. $-4 < x$ 20. $3 < x$ 21. $12 \leq x$ 22. $6 > x$ 23. $x > -11$

Banking A deposit can be considered a positive number. The amount of a check can be considered a negative number.

24. Let A represent the amount of money in an account. The amount of money in the account must always be at least $500. Write an inequality to represent this situation. $A \geq 500$

25. Let M represent the amount of money in another checking account. Suppose that a deposit of $350 is made. The amount in the account is no more than $1240. Write an inequality to represent this situation. $M + 350 \leq 1240$

Tamara had less than $5.00 change after paying her lunch bill.

Consumer Economics Refer to the illustration at the left for Exercises 26 and 27.

26. Write an inequality to represent the amount of money that Tamara could have had before she paid her lunch bill. $0 \le x - 5.45 < 5.00$

27. Solve the inequality you wrote for Exercise 26, and find the amounts of money that Tamara could have had before she paid her lunch bill. $5.45 \le x < 10.45$

For Exercises 28 and 29, write an inequality to represent the situation. Then solve the inequality.

28. Banking Randy had some money in his savings account. After he deposited $50, there was at least $180 in his account. What is the least amount that could have been in the account before his deposit?

29. Savings The cost of a VCR is at least $280. Rhonda has saved $90. What is the least amount that Rhonda still needs to save in order to buy the VCR?

Determine whether the number following each inequality is a solution of the inequality.

30. $x + 3 > -1$; 2 yes   **31.** $x - 4 \ge 2$; 1 no   **32.** $z + 3 \le 5$; 2 yes   **33.** $x + 5 \le 9$; 14 no

34. $y - 2 \le 3$; 5 yes   **35.** $y - 2 < -3$; -1 no **36.** $w + 5 \le 2$; -1 no   **37.** $w - 5 \le 9$; 12 yes

38. $s - 4 \ge -1$; -3 no   **39.** $6 + m < 4$; -2 no **40.** $n - 7 > -2$; 5 no   **41.** $z + 4 < 6$; -5 yes

Use any method to solve each inequality. Graph the solution.

42. $x - 3 < 7$ **43.** $y - 4 > 9$ **44.** $x - 3 < 4$ **45.** $p - 4 < -4$

46. $d - 1 \le 5$ **47.** $y + 6 \ge 3$ **48.** $w + 3 \le -7$ **49.** $z + 5 < -12$

50. $k + 2 < -2$ **51.** $s - 0.5 \ge 5.5$ **52.** $m - 4 < 2$ **53.** $2 \ge x + 9$

Look Back

Simplify each expression. [Lessons 2.2, 2.4]

54. $|-12| + |16|$ 28   **55.** $|-12 + 16|$ 4   **56.** $-16 + (-9) + 32$ 7 **57.** $-154 - (-198)$ $-83$127

Simplify each expression. [Lessons 3.4, 3.5]

58. $\frac{1}{2} + \frac{4}{5}$ $\frac{13}{10}$, or $1\frac{3}{10}$ **59.** $\frac{3}{4} - 1\frac{7}{8}$ $-\frac{9}{8}$, or $-1\frac{1}{8}$ **60.** $\frac{4}{7} \cdot 2\frac{1}{4}$ $\frac{9}{7}$, or $1\frac{2}{7}$ **61.** $\frac{2}{3} \div \frac{3}{8}$ $\frac{16}{9}$, or $1\frac{7}{9}$

Simplify each expression. [Lessons 5.1, 5.2]

62. $(7w - 3) + (8w - 7)$ **63.** $3y - 7y$ **64.** $(12x - 3) - (8x - 12)$
      $15w - 10$ $-4y$ $4x + 9$

Look Beyond

65. Write an inequality to compare the integers -8 and -1. What is the effect of multiplying each side of the inequality by 2? What is the effect of multiplying each side of the inequality by -2? $-8 < -1$; $-16 < -2$ True; $16 < 2$ False

LESSON 5.7 Solving Related Inequalities

Why When the weather report says "The high temperature today will be 75°F," it means that the temperatures throughout the day will be less than or equal to 75°F. This inequality describes a range of values. Inequalities can also be used to determine the items you can purchase when you have a given amount of money.

Hamburger $1.45
Cheeseburger 2.50
Brownie/
Ice Cream 3.00
Fruit 30¢

Michael can spend at most $3.10 for lunch. He buys a hamburger for $1.45. How much can he spend on the rest of his lunch and stay within his spending limit?

Let x be the amount Michael spends on the rest of his lunch. The total amount he spends is then $x + 1.45$. Michael's situation can be modeled by the statement $x + 1.45 \le 3.10$. It is read "x plus 1.45 *is less than or equal to* 3.10." Michael will stay within his limit as long as $x + 1.45$ remains less than or equal to 3.10.

Suppose that Michael buys a piece of fruit. Substitute 0.30 for x. The result is $0.30 + 1.45$, or 1.75. This is within the limit because $1.75 is less than $3.10.

Suppose that Michael buys a brownie with ice cream. Substitute 3.00 for x. The result is $3.00 + 1.45$, or 4.45. This is beyond Michael's limit because $4.45 is *not* less than or equal to $3.10.

Lesson 5.7   Solving Related Inequalities   **313**

For Michael, any amount for x that is less than or equal to 1.65 will make the inequality $x + 1.45 \leq 3.10$ a true statement. The *solution of an inequality* is the set of numbers that make the inequality statement true.

There are several inequality statements used in algebra.

 **Exploration** *Solving an Inequality*

 Tell whether each of the following values makes the inequality $x + 1.25 \leq 2.35$ true or false.
  a. 0.50 True
  b. 1.00 True
  c. 1.50 False
  d. 1.75 False

 Find the largest value of x that makes the inequality true. 1.10

 Describe a method that does not use guessing to determine the largest value of x that makes the inequality true. ❖ Subt 1.25 from each side. The result is $x \leq 1.10$.

The inequality $x + 1.45 \leq 3.10$ can be thought of as two statements.

  a. $x + 1.45 < 3.10$ an inequality
  b. $x + 1.45 = 3.10$ an equation

The equation $x + 1.45 = 3.10$ can be solved by using the Subtraction Property of Equality. Thus, $x + 1.45 = 3.10$ is true when x is 1.65. The inequality $x + 1.45 < 3.10$ is true when x is less than 1.65. This suggests that the properties used for solving inequalities are similar to the ones used for solving equations.

ADDITION PROPERTY OF INEQUALITY

If equal amounts are added to the expressions on each side of an inequality, the resulting inequality is still true.

SUBTRACTION PROPERTY OF INEQUALITY

If equal amounts are subtracted from the expressions on each side of an inequality, the resulting inequality is still true.

EXAMPLE 1

Solve the inequality $8m - 8 \geq 7m + 2$.

Solution ➤

One strategy is to arrange the terms with the variable on one side of the inequality sign and the numbers on the other side. Then combine like terms.

$8m - 8 \geq 7m + 2$	Given
$8m - 8 - 7m \geq 7m + 2 - 7m$	Subtraction Property of Inequality
$8m - 7m - 8 \geq 7m - 7m + 2$	Rearrange terms.
$m - 8 \geq 2$	Combine like terms.
$m - 8 + 8 \geq 2 + 8$	Addition Property of Inequality
$m \geq 10$	Combine like terms.

The solution is $m \geq 10$. Now substitute values equal to 10 and values greater than 10 into the *original* inequality to see if the results are true.

Substitute **10** for m.
$$8(10) - 8 \geq 7(10) + 2$$
$$80 - 8 \geq 70 + 2$$
$$72 \geq 72$$
True, because 72 equals itself

Substitute **11** for m.
$$8(11) - 8 \geq 7(11) + 2$$
$$88 - 8 \geq 77 + 2$$
$$80 \geq 79$$
True, because 80 is greater than 79

When m is 10 the inequality is true.   When m is 11, the inequality is true. ❖

Using the Number Line to Represent an Inequality

EXAMPLE 2

Graph the solution to $x - 2 < 8$.

Solution ➤

First, solve the inequality for x.

$x - 2 < 8$	Given
$x - 2 + 2 < 8 + 2$	Addition Property of Inequality
$x < 10$	Combine like terms.

To graph the result on a number line, shade *all* points to the left of 10 on the number line. This includes all points whose coordinates are less than 10. To indicate that 10 *is not included,* put an *open circle* at 10.

$x < 10$

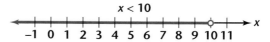

To check, substitute 9 for x.
$$x - 2 < 8$$
$$9 - 2 < 8$$
$$7 < 8 \text{ True}$$

Substitute 10 for x.
$$x - 2 < 8$$
$$10 - 2 < 8$$
$$8 < 8 \text{ False} ❖$$

Lesson 5.7   Solving Related Inequalities   **315**

What numbers do you think are represented by the graph on this number line? Write an inequality that describes these points.

$x \geq 3$

Displaying Solutions

The graph of an inequality involving integers is not the same as the graph involving all numbers. Integers are graphed as shaded circles (●). An inequality involving all numbers is graphed as an

$-2 \leq x < 1$

Using integers

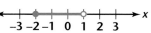

Using all numbers

interval, a ray, or a line. Remember that endpoints that are included are shown as shaded circles, while endpoints that are not included are shown as open circles (○). It is important to consider the reasonable solutions when solving a real-world related inequality.

Graph an inequality to represent the statement "No more than 5 people can fit in this car."

EXAMPLE 3

Graph the solution to $-4 \leq x$ and $x < 2$.

Solution ➤

Since the word *and* is used, the solution set must be true for both inequalities. This includes -4, since $-4 \leq 2$ and $-4 \leq -4$ are both true. It does not include 2, since $2 < 2$ is false.

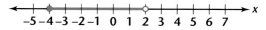

The statement **$-4 \leq x$ and $x < 2$** can also be written as **$-4 \leq x < 2$**. This means that x is between -4 and 2 and includes -4. ❖

Try This   Graph the solution to $-3 < x \leq 2$.

In everyday conversation, certain inequalities have important meanings. Some common examples are the following.

$x \geq 5$
x is at least 5.

$5 < x < 7$
x is between 5 and 7.

$x \leq 5$
x is no more than 5.

$5 \leq x \leq 7$
x is between 5 and 7 inclusive.

No solution; x cannot be ≤ 5 and ≥ 7 at the same time.

What is the solution to $5 \geq x \geq 7$? Explain your answer.

EXERCISES & PROBLEMS

Communicate

Explain how you would determine whether the following inequalities are true or false:

1. $5 \geq 2 + 3$ **2.** $4 \leq 2 + 3$

3. $6 < 7 - 3$ **4.** $8 > 10 - 2$

5. Tell how the properties for inequalities are similar to the properties for equations.

6. Tell the steps necessary to solve the inequality $3x - 4 \leq 2x + 1$. Name the property you would use for each step.

7. Which values of the solution to $x - 4 \leq 9$ represent equality and which values represent inequality?

8. Explain how to graph the solution to $x + 3 < 7$ on a number line.

9. How do you write an inequality that represents the points shown on the graph below?

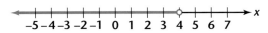

10. Discuss the word *inclusive* and its meaning in mathematics. List the integers between 5 and 10 inclusive. Graph them on a number line.

Practice & Apply

State whether each inequality is true or false.

11. $8 > 9 - 1$
False

12. $-2 \leq 5 - 7$
True

13. $8 \leq 9 - 1$
True

14. $-2 > 5 - 7$
False

Solve each inequality.

15. $x + 8 > -1$ $x > -9$

16. $x - 6 \leq 7$ $x \leq 13$

17. $x + \frac{3}{4} > 1$ $x > \frac{1}{4}$

18. $x + \frac{3}{4} \leq \frac{1}{2}$ $x \leq -\frac{1}{4}$

19. $x + 0.04 > 0.6$
  $x > 0.56$

20. $x - 0.1 < 8$ $x < 8.1$

21. Describe a real-world situation modeled by $x + 10 < 100$.

22. Graph the solution to $x - 4 \geq -1$ on a number line.

23. Graph the solution to $x + 3 < 2$ on a number line.

Write an inequality that describes the points on each number line.

$x \geq -1$

24.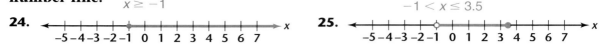

$-1 < x \leq 3.5$

25.

26. List the integers between 57 and 66 inclusive.
57, 58, 59, 60, 61, 62, 63, 64, 65, 66

Temperature Write an equation or an inequality to represent each situation. Use t to represent the variable.

27. all possible daytime temperatures where you live   Answers may vary.

28. a high temperature of 66°F $t \leq 66$

29. the low temperature was 54° $t \geq 54$

30. 54° $t = 54$

31. If t represents the first reading, then after a rise of 5 degrees the temperature was between 70 and 80 degrees inclusive. What is the inequality for this situation? $70 \leq t + 5 \leq 80$

The sports stadium holds 15,000 people.

32. Sports Everyone in the stadium has a seat, but the stadium is not full. Write an inequality that models this situation using P as the number of people. $1 \leq P < 15,000$

33. The school auditorium can seat 450 people for graduation. Graduates will use 74 seats. Write an inequality to describe the number of others who can be seated in the auditorium.
$74 \leq P + 74 \leq 450$

34. From 30 to 50 people attended a party. From 5 to 10 people left early. Write an inequality to represent the possible range for the number of people who did *not* leave early. $20 \leq x \leq 45$

35. A table is supposed to be 42.3 centimeters long. Write an inequality for M, the measure of the table, that allows a possible error of 0.5 centimeters. $M - 0.5 \leq 42.3 \leq M + 0.5$

36. Students in Ms. Ambrose's algebra class earn an A if they average at least 90, and they earn a B if they average at least 80 but less than 90. This can be represented by the following inequalities:

$$A \geq 90 \qquad 80 \leq B < 90$$

Translate the inequality below into words.

$$70 \leq C < 80$$

A student will earn a C if the average is at least 70 and less than 80.

37. Represent the following statement with an inequality:
Students earn a B if they average at least 78, but less than 89. $78 \leq B < 89$

Find each value. **[Lessons 2.1, 2.2, 2.4, 2.5]**

38. $|-4.5|$ 4.5   **39.** $|-9|$ 9   **40.** $|-3+4|$ 1

41. $|-2 + -6|$ 8   **42.** $|(-2)(-6)|$ 12   **43.** $|-4| - |3|$ 1

Simplify each expression. **[Lessons 3.4, 3.5]**   **45.** $\frac{15}{8}$, or $1\frac{7}{8}$

44. $-\frac{2}{3} \cdot -\frac{3}{4}$ $\frac{1}{2}$ **45.** $2\frac{5}{8} + \left(-\frac{3}{4}\right)$ **46.** $2\frac{3}{7} \div \left(-\frac{7}{9}\right)$ $\frac{-153}{49}$, or $-3\frac{6}{49}$

47. $4\frac{1}{8} - 3\frac{2}{5}$ $\frac{29}{40}$ **48.** $9\frac{3}{4} \cdot \frac{8}{3}$ 26   **49.** $\frac{2}{3} \div \frac{2}{3}$ 1

50. The regular price of a coat is discounted by 20%. If the regular price is $120, what is the amount of the discount? **[Lesson 3.7]** $24

51. Dana performs an experiment in which he rolls two number cubes. What is the theoretical probability that he will roll less than 3 on both of the number cubes? **[Lesson 3.9]** $\frac{1}{9}$

52. The legs of a right triangle have lengths of 120 meters and 90 meters. What is the length of the hypotenuse? **[Lesson 4.7]** 150 m

Solve for x. **[Lessons 5.3, 5.4]**

53. $x - 11 = -28$ $x = -17$

54. $x + \frac{1}{2} = \frac{3}{4}$ $x = \frac{1}{4}$

55. $x - 5 = 16$ $x = 21$

56. $\frac{2}{5} - x = \frac{1}{10}$ $x = \frac{3}{10}$

Look Beyond 〜〜

Solve each of the following inequalities. Use several values of x to check your answers.

57. $2x < 8$ $x < 4$ **58.** $4x + 5 \le 16$ $x \le \frac{11}{4}$ **59.** $8x - 3 > 33$ $x > \frac{9}{2}$

60. In how many ways can you make change for a quarter using pennies, nickels, and dimes? 12 ways

61. Place dots on a circle, and connect the dots with line segments. Count the regions that are formed. Be sure that no three lines meet at the same point unless the point is on the circle. What is the maximum number of regions that can be formed with 5 dots? 16

| 1 dot | 2 dots | 3 dots | 4 dots |
| 1 region | 2 regions | 4 regions | 8 regions |

PROJECT CHAPTER 5

Find it FASTER

A database is a tool to organize and manipulate information. A basic example is an address book. A computer database is an electronic version that provides the speed of automatic organization and fast access. Many computer spreadsheets can be used for a database.

Suppose you are planning to continue your education after high school. A database can be constructed and used to help you decide on a college or institute to attend. You can collect information like the college name, region, state, tuition, room and board expenses, and undergraduate student population. You then use a database to store, organize, and manipulate the information.

Examine the terminology of a database shown below. Notice that the first database is arranged alphabetically by school name.

To view one of the other fields in order, select a field, like tuition, and use the SORT command. Notice that the records listed for tuition are now sorted in ascending order.

	A School Name	B Region	C State	D Tuition	E Expenses	F Students
1	Bennett College	Southeast	NC	5600	3095	650
2	Grinnell College	Midwest	IA	15688	4618	133
3	Heidelberg College	East	OH	13500	4400	918
4	Hamline University	Midwest	MN	13252	4194	1362
5	Mills College	West	CA	14100	6000	740
6	St. John's College	Northeast	MD	17430	5720	3932

an entry · *a field* · *a record*

	A School Name	B Region	C State	D Tuition	E Expenses	F Students
1	Bennett College	Southeast	NC	5600	3095	650
2	Hamline University	Midwest	MN	13252	4194	1362
3	Heidelberg College	East	OH	13500	4400	918
4	Mills College	West	CA	14100	6000	740
5	Grinnell College	Midwest	IA	15688	4618	133
6	St. John's College	Northeast	MD	17430	5720	3932

Different spreadsheets and database software have different features for requesting various combinations of information. A request for information is called a **query**. Some common operations for making a query are comparisons that use AND, OR, and NOT.

A query using OR will retrieve all information from the records that fulfills any specification.

Query: Schools in the **Midwest OR** the **West**

Output: Hamline University
Grinnell College
Mills College

A query using AND will give only the information from the records that fulfills all the specifications.

Query: Schools with **tuition less than $14,000 AND** schools in the **Midwest**

Output: Hamline University

Different databases and spreadsheets treat AND and OR operations in different ways.

Answers may vary.

Alternative 1 (With a computer and a spreadsheet or database): Make a database from some data that is of interest to you. It should contain regular lists of information.

Alternative 2 (Without a computer): Create your own database from index cards. As a group project, you can also act out the computer procedure for creating a database. This includes writing, storing, sorting, and retrieving information.

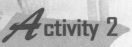

Answers may vary.

Once you have created your database, devise a strategy for sorting various fields and retrieving a record. Form queries that ask for specific combinations of the data. The search should contain a method using OR and a method using AND.

Chapter 5 Review

Vocabulary

Addition Property of Equality	294	like terms	274
Addition Property of Inequality	314	literal equation	295
algebraic expression	274	monomial	299
binomial	299	polynomial	299
coefficient	274	Subtraction Property of Equality	293
constant	274	Subtraction Property of Inequality	314
Distributive Property	275	terms	274
Inequality	308	trinomial	299

Key Skills & Exercises

Lesson 5.1

➤ **Key Skills**

Simplify expressions with several variables by adding like terms.

Simplify $(3r + 7t) + (4r - 8t)$.

$(3r + 7t) + (4r - 8t)$	Given
$(3r + 7t) + (4r + (-8t))$	Definition of Subtraction
$(3r + 4r) + (7t + (-8t))$	Rearrange terms.
$7r - t$	Simplify.

➤ **Exercises**

Simplify.

1. $(6a - 1) + (5a - 4)$ $11a - 5$

2. $(7 - t) + (3t + 4)$ $2t + 11$

3. $\left(\dfrac{x}{3} - 2\right) + \left(\dfrac{x}{2} + 4\right)$ $\dfrac{5x}{6} + 2$

4. $(1.4m - 6.2n) + (2.4m - 5.5n)$ $3.8m - 11.7n$

5. $(3x + 2y + z) + (6x - 4y - 3z)$ $9x - 2y - 2z$

Lesson 5.2

➤ **Key Skills**

Use the definition of subtraction to subtract expressions.

Simplify $(7x + 4y - 2z) - (6x - 5y + z)$.

$(7x + 4y - 2z) - (6x - 5y + z)$	Given
$7x + 4y - 2z - 6x + 5y - z$	Definition of Subtraction
$7x - 6x + 4y + 5y - 2z - z$	Rearrange terms.
$x + 9y - 3z$	Simplify.

➤ **Exercises**

Simplify.

6. $3x - 5x$ $-2x$

7. $7y - (7 - 5y)$ $12y - 7$

8. $(8m - 4) - (6m - 3)$ $2m - 1$

9. $(6d + 3) - (4d - 7) + (3d - 5)$ $5d + 5$

10. $(4a - 3b - c) - (6a + 5b - 4c)$ $-2a - 8b + 3c$

Lesson 5.3

➤ Key Skills

Use algebra tiles to solve addition and subtraction equations.

To solve $x - 3 = -2$ using algebra tiles, set up the model as shown below.

Form neutral pairs by adding 3 positive 1-tiles to each side. Remove neutral pairs.

The result is 1 x-tile on the left side and 1 positive 1-tile on the right side.

➤ Exercises

Solve each equation using algebra tiles. Sketch a picture of each step, and write the solution.

11. $x + 3 = 2$ **12.** $x - 1 = -2$ **13.** $-3 = x + 4$ **14.** $x + 5 = 2$

Lesson 5.4

➤ Key Skills

Solve algebraic equations that contain addition and subtraction.

Solve $x + 15 = 11$.

$$x + 15 = 11$$
$$x + 15 - 15 = 11 - 15$$
$$x = -4$$

Check: $-4 + 15 = 11$ True

Solve $-8 = y - 14$.

$$-8 = y - 14$$
$$-8 + 14 = y - 14 + 14$$
$$6 = y$$

Check: $-8 = 6 - 14$ True

➤ Exercises

Solve.

15. $w + 16 = 25$ $w = 9$ **16.** $r + 26 = 16$ $r = -10$ **17.** $t + 7 = -5$ $t = -12$ **18.** $a + 1.5 = 3.6$ $a = 2.1$

19. $m + \frac{1}{2} = \frac{5}{6}$ $m = \frac{1}{3}$ **20.** $y - 13 = 12$ $y = 25$ **21.** $24 = x - 19$ $x = 43$ **22.** $-6 = g - 17$ $g = 11$

23. $h - \frac{1}{6} = \frac{2}{3}$ $h = \frac{5}{6}$ **24.** $7k - (6k + 5) = 7$ $k = 12$ **25.** $4 - (2 - 3z) = 6z - (4z + 3)$ $z = -5$

Lesson 5.5

➤ Key Skills

Simplify polynomials by combining like terms.

To simplify $6x + 3 + 5x$, first rearrange the order of the terms. Then combine like terms.

$$6x + 3 + 5x = 6x + 5x + 3$$
$$= 11x + 3$$

Add and subtract binomials.

To subtract $4x^2 + 2x$ from $7x^2 - 2$, use the Definition of Subtraction for Expressions, and combine like terms.

$$(7x^2 - 2) - (4x^2 + 2x)$$
$$= (7x^2 - 2) + (-4x^2 - 2x)$$
$$= (7x^2 - 4x^2) + (-2x) + (-2)$$
$$= 3x^2 - 2x - 2$$

➤ Exercises

Simplify each binomial addition expression or binomial subtraction expression.

26. $(5x - 1) + (2x + 3)$ $7x + 2$ **27.** $(x^2 + 1) - (2x^2 - 4)$ $-x^2 + 5$

28. $(-4x^2 - 8) - (7x + 2)$ $-4x^2 - 7x - 10$ **29.** $(3c - 8) + (-6c - 12)$ $-3c - 20$

Lesson 5.6

➤ Key Skills

Use guess-and-check to solve an inequality.
To solve the inequality $x + (-3) < 10$ by the guess-and-check method, substitute a value of x that is too large. Then substitute smaller and smaller values until a value is found that makes the statement true. The statement $13 + (-3) < 10$ is not true but, the statement $12 + (-3) < 10$ is true. The solution is all numbers less than 13, or $x < 13$.

Guessed value of x	Addition process	Simplified value	Less than 10?
20	20 + (−3)	17	no
15	15 + (−3)	12	no
14	14 + (−3)	11	no
13	13 + (−3)	10	no
12	12 + (−3)	9	yes

➤ Exercises

Solve each inequality.

30. $x - 5 < -3$
$x < 2$

31. $x + 4 \geq 7$
$x \geq 3$

32. $x - 7 > -9$
$x > -2$

33. $x + \frac{1}{2} \leq 3\frac{1}{2}$
$x \leq 3$

Lesson 5.7

➤ Key Skills

Use the Addition or Subtraction Property of Inequality to solve an inequality, and show the solution on a number line.
For inequalities involving addition and subtraction, solve as you would an equation. Then graph the solution on a number line.

$$5t - 6 \leq 4t + 1$$
$$5t - 6 - 4t \leq 4t + 1 - 4t$$
$$t - 6 \leq 1$$
$$t - 6 + 6 \leq 1 + 6$$
$$t \leq 7$$

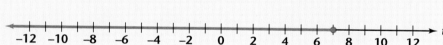

➤ Exercises

Solve each inequality, and graph the solution on a number line.

34. $x + 5 > 10$

35. $n - 15 \leq -3$

36. $y + 0.09 < 3.09$

37. $d - \frac{2}{3} \geq \frac{1}{3}$

Applications

38. Sales Marsha is stocking the art supply cabinet at school. The paintbrushes that she needs to buy cost $1.75 each, and the paint she needs costs $2.45 per jar. Write an expression to show how much she will spend if she buys b paintbrushes and j jars of paint. $1.75b + 2.45j$

39. Hobbies Rita has $15 to spend on new baseball cards. She decides to buy a card that costs $3. Write and solve an inequality to find the most that she can spend on other baseball cards. $x + 3 \leq 15$
$x \leq 12$
$12 or less

Chapter 5 Assessment

Simplify.

1. $(8n - 2) + (6n + 4)$ $14n + 2$

2. $5t - (9t - 6)$ $-4t + 6$

3. $(8 - 9z) - (7z + 15)$ $-16z - 7$

4. $(7 - x) - (8 + 3x) - (6x - 1)$ $-10x$

5. $(17x + 20y - 15z) - (3x + 5y - z)$ $14x + 15y - 14z$

There are 5 boxes of books and 5 additional books in a storage room. Write an algebraic expression to represent each situation. Let *b* represent the number of books in a box.

6. If Mr. Weaver puts 2 more boxes of books and 2 more books in the storage room, in terms of *b*, how many books are in the storage room? $7b + 7$ books

7. If Mrs. Thompson then removes 3 boxes and 1 book from the storage room, in terms of *b*, how many books are left in the storage room? $4b + 6$ books

Solve each equation using algebra tiles. Sketch a picture of each step, and write the solution.

8. $x - 5 = 8$ **9.** $x + 2 = -5$ **10.** $x + 4 = 2$

Solve.

11. $c + 18 = 10$ $c = -8$ **12.** $t - 36 = 19$ $t = 55$ **13.** $8 + y = -14$ $y = -22$

14. $34 = h - 4$ $h = 38$ **15.** $w + \dfrac{2}{5} = \dfrac{7}{10}$ $w = \dfrac{3}{10}$ **16.** $4x - (2x + 2) = 3 + x$ $x = 5$

17. **Geometry** Complementary angles are pairs of angles whose measures have a sum of 90°. Find the measure of an angle that is complementary to an angle with a measure of 48°.   42°

The relationship between total cost of an item, *I*, the cost without sales tax, *C*, and the amount of the sales tax, *T*, is given by the formula *I* = *C* + *T*.

18. Solve $I = C + T$ for C. $C = I - T$

19. Using the formula $I = C + T$, find C when I is \$25.19 and T is \$1.20.
   $C = \$23.99$

Solve each inequality.

20. $x + 6 \geq 4$ $x \geq -2$ **21.** $y - 2.3 < 1.4$ $y < 3.7$ **22.** $t + \dfrac{1}{2} \geq 2$ $t \geq \dfrac{3}{2}$

23. Graph the solution to $x - 8 > -4$ on a number line.

24. Graph the inequality $x \geq 5$ on a number line.

25. There is room for 40 people in an aerobics class. There are 25 people signed up for the class. Write an inequality to describe the number of additional people that can still sign up for the class. $x + 25 \leq 40$

CHAPTER 6

Multiplication and Division in Algebra

Chapter 6 continues the development of solving linear equations. It extends the concept of maintaining balance in an equation when you multiply and divide to find the value of the variable.

This chapter blends traditional methods with the latest technology. It begins with a very old problem written about Diophantus, who is often called the father of algebra. Applications of linear functions to business are stressed. Although modern applications of mathematics to commerce and business are often rather complex, we still use linear relationships to compute percentages, simple interest, and discounts.

PORTFOLIO ACTIVITY

Diophantus lived $\frac{1}{6}$ of his life as a child,

$\frac{1}{12}$ more as a youth,

$\frac{1}{7}$ more before he married, and

5 years more before his son was born.

His son was alive for only $\frac{1}{2}$ of Diophantus's life.

Diophantus then found solace in his studies for 4 more years,
until he too died.

How long did Diophantus live? 84 years

Multiplying and Dividing Expressions

 Businesses use spreadsheets and computer programs to compute employees' earnings. With technology, the operations, such as multiplication and division, are performed automatically.

D2	=6*C2			
	A	**B**	**C**	**D**
1	Month	Day	Hours	Earnings
2	October	1	3	18
3	October	2	2	12
4	October	3	4	24
5	October	4	0	0
6	October	5	6	36

Richard has a part-time job making $6 an hour. His supervisor keeps a record of his earnings on a spreadsheet.

The formula **= 6*C2** at the top of the spreadsheet above computes the earnings for October 1. The value in cell D2 is 6 times the value in cell C2.

Spreadsheet

Just as a spreadsheet uses **6*C2** to mean *6 times the value in cell C2*, you can write $6 \cdot h$, or simply $6h$, to represent 6 times a particular quantity h.

Word Expression	Algebraic Expression
six times the number of hours	$6h$

The algebraic expression $6h$ can be used to find how much Richard earned for working h hours.

To find what Richard earned on October 1, evaluate the expression $6h$ for h equal to 3. That is, substitute 3 for h, and then multiply 6 times 3.

$$6h = 6(3) = 18$$

Richard earned $18 for working 3 hours on October 1. What expression represents Richard's earnings for 5 hours? What are Richard's earnings for 5 hours? for 5.5 hours? 6(5); $30; $33

You can also use a graphics calculator to make a table based on the algebraic expression 6h.

First enter the expression 6X for Y_1.

Then set up a table to substitute values for X.

X	Y_1	
0	0	
.5	3	
1	6	
1.5	9	
2	12	
2.5	15	
3	18	

X=0

Describe the values for X that are substituted in this graphics calculator table. The values for X represent the number of hours worked. The values for Y_1 represent Richard's earnings.

EXAMPLE 1

Travel Suppose you are traveling at a constant rate of 60 miles per hour.

A Write an algebraic expression for this constant rate of travel, using *h* for hours.

B Use substitution to find the distance, in miles, that you would travel in 8 hours.

C Use a graphics calculator to make a table of values for the miles traveled. Include values of *h* beginning at 0 and increasing in increments of 5. Use the table to find out how long you would have to travel (nonstop) in order to go 1500 miles.

Solution ➤

A The word expression 60 *miles per hour* can be written as the algebraic expression 60*h*.

B To find the distance you would travel in 8 hours at a constant rate of 60 miles per hour, substitute 8 for *h* in the expression 60*h*.

$$60h = 60(8) = 480$$

At a constant rate of 60 miles per hour, you would travel 480 miles in 8 hours.

C To make a table for the expression 60*h*, enter the expression 60X for Y_1.

Then set up a table to substitute values for X beginning at 0 and increasing in increments of 5.

From the table, you can see that you would have to travel 25 hours (nonstop) in order to go 1500 miles. ❖

X	Y_1	
0	0	
5	300	
10	600	
15	900	
20	1200	
25	1500	
30	1800	

X=25

EXAMPLE 2

Evaluate $5m - 3n$ when m is 6 and n is -2.

Solution ➤

$5m - 3n$ means 5 times m minus 3 times n. Substitute **m** with **6** and **n** with -2, and then simplify.

$$5\mathbf{m} - 3\mathbf{n} = 5(\mathbf{6}) - 3(-\mathbf{2})$$
$$= 30 - (-6)$$
$$= 30 + 6$$
$$= 36 \;❖$$

Multiplying Expressions

To model $2 \cdot 3$, use 1-tiles. Recall that the area of a rectangle is length times width. Think of $2 \cdot 3$ as the area of a rectangle with dimensions 2 by 3.

The area of the rectangle is $2 \cdot 3 = 6$.

To model $2 \cdot 3x$, use x-tiles. Think of $2 \cdot 3x$ as the area of a rectangle with dimensions 2 by $3x$.

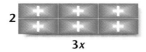

The area of the rectangle is $2 \cdot 3x = 6x$.

To multiply expressions, each of which contains x, a new tile is needed. Think of a tile that has the same height and length as the longer side of an x-tile. The area of the x^2-tile is x^2.

To model $2x \cdot 3x$, use x^2-tiles. Think of $2x \cdot 3x$ as the area of a rectangle with dimensions $2x$ by $3x$.

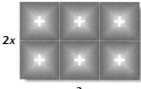

The area of the rectangle is $2x \cdot 3x = 6x^2$.

To check the multiplications, substitute 10 for x.

$$2 \cdot 3x = 2 \cdot 3(10) \qquad\qquad 2x \cdot 3x = 2(10) \cdot 3(10)$$
$$= 60 \qquad\qquad\qquad\qquad\quad = 600$$

$$6x = 6(10) \qquad\qquad\qquad 6x^2 = 6(10^2)$$
$$= 60 \qquad\qquad\qquad\qquad\quad = 600$$

Thus, $2 \cdot 3x = 6x$ is correct.   Thus, $2x \cdot 3x = 6x^2$ is correct.

•Exploration• *Multiplying by − 1*

Create a table of values for the expression $(-1)n$, as shown on the notebook.

Scientific Calculator

1 Choose seven values for *n*: three positive, three negative, and 0. Include at least two fractions and two decimals. List your choices in the first row of the table.

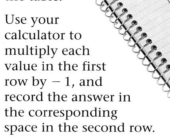

1-2. Answers may vary.

2 Use your calculator to multiply each value in the first row by − 1, and record the answer in the corresponding space in the second row.

3. Their signs are the opposite of those in the first row.

3 What do you notice about the entries in the second row?

4 In relation to the original numbers, what do you call the resulting products of multiplication by − 1?

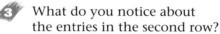

opposites

Multiply an expression by − 1 to find the *opposite* of the expression.

MULTIPLICATIVE PROPERTY OF − 1

For all numbers *a*,

− 1(*a*) = − *a*, or the opposite of *a*.

The Distributive Property Over Subtraction

Which expression below is equal to $-2(12 - 9)$?

a. $-2(12) - 9$ **b.** $-2(12) + 9$
c. $-2(12) - 2(9)$ **d.** $-2(12) - (-2)(9)$

Since $-2(12) - (-2)(9) = -6$ and $-2(12 - 9) = -6$, **d** is the correct choice. Multiplication can be distributed over subtraction. The sign of the number being distributed must be distributed to every term.

DISTRIBUTIVE PROPERTY OVER SUBTRACTION

For all numbers *a*, *b*, and *c*,

$a(b - c) = ab - ac$ and $(b - c)a = ba - ca$.

EXAMPLE 3

Simplify $-2(5a - 4)$.

Solution ➤

An expression is simplified when no more operations can be performed.

$$(-2)(5a - 4) = (-2)(5a) - (-2)(4) \qquad \text{Distributive Property}$$
$$= -10a + 8 \qquad \text{Simplify.} \qquad ❖$$

You can distribute a negative sign across the expressions you are subtracting. That is, if a subtraction sign is in front of an expression, change the sign to addition, and then distribute the negative number to every term of the quantity. This is shown in Example 4.

EXAMPLE 4

Simplify $(5x + 3y - 7) - 3(2x - y)$.

Solution ➤

Use the definition of subtraction to write an addition problem.
$$(5x + 3y - 7) - 3(2x - y) = (5x + 3y - 7) + (-3)(2x - y)$$

Distribute -3 over $(2x - y)$.

$$(5x + 3y - 7) + (-3)(2x - y) = 5x + 3y - 7 + [(-3)(2x) - (-3)(y)]$$
$$= 5x + 3y - 7 + (-6x + 3y)$$
$$= 5x + 3y - 7 - 6x + 3y$$
$$= (5x - 6x) + (3y + 3y) - 7$$
$$= -x + 6y - 7 \quad ❖$$

Try This Simplify $(2b - 5c) - 4(3b + c)$.
$-10b - 9c$

Dividing Expressions

The division problem $\dfrac{2x + 6}{2}$ can be modeled with tiles.

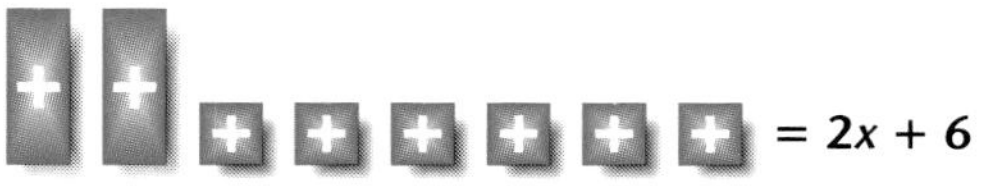

= 2x + 6

Divide the tiles into 2 sets.

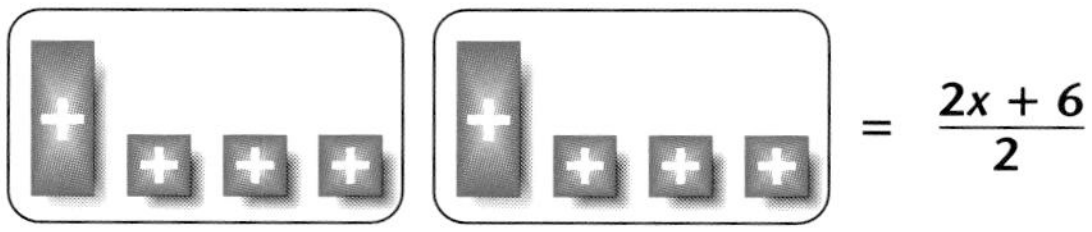

$= \dfrac{2x + 6}{2}$

Since each set contains 1 x-tile and 3 1-tiles, $\dfrac{2x + 6}{2} = x + 3$.

Use tiles to simplify the expression $\frac{8x - 4}{4}$.

When you divide an expression by a number, *each term* is divided by that number.

DIVIDING AN EXPRESSION

For all numbers a, b and c, $c \neq 0$,

$$\frac{a + b}{c} = \frac{a}{c} + \frac{b}{c} \text{ and } \frac{a - b}{c} = \frac{a}{c} - \frac{b}{c}.$$

EXAMPLE 5

Simplify $\frac{8 - 4x}{-4}$.

Solution ➤

Remember that since the quantity $8 - 4x$ is divided by -4, each term must be divided by -4.

$$\frac{8 - 4x}{-4} = \frac{8}{-4} - \frac{4x}{-4}$$

$$= \frac{8}{-4} + \frac{-4x}{-4}$$

$$= -2 + x \quad \diamond$$

 **CRITICAL Thinking** Explain why $\frac{-4x^2}{-4} = x^2$.

Since $\frac{-4x^2}{-4}$ can be rewritten $\frac{-4}{-4}x^2$, then $\frac{-4x^2}{-4} = x^2$.

EXERCISES & PROBLEMS

Communicate

Explain how to write an expression that shows Jan's earnings if she earns $6 an hour and works

1. 4 hours.   **2.** 2.5 hours.   **3.** h hours.

4. Discuss how to evaluate $2j + 3$ when j is 10.

5. Explain how to simplify the expression $3y \cdot 4y$.

6. Discuss how to simplify $-2(4m + 5)$.
How do you know when an expression is simplified?

7. Describe how to simplify the expression $\frac{5p - 15}{-5}$.

Practice & Apply

Evaluate 13r for the following values of r.

8. -4 -52 **9.** 1.5 19.5   **10.** 7 91   **11.** $\frac{1}{2}$ $\frac{13}{2}$, $6\frac{1}{2}$, or 6.5

Evaluate 2t + 1 for the following values of t.

12. 10 21   **13.** 8.5 18   **14.** -6.2 -11.4 **15.** $\frac{3}{4}$ $\frac{5}{2}$, $2\frac{1}{2}$, or 2.5

Simplify each expression.

16. $2 \cdot 6x$ 12x   **17.** $-6x \cdot 2$ $-12x$ **18.** $6x \cdot 2x$ $12x^2$ **19.** $-66x \div 2$ $-33x$

20. $12x \cdot 3x$ $36x^2$ **21.** $-2(6x + 3)$ $-12x - 6$ **22.** $-1.2x \cdot 3x$ $-3.6x^2$ **23.** $-12x \div 3$ $-4x$

24. $7x - (3 - x)$ $8x - 3$ **25.** $-3(7x - 3)$ $-21x + 9$ **26.** $-2(4x - 1) - 8x + 2$ **27.** $3x \cdot 5 + 2x \cdot 2$ 19x

28. $-2 \cdot 8x$ $-16x$ **29.** $8x \cdot 2x$ $16x^2$ **30.** $-8x \div 2$ $-4x$ **31.** $-21x \cdot 3x$ $-63x^2$

32. $2.1x \cdot 3x$ $6.3x^2$ **33.** $-21x \div 7$ $-3x$ **34.** $8x - (2 - 5x)$ $13x - 2$ **35.** $8(x + 1) - (2 - 5x)$ $13x + 6$

36. $(3x + 2y - 9) - 2(x - y)$ $x + 4y - 9$ **37.** $4(x - 5y) - (2x - 2y)$ $2x - 18y$ **38.** $\frac{5w + 15}{-5} - w - 3$

39. $8(x - y) + 5(3x - 3y)$ $23x - 23y$ **40.** $(3x + 5y - 9) - 7(x - y)$ $-4x + 12y - 9$ **41.** $\frac{8 + 16w}{8}$ $1 + 2w$

42. $\frac{-90x + 2.7}{-9}$ $10x - 0.3$ **43.** $-(x - y) - 4(x - y + 9)$ $-5x + 5y - 36$ **44.** $9(2x + y) - 3(3x + 3y)$ 9x

45. $\frac{11 - 33y}{11}$ $1 - 3y$ **46.** $\frac{-10x + 35}{5}$ $-2x + 7$ **47.** $-6(4x - y) - 4(2x - 3y)$ $-32x + 18y$

48. A computer program shows the formula A=$-5*$B+B.
What is the value of A when B is 3? -12

Nicole earns $5.25 per hour at her part-time job.

Wages Find Nicole's earnings for each of the following days:

49. 4 hours on Friday $21.00

50. 6.5 hours on Saturday $34.13

51. total hours for Friday and Saturday $55.13

52. h hours on Monday 5.25h

53. Write an expression in simplified form that shows the total number of hours that Nicole worked for these three days. $10.5 + h$

54. **Geometry** The formula for area is $A = lw$. Find the area of the rectangle by evaluating the formula for the given length and width. 24 sq cm

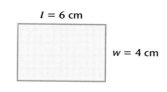
$l = 6$ cm
$w = 4$ cm

55. **Geometry** Evaluate the volume formula, $V = lwh$, for a rectangular prism with the given dimensions. 96 cu in.

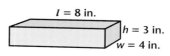
$l = 8$ in.
$h = 3$ in.
$w = 4$ in.

Charlotte is a plumber who charges a fixed service fee of $20 per job, plus $35 an hour.

Small Business How much does Charlotte charge for a job that takes

56. 1 hour? $55   **57.** 3 hours? $125   **58.** h hours?
$C = 35h + 20$

On Sundays, Charlotte doubles her hourly charges. How much does she charge for a job that takes

59. 1 hour?
$90   **60.** 3 hours?
$230   **61.** h hours?
$C = 70h + 20$

Look Back

Simplify. **[Lesson 5.1]**

62. $(3x + 2y) + (3x - 2y) - (3x - 2y)$ $3x + 2y$

63. $(x^2 + 2y + 4) - (2x - 3y - 2)$ $x^2 - 2x + 5y + 6$

Solve each inequality. **[Lessons 5.6, 5.7]**

64. $x - 4 \le 3$
$x \le 7$ **65.** $x + 5 \ge -2$
$x \ge -7$ **66.** $x < -3 + 4$
$x < 1$

Solve each inequality. **[Lesson 5.7]**

67. $x + \frac{1}{3} \le \frac{1}{2}$
$x \le \frac{1}{6}$ **68.** $x - 16 \ge \frac{2}{5}$
$x \ge 16\frac{2}{5}$ **69.** $\frac{4}{5} + x > 7$
$x > 6\frac{1}{5}$

Look Beyond

Simplify.

70. $\dfrac{2x^2 + 4x}{4x}$ $\dfrac{x}{2} + 1$

71. $\dfrac{5y^2 - 15y}{-5y}$ $-y + 3$

72. In a bowling alley, the pins are set up as shown. Move exactly three of the pins to form a triangle like the original one, but pointing away from you.

73. Sixty-four teams are picked to play in the NCAA tournament. Each team plays until it loses one game. How many games are played in order to have a champion?   63 games

Is There ORDER in CHAOS?

Finding Order in Disorder

The science of chaos reveals nature's secrets.

Stock prices lurch and career during weeks of financial mayhem. The number of measles cases inexplicably soars and crashes. A storm unparalleled in more than 40 years slams into England, killing at least 13 people. Such apparently random events have always seemed far beyond the understanding of even the most powerful computers and brilliant researchers. But are they?

A growing and eclectic band of scientists has come to suspect that such chaotic happenings are governed by laws, just as orbiting planets and falling apples are.

The premise of chaos theory is the oldest cliché in science: beneath disorder lurks order. "Chaologists" find that, although it may never be possible to precisely predict the weather, the stock market, or even the path of a roulette ball, one can foresee *patterns* in their behavior. These patterns are the order within the chaos.

Chaos's power to explain diverse phenomena has encouraged researchers to seek it everywhere. Cardiologists studying the human heart find that its normal rhythm is subtly chaotic, rather than regular like a metronome. Climatologists pondering the ice ages find that their timing follows no regular pattern.

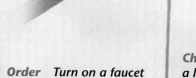

Order Turn on a faucet just a little. The flow of water looks as smooth as glass tube. This is order.

Chaos Turn the handle a little more. The stream changes. Its outer edges flicker in unpredictable turbulence. This is chaos.

Some things, like the stream of water, are sometimes orderly and sometimes chaotic. You can use equations to model such dual behavior. To see how this works, explore a biological system that is easier to model than a stream of water—an imaginary population of insects.

The insects hatch in the spring and die after laying eggs in the fall. The population each year depends on the population of the prior year. The more insects there are, the more eggs there are to hatch next spring. Other factors, such as limited food supply, keep the population from becoming infinitely large.

If you know the size of the insect population one year, you can find the size of the population the next year by using the following equation.

$$P_2 = r \cdot P_1(1 - P_1)$$ P_1 is the size of the population in year 1, P_2 is the size of the population in year 2, and r is some constant.

To find the population in year 3, use $P_3 = r \cdot P_2(1 - P_2)$.

What equation would you use for year 4?

In these equations, all values of P must stay in the range 0 to 1. A value of 0 represents no insects, and a value of 1 represents the final population size that year.

Now you are ready to see how these equations can model behavior that is sometimes orderly and predictable and sometimes chaotic and unpredictable.

Cooperative Learning

1. Copy and complete the chart for $r = 2$ and $r = 4$.

   a. To start, choose any value for P_1 where $0 < P_1 < 1$. Round each answer to three decimal places.

   b. Do you see a pattern in your results for each value of r? Explain.

   c. Compare your results with the results of groups that started with a different value for P_1.

   d. For which values of r can you make a good prediction about the size of the population in year 6 even if your year 1 data were a little off?

2. Copy and complete the chart for $r = 4$. This time start with a value for P_1 that is 0.001 less or 0.001 greater than the value you originally used.

3. In chaos, a tiny difference at the beginning can make a big difference later on. For which value of r resulted in chaos? Explain.

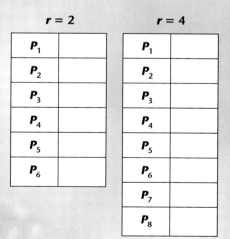

$r = 2$	
P_1	
P_2	
P_3	
P_4	
P_5	
P_6	

$r = 4$	
P_1	
P_2	
P_3	
P_4	
P_5	
P_6	
P_7	
P_8	

Multiplication and Division Equations

why *Carpenters use problem-solving strategies, estimation, and mental computation in their work. They write and solve equations to calculate measurements and to compute the cost of materials and labor.*

8FT.
2×6's
$2.40 EACH

Alice needs to purchase 20 8-foot reinforcement beams. What is the cost per foot?

Multiplication Equations

Carpentry To find the cost per foot, you can draw a picture or use an equation.

Draw a picture of a beam divided into 8 1-foot pieces.

To find the cost of one piece, divide $2.40 by 8. $\quad \frac{2.40}{8} = 0.30$

The cost per foot is $0.30.

Since the cost per foot is represented by x, the equation that models the problem is $8x = 2.40$. To find x, divide each side of the equation by 8.

$$8x = 2.40 \qquad \text{Given}$$
$$\frac{8x}{8} = \frac{2.40}{8} \qquad \text{Divide both sides by 8.}$$
$$x = 0.30 \qquad \text{Simplify.}$$

Each piece costs $2.40 ÷ 8, or $0.30.

You can solve multiplication equations by using division, as shown in Example 1.

> ### DIVISION PROPERTY OF EQUALITY
> If the expressions on each side of an equation are divided by equal nonzero amounts, the expressions remain equal.

EXAMPLE 1

GEOMETRY
Connection

In a regular pentagon all of the angles have equal measures. If the sum of the measures is 540°, find the measure of each angle.

Regular pentagon

Solution ➤

Organize.
Find the information needed to write an equation. The total number of angles is 5. The sum of the measures of the angles is 540°.

Write.
Let x be the measure of each angle.
  Then $5x$ is the sum of the measures of all 5 angles.
  Thus, $5x = 540$.

Solve.

$5x = 540$	Given
$\dfrac{5x}{5} = \dfrac{540}{5}$	Division Property of Equality
$x = 108$	Simplify.

Check.
$5(108) \stackrel{?}{=} 540$

$540 = 540$ True

Each angle has a measure of 108°. ❖

EXAMPLE 2

GEOMETRY
Connection

The formula for the circumference of a circle is $C = \pi d$. Solve for d.

Solution ➤

Divide both sides of the equation by π.

$C = \pi d$	Given
$\dfrac{C}{\pi} = \dfrac{\pi d}{\pi}$	Division Property of Equality
$\dfrac{C}{\pi} = d$, or $d = \dfrac{C}{\pi}$	Simplify. ❖

$C = \pi d$

Try This   The area of a rectangle is $A = lw$. Solve for w.

$w = \dfrac{A}{l}$

Division Equations

Sports Division is used to find batting averages. The number of hits is divided by the number of times at bat. To find the number of hits, a division equation can be used as a model.

$$\frac{\text{Hits}}{\text{Times at bat}} = \text{Batting average}$$

Alex batted 150 times during his varsity career. If his final batting average is 0.300, how many hits did he get?

Let H represent the number of hits. The problem can be modeled by the following division equation:

$$\frac{H}{150} = 0.300$$

To solve the division equation, multiply each side of the equation by the divisor of H, 150.

$$150\left(\frac{H}{150}\right) = 150(0.300)$$
$$H = 45$$

Alex had 45 hits in his varsity career.

The solution to the batting average problem leads to another important property of equality.

MULTIPLICATION PROPERTY OF EQUALITY

If the expressions on each side of an equation are multiplied by equal amounts, the expressions remain equal.

EXAMPLE 3

Solve the equation $\frac{x}{-3} = -6$.

Solution ➤

Use the Multiplication Property of Equality.

$$\frac{x}{-3} = -6 \qquad \text{Given}$$

$$-3\left(\frac{x}{-3}\right) = -3(-6) \qquad \text{Multiplication Property of Equality}$$

$$x = 18 \qquad \text{Simplify.}$$

Check the solution.

$$\frac{18}{-3} \stackrel{?}{=} -6$$

$$-6 = -6 \qquad \text{True} \diamond$$

Compare the Division Property of Equality with the Multiplication Property of Equality. How are they alike? How are they different?

You now have tools to solve addition, subtraction, multiplication, or division equations. Make sure that you apply the correct property when solving an equation.

EXAMPLE 4

Solve. **(A)** $y + \frac{1}{3} = 6$ **(B)** $\frac{x}{0.7} = -56$

Solution ➤

(A) This is an addition equation. Subtract $\frac{1}{3}$ from each side.

$$y + \frac{1}{3} = 6$$

$$\left(y + \frac{1}{3}\right) - \frac{1}{3} = 6 - \frac{1}{3}$$

$$y = 5\frac{2}{3}$$

Check $5\frac{2}{3} + \frac{1}{3} \stackrel{?}{=} 6$

$6 = 6$ True

(B) This is a division equation. Multiply each side by 0.7.

$$\frac{x}{0.7} = -56$$

$$0.7\left(\frac{x}{0.7}\right) = (0.7)(-56)$$

$$x = -39.2$$

Check $\frac{-39.2}{0.7} \stackrel{?}{=} -56$

$-56 = -56$ True ❖

Why is it important to identify what type of equation you are solving before you apply one of the properties of equality?

Try This Solve. **a.** $d - \frac{3}{4} = 8$ $8\frac{3}{4}$ **b.** $\frac{m}{-1.5} = -6$ 9

EXAMPLE 5

Physics In science, **density** is the mass of a material divided by its volume. The formula for density is $\frac{m}{v} = d$. Find the mass, m, in terms of the density, d, and the volume, v. In other words, solve $\frac{m}{v} = d$ for m.

Solution ➤

Multiply each side of the equation by v.

$$\frac{m}{v} = d$$ Given

$$v\left(\frac{m}{v}\right) = v(d)$$ Multiplication Property of Equality

$$m = vd$$ Simplify. ❖

Try This Solve $m = \frac{f}{a}$ for f. $f = ma$

Lesson 6.2   Multiplication and Division Equations   **341**

Rules for solving equations are summarized in the following chart:

SUMMARY FOR SOLVING EQUATIONS	
Type of equation	Operation
Addition	Subtract equal amounts from the expressions on each side.
Subtraction	Add equal amounts to the expressions on each side.
Multiplication	Divide the expressions on each side by equal nonzero amounts.
Division	Multiply the expressions on each side by equal amounts.

EXERCISES & PROBLEMS

Communicate

1. Make up a real-world problem that can be modeled by each equation.
   a. $5x = 100$
   b. $\frac{x}{2} = 10$

Explain how to solve for x.

2. $592x = 812$

3. $x - 246 = 528$

4. $5x = \frac{1}{10}$

5. $x + 10 = 5$ Subtract 10 from each side of the equation.

6.  **Geometry** From the formula for the circumference of a circle, $C = 2\pi r$, explain how to find the radius, r, in terms of the circumference, C, and π.

Describe a real-world problem that could be modeled by each equation.

7. $4x = 40$

8. $\frac{x}{4} = 20$

9. $2.50p = 15$

10. $\frac{x}{20} = 4$

Practice & Apply

Give the property needed to solve each equation. Then solve it.

11. $\frac{y}{3} = -13$

12. $\frac{x}{27} = -26$

13. $x - \frac{1}{3} = 2$

14. $7x = 56$

15. $\frac{b}{-9} = 6$

16. $-12y = 84$

17. $x - \frac{2}{3} = 2$

18. $777x = -888$

19. $5.6v = 7$

20. $7 = -56w$

21. $\frac{x}{-7} = -1.4$

22. $3x = 2$

Solve each equation.

23. $888x = 777$ $\frac{7}{8}$

24. $x + \frac{3}{4} = 12$ $11\frac{1}{4}$

25. $x - 888 = 777$ 1665

26. $-3f = 15$ -5

27. $\frac{x}{7} = -8$ -56

28. $\frac{x}{0.5} = 6$ 3

29. $w + 0 = -22$ -22

30. $-4x = -3228$ 807

31. $4b = -15$ $-3\frac{3}{4}$

32. $\frac{p}{111} = -10$ -1110

33. $\frac{s}{-1} = -40$ 40

34. $c + 7 = 63$ 56

35. $r - \frac{1}{5} = 2$ $2\frac{1}{5}$

36. $2a = 13$ $6\frac{1}{2}$

37. $\frac{x}{10} = -1.9$ -19

38. $d - 7 = 35$ 42

39. $\frac{w}{-1.2} = 10$ -12

40. $7e = -14$ -2

41. $\frac{m}{-9} = 0$ 0

42. $\frac{b}{15} = 1$ 15

43. $0.55x = 0.55$ 1

44. $\frac{p}{-9} = 0.9$ -8.1

45. $-3x = -4215$ 1405

46. $p + 2300 = 890$
-1410

Carpentry   Use the information at the right for Exercises 47–51. Write and solve an equation to describe each situation.

47. How many rolls of masking tape can you buy with $6.00? 5

48. Masking tape comes in packages of 4 rolls for $4.32. What is the cost of one roll in the package? Is the cost of a single roll of masking tape more than the cost of one packaged roll?
$1.08; yes

49. An employer has $19.00 to spend on tape measures for his crew. How many can he buy for that amount? 3

50. What is the price of a single extension cord if you can buy a package of 6 for $7.26? $1.21

51. Suppose that a package of 4 AA batteries costs $2.52. What is the price of each battery? 63¢

Masking Tape $1.15
Hammers $7.99
Tape Measure $4.95

52. Travel   Natalie's family wants to travel 400 miles in 10 hours. They figure that they will stop for a total of 2 hours along the way. At what speed must they travel for the trip? 50 mph

53. Travel   Derrick averaged 50 miles per hour during 3 hours of driving. He has 75 miles more to go. How many miles will he drive in all? 225 miles

54. Maria wants to drive 320 miles in 8 hours. What speed should she average for the trip? 40 mph

Solve each formula for the indicated variable.

55. $r = \frac{d}{t}$ for d $rt = d$ **56.** $A = bh$ for b $\frac{A}{h} = b$ **57.** $V = lwh$ for w $\frac{V}{lh} = w$

 Geometry   In a regular polygon with n sides, the measure, m, of each interior angle is given by the formula $m = \frac{180(n-2)}{n}$. Find the measure of an interior angle in the following regular polygons. HINT: Use the number of sides of each polygon for n.

58. equilateral triangle   60°
59. square   90°
60. pentagon   108°
61. octagon   135°

62. What effect does the *number* of angles in each regular polygon have on the *size* of the angles? As the number of angles increases, the measure of the angle increases.

63. 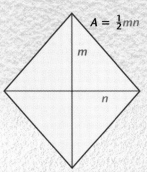 **Geometry**   Solve the formula for the area of a rhombus for the variable n. $\frac{2A}{m} = n$

Rhombus

$A = \frac{1}{2}mn$

64. The diagonals of the rhombus are m and n. If the area, A, is 21 square inches and the longer diagonal, m, is 7 inches, what is the length of the other diagonal, n?   6 in.

Look Back

65. Write a sequence of five terms, starting with 3 and doubling each term to get the next term.   **[Lesson 1.1]**   3, 6, 12, 24, 48

Evaluate each expression if a is 3, b is 2, and c is 0.   [Lesson 1.7]

66. $\frac{a + b \cdot c}{3}$ 1   **67.** $\frac{a \cdot b}{b + c}$ 3   **68.** $\frac{a + b}{b + c}$ $2\frac{1}{2}, \frac{5}{2}$, or 2.5

69. Place parentheses to make $30 \cdot 7 - 4 \cdot 10 \div 2 \div 2 = 515$ true.
[Lesson 1.7] $(30 \cdot 7 - 4) \cdot 10 \div 2 \div 2 = 515$

Find the absolute value.   [Lesson 2.1]

70. $|0|$ 0   **71.** $|-5|$ 5

72. $\left|-\frac{1}{6}\right|$ $\frac{1}{6}$ **73.** $|9|$ 9

Look Beyond

Solve each equation.

74. $3y = 10 + 5y$ -5 **75.** $4x + \frac{1}{2} = 5x$ $\frac{1}{2}$
76. $x + 4 = 3x - 2$ 3   **77.** $6x - 2 = 14 - 2x$ 2

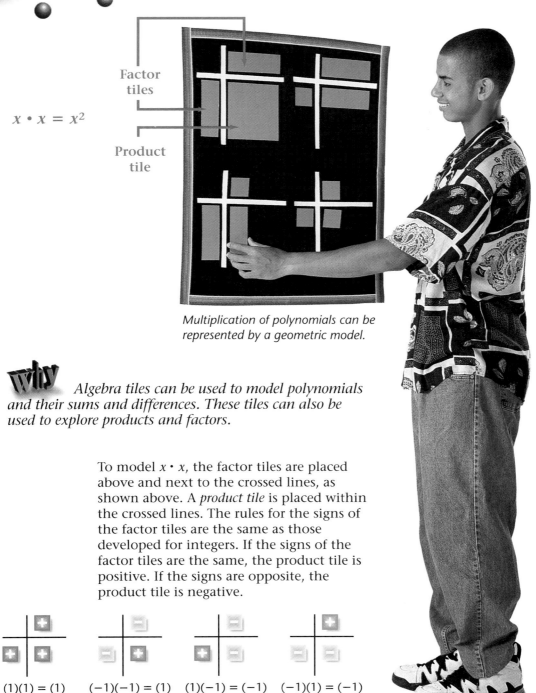

LESSON 6.3

Exploring Products and Factors

$x \cdot x = x^2$

Factor
tiles

Product
tile

*Multiplication of polynomials can be
represented by a geometric model.*

Why Algebra tiles can be used to model polynomials
and their sums and differences. These tiles can also be
used to explore products and factors.

To model $x \cdot x$, the factor tiles are placed
above and next to the crossed lines, as
shown above. A *product tile* is placed within
the crossed lines. The rules for the signs of
the factor tiles are the same as those
developed for integers. If the signs of the
factor tiles are the same, the product tile is
positive. If the signs are opposite, the
product tile is negative.

$(1)(1) = (1)$ $(-1)(-1) = (1)$ $(1)(-1) = (-1)$ $(-1)(1) = (-1)$

345

•Exploration 1 Modeling With Tiles

1 What product is shown in the diagram that models $(x)(-x)$? $-x^2$

2 Use tiles to model $(x)(x)$, $(-x)(x)$, and $(-x)(-x)$. What is the product for each?
x^2; $-x^2$; x^2

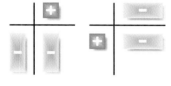

3 The diagrams model $(-x)(1)$ and $(1)(-x)$. What product is shown for both diagrams? $-x$; $-x$

4 Use tiles to model $(x)(-1)$, $(x)(1)$, $(-x)(-1)$, $(1)(x)$, $(1)(-x)$, and $(-1)(-x)$. What is the product for each? ❖ $-x$; x; x; x; $-x$; x

Either tiles or the Distributive Property can be used to find the product of 3 and $2x + 1$.

Algebra-Tile Model

$2x + 1$

Count the tiles for the product.
$6x + 3$

Distributive Property

$$3(2x + 1) = 3(2x) + 3(1)$$
$$= 6x + 3$$

The model shows that the product of 3 and $2x + 1$ is $6x + 3$. The tiles modeling 3 and $2x + 1$ represent two *factors* of $6x + 3$. Notice that applying the Distributive Property results in the same algebraic product.

APPLICATION

Suppose that 3 students each have 2 boxes of pencils and 1 loose pencil. If b represents the number of pencils in a box, then $2b + 1$ represents the number of pencils that each student has. Since there are 3 students, $3(2b + 1)$ represents the total number of pencils. This product can be rewritten as $6b + 3$; that is, $3(2b + 1) = 6b + 3$. ❖

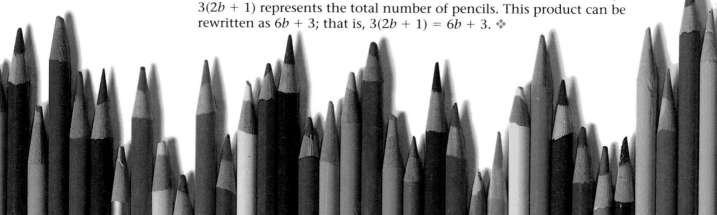

Exploration 2 *Finding Products*

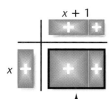

GEOMETRY
Connection

$x + 1$

x

Product rectangle

Examine the product rectangle.

 What are the two factors? $x, x + 1$

 What is the product? $x^2 + x$

 Use the Distributive Property to find the product $x(x + 1)$. $x^2 + x$

 Use tiles to model each product.
  a. $2x(x + 3)$ **b.** $3(3x + 2)$ $9x + 6$
    $2x^2 + 6x$

 Use the Distributive Property to find each product.
  a. $2x(x + 3)$ **b.** $3(3x + 2)$ $9x + 6$
    $2x^2 + 6x$

 Explain how to use the Distributive Property to find the product $2x(x + 1)$. ❖ Multiply $2x$ and x, $2x$ and 1. Add the products. $2x^2 + 2x$.

Rectangles that model products can be built by using positive *and* negative tiles. When you build *product rectangles,* be careful to obey the rules for multiplying positive and negative integers.

Exploration 3 *Using Negative Tiles*

$x - 1$

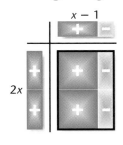

$2x$

Examine the product rectangle.

 What are the two factors? $2x, x-1$

 What is the product? $2x^2 - 2x$

 Use the Distributive Property to find the product $2x(x - 1)$. $2x^2 - 2x$

 Use tiles to model each multiplication problem.
      a. $4(-x + 2)$ **b.** $3(2x - 1)$
        $-4x + 8$ $6x - 3$

 Use the Distributive Property to find each product.
  a. $4(-x + 2)$ **b.** $3(2x - 1)$
    $-4x + 8$ $6x - 3$

 Explain how to use the Distributive Property to find the product $-3(-x + 4)$. ❖ Multiply -3 and $-x$, -3 and 4. Add the products. $3x - 12$

Once a product rectangle is formed, you can tell what two factors have been multiplied to form the product.

Exploration 4 *From Product Rectangles to Factors*

1 Begin with a product consisting of the tiles for $4x^2$ and $6x$. Form a product rectangle with $2x$ as one factor.

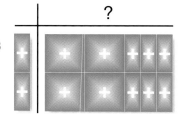

2 What is the second factor? $2x + 3$

3 Start with a product consisting of the tiles for $3x^2$ and $6x$. Form a product rectangle with $3x$ as one factor.

4 What is the second factor? $x + 2$

5 Explain how to use the Distributive Property to complete $3x^2 + 6x = 3x(\underline{\ ?\ } + \underline{\ ?\ })$. Think: $3x$ times what number equals $3x^2$? Think: $3x$ times what number equals $6x$? $3x(x + 2)$.

6 Explain how to use a product rectangle and the Distributive Property to complete $5x^2 + 15x = 5x(\underline{\ ?\ } + \underline{\ ?\ })$. ❖ Think: $5x$ times what number equals $5x^2$? Think: $5x$ times what number equals $15x$? $5x(x + 3)$.

The greatest common factor, or GCF, of $4x^2$ and $6x$ is $2x$. For the polynomial $4x^2 + 6x$, the factored form is $2x(2x + 3)$. Here are some more examples.

Polynomial	GCF	Factored Form
$5x^2 + 10x$	$5x$	$5x(x + 2)$
$3x^2 - 9x$	$3x$	$3x(x - 3)$
$8x^3 - 4x$	$4x$	$4x(2x^2 - 1)$
$-2x - 6x^2$	$-2x$	$-2x(1 + 3x)$
$6x^4 - 9x^2$	$3x^2$	$3x^2(2x^2 - 3)$

CT– No, some polynomials cannot be written as factors; for example $x^2 + 1$ and $x^2 + x + 1$ cannot be arranged in a rectangle.

CRITICAL Thinking

Given the tiles that model a polynomial containing x^2, can you always form a product rectangle from the tiles? Explain.

EXERCISES & PROBLEMS

Communicate

1. Explain how to use algebra tiles to show the area of a rectangle with sides $2x$ and $x + 5$.

2. Explain how to use the Distributive Property to find the area of a rectangle with sides $2x$ and $x + 5$.

3. Draw a tile model to find the sides of a rectangle that has an area of $4x^2 + 2x$.

4. Tell what it means to write the polynomial $x^3 + x$ in factored form.

Practice & Apply

5. Draw a tile model to show the area of a rectangle that has factors of $x + 7$ and x.

6. Draw a tile model to find the sides of a rectangle that has a product of $2x^2 + 2x$.

Geometry Find the factors for the rectangle modeled below with positive tiles. Then write the product shown by the tiles.

7.

$2x(3x + 1); 6x^2 + 2x$

8.

$2x(2x + 1); 4x^2 + 2x$

9.

$2x(x + 5); 2x^2 + 10x$

10.

$3(x + 3); 3x + 9$

11.

$2(3x + 1); 6x + 2$

12.

$4(x + 3); 4x + 12$

33. $7y(y^3 + 7)$
34. $m^2(m^2 - 6)$
35. $-2r(2r + 3)$

Use the Distributive Property to find each product.

13. $4(x + 2)$ $4x + 8$ 14. $6(2x + 7)$ $12x + 42$ 15. $5(y + 10)$ $5y + 50$ 16. $3(m + 8)$ $3m + 24$

17. $x(x + 2)$ $x^2 + 2x$ 18. $2y(y - 4)$ $2y^2 - 8y$ 19. $3r(r^2 - 3)$ $3r^3 - 9r$ 20. $p^2(p + 7)$ $p^3 + 7p^2$

21. $8(3x - 4)$ $24x - 32$ 22. $4y(y - 4)$ $4y^2 - 16y$ 23. $5x(4x + 9)$ 24. $3w(w^2 - w)$ $3w^3 - 3w^2$

25. $2y^2(y^2 + y)$ 26. $2t^2(2t^2 + 8t)$ $20x^2 + 45x$

   $2y^4 + 2y^3$ $4t^4 + 16t^3$ 27. $3y^2(3y - 6)$ 28. $z(11z^2 + 22z)$

                                                              $9y^3 - 18y^2$ $11z^3 + 22z^2$

Write each polynomial in factored form.

29. $3x^2 + 6$ $3(x^2 + 2)$ 30. $5x^2 - 20$ $5(x^2 - 4)$ 31. $y^2 - y^3$ $y^2(1 - y)$ 32. $4p + 12p^3$ $4p(1 + 3p^2)$

33. $7y^4 + 49y$ 34. $m^4 - 6m^2$ 35. $-4r^2 - 6r$ 36. $9n^2 - 27n^4$ $9n^2(1 - 3n^2)$

Geometry Mary wants a rectangular flower bed and lawn in her backyard.

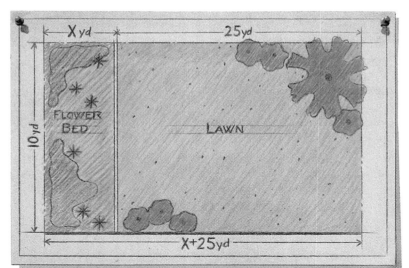

A landscape artist is designing Mary's backyard.

37. Write an expression for the area of the flower bed. 10x

38. Write an expression for the area of the lawn. 25·10

39. Write an expression for the total area of the backyard. 10(x + 25)

40. What conjecture can you make about the areas of the flower bed and the lawn compared with the area of the backyard?

41. Discounts Wholesale Grocery has pieces of dinnerware for sale in the houseware department. Glasses are $1.50, plates are $2.50, cups are $1.25, and saucers are $1.00. A 25% discount is given for purchases over $10.00. If a customer buys x glasses and y plates and receives a discount, write an expression for how much of a discount the customer receives. 0.25(1.50x + 2.50y)

42. **Geometry** Find the perimeter of a rectangle with a width of $2x + 3$ and a length of $3x$. 10x + 6

Look Back

Use the LCD to compare each pair of fractions, using the symbol <, >, or =. **[Lesson 3.2]**

43. $\frac{12}{35}$ and $\frac{18}{40}$ <

44. $\frac{1}{4}$ and $\frac{3}{15}$ >

45. Hobbies Christine is carving plaques from a piece of wood that is 8 feet long. How many plaques can she carve if each plaque requires a piece of wood that is $\frac{5}{6}$ of a foot long? **[Lesson 3.5]** 9 plaques

 Geometry The lengths of the sides of two triangles are given. Use the converse of the "Pythagorean" Right-Triangle Theorem to decide if each triangle is a right triangle. **[Lesson 4.7]**

46. 2, 4, 5 No

47. 10, 24, 26 Yes

Look Beyond

Use tiles to multiply each binomial.

48. $(2x + 1)(x - 5)$

49. $(x + 3)(3x - 4)$

LESSON 6.4 Rational Numbers

why *In Lesson 6.2 you solved equations involving fractions and decimals. Although fractions and decimals are not integers, the properties for solving equations still apply.*

Geno decides to withdraw $25 from his savings account in 2 equal withdrawals.

Banking How can you use negative numbers to express the amount of each withdrawal?

Since Geno is withdrawing $25, use -25 to represent the total withdrawal.

Let x represent each withdrawal. Then $2x$ represents the 2 equal withdrawals. The equation to model the problem is

$$2x = -25.$$

Apply the Division Property of Equality.

$$\frac{2x}{2} = \frac{-25}{2}$$

$$x = -12\frac{1}{2}$$

The result is $x = -12\frac{1}{2}$, or -12.50. Geno should withdraw $12.50 each time for a total withdrawl of $25.

Recall from Chapter 3 that numbers such as $-12.5, -12\frac{1}{2}$, and $\frac{-25}{2}$ are called negative rational numbers. Other examples of rational numbers include $\frac{1}{2}, -\frac{1}{2}$, 0, and 2. A rational number is a number that can be expressed as the ratio of two integers, with 0 excluded from the denominator.

Rational numbers can be graphed on a number line.

How can you show that the integers 0 and -2 are also rational numbers?

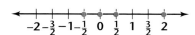

EXAMPLE 1

Stocks   A share of stock increased by $\frac{1}{4}$ of a point on Monday and then decreased by $\frac{5}{8}$ of a point on Tuesday. What was the net effect on the stock?

Solution A ➤

Use a number line to model this problem.

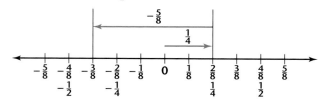

Solution B ➤

Use a common denominator to solve this problem.

$$\frac{1}{4} + \left(-\frac{5}{8}\right) = \frac{2}{8} + \left(-\frac{5}{8}\right)$$

$$= \frac{2 + (-5)}{8}$$

$$= -\frac{3}{8}$$

The net effect is a decrease of $\frac{3}{8}$ of a point. ❖

The set of rational numbers shares all of the properties of the set of integers. However, the *Reciprocal Property* is one additional rational number property that is not shared by the set of integers.

RECIPROCAL PROPERTY

For any nonzero number r, there is a number $\frac{1}{r}$ such that

$$r \cdot \frac{1}{r} = 1.$$

Examine the table of values shown for the equation $y = \frac{1}{x}$. How do the y-values for negative x-values compare with the y-values for positive x-values? They are opposites.

X	Y₁	
-3	$-.3333$	
-2	$-.5$	
-1	-1	
0	ERROR	
1	1	
2	.5	
3	.33333	

X=0

EXAMPLE 2

Write the reciprocal of each rational number.

A $\frac{2}{3}$ **B** 7   **C** $-\frac{4}{5}$ **D** $-1\frac{1}{2}$

Solution ➤

A The reciprocal of $\frac{2}{3}$ is $\frac{3}{2}$ $\left(\text{or } 1\frac{1}{2}\right)$ because $\frac{2}{3} \cdot \frac{3}{2} = \frac{6}{6} = 1$.

B The reciprocal of 7 $\left(\text{or } \frac{7}{1}\right)$ is $\frac{1}{7}$ because $\frac{7}{1} \cdot \frac{1}{7} = \frac{7}{7} = 1$.

C The reciprocal of $-\frac{4}{5}$ is $-\frac{5}{4}$ $\left(\text{or } -1\frac{1}{4}\right)$ because $-\frac{4}{5} \cdot -\frac{5}{4} = \frac{20}{20} = 1$.

D The reciprocal of $-1\frac{1}{2}$ $\left(\text{or } -\frac{3}{2}\right)$ is $-\frac{2}{3}$ because $-\frac{3}{2} \cdot -\frac{2}{3} = \frac{6}{6} = 1$. ❖

Why is -1 its own reciprocal?   Because $\frac{-1}{1} = \frac{1}{-1}$.

Calculator

You can use the $\boxed{x^{-1}}$ or $\boxed{1/x}$ key on a calculator to find reciprocals. The reciprocal of a number is also called its **multiplicative inverse.** How can you find the reciprocal of $-\frac{2}{5}$ on your calculator?

EXAMPLE 3

Find the coefficient of the variable for each expression.

A $-\frac{5}{6}k$ **B** $\frac{3x}{4}$

C $\frac{-y}{8}$ **D** $-\frac{p}{7}$

Solution ➤

A The coefficient of $-\frac{5}{6}k$ is $-\frac{5}{6}$.

B The coefficient of $\frac{3x}{4}$ is $\frac{3}{4}$ because $\frac{3x}{4} = \frac{3}{4}x$.

C The coefficient of $\frac{-y}{8}$ is $-\frac{1}{8}$ because $\frac{-y}{8} = -\frac{1}{8}y$.

D The coefficient of $-\frac{p}{7}$ is $-\frac{1}{7}$ because $-\frac{p}{7} = -\frac{1}{7}p$. ❖

Does the position of a negative sign affect the coefficient of a variable?   No.

The Reciprocal Property can be used to solve equations, as shown in Example 4.

EXAMPLE 4

Solve. **A** $\frac{2}{3}x = 8$ **B** $\frac{-x}{5} = -3$

Solution ➤

Multiply each side of the equation by the reciprocal of the coefficient of x.

A The coefficient of $\frac{2}{3}x$ is $\frac{2}{3}$.

$$\frac{2}{3}x = 8$$

$$\frac{3}{2}\left(\frac{2}{3}x\right) = \frac{3}{2}(8)$$

$$x = \frac{24}{2}$$

$$x = 12$$

B The coefficient of $\frac{-x}{5}$ is $-\frac{1}{5}$.

$$\frac{-x}{5} = -3$$

$$-\frac{1}{5}x = -3$$

$$-5\left(-\frac{1}{5}x\right) = -5(-3)$$

$$x = 15 \ ❖$$

Try This Solve. **a.** $\frac{3}{4}y = -15$ -20 **b.** $\frac{-w}{9} = 8$ -72

What is the reciprocal of the coefficient of x in the expression $\frac{2x}{-3}$? $-\frac{3}{2}$

Remember, a proportion states that two ratios are equal. Example 5 shows two different methods to solve a proportion.

EXAMPLE 5

Solve the proportion $\frac{x}{-5} = \frac{2}{3}$.

Solution ➤

Method A
Multiply each side by the reciprocal of the coefficient of x.

$$\frac{x}{-5} = \frac{2}{3}$$

$$-5\left(\frac{x}{-5}\right) = -5\left(\frac{2}{3}\right)$$

$$x = \frac{-10}{3}$$

$$x = -3\frac{1}{3}$$

Method B
Multiply each side by the least common denominator.

$$\frac{x}{-5} = \frac{2}{3}$$

$$-15\left(\frac{x}{-5}\right) = -15\left(\frac{2}{3}\right)$$

$$3x = -10$$

$$\frac{3x}{3} = -10$$

$$x = \frac{-10}{3}$$

$$x = -3\frac{1}{3}$$

❖

CRITICAL *Thinking*

Compare the two methods shown in Example 5. How are they alike? How are they different?

EXERCISES & PROBLEMS

Communicate

1. Define a rational number. Give three examples.
2. Explain why an integer is also a rational number.
3. Describe how to model $\frac{1}{2} + (-3)$ on the number line.
4. How can you find $\frac{1}{2} - \frac{5}{6}$ using the number line?
5. State the Reciprocal Property. Use the number -3 to give an example of the Reciprocal Property.
6. Give two ways to solve the proportion $\frac{y}{8} = \frac{-2}{3}$.

Practice & Apply

Write the reciprocal of each number.

7. -2 $-\frac{1}{2}$ **8.** $\frac{-1}{9}$ -9 **9.** $4\frac{1}{5}$ $\frac{5}{21}$ **10.** $\frac{10}{3}$ $\frac{3}{10}$ **11.** 1 1

Solve each equation.

12. $\frac{3}{5}x = 3$ 5 **13.** $\frac{1}{-7}y = -6$ 42 **14.** $\frac{-5}{8}q = 10$ -16 **15.** $-\frac{3}{4}x = -9$ 12

16. $\frac{-y}{7} = 11$ -77 **17.** $\frac{p}{4} = -1.2$ -4.8 **18.** $\frac{p}{-11} = -5$ 55 **19.** $-\frac{t}{4} = -5$ 20

20. $-\frac{w}{8} = 9$ -72 **21.** $\frac{-m}{-12} = -10$ -120

Solve each proportion.

22. $\frac{x}{5} = \frac{3}{4}$ $3\frac{3}{4}$ **23.** $\frac{-x}{3} = \frac{2}{5}$ $-1\frac{1}{5}$ **24.** $\frac{x}{4} = \frac{-3}{6}$ -2 **25.** $\frac{p}{-3} = \frac{-4}{7}$ $1\frac{5}{7}$

26. $\frac{x}{16} = \frac{-3}{8}$ -6 **27.** $\frac{x}{-25} = \frac{3}{-5}$ 15 **28.** $\frac{-x}{6} = \frac{5}{3}$ -10 **29.** $\frac{w}{-14} = \frac{6}{7}$ -12

30. $\frac{-x}{6} = \frac{3}{7}$ $-2\frac{4}{7}$ **31.** $\frac{-y}{7.5} = \frac{-2}{4}$ $3\frac{3}{4}$

32. Savings Miguel plans to mow lawns during summer vacation. He spends $\frac{2}{3}$ of his savings to buy a lawn mower. If Miguel paid \$210 for the lawn mower, how much did he have in his savings before buying the lawn mower? $315

33. Sports If Alicia shoots $\frac{3}{4}$ of her free throws by noon, how many total free throws is she supposed to practice? 60

Alicia spent Saturday shooting free throws for basketball. By noon she had practiced shooting 45 free throws.

The cheerleaders sold $\frac{2}{3}$ of their spirit ribbons on the first day.

Set up a proportion and solve each problem.

34. Fund-raising If they sold 450 ribbons on the first day, how many ribbons did the cheerleaders have before they started selling ribbons? 675

35. Sports Mary can run 50 yards in 9 seconds. How many seconds will it take Mary to run 75 yards at the same speed? 13.5 seconds

36. Consumer Economics Miguel went to the store to buy dog food. Six cans of dog food sell for $1.74. He has $5 for his purchase. Does he have enough money to buy 16 cans? yes

37. Cultural Connection: The Americas If 9 crowns were exchanged for 7 pesos, how many pesos were exchanged for 63 crowns? 49 pesos

In the days of colonial Mexico, crowns and pesos were forms of money.

38. **Portfolio Activity** Complete the problem in the portfolio activity on page 327. 84 years

Look Back

39. Evaluate $6p - 3q + r$ if p is 3, q is 4, and r is 10.   **[Lesson 1.7]** 16

Simplify each expression. Write your answers in lowest terms.   **[Lessons 3.4, 3.5]**

40. $\frac{4}{18} - \frac{5}{9}$ $-\frac{1}{3}$ **41.** $\frac{5}{17} \div \frac{10}{17}$ $\frac{1}{2}$ **42.** $\frac{3}{8} + \frac{1}{4} - \frac{5}{16}$ $\frac{5}{16}$

Simplify.   [Lessons 5.1, 5.2]

43. $(3x - 2) + (4x + 7)$ **44.** $2(4q + 3) - (q - 2)$ **45.** $(2a + 3b - 1) - 2(a - 2b - 1)$
     $7x + 5$ $7q + 8$ $7b + 1$

Solve each inequality.   [Lesson 5.6]

46. $x + 4 > -12$ **47.** $y - 0.05 \leq 10.5$ **48.** $w + 1.4 \geq 10.2$
     $x > -16$ $y \leq 10.55$ $w \geq 8.8$

Look Beyond

Find a pair of numbers (x, y) that satisfies both equations.

49. $3x + 2y = 14$ and $6x - 4 = 8$ $(2, 4)$

Exploring Proportion Problems

why *Ratios are used to compare information. They are often used to solve problems and make predictions.*

Pedaling a bicycle is made easier by the front and rear gears, which are connected by a chain. The relationship between the number of revolutions of the front gear and the number of revolutions of the rear gear over a given period of time can be described by using ratios.

•Exploration 1 *Using Proportions*

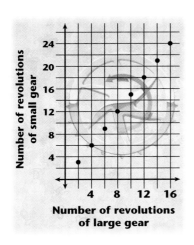

This graph shows the relationship between the number of revolutions of each gear of a bicycle over the same period of time.

1 Copy and complete the following table using the data in the graph.

Number of revolutions of larger gear	2	4	6	?	?	?	?	?
				8	10	12	14	16
Number of revolutions of smaller gear	3	6	?	?	?	?	?	?
				9	12	15	18	21	24

Number of revolutions of small gear

Number of revolutions of large gear

2 Write a sentence to describe the ratio represented by corresponding values in the table.

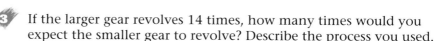 If the larger gear revolves 14 times, how many times would you expect the smaller gear to revolve? Describe the process you used.

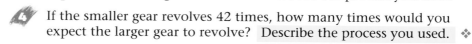 If the smaller gear revolves 42 times, how many times would you expect the larger gear to revolve? Describe the process you used. ❖

In Exploration 1, you used equations called proportions. Recall that a proportion is an equation which states that two ratios are equal. Several ways to write the same proportion are shown below.

$$2:3 = n:27 \quad \text{or} \quad \frac{2}{3} = \frac{n}{27} \quad \text{or} \quad \frac{n}{27} = \frac{2}{3}$$

There are several ways to solve this proportion. In Lesson 3.5, you learned to use multipliers to solve proportions.

Look at the denominators in the proportions. To get from 3 to 27, you multiply by 9, so multiply both the numerator and the denominator of $\frac{2}{3}$ by 9.

$$\frac{n}{27} = \frac{2}{3}$$
$$\frac{n}{27} = \frac{2 \cdot 9}{3 \cdot 9}$$
$$\frac{n}{27} = \frac{18}{27}$$
$$n = 18$$

You can also solve this proportion by multiplying each side of the equation by 27.

$$\frac{n}{27} = \frac{2}{3}$$
$$27 \cdot \frac{n}{27} = 27 \cdot \frac{2}{3}$$
$$n = \frac{54}{3}, \text{ or } 18$$

Notice that a proportion contains four terms. The middle terms are called the **means**. The first and last terms are called the **extremes**.

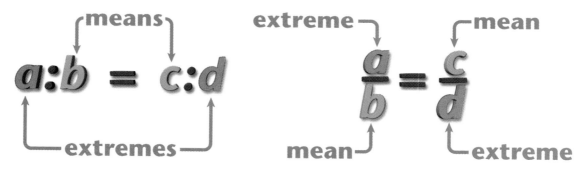

For example, in the proportion from above,

$$2:3 = 18:27 \quad \text{or} \quad \frac{2}{3} = \frac{18}{27},$$

the **means** are **3** and **18**, and the **extremes** are **2** and **27**.

What are the means and extremes in each of the following proportions?

a. $\frac{1}{5} = \frac{6}{30}$ b. $4:12 = 1:3$ c. $\frac{20}{10} = \frac{8}{4}$

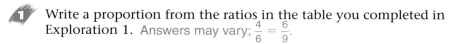

Exploration 2 A Property of Proportions

1 Write a proportion from the ratios in the table you completed in Exploration 1. Answers may vary; $\frac{4}{6} = \frac{6}{9}$.

2 What is the product of the means of this proportion? Answers may vary; 36.

3 What is the product of the extremes of this proportion? Answers may vary; 36.

4 How are the two products related? They are equal.

5 Write a different proportion from the ratios in the table you completed in Exploration 1. Answers may vary; $\frac{2}{3} = \frac{4}{6}$.

6 What is the product of the means of this proportion? Answers may vary; 12.

7 What is the product of the extremes of this proportion? Answers may vary; 12.

8 How are the two products related? They are equal.

9 The product of the means and the product of the extremes are called **cross products**. What appears to be the relationship between the cross products of a proportion? ❖ Cross products of a true proportion are equal.

You can use cross products to determine whether a given statement is a true proportion.

EXTENSION

The cross products for the statement $\frac{42}{54} = \frac{49}{63}$ are

$$42 \cdot 63 = 2646 \quad \text{and} \quad 54 \cdot 49 = 2646.$$

Since the cross products are equal, the statement is a proportion.

The cross products for the statement $\frac{30}{25} = \frac{63}{49}$ are

$$30 \cdot 49 = 1470 \quad \text{and} \quad 25 \cdot 63 = 1575.$$

Since the cross products are *not* equal, the statement is *not* a proportion. ❖

What conclusion can you make if the product of the means does *not* equal the product of the extremes? If the product of the means does not equal the product of the extremes, the statement is not a true proportion.

CRITICAL Thinking

Which is a better buy, 6 items for 75¢ or 8 of the same items for 99¢? Explain how to use a proportion to find the answer.

You can also use cross products to solve proportions.

It is the goal of the Bartlesville City Library to have 2 books for every 5 residents. The recent census showed the population is 12,455. To find the number of books that should be in the library in order to meet the goal, first make a table to organize the information.

Number of books	2	n
Number of residents	5	12,455

Write the proportion shown by the table.

$$\frac{2}{5} = \frac{n}{12{,}455}$$

Then use cross products to solve the proportion. $\frac{2}{5} = \frac{n}{12{,}455}$

$$2 \cdot 12{,}455 = 5n$$
$$24{,}910 = 5n$$
$$\frac{24{,}910}{5} = \frac{5n}{5}$$
$$4982 = n$$

There should be 4982 books in the library. ❖

How can you use a calculator to solve the proportion in one step?

Multiply $\frac{2}{5}$ times 12,455.

Exploration 3 A Geometric Application

1 Recall that similar figures have corresponding sides that are proportional. What is the ratio of *WX* to *AB*? $\frac{1}{2}$

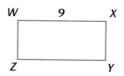

Rectangle WXYZ is similar to rectangle ABCD.

2. Answers may vary.
$\frac{WX}{WZ} = \frac{AB}{AD}$

2 Write a proportion to describe the relationship among *WX*, *WZ*, *AB*, and *AD*.

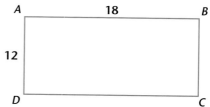

3 Use the proportion you wrote in Step 2 to find *WZ*. *WZ* = 6

4 Find the perimeter of each rectangle. Perim. of *WXYZ* = 30
Perim. *ABCD* = 60

5 What is the ratio of the perimeters of the rectangles? $\frac{1}{2}$

6 What is your conclusion about the ratios of the perimeters of similar rectangles? ❖

EXERCISES & PROBLEMS

Communicate

1. Write a definition of *proportion* in your own words.
2. Describe two different methods that could be used to solve the proportion $\frac{n}{12} = \frac{18}{24}$.
3. Describe two different methods for determining whether the statement $\frac{27}{48} = \frac{18}{32}$ is a proportion.
4. Explain how proportions can be used with similar figures.
5. Write a real-world problem that could be modeled by the proportion $\frac{n}{36} = \frac{2}{3}$.

Practice & Apply

Solve each proportion.

6. $\frac{27}{18} = \frac{42}{n}$ 28

7. $\frac{38}{19} = \frac{n}{20}$ 40

8. $\frac{42}{28} = \frac{36}{n}$ 24

9. $\frac{n}{48} = \frac{72}{96}$ 36

10. $\frac{21.5}{x} = \frac{64.5}{18}$ 6

11. $\frac{x}{37.2} = \frac{16}{24.8}$ 24

12. $\frac{30.8}{112} = \frac{y}{10}$ 2.75

13. $\frac{t}{25} = \frac{473}{15}$ $788\frac{1}{3}$

Determine if each statement is a true proportion.

14. $\frac{15}{9} = \frac{35}{21}$ Yes

15. $\frac{12}{9} = \frac{18}{12}$ No

16. $\frac{56}{24} = \frac{49}{21}$ Yes

17. $\frac{27}{21} = \frac{35}{28}$ No

18. $\frac{18}{8} = \frac{108}{48}$ Yes

19. $\frac{3}{5} = \frac{81}{135}$ Yes

20. $\frac{3}{13} = \frac{10}{65}$ No

21. $\frac{12}{20} = \frac{27}{45}$ Yes

22. **Geometry** Triangle *ABC* is similar to triangle *MNP*. Find the missing side lengths.
MN = 2.52 in.
AC = 1.57 in.

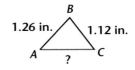

23. Consumer Economics To clean her cedar deck, Kim makes a mixture that will clean 200 square feet. How many ounces of cleaner will she need to clean 600 square feet? 9 oz

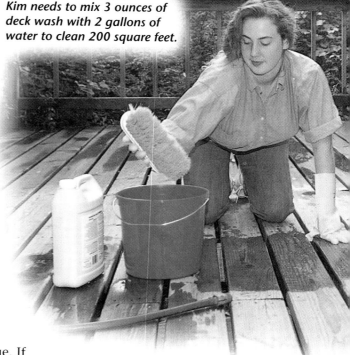

Kim needs to mix 3 ounces of deck wash with 2 gallons of water to clean 200 square feet.

24. Show that $\frac{2}{3} = \frac{6}{9}$ is a proportion. Use the numbers 2, 3, 6, and 9 to write at least three other true proportions.

25. True or false: If two hockey teams have a record of winning 2 out of every 3 games played, then each team played the same number of games. False

26. Student Government The principal of Brookside High School requires that the ratio of boys to girls on the student council be the same as the ratio of boys to girls in the school. If there are 280 boys and 360 girls in the school and the student council has 72 girls, how many boys must be on the council? 56

27. Government Two out of three eligible voters in Kettering are in favor of the school bond issue. If the number of eligible voters in Kettering is 28,950, how many voters are in favor of the bond issue? 19,300

28. Agriculture A manufacturer of fertilizer recommends that you use 4 pounds of fertilizer for every 1500 square feet of lawn. How many pounds of fertilizer are needed for 2400 square feet? 6.4 lbs

37.

Sale Price	Reg. Price
6	8
9	12
12	16
15	20
18	24

Rearrange the numbers to write three more true proportions.

29. $\frac{2}{3} = \frac{24}{36}$ **30.** $\frac{36}{54} = \frac{14}{21}$ **31.** $\frac{48}{27} = \frac{64}{36}$ **32.** $\frac{15}{9} = \frac{10}{6}$

33. $\frac{12}{20} = \frac{27}{45}$ **34.** $\frac{24}{8} = \frac{54}{18}$ **35.** $\frac{2}{13} = \frac{10}{65}$ **36.** $\frac{12}{16} = \frac{18}{24}$

Discounts This graph shows the relationship between the regular price and the sale price of some items at Murray's Department Store.

37. Make a table of values for the points shown on the graph.

38. Find the ratio of the sale price to the regular price for each point. What do you notice about the ratios? 3/4; the ratios all equal 3/4.

39. If the sale price of an item is $56, find the regular price. $74.67

40. If the regular price of an item is $36, find the sale price. $27

41. If the regular price of an item is $60 and the sale price is $48, was the pricing ratio the same as the one described by the graph? Explain.

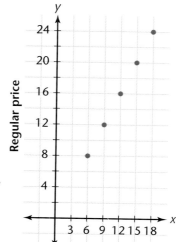

42. Health Frank took his pulse and counted 24 beats in 20 seconds. How many times did his heart beat in one minute? 72 times

43. 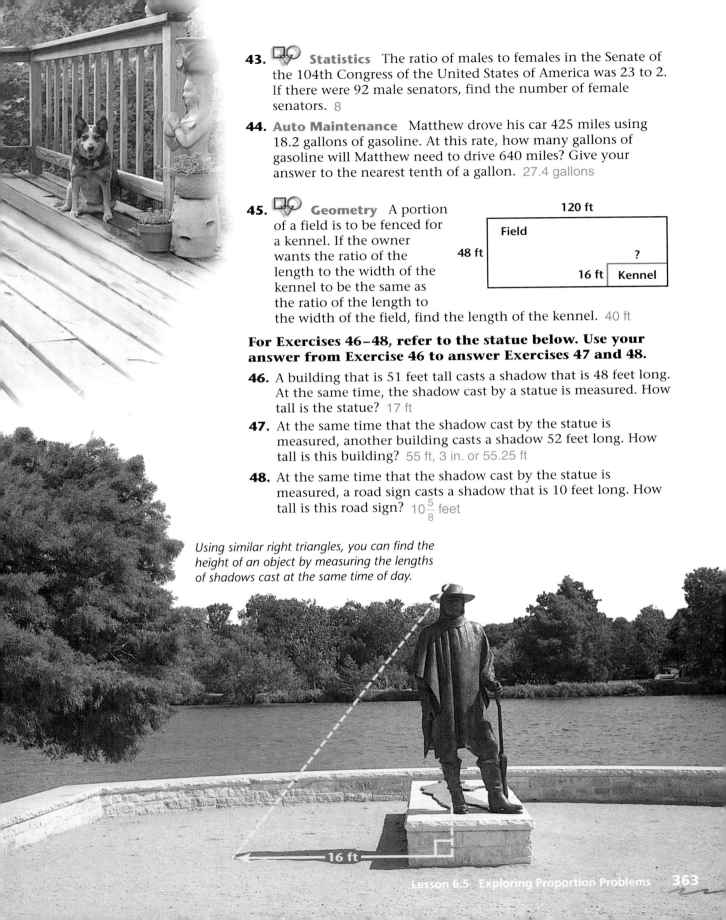 **Statistics** The ratio of males to females in the Senate of the 104th Congress of the United States of America was 23 to 2. If there were 92 male senators, find the number of female senators. 8

44. Auto Maintenance Matthew drove his car 425 miles using 18.2 gallons of gasoline. At this rate, how many gallons of gasoline will Matthew need to drive 640 miles? Give your answer to the nearest tenth of a gallon. 27.4 gallons

45. **Geometry** A portion of a field is to be fenced for a kennel. If the owner wants the ratio of the length to the width of the kennel to be the same as the ratio of the length to the width of the field, find the length of the kennel. 40 ft

120 ft

Field	
48 ft	
	?
16 ft	Kennel

For Exercises 46–48, refer to the statue below. Use your answer from Exercise 46 to answer Exercises 47 and 48.

46. A building that is 51 feet tall casts a shadow that is 48 feet long. At the same time, the shadow cast by a statue is measured. How tall is the statue? 17 ft

47. At the same time that the shadow cast by the statue is measured, another building casts a shadow 52 feet long. How tall is this building? 55 ft, 3 in. or 55.25 ft

48. At the same time that the shadow cast by the statue is measured, a road sign casts a shadow that is 10 feet long. How tall is this road sign? $10\frac{5}{8}$ feet

Using similar right triangles, you can find the height of an object by measuring the lengths of shadows cast at the same time of day.

16 ft

A recipe for 12 servings of chili calls for 3 pounds of ground beef, 8 tomatoes, and various spices.

49. Cooking How many pounds of ground beef are needed to make enough chili to serve 50 people? 12.5 lbs

50. How many pounds of ground beef are needed to make enough chili to serve 65 people? 16.25 lbs

51. Suppose you only have 5 tomatoes. How many servings of chili could you make if you used all the tomatoes? 7.5 servings

52. Suppose you have 4 pounds of ground beef. How many servings of chili could you make? 16 servings

Look Back

53. Evaluate $\frac{1}{12} - \frac{2}{30}$. **[Lesson 3.4]** $\frac{1}{60}$

Change each decimal to a percent. [Lesson 3.7]

54. 0.4 40%   **55.** 1.25 125%   **56.** 0.003 0.3%

Change each fraction to a percent. [Lesson 3.7]

57. $\frac{3}{4}$ 75%   **58.** $\frac{2}{3}$ $66\frac{2}{3}$%   **59.** $\frac{3}{8}$ $37\frac{1}{2}$%

Simplify each expression. [Lessons 5.2, 6.1]

60. $3 \cdot 5x$ $15x$ **61.** $-2y \cdot 3y$ $-6y^2$ **62.** $7(x - 3) - (5 - 4x)$ $11x - 26$

Solve each equation. [Lesson 6.2]

63. $\frac{w}{-12} = 13$ **64.** $-18 = \frac{x}{12}$ **65.** $\frac{12}{100} = \frac{x}{18}$
  $w = -156$ $x = -216$ $x = 2.16$

Look Beyond

66. Recall from Lesson 4.3 that the sum of the measures of the angles in a triangle is 180°. Find m∠B in triangle *ABC*. 26°

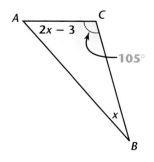

Solve each inequality.

67. $3x - 4 > 2x + 5$ $x > 9$

68. $8x - 16 \le 5x + 92$ $x \le 36$

69. $\frac{2}{3}x + 9 \le 87$ $x \le 117$

LESSON 6.6 Solving Problems Involving Percent

why *The next time you read a newspaper, notice how many times a percent is used. Percents are found in tables, charts, and graphs. They are also in ads, on display signs, and in financial statements.*

THE WEATHER PAGE

WARM FRONT

Palm Beach, FL
Temperature 30°C
Relative Humidity 50%

The relative humidity is expressed as a percent. For example, at 30°C, if 1 cubic meter of air contains 42 liters of water vapor, the air is saturated, and the relative humidity is 100%. If 1 cubic meter of air contains 21 liters of water vapor, or $\frac{1}{2}$ of the saturated amount, at the same temperature, the relative humidity is 50%.

Fractions, Decimals, and Percent

Recall from Lesson 3.7 that a percent is a *ratio* that compares a number with 100. A percent can be expressed in many ways. For example, 25% can be written as a decimal or a fraction. Sometimes you may wish to write the fraction in lowest terms.

$$25\% = \frac{25}{100} = 0.25 \qquad \frac{25}{100} = \frac{1}{4}$$

Suppose 1 cubic meter of air contains $10\frac{1}{2}$ liters of water vapor. Then the relative humidity is $\frac{10\frac{1}{2}}{42}$ or $\frac{1}{4}$ of 100%. The relative humidity is 25%.

EXAMPLE 1

Write each percent as a decimal and as a fraction.

A 75%   **B** 110%   **C** 4.4%

Solution ➤

A $75\% = \dfrac{75}{100} = 0.75$ $\dfrac{75}{100} = \dfrac{3}{4}$

B $110\% = \dfrac{110}{100} = 1.10$ $\dfrac{110}{100} = \dfrac{11}{10}$

C $4.4\% = \dfrac{4.4}{100} = 0.044$ $\dfrac{4.4}{100} = \dfrac{44}{1000} = \dfrac{11}{250}$ ❖

What percent would you write for the decimal 0.08?   8%

Writing and Solving Percent Problems

In a percent equation such as 25% of 40 = 10, 25% is the **percent**, 40 is the **base**, and 10 is the **percentage**. How can you solve an equation if one of the parts is unknown? Explain how each equation is solved using basic properties.

Equation Method

1. The percentage is unknown.
$$25\% \text{ of } 40 = x$$
$$0.25 \cdot 40 = x$$
$$10 = x$$

2. The base is unknown.
$$25\% \text{ of } x = 10$$
$$0.25 \cdot x = 10$$
$$x = 40$$

3. The percent is unknown.
$$x \text{ of } 40 = 10$$
$$x \cdot 40 = 10$$
$$x = 0.25 \text{, or } 25\%$$

Proportion Method

You can also visualize a percent bar and form a proportion.

1. The percentage is unknown.   25% of 40 = x

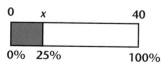

$$\frac{x}{0.25} = \frac{40}{1.00}$$
$$0.25\left(\frac{x}{0.25}\right) = 0.25\left(\frac{40}{1.00}\right)$$
$$x = 10$$

2. The base is unknown.   25% of $x = 10$

$$\frac{10}{0.25} = \frac{x}{1.00} \rightarrow 1.00\left(\frac{10}{0.25}\right) = 1.00\left(\frac{x}{1.00}\right)$$

$$40 = x$$

3. The percent is unknown. x of $40 = 10$

$$\frac{10}{x} = \frac{40}{1.00} \rightarrow \frac{x}{10} = \frac{1.00}{40}$$

$$10\left(\frac{x}{10}\right) = 10\left(\frac{1.00}{40}\right)$$

$$x = 0.25, \text{ or } 25\%$$

CRITICAL
Thinking

Explain how to write an equation and use one step to solve for each variable in the statement $a\%$ of b is c.

EXAMPLE 2

Consumer Economics
Tom plans to buy a CD player for his mother. He sees that the Royal Appliance Store is having a sale. How much will he save at the sale if he buys a $120 CD player?

Solution ➤
The CD player Tom is buying has an original price of $120. The sign says that CD players are on sale for 25% off. You want to find the amount of savings. Let x represent the amount of savings.

Write a percent statement for the problem, and then solve the equation.

$$25\% \text{ of } \$120 \text{ is } \underline{\ ?\ }.$$
$$0.25 \cdot 120 = x$$
$$30 = x$$

Tom will save $30 on the CD player during this sale. ❖

EXAMPLE 3

Discounts The freshman class is sponsoring a trip to see the musical *Cats*. If a group of 20 students buy tickets, there is a 30% discount. How much will each ticket cost at the discounted price?

Students $8.00
Group discounts available

The Rialto Theater
Matinee Wednesday 3:00pm
Daily 8:30pm

Solution A ➤

Proportion method You can visualize the problem using a percent bar. The regular cost, $8, is the length of the bar. Since the tickets will be 30% off, the discounted price will be 70% of the original price. Mark off 70% of the bar, and use *x* to represent the discounted price.

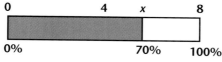

The percent bar indicates that the price is between $4 and $8. Write the proportion $\frac{x}{0.70} = \frac{8}{1.00}$ from the percent bar. Apply the Multiplication Property of Equality, and simplify.

$$\frac{x}{0.70} = \frac{8}{1.00}$$
$$0.70\left(\frac{x}{0.70}\right) = 0.70\left(\frac{8}{1.00}\right)$$
$$x = 5.60$$

The discounted price is $5.60, which agrees with the estimate.

Solution B ➤

Equation method Use *x* to represent the discounted price.

The discounted price, *x*, is 70% of the full price, $8. Model this statement with an equation.

$$x = 0.70 \cdot 8$$
$$x = 5.60$$

The discounted price is $5.60. ❖

Try This Solve the following percent statements.

a. _?_ % of 80 is 15.   18.75
b. 115% of 200 is _?_.   230
c. 35% of _?_ is 45.   about 128.57

A VCR is on sale for $239.40.

EXAMPLE 4

What was the original price of the VCR before the 40% off sale?

Solution ➤

Use the equation method. Let w represent the original price. The sale price of $239.40 is 60% of the original price, w. Model this statement by an equation.

$$239.40 = 0.60w$$

Apply the Division Property of Equality, and simplify.

$$\frac{239.40}{0.60} = \frac{0.60w}{0.60}$$
$$399 = w$$

The original price was $399. ❖

EXAMPLE 5

Sales Tax   Some states have a sales tax for certain items. Linda paid $47.74 for a $45.25 item. Find the percent of tax or the tax *rate*.

Solution ➤

Use the proportion method. First, find the amount of tax. Subtract the amount of the item from the total amount paid.

$$\$47.74 - \$45.25 = \$2.49$$

The tax was $2.49. Use the percent bar to show the proportion.

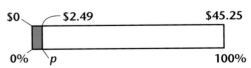

Let p represent percent.

Use the Multiplication Property of Equality to solve the proportion.

$$\frac{2.49}{p} = \frac{45.25}{1.00} \text{ or } \frac{p}{2.49} = \frac{1.00}{45.25}$$

$$2.49\left(\frac{p}{2.49}\right) = 2.49\left(\frac{1.00}{45.25}\right)$$

$$p \approx 0.055, \text{ or } 5.5\% ❖$$

5.5% of $45.25 is $2.49.

What is a percent statement containing "of" and "is" for Example 5?

EXERCISES & PROBLEMS

Communicate

1. Explain the procedure for changing a percent to a fraction. Place the percent over 100 and divide.

Explain how to draw a percent bar to model each problem.

2. 40% of 50   3. 200% of 50

4. 30 is 50% of what number?

5. What percent of 80 is 60?

Estimate each answer as more or less than 50. Explain how you made your estimate.

6. 40% of 50   7. 200% of 50

8. 30 is 50% of what number?

9. What percent of 80 is 60?

Each ounce of pretzels contains 10% of the U.S. Recommended Daily Allowance (U.S. RDA) of iron.

Describe how to set up the equation for the following problem.

10. How many ounces of pretzels would you have to eat to get 100% of the U.S. RDA of iron? Let x represent the number of ounces of pretzels that must be eaten. Since 1 ounce contains 10% of the U.S. RDA, set up the proportion $\frac{1}{0.10} = \frac{x}{1.00}$; $x = 10$.

Practice & Apply

Write each percent as a decimal.

11. 55% 0.55 12. 1.2% 0.012 13. 8% 0.08 14. 145% 1.45 15. 0.5% 0.005

Write each decimal as a percent.

16. 0.47 47% 17. 0.019 1.9% 18. 8.118 811% 19. 0.001 0.1% 20. 9.00 900%

Draw a percent bar to model each problem.

21. 35% of 80   22. 5 is what percent of 25?   23. What number is 10% of 8?

24. 150% of 40   25. What percent of 90 is 40?   26. 18 is 20% of what number?

Estimate each answer as more or less than 50 or 50%.

27. 5% of 80 less   28. 18 is 20% of what number? more   29. 150% of 40 more

30. 2.5% of 100 less   31. What percent of 90 is 40? less   32. What percent of 60 is 120? more

Find each answer.

33. 40% of 50 20   34. 125% of what number is 45? 36   35. 8 is 20% of what number? 40

36. 200% of 50 100   37. 30 is 60% of what number? 50   38. What percent of 80 is 10? 12.5%

39. 35% of 80 28   40. What number is 3.5% of 120? 4.2   41. 3 is what percent of 3000? 0.1%

42. 75% of 900 675   43. 72 is 9% of what number? 800   44. What percent of 60 is 24? 40%

Income Tax Federal income tax is withheld from the paychecks of most people. However, when you prepare your tax return, you need to determine whether the amount that has been withheld from your check throughout the year is the correct amount.

For example, suppose Mandy is single and her taxable income is $23,850. During the year, she had $3,956.78 withheld from her paychecks. Does Mandy need to pay additional income tax, or will she receive a tax refund of the extra tax that was taken?

Single Taxpayers

If taxable income is: over—	but not over—	The tax is:	of the amount over:
$0	$22,100	...15%	$0
$22,100	$53,500	$3,315+28%	$22,100
$53,500	$115,000	$12,107+31%	$53,500
$115,000	$250,000	$31,172+36%	$115,000
$250,000	– – – –	$79,772+39.6%	$250,000

Step 1: Read the table's instructons.
 THINK: *$23,850 is greater than $22,100 and less than $53,500.*

Step 2: Find the total tax.

- Subtract to find out how much more than $22,100 Mandy's income of $23,850 is. $23,850 − $22,100 = $1750

- Multiply to find 28% of $1750. THINK: *28% is 0.28.* 0.28 × $1750 = $490

- Add to find the total tax that should have been paid. $3315 + $490 = $3805

Step 3: Find the difference between the amount Mandy had withheld from her paychecks, $3956.78, and the amount that should have been paid, $3805. $3956.78 − $3805 = $151.78

Since Mandy had more taxes withheld than should be paid, she will recieve a tax refund of $151.78.

Use the tax table above to find the federal income tax for a single taxpayer with each income below.

45. $35,600 $7095 **46.** $24,850 $4085 **47.** $42,950 $9153 **48.** $17,890 $2683.50
49. $55,930 $12,860.30 **50.** $51,865 $11,649.20 **51.** $112,050 $30,257.50 **52.** $53,500 $12,107

Find the tax for a single taxpayer. Then determine whether an amount is still owed or an amount is due to be refunded.

	Taxable income	Amount of tax on income	Amount witheld	Amount owed	or	Amount to be refunded
53.	$65,750	?	$1856			?
54.	$55,930	?	$25,036			?
55.	$43,100	?	$5850			?
56.	$108,426	?	$42,865			?

 **Probability** According to the National Highway Safety Administration, a 0.1% blood alcohol concentration increases the odds of a car accident 7 times, and even one-half of that concentration impairs reflex time and depth perception.

57. Write the percent that is one-half of 0.1%. 0.05%

58. Write a percent statement using "of" and "is" for Exercise 57.
$\frac{1}{2}$ of 0.1% is 0.05%

59.  **Statistics** In a sample of 50 students, 52% opposed the school's new on-campus lunch policy. How many students in the survey opposed the policy? 26

60. Discounts Troy buys a tennis racket for 30% off the original price of $66. What is the sale price of the racket? $46.20

Statistics

61. What percent of the popular vote did Clinton get? ≈43.24%

62. What percent of the electoral vote did Bush get? ≈37.73%

63. What percent of the popular vote did Perot get? ≈19.02%

64. What percent of the popular vote did Clinton and Bush together get? ≈80.98%

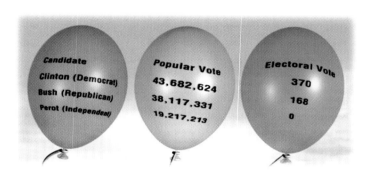

Candidate	Popular Vote	Electoral Vote
Clinton (Democrat)	43,682,624	370
Bush (Republican)	38,117,331	168
Perot (Independent)	19,217,213	0

65. Government The newspaper says that only 42% of the registered voters voted in a school bond election. If 11,960 people voted, how many registered voters are there? about 28,476

66. **Geometry** The dimensions of a rectangle are increased by 10%. By what percent is the perimeter increased? 10%

67. Mr. Arnes assigns 100 points out of 600 points for a group project. What percent is this? about 16.7%

Look Back

Plot the following points on a coordinate plane. **[Lesson 2.6]**

68. $A(7, 8)$ **69.** $B(5, 2)$ **70.** $C(3, 3)$ **71.** $D(5, 7)$

72. If x is an even integer, represent the next two even integers in terms of x.
[Lesson 5.1]

Simplify.   **[Lesson 6.1]**

73. $-2 \cdot 3x$
$-6x$

74. $-8x \div 4$
$-2x$

75. $\frac{10x + 25}{-5}$
$-2x - 5$

76. $\frac{22 - 2y}{2}$
$11 - y$

Look Beyond

77. Solve $4x + 6 = 2x - 2$. -4 **78.** Solve $4x + \frac{1}{2} = 5x - \frac{3}{4}$. $\frac{5}{4}$, $1\frac{1}{4}$, or 1.25

EGYPTIAN
Equation Solving

CHAPTER 6 PROJECT

... *from the Rhind papyrus*

tp-ḥśb, *accurate reckoning*
n, *of*
ḥⁱ·t , *entering*
m, *into*
ḫ·t, *things*
rḫ, *knowledge*
nt·t, *of existing things*
nb·t, *all*

About 4000 years ago, Egyptian mathematicians wrote texts to prepare students for jobs in commerce. In 1858, a Scotsman named Rhind bought one of these texts, which had been copied by a scribe named Ahmes about 1650 B.C.E.

One problem of the Rhind papyrus asks, "What is the amount of meal in each loaf of bread if 2520 *ro* of meal is made into 100 loaves?"

$$100m = 2520$$

The Egyptians solved the equation using the methods of make-a-table and guess-and-check.

They first substituted key values, like different powers of 10, for the variable to find the value of the left side of the equation.

They then used multiples of those values to see how close they could get to 2520. Since 2520 = 2000 + 500 + 20, they asked what value of *m* would produce 20 for 100*m*.

m	$100m$
1	100
10	1000
20	2000
5	500
$\frac{1}{5}$	20

For the equation 100*m* = 2520, they needed to compute $100(20) + 100(5) + 100\left(\frac{1}{5}\right)$ on the left side to get 2520 on the right side.

So, *m* was $20 + 5 + \frac{1}{5}$, or $25\frac{1}{5}$ *ro*.

Activity

Solve these equations as the Egyptians might have.

   a. $40m = 4800$ **c.** $60j = 930$

   b. $24p = 252$ **d.** $10w = 2432$

Make up and solve some equations of your own.

> **HINT:**
> For problem **a** complete this table.
>
> | m | $40m$ |
> |---|---|
> | 1 | 40 |
> | 100 | — |
> | 20 | — |

Chapter 6 Review

Vocabulary

Key Skills & Exercises

Lesson 6.1

➤ **Key Skills**

Use the rules for multiplying and dividing expressions.

$$-4(5x - 6) = (-4)(5x) - (-4)(6)$$
$$= -20x + 24$$

$$(2a - 3b + 1) - 4(a + 6b - 3) = 2a - 3b + 1 - 4a - 24b + 12$$
$$= (2a - 4a) + (-3b - 24b) + (12 + 1)$$
$$= -2a - 27b + 13$$

$$\frac{12 - 8y}{-4} = \frac{12}{-4} + \left(\frac{-8y}{-4}\right)$$
$$= -3 + 2y$$

➤ **Exercises**

Multiply or divide. Simplify when possible.

1. $3 \cdot 9x$ $27x$ **2.** $-33x \div 3$ $-11x$ **3.** $-2(7y - 2)$ $-14y + 4$

4. $-2.4x \cdot 2x$ $-4.8x^2$ **5.** $\dfrac{-30y + 3.6}{-3}$ $10y - 1.2$ **6.** $9r^2 - 8(4 - 3r^2)$ $33r^2 - 32$

Lesson 6.2

➤ **Key Skills**

Solve multiplication equations.

Solve $-9k = 108$.

$$-9k = 108 \quad \text{Given}$$
$$\frac{-9k}{-9} = \frac{108}{-9} \quad \text{Division Property of Equality}$$
$$k = -12 \quad \text{Simplify.}$$

Solve division equations.

Solve $\dfrac{w}{-5} = -2.2$.

$$\frac{w}{-5} = -2.2 \quad \text{Given}$$
$$-5\left(\frac{w}{-5}\right) = -5(-2.2) \quad \text{Multiplication Property of Equality}$$
$$w = 11 \quad \text{Simplify.}$$

Solve each equation.

7. $17x = -85$ -5 **8.** $-4g = -56$ 14 **9.** $-2.2h = 33$ -15

10. $24f = 150$ $6\frac{1}{4}$ **11.** $\frac{w}{-8} = 0.5$ -4 **12.** $\frac{y}{-2.4} = -10$ 24

Lesson 6.3

➤ Key Skills

Use the Distributive Property to find products.

$3x(2x - 3) = 3x(2x) - 3x(3)$
$= 6x^2 - 9x$

Write a polynomial in factored form.
$-15z^2 - 20z^3 = -5z^2(3 + 4z)$

➤ Exercises

Use the Distributive Property to find each product.

13. $5(x - 5)$ $5x - 25$ **14.** $y(y + 4)$ $y^2 + 4y$ **15.** $4t(t^2 + 7)$ $4t^3 + 28t$ **16.** $2r^2(r^2 - 3r)$ $2r^4 - 6r^3$

17. $b(12b^2 + 11b)$ **18.** $4y(y + 5)$ **19.** $5x^2(2x^2 - x)$ $10x^4 - 5x^3$ **20.** $6d^2(d^2 - 1)$ $6d^4 - 6d^2$

17. $12b^3 + 11b^2$
18. $4y^2 + 20y$

Write each polynomial in factored form.

21. $6x^2 + 8$ $2(3x^2 + 4)$ **22.** $5c^3 - 25c$ $5c(c^2 - 5)$ **23.** $n^4 + 2n^3$ $n^3(n + 2)$ **24.** $-9w^2 - 21w^4$ $-3w^2(7w^2 + 3)$

25. $8y^2 - 3y$ $y(8y - 3)$ **26.** $-8p - 14p^2$ $-2p(4 + 7p)$ **27.** $z^4 + 5z^2$ $z^2(z^2 + 5)$ **28.** $16y^5 - 4y^3$ $4y^3(4y^2 - 1)$

Lesson 6.4

➤ Key Skills

Solve an equation using the Reciprocal Property.

Solve $\frac{-2}{5} t = 14$.

$\frac{-2}{5} t = 14$	Given
$\frac{5}{-2}\left(\frac{-2}{5} t\right) = \frac{5}{-2}(14)$	Reciprocal Property
$t = -35$	Simplify.

➤ Exercises

Solve each equation.

29. $\frac{4}{7} x = 4$ 7 **30.** $\frac{1}{-9} y = 2.5$ -22.5 **31.** $\frac{w}{-16} = \frac{3}{-4}$ 12 **32.** $-\frac{3}{5} m = \frac{8}{15}$ $-\frac{8}{9}$

Lesson 6.5

➤ Key Skills

Solve a proportion.

Solve $\frac{n}{32} = \frac{3}{8}$.

Method A

$\frac{n}{32} = \frac{3}{8}$

$32 \cdot \frac{n}{32} = 32 \cdot \frac{3}{8}$

$n = \frac{96}{8}$, or 12

Method B

$\frac{n}{32} = \frac{3}{8}$

$8 \cdot n = 32 \cdot 3$

$8n = 96$

$n = 12$

Use cross products to determine whether a statement is a true proportion.

The cross products for $\frac{72}{81} = \frac{96}{108}$ are

$72 \cdot 108 = 7776$ and $81 \cdot 96 = 7776$.

The statement is a proportion.

The cross products for $\frac{31}{27} = \frac{60}{52}$ are

$31 \cdot 52 = 1612$ and $27 \cdot 60 = 1620$.

Since $1612 \neq 1620$, the statement is *not* a proportion.

Solve each proportion.

33. $\frac{n}{6} = \frac{12}{9}$ 8   **34.** $\frac{13}{n} = \frac{39}{27}$ 9   **35.** $\frac{11.4}{8} = \frac{45.6}{x}$ 32   **36.** $\frac{16}{41} = \frac{x}{820}$ 320

Determine whether each statement is a true proportion.

37. $\frac{17}{11} = \frac{68}{44}$ Yes   **38.** $\frac{7}{19} = \frac{20}{59}$ No   **39.** $\frac{2}{5} = \frac{82}{405}$ No   **40.** $\frac{23}{43} = \frac{45}{87}$ No

Lesson 6.6

➤ *Key Skills*

Solve percent problems.

Find 30% of 15.

Use the proportion method.

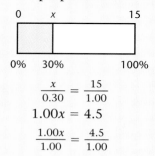

$$\frac{x}{0.30} = \frac{15}{1.00}$$

$$1.00x = 4.5$$

$$\frac{1.00x}{1.00} = \frac{4.5}{1.00}$$

$$x = 4.5.$$

30% of 15 is 4.5.

Use the equation method.

$$30\% \cdot 15 = x$$

$$0.30 \cdot 15 = x$$

$$4.5 = x$$

30% of 15 is 4.5.

What percent of 50 is 75?

Use the proportion method.

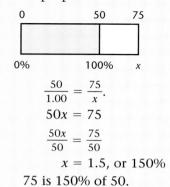

$$\frac{50}{1.00} = \frac{75}{x}.$$

$$50x = 75$$

$$\frac{50x}{50} = \frac{75}{50}$$

$$x = 1.5, \text{ or } 150\%$$

75 is 150% of 50.

Use the equation method.

$$x \cdot 50 = 75$$

$$\frac{x \cdot 50}{50} = \frac{75}{50}$$

$$x = 1.5, \text{ or } 150\%$$

75 is 150% of 50.

➤ *Exercises*

Find each answer.

41. What is 55% of 60? 33   **42.** 28 is 70% of what number? 40   **43.** What is 200% of 40? 80
44. What is 35% of 140? 49 **45.** What percent of 90 is 40.5? 45% **46.** 4 is what percent of 50?
8%

Applications

47. **Geometry** The area of a triangle with a base of 6 centimeters
is 21 square centimeters. What is the height of the triangle? 7 cm
(HINT: $A = \frac{1}{2}bh$)

48. **Statistics** Suppose that of all of the potatoes grown in the
United States, 22% are made into french fries. Out of 50 pounds of
potatoes, how many pounds are made into french fries? 11 lbs

Chapter 6 Assessment

1. Are all of the following expressions equivalent? Explain. $1c$ is different.
   a. $(8x - 32) \div 4$ $2x - 8$
   b. $(8x - 32)(0.25)$ $2x - 8$
   c. $\dfrac{32 - 8x}{4}$ $8 - 2x$
   d. $\dfrac{-32 + 8x}{4}$ $2x - 8$

Perform the indicated multiplication or division.

2. $-5x \cdot 7$ $-35x$

3. $-3(4 - 11y)$ $33y - 12$

4. $\dfrac{-28q}{-14}$ $2q$

5. $\dfrac{-10t + 35}{-5}$ $2t - 7$

Solve each equation.

6. $\dfrac{r}{5} = -22$ -110
7. $-10f = 0.55$ -0.055
8. $\dfrac{z}{-3.5} = -7$ 24.5

Use the Distributive Property to find each product.

9. $8(x - 7)$ $8x - 56$
10. $3a^2(a - 1)$ $3a^3 - 3a^2$
11. $q(2q^2 + q)$ $2q^3 + q^2$

Write each polynomial in factored form.

12. $5y^2 + 10y$ $5y(y + 2)$
13. $m^4 - 5m^3$ $m^3(m - 5)$
14. $36x^5 + 18x^2$ $18x^2(2x^3 + 1)$

15. Are all rational numbers integers? Explain your answer.

16. What is the only rational number that does not have a reciprocal? Explain why it does not.

Solve each equation or proportion.

17. $\dfrac{2}{3}y = -8$ -12
18. $\dfrac{-5}{11}p = -15$ 33
19. $\dfrac{-c}{14} = -\dfrac{1}{7}$ 2

20. $\dfrac{x}{3} = \dfrac{-2}{5}$ $-\dfrac{6}{5}$
21. $\dfrac{-a}{6} = \dfrac{4}{-9}$ $\dfrac{8}{3}$
22. $\dfrac{r}{2.8} = -\dfrac{5}{7}$ -2

Determine whether each statement is a true proportion.

23. $\dfrac{1}{3} = \dfrac{19}{57}$ Yes
24. $\dfrac{2}{9} = \dfrac{7}{26}$ No
25. $\dfrac{113}{29} = \dfrac{1017}{261}$ Yes

26. Paula made $350 in 4 weeks. At this rate, how much will she make in 10 weeks? $875

27. What is 35% of 90? 31.5
28. 3 is what percent of 15? 20%
29. What is 150% of 28? 42

30. Jamie made 80% of his free throws in his game last night. He attempted 15 free throws. How many free throws did he make? 12

Chapters 1–6
Cumulative Assessment

College Entrance Exam Practice

Quantitative Comparison For Questions 1–4, write
A if the quantity in Column A is greater than the quantity in Column B;
B if the quantity in Column B is greater than the quantity in Column A;
C if the two quantities are equal; or
D if the relationship cannot be determined from the information given.

	Column A	Column B	Answers
1.	$x = 6$ $2x + 1$	$4x - 3$	B Ⓐ Ⓑ Ⓒ Ⓓ **[Lesson 1.7]**
2.	$4\frac{1}{3}$	$4\frac{3}{8}$	B Ⓐ Ⓑ Ⓒ Ⓓ **[Lesson 3.2]**
3.	$(3y + 2) + (2y - 5)$	$(2y + 6) - (y + 7)$	D Ⓐ Ⓑ Ⓒ Ⓓ **[Lessons 5.1, 5.2]**
4.	The solutions to the equations $-\frac{3}{4}x = 3$	$-\frac{4}{3}x = -4$	B Ⓐ Ⓑ Ⓒ Ⓓ **[Lesson 6.2]**

5. Which expression is equal to 19?   **[Lesson 1.7]** d
   a. $15 \cdot (6 + 4) - 57$ **b.** $2^3 - 6 \cdot (3 + 2)$
   c. $4^2 - 5^2 \div 5 + 2(0)$ **d.** $9 \div 3 \cdot 2^3 - 5$

6. What is the value of $5\frac{5}{6} \cdot \frac{3}{5}$? **[Lesson 3.5]** c
   a. $5\frac{1}{2}$ **b.** $2\frac{1}{3}$ **c.** $3\frac{1}{2}$ **d.** $\frac{2}{7}$

7. The width of a rectangle is 6.4 inches. The length is 16.05 inches. What
   is the perimeter? **[Lesson 4.5]** d
   a. 22.09 inches   **b.** 44.18 inches   **c.** 22.45 inches   **d.** 44.9 inches

8. Which expression is *not* equal to the others?   **[Lessons 5.1, 5.2, 6.1]** c
   a. $3x - 2(x - 3)$ **b.** $3x + 2(3 - x)$
   c. $3x - 2(x + 3)$ **d.** $3x - 2(-3 + x)$

9. What is the solution to the equation $\frac{q}{139.2} = -58$?   **[Lesson 6.2]**
   a. -2.4 **b.** -8073.6 b
   c. -0.417 **d.** -7551.6

10. What percent is equivalent to 0.5?   **[Lesson 3.7]** b
   a. 0.5%   **b.** 50%
   c. 5%   **d.** 0.05%

Find each sum.   **[Lesson 2.2]**

11. $-25 + (-4)$ -29 **12.** $-36 + 6 + |-6|$ -24 **13.** $452 + (-452)$ 0

14. Use the LCD to compare the fractions $\frac{7}{8}$ and $\frac{13}{16}$ using the symbol $<$, $>$, or $=$.   **[Lesson 3.2]** $>$

15. The complement of an angle is 23°. What is its supplement?
   [Lesson 4.2] 113°

16. The perimeter of a rectangle is 52 feet and its width is 6 feet less than its length. Find the length and the width of the rectangle.
   [Lessons 5.3, 5.4] $l = 16$ ft; $w = 10$ ft

17. Solve $6 - (3 + 2x) = -3x + (2x - 4)$.   **[Lessons 5.3, 5.4]** 7

18. Solve the inequality $y + 4.5 \geq 6$.   **[Lesson 5.7]** $y \geq 1.5$

19. Write the following expression in factored form: $18x^3 - 12x^2$.
   [Lesson 6.3] $6x^2(3x - 2)$

20. Solve the proportion $\frac{15}{n} = \frac{20}{32}$.   **[Lesson 6.5]** 24

Free-Response Grid   The following questions may be answered using a free-response grid commonly used by standardized test services.

21. Misty's team is in a league with 12 other teams. If each of the teams plays each of the other teams once, how many games will be played?   **[Lesson 1.1]** 66

22. Find the sum: $\frac{1}{6} + \frac{1}{2}$.   **[Lesson 3.4]** $\frac{2}{3}$

23. One base of a trapezoid is 6 feet long and the other is 7 feet long. The height of the trapezoid is 8 feet. Find the number of square feet in the area.   **[Lesson 4.6]** 52

24. The shortest side of a triangle is 3 yards long and the longest side is 11 yards long. The shortest side of a similar triangle is 1.5 yards long. What is the length of the longest side, in yards?   **[Lesson 4.8]** 5.5

25. What is the solution to the proportion $\frac{c}{16} = \frac{3}{8}$?
   [Lesson 6.5] 6

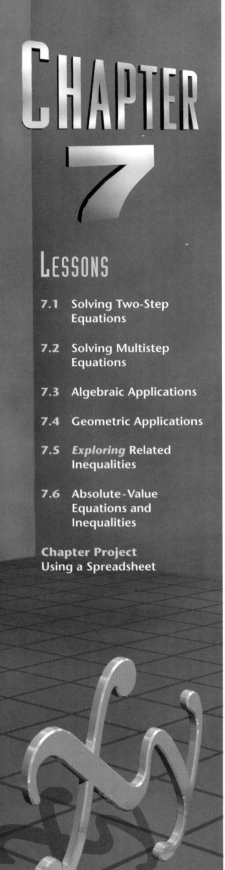

CHAPTER 7

Solving Equations and Inequalities

In many situations, it is necessary to solve equations and inequalities in order to solve larger problems. For instance, in business, equations can be used to determine incomes that involve a commission based on sales and to determine profit based on costs and revenue. Inequalities can be used in quality control to determine the margin of error allowable in measurements.

In mass-produced products, the required accuracy of a measurement can be described by an inequality.

A salesperson may use equations to solve problems involving income that is based on a commission. A small-business owner solves problems involving the cost and revenue required to make a profit.

Small-business owners, such as this yogurt-stand owner, may solve problems involving sales tax, discounts, and inventory.

PORTFOLIO ACTIVITY

In warm weather, outside temperatures are higher than inside temperatures. The following formula is used to estimate the rate at which heat is transferred from the exterior to the interior of a building through the walls and windows. **Heat transfer** is measured in BTUs (British Thermal Units) per hour.

Heat transfer = $A \cdot U(i - o)$, where

- A is surface area, in square feet;
- U is the heat transfer factor;
- i is the inside temperature, in degrees Fahrenheit; and
- o is the outside temperature, in degrees Fahrenheit.

On page 407, you will be asked to use this formula to evaluate the rates of heat transfer through different materials. You may wish to save your work for your portfolio.

LESSON 7.1 Solving Two-Step Equations

Felicia's subtotal at the deli is $5.00. With tax, she pays $5.40. What is the percent sales tax that Felicia paid?

why *It usually takes more than one step to solve a problem about money. To find sales tax, first add the item prices, then multiply by the tax rate. You can solve problems with more than one step by solving equations with more than one step.*

You will be asked in the exercises at the end of this lesson to find the percent sales tax that Felica paid using a two-step equation. Solving two-step equations combines what you learned in Chapters 5 and 6 about solving equations with addition, subtraction, multiplication, and division.

EXAMPLE 1

Solve $2x - 3 = 5$ using algebra tiles.

Solution ➤

First model the equation with algebra tiles.

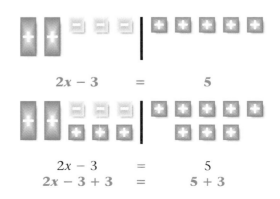

$$2x - 3 \quad = \quad 5$$

To isolate the *x*-tiles on the left side, add 3 positive 1-tiles to each side.

$$2x - 3 \quad = \quad 5$$
$$2x - 3 + 3 \quad = \quad 5 + 3$$

To simplify the model, remove neutral pairs.

$$2x - 3 = 5$$
$$2x - 3 + 3 = 5 + 3$$
$$2x = 8$$

Since there are 2 x-tiles, divide each side into 2 equal parts.

$$2x - 3 = 5$$
$$2x - 3 + 3 = 5 + 3$$
$$2x = 8$$
$$\frac{2x}{2} = \frac{8}{2}$$

The result is 4 1-tiles for each x-tile.

$$x = 4$$

To check your answer, substitute **4** for x in the original equation.

$$2x - 3 = 5$$
$$2(\mathbf{4}) - 3 \stackrel{?}{=} 5$$
$$8 - 3 \stackrel{?}{=} 5$$
$$5 = 5 \quad \text{True}$$

Therefore, the solution is 4. ❖

Try This Solve $5x - 7 = 8$ for x using algebra tiles. $x = 3$

What are the main steps involved in solving two-step equations?
Adding or subtracting, and multiplying or dividing

EXAMPLE 2

Use a graphics calculator to solve the equation $2x - 3 = 5$.

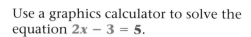

Graphics Calculator

Solution ➤
Graph the equations $Y_1 = 2x - 3$ and $Y_2 = 5$.

Find the point of intersection.

The solution to the equation $2x - 3 = 5$ is the x-value of the point of intersection. The point of intersection is (4, 5), so the solution is 4. ❖

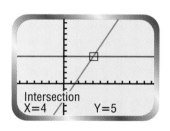

Try This Use a graphics calculator to solve the equation $4x + 5 = 3$. $x = -0.5$

CRITICAL Thinking When solving an equation by graphing, the solution is the *x*-coordinate of the point of intersection. What does the *y*-coordinate represent? The *y*-coordinate represents the value that you get when you substitute the solution back into the original equation.

You can solve some problems by solving two-step equations.

EXAMPLE 3

Consumer Economics

Jim bought 6 glasses and 6 mugs.

Each mug costs $5.98 and each glass costs $2.99. The salesperson tells Jim that his total bill is $47.84.

Jim estimates that 1 mug and 1 glass would cost between $9 and $10, so 6 mugs and 6 glasses should cost between $54 and $60.

Jim thinks that $47.84 is too low. Jim counted 6 mugs as the salesperson entered the prices, but he did not count the number of glasses. How many glasses did the salesperson enter?

Solution ➤

The cost for 6 mugs is $5.98 · 6, or **$35.88**.

Let *g* represent the number of glasses. Then the cost of the glasses is represented by the expression **$2.99g**.

The total bill for the mugs and glasses is **$47.84**.

First write the equation that models the problem.

Cost of mugs + Cost of glasses = Total bill

$$35.88 \quad + \quad 2.99g \quad = \quad 47.84$$

Then solve for g.

$35.88 + 2.99g = 47.84$	Given
$35.88 + 2.99g - 35.88 = 47.84 - 35.88$	Subtraction Property of Equality
$2.99g = 11.96$	Simplify.
$\dfrac{2.99g}{2.99} = \dfrac{11.96}{2.99}$	Division Property of Equality
$g = 4$	Simplify.

So Jim was charged for 4 glasses instead of 6 glasses. ❖

Some computer programs solve equations like the one in Example 3. For example, to solve the equation $35.88 + 2.99g = 47.84$, enter

$$\text{solve}(35.88 + 2.99 \text{*} n = 47.84, n)$$

Computer   The computer will display the answer, 4. As equations become more complicated, you can use technology to solve them.

Simplifying Equations

Sometimes it is possible to simplify an equation before solving it.

EXAMPLE 4

Life Science   Scientists have noticed a relationship between the temperature and the number of times a cricket chirps per minute. For one kind of cricket, this relationship is given by the formula $4.7(t - 50) = c - 92$, where c is the number of chirps per minute and t is the temperature in degrees Fahrenheit. According to this formula, what is the temperature if this type of cricket chirps 186 times per minute?

Solution ➤
Use the Distributive Property to simplify the equation.
Then substitute 186 for c.

$4.7(t - 50) = c - 92$	Given
$4.7t - (4.7)(50) = c - 92$	Distributive Property
$4.7t - 235 = c - 92$	Simplify.
$4.7t - 235 + 235 = c - 92 + 235$	Addition Property of Equality
$4.7t = c + 143$	Simplify.
$4.7t = (186) + 143$	Substitute 186 for c.
$4.7t = 329$	Simplify.
$\dfrac{4.7t}{4.7} = \dfrac{329}{4.7}$	Division Property of Equality
$t = 70$	Simplify.

The temperature is 70° F. ❖

EXAMPLE 5

STATISTICS
Connection

What score does Heather need on her third test so that her test average will be 80?

Heather is looking at her first two test grades.

Solution ➤

Let *s* represent the score that Heather needs on her third test. The average after 3 tests can be modeled by the expression $\frac{72 + 79 + s}{3}$. She wants her average to be 80.

$$\frac{72 + 79 + s}{3} = 80 \qquad \text{Given}$$

$$\frac{151 + s}{3} = 80 \qquad \text{Simplify.}$$

$$3\left(\frac{151 + s}{3}\right) = 3(80) \qquad \text{Multiplication Property of Equality}$$

$$151 + s = 240 \qquad \text{Simplify.}$$

$$151 + s - 151 = 240 - 151 \qquad \text{Subtraction Property of Equality}$$

$$s = 89 \qquad \text{Simplify.}$$

To check your answer, substitute 89 for *s* in the equation $\frac{72 + 79 + s}{3} = 80$.

$$\frac{72 + 79 + s}{3} = 80$$

$$\frac{72 + 79 + 89}{3} \overset{?}{=} 80$$

$$80 = 80 \qquad \text{True}$$

In order to have a test average of 80, Heather needs to score an 89 on her third test. ❖

How do you determine the number to divide by to find an average?
Divide by the number of items to be averaged.

CRITICAL
Thinking

What will happen to Heather's average if she scores above 89 on her third test? What will happen to her average if she scores below 89 on her third test? Heather's average will be > 80; Heather's average will be < 80.

EXERCISES & PROBLEMS

Communicate

1. Describe how algebra tiles can be used to solve $2x + 4 = -6$.

2. Without actually solving it, explain the steps needed to solve $7(8 - 2x) = 86$.

3. **Statistics** Explain how to determine the score that Brian needs on his third history test to average 90 for all three tests.

4. Explain how to decide if 7.2 is the solution to the equation $5x - 6 = 32$.

5. Explain how to solve $6x + 4 = -2$ by graphing.

Brian scored 85 on both of his first two history tests. He has one more history test to take this semester.

Practice & Apply

Solve each equation.

6. $2m + 5 = 17$ 6
7. $9p + 11 = -7$ -2
8. $5x + 9 = 39$ 6
9. $3 + 2x = 21$ 9
10. $6 - 8d = -42$ 6
11. $9 - 14z = 51$ -3
12. $12 = 9x - 6$ 2
13. $16 = 5w - 9$ 5
14. $-4 - 11w = 18$ -2
15. $-7 - 13y = 32$ -3
16. $36 = -3y + 12$ -8
17. $58 = -8z + 18$ -5
18. $2(x - 3) = 14$ 10
19. $4(x - 20) = 16$ 24
20. $5(2x + 3) = 25$ 1
21. $3(4x - 3) = 15$ 2
22. $8 = 2(3x + 4)$ 0
23. $20 = 5(3x - 2)$ 2
24. $3x + 5 + 7x = 25$ 2
25. $9x + 6 + 4x = 45$ 3
26. $8x - 3 - 5x = 21$ 8
27. $17y - 9 - 9y = -41$ -4
28. $56 = -5w + 15 + 3w - 7$ -24
29. $-72 = -8y + 12 + 4y - 16$ 17
30. $5x + \frac{1}{8} = \frac{1}{3}$ $\frac{1}{24}$
31. $3x + \frac{2}{7} = \frac{1}{3}$ $\frac{1}{63}$
32. $3 + \frac{1}{4} + 2x = \frac{5}{8}$ $-1\frac{5}{16}$
33. $4\left(x - \frac{1}{3}\right) = \frac{1}{5}$ $\frac{23}{60}$
34. $\frac{1}{6} + 2x = \frac{7}{24}$ $\frac{1}{16}$
35. $8t = \frac{2}{3} - 3t + 7$ $\frac{23}{33}$
36. $5x + 10 = \frac{5}{11}$ $-1\frac{10}{11}$
37. $\frac{3}{4}x + 1 + \frac{1}{6} + x = \frac{7}{24}$ $-\frac{1}{2}$
38. $239(x + 20) = -956$ -24

39. The sum of three consecutive whole numbers is 48. Write and solve an equation to find the numbers. $x + (x + 1) + (x + 2) = 48$; 15, 16, 17

40. The sum of four consecutive whole numbers is 58. Write and solve an equation to find the numbers. $x + (x + 1) + (x + 2) + (x + 3) = 58$; 13, 14, 15, 16

41. **Statistics** Lynn has scores of 95, 91, and 88 on three tests. Write and solve an equation to find a fourth score that will produce an average of 90 for the four tests. 86 $\frac{95 + 91 + 88 + x}{4} = 90$

42. 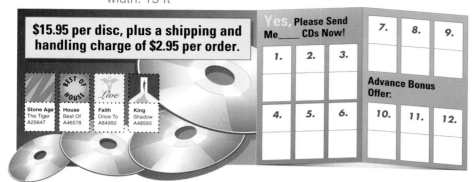 **Geometry** If the perimeter of this rectangle is 180 feet, write and solve an equation to find the length and width. $(x + 60) + x + (x + 60) + x = 180$; length: 75 ft; width: 15 ft

$x + 60$

x

43. Consumer Economics If the total cost, C, was $98.65, how many compact discs did Kara and her friend buy from the advertisement above?
6

Sales Tax Refer to the cost of Felicia's sandwich given at the beginning of the lesson on page 382. Felicia's subtotal at the deli is $5.00. With tax she pays $5.40.

44. What is the percent sales tax that Felicia paid? 8%

45. Dan pays a total of $4.59 for a sandwich. If he pays the same percent sales tax as Felicia pays, what was the subtotal for Dan's sandwich? $4.25

46. Cultural Connection: Africa Find the answer to this problem from ancient Egypt. Fill a large basket $1\frac{1}{2}$ times. Then add 4 *hekats* (a hekat is about half a bushel). The total is 10 *hekats*. How many *hekats* does the basket hold?
4 hekats

Look Back

47. 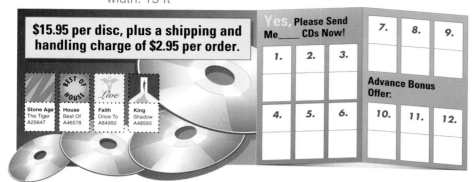 **Geometry** Find the side length of a square with an area of 36 square feet. **[Lesson 4.6]** 6 ft

Write the opposite for each expression. [Lesson 5.2]

48. $6c - 23d$ $-6c + 23d$ **49.** $-s + t$ $s - t$ **50.** $-a - c$ $a + c$

Solve. [Lessons 5.3, 5.4]

51. $5 - x = -12$ 17 **52.** $2.1 + x = -8.3$ -10.4 **53.** $-\frac{2}{3} + x = -\frac{1}{6}$ $\frac{1}{2}$

Look Beyond

Solve.

54. $3x - 4 = 2x + 7$ 11 **55.** $5x + 8 = 2x - 10$ -6 **56.** $-8x + 12 = 2x - 48$ 6

LESSON 7.2 Solving Multistep Equations

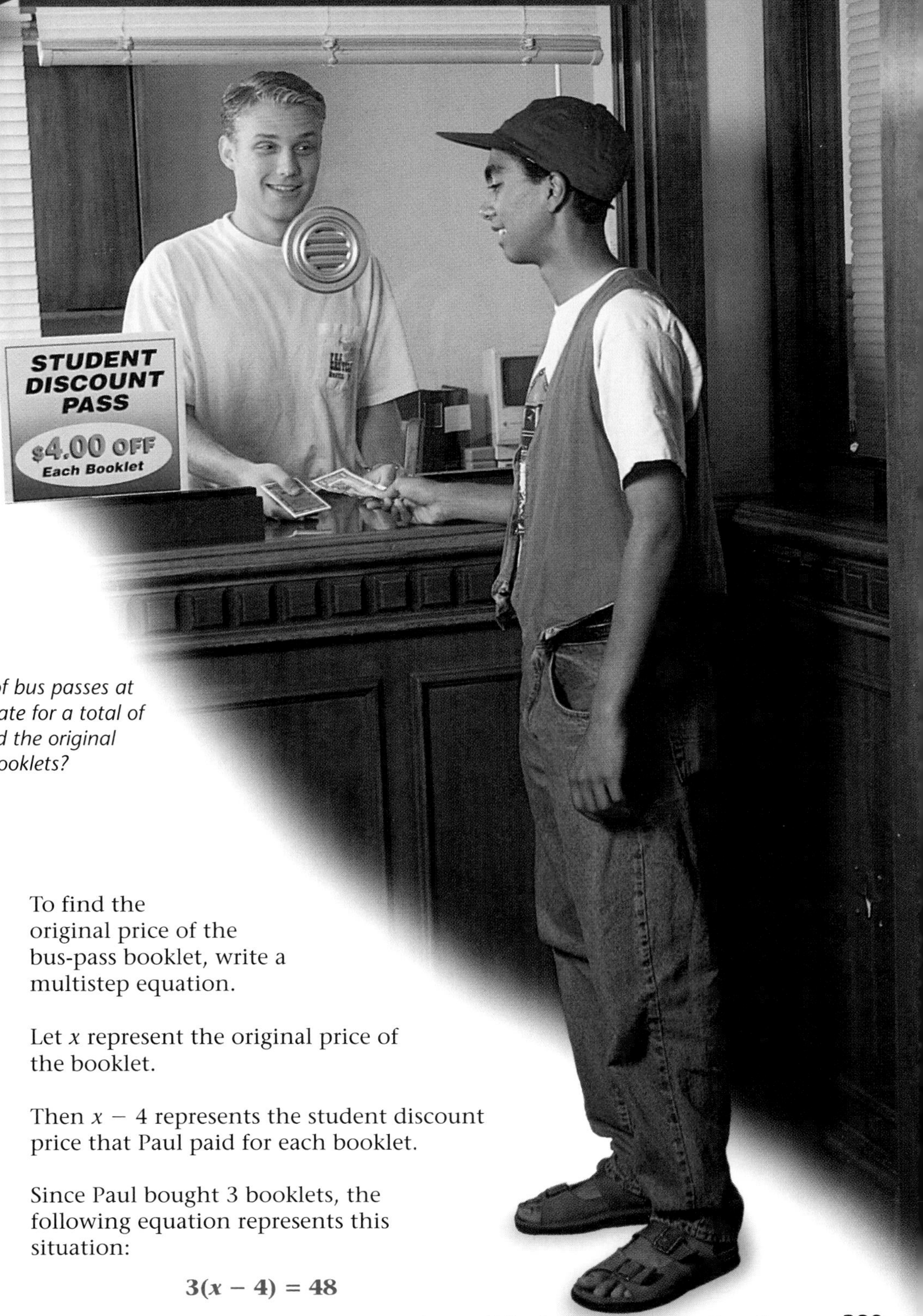

Why

Solutions to real-world problems often have multiple steps. These multiple steps involve various mathematical properties.

STUDENT DISCOUNT PASS

$4.00 OFF
Each Booklet

Paul buys 3 booklets of bus passes at the student discount rate for a total of $48. How can you find the original price for each of the booklets?

To find the original price of the bus-pass booklet, write a multistep equation.

Let x represent the original price of the booklet.

Then $x - 4$ represents the student discount price that Paul paid for each booklet.

Since Paul bought 3 booklets, the following equation represents this situation:

$$3(x - 4) = 48$$

•Exploration *Modeling Equations With Variables on Both Sides*

1 Use algebra tiles to model the equation $4x - 2 = x + 4$.

$$4x - 2 \qquad = \qquad x + 4$$

2 How can you get all of the *x*-tiles on the left side? Do this with your algebra tiles. Subtract 1 positive *x*-tile from each side.

3 Describe your model in Step 2 by completing the equation below.

$$4x - 2 = x + 4$$

$$4x - 2 - (x) \qquad 4x - 2 - (\,?\,) = x + 4 - (\,?\,) \quad x + 4 - (x)$$
$$(3x) - 2 \qquad\qquad (\,?\,) - 2 = 4$$

4 You now have an equation with a variable on only one side. Finish solving the equation using tiles, and describe the step or steps you used. Isolate the *x*-tiles on the left side by adding 2 positive 1-tiles to each side and removing neutral pairs. From equal groupings of 3; $x = 2$.

5 Solve the equation $4x - 2 = x + 4$ without using tiles.

6 Check your result by substituting numbers into the original equation. ❖

CRITICAL
Thinking

Why is it best to check the solution in the *original* equation rather than in an equivalent equation that results as you solve the equation? It is best to check the solution in the original equation in case an error is made in the steps, which could make the wrong answer seem correct.

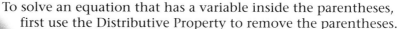
EXAMPLE 1

Solve the equation from page 389, $3(x - 4) = 48$, to find the original price of the bus-pass booklets that Paul bought.

Solution ➤

To solve an equation that has a variable inside the parentheses, first use the Distributive Property to remove the parentheses.

$3(x - 4) = 48$	Given
$3x - 12 = 48$	Distributive Property
$3x - 12 + 12 = 48 + 12$	Addition Property of Equality
$3x = 60$	Simplify.
$\dfrac{3x}{3} = \dfrac{60}{3}$	Division Property of Equality
$x = 20$	Simplify.

To check your answer, substitute 20 for *x* in the *original* equation.

$$3(x - 4) = 48$$
$$3(20 - 4) \stackrel{?}{=} 48$$
$$3(16) \stackrel{?}{=} 48$$
$$48 = 48 \quad \text{True}$$

The original price of the booklet is $20. ❖

EXAMPLE 2

Time   The watch is $4x$ minutes slow. The clock is x minutes fast. What is the actual time?

10:13 A.M.

10:53 A.M.

Solution ➤
First write an expression for the actual number of minutes past 10:00.

If the watch is $4x$ minutes slow, then it should show *more* than 13 minutes past 10:00. Add $4x$ to 13 to get the actual number of minutes past 10:00.

Watch: $13 + 4x$

If the clock is x minutes fast, then it should show *less* than 53 minutes past 10:00. Subtract x from 53 to get the actual number of minutes past 10:00.

Clock: $53 - x$

You can represent the actual number of minutes with an equation.

$$13 + 4x = 53 - x$$

Solve this equation for x.

$13 + 4x = 53 - x$	Given
$13 + 4x - 13 = 53 - x - 13$	Subtraction Property of Equality
$4x = 40 - x$	Simplify.
$4x + x = 40 - x + x$	Subtraction Property of Equality
$5x = 40$	Simplify.
$\dfrac{5x}{5} = \dfrac{40}{5}$	Division Property of Equality
$x = 8$	Simplify.

Substitute 8 for x to find the actual number of minutes past 10:00.

Watch: $13 + 4x$
actual number = $13 + 4(8)$
of minutes $= 13 + 32$, or 45

Clock: $53 - x$
actual number = $53 - (8)$
of minutes $= 45$

Therefore, the actual time is 10:45 A.M. ❖

Try This   According to the information about the clocks shown at the right, what is the actual time?

8:23 P.M.

8:15 P.M.
This clock is
2x minutes too slow.

8:47 P.M.
This clock is
6x minutes too fast.

EXAMPLE 3

Graphics Calculator

Solve $5x + 6 = 2x + 18$ using a graphics calculator.

Solution ➤

Graph the equations $\mathbf{Y}_1 = 5x + 6$ and $\mathbf{Y}_2 = 2x + 18$. Find the intersection point.

The value of the x-coordinate is 4, so the solution is 4. ❖

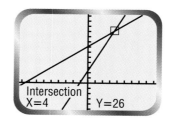

Intersection
X=4   Y=26

Try This   Solve $4x + 7 = 9x - 8$ using a graphics calculator.

In Example 4, many properties of equality are used.

EXAMPLE 4

Solve $4x - 3(2x + 4) = 8x - 25$.

Solution ➤

$4x - 3(2x + 4) = 8x - 25$	Given
$4x - 3(2x) + (-3)(4) = 8x - 25$	Distributive Property
$4x - 6x - 12 = 8x - 25$	Simplify.
$-2x - 12 = 8x - 25$	Simplify.
$-2x - 12 - 8x = 8x - 25 - 8x$	Subtraction Property of Equality
$-10x - 12 = -25$	Simplify.
$-10x - 12 + 12 = -25 + 12$	Addition Property of Equality
$-10x = -13$	Simplify.
$\dfrac{-10x}{-10} = \dfrac{-13}{-10}$	Division Property of Equality
$x = 1.3$	Simplify.

Check the solution. Substitute 1.3 for x in the original equation.

$$4x - 3(2x + 4) = 8x - 25$$
$$4(1.3) - 3[2(1.3) + 4] \stackrel{?}{=} 8(1.3) - 25$$
$$5.2 - 3(6.6) \stackrel{?}{=} 10.4 - 25$$
$$5.2 - 19.8 \stackrel{?}{=} 10.4 - 25$$
$$-14.6 = -14.6 \qquad \text{True ❖}$$

Try This   Solve $4y - 7(y + 6) = 5y - 2$. $y = -5$

EXERCISES & PROBLEMS

Communicate

Describe how algebra tiles can be used to solve each equation.

1. $2x - 3 = 5x + 9$

2. $4x - 6 = -x + 4$

Explain the steps needed to solve each equation.

3. $5x - 3 = 4x + 7$

4. $-2(x - 3) = 3x + 10$

5. Explain how to solve $3x + 6 = 2x + 4$ using a graph. Graph $Y_1 = 3X + 6$ and $Y_2 = 2X + 4$, and find the point of intersection. The point of intersection is $(-2, 0)$; the x-coordinate is the solution; $x = -2$.

Practice & Apply

Solve each equation.

6. $8y = 6y + 24$ 12

7. $12w - 15 = 7w$ 3

8. $4g + 1 = 12 - 8g$ $\frac{11}{12}$

9. $3r - 8 = 5r - 20$ 6

10. $1 - 3x = 2x + 8$ $-1\frac{2}{5}$

11. $15 - 2y = 12 - 8y$ $-\frac{1}{2}$

12. $5 - 3y = 5y + 65$ $-7\frac{1}{2}$

13. $18 + 2w = 7w - 13$ $6\frac{1}{5}$

14. $4(2w + 5) = 12w - 9$ $7\frac{1}{4}$

15. $5x - 7 = 2x + 2$ 3

16. $7m - 2(m - 3) = 3m - 14$ -10

17. $2(y - 3) + 4y + 8 = 3(y + 6)$ $5\frac{1}{3}$

18. $8f - 3(f + 6) = 2f - 16$ $\frac{2}{3}$

19. $4t - 5 + 8t = 7(t + 6)$ $9\frac{2}{5}$

20. During the frozen yogurt special shown below, Franklin served a total of 33 scoops of yogurt. Write and solve an equation to find out how many of those scoops were free. $2x + x = 33$; 11 scoops

21. Andalon ordered 263 cups for the yogurt stand. The number of cups ordered was 8 more than 3 times the number of cones ordered. Write and solve an equation to find out how many cones were ordered. $8 + 3x = 263$; 85 cones

22. Last week Franklin worked 4 days. He earned the same amount of money on each of the first 3 days and earned $32 on the fourth day. His paycheck for that week was $200. How much did Franklin earn on each of the first 3 days? $56

23. Andalon also makes sodas and sundaes at the yogurt stand. Last week the number of sodas she made was 4 fewer than 5 times the number of sundaes. If she made 96 sodas, how many sundaes did she make? 20 sundaes

Buy Two Scoops
Get One **FREE**

Solve each equation.

24. $2(p - 3) = 3(p - 4)$ 6

25. $8.3y = 4.2y + 143.5$ 35

26. $187a + 265 = -456a - 378$ -1

27. $0.95q - 4.56 = 0.35(2.7q + 1.5)$ 1017

28. $2.1(y - 5) = 3(y - 5)$ 5

29. $1.5(k + 4) = 4(k + 0.5)$ 1.6

30. $17.2a + 1291.5 = 14.2a$ -430.5

31. $2.85b + 102.96 = 4.5b$ 62.4

32. $1.4m - 3.7 = 0.9m + 6.3$ 20

33. $0.8n - 4.7 = 0.75n - 4.1$ 12

34. $x + \frac{5}{8} + \frac{3x}{4} = \frac{2}{3} + 5x$ $-\frac{1}{78}$

35. $3\left(x + \frac{1}{3}\right) = 6\left(x + \frac{1}{4}\right)$ $-\frac{1}{6}$

36. 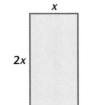 **Geometry** The perimeters of the two rectangles at the right are equal. Find the value of x and the common perimeter. 9; 54

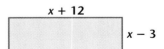

37. Wages Last year Clayton made $6 an hour and worked 20 hours a week. He recently received a 20% raise. How many hours a week must he work now to make the same weekly wage he made last year? About 17 hr

38. Fund raising The Jefferson High Band sold 54 fewer football programs than they sold last week. They sold 192 football programs this week. How many programs did they sell last week? 246

39. Forestry The number of trees planted by C. M. Grow's Nursery in April was 3 more than twice the number of trees planted by the nursery in March. Seventy-one trees were planted in April. Find the number of trees planted in March. 34

40. Student Government In the last student council election, Jill got x number of votes. Morgan got twice as many votes as Jill. Patricia won the election with only 7 more votes than Morgan. If 327 students voted in the election, how many votes did each candidate get?

Jill: 64
Morgan: 128
Patricia: 135

41. Hobbies Alfredo has a coin collection. He has 26 fewer U.S. coins than 3 times the number of his foreign coins. The collection consists of 998 coins. Find the number of each type of coin that Alfredo has.

256 foreign coins; 742 U.S. coins.

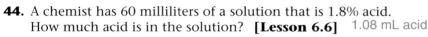 *Look Back*

42. Write 8% as a decimal. **[Lesson 3.7]** 0.08

43. Name two pairs of alternate interior angles in the figure at the right. **[Lesson 4.3]** $\angle 5$ and $\angle 7$, $\angle 6$ and $\angle 8$

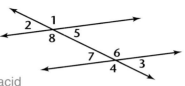

44. A chemist has 60 milliliters of a solution that is 1.8% acid. How much acid is in the solution? **[Lesson 6.6]** 1.08 mL acid

45. 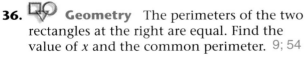 **Statistics** In his first five league games, Ben has scored 24, 18, 27, 32, and 21 points. How many points must he score in the sixth game to have an average of 25 points per game? **[Lesson 7.1]** 28

Look Beyond

46. Solve the inequality $x - 4 \geq 2x + 5$ $-9 \geq x$

LESSON 7.3 Algebraic Applications

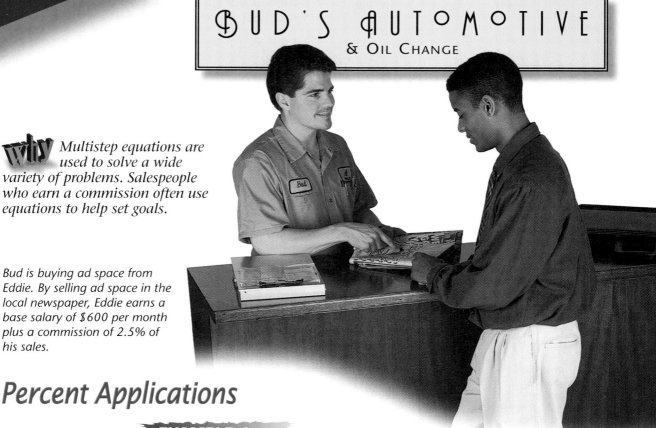

BUD'S AUTOMOTIVE & OIL CHANGE

Why *Multistep equations are used to solve a wide variety of problems. Salespeople who earn a commission often use equations to help set goals.*

Bud is buying ad space from Eddie. By selling ad space in the local newspaper, Eddie earns a base salary of $600 per month plus a commission of 2.5% of his sales.

Percent Applications

EXAMPLE 1

Wages Eddie estimates that he needs a monthly income of $1000 to meet his budget. What must his sales be to meet this goal?

Solution ➤

Write an equation for Eddie's income, I, in terms of his sales, s.

Eddie's income is $600 plus 2.5% of his sales.
$$I \qquad = \quad 600 \quad + \qquad 0.025s$$

In this case, Eddie's desired monthly income is $1000. Substitute 1000 for I.

$I = 600 + 0.025s$	Given
$1000 = 600 + 0.025s$	Substitute 1000 for I.
$1000 - 600 = 600 + 0.025s - 600$	Subtraction Property of Equality
$400 = 0.025s$	Simplify.
$\dfrac{400}{0.025} = \dfrac{0.025s}{0.025}$	Division Property of Equality.
$16,000 = s$	Simplify.

So Eddie must have $16,000 in monthly sales to earn $1000 per month. ❖

Lesson 7.3   Algebraic Applications   **395**

You can also solve the problem in Example 1 by graphing. When you graph the equations $Y_1 = 600 + 0.025X$ and $Y_2 = 1000$, the x-value of the point of intersection gives the amount of sales required to earn $1000. The graph shows that Eddie will need to sell $16,000 worth of ads in a month to earn $1000 in a month.

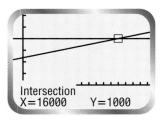

Intersection
X=16000   Y=1000

Graphics Calculator

Use a graphics calculator to find what Eddie's monthly sales must be to earn $1500 in a month. $36,000 in monthly sales

Try This   Jackson is the manager of the men's department at Lovell's Clothing Store. His base salary is $300 per week plus a commission of 1.2% of the weekly sales in the department. What must the total of the weekly sales be if he wants to earn a weekly income of $900? $50,000 in weekly sales

Exploration *Break-Even Point*

Fund-raising   The pep squad at Barton High School is selling pennants to raise money for their activities. They must pay the manufacturer $65.25 for the design of the pennant and $2.15 for each pennant ordered.

The pep squad plans to sell each pennant for $4.50.

 Write a verbal expression to describe the total amount paid to the manufacturer for the pennants. $65.25 plus $2.15 times the number of pennants purchased

2 *Revenue* is the total amount received from the sales. Write a verbal expression to describe the revenue from selling the pennants.
$4.50 times the number of pennants sold

3 Copy and complete the table with amounts for cost and revenue from the given numbers of pennant sales.
Total cost: $76, $86.75, $97.50, $108.25, $119, $129.75

Number of pennants	5	10	15	20	25	30
Total cost	?	?	?	?	?	?
Total revenue	?	?	?	?	?	?

Total rev: $22.50, $45, $67.50, $90, $112.50, $135

 Write an algebraic equation for the total cost in terms of the number of pennants, p, ordered.
$C = 65.25 + 2.15p$

 Write an algebraic equation for the total revenue in terms of the number of pennants, p, sold. $R = 4.5p$

 The point at which the total revenue equals the total cost is the **break-even point**. Write an equation that you could use to determine the number of pennants that must be sold to break even. $65.25 + 2.15p = 4.5p$

 Solve the equation you wrote in Step 6. How many pennants need to be sold to break even? Be sure that your answer is reasonable.
28 pennants

 The *profit* from a sale is the total revenue minus the total cost. Write and solve an equation to determine the number of pennants the pep squad must sell to make a profit of $100. ◆   71 pennants

Mixture Applications

You can use algebra to determine the appropriate amounts of different solutions to mix.

EXAMPLE 2

Chemistry A chemist uses mathematics to determine amounts of different solutions to mix. How much acid is in the new solution? What percent of the new solution is acid?

Karla mixes 120 mL of a 2.4% acid solution with 180 mL of a 3.6% solution.

Solution ➤

First organize the data in a table.

	First solution	Second solution	New solution
Percent acid	2.4% = 0.024	3.6% = 0.036	?
Amount of solution	120 mL	180 mL	120 + 180 = 300 mL
Amount of acid	0.024 (120 mL)	0.036 (180 mL)	0.024(120) + 0.036(180) = 9.36 mL

There is 300 mL of new solution. The amount of acid in the new solution is 9.36 mL. Write an equation to find the percent of acid in the new solution.

9.36 mL is what percent of 300 mL?

$$9.36 = x \cdot 300$$

Solve the equation $9.36 = x \cdot 300$, or $9.36 = 300x$, for x.

$$9.36 = 300x$$
$$\frac{9.36}{300} = \frac{300x}{300}$$
$$0.0312 = x$$

Therefore, 3.12% of the new solution is acid. ❖

CRITICAL Thinking

In Example 2, why is the percent of acid in the new solution closer to the percent of acid in the second solution, 3.6%, than to the percent of acid in the first solution, 2.4%? Because a greater amount of the 3.6% solution was used in the mixture, the percent of acid in the new solution is closer to 3.6% acid.

Try This
0.558 mL;
about 2.15%

Ted has 16 mL of a solution that is 1.8% acid. He mixes that solution with 10 mL of solution that is 2.7% acid. How much acid is in the new solution? Approximately what percent of the new solution is acid?

EXERCISES & PROBLEMS

Communicate

1. Describe the relationships among cost, revenue, and profit.

2. A chemist is mixing 200 milliliters of a 1% acid solution with 50 milliliters of a 5% acid solution. Explain how to find the percent of acid in the resulting mixture.

3. Explain how you would solve the following problem:

   Harry is a carpenter. He earns $16.50 per hour and receives a travel allowance of $90 per week. How many hours per week would he have to work to earn $620, including the travel allowance?

4. Describe how to solve the problem in Exercise 3 using a graph.

Practice & Apply

Solve. Be sure that your answer is reasonable.

5. **Small Business**   Orthea is a painter. She charges $110 for the paint plus $8 per hour to paint a house. How many hours does Orthea work to earn $600 painting a house?   61 hr, 15 min

6. **Small Business**   A plumber charges $32 per job plus $18 per hour. If the total bill for a job was $92, how many hours did the plumber work?   3 hr, 20 min

7. **Wages**   Holly works in a clothing store. She earns $120 per week plus 2.5% of her weekly sales. What must her weekly sales be in order for her to make $200 per week? $3200

8. **Wages**   Denny manages a restaurant. His base salary is $400 per week plus 1.2% of the weekly sales at the restaurant. What must the weekly sales be in order for him to make $700 per week? $25,000

9. **Wages**   Kendall works as a makeup consultant in a department store. She makes $150 per week plus 5.2% of her weekly sales. If she wants to earn $400 per week, what must her weekly sales be? $4807.69

10. **Fund-raising**   The football booster club spent $1240 to print 2000 football programs. They plan to sell each program for $2. How many programs must they sell to make a profit of $600?   920 programs

11. **Theater**   The community-theater manager determined that the cost of costumes, ticket printing, and theater rental for the summer season will be $2400. If each ticket is sold at $7.50, how many tickets must be sold to make a profit of $6000?   1120 tickets

12. Chemistry A chemist has 40 ounces of a solution that is 2% acid. He wants to add pure acid to make a solution that is 2.5% acid. How much pure acid should he add? Approx 0.2 oz

13. Chemistry Abigail is mixing 60 mL of a solution that is 3% acid with pure water to dilute the solution. How much water should she add to obtain a solution that is 2.2% acid? Approx 21.8 mL

14. Chemistry Laura has 48 ounces of a solution that is 48% salt. How much of a solution that is 30% salt should she add to make a new solution that is 38% salt? 60 oz

15. Chemistry Gary plans to mix 2 liters of a solution that is 25% antifreeze with a solution that is 32% antifreeze. How much of the 32% solution should he add to the 25% solution to obtain a new solution that is 30% antifreeze? 5 liters

16. Sports A basketball team played 4 more games this season than last season. Last season, the team won 55% of its games, and this season it won 50% of its games. The team won the same number of games this season as last season. Find the number of games played each season.
Last season: 40 games; This season: 44 games

17. Sports The number of players selected for the girls' tennis team was 75% of the number who tried out. There were 4 more who tried out for the boys' team than for the girls' team, but only 60% of the boys were selected. The two teams had the same number of members. Find how many tried out for each team. 16 girls, 20 boys

18. Small Business In July, Pop's Auto Parts sold 40% of its stock of batteries and 30% of its stock of oil filters. At the beginning of July, there were 40 more oil filters in stock than batteries. An equal number of batteries and oil filters were sold. How many of each item were in stock at the beginning of July? 120 batteries, 160 oil filters

Small Business Jesse makes guinea-pig cages to sell to pet shops. He spent $155 on tools. The materials for each cage cost $4.25. Use this information for Exercises 19–25.

19. Write an equation for Jesse's cost to make the cages in terms of the number of cages, n. $C = 155 + 4.25n$

20. Write an equation for the revenue in terms of the number of cages, n. $R = 12.50n$

21. Write an equation for the profit in terms of the number of cages, n. $P = 8.25n - 155$

22. How many cages must Jesse sell to at least break even? 19

23. How many cages must Jesse sell to make a profit of $450? 74

24. Technology Use a graph to determine the number of cages Jesse must sell to make a profit of $200. 44

25. Technology Suppose Jesse can make 8 cages per day. Use a graph to determine the number of days he will have to work in order to make enough cages for a profit of $200.
5.5 days

Fund-raising The French club is selling wrapping paper. They pay $24 for the advertising brochures and $2.50 for each package of wrapping paper.

26. Write an equation for the cost in terms of the number of packages sold, n. $C = 24 + 2.50n$

27. Write an equation for the revenue in terms of the number of packages sold, n. $R = 4n$

28. Write an equation for the profit in terms of the number of packages sold, n. $P = 1.5n - 24$

29. How many packages must they sell to at least break even? 16

30. How many packages must they sell to make a profit of $300? 216

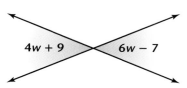

$4w + 9$ $6w - 7$

31. **Geometry** Find the value of w from the angles at the left. What is the measure of each angle? 41°

Look Back

Compare each pair of numbers. Use <, >, or =. [Lessons 2.1, 3.2, 3.3]

32. -8, -3.2 **33.** 12, -17 **34.** $-\dfrac{1}{2}$, $-\dfrac{2}{3}$

32. $-8 < -3.2$ or $-3.2 > -8$
33. $12 > -17$ or $-17 < 12$
34. $-\dfrac{1}{2} > -\dfrac{2}{3}$ or $-\dfrac{2}{3} < -\dfrac{1}{2}$

35. What must be true of two numbers if their product is a positive number? **[Lesson 2.5]** Both positive or both negative

36. **Geometry** Describe the relationship between supplementary angles. **[Lesson 4.2]** The sum of their measures is 180°.

37. **Geometry** Describe the relationships among the angles in a parallelogram. **[Lesson 4.4]**

38. **Geometry** Write an expression for the perimeter of the rectangle at the right. **[Lesson 4.5]**
$2(x + 3) + 2(24)$

39. **Geometry** Write an expression for the area of the rectangle at the right. **[Lesson 4.6]** $24(x + 3)$

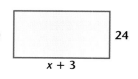

24

$x + 3$

Solve each equation. [Lessons 7.1, 7.2]

40. $3x + 6 + 4x = 8$ $\dfrac{2}{7}$ **41.** $2(x - 8) = 3x + 24$ -40

Look Beyond

Describe a set of numbers that satisfies each inequality.

42. $|x| < 10$ **43.** $|x - 2| < 10$ **44.** $|x| \geq 3$

LESSON 7.4 Geometric Applications

Many applications of mathematics involve geometric figures. Algebraic methods can be used to solve problems about the perimeter and area of different polygons.

Jayne has 396 feet of chicken wire to build a pen for her chickens.

Perimeter

GEOMETRY
Connection

EXAMPLE 1

Jayne is putting a fence around part of her 1-acre lot to fence in her chickens. She wants the length of the pen to be twice the width. Find the dimensions of the pen.

Solution ➤

Draw a diagram of the pen. Since the length is twice the width, label the sides w and $2w$.

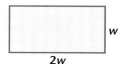

The length of fencing that Jayne has is the perimeter of the pen. Use the formula for the perimeter of a rectangle, $P = 2l + 2w$, to write an equation for a perimeter of 396 feet.

$$P = 2l + 2w \qquad \text{Perimeter formula}$$
$$396 = 2(2w) + 2w \qquad \text{Substitute } 2w \text{ for } l.$$

Solve the equation to find the dimensions.

$$396 = 2(2w) + 2w$$
$$396 = 4w + 2w \qquad \text{Simplify.}$$
$$396 = 6w \qquad \text{Simplify.}$$
$$\frac{396}{6} = \frac{6w}{6} \qquad \text{Division Property of Equality}$$
$$66 = w \qquad \text{Simplify.}$$

The width is 66 feet, so the length, $2w$, is 2(66), or 132 feet. ❖

Graphics Calculator

To find the length and width of the chicken pen in Example 1 using a graphics calculator, use the variable X instead of w.

For the perimeter, $P = 2l + 2w$, enter $Y_1 = 2(2X) + 2X$. Enter a second equation for length, $Y_2 = 2X$. Create a table and find the perimeter of 396 feet in column Y_1. The corresponding number in the Y_2 column, 132 feet, is the length. The number in the X column, 66 feet, is the width.

X	Y_1	Y_2
63	378	126
64	384	128
65	390	130
66	**396**	132
67	402	134
68	408	136
69	414	138

$Y_1=396$

Try This A gymnasium is to be designed with a length that is 20 meters greater than the width. The perimeter must be 480 meters. Find the dimensions of the gymnasium. width: 110 m, length: 130 m

Area

GEOMETRY Connection

Problems involving area can often be solved by substituting the known values into the appropriate formula and then solving for the unknown value.

EXAMPLE 2

Carpentry

Jeffrey is building a table with a tabletop in the shape of a trapezoid. The area of the tabletop is to be 18 square feet. Find the length of the shorter base of the tabletop.

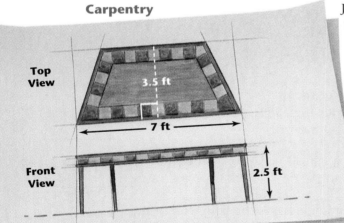

The plans for Jeffrey's table are shown above.

Solution ➤

Recall that the formula for the area of a trapezoid is $A = \frac{1}{2} h(b_1 + b_2)$, where h is the height and b_1 and b_2 are the bases. Substitute 18 for A, 3.5 for h, and 7 for b_1 in the formula. Solve the equation for b_2, the length of the second base.

$A = \frac{1}{2} h(b_1 + b_2)$ Given

$18 = \frac{1}{2} (3.5)(7 + b_2)$ Substitute.

$18 = 1.75(7 + b_2)$ Simplify.

$18 = 1.75(7) + 1.75(b_2)$ Distributive Property

$18 = 12.25 + 1.75b_2$ Simplify.

$18 - 12.25 = 12.25 + 1.75b_2 - 12.25$ Subtraction Property of Equality

$5.75 = 1.75b_2$ Simplify.

$\frac{5.75}{1.75} = \frac{1.75b_2}{1.75}$ Division Property of Equality

$3.2857 \approx b_2$ Simplify.

So the shorter base should be about 3.3 feet. ❖

To find the length of the second base (b_2) in Example 2, use the variable X for b_2 and Y_1 for A. Enter the formula for the area of a trapezoid, with 3.5 substituted for h and 7 substituted for b_1.

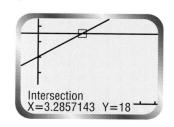

Intersection
X=3.2857143 Y=18

$$Y_1 = \frac{1}{2}(3.5)(7 + X)$$

Graph $Y_2 = 18$, and find the intersection. The x-value of the point of intersection gives the length of the second base.

Try This The base of a triangle is 14.2 yards. If the area of the triangle is 106.5 square yards, what is the height of the triangle? 15 yd

Applying Geometric Properties

Many problems involving the geometric concepts that you studied in Chapter 4 can be solved using algebraic equations.

GEOMETRY
Connection

EXAMPLE 3

Find the measure of each angle in triangle ABC.

Solution ➤

Recall that the sum of the measures of the angles of a triangle is 180°.

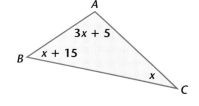

$m\angle A + m\angle B + m\angle C = 180°$	
$(3x + 5) + (x + 15) + x = 180$	Substitute.
$5x + 20 = 180$	Simplify.
$5x + 20 - 20 = 180 - 20$	Subtraction Property of Equality
$5x = 160$	Simplify.
$\frac{5x}{5} = \frac{160}{5}$	Division Property of Equality
$x = 32$	Simplify.

Substitute 32 for x to find the measures of angles A, B, and C.

$m\angle A = 3x + 5$ $m\angle B = x + 15$ $m\angle C = x$
$\quad = 3(32) + 5$ $\quad = 32 + 15$ $\quad = 32°$
$\quad = 101°$ $\quad = 47°$

So $m\angle A$ is 101°, $m\angle B$ is 47°, and $m\angle C$ is 32°. Notice that the sum of the measures is 180°. ❖

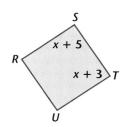

Try This Find the measure of each angle of parallelogram $RSTU$. $m\angle R = m\angle T = 89°$; $m\angle S = m\angle U = 91°$

EXAMPLE 4

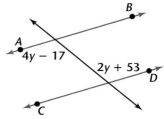

Line *AB* and line *CD* are parallel. Find the value of *y*. Then find the measures of the indicated angles.

GEOMETRY
Connection

Solution ➤

The labeled angles are alternate interior angles. Recall from Lesson 4.3 that alternate interior angles have equal measures when they are formed by parallel lines. Therefore, $2y + 53 = 4y - 17$.

$2y + 53 = 4y - 17$	Given
$2y + 53 - 4y = 4y - 17 - 4y$	Subtraction Property of Equality
$-2y + 53 = -17$	Simplify.
$-2y + 53 - 53 = -17 - 53$	Subtraction Property of Equality
$-2y = -70$	Simplify.
$\dfrac{-2y}{-2} = \dfrac{-70}{-2}$	Division Property of Equality
$y = 35$	Simplify.

Substitute the 35 for *y* in either expression to find the measure of the angles.

$$2y + 53 = 2(35) + 53 \qquad\qquad 4y - 17 = 4(35) - 17$$
$$= 70 + 53 \qquad\text{OR}\qquad = 140 - 17$$
$$= 123° \qquad\qquad\qquad = 123°$$

So the measure of both alternate interior angles is 123°. ❖

CRITICAL
Thinking

Suppose that lines *AB* and *CD* in Example 4 were *not* parallel. Would it be possible to find the value of *y* with the information given? Explain.

No; if lines *AB* and *CD* were not parallel, then $2y + 53 = 4y - 17$ would not be true, and it would not be possible to solve for *y* unless more information were given.

Literal Equations

A **literal equation** is an equation that contains different letters. Many formulas are examples of literal equations.

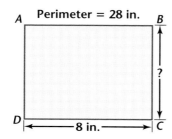

Suppose that you know the perimeter and length of a rectangular object, and you need to find the width. You can solve the perimeter formula, $P = 2w + 2l$, for *w* and use the resulting formula to find the width.

EXAMPLE 5

Solve the formula for the perimeter of a rectangle, $P = 2w + 2l$, for w in terms of P and l.

$P = 2w + 2l$	Perimeter formula
$P - 2l = 2w + 2l - 2l$	Subtraction Property of Equality
$P - 2l = 2w$	Simplify.
$\dfrac{P - 2l}{2} = \dfrac{2w}{2}$	Division Property of Equality
$\dfrac{P - 2l}{2} = w$, or $w = \dfrac{P - 2l}{2}$	Simplify.

So the formula for width in terms of perimeter and length is $w = \dfrac{P - 2l}{2}$. ❖

Use the formula from Example 5 to find the width of rectangle $ABCD$ on page 404. 6 in.

Try This Solve the formula for the circumference of a circle, $C = 2\pi r$, for r. $r = \dfrac{C}{2\pi}$

EXERCISES & PROBLEMS

Communicate

1. Explain how to determine the height of a triangle if the base measures 12 inches and the area is 84 square inches.

2. Explain how to solve the following equation for h:
$$A = \frac{1}{2} h(b_1 + b_2)$$

Describe the relationship between the length and width of each rectangle.

3.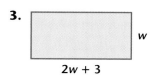
    w
    2w + 3

4.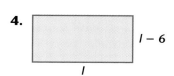
    l − 6
    l

Practice & Apply

Angles 1 and 2 are complementary.
Find m∠1 and m∠2.

5. m∠1 = x + 9; m∠2 = 2x + 3   35°, 55°

6. m∠1 = 5x + 2; m∠2 = 3x   57°, 33°

7. m∠1 = 4x + 17; m∠2 = 5x − 8   53°, 37°

8. m∠1 = $\frac{x}{2}$ + 18; m∠2 = 4x   26°, 64°

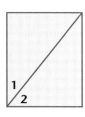

▽ **Geometry**   Find the measures of the angles
in triangle *ABC*.   **9.** 91°, 61°, 28°

9. m∠A = 7x; m∠B = 4x + 9; m∠C = 15 + x

10. m∠A = 6x − 7; m∠B = 4x; m∠C = 2x + 11

11. m∠A = $\frac{x}{3}$; m∠B = $\frac{1}{3}$ + x; m∠C = 6x

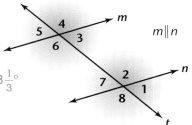

10. 81°, 58$\frac{2}{3}$°, 40$\frac{1}{3}$°

11. 8$\frac{1}{6}$°, 24$\frac{5}{6}$°, 147°

▽ **Geometry**   Use the figure at the right to
find *x* and the measures of the indicated angles
from the values given in Exercises 12–15.

12. m∠1 = 2x + 2; m∠3 = 4x − 5 $\frac{7}{2}$, 9°, 9°

13. m∠6 = 4x + 7; m∠7 = 8x + 9   13$\frac{2}{3}$, 61$\frac{2}{3}$°, 118$\frac{1}{3}$°

14. m∠5 = 4x + 1; m∠6 = 2x + 2   29$\frac{1}{2}$, 119°, 61°

15. m∠4 = 2x + 8; m∠8 = 6x   2, 12°, 12°

16. The area of a triangle is 48 square meters. The base of the
triangle is 24 meters. What is the height of the triangle?   4 m

17. A trapezoid has bases of 15.5 feet and 14.75 feet. If the area is 121 square
feet, what is the height of the trapezoid?   8 ft

18. Two angles are complementary. The larger angle is 4 times the smaller
angle. What are the measures of the angles?   18°, 72°

19. Two angles are supplementary. The larger angle is 60 degrees less
than twice the smaller angle. What are the measures of the angles?
100°, 80°

20. **Entertainment**   The length of a movie screen is 1$\frac{1}{2}$ times the
width. If the perimeter of the screen is 400 feet, what are the
dimensions?   80 ft by 120 ft

21.   ▽ **Geometry**   A sail has a tear that is to be patched with
a piece of cloth shaped like a right triangle. The length of
the tear is 10 centimeters, and the width of the tear is 7
centimeters. The area of the triangular patch will be twice as
large as the area of the tear, and the base of the patch will
be 10 centimeters. What are the dimensions of the patch?
base: 10 cm; height: 14 cm; area: 70 sq cm

Solve each formula for the indicated variable.

22. $A = \frac{1}{2}h(b_1 + b_2)$ for b_1

23. $S = (n - 2)180$ for n

24. $3x + g + 2h = t - y$ for h

25. $x + 4y = r$ for y

26. $\frac{x}{y} = \frac{4}{5}$ for x

27. $\frac{9}{5}C + 32 = F$ for C

Portfolio Activity Refer to the portfolio activity introduced on page 381. The heat-transfer factor, represented by the variable U in the formula heat transfer $= A \cdot U(i - o)$, is a number that varies according to the type of surface through which the heat passes. A negative heat-transfer rate means that the interior gains heat, and a positive heat-transfer rate means that the interior loses heat. **Complete Exercises 28–31.**

Heat Transfer Factors	
Surface	Value of U
Concrete, 6 inches thick	0.58
Glass, single pane	1.13
Brick, 8 inches thick	0.41
Wood, 2 inches thick	0.43

28. Find the rate of heat transfer through a 200-square-foot wall of 6-inch concrete when the indoor temperature is 65°F and the outdoor temperature is 80°F. −1740 BTU

29. Find the rate of heat transfer through a 400-square-foot, 2-inch-thick piece of wood when the indoor temperature is 72°F and the outdoor temperature is 85°F. −2236 BTU

30. Which is greater, heat transfer through a concrete wall 6 inches thick or heat transfer through a brick wall 8 inches thick? Explain.

31. If the outside temperature is 30°F and the inside temperature is 71°F, will there be a heat loss or a heat gain in the interior? Explain.

Insulation is installed in houses to reduce heat transfer. Proper insulation can help keep heat inside in the winter and outside in the summer.

Look Back

Simplify. [**Lesson 2.2**]

32. $|-3| + |7|$ 10   **33.** $|-5 - 8|$ 13   **34.** $|10 - 24|$ 14

Solve each equation. [**Lessons 6.2, 7.1, 7.2**]

35. $0.25x = 7$ 28   **36.** $5x = 3x - 10$ −5   **37.** $9x - 10 = 5x - 16$ $-\frac{3}{2}$

38. **Small Business** Juanita makes and sells figurines. She spends $7 on tools and $0.65 per figurine for the materials. She plans to sell each figurine for $3.25. How many figurines will she need to sell for a profit of $250? [**Lesson 7.3**] 99

Look Beyond

39. Draw a graph on a number line to show all numbers x such that the distance from x to 1 is less than 3.

LESSON 7.5

Exploring Related Inequalities

why *Many real-life situations are not solved by finding a single answer, but by finding a minimum, maximum, or range of answers that satisfy the conditions. For problems of this type, you often need to solve an inequality.*

This ballet company must maintain a certain minimum average attendance to stay in business. Fire codes and health regulations establish a certain maximum number of people allowed in the theater at one time.

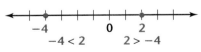

•Exploration 1 *Multiplying and Dividing Inequalities*

Numbers graphed on the number line are in ascending order from left to right.

```
←—+——+——+——+——+——+——+——+——+——→
      −4             0      2
        −4 < 2          2 > −4
```

Complete the pattern by filling in <, =, or >.

  a. $-4 < 2$, so $-4 \cdot 2 < 2 \cdot 2$
b. $-4 < 2$, so $-4 \cdot 1 < 2 \cdot 1$
c. $-4 < 2$, so $-4 \cdot 0$ ___?___ $2 \cdot 0$ =
d. $-4 < 2$, so $-4 \cdot (-1)$ ___?___ $2 \cdot (-1)$ >
e. $-4 < 2$, so $-4 \cdot (-2)$ ___?___ $2 \cdot (-2)$ >

  a. $-4 < 2$, so $-4 \div 2 < 2 \div 2$
b. $-4 < 2$, so $-4 \div 1 < 2 \div 1$
c. $-4 < 2$, so $-4 \div (-1)$ ___?___ $2 \div (-1)$ >
d. $-4 < 2$, so $-4 \div (-2)$ ___?___ $2 \div (-2)$ >

3 What happens to an inequality when the expressions on each side are multiplied or divided by the same *positive* number? It remains true.

4 What happens to an inequality when the expressions on each side are multiplied by 0?

5. It becomes false.

5 What happens to an inequality when the expressions on each side are multiplied or divided by the same *negative* number?

6 **a.** Explain what you must do to the inequality sign when you multiply or divide each side by the same *positive* number. Nothing
b. Explain what you must do to the inequality sign when you multiply or divide each side by the same *negative* number. ❖

Reverse the sign.

•Exploration 2 *Multiplying Inequalities*

To visualize what happens when you multiply each side of an inequality by the same number, complete this exploration.

1 Start with a set of 3 positive tiles and a set of 5 positive tiles. The < sign is used.

2 To multiply each side of the inequality by 2, think of doubling each set of tiles.

Fill in <, =, or >.

3 To multiply each side of the inequality by -2, think of doubling each set *and* changing the sign of each tile.

Fill in <, =, or >.

3 < 5

$2 \cdot 3$ ___?___ $2 \cdot 5$
<

$-2 \cdot 3$ ___?___ $-2 \cdot 5$
>

4 Use a model to show what happens to $-3 > -4$ when each side of the inequality is multiplied by   **a.** 2.   **b.** -2. ❖

Lesson 7.5 Exploring Related Inequalities **409**

Exploration 3 *Dividing Inequalities*

To visualize what happens when you divide each side by the same number, complete this exploration.

1 Start with a set of 4 positive tiles and a set of 8 positive tiles. The < sign is used for the inequality.

4 < 8

2 To divide each side of the inequality by 4, think of dividing each set of tiles by 4.

Fill in <, =, or >.

4 ÷ 4 **?** 8 ÷ 4
    <

3 To divide each side of the inequality by − 4, think of dividing each set by 4 *and* changing the sign of each tile.

Fill in <, =, or >.

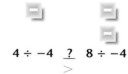

4 ÷ −4 **?** 8 ÷ −4
    >

4 Use a model to show what happens when each side of the inequality − 4 < 2 is divided by **a.** 2. **b.** − 2. ❖

The properties for multiplying and dividing inequalities express the results of multiplying or dividing each side of an inequality by a positive or negative number.

MULTIPLICATION AND DIVISION PROPERTIES OF INEQUALITY

If the expressions on each side of an inequality are multiplied or divided by the same positive number, the resulting inequality is still true.

If the expressions on each side of an inequality are multiplied or divided by the same negative number and the inequality sign is reversed, the resulting inequality is still true.

CRITICAL *Thinking*

A famous inequality states that $|a| + |b| \geq |a + b|$. Explain why this inequality is true for all numbers a and b.

Special women's ballet slippers cost about $50 a pair and wear out very quickly.

Ballet A dancer rehearsing and performing the role of Clara during one run of *The Nutcracker* can typically use 15 to 25 pairs of these ballet slippers. If a ballet company has no more than $1000 to spend, how many pairs of these special ballet slippers can the company buy?

Use an inequality to model the problem.

Let x equal the number of ballet slippers the company buys. Then $50x$ equals the amount they spend.

The amount must be less than or equal to 1000. Thus, $50x \leq 1000$.

Divide each side of the inequality by 50.

$$\frac{50x}{50} \leq \frac{1000}{50}$$

$$x \leq 20$$

The company can buy no more than 20 pairs of special ballet slippers. ❖

EXERCISES & PROBLEMS

Communicate

Tell whether each statement is true or false. Explain your reasoning.

1. $7 < 8$ T   **2.** $7 < 7$ F; $7 = 7$ **3.** $7 \leq 7$ T; $7 = 7$ **4.** $7 \neq 7$ F; $7 = 7$

Describe the steps needed to solve each inequality.

5. $x + 1 > 4$ **6.** $x - 3 \leq 13$ **7.** $-3p < 12$ **8.** $4x - 2 \geq 2x + 3$

Practice & Apply

Write an inequality that corresponds to each statement.

9. L is greater than W. $L > W$

10. r is greater than or equal to 4. $r \geq 4$

11. V is between 3.1 and 3.2, inclusive. $3.1 \leq V \leq 3.2$

12. x cannot equal 0. $x \neq 0$

13. m is positive. $m > 0$

14. y is not negative. $y \geq 0$

Tell whether each statement is true or false.

15. $4.2 \geq 4.2$ T

16. $9.22 \leq 9.22$ T

17. $3.1 < 3.01$ F

18. $8.55 > 8.505$ T

19. $\frac{1}{7} \geq \frac{1}{6}$ F

20. $\frac{3}{4} \leq \frac{4}{5}$ T

21. $-8 < -4$ T

22. $0 \geq -3$ T

Solve each inequality.

34. $x > \frac{11}{3}$ or $3\frac{2}{3}$

23. $x + 8 \geq 11$ $x \geq 3$

24. $x - 11 < -20$ $x < -9$

25. $G - 6 \leq 9$ $G \leq 15$

26. $8 - H > 9$ $H < -1$

27. $6 - x > -1$ $x < 7$

28. $5 - y \geq 2$ $y \leq 3$

29. $\frac{x}{8} < 1$ $x < 8$

30. $\frac{u}{-3} \geq 21$ $u \leq -63$

31. $5b > 3$ $b > \frac{3}{5}$

32. $9c \geq -21$ $c \geq -\frac{7}{3}$ or $-2\frac{1}{3}$

33. $2d + 1 < 5$ $d < 2$

34. $5x - 2 > 2x + 9$

35. $\frac{x}{3} + 4 < 10$ $x < 18$

36. $\frac{-x}{5} - 1 < 3$ $x > -20$

37. $8x - 3 \leq 9$ $x \leq \frac{3}{2}$ or $1\frac{1}{2}$

38. $15 + \frac{y}{4} > 10$ $y > -20$

39. Ballet Express the cost, C, of shoes used in a year (for one man) as an inequality. $255 \leq C \leq 300$

A man uses about 15 pairs of ballet shoes a year, and the price ranges from $17 to $20 per pair.

Cultural Connection: Europe A Russian mathematician named Chebyshev (1821–1894) proved that the next prime number after p is less than $2p$. For example, 7 is a prime, so according to Chebyshev, the next prime after 7 is less than 14, which is true because 11 is the next prime and $11 < 14$.

Substitute each number for p to show that Chebyshev's statement is true.

40. 2 $3 < 4$

41. 3 $5 < 6$

42. 5 $7 < 10$

43. 89 $97 < 178$

Geometry The length, *l*, of a rectangle is at least 5 centimeters more than the width, *w*.

44. Express the statement as an inequality. $l \geq 5 + w$

45. Write an inequality for the length if the width is 20 centimeters.
$$l \geq 5 + 20$$
$$l \geq 25$$

$0 \leq 14.99v + 3.95 \leq 50; 0 \leq v \leq 3.07$
Consumer Economics Robin has $50 to spend on videotapes that are on sale in a catalog.

46. Write *two* inequalities that express the possible number of videotapes she can buy.
   a. the maximum number $v \leq 3$
   b. the minimum number $v \geq 0$

47. List all the possibilities for the number of videotapes Robin can buy. 0, 1, 2, 3

48. How many videotapes can Robin and two friends buy if they have $80 to spend?
   5 or less

Look Back

Evaluate.   **[Lesson 2.4]**

49. $89 - (-14)$ 103   **50.** $400 - (-111)$ 511   **51.** $-16 - (-3)$ -13 **52.** $-674 - 9(-900)$
   7426

Determine whether the value is negative or positive.   **[Lesson 2.5]**

53. $(-3)(-3)(-1)(-1)$ pos   **54.** $(-1)(1)(-1)(-1)$ neg   **55.** $\frac{-22}{2}$ neg   **56.** $\frac{-16}{-4}$ pos

Use the Distributive Property to simplify each expression.   **[Lesson 6.1]**

57. $(3x - 2y + 1) - 3(x + 2y - 1)$
   $-8y + 4$

58. $3(a + b) - 2(a - b)$ $a + 5b$

Look Beyond

Sometimes inequalities are used to indicate values that make sense in an equation, such as in these examples.

If $y = \frac{1}{x}$, then $x \neq 0$.   If $y = \sqrt{x}$, then $x \geq 0$.

Complete each statement.

59. If $y = \frac{1}{x - 2}$, then $x \neq \underline{\ ?\ }$. 2   **60.** If $y = \sqrt{x + 3}$, then $x \geq \underline{\ ?\ }$. -3

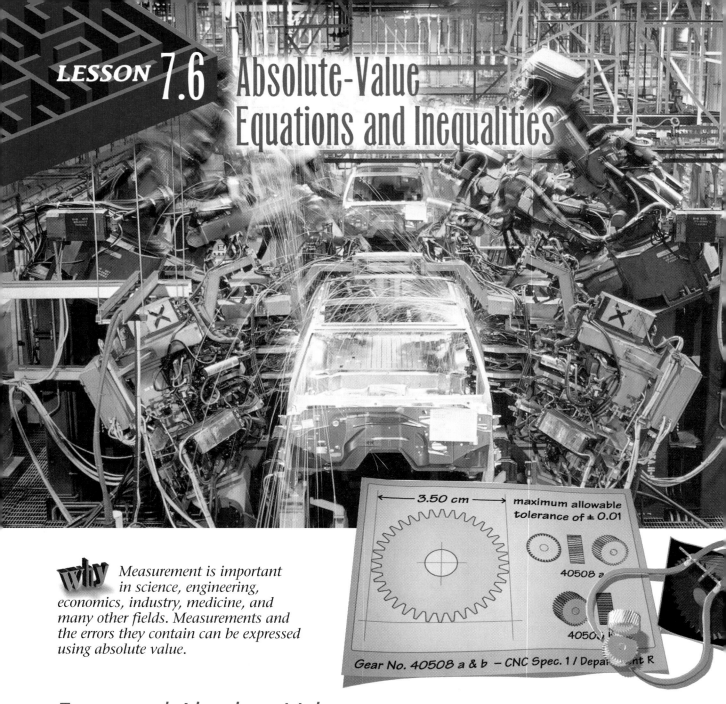

Absolute-Value Equations and Inequalities

why *Measurement is important in science, engineering, economics, industry, medicine, and many other fields. Measurements and the errors they contain can be expressed using absolute value.*

← 3.50 cm →

maximum allowable tolerance of ± 0.01

40508 a

4050

Gear No. 40508 a & b — CNC Spec. 1 / Department R

Error and Absolute Value

A company manufactures a small gear for a car. If the gear is made too large, it will not fit. If it is made too small, the car will not run properly. How accurate must the measurements be?

Manufacturing

A gear is designed with a specification of 3.50 centimeters for the diameter. It will work if it is within ± 0.01 centimeter of the specified measurement. The **absolute error** is the absolute value of the difference between the actual measure, x, and the specified measure, 3.50 centimeters. This is written as $|x - 3.50|$.

MAXIMUM MINIMUM *Connection*

If the maximum error permitted is 0.01 centimeter, the acceptable diameters can be shown with an absolute-value inequality.

$$|x - 3.50| \leq 0.01$$

To find the maximum and minimum diameters, solve the equation part of the inequality.

Recall the definition for absolute value.

Case 1 $|x| = x$ if x is positive or zero.
Case 2 $|x| = -x$ (the opposite of x) if x is negative.

To solve absolute-value equations, you must consider two cases.

Case 1
Consider the quantity within the absolute-value sign to be positive or zero.

$$|x - 3.50| = 0.01$$
$$x - 3.50 = 0.01$$
$$x - 3.50 + 3.50 = 0.01 + 3.50$$
$$x = 3.51$$

Case 2
Consider the quantity within the absolute-value sign to be negative.

$$|x - 3.50| = 0.01$$
$$-(x - 3.50) = 0.01$$
$$-x + 3.50 = 0.01$$
$$-x + 3.50 - 3.50 = 0.01 - 3.50$$
$$x = 3.49$$

The maximum and minimum allowable diameters for the gears are 3.51 centimeters and 3.49 centimeters.

EXAMPLE 1

Solve $|3x - 2| = 10$.

Solution ➤

Case 1
Consider the quantity within the absolute-value sign to be positive or zero.

$$|3x - 2| = 10$$
$$3x - 2 = 10$$
$$3x - 2 + 2 = 10 + 2$$
$$3x = 12$$
$$x = 4$$

Case 2
Consider the quantity within the absolute-value sign to be negative.

$$|3x - 2| = 10$$
$$-(3x - 2) = 10$$
$$-3x + 2 = 10$$
$$-3x + 2 - 2 = 10 - 2$$
$$-3x = 8$$
$$x = -\frac{8}{3}$$

Check

$$|3x - 2| = 10$$
$$|3(4) - 2| \stackrel{?}{=} 10$$
$$|10| = 10 \quad \text{True}$$

Check

$$|3x - 2| = 10$$
$$|3\left(-\frac{8}{3}\right) - 2| \stackrel{?}{=} 10$$
$$|-10| = 10 \quad \text{True} \; \diamond$$

Try This   Solve $|2x - 4| = 8$.   *$x = 6$ or $x = -2$*

Distance and Absolute Value

GEOMETRY
Connection

Absolute value can be used to describe the distance between any two points on a number line. The number line and geometry can help you visualize the meaning of the absolute-value equation and its solution.

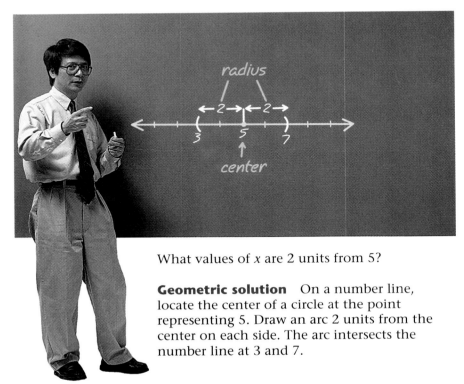

What values of x are 2 units from 5?

Geometric solution On a number line, locate the center of a circle at the point representing 5. Draw an arc 2 units from the center on each side. The arc intersects the number line at 3 and 7.

Algebraic solution Express this relationship as an absolute-value equation. The expression $|x - 5|$ represents the distance between 5 and each of the two points marked on the number line.

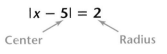

Then solve the absolute-value equation $|x - 5| = 2$.

C.T. The absolute value of a number is defined as the distance of that number from zero, and distance is always positive.

Case 1
The quantity within the absolute-value sign is positive or zero.

$$|x - 5| = 2$$
$$x - 5 = 2$$
$$x - 5 + 5 = 2 + 5$$
$$x = 7$$

Case 2
The quantity within the absolute-value sign is negative.

$$|x - 5| = 2$$
$$-(x - 5) = 2$$
$$-x + 5 = 2$$
$$-x + 5 - 5 = 2 - 5$$
$$-x = -3$$
$$x = 3$$

These are the points where the arcs of the circle cross the number line. They are each 2 units from 5.

CRITICAL
Thinking

Explain why absolute value is always a non-negative number. Why is absolute value a good method for finding distance?

Absolute Values and Inequalities

An artist must fit a square piece of stained glass in a 48.00-by-48.00-centimeter space in a wood block.

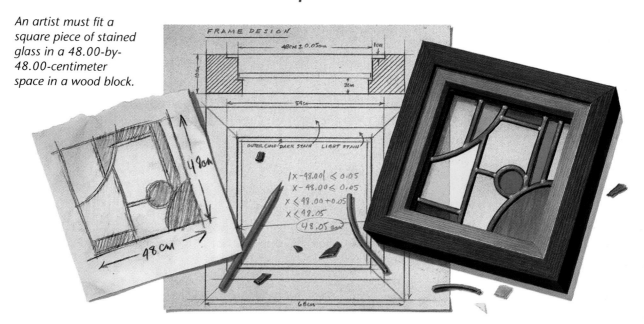

EXAMPLE 2

Crafts The metal square will fit properly in the block if the amount of error in the length of the sides is within 0.05 centimeter. Use absolute value to express the amount of error allowed for the length of the stained-glass square's side.

Solution ➤

The absolute error can be expressed as $|x - 48.00|$. The allowable error is ± 0.05 centimeter. Since the allowable error includes all the lengths within 0.05 centimeter of the specified measure, use an inequality.

Solve $|x - 48.00| \leq 0.05$.

Case 1
Consider the quantity within the absolute-value sign to be positive or zero.

$$|x - 48.00| \leq 0.05$$
$$x - 48.00 \leq 0.05$$
$$x - 48.00 + 48.00 \leq 0.05 + 48.00$$
$$x \leq 48.05$$

Case 2
Consider the quantity within the absolute-value sign to be negative.

$$|x - 48.00| \leq 0.05$$
$$-(x - 48.00) \leq 0.05$$
$$-x + 48.00 \leq 0.05$$
$$-x + 48.00 - 48.00 \leq 0.05 - 48.00$$
$$-x \leq -47.95$$
$$x \geq 47.95$$

In Case 2, recall that the inequality reverses when you multiply the expressions on each side of the inequality by a negative value.

The allowable lengths for the sides of the stained-glass square are all of the measures between 48.05 centimeters and 47.95 centimeters, inclusive. Write $47.95 \leq x \leq 48.05$. ❖

EXAMPLE 3

What values of x are less than or equal to 2 units from 5? Solve the inequality $|x - 5| \le 2$.

Solution ➤

Case 1
The quantity within the absolute-value sign is positive or zero.

$$|x - 5| \le 2$$
$$x - 5 \le 2$$
$$x - 5 + 5 \le 2 + 5$$
$$x \le 7$$

Case 2
The quantity within the absolute-value sign is negative.

$$|x - 5| \le 2$$
$$-(x - 5) \le 2$$
$$-x + 5 \le 2$$
$$-x + 5 - 5 \le 2 - 5$$
$$-x \le -3$$
$$x \ge 3$$

The inequality is true when x is less than or equal to 7 and greater than or equal to 3. Write $x \le 7$ *and* $x \ge 3$, or $3 \le x \le 7$.

This can be represented on the number line.

$3 \le x \le 7$

EXAMPLE 4

Solve $|x - 6| > 2$.

Solution ➤

Consider two cases.

Case 1
The quantity $x - 6$ is positive or zero.

$$|x - 6| > 2$$
$$x - 6 > 2$$
$$x - 6 + 6 > 2 + 6$$
$$x > 8$$

Case 2
The quantity $x - 6$ is negative.

$$|x - 6| > 2$$
$$-(x - 6) > 2$$
$$-x + 6 > 2$$
$$-x + 6 - 6 > 2 - 6$$
$$-x > -4$$
$$x < 4$$

The inequality is true when $x > 8$. The inequality is also true when $x < 4$. Write $x > 8$ *or* $x < 4$.

Check by testing numbers from the solution in the original inequality. ❖

CRITICAL *Thinking*

What problem occurs when you try to solve $|x - 5| < -1$? Test some values to see. Why does this happen?

EXERCISES & PROBLEMS

Communicate

1. What is the meaning of the specification 45 ± 0.001 centimeters?

2. Explain how to write an absolute-value equation that can be used to express 45 ± 0.001 centimeters.

3. Why must you consider two cases when you solve absolute-value equations and inequalities?

4. Describe how to use an absolute-value inequality to represent all the values on the number line that are within 3 units of the number −7.

5. Explain how to check the values of an absolute-value inequality to see if they are inside or outside the boundary.

Practice & Apply

Find the values of x that solve each absolute-value equation. Check your answer.

6. $|x - 5| = 3$ 8, 2

7. $|x - 1| = 6$ 7, −5

8. $|x - 2| = 4$ 6, −2

9. $|x - 8| = 5$ 13, 3

10. $|5x - 1| = 4$ 1, $-\frac{3}{5}$

11. $|2x + 4| = 7$ $1\frac{1}{2}, -5\frac{1}{2}$

12. $|4x + 5| = 1$ −1, $-1\frac{1}{2}$

13. $|-1 + x| = 3$ 4, −2

Find the values of x that solve each absolute-value inequality. Graph the answer on a number line. Check your answer.

14. $|x - 3| < 7$

15. $|x + 4| > 8$

16. $|x - 8| \le 4$

17. $|x - 5| \ge 2$

18. $|x - 2| > 6$

19. $|x - 2| \le 10$

20. $|x + 1| < 5$

21. $|x - 4| > 2$

Health According to a height and weight chart, Margo's ideal weight is 118 pounds. She wants to stay within 5 pounds of her ideal weight.

22. Draw a diagram. Identify boundary values, acceptable weights, and unacceptable weights on your diagram.

23. Write an absolute-value equation to describe the boundary values. What is the solution of your equation? $|w - 118| = 5; w = 113$ or $w = 123$

24. Write an absolute-value inequality to describe the acceptable weights. What is the solution of your inequality? $|w - 118| \le 5; 113 \le w \le 123$

The distance between x and 2 is 7.

25. Draw a number-line diagram to illustrate the given sentence.

26. Translate the given sentence into an absolute-value equation. $|x - 2| = 7$

Let $|x - 3| < 4$.

27. Write the absolute-value inequality as a sentence that begins, "The distance between . . ." . . . *x* and 3 is less than 4 units.

28. Draw a number-line solution for the inequality.

29. Describe the solution in words.

30. Give 5 specific numbers that satisfy the inequality.
Answers may vary. 0, 1, 2, 3, 4

Let $|x - 3| > 4$.

31. Write the absolute-value inequality as a sentence that begins, "The distance between . . ." . . . *x* and 3 is greater than 4 units.

32. Draw a number-line solution for the inequality.

33. Describe the solution in words.

34. Give 5 specific numbers that satisfy the inequality.
Answers may vary. −2, −3, −4, 8, 9

Government In a recent voter preference poll between two candidates, respondents gave Candidate A 50% of the vote. The polling technique used gives results that are accurate within 3 percentage points.

35. What are the upper and lower boundaries for the actual percent of voters who favor Candidate A? 53%; 47%

36. Describe the upper and lower boundaries using an absolute-value equation. $|A - 50| \leq 3$

37. Is it possible that Candidate A will lose the election? Explain.
Yes, candidate A could get between 47% and 50% of the vote.

Technology Use technology to compare each pair of graphs.

38. $y = 2x - 6$
$y = |2x - 6|$

39. $y = -x + 5$
$y = |-x + 5|$

Look Back

Evaluate each expression if p is 4, q is 1, and r is 2.   [Lesson 1.7]

40. $pqr - q$ 7

41. $\frac{pq}{r}$ 2

42. $\frac{pqr}{q} + pqr$ 16

Simplify.   [Lesson 5.5]

43. $-5(8c + 3)$
$-40c - 15$

44. $9(7b + 2)$
$63b + 18$

45. $-4(-5k + 8)$
$20k - 32$

Solve each inequality.   [Lesson 5.6]

46. $4x + 5 \leq 25$
$x < 5$

47. $6y - 10 > 5$

48. $9m - 8 < 4 + 8m$

47. $y > \frac{5}{2}$ or $2\frac{1}{2}$

48. $m < 12$

49. What percent of 480 is 60?   **[Lesson 6.6]**   12.5%

Solve.   [Lessons 7.1, 7.2]

50. $3x - 4 = 5$ 3

51. $8x - 7 = 25$ 4

52. $4(2 - x) = 20$ −3

Look Beyond

53. Subtract $x^2 + 4$ from $x^2 - x + 5$. $-x + 1$

54. Subtract $x^2 + 2x + 3$ from $x^2 + 10$. $-2x + 7$

Using a Spreadsheet

Spreadsheets allow you to organize lists or tables of data. The spreadsheet below has been created to show the perimeters and areas of different rectangles. Each row gives the information for a different rectangle. Use this Spreadsheet for Activity 1.

ACTIVITY 1

1. Look at the length and the width shown for each rectangle. What appears to be the relationship between the length and the width of each rectangle?

2. Let *l* represent the length of a rectangle and *w* represent the width. Write an equation to express the width in terms of the length.

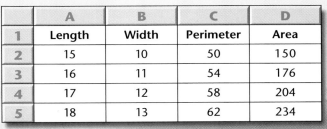

	A	B	C	D
1	Length	Width	Perimeter	Area
2	15	10	50	150
3	16	11	54	176
4	17	12	58	204
5	18	13	62	234

3. What formula has been used to compute the data in column C?

4. Describe the relationship among the entries in column A.

5. The spreadsheet below shows some of the formulas that were entered into the cells to compute the data. Copy and complete the spreadsheet by filling in the remaining formulas.

	A	B	C	D
1	Length	Width	Perimeter	Area
2	15	=A2–5	=2*A2+2*B2	=A2*B2
3	=A2+1	=A3–5	=2*A3+2*B3	?
4	?	?	?	?
5	?	?	?	?

6. Write an algebraic equation that can be used to determine the dimensions of a rectangle from this spreadsheet that has a perimeter of 114.

ACTIVITY 2

A city park is 500 feet by 250 feet. The city council has asked the high-school garden club to design a water garden in the park. They want the length-to-width ratio of the garden to have the same length-to-width ratio of the park. Create a spreadsheet to show possible perimeters and areas for the garden. Use your spreadsheet to determine the dimensions of a garden with a perimeter of 720 feet. What is the area of this garden?

Chapter 7 Review

Vocabulary

Key Skills & Exercises

Lesson 7.1

➤ Key Skills

Solve two-step equations.

To solve the equation $2(5x - 3) = 56$, first simplify the equation.

$2(5x - 3) = 56$	Given
$10x - 6 = 56$	Distributive Property
$10x - 6 + 6 = 56 + 6$	Addition Property of Equality
$10x = 62$	Simplify.
$\frac{10x}{10} = \frac{62}{10}$	Division Property of Equality
$x = 6.2$	Simplify.

➤ Exercises

Solve each equation.

1. $3x + 7 = 31$ 8

2. $7n - 6 = 29$ 5

3. $34 = -6 + 8m$ 5

4. $7(2y + 18) = 28$ -7

5. $8\left(z - \frac{1}{3}\right) = 4$ $\frac{5}{6}$

6. $\frac{1}{3} + 2t = 1$ $\frac{1}{3}$

Lesson 7.2

➤ Key Skills

Solve multistep equations.

To solve an equation with the variable on both sides of the equation, first obtain an equation with the variable on only one side. Then solve.

$5(x + 4) = 2x + 11$	Given
$5x + 20 = 2x + 11$	Distributive Property
$5x + 20 - 2x = 2x + 11 - 2x$	Subtraction Property of Equality
$3x + 20 = 11$	Simplify.
$3x + 20 - 20 = 11 - 20$	Subtraction Property of Equality
$3x = -9$	Simplify.
$\frac{3x}{3} = \frac{-9}{3}$	Division Property of Equality
$x = -3$	Simplify.

➤ Exercises

Solve each equation.

7. $10x = 7x - 12$ -4

8. $11t + 9 = 61 - 2t$ 4

9. $4s - 4 = 6s - 21$ $\frac{17}{2}$ or $8\frac{1}{2}$

10. $8(d - 1) = 15d + 6$ -2

11. $5(y - 1) = 4(10 - y)$ 5

12. $0.6z + 1.4 = -6.7 - 0.3z$ -9

Lesson 7.3

➤ Key Skills

Use equations to solve algebraic applications.

A salesperson needs to make $1200 each month. He makes a 3.5% commission on his monthly sales and also receives a base salary of $500 per month. Find the amount of monthly sales that he needs to make.

The needed income is $500 plus 3.5% of the sales.

$$1200 = 500 + 0.035 \cdot s$$

Solve for s.

$$1200 = 500 + 0.035s$$
$$1200 - 500 = 500 + 0.035s - 500$$
$$700 = 0.035s$$
$$\frac{700}{0.035} = \frac{0.035s}{0.035}$$
$$20{,}000 = s$$

➤ Exercises

13. Janice made $3400 last month as a real-estate salesperson. Her base salary was $1000, and she received a 4% commission on her sales. What were her total sales for the month? $60,000

14. Thirty milliliters of a solution that is 3% acid are mixed with 10 milliliters of a solution that is 5% acid. How much acid is in the new solution? What percent of the new solution is acid?   1.4 mL; 3.5%

Lesson 7.4

➤ Key Skills

Use equations to solve geometric applications.

The area of triangle ABC is 13.5 square feet. To find the length of the base, substitute the known values into the formula for the area of a triangle, $A = \frac{1}{2}bh$, and solve for the unknown variable, b.

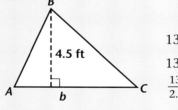

$$A = \frac{1}{2}bh$$
$$13.5 = \frac{1}{2}b(4.5)$$
$$13.5 = 2.25b$$
$$\frac{13.5}{2.25} = \frac{2.25b}{2.25}$$
$$6 = b$$

The length of the base is 6 feet.

➤ Exercises

Lines s and r are parallel. Find the value of x and the measures of the angles in each exercise below.

15. $m\angle 1 = 3x - 1$; $m\angle 3 = x + 2$

16. $m\angle 5 = 6x - 1$; $m\angle 6 = 3x + 1$

17. $m\angle 1 = 8x - 2$; $m\angle 7 = 4x$

18. $m\angle 2 = x + 5$; $m\angle 3 = 6x - 7$

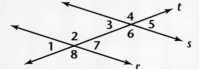

15. $\frac{3}{2}$; $3\frac{1}{2}°$; $3\frac{1}{2}°$
16. 20; 119°, 61°
17. $\frac{1}{2}$; 2°, 2°
18. 26; 31°; 149°

19. Two angles are supplementary. The larger angle is 20° greater than 3 times the smaller angle. Find the measures of both angles.   40°, 140°

20. The area of a triangle is 56 square yards. The base of the triangle is 16 yards long. What is the height of the triangle?   7 yd

Lesson 7.5

➤ Key Skills

Solve inequalities.

Solve $14 - x \leq -6$.

$$
\begin{aligned}
14 - x &\leq -6 & &\text{Given} \\
14 - x - 14 &\leq -6 - 14 & &\text{Subtraction Property of Inequality} \\
-x &\leq -20 & &\text{Simplify.} \\
\text{Reverse the inequality sign.} \quad -1(-x) &\geq -1(-20) & &\text{Multiplication Property of Inequality} \\
x &\geq 20 & &\text{Simplify.}
\end{aligned}
$$

➤ Exercises

Solve each inequality.

21. $x + 4 < 6$ $x < 2$ **22.** $8 - y \geq 7$ $y \leq 1$ **23.** $5r > -60$ $r > -12$

24. $\frac{-p}{8} \leq -3$ $p \geq 24$ **25.** $x + 3 \geq 9 - x$ $x \geq 3$ **26.** $t + \frac{1}{2} < \frac{t}{4} + 2$ $t < 2$

Lesson 7.6

➤ Key Skills

Solve absolute-value equations and inequalities.

Solve $|x - 2| = 3$.

Case 1	**Case 2**
The quantity $x - 2$ is positive or zero.	The quantity $x - 2$ is negative.

$$
\begin{aligned}
|x - 2| &= 3 \\
x - 2 &= 3 \\
x &= 5
\end{aligned}
$$

$$
\begin{aligned}
|x - 2| &= 3 \\
-(x - 2) &= 3 \\
-x + 2 &= 3 \\
-x &= 1 \\
x &= -1
\end{aligned}
$$

$$x = 5 \ or \ x = -1$$

Solve $|x - 2| < 3$.

Case 1	**Case 2**
The quantity $x - 2$ is positive or zero.	The quantity $x - 2$ is negative.

$$
\begin{aligned}
|x - 2| &< 3 \\
x - 2 &< 3 \\
x &< 5
\end{aligned}
$$

$$
\begin{aligned}
|x - 2| &< 3 \\
-(x - 2) &< 3 \\
-x + 2 &< 3 \\
-x &< 1 \\
(-1)(-x) &> (-1)(1) \\
x &> -1
\end{aligned}
$$

$$-1 < x < 5$$

➤ Exercises

Solve each absolute-value equation or inequality. Graph each inequality on a number line.

27. $|x - 4| = 8$ **28.** $|3x + 2| = 8$ **29.** $|x - 3| \leq 8$ **30.** $|x - 6| < 8$

   $x = 12 \ or \ x = -4$ $x = 2 \ or \ x = -\frac{10}{3}$

Applications

31. Chemistry How many ounces of an 80% glucose solution should be added to 20 ounces of a 16% glucose solution to obtain a solution that is 30% glucose?   5.6 oz

32. Government In a poll to predict a school-board election, it was found that 51% of the respondents seemed to prefer Mr. Greene. It is known that the actual value of Mr. Greene's percent is within 3.5 percentage points of the poll's results. Describe the upper and lower boundaries using an absolute-value inequality. Is it possible that Mr. Greene will lose the election? Explain your reasoning.

Chapter 7 Assessment

Solve each equation.

1. $2x - 9 = 23$ 16

2. $53 = -7 + 2s$ 30

3. $4(15 - 5r) = 90$ $-1\frac{1}{2}$

4. $-16x + 3 = 37 + x$ -2

5. $2\left(t + \frac{1}{4}\right) = 4t - 1$ $\frac{3}{4}$

6. $1.1y - 8 = 0.5 + 0.25y$ 10

7. $4v - 5(v - 2) = 6v - 46$ 8

8. $-2z + 4(z - 7) = 9z + 7(-2 + z)$ -1

9. **Statistics** Alberto received a grade of 100 on one of his English tests and 75 on another. What grade does he have to receive on the next test in order to average 85 on all three tests?   80

The perimeters of the two triangles below are equal.

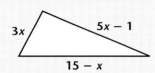

3x   5x − 1   15 − x

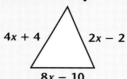

4x + 4   2x − 2   8x − 10

10. Find the value of x. $3\frac{1}{7}$

11. Find the common perimeter.   36

12. **Sports** The high-school baseball team played 6 fewer games this year than last year but won 60% of the games played, compared with 50% of its games last year. The team won 1 more game this year than last year. How many games did the team play each year?   46 last year; 40 this year

Find the value of x and the measures of the angles of triangle ABC.

13. $m\angle A = 5x$; $m\angle B = 3x + 11$; $m\angle C = 29 - x$

14. $m\angle A = 2x - 10$; $m\angle B = 5x - 100$; $m\angle C = x + 10$

15. $m\angle A = \frac{x}{8} + 25$; $m\angle B = \frac{x}{2} + 80$; $m\angle C = 90 - x$

16. Solve the formula $s = mph$ for m. $m = \frac{s}{ph}$

17. Solve the formula $A = \frac{x}{2}h(b_1 + b_2)$ for h. $h = \frac{2A}{x(b_1 + b_2)}$

Solve each inequality.

18. $x - 8 < 15$ $x < 23$

19. $7 + y \geq 23$ $y \geq 16$

20. $4t \leq -16$ $t \leq -4$

21. $-5z > 45$ $z < -9$

Solve each absolute-value equation or inequality. Graph each inequality on a number line.

22. $|x - 8| = 14$ $x = 22$ or $x = -6$

23. $|4x + 3| = 7$ $x = 1$ or $x = -2\frac{1}{2}$

24. $|x - 6| \leq 17$

25. $|x + 2| > 4$

26. **Geometry** The formula for the area of a triangle is $A = \frac{1}{2}bh$. If the area of a triangle is 336 square centimeters and the base is 24 centimeters long, what is the height of the triangle?   28 cm

CHAPTER 8

LESSONS

Linear Functions

Pyramids are part of the heritage of people who trace their roots to Africa or to Central America. The Great Pyramid of Giza in Egypt was built over a 30-year period around 2900 B.C. E. The builders used blocks averaging 2.5 tons, but some weighed as much as 54 tons. The blocks were transported from a quarry 600 miles away. The pyramid has a square base, and the error in constructing the right angles of the base was only one part in 14,000.

In order to build the pyramids, people needed to solve problems about slope. One problem asks to find the *seked* (a form of slope) of a pyramid with a base of 360 cubits (618 feet) on each side and a height of 250 cubits (429 feet).

The Rhind papyrus

PORTFOLIO ACTIVITY

How steep are the sides of the pyramids of Egypt? The use of mathematics allows us to measure steepness. One measure of steepness is the slope.

1. Use cubit measures to find the slope of the triangular face of the pyramid.

2. Repeat the process using measurement in feet.

3. What do you notice about the slope measured in these two ways?

You may wish to include your calculations and explanation in your portfolio.

250 cubits
429 feet

360 cubits
618 feet

Representing Linear Functions by Graphs

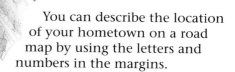

Why *Sometimes it is difficult to recognize patterns in a numerical problem. It is often easier to see the pattern from a picture. In algebra, a graph is used to display a picture of a mathematical pattern.*

You can describe the location of your hometown on a road map by using the letters and numbers in the margins.

To find Northfield, Minnesota, on a map, for instance, you look up from the letter K on the bottom and across from the number 18 in the margin of a Minnesota road map. The region where these strips overlap contains Northfield.

Graphing With Rectangular Coordinates

Cultural Connection: Africa The idea of graphing has existed for over 4500 years. Around 2650 B.C.E., Egyptians sketched curves using pairs of numbers to locate points. This idea gave rise to the use of coordinates.

Recall that the set of real numbers can be represented on a number line.

- To construct the number line, mark a point on the line as zero.

- Mark off equal spaces to the right of zero, and number them 1, 2, 3, 4, . . .

- Then mark off corresponding spaces to the left of zero, and number them − 1, − 2, − 3, − 4, . . .

Where is $\frac{1}{2}$ on this number line? Where is $-\frac{1}{2}$?

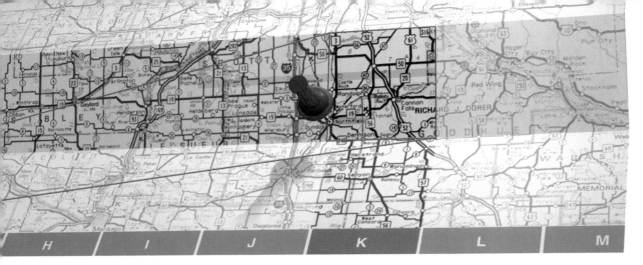

H   I   J   K   L   M

Cultural Connection: Europe In the seventeenth century, Rene Descartes, a French mathematician and philosopher, used a horizontal and vertical number line to divide a plane into four regions, called **quadrants**.

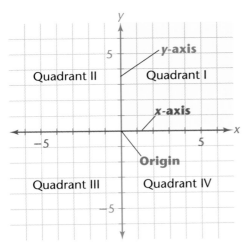

The rectangular, or Cartesian, coordinate system was named for Rene Descartes.

The horizontal line is called the **x-axis**, and points on this axis are called **x-coordinates**.

The vertical line is called the **y-axis**, and points on this axis are called **y-coordinates**.

Recall that the x- and y-axes intersect at the **origin**.

Also recall that letters other than x and y can represent the axes. For example, you might see the x-axis shown as the t-axis when t refers to time in a problem.

Coordinates give the address of a point. They are written as an ordered pair, indicated by two numbers in parentheses, (x, y). The number from the x-axis will appear first in an ordered pair.

To locate the point shown on the graph, start at the origin, (0, 0). Move to the right 10 units along the x-axis, then move up 5 units to the point. The ordered pair is (**10, 5**).

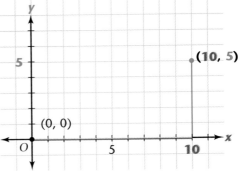

CRITICAL Thinking Find a way to locate a point on a plane without using the rectangular coordinate system.

Linear Functions

Once you organize the data from a problem, you can form ordered pairs and graph the data.

Sabrina normally rides her bicycle at a constant rate of 0.2 miles per minute.

EXAMPLE 1

On Wednesday, Sabrina left school and rode her bicycle for 15 minutes in a direction away from home. If Sabrina lives 2 miles from school, how far is Sabrina from home?

Solution ➤

The equation that models Sabrina's distance from home with respect to time spent bicycling is based on the following relationship:

Distance = rate • time + miles from home to school.

Replace *distance* with d, *rate* with 0.2, *time* with t, and *miles from home to school* with 2.

The equation is now written as $d = 0.2t + 2$.

To make a table of values, substitute values for t in minutes, and find the corresponding values for d in miles. The ordered pairs are then plotted and connected by a line.

t	$0.2t + 2.0$	d	(t, d)
0	$0.2(0) + 2.0$	2.0	(0, 2.0)
1	$0.2(1) + 2.0$	2.2	(1, 2.2)
2	$0.2(2) + 2.0$	2.4	(2, 2.4)
3	$0.2(3) + 2.0$	2.6	(3, 2.6)
4	$0.2(4) + 2.0$	2.8	(4, 2.8)
5	$0.2(5) + 2.0$	3.0	(5, 3.0)
⋮	⋮	⋮	⋮
15	$0.2(15) + 2.0 =$	5.0	(15, 5.0)

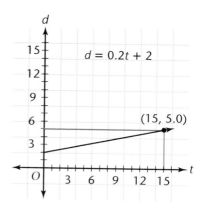

The table and the graph indicate that after 15 minutes, Sabrina is 5 miles from home. ❖

The graph of the equation $d = 0.2t + 2$ is a straight line. For this reason the equation is called a linear equation. This equation generates a set of ordered pairs, (t, d). We say that d is a function of t or that d is dependent on t. Thus, $d = 0.2t + 2$ is a **linear function.** In the ordered pairs, the set of first values, t, is the **domain** of the function. The set of second values, d, is the **range** of the function.

In the function $d = 0.2t + 2$, t is called the **independent variable**. The d is called the **dependent variable** because its value depends on the value chosen for t.

The domain and range for the function $d = 0.2t + 2$ are all real numbers. However, all real numbers do not make sense for the situation in Example 1. Since time, t, and distance, d, cannot be negative numbers, the reasonable domain and range for this situation are all positive real numbers and zero.

EXAMPLE 2

Health   Represent maximum heart rate by r and age by a. Write a function to express the maximum heart rate in terms of a. Graph the function for ages 10 to 50.

Solution ➤

Since the rate, r, equals 220 beats per minute minus the age, write the equation $r = 220 - a$. Values for a are elements of the domain, and values for r are elements of the range. When graphing a function, it is customary to represent the domain on the x-axis and the range on the y-axis.

When the values 10, 20, 30, 40, and 50 are in the domain, the table shows how substitution allows you to find the corresponding values in the range.

To graph the function, enter **220 − x** after Y = on a graphics calculator. Set the limits of the domain and range in the viewing window. Then press GRAPH.

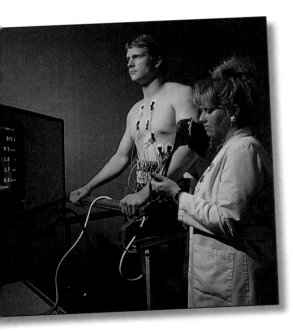

A reasonable estimate for the maximum heart rate during exercise should be no more than 220 beats per minute minus the person's age.

a	$220 - a$	r
10	220 − 10	210
20	220 − 20	200
30	220 − 30	190
40	220 − 40	180
50	220 − 50	170

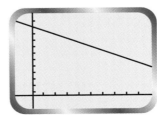

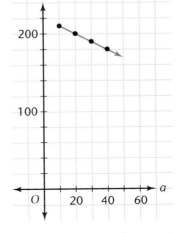

EXERCISES & PROBLEMS

Communicate

1. Explain how you find the coordinates of a given point from a graph.

2. Is (6, 7) the same point as (7, 6)? Explain.

3. Describe the steps for plotting the point with the coordinates (7, 3) on a graph.

4. Describe the relationship between the coordinates of an ordered pair and the domain and range of a function.

5. Which axis in the coordinate plane represents the dependent variable?

6. How would you make a table of the values for x and y for the equation $y = 2x + 5$?

Practice & Apply

Graph each list of ordered pairs. State whether they lie on a straight line.

7. (1, 3), (2, 6), (3, 9)

8. (1, 5), (2, 4), (3, 1)

9. (1, 10), (2, 7), (3, 2)

10. (1, − 3), (2, − 6), (3, − 9)

11. (5, 2), (7, 2), (9, 2)

12. (4, 1), (4, 5), (4, 9)

What are the coordinates of the given points?

13. A (0, 7)

14. B (7, 7)

15. C (5, 4)

16. D (6, 0)

Graph and compare the following two linear functions. What effect does the operation before the x have on the graph of the line?

17. $y = 3 + x$

18. $y = 3 - x$

Graph each of the following functions on the same coordinate plane, and compare the graphs. Explain your conclusions.

19. $y = x + 7$

20. $y = x - 7$

21. $y = 7 - x$

22. $y = -7 - x$

Find the values for y by substituting 1, 2, 3, 4, and 5 for x. Make a table.

23. $y = x + 3$

24. $y = x + 4$

25. $y = 2x$

26. $y = 2x + 5$

Don walks at a rate of 3 miles per hour. You can determine the distance that he walks by multiplying the rate by the number of hours that he walks.

27. Write a function for distance, *d*, as a function of the number of hours, *h*. $d = 3h$

28. Make a table to show how far Don walks in 0, 1, 2, and 3 hours. Graph the function.

29. Determine the reasonable domain and range for this situation. $h \geq 0;\ d \geq 0$

30. Which is the dependent variable? Which is the independent variable?
d is the dependent variable; *h* is the independent variable.

Suppose a T-shirt company charges $8 per T-shirt and a $3 handling fee per order.

31. How much does an order of 2 T-shirts cost? $19

32. How much does an order of 5 T-shirts cost? $43

33. Make a set of ordered pairs from the given information, and plot them as points on a graph. Do the points lie on a straight line?

34. Determine the reasonable domain and range for this situation. domain: $x \geq 0$; range: $y \geq 0$

8^{00}

Look Back

Many examples of the Fibonacci sequence are found in nature.

35. What are the next three terms in the 31, 46, 64
sequence 1, 1, 4, 10, 19, . . . ? **[Lesson 1.1]**

Simplify.   [Lesson 1.7]

36. $2 \cdot 14 \div 2 + 5$ 19   **37.** $6 + 12 \div 6 - 4$ 4

38. $4[(12 - 3) \cdot 2] \div 11$ $6\frac{6}{11}$ **39.** $[3(4) - 6] - [(15 - 7) \div 4]$ 4

40. How many differences are needed to reach a constant for the sequence 1, 2, 6, 15, 31, . . . ? What is the constant difference for the sequence? **[Lesson 2.7]**
3rd differences; 2

41. The sequence 1, 1, 2, 3, 5, 8, . . . is a famous sequence named for the Italian mathematician Fibonacci. Find the next two terms. What kind of pattern do the differences show? **[Lesson 2.7]**

42. Solve the equation $3x + 406 = 421$. **[Lesson 7.1]** $x = 5$

43. Solve the equation $7x + 4 = 40$. **[Lesson 7.1]** $x = 5\frac{1}{7}$

Look Beyond

44. The graph of the path of a rocket is a parabola. The highest point of this curve is called the vertex. Graph the set of points, and determine which point is the vertex.

x	0	1	2	3	4	5	6	7	8	9	10
y	0	9	16	21	24	25	24	21	16	9	0

LESSON 8.2
Exploring Slope

why *Kara and Rich hike up the steep slope of a mountain trail. They remember how easy it was walking along the nearly flat streets back home. If you want to compare the steepness of a mountain trail with a level street, you can compare their slopes. You can also measure the steepness of lines on a coordinate plane in the same way.*

Slope is the ratio of the vertical rise to the horizontal run. The slope of the hill is the same for both Kara and Rich.

To calculate the slope of the hill, find the ratio of rise to run for each hiker. The horizontal length of Rich's step (the run) is 35 centimeters, and the vertical length (the rise) is 14 centimeters. The run for Kara's step is 30 centimeters, and the rise is 12 centimeters.

<table>
<tr><td align="center">Rich</td><td align="center">Kara</td></tr>
<tr><td align="center">slope $= \dfrac{\text{rise}}{\text{run}} = \dfrac{14}{35} = \dfrac{2}{5}$</td><td align="center">slope $= \dfrac{\text{rise}}{\text{run}} = \dfrac{12}{30} = \dfrac{2}{5}$</td></tr>
</table>

The slope of the hill is $\dfrac{2}{5}$.

GEOMETRIC OR GRAPHIC INTERPRETATION OF SLOPE

Slope measures the steepness of a line by the formula

$$\text{slope} = \frac{\text{rise}}{\text{run}}.$$

Sign of the Slope

The slope of the mountain trail was found from an upward rise left to right and was calculated to be positive $\frac{2}{5}$. In Exploration 1, you will investigate a situation in which a line slopes downward from left to right.

 **Exploration 1** *Negative Slope*

For this exploration, you will need 3 meter sticks and graph paper.

Maya and her class are studying the effects of friction. To model the concept, they lean a meter stick against a wall. The object is to see how steep the incline must be to keep the ruler from slipping.

Use a flat tabletop pushed against a wall. Attach one meter stick to the wall and one meter stick to the tabletop. Make sure the meter sticks are perpendicular to the tabletop and to the wall. Then lean the third meter stick against the wall.

1 What is the distance from the top of the meter stick straight down to the tabletop? from the bottom of the meter stick to the wall? Answers may vary.

2 On a graph, the wall represents the _?_ -axis and the tabletop represents the _?_ -axis. The intersection of the table and the wall represents the _?_ . *y; x; origin*

3 Graph the results of Step 2 on your graph so that the run is measured toward the right and the rise is measured downward. How is this slope different from the rise and run of the mountain trail? This slope goes downward from left to right.

4 Is the run positive or negative? Is the rise positive or negative? *positive run; negative rise*

5 Find the slope of the line formed by the ruler. Is the slope positive or negative? Answers may vary; negative.

6 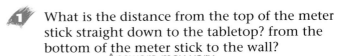 What makes the slope of a line positive? What makes the slope of a line negative? ❖

DIRECTIONS AND SIGNS
The positive directions (+) on the graph are toward the right and
up. The negative directions (−) on the graph are toward the left
and down.

EXTENSION

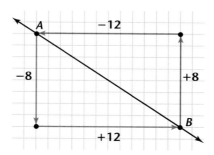

You can find the slope of line AB in more than one way.

From point A to point B, the rise is -8 and the run is $+12$.

$$\text{slope} = \frac{\text{rise}}{\text{run}} = \frac{-8}{12} = \frac{-2}{3}$$

From point B to point A, the rise is $+8$ and the run is -12.

$$\text{slope} = \frac{\text{rise}}{\text{run}} = \frac{8}{-12} = \frac{2}{-3} \ \diamondsuit$$

What combinations of positive and negative values for rise and run produce a negative slope? a positive slope? For a negative slope, either the rise is positive and the run is negative, or the rise is negative and the run is positive. For a positive slope, the rise and run must be either both positive or both negative.

•Exploration 2 Finding Slope From Two Points

CT− The line with slope −2 falls sharply from left to right. As the value of the slope increases, the lines continue to fall but less steeply. The line with slope 0 is horizontal. The line with slope $\frac{1}{3}$ rises gradually from left to right. As the slope continues to increase, the lines rise more and more sharply.

1 Plot the points (1, 2) and (3, 5) on a coordinate grid. Draw a line through the two points.

2 Use the ratio of rise to run to determine the slope of the line. $\frac{3}{2}$

3 Find the difference between the y-coordinate of the first point and the y-coordinate of the second point. What does this difference represent? −3; represents the rise

4 Find the difference between the x-coordinate of the first point and the x-coordinate of the second point. What does this difference represent? −2; represents the run

5 Write the ratio $\frac{\text{difference in } y\text{-coordinates}}{\text{difference in } x\text{-coordinates}}$. How does this ratio compare with the slope you found in Step 2? $\diamondsuit$ $\frac{3}{2}$; same

**CRITICAL
Thinking**

Suppose seven different lines have the slopes -2, $-\frac{3}{4}$, $-\frac{1}{3}$, 0, $\frac{1}{3}$, $\frac{3}{4}$, and 2. Describe the graph of each line as the slope increases from negative to positive numbers.

Exploration 3 *Using a Graph*

Jeff leaves his home to ride his new scooter. This graph shows the distance he has traveled for the first 6 minutes.

 How many miles did Jeff travel in 6 minutes? 4 mi

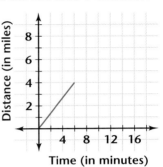

The rate at which Jeff traveled is $\frac{\text{the change in distance}}{\text{the change in time}}$.
Jeff's rate is _?_ miles per minute.

$\frac{4}{6} = \frac{2}{3}$ mi/min

Find the slope of the graph. How does the rate in Step 2 compare with the slope of the graph? $\frac{2}{3}$; same

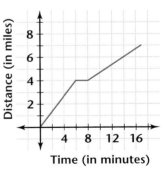

Jeff stopped at a traffic light for 2 minutes and then started again. When Jeff stopped, he traveled 0 miles in 2 minutes. The graph at the right shows Jeff's stop and the distance he traveled for 9 minutes the second time. What was his rate while he was stopped at the traffic light? What was his rate after he started again? 0 mi/min; $\frac{1}{3}$ mi/min

 Suppose Jeff stops again for 3 minutes. Copy and extend the graph to show this stop.

 Suppose Jeff starts again and travels 2 miles in 4 minutes. Show this on your graph. What is Jeff's rate of speed on this part of his trip?

 Complete the following sentence. For each line segment on your graph, the number of units the line segment extends in the _?_ direction divided by the number of units it extends in the _?_ direction determines the rate of speed. ❖ vertical, horizontal

Lesson 8.2 Exploring Slope **437**

EXERCISES & PROBLEMS

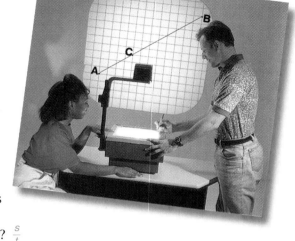

Communicate

1. Explain how to draw a line with a rise of 4 and a run of 3.

2. Describe the slope of a line with negative rise and positive run. negative slope; falls downward from left to right

3. Explain how to find the slope of line *AC*.

4. Explain why the slope of line *AC* is the same as the slope of line *AB*.

5. What is the slope of a line with rise *s* and run *t*? $\frac{s}{t}$

6. Point *R*(5, −3) and point *N*(9, −4) are on line *k*. Explain how to find the slope of line *k*.

Practice & Apply

Examine the graphs below. Which line has a positive slope? Which has a negative slope? Which has neither?

7. pos
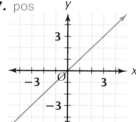

8. neither
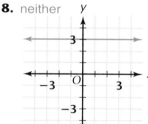

9. neg
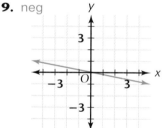

10. **Recreation** Suppose the slopes of two ski hills are negative. On graph paper, draw a line to represent a beginner's ski hill and another to represent an advanced ski hill. Indicate each slope.

Find the slope for each of the given lines.

11. rise 6, run 2 3

12. rise 1, run 7 $\frac{1}{7}$

13. rise −1, run 7 $-\frac{1}{7}$

14. rise 0, run 5 0

15. rise 5 − 3, run −3 − 1 $-\frac{1}{2}$

16. rise −7 + 2, run 3 − 1 $-\frac{5}{2}$

17. Draw coordinate axes on graph paper. Place a point somewhere in the upper left region of your graph. From that point, move 6 spaces to the right, and then down 4 spaces. Place a point there. Move again 6 spaces right, and then down 4 spaces. Repeat this at least four times. What shape appears when you connect the points? What is the slope of the line connecting the points? Try this exercise with patterns of your own.

18. Using the table at the right, find the difference for several pairs of *x*-coordinates and the difference for corresponding pairs of *y*-coordinates. Next, plot the values on a graph, and find the slope. Explain the relationships that you find.

x	1	2	3	4	8
y	3	6	9	12	24

19. 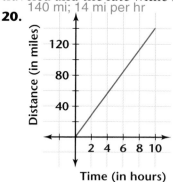 **Geometry** Find at least four pictures of houses that have roofs with different slopes. Use a protractor to measure the angle for the slope of each roof. Use a ruler to find the slopes. Make a table to match the measure of each angle to each slope.

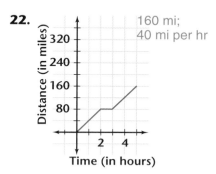

Travel Each graph below represents the distance a driver traveled in terms of the number of hours since the driver left home. In each case, the driver traveled at a constant rate. Determine the distance traveled and the rate while moving for each graph.

20. 140 mi; 14 mi per hr

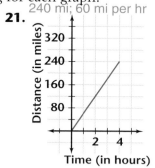

21. 240 mi; 60 mi per hr

22. 160 mi; 40 mi per hr

Travel Jarel left his grandmother's house riding his bicycle at a constant rate. The graph at the right shows his distance from his house in terms of the number of minutes since he left his grandmother's house.

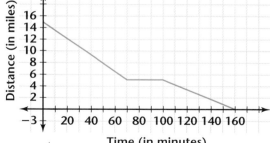

23. How far is his grandmother's house from his house? 15 mi

24. Did Jarel stop and rest on the way home? If so, for how long? yes; 30 min

25. How many different rates did Jarel travel? 3
What were each of those rates? $\frac{1}{7}$ mi per min; 0 mi per min; $\frac{1}{12}$ mi per min

26. How long did it take Jarel to get home? 160 min = 2 hr 40 min

27. Use the information in the table to find the average amount you climb per mile from Kansas City to Denver. $7\frac{14}{25}$ ft per mi

Denver, Colorado, elevation 1 mile

City	Miles from Kansas City	Elevation in feet
Kansas City, KS	0	744
Denver, CO	600	5280

28. The elevation of Loveland Pass, 70 miles west of Denver near Interstate 70, is 12,992 feet. Find the average slope of the land from Denver to Loveland Pass. $110\frac{6}{35}$ ft per mi

There are 91 steps in each of the four stairways on the Mayan/Toltec pyramid at Chichen Itza, in the Yucatan.

Cultural Connection: Americas
Mexican pyramids are somewhat different in concept from African pyramids.

29. If the height (rise) of the stairway is 27 meters and the length (run) is 25 meters, what is the slope of the stairway of the pyramid? $\frac{27}{25}$, or 1.08

30. The average rise of a step is 29.7 centimeters, and the average run is 27.5 centimeters. Find the slope of a step, and compare it with the slope of the pyramid's stairway.
$\frac{29.7}{27.5}$, or 1.08; same

Determine the slope of each line from the graph.

31. $\frac{1}{2}$

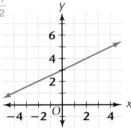

32. -2

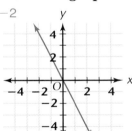

33. -1

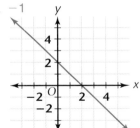

34. 3

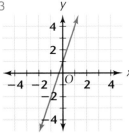

35. $\frac{1}{3}$

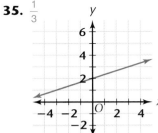

36. $-\frac{3}{4}$

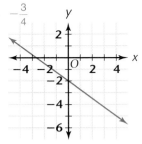

37. 0

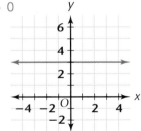

38. undefined

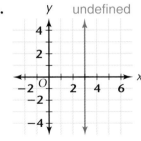

39. $\frac{3}{2}$

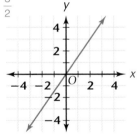

Each pair of points is on a line. What is the slope of each line?

40. $M(9, 6)$, $N(1, 4)$ $\frac{1}{4}$

41. $M(1, 3)$, $N(1, 5)$ undef

42. $M(-3, 1)$, $N(2, 6)$ 1

43. $M(0, 2)$, $N(3, 0)$ $-\frac{2}{3}$

44. $M(-3, 5)$, $N(7, 6)$ $\frac{1}{10}$

45. $M(2, -4)$, $N(5, 9)$ $\frac{13}{3}$

46. $M(3.2, 8.9)$, $N(9.1, 7.2)$ $-\frac{17}{59}$

47. $M(8, 10)$, $N(8, 7)$ undef

48. $M(10, 4)$, $N(7, 4)$ 0

Engineering Water is being pumped into a tank. The graph shows the number of gallons of water in the tank relative to the number of minutes.

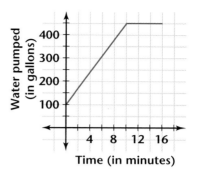

49. How much water was in the tank when the pump was turned on? 100 gal

50. How long was the pump on before it stopped? 10 min

51. At what rate was the water pumped into the tank from 0 to 10 minutes? 35 gal per min

52.  **Geometry** Use a string to measure several round objects. Record their diameters and circumferences in a table. Then graph the data with diameters on the *x*-axis, and fit the points with a straight line. How well does the line fit the points? Find the slope of the line. What conjectures can you make from this information?

Look Back

Evaluate. **[Lesson 2.2]**

53. $|-7|$ 7

54. $|50|$ 50

55. $-|-9|$ -9

56. $-|99|$ -99

57. Fund-raising If there are 8 people contributing to a charity, how much would each person have to contribute to raise $98? $12.25
[Lesson 6.2]

Solve. **[Lesson 7.1]**

58. $3x + 16 = 19$ 1

59. $28 = -4 + 4x$ 8

Solve. **[Lesson 7.2]**

60. $2x - 7 = 4x - 9$ 1

61. $7(x - 1) = 2(3 - 3x)$ 1

62. $\frac{2}{5} + x = \frac{1}{3} - 4x$ $-\frac{1}{75}$

63. Solve the equation $2x - (x + 3) = 3x - 4$. **[Lesson 7.2]** $x = \frac{1}{2}$

64. Solve this equation for *n*: $2m + n = 10$. **[Lesson 7.4]** $n = 10 - 2m$

65. Solve this equation for *p*: $3q - p = 4$. **[Lesson 7.4]** $p = 3q - 4$

Look Beyond

66. Graph $y = 4x - 2$, $y = 4x$, and $y = 4x + 3$ on the same coordinate plane. Find the slope of each function. What do you notice about the lines?

67. On graph paper, plot the point $(0, 0)$. From that point, move right 5 and up 3 to plot the second point. Then move right 5 and up 6 to plot the third point. Next move right 5 and up 9 for the fourth point. Continue until you have plotted six points. Connect the points with a smooth curve. Describe this curve.

Watch Your Step!

Atlanta Architect Steps Up Quest for Safe Staircases

Numerous Accidents, Deaths Push Researcher into Career

By Lauran Neergaard
THE ASSOCIATED PRESS

ATLANTA – ...Every year, 1 million Americans seek medical treatment for falls on staircases. About 50,000 are hospitalized and 4,000 die.

Templer's interest began as a student at Columbia University, when someone asked him why people were always falling down the steps outside Lincoln Center. Templer discovered there was no research on stair safety. Nobody even counted falls.

Curious, he visited Lincoln Center with his family, and his sister-in-law tripped on the steps.

Templer called his mentor to propose the topic as a thesis and learned that the man had just broken his leg falling down a stairwell of a subway station.

A career was born.

"All stairs are dangerous, it's a matter of degree," Templer said. "There are ways to mitigate that danger, if we could get that message to people."

Stairs evolved from ladders and first were used for defense. Narrow winding staircases, for instance, hampered intruders. Europeans wrought stairs into works of art, building grand palace staircases that gradually steepened, forcing visitors into a slow, stately pace as they approached royalty.

Most stairs now have 9-inch treads and $8\frac{1}{2}$-inch risers, a size determined around 1850, Templer said. But people today have bigger feet that hang over the edges of stairs, throwing them off balance, he said.

Stairs also are too high, Templer concluded after experiments in which he forced volunteers to trip on collapsible stairs. They were harnessed so they didn't tumble all the way to the floor, but Templer used videos to simulate how they would have landed.

He wants building codes revised for stairs with 11-inch treads and 7-inch risers. His proposal prompted a lobbying blitz from the National Association of Home Builders, which contends that larger stairs would add at least 150 square feet and $1,500 in costs to a house.

NAHB's Richard Meyer dismissed Templer's work, saying people fall when stairs are improperly lighted, have loose carpeting or have objects placed in the way.

That's true, too, Templer said. But he said his experiments, funded by the National Science Foundation, prove stair shape is a large problem.

Cutaway Diagram of Steps of Stairway

Tread Width (run)

Riser Height (rise)

TREAD

RISER

Count Your Way to Stair Success

By Karol V. Menzie and Randy Johnson

Baltimore Sun

The riser height (rise) and the tread width (run) determine how comfortable the stairs will be to use. If the rise and run are too great, the stairs will strain your legs and be hard to climb. If the rise and run are too small, you may whack your toe on the back of each step.

Over the years, carpenters have determined that tread width times riser height should equal somewhere between 72 to 75 when the measurements are in inches.

On the main stair, the maximum rise should be no more than $8\frac{1}{4}$ inches and the minimum run should be no less than nine inches.

To determine how many steps, or treads, you need, measure from the top of the finished floor on the lower level to the top of the finished floor on the upper level.

To figure the rise and run in a house with eight-foot ceilings, for instance, start by figuring the total vertical rise. By the time you add floor joists, subfloor and finish floor, the total is about 105 inches.

A standard number of treads in a stair between first and second floors is 14. One hundred five divided by 14 equals $7\frac{1}{2}$. That means the distance from the top of each step to the top of the next step will be $7\frac{1}{2}$ inches.

With a riser height of $7\frac{1}{2}$ inches, tread width (run) should be at least nine inches. Ten inches is a more comfortable run; when you multiply $7\frac{1}{2}$ inches by 10 inches, you get 75 – within the conventional *ratio* of 72 to 75. With fourteen 10-inch treads, the total run of the stairs will be 140 inches. In other words, the entire stair will be 105 inches tall and 140 inches deep.

You can alter the rise and run to some extent. If you used 15 risers instead of 14, for instance, the rise would be 7 inches, and the tread width would be $10\frac{1}{2}$ inches (7 times $10\frac{1}{2}$ equals 73.5, within the rise and run guidelines).

Cooperative Learning

1. In the news article above, is the underlined term *ratio* used correctly? Explain.

2. Find the angle of each of the following staircases. Draw the triangle and measure the angle.

   a. the 105-inch by 140-inch staircase described in the article above about 37°

   b. the same staircase changed as described in the article to include 15 risers instead of 14 about 34°

   c. the same staircase changed to meet Templer's preference for 7-inch risers and 11-inch treads about 32.5°

3. Which of the three staircases in activity 2 would take up the most room? Explain.

4. Suppose you did not want the building code for stairs to change. Write an argument (with numbers and a diagram) to show that larger stairs would add at least 150 square feet to a typical house. Make up your own estimates of the size of a typical house.

Project

Measure the rise and run on staircases in your school or home. Do they match the information in the two news articles about standard stairs?

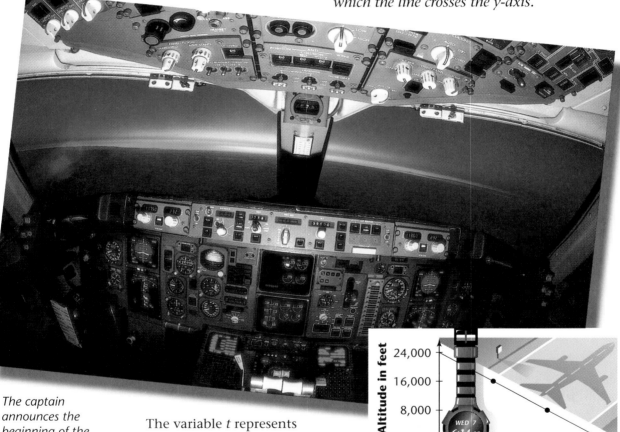

Exploring Graphs of Linear Functions

why *A linear function can be used to model the descent of an airplane. When you model situations using the graph of a linear function, you need to know the slope and the point at which the line crosses the y-axis.*

The captain announces the beginning of the descent from 24,000 feet for a landing in Detroit in 30 minutes.

The variable t represents the time in minutes after the airplane begins its descent. The starting condition is 24,000 feet. The constant rate of descent is -800 feet per minute because the plane must descend 24,000 feet in 30 minutes. The function $y = 24,000 - 800t$, or $y = -800t + 24,000$, models the descent.

Recall from Lesson 8.1 that the equation $y = -800t + 24,000$ is a linear function because its graph is a line. In fact, any equation of the form $y = mx + b$ is a linear function. When b is 0, $y = mx + b$ becomes $y = mx$, which is a line that passes through the origin.

Graphics Calculator

Graph the point $A(3, 6)$ on a coordinate plane. Start with the equation $y = mx$. Guess and check numbers for m by graphing the equations until you find a line that intersects point A.

If you use a graphics calculator, enter point $A(3, 6)$, and let your first guess for m be 4. Enter **Y = 4X**, and then graph the function. See if the line passes through $A(3, 6)$.

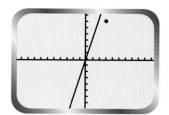

If you use graph paper, substitute values for x in the equation $y = 4x$ to locate two points on the line, such as $(0, 0)$ and $(2, 8)$. Then draw the line that connects the points. See if the line passes through $A(3, 6)$.

When m is 4, the line $y = 4x$ is too steep to go through point A.

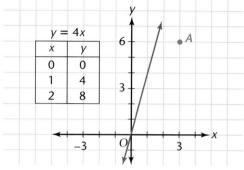

$y = 4x$

x	y
0	0
1	4
2	8

Repeat the process with different values for m until you find an equation for the line that intersects point A.

•Exploration 1 *Fitting the Line to a Point*

Repeat the process above for points B through F.

$B(2, 8)$ $C(6, 3)$ $D(3, -6)$ $E(2, -8)$ $F(4, 7)$

What is the connection between the coordinates of the given target point and the slope, m? ❖

•Exploration 2 *Changing the Slope*

Draw the graph of each equation on the same coordinate plane. Compare the graphs you get by varying the slope, m.

$$y = x \qquad y = 5x \qquad y = -2x \qquad y = \frac{1}{2}x \qquad y = -\frac{1}{3}x$$

1 How are the graphs of these lines alike? All pass through (0, 0).

2 How are the graphs of these lines different? Their slopes are different.

3 Guess what the graph of $y = 3x$ looks like. Passes through (0, 0) with slope 3.

4 Check your guess by drawing the graph.

5 Make a conjecture about how m, the coefficient of x, affects the graph of $y = mx$. ❖

•Exploration 3 *Introducing a Constant*

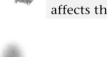

Draw the graphs of the next set of equations on the same axes.

$$y = x + 3 \qquad y = 2x + 3 \qquad y = 5x + 3 \qquad y = -2x + 3$$

1 How are the graphs of these lines alike? All cross *y*-axis at (0, 3)

2 How are the graphs of these lines different?
Their slopes are different.

3 Guess what the graph of $y = 4x + 3$ looks like.
Passes through (0, 3) with slope 4.

4 Check your guess by graphing.

5 Make a conjecture about how adding 3 affects the graph of $y = mx$.

Draw the graphs of the next set of equations on the same axes.

$$y = 2x \qquad y = 2x + 3 \qquad y = 2x + 5 \qquad y = 2x - 3$$

6 How are the graphs of these lines alike? All have the same slope.

7 How are the graphs of these lines different?
Their *y*-intercepts are different.

8 Guess what the graph of $y = 2x - 4$ looks like.
Parallel to other four lines, but with *y*-intercept at (0, −4).

9 Check your guess by graphing.

10 Make a conjecture about how the value of *b* The value of *b* is the
affects the graph of $y = 2x + b$. ❖ *y*-coordinate of the point where
the line crosses the *y*-axis. It raises or lowers the line.

Describe the graph of each equation below without graphing. Check the accuracy of your descriptions by graphing each equation.

$$y = -3x \qquad y = 5x - 6$$
$$y = 5 + 6x \qquad y = \frac{1}{2}x + 3$$

How do the values of *m* and *b* affect the graph of $y = mx + b$? Discuss the results of what you have discovered from the explorations. *m* affects the steepness of the graph; *b* affects where the graph will cross the *y*-axis

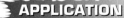

APPLICATION

An orange carton weighs 3 pounds when empty, and each orange you place in it weighs an average of $\frac{1}{2}$ pound. If you know the number of oranges, what is the total weight of the carton and the oranges? Let *n* equal the number of oranges. Write an equation for the total weight, *w*, of a carton containing *n* oranges. Graph the equation with *n* on the horizontal axis and *w* on the vertical axis.

total weight = weight of oranges + weight of carton
$$w \qquad = \qquad \frac{1}{2}n \qquad + \qquad 3$$

The weight and the number of oranges cannot be negative. Thus, the graph will include only points where n and w are greater than or equal to 0.

Try $n = 0$. When there are 0 oranges, the weight is only that of the carton, or 3 pounds. Plot the first point (0, 3). Each time you add an orange, the weight increases by $\frac{1}{2}$ pound. This is the **rate of change** of the function $w = \frac{1}{2}n + 3$. How does the rate of change compare with the slope of a line? They are the same.

You can add new points to the graph by moving to the right 1 unit and up $\frac{1}{2}$ unit from each point. If you draw a line through the points, the line will have a rate of change, or slope, of $\frac{1}{2}$.

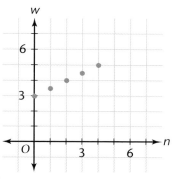

The individual points represent the weight of whole oranges.

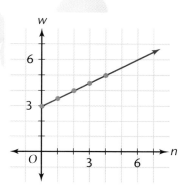

The solid line also includes the weight of partial oranges. ❖

What happens to the function that models the airplane descent, $y = -800t + 24,000$, if the starting altitude or rate of descent changes? Find the new equation if the starting condition is 30,000 feet and the rate of change is -1000 feet per minute. If the starting altitude changes, the constant term 24,000 changes. If the rate of descent changes, the slope, -800, changes.
$y = -1000t + 30,000$

EXERCISES  PROBLEMS

Communicate

1. Summarize your discoveries from each of the explorations.

Give the equation of the line that passes through the origin and each of the following points. Check by drawing the graph.

2. $G(8, 4)$ $y = \frac{1}{2}x$ **3.** $H(4, 8)$ $y = 2x$ **4.** $I(4, 9)$ $y = \frac{9}{4}x$ **5.** $J(6, -4)$ $y = -\frac{2}{3}x$

6. Line l passes through the origin (0, 0) and $A(3, 6)$. Explain how to write an equation for line l.

7. If the slope of a line that passes through the origin and through a given point is 6, what is the equation of the line? $y = 6x$

Practice & Apply

Draw each *pair* of graphs on the same coordinate plane. In each case, tell how the graphs are alike and how they are different.

8. $y = 6x$; $y = -6x$

9. $y = 8x$; $y = -8x$

10. $y = \frac{1}{2}x$; $y = -\frac{1}{2}x$

11. $y = \frac{4}{3}x$; $y = -\frac{4}{3}x$

12. $y = 4x - 7$; $y = -4x - 7$

13. $y = 8x - 7$; $y = -8x - 7$

14. How does changing $y = mx$ to $y = -mx$ affect the line that is graphed? The line $y = mx$ has the opposite slope as the line $y = -mx$, but the y-intercept will be the same.

15. Chemistry From the information shown, write a function for the total mass, M, of a beaker containing w milliliters of water. Substitute values for w, and graph the function.

Empty beaker— 30 grams

1 milliliter of water = 1 gram

Find the equation of a line that passes through the origin and each of the following points.

16. $(2, 5)$ $y = \frac{5}{2}x$

17. $(5, 8)$ $y = \frac{8}{5}x$

18. $(1, 9)$ $y = 9x$

19. $(4, 2)$ $y = \frac{1}{2}x$

20. $(7, 3)$ $y = \frac{3}{7}x$

Draw each *pair* of graphs on the same axes. In each case, tell how the graphs are alike and how they are different.

21. $y = \frac{1}{4}x$ and $y = \frac{1}{4}x + 6$

22. $y = \frac{1}{4}x + 8$ and $y = \frac{1}{4}x - 2$

23. Describe the effect on the lines that are graphed if only the values of b change in the equation $y = mx + b$. The value of b is where the line crosses the y-axis. The lines will be parallel: they have the same slope but different y-intercepts.

The normal water level of the river is 32 feet.

24. Ecology The water level of a river is rising 0.8 foot per day. Write a linear function for the water level after d days. Graph the function.

25. Suppose the water level of the river is 34 feet and is receding 0.5 feet per day. Write a linear function for the water level after d days. What is the rate of change for this function? Graph the function. In how many days will the water level be 26 feet?

Describe the graph of each line. Then check your description by graphing.

26. $y = -5x$
27. $y = -6x + 3$
28. $y = 7$
29. $y = -6$

30. $y = -5x - 1$
31. $y = -2$
32. $y = -x - 3$
33. $y = 2x + 3$

34. $y = 3x + 7$
35. $y = 7 - x$
36. $y = -3 + \frac{1}{2}x$
37. $y = x + 4$

38. **Portfolio Activity** Complete the problem in the portfolio activity on page 427.

Look Back

39. If Calvin earns $8 per hour, h, find his total earnings, t, for 33 hours. Write an equation and find the solution. **[Lesson 2.5]** $t = 8h; t = 8(33) = \$264$

40. Dan earns 15% of the price of each $20 box of candles that he sells. How many boxes will he have to sell to earn $100? **[Lesson 6.6]** 34

41. Solve the equation $3x + 4y = 24$ for y. **[Lesson 7.1]** $y = -\frac{3}{4}x + 6$

42. Suppose that in the orange carton problem on page 446 the total weight is 51 pounds. How many oranges are in the carton? **[Lesson 7.1]** 96

Is the slope of the line negative, positive, or neither? **[Lesson 8.2]**

43. neither
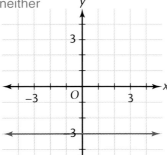

44. negative
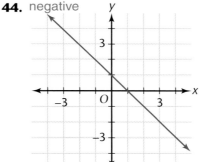

45. positive
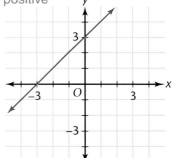

46. Cultural Connection: Africa The Rhind papyrus that describes the mathematics of building the ancient pyramids of Egypt refers to a *seked*. Consider a square pyramid with a base edge of 360 cubits and a height of 250 cubits. Scribes took half of the base as the run and the height as the rise. They calculated the *seked* to be $\frac{180}{250}$. Compare the calculation of a *seked* with the calculation of the slope for the same pyramid. Explain the relationship between the *seked* and the slope. **[Lesson 8.2]**

$seked = \frac{run}{rise} = \frac{180}{250}$; $slope = \frac{rise}{run} = \frac{250}{180}$; reciprocals

Look Beyond

47. Do the graphs of the lines $y = 0.5x + 3$ and $y = \frac{3}{6}x + 4$ intersect? No, lines are parallel.

LESSON 8.4 The Slope-Intercept Form

Why *Once you understand how to interpret the components of a linear function y = mx + b, you can sketch the graph. When you can identify the slope and the y-intercept of a line, you can write its equation.*

Skaters practice several hours a week. Kim pays $50 once a year for club fees and $3 an hour to practice.

You can use a graph to see the pattern of Kim's skating expenses. How would you draw this graph?

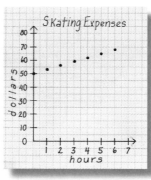

Examine the data. Let x be the number of hours Kim practices for the year. Let y be the total cost of practice time for the year. Calculate the first differences to determine the slope from a table of x- and y-values.

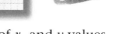

Hours	x	0	1	2	3	4	5	6	7	8	9	10
Cost	y	50	53	56	59	62	65	68	71	74	77	80

First differences → 3 3 3 3 3 3 3 3 3 3

Since the first differences are 3, the table represents a linear function in the form $y = mx + b$. How can you find the slope of the line that represents the function? The table shows that the change in the y-variable is 3. Since the x-values are consecutive, the change in the x-variable is 1.

$$\text{slope} = m = \frac{\text{change in } y}{\text{change in } x} = \frac{3}{1} = 3$$

The slope indicates that Kim pays $3 for every hour that she practices.

The table also shows that when x is 0, y is 50. Thus, (0, 50) is on the graph. The y-value, 50, of the point where the line crosses the y-axis, is the **y-intercept**. It means that Kim pays a $50 fixed fee per year in addition to the hourly rate.

Recall from Lesson 8.3 that the graph of an equation in the form $y = mx + b$ has slope m and crosses the y-axis at b. If you substitute 3 for m and 50 for b, the equation for Kim's practice costs becomes $y = 3x + 50$.

SLOPE-INTERCEPT FORM

The slope-intercept form for a line with slope **m** and y-intercept **b**, is **$y = mx + b$**.

EXAMPLE 1

The equation of the line that represents Kim's skating costs is $y = 3x + 50$. How can you use the slope-intercept formula to construct the graph?

Solution ➤

The formula is $y = \mathbf{mx + b}$. Since **b** is 50, measure 50 units up from the origin (0, 0) on the y-axis, and graph the point.

From that point, measure the run and the rise of the slope to locate a second point.

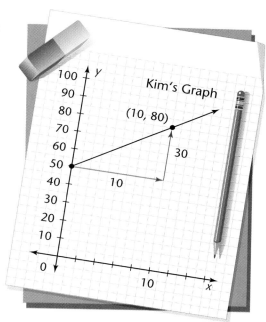

Kim's Graph

(10, 80)

The slope, **m**, is 3. You can use any equivalent of $\frac{3}{1}$, such as $\frac{6}{2}$ or $\frac{30}{10}$ to measure the run and rise. The axes are in intervals of 10 units, so use $\frac{30}{10}$. Move right 10, then up 30.
Graph the second point. Draw a line through the points. ❖

CRITICAL
Thinking

Why do you think the line in Kim's graph does not extend to the left of the y-axis? How far do you think the line would realistically extend in the positive direction? What are the reasonable domain and range? The x-values to the left of the y-axis represent negative numbers of practice hours. The graph may extend as far as (624, 1922). 624 is the number of hours Kim would skate in one year if she skated 12 hours per week for 52 weeks.

From Two Points to an Equation

When you know two points on a line, you can write the equation for that line. First, calculate the slope, m, using the slope formula. Then calculate b from the slope-intercept formula and one of the points.

SLOPE FORMULA

Given two points with coordinates (x_1, y_1) and (x_2, y_2), the formula for the slope is

$$m = \frac{\text{change in } y}{\text{change in } x} = \frac{\text{difference in } y}{\text{difference in } x} = \frac{y_2 - y_1}{x_2 - x_1}.$$

EXAMPLE 2

Sports

Kim's notes show that after the first 7 hours of practice, the total cost for her skating that year is $71. Earlier in her notes it shows that after 5 hours of practice the total cost for the year was $65. How can you write an equation for a line knowing only this information?

Solution ➤

You can represent the data as points by writing the hours and cost as ordered pairs, (5, 65) and (7, 71). To write the equation of a line from these two points, substitute the values into the slope formula.

$$m = \frac{\text{difference in } y}{\text{difference in } x}$$

$$m = \frac{71 - 65}{7 - 5} = \frac{6}{2} = 3$$

Substitute 3 for m in $y = mx + b$.

$$y = 3x + b$$

Next, choose either point, and substitute the coordinates for x and y into the equation. If you use the point (5, 65), substitute 5 for x and 65 for y. Then solve for b.

$$y = 3x + b$$
$$65 = 3(5) + b$$
$$65 = 15 + b$$
$$50 = b$$

Now substitute 3 for m and 50 for b in $y = mx + b$. The equation for the line is $y = 3x + 50$. ❖

Why is the equation in Example 2 the same as the equation in Example 1?
The points (7, 71) and (5, 65) are on the same line as the line in Example 1.

Try This Write an equation for a line passing through points (3, 3) and (5, 7).
$$y = 2x - 3$$

EXAMPLE 3

Auto Racing

For decades the Indianapolis 500 automobile race has attracted the attention of millions of enthusiasts. Since 1911 the average speed for the race has increased from 74.59 miles per hour to 185.987 miles per hour in 1990. How steadily has this average increased? How can you find an equation for the line of best fit that shows the trend in average speed? Use the data given for a sample of speed averages from 1915 to 1975 in 5-year intervals. Consider 1900 as year 0.

Year	Average Speed in MPH	Year	Average Speed in MPH	Year	Average Speed in MPH	Year	Average Speed in MPH
1915	89.84	1935	106.240	1955	128.209	1975	149.213
1920	88.62	1940	114.277	1960	138.767	1980	?
1925	101.13	1945	none	1965	150.686	1985	?
1930	100.448	1950	124.002	1970	155.749	1990	?

Solution ➤

One way to find the equation for the line of best fit is to plot the points on a graph and estimate the location of the line using a clear ruler.

Find the slope and y-intercept. Then substitute the numbers into the formula for a line, $y = mx + b$.

Graphics Calculator

The developments in calculator technology have made graphing a scatter plot much easier. On a graphics calculator, the line of best fit is called the **regression line**. The calculator will automatically calculate the slope, the y-intercept, and the equation, based on the line of best fit.

L1	L2	L3
15.00	89.84	-------
20.00	88.62	
25.00	101.13	
30.00	100.45	
35.00	106.24	
40.00	114.28	
50.00	124.00	

L2={89.84,88.62...

LinReg
y=ax+b
a=1.16
b=68.24
r=.98

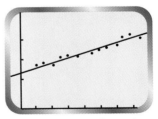

Place the information from the table into a 2-variable (x, y) data table.

Use the linear regression feature to find the line of best fit. Here, the slope for the line of best fit is a, and the y-intercept is b. The r is the correlation coefficient.

*The calculator will graph the scatter plot and draw the line of best fit when you enter the equation for the regression line after **Y =**.*

The rate of change in average speed is the slope, about 1.16 miles per hour per year. The line of best fit (regression line) has the approximate equation $y = 1.16x + 68.24$. ❖

Use the line of best fit to estimate the average speeds in 1980, 1985, 1990, and 1995. Compare your estimates with the actual speeds achieved in those years. What might explain the variation?

Direct Variation

Springs are used in science to measure mass. Some springs are thin and can be stretched easily. Others are much stiffer and seemingly cannot be stretched at all. The relationship between the amount of stretch and the mass on the spring is defined by a law of physics called Hooke's law.

Hooke's law is an example of direct variation. It states that the distance a spring stretches varies directly as the force applied.

EXAMPLE 4

Jill works as a cable technician and charges by the hour. Her records show the hours she works and pay she receives for various jobs.

Hours on the job	3	5	7	10	15
Pay for the job	$28.80	$48	$67.20	$96	$144

Write an equation that will calculate the hourly wage that Jill earns.

Solution ➤

In looking at Jill's records, as the number of hours she works increases, the pay she earns also increases at the same rate. This is an example of *direct variation*. The pay that Jill earns varies directly as the number of hours she works. The hourly wage that Jill earns is determined by

$$\frac{\text{Pay for the job}}{\text{Hours on the job}} = \text{Hourly wage, so } \frac{48}{5} = \$9.60.$$

The hourly wage, $9.60, is a constant.

Let p represent the pay for the job. Let h represent the hours on the job.

$$\frac{p}{h} = 9.6$$

If you solve for p, you find that $p = 9.6h$. ❖

The equation $\frac{p}{h} = 9.6$ models direct variation. The expression $\frac{p}{h}$ is a ratio, and 9.6 is the constant of variation.

DIRECT VARIATION

If y varies directly as x, then $\frac{y}{x} = k$, or $y = kx$. The k is the **constant of variation.**

The direct-variation equation written in the form $y = kx$ is a linear function. To find the constant of variation, find the ratio $\frac{y}{x}$. How does this compare with finding the slope of a line? The ratio is the same as the ratio used to calculate the slope of the line through the origin.

EXAMPLE 5

Physics A force of 5 pounds stretches a spring a distance of 17.5 inches. How far will a spring stretch if an 8-pound weight is applied?

Solution ➤

The amount of stretch varies directly as the amount of force applied.

Let d represent the distance that the spring stretches.

Let f represent the force of the weight applied.

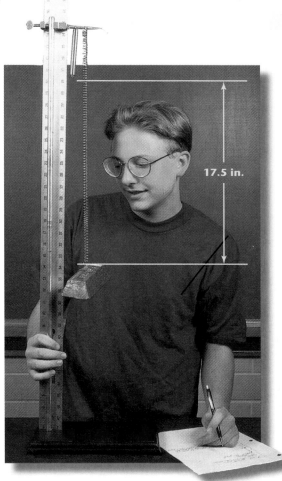

17.5 in.

$$\frac{d}{f} = k$$ From the definition of direct variation, identify the value of k, the constant of variation.

$$\frac{17.5}{5} = k$$ The ratio of 17.5 inches to a 5-pound weight simplifies to the constant $k = 3.5$. The constant k is the constant of variation.

$$\frac{d}{f} = 3.5$$ Once you know the constant of variation, you can find the distance of a given stretch, d.

$d = 3.5f$ Solve for d to form a linear function in terms of f.
$d = 3.5(8)$ Substitute 8 for the weight.

$d = 3.5(8)$ The distance the spring will stretch is 28 inches.❖
$\quad = 28$ inches

If you graph a direct-variation equation, you will find that direct variation represents a linear function in the form $y = mx + b$. The y-intercept is 0 and the slope is k, the constant of variation.

EXERCISES & PROBLEMS

Communicate

1. Explain what effect a change in b represents on the graph of the equation $y = mx + b$.
changes the y-intercept

2. Explain what effect a change in m represents on the graph of the equation $y = mx + b$.
changes the slope

3. Describe how to find the slope, m, of a line passing through the points (3, 5) and (7, 2).

4. How do you determine the y-intercept for the equation $y = 5x - 1$? *Substitute 0 for x, and solve for y.*

5. How do you determine the slope for the equation $y = 10 - x$?

6. Tell how to graph the equation $y = 2x - 3$ without plotting points.

7. Explain how to write the equation of a line passing through points $A(-2, 4)$ and $B(1, 5)$.

Practice & Apply

Make a table from the given data points, and find the first differences of the x- and y-values.

(0, 4), (2, 10), (4, 16), (6, 22), (8, 28), (10, 34)

8. According to the differences, what is the slope? *3*

9. At what point does the line passing through these points cross the y-axis? *(0, 4)*

10. Write the equation for this line, and graph the line.

Give the coordinates of the point where the line for each equation crosses the y-axis.

11. $y = 4x + 5$ *(0, 5)*   **12.** $y = 8x - 1$ *(0, -1)*

13. $y = -3x + 7$ *(0, 7)*   **14.** $y = -5x - 9$
(0, -9)

Examine the lines on the graph.

15. Which line has a y-intercept of -2?
line k

16. Which line has a positive slope?
line n

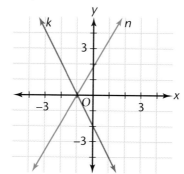

Find the slope of the line passing through each pair of points.

17. $(0, 6)$, $(5, 0)$ $-\dfrac{6}{5}$ **18.** $(3, 4)$, $(-1, -2)$ $\dfrac{3}{2}$ **19.** $(1, 7)$, $(5, 3)$ -1

20. $(7, 2)$, $(-4, -2)$ $\dfrac{4}{11}$ **21.** $(-5, 0)$, $(8, -4)$ $-\dfrac{4}{13}$ **22.** $(7, -7)$, $(-4, -3)$ $-\dfrac{4}{11}$

23. $(-4, -3)$, $(-2, -6)$ $-\dfrac{3}{2}$ **24.** $(6, 6)$, $(-2, -2)$ 1 **25.** $(-1, 1)$, $(5, -7)$ $-\dfrac{4}{3}$

Write the equation, in slope-intercept form, for each line.

26. with slope -1 and y-intercept 0 $y = -x$ **27.** through $(0, -4)$ and with slope -4 $y = -4x - 4$

28. with slope 11 and y-intercept 15 $y = 11x + 15$ **29.** with slope -5 and y-intercept 7 $y = -5x + 7$

30. through $(0, 5)$ and with slope 1 $y = x + 5$ **31.** with slope -3 and y-intercept -1 $y = -3x - 1$

32. through $(0, 3)$ and with slope 3 $y = 3x + 3$ **33.** with slope $\dfrac{2}{3}$ and y-intercept 2 $y = \dfrac{2}{3}x + 2$

Write an equation, in slope-intercept form, for the line passing through each pair of points.

34. $(-1, 0)$, $(0, 3)$ $y = 3x + 3$ **35.** $(7, 9)$, $(3, -2)$ $y = \dfrac{11}{4}x - \dfrac{41}{4}$ **36.** $(4, 5)$, $(-1, -2)$ $y = \dfrac{7}{5}x - \dfrac{3}{5}$

37. $(3, 3)$, $(-2, -6)$ $y = \dfrac{9}{5}x - \dfrac{12}{5}$ **38.** $(-8, -3)$, $(6, -2)$ $y = \dfrac{1}{14}x - \dfrac{17}{7}$ **39.** $(-1, -4)$, $(-3, -5)$ $y = \dfrac{1}{2}x - \dfrac{7}{2}$

40. Consumer Economics A catalog company charges a fixed fee of $2 per order plus $0.50 per pound for shipping. Write an equation for the total shipping cost, c, in terms of the number of pounds, p. What is the form of your equation? $c = 0.50p + 2$; slope-intercept form

Write an equation, in slope-intercept form, for the graph of each line.

41. $y = 3x + 4$

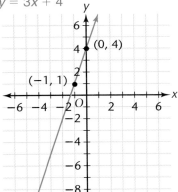

42. $y = -x - 3$

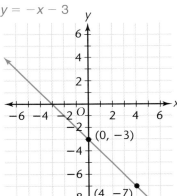

Josh sold 20 animal figures for $6.50 each. The cost of materials was $43.80. **46.** $\dfrac{d}{t} = 65$; 195 mi

43. Write an equation, in slope-intercept form, for the amount of money Josh will have after each sale. $y = 6.50x - 43.80$

44. How many figures must Josh sell before the cost of materials is less than the income from sales? 7

45. Graph the equation for Josh's business.

46. The distance a car travels at a constant rate of 65 miles per hour varies directly as the time in hours. Write an equation to express this direct variation. Then find the distance if the time is 3 hours.

$ 6.50 each

47. Physics Find a sturdy rubber band or spring, and suspend it so that you can measure its length. Measure the length without any weight on it. Find several weights that the rubber band or spring will hold without breaking. Hang the weights on the rubber band or spring, and measure the length of the stretch. Make a table of the weights and the stretch of the rubber band or spring. Make a scatter plot, and find the line of best fit. Use a graphics calculator if possible to find the equation of the regression line. Write a description of your findings.

48. **Geometry** If the base area of a container is kept constant, the volume of the container varies directly as the height. The table shows the volumes and heights. Find the constant of variation. Then find the volume if the height is 12 centimeters. $k = 9$; 108 cu cm

V	27	36	45	54
h	3	4	5	6

Look Back

Technology The population of Cooperstown is 12,345 and is growing by 678 people per year.

49. Make a table or use the repeating operations feature of a calculator to determine when the population of Cooperstown will reach 20,000. **[Lesson 1.4]** between 11 and 12 years

50. Write an equation for the population, p, in terms of y, the number of years the population increases at this rate. **[Lesson 1.4]** $p = 678y + 12,345$

51. Technology Plot a graph (using a graphics calculator if possible) to find when the population will reach 20,000. **[Lesson 1.4]**

Simplify. [Lessons 5.1, 5.2]

52. $(3a - 2) + (2a - 2)$
$5a - 4$

53. $(4x + 3y) - (3x - 2y)$
$x + 5y$

Solve each inequality. **[Lesson 5.6]**

54. $x + 7 \le -2$
$x \le -9$

55. $x - 4 \ge 6$
$x \ge 10$

56. $x - 5 < 21$
$x < 26$

Look Beyond

57. If a line is horizontal, what is the rise when you compute the slope? 0

58. If a line is vertical, what is the run when you compute the slope? What does this indicate about the slope? 0; the slope is undefined

LESSON 8.5 Other Forms for Equations of Lines

Wynton Marsalis studied at the Julliard School of Music and played trumpet with the New Orleans Philharmonic Orchestra before gaining fame as a jazz musician. He was born in New Orleans on October 16, 1961.

Why *The equation of a line can be written in different forms. If you know which form to use, you can often save yourself time and effort.*

You can write the birth date of Wynton Marsalis many ways. Each way has its advantages. The long form is easy to read. The short form is quick and easy to write. The computer form is easy to sort by year, month, and day.

October 16, 1961	10-16-61	611016
long form	**short form**	**computer form**

You have been using the slope-intercept form of a line, $y = mx + b$, to solve problems. As with dates, there are other forms for equations of lines. Each form has its advantages. The standard form provides a simple method for graphing the equation of a line.

Standard Form

Consider an equation written in the form $Ax + By = C$. After values for A, B, and C are determined, a solution to this equation is any ordered pair of numbers (x, y) that makes the equation true. When these ordered pairs are graphed, they form a straight line.

> **STANDARD FORM**
> An equation in the form $Ax + By = C$ is in standard form when
> - A, B, and C are integers,
> - A and B are not both zero, and
> - A is not negative.

EXAMPLE 1

Fund-raising Jackie is in charge of selling tickets for the school jazz concert. She hopes the total ticket sales will be about $588. This will cover expenses and make a modest profit.

A The ticket prices are shown on the poster. Write an equation in standard form that models this situation.

B Graph the equation.

Solution ➤

A The equation must show that $4 times the number of adult tickets plus $2 times the number of student tickets is $588.

Let x represent the number of adult tickets sold. Let y represent the number of student tickets sold. The equation that models the problem is $4x + 2y = 588$. Notice that the equation is in standard form $Ax + By = C$.

B Graph the number of adult tickets, x, on the horizontal axis and the number of student tickets, y, on the vertical axis. The points (x, y) represent (adult tickets, student tickets).

To graph an equation in standard form, substitute 0 for x, and solve for y. Then substitute 0 for y, and solve for x. This gives the intercepts $(0, y)$ and $(x, 0)$, which are the points where the line crosses each axis. Draw the line connecting these two points.

Start with the equation $4x + 2y = 588$.

First substitute 0 for x.
$$4x + 2y = 588$$
$$0 + 2y = 588$$
$$y = 294$$
$(0, 294)$ is on the graph.

Then substitute 0 for y.
$$4x + 2y = 588$$
$$4x + 0 = 588$$
$$x = 147$$
$(147, 0)$ is on the graph.

Draw the line connecting the points. The coordinates of any point on the line will solve the equation. ❖

To solve the jazz-concert problem, however, the values for x and y must be whole numbers. Why? A fraction of a ticket cannot be sold.

Sale of 97 adult tickets and 100 student tickets. ▶ What does the point $(97, 100)$ represent with respect to the problem?

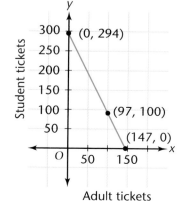

EXAMPLE 2

Graphics Calculator

Use a graphics calculator to graph the equation $4x + 2y = 588$.

Solution ➤

To enter an equation in a graphics calculator, first rewrite the equation in slope-intercept form, $y = mx + b$.

$$4x + 2y = 588$$
$$2y = 588 - 4x$$
$$y = 294 - 2x$$
$$y = -2x + 294$$

The equation can now be entered into the calculator.

$$\mathbf{Y = -2X + 294} \ \ \diamond$$

CRITICAL Thinking

If the equation of a line is written in standard form, $Ax + By = C$, the slope is $-\frac{A}{B}$ and the y-intercept is $\frac{C}{B}$. Explain why. Solving $Ax + By = C$ for y gives the slope-intercept form $y = -\frac{A}{B}x + \frac{C}{B}$.

Point-Slope Form

If you know the slope of a line and the coordinates of one of its points, you can write a third form of an equation for a line.

Remember that the value for the slope **m** can be calculated from any two points (x_1, y_1) and (x_2, y_2) and the formula $m = \frac{y_2 - y_1}{x_2 - x_1}$.

Multiply both sides of the slope formula by $x_2 - x_1$, and simplify.

If you replace the specific point (x_2, y_2) with a general point on the line, represented by (x, y), the result is $y - y_1 = m(x - x_1)$.

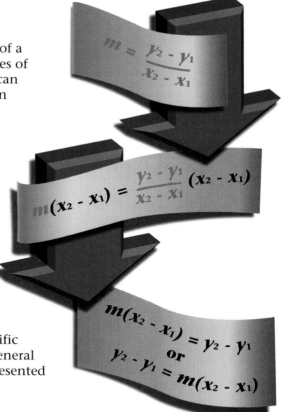

$$m = \frac{y_2 - y_1}{x_2 - x_1}$$

$$m(x_2 - x_1) = \frac{y_2 - y_1}{x_2 - x_1}(x_2 - x_1)$$

$$m(x_2 - x_1) = y_2 - y_1$$
or
$$y_2 - y_1 = m(x_2 - x_1)$$

POINT-SLOPE FORM

The form $y - y_1 = m(x - x_1)$ is the **point-slope form** for the equation of a line. The coordinates x_1 and y_1 are taken from a given point (x_1, y_1), and the slope is **m**.

EXAMPLE 3

A line with slope 3 passes through point (2, 7). Write the equation of the line in point-slope form.

Solution ➤

Let $m = 3$, $x_1 = 2$, and $y_1 = 7$. Substitute the given values into the point-slope equation.

$$y - y_1 = m(x - x_1)$$
$$y - 7 = 3(x - 2) ❖$$

EXAMPLE 4

Janet's class is ordering T-shirts. Write the point-slope equation that models the information on the order form.

The shipping and handling charge at Shirts, inc. is $5.

T-Shirts Inc.
506 Front Street

T-Shirts $9.00 ea.
* Plus Shipping and Handling

COLOR	S	M	L	TOTAL
Red	2			2
Blue	1	4	5	10
White		1	3	4
TOTAL				16
TOTAL COST				$149.00

Solution ➤

Each time a shirt is ordered, the cost increases by $9, so 9 is the rate of change, or slope. Since 16 shirts cost $149, the point (16, 149) represents a point on the graph. Substitute 9 for m, 16 for x_1, and 149 for y_1 into $y - y_1 = m(x - x_1)$. The result is $y - 149 = 9(x - 16)$. ❖

EXAMPLE 5

Change $y - 149 = 9(x - 16)$ to an equation

A in slope-intercept form.   **B** in standard form.

Solution ➤

A

$y - 149 = 9(x - 16)$	Given
$y - 149 = 9x - 144$	Distributive Property
$y = 9x + 5$	Addition Property of Equality

The equation is now in slope-intercept form. Note that the shipping and handling charge is $5.

B Next, change the slope-intercept form to standard form, $Ax + By = C$.

$y = 9x + 5$	Given
$y - 9x = 5$	Subtraction Property of Equality
$-9x + y = 5$	Rearrange terms.
$9x - y = -5$	Multiply each side by -1. ❖

EXAMPLE 6

Find an equation in point-slope form for the graph of a line that passes through the points $(-1, 10)$ and $(5, 8)$.

Solution ➤

Let $(x_1, y_1) = (-1, 10)$ and $(x_2, y_2) = (5, 8)$. Find the slope.

$$\text{Slope} = \frac{y_2 - y_1}{x_2 - x_1} = \frac{8 - 10}{5 - (-1)} = \frac{-2}{6} = -\frac{1}{3}$$

Use the point $(-1, 10)$ for (x_1, y_1). Substitute the slope and the coordinates of the point into the equation $y - y_1 = m(x - x_1)$. Then $y - 10 = -\frac{1}{3}(x - (-1))$, or $y - 10 = -\frac{1}{3}(x + 1)$. ❖

Try This Use the point-slope form to find the equation of the line that passes through the points $(5, 65)$ and $(7, 71)$. $y - 65 = 3(x - 5)$ or $y - 71 = 3(x - 7)$

SUMMARY OF THE FORMS FOR LINEAR EQUATIONS		
Name	**Form**	**Example**
Slope-intercept	$y = mx + b$	$y = 3x + 5$
Standard	$Ax + By = C$	$3x - y = -5$
Point-slope	$y - y_1 = m(x - x_1)$	$y - 11 = 3(x - 2)$

EXERCISES & PROBLEMS

Communicate

1. Explain how to write the equation $5y - 2 = -3x$ in standard form.

2. Tell how you would find the intercepts for the equation $3x + 6y = 18$.

3. Describe how to graph $2x + 3y = 12$ by finding the intercepts.

4. Explain how to change $x - 3y = 9$ into slope-intercept form.

5. How would you use the point-slope formula to write the equation of the line that passes through points $(-2, 4)$ and $(4, -8)$?

6. The equation of a line is $5x + 2y = 40$. Tell how you would find the slope of this line.

Practice & Apply

Write each equation in standard form.

$$4x + 3y = 24$$
7. $4x = -3y + 24$

$$5x + 7y = -35$$
8. $7y = -5x - 35$

$$6x + 4y = -12$$
9. $6x + 4y + 12 = 0$

$$2x - 4y = 0$$
10. $2x = 4y$

11. $6x - 8 = 2y + 6$
$$6x - 2y = 14$$

12. $x = \frac{2}{3}y + 6$
$$3x - 2y = 18$$

13. $2 + 7x + 14y = 3x - 10$
$$4x + 14y = -12$$

14. $5 = y - x$
$$x - y = -5$$

Find the intercepts for each equation.

15. $x + y = 10$
$(0, 10); (10, 0)$

16. $3x - 2y = 12$
$(0, -6); (4, 0)$

17. $5x + 4y = 20$
$(0, 5); (4, 0)$

18. $x = 2y$
$(0, 0); (0, 0)$

19. Draw a graph of a line that intercepts the axes at $(3, 0)$ and $(0, 7)$.

20. Graph the equation $2x + 6y = 18$ by finding the intercepts.

21. The equation of a given line is $6x + 2y = 40$. What is the slope of the line? -3

22. Compare the graphs of $4a + 2s = 588$ and $2a + s = 294$. What do you find? Both represent the same line.

23. Fund-raising Suppose that in Example 1, on page 460, the number of student tickets sold, y, is 50. If student tickets are $2.00 each and adult tickets are $4.00 each, how many adult tickets, x, must be sold for the total ticket sales to be $588? 122

Write the equation, in slope-intercept form, of each line.

24. through $(5, 2)$ and with slope -8 $y = -8x + 42$

25. crossing the x-axis at $x = 7$ and the y-axis at $y = 2$ $y = -\frac{2}{7}x + 2$

26. through $(5, 2)$ and with slope 8 $y = 8x - 38$

27. through $(5, 2)$ and with slope 0 $y = 2$

28. crossing the x-axis at $x = 1$ and the y-axis at $y = \pi$ $y = -\pi x + \pi$

29. through $(2, 3)$ and $(8, -3)$ $y = -x + 5$

30. through $(5, 9)$ and $(10, 9)$ $y = 9$

31. Technology On some graphics calculators, equations must be in function form, $y = f(x)$, to draw graphs. The symbol $f(x)$ indicates a function. Which of the three forms for the equation of a line given in the summary at the end of the lesson is (or are) in the form $y = f(x)$? slope-intercept form
$$y = mx + b$$

Rewrite each equation in the form $y = f(x)$, where the equation is solved for y.

32. $2x + 3y = 12$
$$y = -\frac{2}{3}x + 4$$

33. $y - 2 = 3(x - 7)$
$$y = 3x - 19$$

34. $2x + 2y + 5 = 0$ $y = -x - \frac{5}{2}$

35. $3x + 4y = 6 + 5y$
$$y = 3x - 6$$

36. $\frac{x}{2} + \frac{y}{3} = 18$
$$y = -\frac{3}{2}x + 54$$

37. $\frac{5x - y}{7} = 14$ $y = 5x - 98$

38. Write an equation which shows that the value of n nickels and d dimes is $5. HINT: Be careful about the units. $0.05n + 0.10d = 5$ (in dollars)

Fund-raising Write an expression for the sales, in dollars, of

39. 40 adult tickets. 5(40)

40. 20 student tickets and 37 adult tickets. 3(20) + 5(37)

41. s student tickets and a adult tickets. $3s + 5a$

42. Write an equation that says the total sale of s student tickets and a adult tickets is $700. Which form did you choose for your equation? $3s + 5a = 700$; standard form

43. Find the number of adult tickets sold if 90 student tickets are sold. 86

44. Find the number of student tickets sold if 80 adult tickets are sold. 100

Social Studies Complete this table which shows various forms of the same date.

	Long form	Short form	Computer form	Importance
45.	?	?	081001	1st Model-T Ford
46.	January 31, 1958	?	?	1st U.S. satellite
47.	?	7-20-69	?	1st moonwalk

48. Why do you think the computer form uses 07 instead of 7 for July?
Two digit spaces must be assigned to allow for all possible months: 01, 02, . . . , 11, 12

Look Back

Evaluate each expression if x is 1, y is 1, and z is 2. [Lesson 1.4]

49. $x^2 + y + z^2$ 6

50. $x - y + z$ 2

51. $x + y - z$ 0

52. $-(x + y + z)$ -4

53. Simplify $\frac{12x - 18}{-6}$. **[Lesson 6.1]** $-2x + 3$

Solve each equation. [Lesson 6.2]

54. $-5y = 30$
$y = -6$

55. $3x = 420$
$x = 140$

56. $\frac{y}{9} = 36$
$y = 324$

57. $\frac{x}{2} = 108$
$x = 216$

Find the slope of the line containing the origin and the given point. Then give the equation of the line passing through the origin and the given point. [Lesson 8.2]

58. $A(3, 6)$
$m = 2; y = 2x$

59. $B(2, 8)$
$m = 4; y = 4x$

60. $C(6, 3)$
$m = \frac{1}{2}; y = \frac{1}{2}x$

61. $D(-5, -7)$
$m = \frac{7}{5}; y = \frac{7}{5}x$

Look Beyond

62. If two lines intersect, how many points do they have in common? 1

63. If two lines are parallel, how many points do they have in common? 0

64. Michael has a 5-liter and a 3-liter bottle. Neither have markings. He has a supply of water. How can he measure exactly 1 liter?

Vertical and Horizontal Lines

why *A line along the diving tower, seen from the side, is horizontal. A line down the tower is vertical. The slopes and equations for such horizontal and vertical lines have special significance in algebra.*

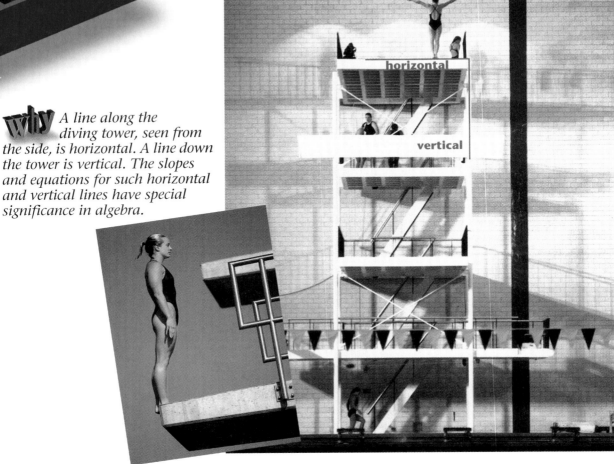

horizontal

vertical

A pool offers two options. Members can pay either $2 per visit or $25 for a season pass. How many times do you have to use the pool to make the season pass the better option?

Recreation   The data points showing the number of visits to the pool, x, and the total cost, y, can be modeled for both options. The graphs show only the first quadrant because the number of visits and the amount of money spent will always be 0 or greater.

Per-visit option For 0 visits the cost is $0. Thus, the y-intercept is 0, and (0, 0) is one point on the graph. Since the rate of increase is $2 per visit, the slope is 2.

The first graph is a line drawn through **(0, 0)** with slope 2. The line represents a typical equation in slope-intercept form, $y = 2x + 0$. You can check this by seeing that it contains the point **(10, 20)**. Ten visits cost $20.

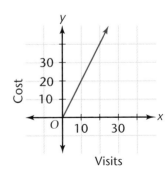

Season-pass option The cost is a constant $25. If you go to the pool twice, the cost is $25. Thus, **(2, 25)** is a point on the graph. If you go to the pool 40 times, the cost is still $25. The point **(40, 25)** is on the graph.

The second graph is a horizontal line consisting of all y-values equal to 25. The line drawn through the two points is a special equation in slope-intercept form, $y = 0x + 25$, or $y = 25$.

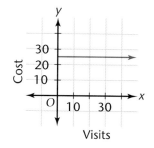

Visits

Now draw both graphs on one coordinate plane.

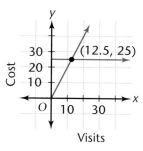

Visits

The graph shows that for $12\frac{1}{2}$ visits the cost is $25. Since you cannot make $\frac{1}{2}$ a visit, the season-pass option is better for 13 or more visits.

Notice that the graph of $y = 25$ is a horizontal line. The graph of $y = b$ is a horizontal line that crosses the y-axis at the point $(0, b)$. The equation $y = b$ is a **constant function**, where b is the **constant**. It can be written as $y = mx + b$, where the slope is zero. In standard form the equation is written $0x + 1y = b$.

EXAMPLE 1

Compare the graphs of $y = 4$ and $x = 4$.

Solution ➤

The graph $y = 4$ consists of all points with a y-coordinate of 4. The graph $x = 4$ consists of all points with an x-coordinate of 4.

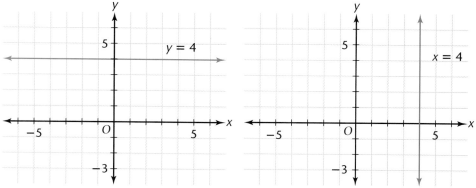

The equation $y = 4$ is a horizontal line 4 units above the origin.

The equation $x = 4$ is a vertical line 4 units to the right of the origin. ❖

Try This Compare the graphs of $y = -3$ and $x = -3$.

EXAMPLE 2

Find the slope of **A** $y = 5$. **B** $x = -3$.

Solution ➤

A The graph of $y = 5$ has the same value of y for any value of x. Use any two points with a **y-coordinate of 5** to find the slope. The points (3, **5**) and (7, **5**) will work.

$$\text{slope} = \frac{y_2 - y_1}{x_2 - x_1} = \frac{5 - 5}{7 - 3} = \frac{0}{4} = 0$$

B The graph of $x = -3$ has the same value of x for any value of y. Use any two points with an **x-coordinate of -3** to find the slope. The points (-3, 3) and (-3, 7) will work.

$$\text{slope} = \frac{y_2 - y_1}{x_2 - x_1} = \frac{7 - 3}{(-3) - (-3)} = \frac{4}{0}$$

The run is 0. Since dividing by 0 is impossible, the *slope* is *undefined*. ❖

The graph of an equation in the form $x = a$ is a vertical line that crosses the x-axis where x equals a. The equation cannot be written in slope-intercept form, but it can be written in standard form.

$$1x + 0y = a$$

HORIZONTAL AND VERTICAL LINES

The equation for a horizontal line is written in the form $y = b$.
The slope is 0.

The equation for a vertical line is written in the form $x = a$.
The slope is undefined.

EXAMPLE 3

Compare the graphs of $y = 3x$ and $x = 3y$.

Solution ➤

Write the equations in slope-intercept form. Graph the lines.

The equation $y = 3x$ is in slope-intercept form, and its graph is a line with slope 3 and y-intercept 0.

Change $x = 3y$ to slope-intercept form by solving for y. Since $x = 3y$ is equivalent to $3y = x$,

$$3y = x \;\rightarrow\; y = \frac{x}{3} \;\rightarrow\; y = \frac{1}{3}x + 0$$

The equation $x = 3y$ is written $y = \frac{1}{3}x + 0$ in slope-intercept form. Its graph is a line with slope $\frac{1}{3}$ and y-intercept 0.

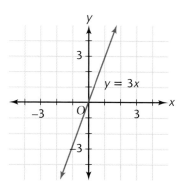

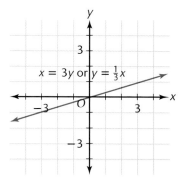

Recall that numbers are reciprocals when their product equals 1. Since $3 \cdot \frac{1}{3} = 1$, the slopes of $y = 3x$ and $y = \frac{1}{3} x$ are reciprocals.

The graphs of $y = 3x$ and $x = 3y$ are lines with reciprocal slopes and a common point at the origin, $(0, 0)$. ❖

Draw the graph of $y = x$. On the same set of axes, draw the graphs of $y = 2x$ and $x = 2y$. Try other pairs of lines in the form $y = mx$ and $x = my$ with different m-values. What relationships do you see? These pairs of lines will have reciprocal slopes and pass through the common point $(0, 0)$.

EXERCISES & PROBLEMS

Communicate

1. Describe how to find the slope of $y = 3$. Is the line vertical or horizontal?

2. Explain how to find the slope of $x = 1$. Is the line vertical or horizontal?

3. Discuss how to find the slope of $y = 5x$. Choose your own points to graph the equation. At what point does the line cross the y-axis?

4. Tell how to find the slope of $x = 5y$. Choose your own points to graph the equation. At what point does the line cross the y-axis?

5. Explain why the graph of $y = 15$ is a constant function.

6. Tell why the slope for a vertical line is unusual.

7. Tell why the slope for a horizontal line is 0.

Practice & Apply

Find the slope for each equation. Plot points, graph the line for each equation, and indicate if the line is vertical or horizontal.

m is undef; vertical

8. $y = 5$ **9.** $x = 7$ **10.** $y = 9x$ $m = 9$ **11.** $x = 5y$ $m = \frac{1}{5}$ **12.** $y = \frac{1}{3}x$ $m = \frac{1}{3}$

$m = 0$; horizontal

13. Language Arts Where do you think the word *horizontal* comes from?
Answers may vary. Sample: from the word *horizon*

> **Match each equation with the appropriate description.**
>
> E **14.** $x + y = 9$ **A.** a horizontal line 9 units above the origin
>
> F **15.** $xy = 9$ **B.** a vertical line 9 units to the right of the origin
>
> B **16.** $x = 9$ **C.** a line through the origin with slope 9
>
> A **17.** $y = 9$ **D.** a line through the origin with slope $\frac{1}{9}$
>
> C **18.** $y = 9x$ **E.** a line with slope -1 and y-intercept 9
>
> D **19.** $x = 9y$ **F.** something other than a straight line

Coordinate Geometry The equation $x = 4$ cannot be written in slope-intercept form because the slope is undefined. It can, however, be written in standard form as $1x + 0y = 4$. Copy and complete the table. Write an equivalent form of each equation, writing "undefined slope" when appropriate for slope-intercept form.

	Given	Slope-intercept form	Standard form
20.	$x = 1$	undefined ?	$1x + 0y = 1$?
21.	$y = 4$	$y = 4$?	$0x + 1y = 4$?
22.	$x + y = 5$	$y = -x + 5$?	$1x + 1y = 5$?
23.	$y = 4x$	$y = 4x$?	$4x - 1y = 0$?
24.	$x = 4y$	$y = \frac{1}{4}x$?	$1x - 4y = 0$?

Determine whether each line is vertical or horizontal.

25. $y = 8$ **26.** $x = -2$ **27.** $y = -4$ **28.** $x = 9$
horizontal   vertical   horizontal   vertical

29. Technology Find how to graph a vertical line on a graphics calculator. Answers may vary.

30. Find the slopes for the equations $y = 6x$ and $x = 6y$. What point do these lines have in common? $6; \frac{1}{6}; (0, 0)$

31. Write the reciprocal of 0.25. 4

32. Write the reciprocal of $-4\frac{1}{2}$. $-\frac{2}{9}$

33. Entertainment Saul's father is thinking of buying him a six-month movie pass for $40. If matinees are $3.00 each, how many times must Saul attend before it would benefit his father to buy the pass? 14

In the swimming-pool example from page 466, which option is cheaper if you visit the pool

34. 8 times?
per-visit

35. 29 times?
season pass

36. **Recreation** Repeat the swimming-pool problem, but change the pool costs to those shown at the right. After how many visits is it less expensive to buy a season pass?
12

37. Technically, should the graphs of the problem consist of points or lines? For example, does it make sense to visit the pool $4\frac{2}{3}$ times? For ease in making and reading graphs, lines are often used instead of points. What is another example in which this might happen?

Season Opening
at the
Community Pool!
Please note our new rates

	old	new
Daily Pass	$ 2.00	$ 3.00
Season Pass	$25.00	$35.00

Lifeguards always on duty!
hours: 9:00am to 9:00pm
Bring the whole family!

Look Back

Use the order of operations to simplify each expression. [Lesson 1.7]

38. $30 \cdot 2 \div 1 + 10$ 70

39. $[2(105 - 5)] \div 2$ 100

40. $\frac{(3 + 9)^2 - 4}{2^2 \cdot 15 - 10}$ $\frac{14}{5}$ or $2\frac{4}{5}$

Simplify.   [Lesson 5.1]

41. $(6a + 3) + (2a + 6)$
$8a + 9$

42. $(x + y + z) + (3x + y + 4z)$
$4x + 2y + 5z$

43. $-7 + 3p - 9 - 7p$
$-4p - 16$

44. Mary needs $1\frac{1}{3}$ yards of fabric to make a skirt. A yard of fabric costs $4.58. How many skirts can she make from 5 yards of fabric?   3 skirts
[Lesson 6.2]

Lunch cost $15.50. There is a 7% sales tax, and the customary tip is 15% of the total.   [Lesson 6.6]

45. How much is the tax on the lunch? $1.09

46. How much is the tip? $2.49

47. Given points (3, 5) and $(-4, -2)$, write the slope-intercept equation for the line.   **[Lesson 8.4]** $y = x + 2$

Look Beyond

48. **Geometry** What kind of angle is formed when a horizontal line and a vertical line intersect?   right angle (90°)

49. Will two different vertical lines ever intersect?   no

50. Will two different horizontal lines ever intersect? no

Parallel and Perpendicular Lines

Why *The graphs of parallel and perpendicular lines are related by their slopes. Algebra can help you solve problems in geometry.*

Parallel lines often look like they meet, but they never do.

Temperature Examine the three different thermometer readings in the table below. It is possible to derive a linear equation for changing degrees in one scale to degrees in another. Notice how slope is used in each solution.

EXAMPLE 1

Use the data in the table to find the formula that changes

A degrees Celsius (°C) to degrees Fahrenheit (°F).

B kelvins to degrees Fahrenheit.

	°F	°C	K
Water boils	212	100	373
Water freezes	32	0	273
Absolute 0	−459	−273	0

Solution ➤

Ⓐ The relationship between any two of the temperature scales is linear. Compare the Fahrenheit and Celsius scales. In the equation $y = mx + b$, let the y-values be Fahrenheit and the x-values be Celsius. Use the data from the table to calculate the slope and y-intercept.

$$\text{slope} = \frac{\text{change in Fahrenheit}}{\text{change in Celsius}} \rightarrow m = \frac{212 - 32}{100 - 0} = \frac{180}{100} = \frac{9}{5}$$

Notice that when C is 0, F is 32. The y-intercept, b, is 32. Find the equation that changes Celsius to Fahrenheit by substituting $\frac{9}{5}$ for m and 32 for b in $y = mx + b$. The equation $y = \frac{9}{5}x + 32$ gives the formula $F = \frac{9}{5}C + 32$.

Ⓑ You can change from kelvins to degrees Fahrenheit in much the same way. The equation for this change is $y = \frac{9}{5}x - 459$. The formula is $F = \frac{9}{5}K - 459$. ❖

Parallel Lines

Graphics Calculator

Graph the equations $F = \frac{9}{5}C + 32$ and $F = \frac{9}{5}K - 459$. Let y represent degrees Fahrenheit and x represent either degrees Celsius or kelvins. On a graphics calculator, enter the Celsius formula as the function $\mathbf{Y_1 = (9/5)X + 32}$. Then enter the Kelvin formula as the function $\mathbf{Y_2 = (9/5)X - 459}$. Place both graphs on the same axes.

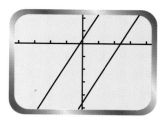

GEOMETRY *Connection*

If you use graph paper, use the y-intercept to locate the first point for one equation. Then use the run and rise of the slope for the second point. Once you have determined 2 or 3 points for the equation, use a straightedge to connect the points. Repeat the process for the second equation.

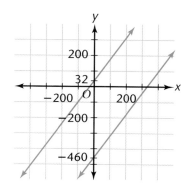

How are the slopes of the two lines related?

PARALLEL LINES

If two different lines have the same slope, the lines are **parallel**.

If two non-vertical lines are parallel, they have the same slope. Two parallel, vertical lines have undefined slopes.

Perpendicular Lines

GEOMETRY
Connection

Recall that *perpendicular lines* form right angles. In Exploration 1, you will explore the algebraic relationship between two lines that are perpendicular. The symbol for perpendicular is ⊥.

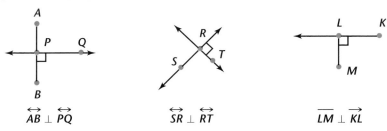

$$\overleftrightarrow{AB} \perp \overleftrightarrow{PQ} \qquad \overleftrightarrow{SR} \perp \overleftrightarrow{RT} \qquad \overline{LM} \perp \overrightarrow{KL}$$

•Exploration 1 *Slopes and Perpendicular Lines*

On a piece of graph paper, draw a line that has a run of 5 squares and a rise of 3 squares, as shown on the grid. Make a copy on clear plastic or tracing paper. Rotate the copy of the graph 90 degrees clockwise, as shown. When one graph is placed on the other, you can see the relationship between the two lines.

The lines are perpendicular, because the line on the second graph was rotated 90 degrees. Where the lines on the two graphs intersect, right angles are formed.

1 What is the slope of the original line? $\frac{3}{5}$

2 Find the slope of the rotated line. Count the squares to find the rise over the run. Notice, for example, that the new line shows a run of 3. Consider the direction of the line. $-\frac{5}{3}$

3 What is the sign for the slope of the perpendicular line? negative

4 What is the relationship between the slope of the perpendicular line and the slope of the original line? The slope of the perpendicular line is the negative reciprocal of the slope of the original line.

5 What happens when this exploration is repeated beginning with a negative slope? ❖ The result is the same. The slope of the perpendicular line is positive, and is the negative reciprocal of the slope of the original line.

•Exploration 2 *Perpendicular Slope on a Calculator*

Graphics Calculator

You can use a graphics calculator to find the equation of a line perpendicular to a given line. Be sure to select a *square window* when you set the viewing window on your graphics calculator. Then the right angles will have the correct appearance on the screen.

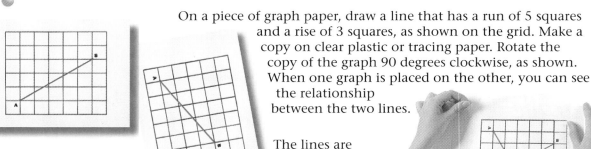
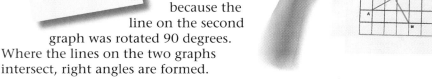

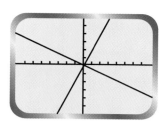

 Enter the graph for the line **Y = 2X**.

 On the same axes, graph new lines in the form $y = mx$ by trying different values for m until the new line looks perpendicular to the original line. Keep a record of your guesses. Graphs may vary.

3 Decide on your best guess for the slope. Using your slope, write an equation for a line perpendicular to $y = 2x$. ❖ $y = -\frac{1}{2}x$

On the same set of axes, graph $y = 2x$ and the line perpendicular to $y = 2x$ that passes through (0, 0). Find the slope of each line. Compare the signs. Compare the absolute values. Multiply the slopes. What is the result?

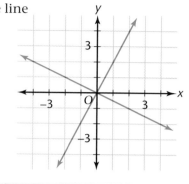

Suppose a given line has positive slope m. Make a conjecture about the product of m and the slope of a line perpendicular to the given line. The product will be -1. The slopes are m and $-\frac{1}{m}$.

PERPENDICULAR LINES

If the slopes of two lines are m and $-\frac{1}{m}$, the lines are perpendicular.

If the slope of a line is m, then the slope of a line perpendicular to it is $-\frac{1}{m}$.

EXAMPLE 2

Find an equation for the line that contains point (4, 5) and is

A parallel to the line $2x + 3y = 7$.

B perpendicular to the line $2x + 3y = 7$.

Solution ➤

A First, write the equation of the given line in slope-intercept form.

$$2x + 3y = 7$$
$$3y = -2x + 7$$
$$y = -\frac{2}{3}x + \frac{7}{3}$$

The slope of the line is $-\frac{2}{3}$. Any line parallel to this line must also have a slope of $-\frac{2}{3}$. The coordinates of the given point are (4, 5), and the slope is $-\frac{2}{3}$. Now use the *point-slope form* to write an equation for the parallel line through the given point. Substitute $-\frac{2}{3}$ for m and (4, 5) for (x_1, y_1) into the point-slope form of the equation.

$$y - y_1 = m(x - x_1)$$
$$y - 5 = -\frac{2}{3}(x - 4)$$

B When a line has slope $-\frac{2}{3}$, any line perpendicular to that line has slope $\frac{3}{2}$. Since the line also contains the point $(4, 5)$, you can substitute $\frac{3}{2}$ for m and $(4, 5)$ for (x_1, y_1) in $y - y_1 = m(x - x_1)$.

$$y - 5 = \frac{3}{2}(x - 4)$$

Graphics Calculator

As a check, change the equations $y - 5 = -\frac{2}{3}(x - 4)$ and $y - 5 = \frac{3}{2}(x - 4)$ to slope-intercept form. Graph the lines using a square window. The lines will appear perpendicular. Both lines will also contain the point $(4, 5)$ where they intersect. ❖

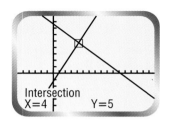

Intersection
X=4   Y=5

Try This Find an equation for a line that contains the point $(4, 1)$ and is perpendicular to $y = 2x - 7$. $y = -\frac{1}{2}x + 3$

GEOMETRY
Connection

If line l has slope $\frac{2}{5}$ and line m has slope $-\frac{5}{2}$, then $l \perp m$.

If $r \perp s$ and r has slope $\frac{2}{5}$, then s has slope $-\frac{5}{2}$.

EXERCISES & PROBLEMS

Communicate

1. Explain how to write an equation for a line parallel to $y = 4x + 3$.

2. The slope of a line is $\frac{3}{2}$. Explain how to find the slope of a line that is perpendicular to that line.

3. How would you find the slope of a line perpendicular to the line $y = \frac{1}{3}x + 2$?

4. Describe how to find an equation for a line perpendicular to $y = 4x + 3$.

5. Explain how to write the standard form of an equation for a line with y-intercept -4 that is parallel to a line with slope 3.

6. Tell how to write an equation in standard form for a line perpendicular to the line with $m = -6$ and $b = 12$.

7. Discuss how to write the equation for a line that passes through the point $(0, 0)$ and is perpendicular to $x - 5y = 15$.

Practice & Apply

Graph the line $y = 5x$.

8. Sketch a graph that is parallel to $y = 5x$. Write its equation.

9. Sketch a graph that is perpendicular to $y = 5x$. Write its equation.

10. What relationship is there between the graph of $y = 3x + 2$ and $y = -3x + 2$? Are these lines perpendicular?

What is the slope of a line

11. parallel to a horizontal line? 0

12. perpendicular to a horizontal line? undefined

13. parallel to a vertical line? undefined

14. perpendicular to a vertical line? 0

Write an equation for a line according to the instructions.

	Contains:	Is parallel to:
15.	$(3, -5)$	$5x - 2y = 10$
17.	$(-2, 7)$	$y = 3x - 4$
19.	$(2, 4)$	$y = 7$

	Contains:	Is perpendicular to:
16.	$(3, -5)$	$5x - 2y = 10$
18.	$(-2, 7)$	$y = 3x - 4$
20.	$(2, 4)$	$y = 7$

Write the slope of a line that is parallel to each given line.

21. $y = 3x + 14$ 3

22. $2x + y = 6$ -2

23. $8 = -4x + 2y$ 2

24. $-2x + \frac{1}{2}y = 16$ 4

Write the slope of a line that is perpendicular to each given line.

25. $y = -\frac{1}{3}x + 10$ 3

26. $-\frac{1}{2}x - y = 20$ 2

27. $13 = -x + y$ -1

28. $3x + 12y = 12$ 4

Geometry Write equations of four lines that meet to form a square whose sides are Answers may vary. Samples are given.

29. *parallel* to the axes.
   $x = 2$, $x = 4$, $y = 2$, and $y = 4$

30. *not parallel* to the axes.
   $y = x$, $y = x - 2$, $y = -x$, and $y = -x + 2$

Look Back

31. Place parentheses to make $2 \cdot (7 + 35) \div 7 - 10 = 2$ true. **[Lesson 1.7]**

Evaluate each expression. [Lessons 2.2, 2.4]

32. $-4 + (-3) + 1$ -6

33. $-2 + 3 + (-7) + 3$ -3

34. $-12 + 4 - (-4)$ -4

Simplify. [Lesson 5.1]

35. $2x^2 + 3y + 4y + 3x^2$
   $5x^2 + 7y$

36. $3x + 2 + 4y + 2 + 3y$
   $3x + 7y + 4$

37. $2x + 3xy + 5x^2 + 7xy$
   $5x^2 + 10xy + 2x$

Look Beyond

How many ordered pairs will satisfy both equations simultaneously if two linear equations have graphs

38. that are parallel? 0

39. that are perpendicular? 1

Exploring Linear Inequalities

why *Limitations, such as a budget, can be described using inequalities. Graphs are used to show the solutions to inequalities.*

Jason can spend no more than $12 of his weekly earnings on collecting card-game packages.

Jason collects card-game packages. The cost of each card package for Game A is $3, and the cost of each card package for Game B is $2. How many card packages for each game can Jason buy? The answer to this question can be found using the graph of a *linear inequality*.

•Exploration 1 *Graphing a Linear Inequality*

1 Let *x* represent the number of card packages for Game A, and let *y* represent the number of card packages for Game B. Write an inequality to represent the amount of money Jason can spend each week on card packages. $3x + 2y \leq 12$

2 Replace the inequality symbol with an equal sign in the inequality you wrote in Step 1. Solve for *y*, and graph the equation on a coordinate plane. This line is called the **boundary line** of the related inequality.

3 The following table shows how you can use substitution to find some points that satisfy the inequality. Copy and complete the table.

Point	Substitute: $3x + 2y \leq 12$	Simplify.	Is the inequality true?
(0, 0)	$3(0) + 2(0) \leq 12$	$0 \leq 12$	yes
(1, 1)	$3(1) + 2(1) \leq 12$	?	?
(2, 1)	?	?	?
(2, 2)	?	?	?
(3, 1)	?	?	?
(3, 2)	?	?	?
(3, 3)	?	?	?
(4, 4)	?	?	?

4 Which points in the table make the inequality true? Plot them on the graph you drew in Step 2. (0, 0), (1, 1), (2, 1), (2, 2), (3, 1)

5 Shade the region on the side of the line containing the points you plotted in Step 4. What is true about all of the points in this region? They all make the inequality $3x + 2y \leq 12$ true.

6 Are all of the points in the shaded region solutions to the inequality? What are the reasonable domain and range for this real-world situation? yes; $0 \leq x \leq 4$; $0 \leq y \leq 6$

7 Can you list *all* of the possible numbers of card packages from Games A and B that Jason can buy in one week? Explain. ❖

CRITICAL
Thinking

Why is the ordered pair $\left(2\frac{1}{2}, 1\frac{1}{2}\right)$ a solution to the inequality but not a real answer to the problem in Exploration 1? These coordinates produce a true statement when substituted in $3x + 2y \leq 12$, but this is not a solution because you cannot buy partial packages.

APPLICATION

Carlos wants to make a birthday bouquet of lilies and roses for his mother. Carlos can spend no more than $25 on the bouquet. To draw a graph to show the possible number of each type of flowers he can purchase, first let x represent the number of lilies and y represent the number of roses.

Since lilies cost $2 each, an expression for the cost of lilies is $2x$.

Since roses cost $4 each, an expression for the cost of roses is $4y$.

The total cost of the bouquet must be less than or equal to $25.

$$2x + 4y \leq 25$$

Solve the inequality $2x + 4y \leq 25$ for y.

$2x + 4y \leq 25$	Given
$2x + 4y - 2x \leq 25 - 2x$	Subtraction Property of Inequality
$4y \leq 25 - 2x$	Simplify.
$\dfrac{4y}{4} \leq \dfrac{25 - 2x}{4}$	Division Property of Inequality
$y \leq \dfrac{25}{4} - \dfrac{2x}{4}$	Simplify.
$y \leq 6\dfrac{1}{4} - \dfrac{x}{2}$	Simplify.
$y \leq -\dfrac{x}{2} + 6\dfrac{1}{4}$	Commutitive Property
$y \leq -\dfrac{1}{2}x + 6\dfrac{1}{4}$	Simplify.

Graph the boundary line
$y = -\dfrac{1}{2}x + 6\dfrac{1}{4}$.

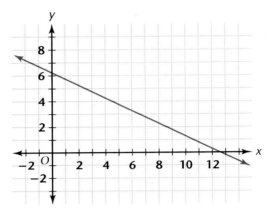

Choose any point above the line, such as $(0, 8)$. Test the coordinates of this point in the inequality.

$$2x + 4y \leq 25$$
$$2(0) + 4(8) \leq 25$$
$$32 \leq 25 \quad \text{False}$$

Since $32 \leq 25$ is a *false* statement, the point $(0, 8)$ is *not* in the solution set of the inequality.

Choose any point below the line, such as $(0, 0)$. Test the coordinates of this point in the inequality.

$$2x + 4y \leq 25$$
$$2(0) + 4(0) \leq 25$$
$$0 \leq 25 \quad \text{True}$$

Since $0 \leq 25$ is a *true* statement, the point $(0, 0)$ *is* in the solution set of the inequality.

So the points in the solution set for the inequality $2x + 4y \leq 25$ must be below the boundary line.

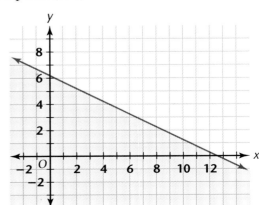

However, since the graph is supposed to show the possible numbers of each type of flower that Carlos can purchase, not *all* of the points below the line should be graphed. The number of lilies, the domain x, and the number of roses, the range y, can be only positive numbers or zero. So the shaded region needs to be contained within the first quadrant.

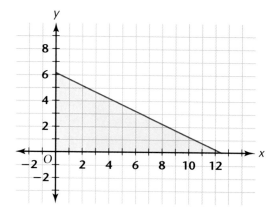

Also, since people do not buy fractions of a flower, the domain and range can only be positive integers or zero.

The final graph that shows the possible numbers of each type of flower that Carlos can purchase is shown at the right. ❖

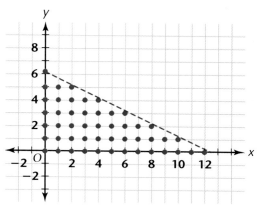

Exploration 2 · A Greater-Than Inequality

1 What is the equation of the boundary line for the inequality $y > 2x - 4$? $y = 2x - 4$

2 Graph the boundary line. If an inequality contains the symbol < or >, the boundary line *is not* part of the graph and is shown as a *dashed* line. If the inequality contains the symbol ≤ or ≥, the boundary line *is* part of the graph and is shown as a *solid* line. Is the boundary line in the inequality $y > 2x - 4$ solid or dashed?

3 Copy and complete the following table to find points that make the inequality true.

Point	Substitute: $y > 2x - 4$	Simplify.	Is the inequality true?
(5, 3)	$3 > 2(5) - 4$	$3 > 6$	no
(2, 3)	?	?	?
(0, 5)	?	?	?
(1, 6)	?	?	?
(5, 6)	?	?	?
(6, 2)	?	?	?
(3, 2)	?	?	?
(5, 1)	?	?	?

4 Plot the points from the table that satisfy the inequality. Do the points lie above or below the boundary line? Shade the region containing these points.

5 **Explain how to use the inequality to determine which side of the line to shade.** ❖ Substitute the coordinates of different points into the inequality. The points that make the inequality true will all be on the side of the boundary line that is to be shaded.

EXTENSION

To graph the inequality $3x + y < 5$, first solve the inequality for y.

$$3x + y < 5$$
$$3x + y - 3x < 5 - 3x$$
$$y < 5 - 3x$$
$$y < -3x + 5$$

Graph the boundary line $y = -3x + 5$. The line is graphed as a dashed line because the inequality contains the symbol $<$ and points on the line do not satisfy the inequality.

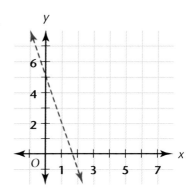

Choose any point above the line, such as $(3, 0)$. Test the point in the original inequality.

$$3x + y < 5$$
$$3(3) + 0 < 5$$
$$9 < 5 \quad \text{False}$$

Since $9 < 5$ is a *false* statement, the point $(3, 0)$ is *not* in the solution set of the inequality.

Choose any point below the line, such as $(0, 0)$. Test the point in the original inequality.

$$3x + y < 5$$
$$3(0) + 0 < 5$$
$$0 < 5 \quad \text{True}$$

Since $0 < 5$ is a *true* statement, the point $(0, 0)$ *is* in the solution set of the inequality. Shade the region below the line. The entire shaded region is the solution to the inequality. ❖

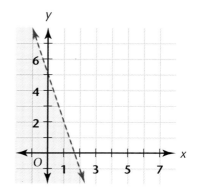

Exercises & Problems

Communicate

1. Explain how to graph the linear inequality $2x + y \le 5$.

2. How do you know which side of the boundary line to shade when graphing an inequality?

3. How do you know whether to include the boundary line in the graph of an inequality?

4. Describe a situation in which all of the shaded area that shows the solution to the inequality might not be included in the solution to the problem.

5. How can you tell whether to use an inequality or an equation to describe a situation?

Practice & Apply

Graph each inequality.

6. $y < 2x - 1$

7. $y \le \frac{-3}{4}x - 3$

8. $y \ge \frac{1}{3}x - 3$

9. $y > \frac{-2}{5}x + 3$

10. $2x + y < 8$

11. $5x - 3y \le 6$

12. $x + y \ge 5$

13. $2x + y < -2$

14. $x + 4y \le 4$

15. $3x + 2y \le 2$

16. $3x + y > -3$

17. $4x + 5y < -20$

18. $2y > 10$

19. $y \le -3$

20. $x \ge 3$

21. $2x < -6$

22. $-3x + y - 5 > 0$

23. $2x + y + 3 < 0$

24. $4x + y \le 6$

25. $-5x + y \ge 2$

26. $x - y \le -3$

27. $\frac{1}{3}y < x - 5$

28. $-\frac{1}{5}y \le -x + 2$

29. $2x - 3y > 6$

30. $y > \frac{1}{3}x + 8$

31. $y \ge \frac{5}{2}x - 2$

32. $-9x - y \le 3$

33. $y < 3x - 1$

34. $y < -2x + 2$

35. $y > -x + 3$

36. $y > 3$

37. $x > 3$

38. $y \ge \frac{1}{2}x - 3$

39. $-3x + 2y - 2 > 0$

40. $-\frac{1}{3}x + y \le 1$

41. $x \le -y - 3$

Technology You can use the shading function of a graphics calculator to create graphs of linear inequalities. Graph each of the following inequalities, and copy the graph onto graph paper.

42. $y \geq x$ **43.** $y \leq \frac{-3}{5}x + 8$ **44.** $y \geq \frac{1}{2}x - 4$ **45.** $y > \frac{2}{3}x + 5$

Write the inequality for each graph.

46.
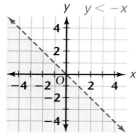
$y \quad y < -x$

47.
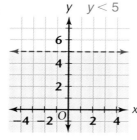
$y \quad y < 5$

48.
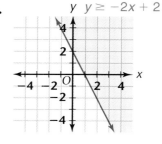
$y \quad y \geq -2x + 2$

49.
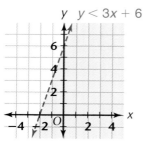
$y \quad y < 3x + 6$

50.
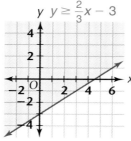
$y \quad y \geq \frac{2}{3}x - 3$

51.
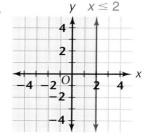
$y \quad x \leq 2$

52. Recreation A diver finds a treasure chest that contains silver and gold coins. All together, the coins weigh at least 25,000 grams, or 25 kilograms. Graph the solution to the inequality. Identify the reasonable domain and range for this situation.

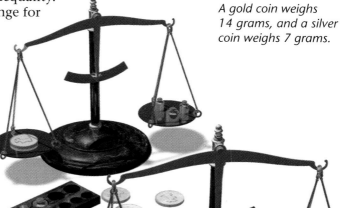

A gold coin weighs 14 grams, and a silver coin weighs 7 grams.

53. Rental The cost to rent a car is $25 per day plus $0.30 per mile. Michelle wants to pay less than $140 to rent a car. Write and graph an inequality to describe this situation. Identify the reasonable domain and range for this situation.

54. Academics Alice is taking a test. To keep students from guessing, Alice's teacher gives 5 points for every correct answer and deducts 2 points for every incorrect answer. No points are added or subtracted if the problem is skipped. Alice wants to earn at least an 80 on the test. Write and graph an inequality to describe this situation. Use your graph to determine whether Alice will earn at least an 80 if she gets 18 correct answers and 4 incorrect answers.

55. Fund-raising The senior class is selling tickets to the Faculty Follies. They plan to charge $4 per student and $5 per adult. Write and graph an inequality to show that they want to sell at least $2000 worth of tickets. If they sell 200 student tickets and 90 adult tickets, will they meet their goal?

56. Entertainment Will is planning a party. He is going to serve pizzas that cost $8 each and soft drinks that cost $2 per bottle. The cost of the party must be no more than $60. Write and graph an inequality for the cost. If Will buys 5 pizzas and 4 bottles of soda, will he spend $60 or less?

57. Wages Jenna is a seamstress. She earns $5 for each blouse she makes and $4 for each shirt she makes. She wants to earn at least $60. Write and graph an inequality to describe this situation. Use the graph to determine whether Jenna can earn at least $60 by making 8 blouses and 7 shirts.

58. Sports Mike scored at least 24 points in every basketball game this season by making 2-point and 3-point field goals. Write and graph an inequality to show the number of points he scored. Use your graph to decide if he could have made five 2-point field goals and four 3-point field goals in one game.

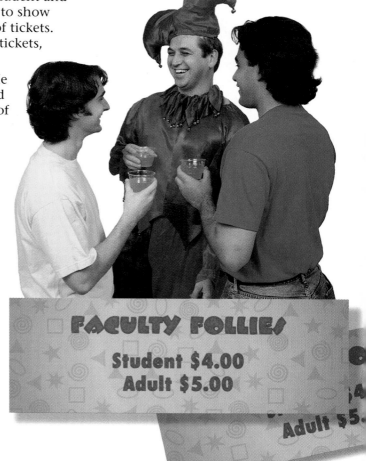

FACULTY FOLLIES
Student $4.00
Adult $5.00

Look Back

59. **Geometry** Quadrilateral $XYZW$ is a parallelogram. If the measure of angle X is 15° less than twice the measure of the angle Y, what are the measures of each angle of the parallelogram? **[Lesson 4.4]** 115°, 65°, 115°, 65°

Solve each proportion. **[Lesson 6.5]**

60. $\frac{7}{18} = \frac{21}{x}$ $x = 54$ **61.** $\frac{15}{32} = \frac{x}{20}$ $x = 9\frac{3}{8}$ **62.** $\frac{3}{x} = \frac{14}{9}$ $x = 1\frac{13}{14}$

63. Chemistry Carter is a chemist who wants to make a 48-milliliter solution that is 2.4% acid. To make the solution, he will mix one solution that is 2% acid and another solution that is 3% acid. How much of each should he use? **[Lesson 7.3]**
28.8 mL of 2% solution and 19.2 mL of 3% solution

Look Beyond

Graph the set of points that satisfies each system of inequalities.

64. $\begin{cases} x < 2 \\ y < 2x - 4 \end{cases}$ **65.** $\begin{cases} y > -2 \\ 2x - y < 5 \end{cases}$

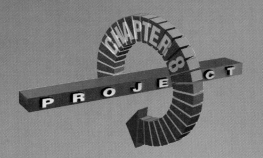

PROJECT 9 · CHAPTER 8

Diophantine EQUATIONS

Donna sees only the legs of horses and the legs of the riders walking the horses. She counts 22 legs in all. How many horses and how many riders does she see?

This is a version of a classic algebra problem. The numbers of legs must be whole numbers, so the number of solutions is restricted.

Equations that require the condition that *the solutions must be integers* are called **Diophantine equations**. The equations are named for Diophantus, who lived about 2000 years ago in the city of Alexandria in Egypt.

Using a Table Solution

Find all the solutions to the horse-and-rider problem.

Let h be the number of horses and r be the number of riders. Since each horse has 4 legs and each rider has two legs, $4h + 2r = 22$. Notice that all the coefficients are even numbers, so you can divide both sides of $4h + 2r = 22$ by 2 to get $2h + r = 11$.

Make a table to find the solutions. Remember, all solutions must be whole numbers. The first solution can be found by letting $h = 0$. When you substitute, you will find that $r = 11$. To find other solutions, note that each time there is one more horse, there are 4 more legs. That means there are 2 fewer riders. The table shows the solutions.

Horses	0	1	2	3	4	5
Riders	11	9	7	5	3	1

It is impossible to have more than 5 horses, because there would then be more than 22 legs.

Using a Graph Solution

Find all of the whole-number solutions to the equation $4h + 2r = 22$.

If $r = 0$, the equation becomes $4h = 22$, which has no whole-number solution. If $r = 1$, the equation becomes $4h + 2 = 22$, or $4h = 20$, which does have a solution, $h = 5$. So $(1, 5)$ is one solution to the Diophantine equation. Now look at the Diophantine equation in slope-intercept form. Solve for h.

$4h + 2r = 22$ becomes $h = -\frac{2}{4}r + \frac{22}{4}$.

The slope of the equation in lowest terms is $-\frac{1}{2}$.

Start with $(1, 5)$. Use a run of 2 and a rise of -1 to find any other solutions. The second solution is $(3, 4)$. After $(11, 0)$, another run of 2 and rise of -1 would place you at $(13, -1)$. But -1 is not a whole number.

How can you be sure that no other whole-number solutions exist between $(1, 5)$ and $(11, 0)$?

Try whole-number values for r between 1 and 11 to be sure.

Notice that if you had used $-\frac{2}{4}$ for the slope from $(1, 5)$, you would have missed the point $(3, 4)$. Why must the slope be in lowest terms?

Activity 1

If flour comes in 2-pound bags and in 5-pound bags, list all the ways to buy exactly 18 pounds of flour. Draw a graph, and locate the solutions on the graph. What can you say about the solutions that appear on the graph?

Activity 2

Ms. Smiley has written a math exam with two parts. Part I has 9 questions, and part II has 8 questions. She wants to know how to assign points to each part so that the total will be exactly 100 points. Find a Diophantine equation that represents this situation and solves Ms. Smiley's problem.

Make a table to represent tests that have 9 questions in part I, 8 questions in part II, and 200 points in all.

Make tables to represent tests that have different numbers of questions but have 100 total points.

Activity 3

How can you tell that $4x + 6y = 125$ has no whole-number solution?

How can you tell that $5x + 10y = 112$ has no whole-number solution?

How can you tell whether $6x + 9y = 100$ has whole-number solutions?

Chapter 8 Review

Vocabulary

boundary line	478	independent variable	431	rate of change	447
constant	467	linear function	431	regression line	453
constant function	467	parallel lines	473	slope	435
constant of variation	454	perpendicular lines	475	slope-intercept form	451
dependent variable	431	point-slope form	461	standard form	459
direct variation	454	quadrants	429	y-intercept	451
domain	431	range	431		

Key Skills & Exercises

Lesson 8.1

➤ **Key Skills**

Graph points from ordered pairs.

Graph the ordered pairs $(-4, 3)$, $(-4, -2)$, $(2, -3)$ and $(3, 2)$. Determine whether they lie on a straight line.

The points do not lie on a straight line.

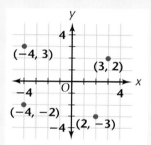

➤ **Exercises**

Graph each list of ordered pairs. Determine whether they lie on a straight line.

1. $(-1, 0)$, $(-3, -2)$, $(1, 2)$ **2.** $(1, 3)$, $(1, 4)$, $(1, -2)$ **3.** $(0, -1)$, $(-1, -2)$, $(1, 6)$

Lesson 8.2

➤ **Key Skills**

Find the slope of a line passing through two given points.

Find the slope, m, of a line passing through the points $A(-2, 4)$ and $B(3, 5)$.

$$m = \frac{\text{difference in } y\text{-values}}{\text{difference in } x\text{-values}} = \frac{5 - 4}{3 - (-2)} = \frac{1}{5}$$

➤ **Exercises**

Each pair of points is on a separate line. Find the slope of each line.

4. $A(-3, 2)$, $B(2, 3)$ $\frac{1}{5}$ **5.** $A(-5, 4)$, $B(1, 4)$ 0   **6.** $A(-3, -1)$, $B(-3, 3)$ undefined

Lesson 8.3

➤ **Key Skills**

Find the equation of a line passing through the origin and a given point.

Find the equation of a line passing through the origin and the point $(2, 4)$. Since the slope is $\frac{4}{2}$, or 2, the equation is $y = 2x$.

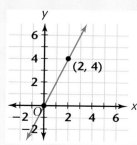

➤ **Exercises**

Find the equation of a line passing through the origin and

7. $(-3, 2)$.
$y = -\frac{2}{3}x$

8. $(2, 3)$.
$y = \frac{3}{2}x$

9. $(-2, -5)$.
$y = \frac{5}{2}x$

10. $(-5, 4)$.
$y = -\frac{4}{5}x$

Lesson 8.4

➤ *Key Skills*

Find the equation, in slope-intercept form, of a line through two points.

Write the equation for the line passing through $(-2, 3)$ and $(-1, 5)$.

$$\text{Slope} = m = \frac{5 - 3}{-1 - (-2)} = \frac{2}{1} = 2$$

Substitute 2 for m, -2 for x, and 3 for y.

$$y = mx + b$$
$$3 = 2(-2) + b$$
$$7 = b$$

Substitute 2 for m and 7 for b.

$$y = mx + b$$
$$y = 2x + 7$$

Graph a linear equation in slope-intercept form.

Graph the equation $y = \frac{1}{2}x + 1$.

The slope is $\frac{1}{2}$ and the y-intercept is 1.

Locate the y-intercept $(0, 1)$. From that point, move 1 unit up and 2 units to the right to the point $(2, 2)$. Draw the line.

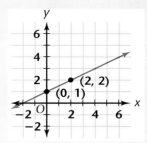

➤ *Exercises*

Write and graph an equation in slope-intercept form for the described line.

11. Slope: 2, y-intercept: 1

12. Contains the points $(2, -1)$ and $(-3, -1)$

Lesson 8.5

➤ *Key Skills*

Write an equation in standard form.

Write $y + 10 = -8x - 22$ in standard form.

$$y + 10 = -8x - 22$$
$$y + 10 + 8x = -8x - 22 + 8x$$
$$y + 10 + 8x - 10 = -22 - 10$$
$$8x + y = -32$$

Write equations in point-slope form.

Write an equation in point-slope form for the line passing through points $(4, 5)$ and $(2, 1)$.

$$m = \frac{1 - 5}{2 - 4} = \frac{-4}{-2} = 2$$

Use $(4, 5)$ for (x_1, y_1).

$$y - y_1 = m(x - x_1)$$
$$y - 5 = 2(x - 4)$$

➤ *Exercises*

13. $4x - y = 17$ **14.** $x + y = 5$

Write each linear equation in standard form.

13. $y + 9 = 4x - 8$ **14.** $y - 4 = -x + 1$ **15.** $y - 13 = 2x + 4$ **16.** $3x + y + 6 = 9$

$2x - y = -17$ $3x + y = 3$

For each line described, write the equation in point-slope form.

17. Slope: 2, y-intercept: -1
$y + 1 = 2(x - 0)$

18. Contains the points $(0, 4)$ and $(1, 2)$
$y - 4 = -2(x + 0)$ or $y - 2 = -2(x - 1)$

Lesson 8.6

➤ *Key Skills*

Recognize vertical and horizontal lines, and find the slope of horizontal lines.

$y = -8$ is an equation of a horizontal line. The slope is 0.
$x = 7$ is an equation of a vertical line. The slope is undefined.

Identify each line as vertical or horizontal, and state the slope (if possible).

vertical; undefined   vertical; undefined
19. $y = -6$ **20.** $x = 2$ **21.** $y = \frac{1}{2}$ **22.** $x = -\frac{1}{3}$
horizontal; 0   horizontal; 0

Lesson 8.7

➤ **Key Skills**

Recognize lines that are parallel and perpendicular to given lines, and write their equations.

The lines $y = -\frac{2}{3}x + 7$ and $y = -\frac{2}{3}x - 6$ are parallel because their slopes, $-\frac{2}{3}$, are equal.

The lines $y = -\frac{2}{3}x + 7$ and $y = \frac{3}{2}x - 6$ are perpendicular because their slopes, $-\frac{2}{3}$ and $\frac{3}{2}$, are negative reciprocals.

➤ **Exercises**

Write the slope of a line that is parallel to the given line. Write the slope of a line that is perpendicular to the given line.

23. $y = \frac{2}{3}x + 4$ $\frac{2}{3}; -\frac{3}{2}$ **24.** $y = -7x - 1$ $-7; \frac{1}{7}$

Write an equation for a line that contains the point (1, 5) and is

25. parallel to the line $2x + y = -1$.   **26.** perpendicular to $2x + y = -1$.
$y = -2x + 7$ $y = \frac{1}{2}x + \frac{9}{2}$, or $y = \frac{1}{2}x + 4\frac{1}{2}$

Lesson 8.8

➤ **Key Skills**

Graph linear inequalities.
To graph the inequality $y > \frac{1}{2}x - 1$, first graph the *dashed* boundary line $y = \frac{1}{2}x - 1$. Then test the point (0, 0).

$$y > \frac{1}{2}x - 1$$
$$0 > \frac{1}{2}(0) - 1$$
$$0 > -1 \qquad \text{True}$$

Shade the side of the boundary that includes the point (0, 0).

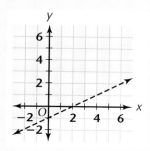

➤ **Exercises**

Graph each inequality.

27. $y > x + 1$ **28.** $y \geq -\frac{1}{2}x - 1$ **29.** $x + y < 4$ **30.** $y \leq 3$

Applications

Travel Marie had driven 110 miles when she stopped at a gas station. After she left the station, she recorded the number of hours she had driven. She averaged 50 miles per hour for the rest of the trip. Write a linear equation, and find the distance that she had driven after $d = 50h + 110$

31. 1.6 hours. 190 mi   **32.** 2.9 hours. 255 mi   **33.** 3.3 hours. 275 mi

Chapter 8 Assessment

1. Graph the ordered pairs (2, 1), (3, 2), and (5, 5). Determine whether they lie on a straight line.

Write each linear equation in standard form.

2. $y + 3 = -5x + 6$
   $5x + y = 3$

3. $4x + 2y + 8 = 10$
   $4x + 2y = 2$

Write each linear equation in slope-intercept form.

4. $2x + y = 10 - 3x$
   $y = -5x + 10$

5. $3x - y = 2y - 5$
   $y = x + 1\frac{2}{3}$

Find the slope and y-intercept for each linear equation.

6. $3x + 6y = 12$ $-\frac{1}{2}; 2$

7. $2x + 3y = -3$ $-\frac{2}{3}, -1$

8. On graph paper, sketch the graph of $y = 3x + 1$.

9. Find the slope of the line that passes through the points (5, 3) and (8, −2). $-\frac{5}{3}$

Write the slope-intercept form of an equation for a line that contains the point (2, 4) and is

10. parallel to the graph of $y = -3x - 7$. $y = -3x + 10$

11. perpendicular to the graph of $2x + y = 13$. $y = \frac{1}{2}x + 3$

12. On graph paper, graph the line that has slope −4 and y-intercept 1.

13. Write the equation in slope-intercept form for the line with a slope of $-\frac{2}{3}$ and y-intercept 5. $y = -\frac{2}{3}x + 5$

14. Write the equation in point-slope form for the line that has a slope of −3 and passes through the point (2, −3). $y + 3 = -3(x - 2)$

15. Write the equation in slope-intercept form for the line that passes through the points (−5, −1) and (1, 5). $y = x + 4$

16. What is the slope-intercept form of the equation for the graph at the right? $y = \frac{2}{3}x + 2$

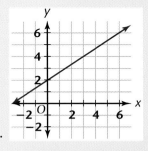

17. Give the slope of a line that passes through the points (1, 8) and (−2, 8). 0

Graph each linear inequality.

18. $y > 2x - 1$

19. $y \leq 3x + 2$

20. $x - y \geq 4$

21. $y < \frac{1}{2}x$

22. **Consumer Economics** Steve can rent a small lawn mower for a daily rate of $15, or he can rent a mower for 10 days at a flat rate of $100. After how many days of renting the mower will the flat rate be cheaper than the daily rate? 7 days

Chapters 1 – 8
Cumulative Assessment

College Entrance Exam Practice

Quantitative Comparison For Questions 1–4, write
A if the quantity in Column A is greater than the quantity in Column B;
B if the quantity in Column B is greater than the quantity in Column A;
C if the two quantities are equal; or
D if the relationship cannot be determined from the information given.

	Column A	Column B	Answers
1. C	$-18 + 4$	$-10 + (-4)$	Ⓐ Ⓑ Ⓒ Ⓓ **[Lesson 2.2]**
2. B	0.37	$\frac{37}{10}$	Ⓐ Ⓑ Ⓒ Ⓓ **[Lesson 3.3]**
3. A	The smallest integer solution to $x + 3 \geq 4$	The largest integer solution to $x - 7 < -6$	Ⓐ Ⓑ Ⓒ Ⓓ **[Lesson 7.5]**
4. C	The slope of the line passing through the points $(1, 2)$ and $(0, 3)$	The slope of the line $4x + 4y = 12$	Ⓐ Ⓑ Ⓒ Ⓓ **[Lesson 8.4]**

5. What are the next three terms of the sequence 5, 8, 11, 14, 17, . . . ?
a **[Lesson 1.1]**

    a. 20, 23, 26   **b.** 14, 11, 8   **c.** 22, 27, 32   **d.** 19, 21, 23

6. What is the solution to $16y = -120$? **[Lesson 6.2]**
d   **a.** 1920   **b.** -1920 **c.** 7.5   **d.** -7.5

7. What is the solution to the inequality $-2x + 3 \leq 7$? **[Lesson 7.5]**
c   **a.** $x \leq 2$ **b.** $x \leq 5$ **c.** $x \geq -2$ **d.** $x > -2$

8. Which of the following is the standard form of the equation
d $y + 4 = 2x - 7$? **[Lesson 8.5]**

    a. $y = 2x - 11$ **b.** $2x + y = -11$ **c.** $-2x + y = 11$ **d.** $2x - y = 11$

9. Which of the following is the equation of a vertical line? **[Lesson 8.6]**

b
  a. $y = -5$ **b.** $x = 3$ **c.** $y = -3x + 1$ **d.** $y = 2x - 2$

10. What is the equation of the line passing through $(-1, 2)$ and parallel to
c the line with the equation $y = 2x - 1$? **[Lesson 8.7]**

  a. $y = -\frac{x}{2} + 4$ **b.** $y = -\frac{x}{2} + 2$ **c.** $y = 2x + 4$ **d.** $y = -x - 1$

11. Which of the following polygons always has at least one pair of parallel
c sides? **[Lesson 4.4]**

  a. quadrilateral   **b.** triangle   **c.** trapezoid   **d.** pentagon

12. Compare the fractions $\frac{9}{12}$ and $\frac{5}{6}$ using the symbols <, >, or =. $\frac{9}{12} < \frac{5}{6}$
    [Lesson 3.2]

13. Simplify $(7x^2 + 2) - (3x - 2x^2)$. **[Lesson 5.5]** $9x^2 - 3x + 2$

Simplify each expression. [Lesson 6.1]

14. $(9w - 5) - (3w - 4)$ $6w - 1$ **15.** $(-54s) \div (6s)$ -9 **16.** $4x^2 - 4(3x^2 + 7)$ $-8x^2 - 28$

17. Solve the proportion $\frac{x}{12} = \frac{-3}{20}$. **[Lesson 6.5]** $x = -\frac{9}{5}$, or $-1\frac{4}{5}$

18. Solve $0.4x + 5.4 = -(3.2 - x)$. **[Lesson 7.2]** $x = \frac{43}{3}$, or $14\frac{1}{3}$

19. What is the y-intercept of the line that is parallel to the line $2x + 3y = 4$
    and contains the point $(3, -1)$? **[Lesson 8.7]** 1, or the point (0, 1)

20. Graph the linear inequality $y \le \frac{1}{2}x + 3$. **[Lesson 8.8]**

Free-Response Grid   The following questions may be
answered using a free-response grid commonly used by
standardized test services.

21. Simplify $4^2 \div 8 + 5(8 - 2) \cdot 2$. **[Lesson 1.7]** 62

22. Solve $\frac{5}{8} - x = \frac{1}{2}$. **[Lesson 5.4]** $\frac{1}{8}$

23. Maurice has scores of 88, 90, and 80 on three tests.
    What score must he make on his next test to have a
    test average of 85? **[Lesson 7.1]** 82

24. The area of a triangle is 27 square meters. The base of
    the triangle is 10 meters. Find the height.
    [Lesson 7.4] 5.4

25. Find the slope of the line that passes through the
    origin and $(4, 2)$. **[Lesson 8.2]** $\frac{1}{2}$

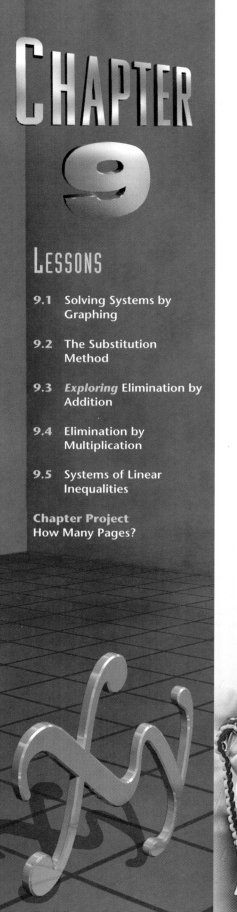

Systems of Equations and Inequalities

The athletic director hears the band playing the fight song and the loud cheers from the crowd. There is great excitement at the first football game of the season. The stadium is filled to capacity, and the team is ready to play. The ticket-booth manager is counting the money from the ticket sales.

A **system of equations** is a set of two or more equations considered at the same time. A **system of inequalities** is a set of two or more inequalities considered at the same time. The solution to a system must satisfy each equation or inequality in the system. Systems of equations or inequalities can be used to solve many different real-world problems, such as determining business profits.

Students $2.00 Adults $5.00

TiCKeTs

PORTFOLIO ACTIVITY

Students pay $2 per ticket and adults pay $5 per ticket. Although the stadium is filled to its capacity of 2000, 254 of those seats are given free of charge to band members, pep squad members, and faculty members. The ticket-booth manager reports to the athletic director that the total income from the sale of tickets for the current game is $5766.

With this information, you can find the number of adult tickets sold and the number of student tickets sold.

Let *a* represent the number of adult tickets sold.

Let *s* represent the number of student tickets sold.

In this chapter you will be asked to write and solve a system of equations to find the number of each type of ticket sold using different methods. You may wish to save your work for your portfolio.

Solving Systems by Graphing

why *Graphs make mathematical data easy to understand. They are used to show comparisons and changes in such fields as business, agriculture, and science.*

Mato makes wooden jewelry boxes in two sizes to sell at craft shows. The material for the small jewelry box costs $1, and the material for the large jewelry box costs $3.

From the graph of two linear equations, you can see where the two lines intersect. The point of intersection can have a very important meaning in the real world.

EXAMPLE 1

Small Business

Mato wants to make the best use of his time and money. He has determined that he has time to make 5 jewelry boxes per week. He can only spend $9 per week on materials. Write a system of two linear equations to find the number of boxes of each size that Mato should plan to make each week. Graph the system, and find the number of small boxes per week and the number of large boxes per week that Mato can make.

Solution ➤

Choose variables to represent the unknowns in this problem.

Let x represent the number of small jewelry boxes that Mato can make each week.

Let y represent the number of large jewelry boxes that Mato can make each week.

It is often helpful to make a table to organize your information.

	Small boxes	Large boxes	Total
Number	x	y	$x + y = 5$
Cost (in dollars)	$1x$	$3y$	$1x + 3y = 9$

The system of equations to represent the situation is shown below.

$$\begin{cases} x + y = 5 \\ x + 3y = 9 \end{cases}$$

To graph this system, first solve each equation for y.

$x + y = 5$	Given
$x + y - x = 5 - x$	Subtraction Property of Equality
$y = 5 - x$, or $y = -x + 5$	Simplify.

$x + 3y = 9$	Given
$x + 3y - x = 9 - x$	Subtraction Property of Equality
$3y = 9 - x$	Simplify.
$\dfrac{3y}{3} = \dfrac{9 - x}{3}$	Division Property of Equality
$y = \dfrac{9}{3} - \dfrac{x}{3}$	Simplify.
$y = 3 - \dfrac{x}{3}$	Simplify.
$y = -\dfrac{x}{3} + 3$, or $y = -\dfrac{1}{3}x + 3$	Commutative Property

Graph both lines on the same coordinate plane. The lines intersect at the point (3, 2).

Since the coordinates of (3, 2) satisfy both equations, the solution to the system of equations is (3, 2).

Mato should make 3 small jewelry boxes and 2 large jewelry boxes each week to make the best use of his time and money. ❖

You can use a graphics calculator to solve the system of equations in Example 1 by graphing. Enter both equations, graph them, and find the point of intersection.

You can also use a table to find the common solution. Notice that when X is 3, both Y_1 and Y_2 are equal to 2. This is the only point at which $Y_1 = Y_2$ for any x-value.

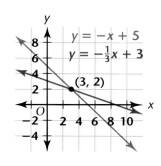

X	Y_1	Y_2
0	5	3
1	4	2.6667
2	3	2.3333
3	**2**	**2**
4	1	1.6667
5	0	1.3333
6	−1	1
X=3		

EXAMPLE 2

Solve the system $\begin{cases} 3x - y = 2 \\ x - 2y = -2 \end{cases}$ by graphing.

Solution ➤

First solve each equation for y.

$3x - y = 2$	Given
$3x - y - 3x = 2 - 3x$	Subtraction Property of Equality
$-y = 2 - 3x$	Simplify.
$-1(-y) = -1(2 - 3x)$	Multiplication Property of Equality
$y = -2 + 3x$	Simplify.
$y = 3x - 2$	Commutative Property

$x - 2y = -2$	Given
$x - 2y - x = -2 - x$	Subtraction Property of Equality
$-2y = -2 - x$	Simplify.
$\dfrac{-2y}{-2} = \dfrac{-2 - x}{-2}$	Division Property of Equality
$y = \dfrac{-2}{-2} + \dfrac{-x}{-2}$	Simplify.
$y = 1 + \dfrac{x}{2}$	Simplify.
$y = \dfrac{1}{2}x + 1$	Commutative Property

Graph both equations on the same coordinate plane. The lines intersect at a point that has rational-number coordinates. From the graph, you can only estimate the coordinates of the point of intersection.

The x-coordinate is about 1.1, and the y-coordinate is about 1.5, so the point of intersection is about (1.1, 1.5). ❖

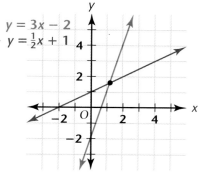

$y = 3x - 2$
$y = \frac{1}{2}x + 1$

Graphics Calculator

Most graphics calculators can calculate the exact coordinates of the point of intersection. The system from Example 2 is graphed here.

The exact coordinates are (1.2, 1.6).

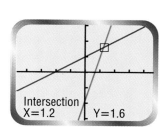

Intersection
X=1.2   Y=1.6

Approximate Solution

Real-world applications often involve solving problems that do not have whole-number solutions. It is sometimes difficult to find an exact solution from a graph. A reasonable estimate for a point of intersection is called an **approximate solution.**

EXAMPLE 3

Fund-raising   The drama club at Wilson High School sponsored a variety show to raise money for a trip to New York City. Student tickets sold for $2.25 each, and adult tickets sold for $4.60 each. All 200 seats were sold. How many student tickets and how many adult tickets were sold?

Members of the drama club reached the goal of earning barely $700 from the sale of tickets. Now they are on their way to New York City.

Solution ➤

Let x represent the number of student tickets that were sold. Let y represent the number of adult tickets that were sold. Make a table to organize the information.

	Student tickets	Adult tickets	Total
Number	x	y	200
Earnings	$2.25x	$4.60y	$700

The system of equations is: $\begin{cases} x + y = 200 \\ 2.25x + 4.60y = 700 \end{cases}$

Solve the first equation, $x + y = 200$, for y.

$$x + y = 200 \qquad \text{Given}$$
$$x + y - x = 200 - x \qquad \text{Subtraction Property of Equality}$$
$$y = -x + 200 \qquad \text{Commutative Property}$$

Solve the second equation, $2.25x + 4.60y = 700$, for y.

$2.25x + 4.60y = 700$	Given
$2.25x + 4.60y - 2.25x = 700 - 2.25x$	Subtraction Property of Equality
$4.6y = 700 - 2.25x$	Simplify.
$\dfrac{4.6y}{4.6} = \dfrac{700 - 2.25x}{4.6}$	Division Property of Equality
$y = \dfrac{700}{4.6} - \dfrac{2.25x}{4.6}$	Simplify.
$y = -\dfrac{2.25x}{4.6} + \dfrac{700}{4.6}$	Commutative Property
$y \approx -0.5x + 152$	Approximation

Graph the equations on one coordinate plane, and estimate the coordinates of the point of intersection.

One reasonable estimate is (95, 105).

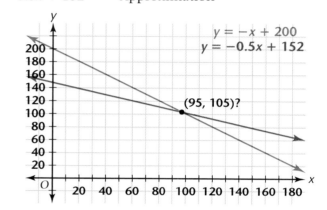

Check this approximate solution in the original equations. Substitute 95 for x and 105 for y.

$x + y = 200$	$2.25x + 4.6y = 700$
$95 + 105 \stackrel{?}{=} 200$	$2.25(95) + 4.6(105) \stackrel{?}{=} 700$
$200 = 200$	$213.75 + 483 \stackrel{?}{=} 700$
	$696.75 \neq 700$

Since 696.75 is less than 700, try estimating different values for x and y that have a sum of 200.

Try (94, 106). Substitute 94 for x and 106 for y.

$$x + y = 200$$
$$94 + 106 \stackrel{?}{=} 200$$
$$200 = 200$$

$$2.25x + 4.6y = 700$$
$$2.25(94) + 4.6(106) \stackrel{?}{=} 700$$
$$211.5 + 487.6 \stackrel{?}{=} 700$$
$$699.1 \neq 700$$

The approximate solution (94, 106) yields earnings that are closer to $700, but are still less than $700.

Try (93, 107). Substitute 93 for x and 107 for y.

$$x + y = 200$$
$$93 + 107 \stackrel{?}{=} 200$$
$$200 = 200$$

$$2.25x + 4.6y = 700$$
$$2.25(93) + 4.6(107) \stackrel{?}{=} 700$$
$$209.25 + 492.2 \stackrel{?}{=} 700$$
$$701.45 \neq 700$$

The point (93, 107) yields earnings that barely reach the goal of $700. ❖

CRITICAL Thinking

Explain why (93, 107) is the best approximation for the situation in Example 3?

Graphics Calculator

You can use the table feature on a graphics calculator to find a more exact solution to a system of equations that has rational-number solutions. On some graphics calculators you can even find the exact coordinates of the point of intersection.

For example, to find the exact solution for the system of equations in Example 3, you can graph the system and find the exact coordinates of the point of intersection, or you can use the table feature.

```
Y₁=-X+200
Y₂=-2.25X/4.6+700/4.6
Y₃=
Y₄=
Y₅=
Y₆=
Y₇=
```

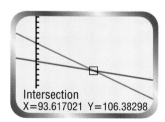

Intersection
X=93.617021 Y=106.38298

X	Y₁	Y₂
93.57	106.43	106.41
93.58	106.42	106.4
93.59	106.41	106.4
93.6	106.4	106.39
93.61	106.39	106.39
93.62	106.38	106.38
93.63	106.37	106.38

X=93.62

Why must you be sure to enter equations that are *equivalent* to the original equations, not an approximation, when finding an exact solution?

EXERCISES & PROBLEMS

Communicate

1. Explain how to determine the solution to this system of equations. $\begin{cases} 2x - y = 5 \\ y = 3x - 7 \end{cases}$

2. How can you check your solution to a system of equations?

3. Describe how a graphics calculator can be used to solve a system of equations.

Practice & Apply

Estimate the solution to each system of equations.

4. (4, 1)
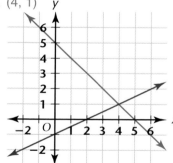

5. (0, 6)
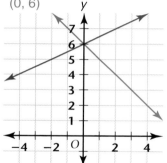

6. (−2, −3)
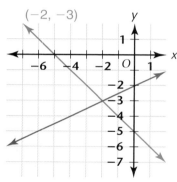

7. (−2, 3)
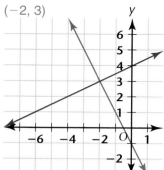

8. $\left(5, \dfrac{1}{2}\right)$
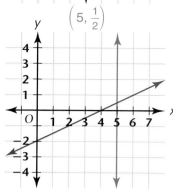

9. (−2.3, −0.4)
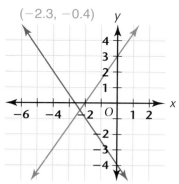

For Exercises 10–21, solve each system of equations by graphing. If necessary, give an approximate solution.

10. $\begin{cases} 2x + y = 1 \\ 3x + y = 2 \end{cases}$ (1, −1)

11. $\begin{cases} x + y = 2 \\ 2x + y = 6 \end{cases}$ (4, −2)

12. $\begin{cases} 3x - y = 4 \\ 2x + y = 6 \end{cases}$ (2, 2)

13. $\begin{cases} 2x + y = 7 \\ x - y = 2 \end{cases}$ (3, 1)

14. $\begin{cases} 3x - y = 6 \\ y = 4x - 4 \end{cases}$ (−2, −12)

15. $\begin{cases} 3x - 2y = 10 \\ y = x - 4 \end{cases}$ (2, −2)

16. $\begin{cases} 3x - y = 6 \\ y = 4 \end{cases}$ $(\approx 3.3, 4)$ **17.** $\begin{cases} 3x - y = 8 \\ x = -2 \end{cases}$ $(-2, -14)$ **18.** $\begin{cases} x = -4 \\ y = 5 \end{cases}$ $(-4, 5)$

19. $\begin{cases} y = -\frac{1}{2}x + 1 \\ y = 2x + 4 \end{cases}$ $(-1.2, 1.6)$ **20.** $\begin{cases} y = \frac{3}{4}x + 2 \\ y = x + 1 \end{cases}$ $(4, 5)$ **21.** $\begin{cases} y = -\frac{1}{3}x + 2 \\ y = x + 2 \end{cases}$ $(0, 2)$

22. Agriculture On a ranch, about 50 calves are born. Of these calves, there are about twice as many bulls as there are heifers. How many bull calves and heifer calves are born? Write a system of equations for this situation, and use a graph to solve the system.

Write a system of equations to solve each problem. Solve the system by graphing. If necessary, give an approximate solution.

23. Sales A health-food store sells a mixture of peanuts and raisins. The mixture weighs $2\frac{1}{2}$ pounds and sells for $6. How many pounds of each should be used in the mixture? $1\frac{1}{2}$ lb raisins, 1lb peanuts

24. Nutrition There are 15 grams of fat in a package of peanut butter crackers. They contain 5 times as much unsaturated fat as saturated fat. How many grams of unsaturated and saturated fat are in the crackers?
2.5 g sat fat, 12.5 unsat fat

25. 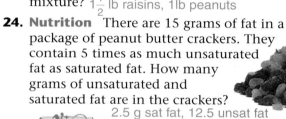 **Portfolio Activity** Write a system of equations for the problem in the portfolio activity on page 495. Solve this system by graphing.

Look Back

26. Chemistry A mixture is 2.8% acid. If the total mixture is 420 milliliters, how many milliliters of acid are in the mixture? **[Lesson 3.7]** 11.76 mL

Solve each inequality. [Lesson 7.5] 27. $x > -3$ 28. $z \geq -15$ 29. $a > \frac{3}{4}$

27. $3x + 5 > -4$ **28.** $3 - z \leq 18$ **29.** $18 < 15 + 4a$

30. What is the slope of a vertical line? a horizontal line? **[Lesson 8.6]** undefined; 0

31. Does the ordered pair $(-2, 5)$ satisfy the inequality $x + 5y < 12$? no **[Lesson 8.8]**

32. Does the ordered pair $(0, 0)$ satisfy the inequality $3x - 4y < 10$? yes **[Lesson 8.8]**

Look Beyond

33. If $m = 3$ and $3m + 4v = 23$, what is the value of v? $v = 3.5$

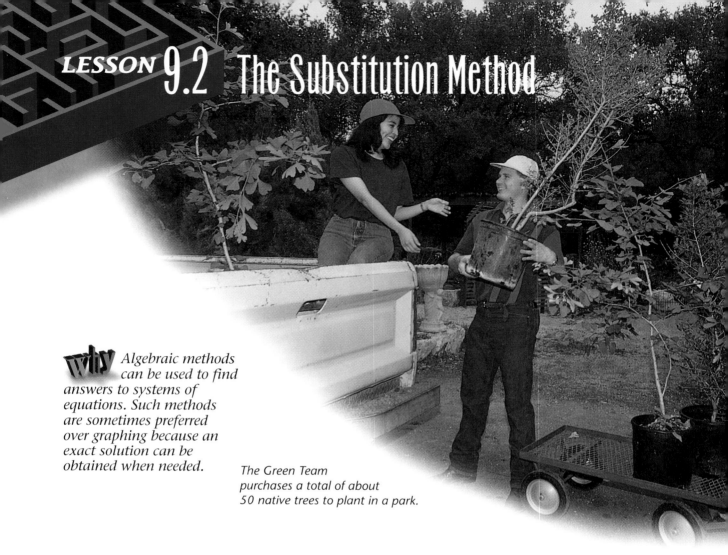

why *Algebraic methods can be used to find answers to systems of equations. Such methods are sometimes preferred over graphing because an exact solution can be obtained when needed.*

The Green Team purchases a total of about 50 native trees to plant in a park.

number of large trees	**and**	number of small trees	**equals**	50
x	**+**	*y*	**=**	**50**

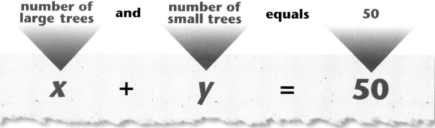

number of small trees	**is**	twice the number of large trees
y	**=**	**2x**

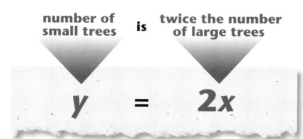

The Green Team will plant large four-year-old trees and smaller two-year-old trees. Since the smaller trees cost less, they will plant twice as many of the smaller trees. How many trees of each size will they plant?

You can find the answer by writing and solving a system of equations. Let *x* represent the number of large trees. Let *y* represent the number of small trees.

The system of equations is shown below.

$$\begin{cases} x + y = 50 \\ y = 2x \end{cases}$$

You can find the solution, or an approximate solution, to a system of equations by graphing. From this graph, an exact solution to the system cannot be determined. A reasonable approximate solution is (17, 34).

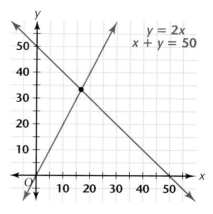

To find the exact solution to this system of equations, you can use the algebraic method of substitution.

You will investigate the algebraic method of solving this system in the following exploration.

•Exploration *The Substitution Method*

Refer to the system of equations shown below.

$$\begin{cases} x + y = 50 \\ y = 2x \end{cases}$$

1 Substitute $2x$ for y in the equation $x + y = 50$, and solve for x. $x = 16\frac{2}{3}$

2 Now substitute your value for x into the equation $y = 2x$, and simplify. $y = 33\frac{1}{3}$

3 What is the resulting value for y? Write the exact solution as an ordered pair. $\left(16\frac{2}{3}, 33\frac{1}{3}\right)$

4 According to the exact solution, the Green Team should order how many large trees and how many small trees? Is your answer reasonable? Explain.

5 What are some reasonable approximations that solve this problem? Do these approximate solutions satisfy both equations? Explain.

6 Which reasonable approximation best solves this problem? Explain. ❖

EXAMPLE 1

Use the substitution method to solve this system of equations. $\begin{cases} -x + y = 4 \\ 3x + y = 36 \end{cases}$

Solution ➤

First solve $-x + y = 4$ for y to get $y = x + 4$. Then substitute $x + 4$ for y in the second equation, $3x + y = 36$.

$3x + y = 36$	Given
$3x + (x + 4) = 36$	Substitution
$4x + 4 = 36$	Simplify.
$4x + 4 - 4 = 36 - 4$	Subtraction Property of Equality
$4x = 32$	Simplify.
$\dfrac{4x}{4} = \dfrac{32}{4}$	Division Property of Equality
$x = 8$	Simplify.

Now substitute 8 for x in the first equation, $y = x + 4$.

$y = x + 4$	Given
$y = 8 + 4$	Substitution
$y = 12$	Simplify.

So the solution is (8, 12). Always check your answers by substituting the x- and y-values into both of the original equations. ❖

Try This   Use the substitution method to solve this system of equations.
$\begin{cases} 3x - y = 10 \\ 5x + 2y = 24 \end{cases}$ (4, 2)

EXAMPLE 2

Biology   In 1994, the whooping-crane population totaled 291. The number of captive whooping cranes was about $\frac{2}{3}$ the number of wild whooping cranes. How many whooping cranes were in captivity and how many were in the wild?

Solution ➤

Let c represent the number of whooping cranes living in captivity, and let w represent the number of whooping cranes living in the wild.

First equation

Whooping-crane population	totaled	291
$c + w$	$=$	291

Second equation

Number of captive whooping cranes	was	$\frac{2}{3}$ the number of wild whooping cranes
c	$=$	$\frac{2}{3}w$

So the system of equations is:
$$\begin{cases} c + w = 291 \\ c = \frac{2}{3}w \end{cases}$$

Since c is equal to $\frac{2}{3}w$ in the second equation, substitute $\frac{2}{3}w$ for c in the first equation, $c + w = 291$.

$c + w = 291$	Given
$\left(\frac{2}{3}w\right) + w = 291$	Substitution
$\frac{2}{3}w + \frac{3}{3}w = 291$	Identity Property
$\frac{5}{3}w = 291$	Simplify.
$\frac{3}{5}\left(\frac{5}{3}w\right) = \frac{3}{5} \cdot 291$	Multiplication Property of Equality
$w = \frac{873}{5}$, or $174\frac{3}{5}$	Simplify.

Since w represents the number of whooping cranes in the wild, and the number of these whooping cranes must be a whole number, round $174\frac{3}{5}$ up to 175.

Substitute 175 for w in the second equation, $c + w = 291$.

$$c + w = 291$$
$$c + 175 = 291$$
$$c + 175 - 175 = 291 - 175$$
$$c = 116$$

So in 1994, there were 175 whooping cranes in the wild and 116 in captivity. ❖

Try This Patrick plans to fence a portion of his 1-acre lot. He wants the length of the fenced yard to be 3 times the width. He can afford to purchase 1200 feet of fencing. Write and solve a system of equations to find the dimensions of the fenced yard.

EXERCISES & PROBLEMS

Communicate ~~~~

1. Why is solving by substitution sometimes preferable to solving by graphing?

2. Explain how to solve this system of equations using the substitution method.
$$\begin{cases} x + y = 7 \\ 2x - y = 12 \end{cases}$$

3. Explain how to check your solution to a system of equations.

Practice & Apply ~~~~

Use the substitution method to solve each system of equations. Check your answer by substituting your solution in the original equations.

4. $\begin{cases} y = 4x \\ x + y = 20 \end{cases}$

5. $\begin{cases} x = 5y \\ x + 3y = 24 \end{cases}$

6. $\begin{cases} x = y - 3 \\ x + 3y = 13 \end{cases}$

7. $\begin{cases} m = n + 7 \\ m + 5n = 37 \end{cases}$

8. $\begin{cases} y = 9 - x \\ x - y = 5 \end{cases}$

9. $\begin{cases} t = 11 - z \\ z - t = 15 \end{cases}$

10. $\begin{cases} w = 5 - 2y \\ y - 3w = 23 \end{cases}$

11. $\begin{cases} c = 9 + 3d \\ 3d - 4c = 36 \end{cases}$

12. $\begin{cases} 2x + y = 11 \\ x + y = 14 \end{cases}$

13. $\begin{cases} 3w + 5z = 19 \\ w - z = 3 \end{cases}$

14. $\begin{cases} m - 2n = 20 \\ 2m + n = 5 \end{cases}$

15. $\begin{cases} x + 2y = 11 \\ 3x + y = 3 \end{cases}$

16. $\begin{cases} x + y = 13 \\ x - 2y = -2 \end{cases}$

17. $\begin{cases} m - n = 16 \\ m + 3n = 12 \end{cases}$

18. $\begin{cases} z - 2h = 18 \\ z + 3h = -2 \end{cases}$

19. $\begin{cases} 2z - 3w = 15 \\ z - 2w = 16 \end{cases}$

20. $\begin{cases} y = x - 4 \\ 2x - 5y = 2 \end{cases}$

21. $\begin{cases} y = -x + 2 \\ 2x - y = 1 \end{cases}$

22. $\begin{cases} x = 2y - 1 \\ y = -2x + 3 \end{cases}$

23. $\begin{cases} 4x - 8 = y \\ 5x - 10 = -3y \end{cases}$

24. $\begin{cases} x - y = 4 \\ 2x - 3y = -2 \end{cases}$

25. $\begin{cases} 5x + y = -15 \\ x - y = 3 \end{cases}$

26. $\begin{cases} x = y - 4 \\ 2y - 5x = 2 \end{cases}$

27. $\begin{cases} x = -y + 2 \\ 2y - x = 1 \end{cases}$

28. $\begin{cases} 4x - 8 = -y \\ 5x - 3 = -3y \end{cases}$

29. $\begin{cases} 3x - 16 = -2y \\ 7x - 19 = -y \end{cases}$

30. $\begin{cases} x - 2y = -2 \\ 2x - 3y = 2 \end{cases}$

31. $\begin{cases} 2x + y = 5 \\ 8x - y = 45 \end{cases}$

32. $\begin{cases} x + 2y = 5 \\ -3x - 2y = -3 \end{cases}$

33. $\begin{cases} y = 2.4x - 1.6 \\ y = -3.6x + 1.4 \end{cases}$

34. $\begin{cases} -5.7x + 1.8y = 9 \\ y = 1.5x + 3 \end{cases}$

35. $\begin{cases} y = \dfrac{5}{2}x - 2 \\ 11x - 4y = 8 \end{cases}$

36. $\begin{cases} x = -\dfrac{3}{2}y + 2 \\ 2x + 4y = 4 \end{cases}$

37. $\begin{cases} y - 3 = \dfrac{1}{4}x - 3 \\ y + 2 = -x + 3 \end{cases}$

38. $\begin{cases} x + y = 320 \\ 0.1x + 0.18y = 0.15 \end{cases}$

39. $\begin{cases} x = \dfrac{1}{2}y - 3 \\ y = -4x + 3 \end{cases}$

40.  **Geometry** Two angles are supplementary. The measure of the larger angle is 3 times the measure of the smaller angle. Find the measures of the angles.   45°, 135°

41. **Geometry** Two angles are complementary. The larger angle measures 10 degrees more than the smaller angle. Find the measures of the angles.   40°, 50°

42. **Investments** Margaret has $4500 to invest. She plans to invest part of the money at 5% and part at 7%. She wants her total interest to be $290. How much should she invest at each rate? $1250 at 5% and $3250 at 7%

43. **Chemistry** A chemist wants to combine a solution that is 30% acid with pure acid. He wants to produce 42 milliliters of a solution that is 50% acid. How much of the 30% solution and how much of the pure acid should the chemist use?   30 mL of 30% solution, 12 mL pure acid

44. **Fund-raising** The baseball booster club earned total of $1422 by selling thermal mugs and baseball caps at the games. If they sold twice as many caps as thermal mugs, how many of each did they sell?   79 mugs, 158 caps

Each thermal mug sold for $2, and each cap sold for $8.

45. The sum of two numbers is 17. The smaller number is 33 less than the greater number. Find the two numbers.   25 and −8

46. The sum of two numbers is 8. Four times the greater number is 2 more than 4 times the smaller number. Find the two numbers.
4.25 and 3.75

47. The sum of two numbers is 28. The difference of the two numbers is 3. Find the two numbers.   12.5 and 15.5

48. **Portfolio Activity** Use substitution to solve a system of equations for the problem in the portfolio activity on page 495.

Look Back

For Exercises 49–52, write an equation for the line described. [Lessons 8.4, 8.5] $y = -3x - 1$ $y = \frac{1}{5}x + 7\frac{2}{5}$ or $y = 0.2x + 7.4$

49. With slope −3 and through the point (−2, 5)   **50.** Through the points (−2, 7) and (3, 8)

51. Through the points (−3, 4) and (−3, 7) $x = -3$ **52.** Through the points (5, 2) and (−1, 2)
$y = 2$

53. Write an equation of a line perpendicular to $2x - 3y = 5$.   **[Lesson 8.7]**

Answers may vary. A sample answer is $y = -\frac{3}{2}x$ or $y = -1.5x$

Look Beyond

Find the least common multiple for each pair of monomials.

54. $2xy$ and $4x^2$ $4x^2y$ **55.** $3xz$ and $5x^2y$ $15x^2yz$

Exploring Elimination by Addition

why *You have solved systems of equations by graphing and by substitution. Another method of solving systems of equations is called the elimination method. It can be modeled using algebra tiles.*

Janet and Matthew design a game to play for a math project.

Janet wants to use a scoring system in which the player moves 2 spaces forward for each correct answer but moves only 1 space forward for each incorrect answer. Matthew wants to use a scoring system in which the player moves 3 spaces forward for each correct answer and moves 1 space backward for each incorrect answer. When Henry plays the game, he moves 10 spaces forward regardless of which scoring system he uses. How many correct and incorrect answers did Henry have?

x-tiles   **y-tiles**

For Exploration 1, you will need positive and negative *y*-tiles to use with your positive and negative *x*-tiles. You can use blue ovals for positive *y*-tiles and yellow ovals to model negative *y*-tiles.

•Exploration 1 *Using the Addition Property of Equality*

1 Let *x* represent the number of correct answers. Let *y* represent the incorrect answers. Write a system of equations.

Janet's scoring system	**Matthew's scoring system**
2 spaces *forward* for a correct answer	3 spaces *forward* for a correct answer
+ 1 space *forward* for an incorrect answer	+ 1 space *backward* for an incorrect answer
10 spaces forward	10 spaces forward

$$\underset{2x}{\underline{\;?\;}} + \underset{y}{\underline{\;?\;}} = 10 \qquad\qquad \underset{3x}{\underline{\;?\;}} + \underset{(-y)}{\underline{\;?\;}} = 10$$

2 Rewrite the equation representing Matthew's scoring system using a minus sign. Model this system of equations using the *x*-tiles and *y*-tiles.

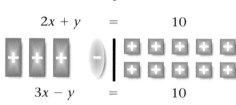

$$2x + y = 10$$

$$3x - y = 10$$

3 Combine the two models. Remove neutral pairs, and write the new equation. $5x = 20$

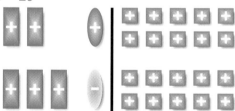

$$(2x + y) + (3x - y) = 10 + 10$$

4 Which variable was eliminated? Why? Solve the resulting equation for the remaining variable. *y*-variable; neutral pairs; *x* = 4

5 Use substitution in either of the original equations to find the value of the variable that was eliminated by removing neutral pairs. The solution to the system is (_?_, _?_). (4, 2)

6 Henry answered _?_ questions correctly and _?_ questions incorrectly. ❖ 4 correctly, 2 incorrectly

CT– Answers may vary. A sample answer is given.

$$\begin{cases} 2x + y = 10 \\ 3x - 2y = 10 \end{cases}$$

CRITICAL Thinking

In Exploration 1, one of the variables in the system of equations $\begin{cases} 2x + y = 10 \\ 3x - y = 10 \end{cases}$ was eliminated by removing neutral pairs. Write a system of equations in which one of the variables would *not* be eliminated using this method.

Use the elimination-by-addition method to solve $\begin{cases} 3x - y = 5 \\ x + y = 7 \end{cases}$.

Since the coefficients of y are opposites in the two equations, you can solve this system using the elimination-by-addition method. Use the Addition Property of Equality to add terms on corresponding sides of the equation.

$$\begin{aligned} 3x - y &= 5 \\ \underline{x + y} &= \underline{7} \\ 4x &= 12 \end{aligned}$$

The y-variable is eliminated.

Solve the resulting equation, $4x = 12$, for x.

$$4x = 12$$
$$\frac{4x}{4} = \frac{12}{4}$$
$$x = 3$$

To find the value of y, substitute 3 for x in one of the *original* equations.

$$x + y = 7$$
$$3 + y = 7$$
$$3 + y - 3 = 7 - 3$$
$$y = 4$$

The solution is (3, 4). Check by substituting 3 for x and 4 for y in both of the original equations. ❖

Another way to use elimination by addition to solve a system is to use the Subtraction Property of Equality.

Exploration 2 *Using the Subtraction Property of Equality*

1 Use the x-tiles and y-tiles to model the system. $\begin{cases} 4x + y = 10 \\ x + y = 4 \end{cases}$

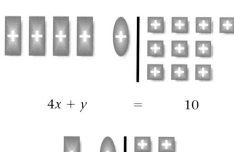

$$4x + y \qquad = \qquad 10$$

$$x + y \quad = \quad 4$$

2 Combine the two models, and remove any neutral pairs. Is a variable eliminated? Why or why not? $5x + 2y = 14$; no variables are eliminated because neither variable combines with an equal number of opposites.

3 Use tiles to model the system of equations again.
Tiles shown on page 512.

4 You can eliminate the y-variable in this system by using the Subtraction Property of Equality. Recall that to subtract, you add the opposite. First model the *opposite* of the equation $x + y = 4$.
Tiles shown below.

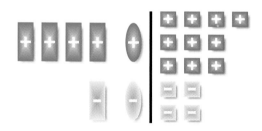

$$x + y = 4$$

$$-1(x + y) = -1(4)$$
$$-x - y = -4$$

5 Now subtract by adding the opposite. Combine the model of the *opposite* of $x + y = 4$ with the model of $4x + y = 10$, and remove neutral pairs. What is the resulting equation? What variable was eliminated? $3x = 6$, y-variable

$$(4x + y) + (-x - y) = 10 + (-4)$$

6 Use the model to solve the resulting equation for x. $x = 2$

7 Substitute the solution for x into one of the original equations, and solve for y. $y = 2$

8 Check your solution in both of the original equations.
$2 + 2 = 4$ (True) and $8 + 2 = 10$ (True)

9 Describe how to solve a system of equations using the Subtraction Property of Equality. ❖

CRITICAL Thinking

Once you have used elimination and found the value of one variable for a system, does it matter in which equation you substitute the value to find the value of the second variable? Explain.

APPLICATION

Investments   Emily is going to invest her savings. She has two options.

> **Option 1**   She can invest part of her savings at 4% and the remaining part at 5%. This option results in a yearly earning of $250.

> **Option 2**   She can invest the same amount of money at 4% and the remaining amount at 6%. This option results in a yearly earning of $280. However, this option is riskier.

What amount of money is she considering investing at each rate?

Write an equation for each option. Emily will divide her savings into two parts, x and y. The x part is the amount she will invest at the 4% rate in either option.

Option 1

Invest the x part at 4% and the y part at 5% resulting in $250.

$$0.04x \quad + \quad 0.05y \quad = \quad 250$$

Option 2

Invest the x part at 4% and the y part at 6% resulting in $280.

$$0.04x \quad + \quad 0.06y \quad = \quad 280$$

Solve this system of equations. $\begin{cases} 0.04x + 0.05y = 250 \\ 0.04x + 0.06y = 280 \end{cases}$

Use the Subtraction Property of Equality.

$$\begin{aligned}
0.04x + 0.06y &= 280 \\
- (0.04x + 0.05y) &= -(250) \\
\hline
0.01y &= 30 \\
\frac{0.01y}{0.01} &= \frac{30}{0.01} \\
y &= 3000
\end{aligned}$$

Substitute 3000 for y in one of the original equations. Solve for x.

$$\begin{aligned}
0.04x + 0.05y &= 250 \\
0.04x + 0.05(3000) &= 250 \\
0.04x + 150 &= 250 \\
0.04x + 150 - 150 &= 250 - 150 \\
0.04x &= 100 \\
\frac{0.04x}{0.04} &= \frac{100}{0.04} \\
x &= 2500
\end{aligned}$$

Emily will invest $2500 at 4% and $3000 at either 5% or 6%. ❖

514   CHAPTER 9

EXERCISES & PROBLEMS

Communicate

1. Describe how algebra tiles can be used to model the solution to this system of equations. $\begin{cases} 3x + y = 6 \\ x - y = 2 \end{cases}$

2. Explain how to decide whether to use the Addition Property of Equality to eliminate one variable in a system of equations.

3. How would you determine whether (5, 3) is a solution to this system of equations? $\begin{cases} 2x - y = 7 \\ 2x + y = 10 \end{cases}$

4. How could you determine by graphing whether (5, 3) is a solution to the system of equations in Exercise 3?

Practice & Apply

Model each system of equations using algebra tiles. Then solve the system using elimination by addition.

5. $\begin{cases} 3x + y = 7 \\ 2x + y = 3 \end{cases}$ (4, −5) **6.** $\begin{cases} 3x + y = 6 \\ x - y = -2 \end{cases}$ (1, 3) **7.** $\begin{cases} 4x - 2y = 10 \\ 3x + 2y = 4 \end{cases}$ (2, −1)

8. $\begin{cases} 3x - y = 4 \\ 2x - y = 2 \end{cases}$ (2, 2) **9.** $\begin{cases} 3x + y = 10 \\ 2x + y = 8 \end{cases}$ (2, 4) **10.** $\begin{cases} 5x + 7y = 11 \\ 5x + 3y = 3 \end{cases}$ $\left(-\frac{3}{5}, 2\right)$

Solve each system of equations using elimination by addition. Check your solutions.

(−1, 5)

11. $\begin{cases} x - y = 4 \\ x + y = 2 \end{cases}$ (3, −1) **12.** $\begin{cases} 2x + y = 5 \\ x + y = -2 \end{cases}$ (7, −9) **13.** $\begin{cases} 4x + y = 2 \\ x - y = 3 \end{cases}$ (1, −2) **14.** $\begin{cases} 3x - 2y = -13 \\ 3x + y = 2 \end{cases}$

15. $\begin{cases} x + 2y = 3 \\ 3x - 2y = 5 \end{cases}$ $\left(2, \frac{1}{2}\right)$ **16.** $\begin{cases} x - y = 3 \\ 2x - y = 2 \end{cases}$ (−1, −4) **17.** $\begin{cases} x + 4y = 10 \\ x + 3y = 13 \end{cases}$ (22, −3) **18.** $\begin{cases} 7a + 5c = 37 \\ 2a - 5c = 8 \end{cases}$

$a = 5, c = \frac{2}{5}$

Solve each system of equations using elimination by addition.

19. $\begin{cases} 4w - 2z = 15 \\ 3w + 2z = 13 \end{cases}$ $w = \frac{4}{1}$ $z = \frac{1}{2}$ **20.** $\begin{cases} 4w + 5c = 12 \\ 4w + 6c = 16 \end{cases}$ $w = -2$ $c = 4$ **21.** $\begin{cases} 2m - 3n = 16 \\ 5m - 3n = 13 \end{cases}$ $m = -1$ $n = -6$

22. $\begin{cases} 4p - 5q = 11 \\ 2p - 5q = 17 \end{cases}$ **23.** $\begin{cases} 3x - y = 12 \\ y = -3x + 5 \end{cases}$ $\left(2\frac{5}{6}, -3\frac{1}{2}\right)$ **24.** $\begin{cases} 4m + 3n = 15 \\ 4m = -2n + 10 \end{cases}$ $m = 0$ $n = 5$

$(p, q) = \left(-3, -4\frac{3}{5}\right)$

25. $\begin{cases} \dfrac{4}{5}x + 2y = 6 \\ \dfrac{4}{5}x + 5y = 21 \end{cases}$ (−5, 5) **26.** $\begin{cases} x + 3y = \dfrac{4}{5} \\ x - 3y = -\dfrac{1}{5} \end{cases}$ $\left(\dfrac{3}{10}, \dfrac{1}{6}\right)$ **27.** $\begin{cases} 5x - y = -\dfrac{3}{5} \\ 2x - y = \dfrac{3}{5} \end{cases}$ $\left(-\dfrac{2}{5}, -\dfrac{7}{5}\right)$

28. $\begin{cases} 5y = 2x - 1 \\ y = -2x + 3 \end{cases}$ $\left(\dfrac{4}{3}, \dfrac{1}{3}\right)$ **29.** $\begin{cases} 1.3x = 2.5y - 1 \\ y = -1.3x + 3 \end{cases}$ $\left(\dfrac{10}{7}, \dfrac{8}{7}\right)$ **30.** $\begin{cases} 0.2m - 0.3n = 1.4 \\ 0.2m - 1.5n = 6.2 \end{cases}$ $m = 1$ $n = -4$

Use a system of equations to solve each problem.

31. Theater Drama club members sold tickets to an afternoon children's program. They charged $1 for each child's ticket and $2 for each adult ticket. They sold $656 worth of tickets. If they sold a total of 400 tickets, how many tickets of each kind did they sell?   144 child, 256 adult

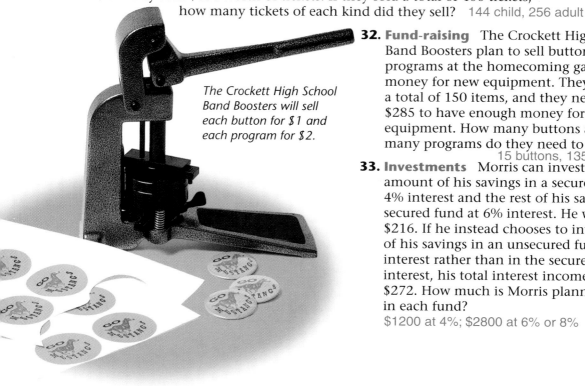

The Crockett High School Band Boosters will sell each button for $1 and each program for $2.

32. Fund-raising The Crockett High School Band Boosters plan to sell buttons and programs at the homecoming game to raise money for new equipment. They plan to sell a total of 150 items, and they need to make $285 to have enough money for the equipment. How many buttons and how many programs do they need to sell?
15 buttons, 135 programs

33. Investments Morris can invest a certain amount of his savings in a secured fund at 4% interest and the rest of his savings in a secured fund at 6% interest. He will earn $216. If he instead chooses to invest the rest of his savings in an unsecured fund at 8% interest rather than in the secured fund at 6% interest, his total interest income will be $272. How much is Morris planning to invest in each fund?
$1200 at 4%; $2800 at 6% or 8%

Look Back

Solve each equation for k.   **[Lesson 7.4]**

34. $d + h - k = 3l$
$k = d + h - 3l$

35. $kxh = r$
$k = \dfrac{r}{xh}$

36. $mk - 4t = 8s$
$k = \dfrac{8s + 4t}{m}$

37. $-k + m = x$
$k = m - x$

Solve each inequality.   **[Lesson 7.5]**

38. $2x - 5 < 8x - 4$
$x > -\dfrac{1}{6}$

39. $3(x - 4) \geq 4(x + 7) - 3$ $x \leq -37$

Solve each system of equations by two methods, graphing and substitution.   **[Lessons 9.1, 9.2]**

40. $\begin{cases} 2x - y = 6 \\ y = 3x - 4 \end{cases}$ $(-2, -10)$

41. $\begin{cases} 2x + 3y = 7 \\ x + y = 2 \end{cases}$ $(-1, 3)$

42. $\begin{cases} y = -5x + 9 \\ 3x - 4y = 10 \end{cases}$ $(2, -1)$

Look Beyond

43. A rectangular box has a base with a width of 5 inches and a length of 11 inches. The height of the box is 6 inches. What is the area of each rectangle that forms a side of the box?

44. Graph the equation $y = 2x - 3$. Use your graph to find a point that satisfies the inequality $y < 2x - 3$.

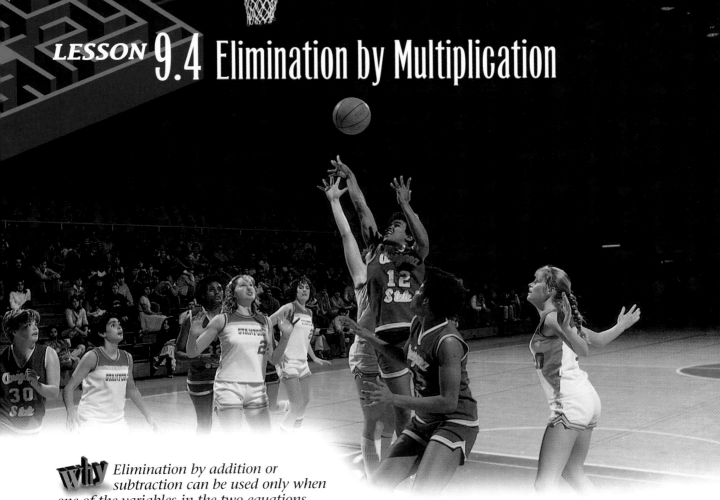

Elimination by Multiplication

why *Elimination by addition or subtraction can be used only when one of the variables in the two equations has the same or opposite coefficient. When no such coefficient exists, you can use the elimination-by-multiplication method of solving systems.*

Lana made 4 baskets for a total of 11 points in the first basketball game.

Using Multiplication in One Equation

Lana did not make any 1-point goals, but she made several 2-point and 3-point goals. How many goals of each type did Lana make? In the following exploration, you will write and solve a system of equations to find out the answer.

•Exploration Solving Systems Using Multiplication

1 What is the total number of goals that Lana made? Let x represent the number of 2-point goals, and let y represent the number of 3-point goals. Write an equation to describe this situation. $4; x + y = 4$

2 How many total points did Lana score? Write an equation using the variables x and y to describe this situation. $11; 2x + 3y = 11$

3 Model each equation using tiles.

4 Combine your models from Step 3 as you did in Lesson 9.3. Was one variable eliminated? Why or why not? No; no neutral pairs were formed.

5 Model the equations again, and subtract the terms of the second equation from the terms of the first equation. Was one variable eliminated? Why or why not?

6 How can you use multiplication to change one of the equations so that addition or subtraction will eliminate one of the variables? Multiply each side of the first equation by -2, as shown below. Simplify this equation, and model the equation with tiles.

$$-2(x + y) = -2(4)$$

7 Combine the model of the new first equation with the model of the original second equation. What variable is eliminated?

8. (1, 3) Lana made 1 2-point goal and 3 3-point goals.

8 Solve the system, and check your solution. How many 2-point and 3-point goals did Lana make?

9 Explain why the first equation was multiplied by -2. Could you have eliminated a variable by adding if you had multiplied both sides of the first equation by 2 instead of -2? Explain. ❖

Graphics Calculator

You can use a graphics calculator to check your solution to a system of equations by graphing. To check your solution in the exploration, first solve each equation in the system for y, and then graph them. If the point of intersection is (1, 3), then the solution is (1, 3).

Intersection
X=1   Y=3

What multiplication can you perform in the following system of equations to make elimination by addition possible?

Multiply the expressions on each side of $9x + 2y = 4$ by 2.

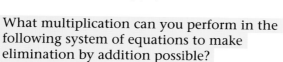
$$\begin{cases} 9x + 2y = 4 \\ 2x - 4y = 1 \end{cases}$$

EXAMPLE 1

Chemistry A chemist will mix the two salt solutions shown at the right. She wants to make 42 ounces of a new solution that contains 11 ounces of salt. How many ounces of each solution should she use?

20%
Salt Solution

Solution ➤

Let f represent the number of ounces of the first solution. Let s represent the number of ounces of the second solution. The total number of ounces must be 42.

$$f + s = 42$$

The first solution has 20% salt, so an expression for the amount of salt in the first solution is $0.2f$. The second solution has 30% salt, so an expression for the amount of salt in the second solution is $0.3s$. The total amount of salt in the combination is to be 11 ounces.

$$0.2f + 0.3s = 11$$

The system to solve for f and s is shown.
$$\begin{cases} f + s = 42 \\ 0.2f + 0.3s = 11 \end{cases}$$

Notice that the variables in the two equations have different coefficients. Multiply the first equation by -0.2 to eliminate f by adding.

$$\begin{cases} -0.2(f + s) = -0.2(42) \\ 0.2f + 0.3s = 11 \end{cases} \Rightarrow \begin{cases} -0.2f - 0.2s = -8.4 \\ 0.2f + 0.3s = 11 \end{cases}$$

Add the terms on corresponding sides of the equations to eliminate f. Then solve for s.

$$
\begin{aligned}
-0.2f - 0.2s &= -8.4 \\
\underline{0.2f + 0.3s} &= \underline{11} \\
0.1s &= 2.6 \\
\frac{0.1s}{0.1} &= \frac{2.6}{0.1} \\
s &= 26
\end{aligned}
$$

   Addition Property of Equality
   Simplify.

   Division Property of Equality

   Simplify.

Substitute 26 for s in the first original equation, and solve for f.

$$
\begin{aligned}
f + s &= 42 \\
f + 26 &= 42 \\
f + 26 - 26 &= 42 - 26 \\
f &= 16
\end{aligned}
$$

   Given
   Substitution Property
   Subtraction Property of Equality
   Simplify.

So the solution, (f, s), appears to be $(16, 26)$. Check the solution in both of the original equations.

$$
\begin{aligned}
f + s &= 42 \\
16 + 26 &\overset{?}{=} 42 \\
42 &= 42 \\
&\text{True}
\end{aligned}
\qquad
\begin{aligned}
0.2f + 0.3s &= 11 \\
0.2(16) + 0.3(26) &\overset{?}{=} 11 \\
3.2 + 7.8 &\overset{?}{=} 11 \\
11 &= 11 \\
&\text{True}
\end{aligned}
$$

Since f is 16, the chemist should use 16 ounces of the first solution (20% solution); and since s is 26, the chemist should use 26 ounces of the second solution (30% solution). ❖

Describe how to solve the system in Example 1 by eliminating s first. Would the solution be the same? Explain. Multiply $f + s = 42$ by -0.3, and add the result to $0.2f + 0.3s = 11$. Yes, both methods give the solution $(16, 26)$.

Using Multiplication in Both Equations

Sometimes you need to use multiplication in both equations in order to eliminate a variable.

EXAMPLE 2

Solve this system of equations using elimination by multiplication and addition. $\begin{cases} 2z + 5w = 9 \\ 3z + 4w = 6.5 \end{cases}$

Solution ➤

To eliminate the z, multiply the first equation by 3 and the second equation by -2.

$$\begin{cases} 3(2z + 5w) = 3(9) \\ -2(3z + 4w) = -2(6.5) \end{cases} \Rightarrow \begin{cases} 6z + 15w = 27 \\ -6z - 8w = -13 \end{cases}$$

Add the terms on corresponding sides of the equations to eliminate z. Then solve for w.

$$
\begin{array}{rl}
6z + 15w = & 27 \\
-6z - 8w = & -13 \\
\hline
7w = & 14 \\
\dfrac{7w}{7} = & \dfrac{14}{7} \\
w = & 2
\end{array}
$$

Addition Property of Equality
Simplify.
Division Property of Equality
Simplify.

Substitute 2 for w in the first *original* equation, and solve for z.

$$
\begin{aligned}
2z + 5w &= 9 \\
2z + 5(2) &= 9 \\
2z + 10 &= 9 \\
2z + 10 - 10 &= 9 - 10 \\
2z &= -1 \\
\dfrac{2z}{2} &= \dfrac{-1}{2} \\
z &= -\dfrac{1}{2}
\end{aligned}
$$

Check the values by substituting in both of the original equations. ❖

Describe how to solve the system of equations in Example 2 by eliminating w first. Multiply $2z + 5w = 9$ by 4, and multiply $3z + 4w = 6.5$ by -5. Then add.

Try This Solve this system of equations using elimination by multiplication and addition. $\begin{cases} 6a + 7c = 9 \\ 4a = 3c - 17 \end{cases}$ $a = -2, c = 3$

EXERCISES & PROBLEMS

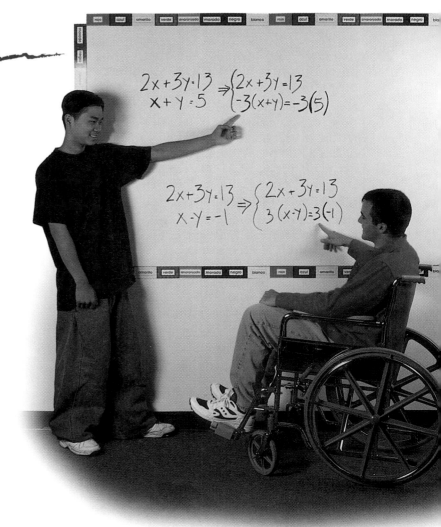

Communicate

1. When do you need to use multiplication to solve a system of equations by elimination?

2. Explain how to decide whether to multiply by a positive or a negative number when using multiplication and addition to solve a system of equations.

3. Describe how to use algebra tiles to solve this system of equations.

$$\begin{cases} 3x + 2y = 8 \\ 2x - y = 3 \end{cases}$$

Explain how you would use elimination to solve each system.

4. $\begin{cases} 5x - 4y = 1 \\ x + 2y = 8 \end{cases}$

5. $\begin{cases} 3x - y = 10 \\ 5x - 7y = 6 \end{cases}$

6. $\begin{cases} 1.6x - 2.4y = 2 \\ 0.8x - 1.2y = 7 \end{cases}$

Practice & Apply

Use algebra tiles to solve each system of equations.

7. $\begin{cases} x + 2y = 7 \\ 2x - y = 4 \end{cases}$ $(3, 2)$

8. $\begin{cases} 2x + y = 5 \\ 3x + 2y = 4 \end{cases}$ $(6, -7)$

9. $\begin{cases} 5x - 4y = 6 \\ 3x + y = 7 \end{cases}$ $(2, 1)$

10. $\begin{cases} 20x + 9y = 10 \\ 5x + 2y = 3 \end{cases}$ $\left(\frac{7}{5}, -2\right)$

11. $\begin{cases} 4x + 5y = 7 \\ -6x + 3y = 4 \end{cases}$ $\left(\frac{1}{42}, \frac{29}{21}\right)$

12. $\begin{cases} 2a - 4c = 5 \\ a - 3c = 7 \end{cases}$ $a = -6\frac{1}{2}$ $c = -4\frac{1}{2}$

Solve each system using elimination by multiplication. Check your solution either by graphing or by substitution.

13. $\begin{cases} 2x + 3y = 6 \\ 4x + y = 2 \end{cases}$ (0, 2) **14.** $\begin{cases} 3x + y = 2 \\ x - 2y = 10 \end{cases}$ (2, −4) **15.** $\begin{cases} y = 2x + 3 \\ 6x - 12y = 6 \end{cases}$ $\left(-2\frac{1}{3}, -1\frac{2}{3}\right)$

16. $\begin{cases} 2a - 4c = 5 \\ 2a - 2c = 7 \end{cases}$ $\begin{array}{l} a = 4\frac{1}{2} \\ c = 1 \end{array}$ **17.** $\begin{cases} 5x - 4y = 1 \\ x + 2y = 8 \end{cases}$ $\left(2\frac{3}{7}, 2\frac{11}{14}\right)$ **18.** $\begin{cases} 2x + y = 9 \\ 3x - 4y = 8 \end{cases}$ (4, 1) **20.** $\left(\frac{1}{24}, \frac{13}{16}\right)$

19. $\begin{cases} 3x - 5y = 11 \\ 2x - 3y = 1 \end{cases}$ (−28, −19) **20.** $\begin{cases} 9x + 2y = 2 \\ -21x + 6y = 4 \end{cases}$ **21.** $\begin{cases} 1.6x - 2.4y = 2 \\ 0.2x - 2.4y = 9 \end{cases}$ $\left(-5, -4\frac{1}{6}\right)$

22. Chemistry A solution that is 90% glucose is to be mixed with a solution that is 40% glucose to make a 4-liter mixture that has 2.1 liters of glucose. How many liters of each solution should be used? 1 liter of 90%, 3 liters of 40%

One 8-serving jug of the house blend at Adrian's Juice Bar sells for $8.50.

23. Chemistry To obtain 40 ounces of a 40% antifreeze solution, pure antifreeze is added to a 25% antifreeze solution. How much pure antifreeze should be added? 8 oz pure, 32 oz of 25%

24. Sales Adrian's Juice Bar sells fresh-squeezed juices. The house blend contains a mixture of cranberry juice, which sells for $1.20 per serving, and apple juice, which sells for $1.00 per serving. How many servings of each juice are used to make the house blend?

2.5 serv cranberry
5.5 serv apple

25. Sales The owner of a health-food store sells a mixture of seeds and nuts. The seeds cost 20¢ per ounce, and the nuts cost 30¢ per ounce. Forty-two ounces of the mixture has a value of $11.00. How many ounces of seeds and nuts did the owner use in the mix? 16 oz seeds, 26 oz nuts

26. Chemistry A 10% salt solution and an 18% salt solution are mixed to obtain 320 grams of a 15% salt solution. How many grams of the 10% solution and the 18% solution are needed? 120 g of 10%, 200 g of 18%

Solve each system of equations by elimination. Check your answers.

27. $\begin{cases} 6m - 3n = -12 \\ 5m + 3n = 1 \end{cases}$ **28.** $\begin{cases} 3a + 4c = 5 \\ 5a + 2c = 13 \end{cases}$ **29.** $\begin{cases} 4w + 5z = 7 \\ 2w + z = -1 \end{cases}$ **30.** $\begin{cases} 7x = 4y + 22 \\ 3x + 2y = 2 \end{cases}$

31. $\begin{cases} 5w + 7z = 10 \\ w + z = 4 \end{cases}$ **32.** $\begin{cases} 2a - 3c = 9 \\ a - c = 7 \end{cases}$ **33.** $\begin{cases} 3m = 8 + n \\ 10m - 4n = 7 \end{cases}$ **34.** $\begin{cases} 5a - 3c = 13 \\ 3a + 2c = 4 \end{cases}$

35. $\begin{cases} 3x + 4y = 6 \\ 5x + 3y = -1 \end{cases}$ **36.** $\begin{cases} 2x - y = 6 \\ 4x - 5y = 2 \end{cases}$ **37.** $\begin{cases} 3x - 5y = 7 \\ 4x - 3y = 2 \end{cases}$ **38.** $\begin{cases} 2y = 3x + 1 \\ -4y = 2x + 1 \end{cases}$

39. $\begin{cases} y = 2x - 8 \\ x - 5y = 9 \end{cases}$ **40.** $\begin{cases} 3g - 2h = 4 \\ 5h = -3g + 2 \end{cases}$ **41.** $\begin{cases} 0.5x - 2y = 9 \\ -\frac{1}{4}x - \frac{1}{2}y = 2 \end{cases}$ **42.** $\begin{cases} 5.2w + 0.5c = -19.3 \\ 3.1w + 2.5c = -4.9 \end{cases}$

Use a system of equations to solve each problem. Use any method to solve.

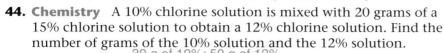

John is 56 years younger than his grandfather.

43. Age The sum of John's age and his grandfather's age is 78. Find their ages. *John: 11 yr; grandfather: 67 yr*

44. Chemistry A 10% chlorine solution is mixed with 20 grams of a 15% chlorine solution to obtain a 12% chlorine solution. Find the number of grams of the 10% solution and the 12% solution. *30 g of 10%; 50 g of 12%*

45. Chemistry A student in a chemistry class said, "When a chemist mixes a 40% solution with a 30% solution, the result is a 35% solution." Give an example to explain why this statement may not be true.

46. Investments Becky invests her savings of $3000 in two separate accounts. One account yields 5% interest and the second account yields 8% interest. If her money earns $202.38 interest, how much did she invest in each account? *$1254 at 5%, $1746 at 8%*

47. Sports Brandi played 4 fewer innings in a softball game than Denise. Twice the number of innings that Brandi played plus 2 equals the number of innings that Denise played. How many innings did each girl play? *Brandi: 2; Denise: 6*

48. 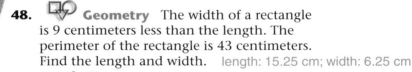 **Geometry** The width of a rectangle is 9 centimeters less than the length. The perimeter of the rectangle is 43 centimeters. Find the length and width. *length: 15.25 cm; width: 6.25 cm*

49. 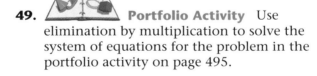 **Portfolio Activity** Use elimination by multiplication to solve the system of equations for the problem in the portfolio activity on page 495.

Look Back

Solve for *t*. [Lessons 7.2, 7.4]

50. $5t + 7t = 9$ $t = \frac{3}{4}$ **51.** $8t - 9 = 2t + 1$ $t = \frac{5}{3}$ or $1\frac{2}{3}$

52. $\frac{1}{5}t + 1 = 7(t + 1)$ $t = -\frac{15}{17}$

53. $3t + 5k = r$ $t = \frac{r - 5k}{3}$

Solve for *x*. [Lesson 7.6]

54. $|x + 7| = 8$ $x = -15$ or $x = 1$

55. $|x - 8| \le 4$ $4 \le x \le 12$

56. $|x - 2| > 10$ $x < -8$ or $x > 12$

Look Beyond

57. Make a table of values for the equation $y = x^2$. Graph the set of points. Make a table of values for the equation $y = 2^x$. Graph the set of points on the same coordinate plane.

58. Examine the graphs from Exercise 57. How are the graphs of the two equations alike? How are they different?

Systems of Linear Inequalities

Applications of algebra often require several conditions to be met in order to solve the problem. Systems of inequalities can often be used to solve such problems.

Libby is making a window frame for etched glass. The frame will be for a window that is square on the bottom with an isosceles triangle on top. The perimeter of the window must be no more than 15 feet. What are some possible dimensions of the window?

Exploration Solving a System of Inequalities

1 Let x represent the length of each side of the square, and let y represent the length of one of the two congruent sides of the isosceles triangle.

2 The perimeter of the window must be no more than 15 feet. Write an inequality for the perimeter using the variables x and y. $3x + 2y \leq 15$

3 The sum of two sides of a triangle is always greater than the third side. Use this fact to write the second inequality. $2y > x$

4 Write the system of inequalities. Solve each inequality for y.

Recall that you can use a graph to find the solutions. All of the points that lie in the solution region of both inequalities are in the solution of the system. Remember that the dimensions of the window must be positive, so the reasonable domain and range are contained in the first quadrant.

5 Graph the boundary lines. Is one or both of the boundary lines solid or dashed? Why?

6 Notice that the boundary lines divide the first quadrant into four regions. Points A, B, C, and D are each placed in one of these four regions. In the following table, the coordinates of these points are given. Substitute the coordinates of each point in both inequalities, and complete the table.

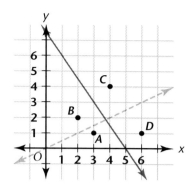

Point	$3x + 2y \leq 15$ Is the inequality true?	$2y > x$ Is the inequality true?	Are both inequalities true?
$A(3, 1)$	$3(3) + 2(1) \leq 15$; Yes	$2(1) > 3$; No	No
$B(2, 2)$	?	?	?
$C(4, 4)$	?	?	?
$D(6, 1)$	?	?	?

7 On your graph, shade the region containing the point that makes both inequalities true. Do either of the boundary lines contain points that are solutions to this system of linear inequalities? Explain. ❖

Adult admission to the Natural Science Exploratorium is $10 per day, and student admission is $5 per day.

EXAMPLE 1

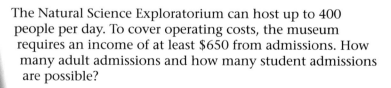

The Natural Science Exploratorium can host up to 400 people per day. To cover operating costs, the museum requires an income of at least $650 from admissions. How many adult admissions and how many student admissions are possible?

Solution ➤

Let x represent the number of adult tickets, and let y represent the number of student tickets. The maximum number of tickets that can be sold is 400.

$$x + y \leq 400$$

Adult tickets cost $10, and student tickets cost $5. At least $650 worth of tickets must be sold.

$$10x + 5y \geq 650$$

Write the system of inequalities. $\begin{cases} x + y \leq 400 \\ 10x + 5y \geq 650 \end{cases}$

Solve each inequality in the system for y.

$$x + y \leq 400 \qquad\qquad 10x + 5y \geq 650$$
$$x + y - x \leq 400 - x \qquad 10x + 5y - 10x \geq 650 - 10x$$
$$y \leq 400 - x \qquad\qquad 5y \geq -10x + 650$$
$$y \leq -x + 400 \qquad\qquad \frac{5y}{5} \geq \frac{-10x}{5} + \frac{650}{5}$$
$$y \geq -2x + 130$$

Graph the boundary lines,
$y = -x + 400$ and $y = -2x + 130$.

Since x represents the number of adult tickets and y represents the number of student tickets, values of x (the domain) and values of y (the range) cannot be negative.

The boundary lines divide the first quadrant into three regions. Select a point from each region, and substitute its coordinates in both of the original inequalities.

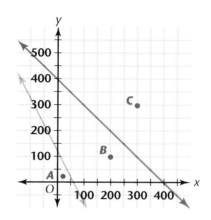

Point	$x + y \leq 400$ True or false?	$10x + 5y \geq 650$ True or false?	Both true?
$A(25, 25)$	$25 + 25 \leq 400$; True	$10(25) + 5(25) \geq 650$; False	No
$B(200, 100)$	$200 + 100 \leq 400$; True	$10(200) + 5(100) \geq 650$; True	Yes
$C(300, 300)$	$300 + 300 \leq 400$; False	$10(300) + 5(300) \geq 650$; True	No

Therefore, the solution region is the one containing point B.

Any point with whole-number coordinates in the shaded region will satisfy the system of inequalities
$$\begin{cases} x + y \leq 400 \\ 10x + 5y \geq 650 \end{cases}$$
and will satisfy the conditions of the problem.

For example, the point (200, 100) represents 200 adult tickets and 100 student tickets. This is a total of 300 tickets, which is less than the 400 person capacity, and which will bring in $2500, greater than the minimum amount that the museum needs. ❖

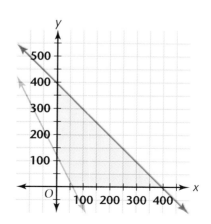

Try This The math booster club is raising money by selling gift-wrapping paper for $4 per roll and greeting cards for $5 per box. Each member has been asked to sell at least $100 worth of items. The goal is that each member will not have to sell more than 30 items. Write and graph a system of inequalities to describe this situation. Determine the reasonable domain and range for this real-world situation.

CRITICAL
Thinking

Examine the graph at the right. What is the system of inequalities graphed?

$$\begin{cases} y \le 2x - 4 \\ y \ge -3x + 3 \end{cases}$$

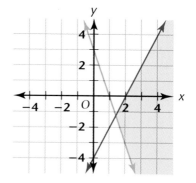

EXAMPLE 2

Graphics Calculator

A chemist will mix a solution that is 2.5% acid with a solution that is 4.8% acid. The total mixture must be at least 45 milliliters and must contain no more than 18 milliliters of acid. How much of each solution will need to be mixed? Write a system of inequalities for this situation. Then use a graphics calculator to graph the boundary lines and to find at least four possible solutions.

Solution ➤

Let x represent the amount of 2.5% solution, and let y represent the amount of 4.8% solution. The total mixture must be at least 45 milliliters.

$$x + y \ge 45$$

Since 2.5% = 0.025, an expression for the amount of acid in the 2.5% solution is 0.025x. Since 4.8% = 0.048, an expression for the amount of acid in the 4.8% solution is 0.048y. The amount of acid in the mixture must not contain more than 18 milliliters.

$$0.025x + 0.048y \le 18$$

Solve each inequality in the system for y. $\begin{cases} x + y \ge 45 \\ 0.025x + 0.048y \le 18 \end{cases}$

$$x + y \ge 45 \qquad\qquad\qquad 0.025x + 0.048y \le 18$$
$$x + y - x \ge 45 - x \qquad 0.025x + 0.048y - 0.025x \le 18 - 0.025x$$
$$y \ge -x + 45 \qquad\qquad\qquad\quad 0.048y \le -0.025x + 18$$
$$\frac{0.048y}{0.048} \le \frac{-0.025x}{0.048} + \frac{18}{0.048}$$
$$y \le \frac{-0.025x}{0.048} + \frac{18}{0.048}$$

Lesson 9.5   Systems of Linear Inequalities   **527**

Graph the boundary lines. Select a point in each of the four regions created by the lines. Test each point in both original inequalities to determine the solution region.

Graphics Calculator

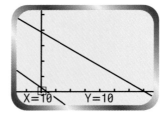

X=10   Y=10

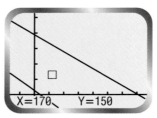

X=170   Y=150

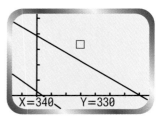

X=340   Y=330

The coordinates of the point (170, 150) make both inequalities true, so it is in the solution region. Therefore, one possible solution is to mix 170 milliliters of the 2.5% solution with 150 milliliters of the 4.8% solution.

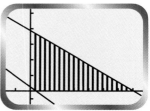

Since x and y represent amounts of a solution, the reasonable domain and range for this situation cannot include negative numbers. The solution region for this problem is contained only in the first quadrant. Any point in the solution region will give the amounts of each solution needed to make the required mixture. ❖

The reasonable domain for the problem in Example 2 is all real numbers between 0 and 720, or $0 < x < 720$. Explain why this reasonable domain does not include 0 and 720.

The reasonable domain for the problem in Example 2 does not include 0 or 720 because neither solution can be used alone to make the mixture.

EXERCISES & PROBLEMS

Communicate

1. How can you determine the region that satisfies two inequalities?

2. What words in a problem indicate that the $\leq$ sign needs to be used?

3. What words in a problem indicate that the $\geq$ sign needs to be used?

4. Describe a situation in which some, but not all, of the points in the solution to a system of inequalities would solve a problem.

5. How can you use a graphics calculator to help you solve a system of linear inequalities?

Practice & Apply

Determine which of the given points are solutions to the system of inequalities.

6. $\begin{cases} 2x - 3y > 10 \\ x + 4y < 6 \end{cases}$ b   **a.** $(2, -1)$ **b.** $(4, -7)$ **c.** $(-3, 5)$

7. $\begin{cases} x + 2y \le 12 \\ x - y > 5 \end{cases}$ b   **a.** $(4, 1)$ **b.** $(6, -5)$ **c.** $(-10, 6)$

8. $\begin{cases} 18x - 12y > 80 \\ 12x - 11y > 12 \end{cases}$ a, b   **a.** $(18, -20)$ **b.** $(15, -10)$ **c.** $(-8, 30)$

9. $\begin{cases} 16x - 24y > 62 \\ 21x + 17y < 48 \end{cases}$ none   **a.** $(17, -18)$ **b.** $(-14, 16)$ **c.** $(22, -11)$

Determine each system of inequalities graphed.

10. $\begin{cases} y \le x + 4 \\ y \ge x - 6 \end{cases}$

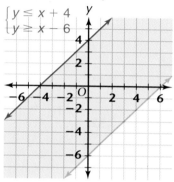

11. $\begin{cases} y < -\frac{1}{2}x + 3 \\ y > 2x + 3 \end{cases}$

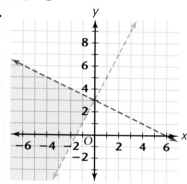

12. $\begin{cases} y \le 4 \\ y > -x - 3 \end{cases}$

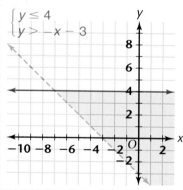

13. $\begin{cases} y \le -\frac{3}{5}x + 3 \\ y > \frac{2}{3}x - 3\frac{1}{3} \end{cases}$

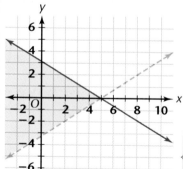

14. Sales At a pet store, you can buy your own mix of fish-tank gravel. How much of each type of gravel could you use to create a mixture that weighs no more than 10 pounds and costs at least $7.50? Write a system of inequalities, and find at least two possible answers to the problem. Then determine the reasonable domain and range for this system of inequalities.

Large-sized gravel costs $1.75 per pound, and small-sized gravel costs $1.25 per pound.

Write a system of inequalities to describe the situation in each problem. Graph the system, and determine the reasonable domain and range for the situation.

15. Crafts Janice makes home decorations. It takes her 2 hours to make a wreath and 1 hour to make a basket. She can work no more than 40 hours per week. The cost to make one wreath is $3 and the cost to make one basket is $2. She can afford to spend no more than $72 per week. How many wreaths and baskets can she make each week?

16. Chemistry A salt solution is to be made with one solution that is 22% salt and another solution that is 30% salt. The solution must be at least 48 ounces and must contain at least 12 ounces of salt. How many ounces of each solution should be used?

Graph each system of inequalities.

17. $\begin{cases} y > 2 \\ y > 3x - 2 \end{cases}$

18. $\begin{cases} y < 1 \\ y \geq \frac{2}{3}x + 1 \end{cases}$

19. $\begin{cases} x < 2 \\ y \leq -\frac{1}{2}x + 3 \end{cases}$

20. $\begin{cases} x + 2y \leq 6 \\ 2x + 4y \geq 4 \end{cases}$

21. $\begin{cases} y + 2x > 3 \\ 3x - 4y < 8 \end{cases}$

22. $\begin{cases} 2x + y < 3 \\ 3x - 2y \geq 4 \end{cases}$

23. $\begin{cases} 4x + 3y \leq 6 \\ x - 2y \geq 4 \end{cases}$

24. $\begin{cases} y \leq 3x + 4 \\ 2x + y \geq 1 \end{cases}$

25. $\begin{cases} y - 7 \leq 2x \\ 3y + 6 \leq 6x \end{cases}$

26. $\begin{cases} y + 2 > 3x \\ 3x - y < 7 \end{cases}$

27. $\begin{cases} 3x + y < 3 \\ 6x + 2y \geq 12 \end{cases}$

28. $\begin{cases} x + y < 4 \\ x - y < -4 \end{cases}$

29. $\begin{cases} y \geq -2 \\ x < -3 \end{cases}$

30. $\begin{cases} 6x + 5y \leq 10 \\ x + 4y > 5 \\ y < 2 \end{cases}$

31. $\begin{cases} y \leq 5 \\ x + y \geq -2 \\ x + y < 8 \end{cases}$

32. $\begin{cases} 2x > 6 \\ x - 5y < 15 \\ 2x + 3y > 6 \end{cases}$

Look Back

Evaluate each expression.   [Lesson 1.7]

33. $4 \cdot 2 + 3 - 6$ 5

34. $9 - 6 \div 2 + 7$ 13

35. $10 \div 5 \cdot 3 + 6$ 12

Use the substitution method to solve each system of equations. Use two different methods to check each solution.   [Lesson 9.2]

36. $\begin{cases} y = 3x - 4 \\ 2x - y = 12 \end{cases}$ $(-8, -28)$

37. $\begin{cases} x - y = 14 \\ 2x + 3y = 12 \end{cases}$ $(10.8, -3.2)$

Use elimination to solve each system of equations. Check each solution using two different methods.   [Lessons 9.3, 9.4]

38. $\begin{cases} 4a - 3c = 12 \\ 2a + c = 10 \end{cases}$ $a = 4.2, c = 1.6$

39. $\begin{cases} 0.5m + 0.3n = 1.4 \\ 0.7m + 0.3n = 2.8 \end{cases}$ $m = 7, n = -7$

Look Beyond

40. Graph this system of inequalities. What is the shape of the solution region? Find the vertices of the solution region. $\begin{cases} 2x + y \geq 1 \\ x - 2y > 1 \\ x < 6 \end{cases}$

CHAPTER 9 PROJECT How Many Pages?

The yearbook staff is deciding how many color pages to include in the yearbook. The yearbook needs to have at least 72 pages. From previous experience, the yearbook staff know that the number of black-and-white pages must be no more than twice the number of color pages and that there can be no more than 40 color pages in the yearbook.

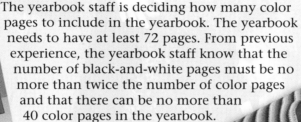

The printer charges $200 for each color page and $150 for each black-and-white page. How many black-and-white pages and how many color pages should be included to minimize the cost of producing the year book?

ACTIVITY

Write a system of inequalities, and find the solution region, which contains all of the possible answers to this problem. Find the coordinates of the vertices of this region by finding the coordinates of the points of intersection of the three boundary lines.

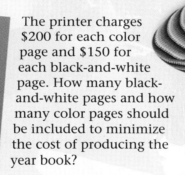

Write the equation to describe the total cost using the variables x and y. The cost equation is called the **optimization equation.**

The cost is equal to $200 per color page plus $150 per black-and-white page.

Substitute the values of x and y at each of the vertices of the solution region into your cost equation to find the costs for these values. Which of these costs is the minimum? Can you find any other point in the solution region that results in a lower cost? Solve the following problem: How many black-and-white pages and how many color pages should be included to minimize the production cost?

Chapter 9 Review

Vocabulary

Key Skills & Exercises

Lesson 9.1

➤ Key Skills

Solve a system of linear equations by graphing.

To solve the system $\begin{cases} 2x + y = 1 \\ x - y = -4 \end{cases}$ by graphing, solve each equation for y. Then graph both equations on the same coordinate plane.

The two lines intersect at the point $(-1, 3)$. Since $x = -1$ and $y = 3$ satisfies both linear equations, the solution is $(-1, 3)$.

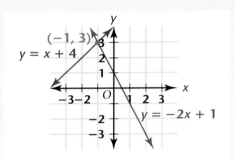

➤ Exercises

Solve each system of linear equations by graphing. If necessary, give an approximate solution.

1. $\begin{cases} 2x - y = 3 \\ x + 2y = 4 \end{cases}$ (2, 1) 2. $\begin{cases} 4x + y = 5 \\ 3x - y = 9 \end{cases}$ (2, −3)

3. $\begin{cases} y = -2x \\ x - y = -6 \end{cases}$ (−2, 4) 4. $\begin{cases} x + y = 4 \\ 2x - 2y = 10 \end{cases}$ (4.5, −0.5)

Lesson 9.2

➤ Key Skills

Use the substitution method to solve a system of linear equations.

To solve the system $\begin{cases} 3x - 7y = 5 \\ 2x - y = 7 \end{cases}$ using the substitution method, first solve the equation $2x - y = 7$ for y to get $y = 2x - 7$.

Next, substitute $2x - 7$ for y in the equation $3x - 7y = 5$. Solve for x.

$$3x - 7y = 5$$
$$3x - 7(2x - 7) = 5$$
$$3x - 14x + 49 = 5$$
$$-11x + 49 - 49 = 5 - 49$$
$$-11x = -44$$
$$\frac{-11x}{-11} = \frac{-44}{-11}$$
$$x = 4$$

Then substitute 4 for x in either original equation.

$$y = 2x - 7$$
$$y = 2(4) - 7$$
$$y = 8 - 7$$
$$y = 1$$

The solution is (4, 1).

Check by substituting 4 for x and 1 for y in both of the original equations.

➤ **Exercises**

Use the substitution method to solve each system of linear equations. Check.

5. $\begin{cases} y = 3x \\ x + y = 12 \end{cases}$ (3, 9)

6. $\begin{cases} y = 2x - 6 \\ x - y = 4 \end{cases}$ (2, −2)

7. $\begin{cases} 3x + 2y = 14 \\ 4 - 2x = y \end{cases}$ (−6, 16)

8. $\begin{cases} x - y = 16 \\ x + y = 74 \end{cases}$

(45, 29)

Lesson 9.3

➤ **Key Skills**

Use elimination by addition to solve a system of linear equations.

To solve the system $\begin{cases} 3x - 2y = 2 \\ 7x + 2y = 18 \end{cases}$, use the Addition Property of Equality to add the expressions on each side of the equal signs.

$$(3x - 2y) + (7x + 2y) = 2 + 18$$
$$10x = 20$$
$$x = 2$$

Next, substitute 2 for x in either of the two original equations.

$$7x + 2y = 18$$
$$7(2) + 2y = 18$$
$$14 + 2y - 14 = 18 - 14$$
$$2y = 4$$
$$y = 2$$

The solution is (2, 2).

Check by substituting 2 for x and 2 for y in both of the original equations.

➤ **Exercises**

Solve each system of linear equations using elimination by addition. Check.

9. $\begin{cases} x + y = 6 \\ x - y = 12 \end{cases}$ (9, −3)

10. $\begin{cases} 5x - 2y = 30 \\ x + 2y = 6 \end{cases}$ (6, 0)

11. $\begin{cases} 4x - 3y = 14 \\ 5x + 3y = 31 \end{cases}$ (5, 2)

12. $\begin{cases} 4x - 5y = 44 \\ 4x + 9y = -12 \end{cases}$ (6, −4)

Lesson 9.4

➤ Key Skills

Use elimination by multiplication to solve a system of linear equations.

To solve the system $\begin{cases} x + y = 4 \\ 2x + 3y = 9 \end{cases}$, multiply each side of the first

equation by -3 to get $\begin{cases} -3x - 3y = -12 \\ 2x + 3y = 9 \end{cases}$.

Then use the Addition Property of Equality to add the expressions on each side of the equal signs.

$$\begin{array}{r} -3x - 3y = -12 \\ 2x + 3y = 9 \\ \hline -x = -3 \\ x = 3 \end{array}$$

To find the value of y, substitute 3 for x in either of the original equations.

$$\begin{array}{r} x + y = 4 \\ (3) + y = 4 \\ y = 1 \end{array}$$

The solution is (3, 1). Check your solution.

➤ Exercises

Solve and check each system of linear equations using elimination by multiplication. Check.

13. $\begin{cases} 4x - 3y = 1 \\ 2x + y = 3 \end{cases}$
(1, 1)

14. $\begin{cases} 5x - y = 14 \\ 3x + 3y = 3 \end{cases}$
(2.5, −1.5)

15. $\begin{cases} 3x + 2y = 13 \\ 2x + 5y = 16 \end{cases}$
(3, 2)

16. $\begin{cases} -3x - 3y = 6 \\ x - 2y = 7 \end{cases}$
(1, −3)

Lesson 9.5

➤ Key Skills

Graph a system of linear inequalities.

To graph the system $\begin{cases} 3x - 5y < 10 \\ 2x + 3y \leq 3 \end{cases}$, first graph the
two boundary lines. The boundary line for
$3x - 5y < 10$ is a dashed line to indicate that the
boundary line is *not* part of the graphed region.
The boundary line for $2x + 3y \leq 3$ is a solid line to
indicate that the boundary line is part of the
graphed region. Test a point, such as (0, 0), in both
inequalities.

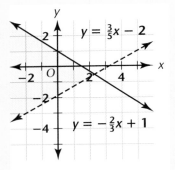

$3(0) - 5(0) < 10$ (True) $2(0) + 3(0) \leq 3$ (True)

Shade the region that includes (0, 0).

➤ Exercises

Graph each system of linear inequalities.

17. $\begin{cases} x < 4 \\ y \geq 3 \end{cases}$

18. $\begin{cases} y > x + 4 \\ x \leq 1 \end{cases}$

19. $\begin{cases} y \leq 2x - 4 \\ y > \frac{1}{2}x - 5 \end{cases}$

20. $\begin{cases} 3x - 4y \leq 12 \\ x - y > 1 \end{cases}$

Application

21. Sports The cost of an adult ticket to a football game is $5.50, and
the cost of a student ticket is $3.50. The receipts for a game attended
by 700 adults and students was $2850. What were the number of
students and the number of adults who attended the game? 200 adults, 500 students

Chapter 9 Assessment

Solve each system of linear equations by graphing.

1. $\begin{cases} 3x - 2y = 6 \\ x + y = 2 \end{cases}$ (2, 0)

2. $\begin{cases} x + 2y = 5 \\ 5x + 3y = -3 \end{cases}$ (−3, 4)

3. $\begin{cases} 3x - 2y = 4 \\ 2x + y = 5 \end{cases}$ (2, 1)

Use the substitution method to solve each system of linear equations. Check.

4. $\begin{cases} y = 2x \\ x + y = 12 \end{cases}$ (4, 8)

5. $\begin{cases} x + y = 15 \\ y = x + 3 \end{cases}$ (6, 9)

6. $\begin{cases} 2x + 2y = 4 \\ x - y = 6 \end{cases}$ (4, −2)

Solve each system of linear equations using elimination by addition. Check.

7. $\begin{cases} 2x + 9y = 24 \\ -2x + 5y = 4 \end{cases}$ (3, 2)

8. $\begin{cases} 3x + 2y = 13 \\ 5x - 2y = 11 \end{cases}$ (3, 2)

9. $\begin{cases} x + y = 32 \\ x - y = 4 \end{cases}$ (18, 14)

Solve each system of linear equations using elimination by multiplication. Check.

10. $\begin{cases} 5x - 2y = 3 \\ 2x + 7y = 9 \end{cases}$ (1, 1)

11. $\begin{cases} 5x + 2y = 9 \\ 3x + 7y = 17 \end{cases}$ (1, 2)

12. $\begin{cases} 10x - 5y = 50 \\ 6x + 2y = 28 \end{cases}$ (4.8, −0.4)

13. **Sales** The "Nuts 4 U" store has found that customers are willing to pay $21 for a package of nuts that contains 4 times as many almonds as pecans. Because of the popularity of the mixture, the store manager does not charge extra for the expense of preparing the mixture. Instead, the manager bases the price on what he would charge if the nuts were purchased separately: $4.50 per pound for the almonds and $3.00 per pound for the pecans. How many pounds of almonds and how many pounds of pecans are in a package of mixed nuts?   1 lb pecans, 4 lb almonds

14. Alex has 95 cents in dimes and nickels. The total number of coins is 1 more than twice the number of dimes. How many coins of each type does Alex have?   6 dimes, 7 nickels

Graph each system of linear inequalities.

15. $\begin{cases} x < 2 \\ y \geq -1 \end{cases}$

16. $\begin{cases} 3x - 5y < 10 \\ 2x + 3y \leq 3 \end{cases}$

17. $\begin{cases} 3x - 2y < 2 \\ 2y - 3x < 4 \end{cases}$

Fund-raising At the annual school fair for the scholarship fund, the senior class sold cookies for $6 per box and brownies for $8 per box. The goal for each senior was to sell at least $120 worth of cookies and brownies. However, it was hoped that no one would have to sell more than 50 boxes.

18. Write and graph a system of linear inequalities to describe this situation.

19. What are the reasonable domain and range for this situation?

20. Find at least three possible solutions to the system.

CHAPTER 10

A Preview of Functions

Many activities that people take for granted are rich in mathematics. Functions are particularly valuable because they let you see the way numerical relationships behave.

Linear functions appeared in Chapter 8. In this chapter, you will have the opportunity to examine other common functions that are part of the mathematician's toolbox. These functions include those that appear on standard scientific and graphics calculators. You will be able to study these functions in more detail in later chapters.

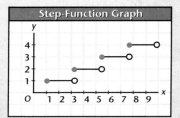

Step-Function Graph

Paula knows that the cost for postage depends on the weight of the package. A function can be used to show this relationship.

By moving his fingers along the neck of the guitar, Alex changes the length of the vibrating part of a guitar string. A function can be used to show how the number of vibrations per second depends on the length of the string.

Reciprocal-Function Graph

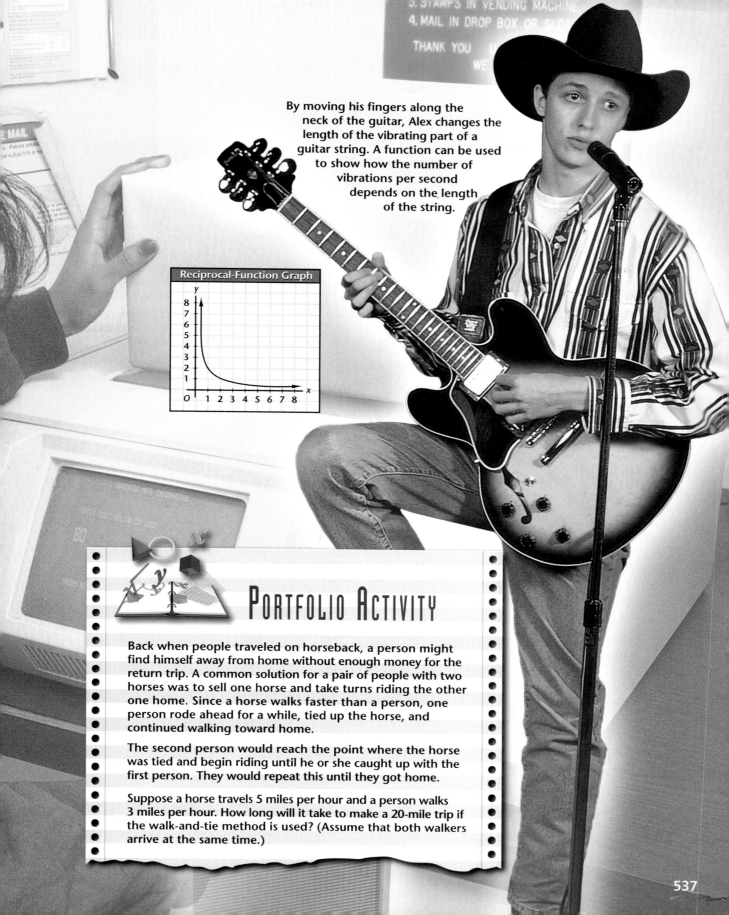

PORTFOLIO ACTIVITY

Back when people traveled on horseback, a person might find himself away from home without enough money for the return trip. A common solution for a pair of people with two horses was to sell one horse and take turns riding the other one home. Since a horse walks faster than a person, one person rode ahead for a while, tied up the horse, and continued walking toward home.

The second person would reach the point where the horse was tied and begin riding until he or she caught up with the first person. They would repeat this until they got home.

Suppose a horse travels 5 miles per hour and a person walks 3 miles per hour. How long will it take to make a 20-mile trip if the walk-and-tie method is used? (Assume that both walkers arrive at the same time.)

Previewing Exponential Functions

Prize A
Start with $100,
add $100 each
day, for 20 days!

Prize B
Start with 1¢, double
your money each
day, for 20 days!

why *What happens each time you double your money? An exponential function can be used to determine the total amount of money at any time during the doubling process.*

Students at Lincoln High have been playing the Math Know-How game. After the game, the winners will have 10 seconds to decide which prize to choose. They get to keep the money from the 20th day.

Exploration 1 *Evaluating Your Options*

Calculator

To help the winners decide, try this exploration with a partner. One person finds the amount of Prize A, and the other finds the amount of Prize B. If calculators are available, use them to perform repeated calculations.

Prize A
Start with 100. Repeatedly *add* 100 to the previous day's total.

Prize B
Start with 0.01. Repeatedly *multiply* the previous day's amount by 2.

	Prize A			Prize B
Day	Total Amount		Day	Total Amount
1	$100		1	$0.01
2	$200		2	$0.02
3	$300		3	$0.04
4	$400		4	$0.08

1 After four days, which prize would you choose? A

2 Do you think Prize B is ever greater than Prize A?
After 4 days, it doesn't seem so.

3 Continue each table through 20 days.

4 Is Prize B ever greater than Prize A? If so, when? Yes, after Day 19

5 Extend each table to 30 days. What is each prize worth? ❖

Introducing Exponential Functions

In Chapter 8 you worked with linear functions. In Example 1 below, you can see that not all functions are linear functions.

EXAMPLE 1

Spreadsheet

The population of Parksburg grew from 10,000 to 11,000 in one year. Calculate and graph the population growth for 10 years if

A the population grows by a *fixed amount* each year.

B the population grows at a *fixed rate* each year.

Solution ➤

Demographics

A The first year's growth is 1000 people. If the population grows by a *fixed amount*, it will gain 1000 people each year. You can use a spreadsheet to display this information. Start with 10,000 and repeatedly *add* 1000 to the previous number. The results appear in column B of the spreadsheet.

	A	B	C
	Year	Fixed amount	Fixed rate
1			
2	0	10,000	10,000
3	1	11,000	11,000
4	2	12,000	12,100
5	3	13,000	13,310
6	4	14,000	14,641
7	5	15,000	16,105
8	6	16,000	17,716
9	7	17,000	19,487
10	8	18,000	21,436
11	9	19,000	23,579
12	10	20,000	25,937

B The first year's growth is 1000 people. To figure the *fixed rate* of growth for each year, divide 1000 by 10,000.

$$fixed\ rate = \frac{first\ year\ change}{original\ population} = \frac{1000}{10,000} = 10\%$$

The new population after one year is 100% of the original population *plus* an additional 10%. Each year, the population is 110% of the previous year's population. Enter this on the spreadsheet by starting with 10,000 and repeatedly *multiplying* by 110%, or 1.1. The results appear in column C of the spreadsheet above.

The graph of each growth pattern over 40 years is shown below.

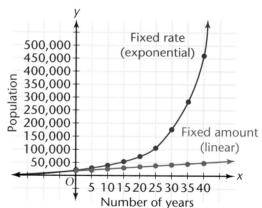

When population grows by a *fixed amount,* the points lie along a line. This relationship is **linear**. When the population grows at a *fixed rate,* the relationship is *exponential.* You can see from the table and from the graph how much faster the population grows at a fixed rate than by a fixed amount. The population growth at a fixed rate is an example of an **exponential function**.

Comparing Exponential Growth and Decay

When you found the population of Parksburg in Example 1 by repeatedly multiplying by 1.1, the population grew exponentially. The function modeled **exponential growth**.

EXAMPLE 2

Calculator

What would happen to Parksburg's population in Example 1 if you repeatedly multiplied it by 0.9 instead of by 1.1?

Solution ➤

If you use a calculator to perform repeated calculations, first enter 10,000, and then multiply by 0.9 repeatedly. Rounded to whole numbers, the outputs for every 10th entry are 3487, 1216, 424, 148, 52, . . . This output shows the decrease in population every 10 years. You can graph this output to see how fast the population decreases. The function models **exponential decay**.

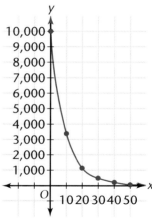

Compound Interest

Rising costs are making it necessary to save for years to cover college expenses. Compound interest accounts earn increasing amounts of interest over long periods of time.

Investments The money paid or earned on a given amount of money often involves **compound interest**. Compound interest is an example of exponential growth.

EXAMPLE 3

When Jill was eight years old, her parents invested $10,000 in a college tuition fund. The interest rate is 8% compounded once at the end of each year. How much will Jill have for tuition after 10 years?

Solution ➤

The interest for the *first year* is 8% of $10,000, which is

$$0.08 \cdot 10,000 = \$800.$$

After one year the total amount is $10,000 + $800 = $10,800.

The interest for the *second year* is 8% of $10,800, which is

$$0.08 \cdot \$10,800 = \$864.$$

After two years the total amount is $10,800 + $864 = $11,664.

Since you are adding 8% to 100% for each year, you can repeatedly multiply Jill's balance by 1.08.

Spreadsheet

You can use a calculator or spreadsheet to help you find what the investment is worth after 10 years. Start with $10,000 and multiply repeatedly by 1.08.

From the spreadsheet, you can see that after 10 years Jill will have $21,589.25 for tuition.

	A	B
	Year	**Cmp. interest**
1		
2	0	10,000.00
3	1	10,800.00
4	2	11,664.00
5	3	12,597.12
6	4	13,604.89
7	5	14,693.28
8	6	15,868.74
9	7	17,138.24
10	8	18,509.30
11	9	19,990.05
12	10	21,589.25

An equation such as $y = 2^x$ is an example of an **exponential function**. The base, 2, is the number that is repeatedly multiplied. The exponent, x, represents the number of times that the base, 2, occurs in the multiplications.

•Exploration 2 *Growth and Decay*

 Copy and complete the table for $y = 2^x$.

x	1	2	3	4	5
y	2	4	8	?	?

16 32

 Graph the function $y = 2^x$ for the values in Step 1.

 Describe the pattern in the y-values for $y = 2^x$? The y-value doubles with each increase by 1 of the x-value.

 Make a table of x- and y-values for $y = \left(\frac{1}{2}\right)^x$. Use the same x-values that you used for Step 1. Then graph the function.

 Compare the graphs of $y = 2^x$ and $y = \left(\frac{1}{2}\right)^x$. How are they different? ❖

Exercises & Problems

Communicate

Describe the pattern for each sequence.

1. 10, 20, 40, 80, 160, . . .   **2.** 20, 10, 5, 2.5, 1.25, . . .

3. 10, 20, 30, 40, 50, . . .   **4.** 160, 140, 120, 100, 80, . . .

5. How can you tell if any of the sequences in Exercises 1–4 are exponential?

6. Explain what is meant by *exponential decay*. Explain what is meant by *exponential growth*.

7. In exponential decay, is the number that is multiplied repeatedly greater than 1 or less than 1?

8. How do you determine which number multiplies the principal amount when you solve the following problem? How much will Sabrina have to pay if she borrows $1000 at 10% interest, compounded once each year for 4 years?

Practice & Apply

9. Write the first five terms of a sequence that starts with 5 and doubles the previous number. 5, 10, 20, 40, 80

10. 24,300; 72,900; 218,700   **11.** 1100, 1300, 1500

What are the next three terms of each sequence?

10. 100, 300, 900, 2700, 8100, . . .   **11.** 100, 300, 500, 700, 900, . . .

12. 100, 10, 1, 0.1, 0.01, . . .   **13.** 40, 35, 30, 25, 20, . . .
        0.001, 0.0001, 0.00001   15, 10, 5

14. Which of the sequences in Exercises 10–13 show exponential growth? Ex. 10

15. Which of the sequences in Exercises 10–13 are linear? Ex. 11, 13

16. Is the sequence 64, 16, 4, 1, $\frac{1}{4}$, . . . linear or exponential? Exponential

17. Generate a table for the function $y = 3^x$ when x is 1, 2, 3, 4, and 5.

x	1	2	3	4	5
y	3	9	27	81	243

18. **Cultural Connection: Africa** An ancient Egyptian papyrus of mathematical problems includes the statement, "Take $\frac{1}{2}$ to infinity." Explore the sequence you get by starting with 0.5 and repeatedly multiplying by 0.5. What happens to each successive product? Gets closer and closer to 0

19. **Investments** Jacy invests $1000. His investment earns 5% interest per year and is compounded once each year. How long will it take Jacy to double his money? Between 14 and 15 years; 15 years, if compounded at end of year

Physics Suppose a ball bounces to $\frac{1}{2}$ of its previous height on each bounce. Find how high the ball bounces on the

20. second bounce. 5 ft **21.** third bounce. 2.5 ft

22. fourth bounce.   **23.** fifth bounce.
      1.25 ft   0.625 ft

24. Write the sequence you get if you start with 1 and keep tripling the previous number.
             1, 3, 9, 27, 81, 243, . . .

25. Is the sequence 100, 90, 80, 70, 60, . . . linear or exponential?
             Linear

26. Define *fixed rate*.

27. Define *fixed amount*.

28. Define *compound interest*.

29. Start with the number 8, and double the amount 5 times. What number do you get? 256

30. **Banking** Consider the terms shown here for a First Money credit card. The company has a promotion in which interest is not charged for the first 6 months. What is the balance to the nearest cent after the sixth payment on $1000 if only minimum payments are made?

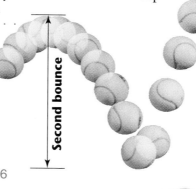

Second bounce

First bounce 10 feet

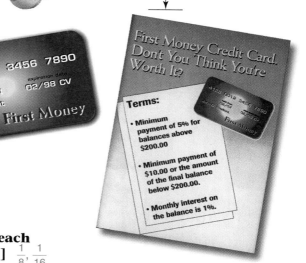

4128 0012 3456 7890

valid from 02/94   expiration date 02/98 CV

John Q. Student   First Money

First Money Credit Card. Don't You Think You're Worth It?

Terms:
- Minimum payment of 5% for balances above $200.00
- Minimum payment of $10.00 or the amount of the final balance below $200.00.
- Monthly interest on the balance is 1%.

Look Back

Write the rule used to find the next term in each pattern. Find the missing terms. [Lesson 1.1]

31. 3, 7, 11, 15, 19, _?_, _?_ 23, 27   **32.** 4, 2, 1, $\frac{1}{2}$, $\frac{1}{4}$, _?_, _?_ $\frac{1}{8}, \frac{1}{16}$ **33.** 33, 30, 27, 24, 21, _?_, _?_ 18, 15
Add 4 to each term.   Divide each term by 2.   Subtract 3 from each term.

Find the next two terms for each sequence. [Lesson 1.1]

34. 10, 21, 34, 49, 66, _?_, _?_   **35.** 99, 78, 72, 51, 45, _?_, _?_
      85, 106   24, 18

Plot each point on a coordinate plane. [Lesson 2.6, 8.1]

36. $A(3, 5)$ **37.** $B(5, 6)$ **38.** $C(6, 5)$ **39.** $D(4, 5)$

Look Beyond

40. **Technology** Explore the sequence you find when you begin with 100 and repeatedly multiply by -2. Do the numbers get close to any particular number? Some calculators have a special key for "negative," such as $\boxed{(\text{-})}$. On other calculators you enter 2 and then $\boxed{+/-}$ to get -2.

41. Explore the sequence you find when you begin with 100 and repeatedly multiply by -0.5. Do the numbers get close to any particular number?

Exploring Quadratic Functions

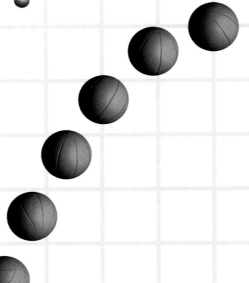

why *The path of a softball, basketball, or volleyball is often a parabola. This path can be modeled by a quadratic function. In geometry, quadratic functions model problems that relate to area.*

•Exploration 1 *Edges and Surface Area*

GEOMETRY
Connection

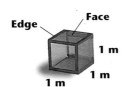

Suppose you are going to paint all six faces of a cube. Guess how much paint will be needed if you decide to double the length of each edge. You might be surprised to find that it will take more than twice the amount to paint the faces of the larger cube.

Suppose the edge of a cube is 1 meter. Since the area of each face is 1 square meter (1 m²), the surface area of the six faces is 6 m². Examine how the surface area of the six faces grows as the length of each edge is increased.

1 Compare the surface area of a 1-meter cube with the surface area of a 2-meter cube. What happens to the surface area of the 6 faces when the length of the edge doubles?

2 Examine a cube whose edge is triple the length of the first cube's edge. What happens to the surface area of the 6 faces when the length of the edge triples?

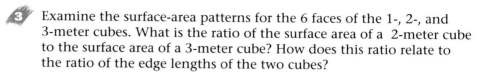

3 Examine the surface-area patterns for the 6 faces of the 1-, 2-, and 3-meter cubes. What is the ratio of the surface area of a 2-meter cube to the surface area of a 3-meter cube? How does this ratio relate to the ratio of the edge lengths of the two cubes?

4 Extend the following table to include an edge length of 5 meters. Notice the pattern.

Length of each edge in meters	1	2	3	4	5	...	e
Area of each face in square meters	1	4	9	16	_?_ 25	...	e^2
Formula for the total surface area of the cube, S	6(1) 6(1²)	6(4) 6(2²)	6(9) 6(3²)	6(16) 6(4²)	_?_ 6(25) ... _?_ 6(5²) ...		$S = \dfrac{?}{6(e^2)}$
Total surface area of the cube	6 sq m	24 sq m	54 sq m	96 sq m	_?_ sq m 150		

5 Write a formula for the total surface area of the 6 faces of a cube based on the pattern that you see in the table. ❖ $S = 6e^2$

In Exploration 1, the relationship involves the *square* of the length of an edge. Functions that involve *squaring*, such as $y = 6e^2$ or $y = x^2$, are called **quadratic functions**.

CRITICAL Thinking How can you use this pattern to find how much paint you need to paint cubes of different sizes? Find out how much paint is needed to cover 1 sq unit, and multiply that amount of paint by the surface area, which is $6e^2$.

These students are painting a portion of the set for a school play.

APPLICATION

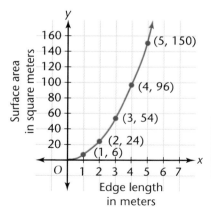

If you want to get a picture of the surface-area data in the table on page 545, you can draw a graph. Start by writing five ordered pairs that relate edge length to surface area as ordered pairs (edge length, surface area). These ordered pairs are (1, 6), (2, 24), (3, 54), (4, 96), and (5, 150). You can plot the points and connect them with a smooth curve like the one at the left.

Only part of the quadratic curve appears. Why does the graph show only the positive *x*- and *y*-values of the quadratic function? ❖

Since *x* represents edge length, the values can only be positive.

APPLICATION

Physics The following spreadsheet shows the data for the flight of a small rocket. The graph of this data, which has a quadratic relationship, is called a **parabola**.

Rocket Data

Time (sec)	Height (ft)
0	0
1	208
2	384
3	528
4	640
5	720
6	768
7	784
8	768
9	720
10	640
11	528
12	384
13	208
14	0

Use the data from the spreadsheet to draw a graph. The curve is a parabola that opens downward. You can see that the rocket reaches its maximum height of 784 feet, after 7 seconds.

A parabola may open either *downward* or *upward*. The point where the curve changes direction is the **vertex**. ❖

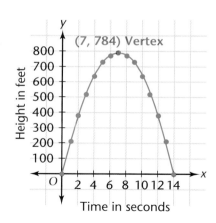

●Exploration 2 *Graphing Sequences*

Use the method of differences to discover the next two terms in the sequence 1, 4, 9, 16, 25, . . .

1 Calculate the first differences. Are they constant? If not, what are the second differences? no; 2

2 Work backward using the pattern, and predict the missing terms.

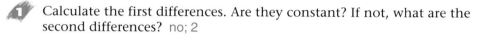

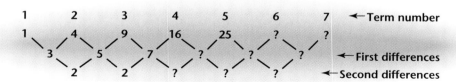

3 What do you notice about the second differences of this sequence?
They are constant.

4 Graph the sequence. Use the term number for *x* and the value of the term for *y*. For example, the first three points will be (1, 1), (2, 4), and (3, 9). What kind of curve appears when you connect the points?

5 Find the first differences for the sequence 4, 8, 12, 16, 20. Are the differences constant? +4; yes

6 Graph the sequence. For example, the first three points will be (1, 4), (2, 8), and (3, 12). What kind of curve appears when you connect the points?

7 How does the graph in Step 4 differ from the graph in Step 6? Which differences were constant for each sequence? ❖
In Step 4, the second differences are constant and the graph is a half parabola. In Step 6, the first differences are constant and the graph is a line.

EXERCISES & PROBLEMS

Communicate

1. Explain what happens to the area of a square if you double the length of each side.

2. What clues allow you to recognize a quadratic relationship? List at least two clues.

3. Explain how you could use a pattern to determine the type of relationship between the given pairs.

x	1	2	3	4	5
y	4	0	−4	0	4

How would you decide which graph matches which relationship?

a. Linear
b. Exponential
c. Quadratic
d. None of these

4.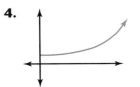

5.

6.

7.

8.

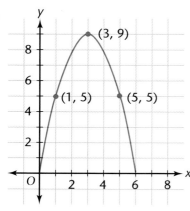

Explain how to determine the vertex of the parabola shown. Explain how to determine the vertex from a table.

9. Discuss how differences can be used to determine sequences and how the differences relate to quadratic relationships. What is the importance of constant differences?

Practice & Apply

Match each table with the correct type of relationship.

a. Linear   **b.** Exponential   **c.** Quadratic   **d.** None of these

10.

x	1	2	3	4	5
y	3	6	11	18	27

c

11.

x	1	2	3	4	5
y	1	2	4	8	16

b

12.

x	1	2	3	4	5
y	9	8	7	6	5

a

13.

x	1	2	3	4	5
y	4	6	8	10	12

a

14.

x	1	2	3	4	5
y	10	17	26	37	50

c

15.

x	1	2	3	4	5
y	80	40	20	10	5

b

Geometry What happens to the area of a square if you

16. triple the length of each side?

17. multiply the length of each side by 10?

18. take $\frac{1}{3}$ of the length of each side?

19. **Geometry** What is the area of each pizza shown? Use 3.14 for π.

20. Consumer Economics Which pizza is the better buy?

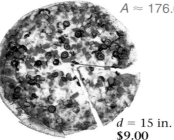

$A \approx 176.625$ sq in.

$A \approx 78.5$ sq in.

15" pizza

$d = 15$ in.
$9.00

$d = 10$ in.
$4.50

Match each relationship with its graph.

a. Linear   **b.** Exponential   **c.** Quadratic   **d.** None of these

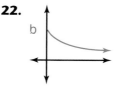

21. c

22. b

23. a

24. d

25. How many times larger is the side of the larger square than the side of the smaller square? 2 times larger

26. A parabola has a vertex at $(8, -2)$ and contains the point $(5, 3)$. Give the coordinates of another point on the parabola. Answers may vary. (11, 3)

27. Tell whether the parabola described in Exercise 26 opens upward or downward. upward

16 square inches

4 square inches

Physics Find the height of each waterfall if the time for a float in the water to fall is

400 ft

28. 1 second. 16 ft **29.** 2 seconds. 64 ft **30.** 5 seconds.

The height, h, (in feet) of a waterfall can be determined by the time, t, (in seconds) that it takes a float in the water to fall from the top to the bottom according to the formula $h = 16t^2$. For example, if the float takes 3 seconds to fall, the height is $16 \cdot 3^2 = 144$ feet.

31. Graph the points you found for the waterfalls in Exercises 28–30. What are the reasonable domain and range values?

32. Use your graph from Exercise 31 to estimate how long it would take a float in the water to fall from a waterfall with a height of 600 feet. Just over 6 seconds

Look Back

Write the next term in each pattern. Leave answers in fraction form. [Lesson 1.1]

33. 11, 33, 99, 297, 891, ? 2673

34. 3125, 625, 125, 25, ? 5

35. $\frac{4}{5}, \frac{8}{10}, \frac{16}{20}, \frac{32}{40}, \frac{?}{}$ $\frac{64}{80}$

Evaluate each expression if x is 2, y is 1, and z is 4. [Lesson 1.4]

36. $3xy$ 6 **37.** $4z$ 16 **38.** $21yz$ 84 **39.** xyz 8

40. If each package of notebook paper costs $1.29, find the cost of 5 packages. [Lesson 1.8]
$6.45

Substitute some values for x to find some values of y. Form ordered pairs, and plot the points. [Lessons 2.6, 8.1]

41. Graph the line for the equation $y = 4x + 6$.

42. Graph the line for the equation $y = 3x + 2$.

43. Which variable, h or t, is the independent variable for the equation $h = 16t^2$ in Exercise 31? [Lesson 8.1]
t is the independent variable.

h

Look Beyond

Identify the coordinates for the vertex of each function.

44. $y = (x - 2)^2 + 7$ (2, 7) **45.** $y = x^2 + 2x + 1$ (−1, 0)

LESSON 10.3 Previewing Reciprocal Functions

why At $5 per car, how many cars must the students wash to meet their goal? If they decide to charge $4 per car, how many cars will they have to wash? As the price per car decreases, what happens to the number of cars they have to wash? A reciprocal function models this relationship.

The carwash was successful. The students met their goal of $500. The students have a new goal of $1000 to help a needy family. If the students find 10 contributors, how much money will each person need to contribute to meet the goal?

For 10 contributors, divide the goal of $1000 by 10. Each of the 10 contributors will need to pay $100. Look at what happens when a greater number of contributors donate. For 20 contributors, divide $1000 by 20. What do you get? Calculate the amount donated per person when there are 50, 100, and 1000 contributors. Compare your results with the data in the table. Results are the same.

Let x be the number of contributors.	10	20	50	100	1000
Let y be the dollars given per person.	100	50	20	10	1

Graphing Reciprocal Functions

The function $y = \dfrac{1000}{x}$ models the relationship between the number of contributors, x, and the number of dollars given per person, y. To graph this function, plot the ordered pairs from the table above.

Plot the ordered pairs (10, 100), (20, 50), (50, 20), and (100, 10), and connect them with a curve. The point (1000, 1) is omitted because it would make the scale too large to see most of the other points. **As the number of contributors increases, what happens to the number of dollars donated by each contributor?** decreases

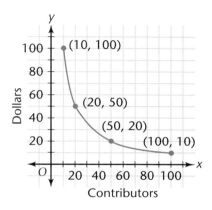

Examine the ordered pairs on the graph. Notice that in each case, the y-value is found by dividing 1000 by x. Functions such as $y = \frac{1000}{x}$ and $y = \frac{1}{x}$ are examples of **reciprocal functions**.

Try This Copy and complete the table that shows the amount, y, that each contributor must donate to raise $500 when x represents the number of contributors.

x	1	2	5	10	20	25	50	100	250	500
y	500	250	100	50	25	?	?	?	?	?
						20	10	5	2	1

EXAMPLE 1

Make a table of the values for the reciprocal function $y = \frac{1}{x}$. Then graph the function.

Solution ➤

Substitute values for x in the equation $y = \frac{1}{x}$ to make a table of ordered pairs. The y-values represent the reciprocals of the values that you choose for x.

x	1	2	4	5	$\frac{1}{2}$	$\frac{1}{4}$	$\frac{1}{5}$
y	1	$\frac{1}{2}$	$\frac{1}{4}$	$\frac{1}{5}$	2	4	5

Plot the ordered pairs on the graph, and connect the points. Examine the curve formed from the values of the function. **How can you use a graph of $y = \frac{1}{x}$ to find the reciprocal of $\frac{1}{8}$?**

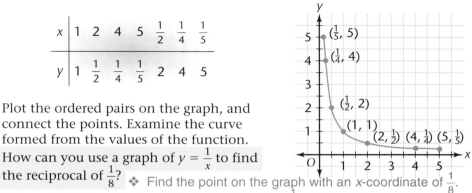

❖ Find the point on the graph with an x-coordinate of $\frac{1}{8}$. The y-coordinate of that point is the reciprocal of $\frac{1}{8}$, which is 8.

Graphics Calculator

You can use a graphics calculator to display the graph of $y = \frac{1}{x}$. Enter x, and press the ⬚ x^{-1} key to enter the function after **Y=**. Draw the graph. Use the trace feature to find the reciprocal of $\frac{1}{8}$.

Examine the graph of $y = \frac{1}{x}$. What happens to the y-value of the function as the x-value decreases from 1 to 0? What is y when x is 0?

As the x-value comes closer to 0, the y-value gets larger and larger. The function is not defined when x is 0 because the y-values approach infinity. Notice that division by 0 is not defined.

Reciprocals and Music

The frequency of the open string is 98 vibrations per second (vps). Stopping the string halfway produces a frequency of 196 vps.

Cultural Connection: Europe   A famous example of a reciprocal relationship occurs in music. Every culture that had stringed instruments probably discovered that the longer the string, the lower the frequency, or pitch, of the note it produces. By the time of the Greek scholar Aristotle, it was known that *doubling the length of a musical string produces half of the frequency* in vibrations per second. In general, the relationship between the length of a musical string and the frequency it produces is reciprocal.

EXAMPLE 2

Suppose a musical string 48 centimeters long produces a frequency of 440 vibrations per second (vps). This frequency produces the musical note A, the usual tuning note for an orchestra. What happens to the frequency when the length of the vibrating string is 96 centimeters?

Solution ➤

To determine the frequency for the 96-centimeter string, find the relationship of the new length to the original 48-centimeter string.

$$96 = 2 \cdot 48$$

The string is 2 times as long as the 48-centimeter string. Based on the reciprocal relationship, *doubling the length produces half of the frequency.* The frequency of the 96-centimeter string is $\frac{1}{2}$ of the 440 vps you get from the 48-centimeter string.

$$\frac{1}{2} \cdot 440 = 220 \text{ vps} ❖$$

Toccata and Fugue in D minor by Johann Sebastian Bach

Note	A	G	A		D	A	G	A
VPS	880	784	880		588	440	392	440

EXAMPLE 3

Suppose a 15-inch organ pipe produces a frequency of 440 vps. What happens to the frequency when the length of the pipe is

A 5 inches?

B 60 inches?

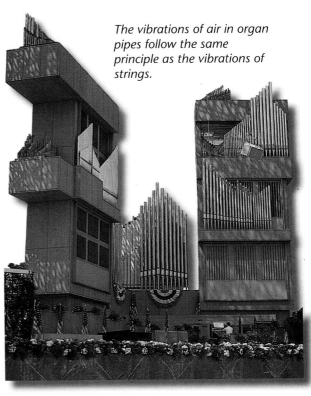

The vibrations of air in organ pipes follow the same principle as the vibrations of strings.

Solution ➤

A To determine the frequency of a 5-inch pipe, find the relationship of the new length to the original 15-inch length.

$$5 = \frac{1}{3} \cdot 15$$

The 5-inch pipe is $\frac{1}{3}$ as long as the 15-inch pipe.

Based on the reciprocal relationship, $\frac{1}{3}$ of the length has 3 times the frequency. The frequency of the 5-inch pipe is 3 times the 440 vps produced by the 15-inch pipe.

$$3 \cdot 440 = 1320 \text{ vps}$$

B To determine the frequency of a 60-inch pipe, find the relationship of the new length to the original 15-inch length.

$$60 = 4 \cdot 15$$

The 60-inch pipe is 4 times as long as the 15-inch pipe. Based on the reciprocal relationship, 4 times the length has $\frac{1}{4}$ of the frequency. The frequency of the 60-inch pipe is $\frac{1}{4}$ of the 440 vps produced by the 15-inch pipe.

$$\frac{1}{4} \cdot 440 = 110 \text{ vps} \; ❖$$

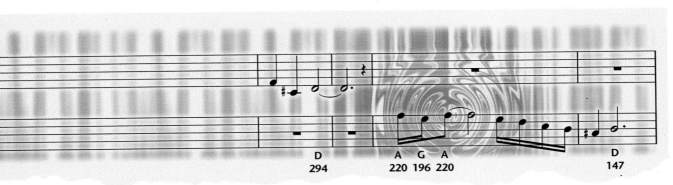

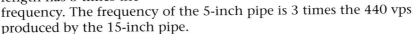

D 294   A 220   G 196   A 220   D 147

EXERCISES & PROBLEMS

Communicate

1. Describe a reciprocal function using the terms *increasing* and *decreasing*. Use examples to describe the relationship.

Tell how to find the reciprocal of each number.

2. 5   **3.** 100   **4.** $\frac{1}{4}$ **5.** $\frac{1}{6}$

Discuss how you would find the amount that each person should equally contribute to reach a goal of $1000 if there were

6. 4 people.   **7.** 200 people.

8. Describe how you would set up the following problem. Jane's co-workers are contributing money to buy her a birthday present. How much must each contribute if they expect to spend $30 and there are 5 co-workers?

Practice & Apply

9. Complete the table. Let each y-value be the reciprocal of each x-value. Graph the function.

x	6	5	3	1	$\frac{1}{3}$	$\frac{1}{5}$	$\frac{1}{6}$
y	?	?	?	?	?	?	?

10. Use variables to set up the following problem. Suppose a trip takes 4 hours if you average 45 miles per hour. How long will the same trip take if you average 60 miles per hour? 3 hr

Find the reciprocal of each number.

11. $7\frac{1}{7}$ **12.** $5\frac{1}{5}$ **13.** $25\frac{1}{25}$ **14.** $\frac{1}{4}$ 4   **15.** $\frac{1}{10}$ 10   **16.** $\frac{1}{7}$ 7

17. Explain why 0 has no reciprocal.   **18.** Explain why 1 is its own reciprocal.

Suppose a string on a guitar is 60 centimeters long and produces a frequency of 300 vibrations per second.

Music   Given the string at the left, what frequencies will be produced by similar guitar strings with the following lengths?

19. 30 cm 600 vps   **20.** 20 cm 900 vps   **21.** 90 cm 200 vps

Music   Given that a string 32 centimeters long produces a frequency of 660 vibrations per second, find the frequencies produced by similar strings with the following lengths.

22. 96 cm 220 vps   **23.** 16 cm 1320 vps   **24.** 24 cm 880 vps

Plot several ordered pairs for each function, and connect them with a smooth curve.

25. $y = \frac{12}{x}$ **26.** $y = \frac{20}{x}$ **27.** $y = \frac{24}{x}$

Fund-raising How much will each person have to contribute to reach a goal of $100 if there are

28. 10 people? $10 **29.** 20 people? $5 **30.** 100 people? $1 **31.** 1000 people? $0.10

Travel Sixty students and teachers from two algebra classes are going on a field trip to a museum. How many cars are needed if each car holds the indicated number of passengers?

32. 3 passengers 20 **33.** 4 passengers 15 **34.** 5 passengers 12

How fast would you have to drive to complete a 240-mile trip in

35. 6 hours? 40 mph **36.** 5 hours? 48 mph **37.** 4 hours? 60 mph

38. 3 hours? 80 mph **39.** 1 hour? 240 mph **40.** $\frac{1}{2}$ hour? 480 mph

41. Suppose a trip takes 4 hours if you average 50 mph. How long will it take if you average 40 mph? 5 hr

42. If other factors are held constant, the frequency of a musical note is a constant multiplied by the reciprocal of the length. If a musical string 48 centimeters long produces a frequency of 440 vps, find the constant. 21,120 cm per sec

Look Back

Geometry In Exercises 43 and 44, angles 1 and 2 are complementary. Find x. **[Lessons 4.2, 7.4]**

43. $m\angle 1 = 2x + 9$; $m\angle 2 = \frac{3}{4} + x$ 26.75 **44.** $m\angle 1 = 1.7x - 0.98$; $m\angle 2 = 1.4x + 7.9$ 26.8

45. Fund-raising The PTA sold hard cover and paperback books. The hard cover books sold for $7 each and the paperbacks sold for $3 each. Each member was asked to sell $48 worth of books. The number of paperbacks sold was 3 times the number of hard cover books sold. How many books of each type did each member sell? Write and solve a system of equations to solve this problem. **[Lessons 9.1, 9.2, 9.3, 9.4]** 3 hard cover, 9 paperback

46. Chemistry A 1.5% acid solution is mixed with a 3% acid solution to obtain 150 grams of a 1.8% acid solution. How many grams of the 1.5% solution and of the 3% solution are needed? Write and solve a system of linear equations to solve this problem. **[Lessons 9.1, 9.2, 9.3, 9.4]** 120 g of 1.5% solution; 30 g of 3% solution

Graph each system of linear inequalities. [Lesson 9.5]

47. $\begin{cases} 3x + y < -3 \\ x - y \geq 2 \end{cases}$ **48.** $\begin{cases} y > -x - 2 \\ y > x + 2 \end{cases}$ **49.** $\begin{cases} y \leq -2x + 1 \\ y \leq 2x - 3 \end{cases}$

Look Beyond

50. Graph $y = \frac{3}{x}$ for x-values 1, 2, 4, 5, $\frac{1}{2}$, $\frac{1}{4}$, $\frac{1}{5}$, -1, -2, -4, -5, $-\frac{1}{2}$, $-\frac{1}{4}$, and $-\frac{1}{5}$.

Are the DINOSAURS Breeding?

In the novel *Jurassic Park*, dinosaurs created from preserved DNA roam an island theme park. New batches of these cloned creatures are added to the population every six months. During a pre-opening tour, Dr. Malcolm, a mathematician, and Mr. Gennaro, a lawyer, discuss a graph of the heights of one of the dinosaur species on the island.

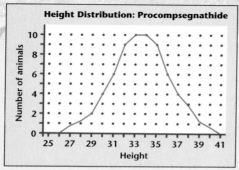

"Yes," Malcolm said. "Look here. The basic event that has occurred in Jurassic Park is that the scientists and technicians have tried to make a new, complete biological world. And the scientists in the control room expect to see a natural world. As in the graph they just showed us. Even though a moment's thought reveals that nice, normal distribution is terribly worrisome on this island."

"It is?"

"Yes. Based on what Dr. Wu told us earlier, one should never see a population graph like that."

"Why not?" Gennaro said.

"Because that is a graph for a normal biological population. Which is precisely what Jurassic Park is not. Jurassic Park is not the real world. It is intended to be a controlled world that only imitates the natural world."

Later, Dr. Malcolm discusses the graph with one of the park's scientists, Dr. Wu. They debate whether the supposedly all-female population is somehow breeding.

"Notice anything about it?" Malcolm said.

"It's a Poisson distribution," Wu said. "Normal curve."

"But didn't you say you introduced the compys in three batches? At six-month intervals?"

"Yes . . ."

"Then you should get a graph with peaks for each of the three separate batches that were introduced," Malcolm said, tapping the keyboard. "Like this."

"But you didn't get this graph," Malcolm said.

"The graph you actually got is a graph of a breeding population. Your compys are breeding."

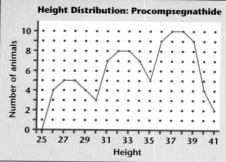

Wu shook his head. "I don't see how."

"You've got breeding dinosaurs out there, Henry."

"But they're all female," Wu said. "It's impossible."

Cooperative Learning

To help settle the breeding debate, you can model
dinosaur populations with random numbers.

1. Model a breeding population by following these steps.
   a. Represent 15 baby dinosaurs with 15 letters, A–O, as shown.

A	B	C	D	E	F	G	H	I	J	K	L	M	N	O

   b. Use a random number table to generate digits between 0 and 9.
      Let each digit represent a week's growth for one dinosaur.

A	B	C	D	E	F	G	H	I	J	K	L	M	N	O	
8	1	0	1	7	4	9	0	2	7	7	9	0	3	1	(week 1)
5	0	9	1	2	0	9	3	9	9	2	3	5	0	1	(week 2)
2	2	6	4	2	6	3	0	8	1	0	8	1	9	1	(week 3)

   c. To simulate breeding population, add one new letter
      (for a new dinosaur) every fourth week.

A	B	C	D	E	F	G	H	I	J	K	L	M	N	O	P	
8	1	0	1	7	4	9	0	2	7	7	9	0	3	1		(week 1)
5	0	9	1	2	0	9	3	9	9	2	3	5	0	1		(week 2)
2	2	6	4	2	6	3	0	8	1	0	8	1	9	1		(week 3)
8	9	4	2	0	6	7	8	0	0	5	5	1	3	7	5	(week 4)

   d. Run the simulation for about 15–20 weeks. Then add the
      digits for each dinosaur to find its height.

   e. Display the dinosaur heights in a frequency distribution
      graph. Which of the graphs on page 556 is shaped most
      like your graph?

2. Now model an artificially controlled population.

   a. Start with 10 dinosaurs, labeled A–J, and run the simulation
      for 8 weeks. To model a population that is not breeding, do
      not add a new dinosaur every 3 weeks.

   b. After the first 8 weeks, introduce a new batch of 10 dinosaurs
      by adding the letters K–T.

   c. Run the simulation for 8 more weeks. Now you will need
      20 digits for each week.

   d. Find the height of each dinosaur. Display the heights on a
      frequency distribution graph. How is your graph like the
      bottom one on page 556? How is it different? Why does the
      difference make sense?

3. Do you agree with Dr. Malcolm or Dr. Wu? Why?

Previewing Other Functions

wky *The absolute-value function and greatest-integer function model special relationships. The graph of each of these functions has a distinctive appearance. One looks like a V, and the other looks like steps.*

These students are participating in an experiment in which some of the students estimate elapsed time.

STATISTICS
Connection

Students work in pairs. Half of the students close their eyes and raise their hands. They are told to put their hands down when they think a minute has passed. The other half of the students act as timers. Later they switch roles.

Data from the first group is entered in a spreadsheet with the students' names in column A and their estimates of the time elapsed in column B.

Spreadsheet

Compute the error (the difference between a guess and the actual time) *in seconds* in column C by subtracting 60 from each student's time.

In column C, a negative number means the guess was *under* 1 minute, and a positive number means the guess was *over* 1 minute. The absolute error in column D shows only the *amount* of the error, or the absolute value of the error.

D2		=ABS(C2)		
	A	B	C	D
1	Student	Time	Error	Abs. error
2	Tricia	49	−11	11
3	Keira	59	−1	1
4	Tom	51	−9	9
5	Louise	65	5	5
6	James	68	8	8
7	Sakeenah	77	17	17
8	Hong	66	6	6
9	Louis	54	−6	6
10	Mary	67	7	7
11	Maria	46	−14	14
12	Marcus	62	2	2
13	Shamar	73	13	13
14	Lois	61	1	1
15	Dianne	53	−7	7
16	Suzanne	64	4	4

For example, row 10 shows that Mary thought 1 minute had elapsed after 67 seconds. Her error was 67 − 60 = 7, indicating that her answer was 7 seconds *over* 1 minute. Row 15 shows that Dianne's time was 53 seconds, so her error was 53 − 60 = −7, or 7 seconds *under* 1 minute.

Look carefully at column D in the spreadsheet. Compare the numbers in column C with those in column D. What is the relationship between each number in column D and the corresponding number in column C?

Absolute-Value Function

Recall from Lesson 2.1 that the absolute value of a number can be shown on a number line. The absolute value of a number is its distance from zero.

A function written in the form $y = |x|$ or $y = \text{ABS}(x)$ is an example of an **absolute-value function**.

7 is 7 units from 0, so the absolute value of 7 is 7.

-7 is 7 units from 0, so the absolute value of -7 is 7.

Thus, $|7| = 7.$

Thus, $|-7| = 7.$

•Exploration How Far From Zero?

1 Complete the table for $y = |x|$.

x	−4	−3	−2	−1	0	1	2	3	4
y	4	? 3	? 2	? 1	0	? 1	? 2	? 3	4

2 Graph the function $y = |x|$ for the values in Step 1.

3 Describe the graph for the absolute-value function you graphed in Step 2. ❖ V-shaped

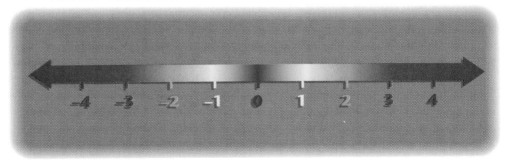

EXAMPLE 1

Graph the function $y = |x|$ using integer x-values from -3 to 3.

Solution ➤

Make a table of ordered pairs. For every value of x in the table, write the absolute value of x as the y-value. For example, if x is -2, then y is $|-2|$, or 2.

x	-3	-2	-1	0	1	2	3
y	3	2	1	0	1	2	3

COORDINATE GEOMETRY *Connection*

Plot the ordered pairs, and connect the points.

The graph of the absolute-value function has a distinctive V shape. The tip of the V shape is at the origin, $(0, 0)$. ❖

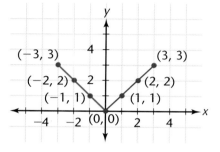

CRITICAL *Thinking*

Compare these graphs. How are they alike? How are they different?

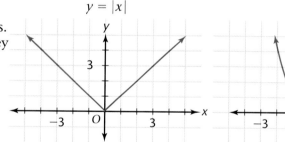

Both have only positive or zero y-values; both contain the points $(0, 0)$, $(1, 1)$, and $(-1, 1)$; and both continue upward in the 1st and 2nd quadrants. One graph is V-shaped and the other is a parabola.

Step Functions

Graphics Calculator

Another function that is especially important when using technology is the *step function*, the most common of which is the *greatest-integer function*. It appears on some calculators as a menu item (INT) or as a key ⟨ INT ⟩. When given a number, the *greatest-integer function* rounds down to the next integer. The following are some examples.

$$\text{INT}(1) = 1 \qquad \text{INT}(2) = 2$$
$$\text{INT}\left(\frac{5}{3}\right) = 1 \qquad \text{INT}(-2.1) = -3$$
$$\text{INT}(1.9) = 1 \qquad \text{INT}(-3.9) = -4$$

It is important to note that the greatest integer function always rounds *down*, and not *up*, to the next integer. Notice that INT(1.9) is rounded *down* to 1, even though it is nearer to 2. You can also represent these examples using ordered pairs, such as $(1, 1)$, $\left(\frac{5}{3}, 1\right)$, $(1.9, 1)$, $(2, 2)$, $(-2.1, -3)$, and $(-3.9, -4)$.

EXAMPLE 2

If you have $8.00 and want to buy movie tickets that cost $3.00 each, how many movie tickets can you buy?

Solution ➤

You can buy only 2 tickets, even though you have enough money for $2\frac{2}{3}$ tickets. You cannot buy $\frac{2}{3}$ of a ticket. In this situation, $\frac{2}{3}$ tickets rounds down to 2 tickets. ❖

Try This How many $3.00 movie tickets can you buy if you have $7.00? if you have $8.50? if you have $7.75? Remember, the integer function *rounds down* to the *next lower integer.* You can only buy 2 tickets with any three of the amounts.

EXAMPLE 3

Marilyn is ordering raffle tickets for her 12 friends. The tickets cost $1.50 each. Her friends have $2.00, $2.50, $2.75, $3.50, $3.75, $4.00, $4.50, $5.00, $5.25, $6.00, $6.25, and $6.50, respectively. Draw a graph representing the amount of money each friend has and the number of tickets that each friend can purchase.

Amount	Number of tickets
$2.00	1
$2.50	1
$2.75	1
$3.50	2
$3.75	2
$4.00	2
$4.50	3
$5.00	3
$5.25	3
$6.00	4
$6.25	4
$6.50	4

Solution ➤

Make a table to show the amount of money each has and the number of tickets that amount will purchase. Plot the pairs of numbers on a graph.

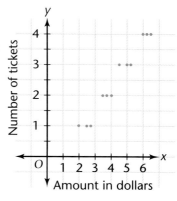

You can see that the *y*-value of the function stays level until the *x*-value reaches a certain point. Then the *y*-value *jumps*. This is the graph of a step function. ❖

EXERCISES & PROBLEMS

Communicate

Explain how to find the absolute value (ABS) of each number.

1. 29   **2.** − 34   **3.** 7.99   **4.** − 3.44

Explain how to find the greatest integer (INT) of each number.

5. 7.99   **6.** $-\frac{34}{5}$ **7.** 29   **8.** − 3.44

9. Describe the graph of an absolute-value function. V-shape

10. Describe the graph of a step function. steps of a staircase

Practice & Apply

In the minute-timing experiment, one student's estimate of a minute was 52 seconds.

11. Was the student over or under the correct amount? under

12. What was the student's error? − 8 sec

13. What was the absolute value of the student's error? 8 sec

Find the absolute value (ABS) of each number.

14. 17 17

15. − 33 33

16. 8.67 8.67

17. − 7.11 7.11

Evaluate.

18. |4.8| 4.8

19. |− 3.2| 3.2

20. INT(5.8) 5

21. INT(11/2) 5

22. ABS(5.8) 5.8

23. ABS(11/2) $\frac{11}{2}$, or 5.5

Find the greatest integer (INT).

24. 17 17

25. $\frac{33}{4}$ 8

26. − 8.67 −9

27. − 7.11 −8

Evaluate.

28. |8| 8

29. |−8| 8

30. −|8| − 8

31. −|− 8| − 8

32. Consumer Economics The integer function is one type of step function. The table at the right shows postal costs as a function of package weights. Paula often mails several packages and needs to know the cost for the various weights. Make a graph of the weights and costs from the table. Explain why this function is a step function, but *not* the greatest-integer function.

Crafts Melissa is making place mats, each requiring 1 yard of fabric. How many place mats can she make from the following amounts of fabric?

33. $6\frac{3}{4}$ yards 6

34. $3\frac{1}{2}$ yards 3

35. $5\frac{1}{4}$ yards 5

36. $\frac{9}{10}$ yard 0

Technology After collecting the data shown for the minute-timing experiment, students switched roles. This time, the students estimating the time were distracted by having conversations at the same time. The spreadsheet shows the data for this group.

37. On your own paper, write the errors for column C.

38. Write the absolute errors for column D.

39. Life Science Perform both parts of the minute-timing experiment in your class. Give the estimate, the error, and the absolute value of the error for each student. Use a spreadsheet to display the results for the two groups. Answers may vary.

D3		=ABS(C3)	
	A	B	C	D
1	Student	Time	Error	Abs. error
2	Buster	57	-3	3
3	Tony	74	14	14
4	Charlotte	59		
5	Amy	60		
6	Gerti	71		
7	Nathan	56		
8	Jamie	58		
9	Marni	70		
10	Lee	58		
11	Wei	62		
12	Ned	68		
13	Debbie	74		
14	Wynton	70		
15	Bob	73		
16	Jill	65		

37. −3, 14, −1, 0, 11, −4, −2, 10, −2, 2, 8, 14, 10, 13, 5

38. 3, 14, 1, 0, 11, 4, 2, 10, 2, 2, 8, 14, 10, 13, 5

40.  **Portfolio Activity** Complete the problem in the portfolio activity on page 537.

Look Back

Simplify each polynomial expression.   [Lesson 5.5]

41. $x^2 + 3x - 2x^2 + 4x$ $-x^2 + 7x$ **42.** $2x^2 - (-4x + 3)$ $2x^2 + 4x - 3$

43. $(2x + 5) + (3x + 4)$ $5x + 9$ **44.** $(4x + 3) - (2x - 5)$ $2x + 8$

Use the Distributive Property to find each product. [Lesson 6.3]

45. $-4x(3x - 2)$ $-12x^2 + 8x$ **46.** $2y^2(-5y + 1)$ $-10y^3 + 2y^2$

47. $6(-x^2 + 4x)$ $-6x^2 + 24x$ **48.** $3x(x^2 - x)$ $3x^3 - 3x^2$

Write each polynomial in factored form.   [Lesson 6.3]

49. $3m + 24$ $3(m + 8)$ **50.** $6x^4 + 36x$ $6x(x^3 + 6)$

51. $4u - 20u^3$ $4u(1 - 5u^2)$ **52.** $9x^2 - 36x^3$ $9x^2(1 - 4x)$

53. Use the following data to make a graph. Which point is the vertex? **[Lesson 10.2]**

x	1	2	3	4	5
y	4	3	4	7	12

Look Beyond

Choose positive and negative values for x, and graph each function.

54. $y = |x| + 3$ **55.** $y = |x + 3|$

Identifying Types of Functions

Acceleration can be modeled with the graph of a function. Many types of functions can be identified by the shapes of their graphs.

Have you ever heard the phrase "zero to sixty in eight seconds?" For each second that an automobile accelerates, the distance that it travels per second increases.

Quadratic Functions

The following table shows the relationship between time and distance traveled as a car accelerates its speed.

Time in seconds	0	1	2	3	4	5	6	7	8
Distance in feet	0	5.5	22	49.5	88	137.5	198	269.5	352

When you graph the ordered pairs (time, distance), the results appear to show a quadratic relationship.

You can use the method of differences to verify that the data have a quadratic relationship

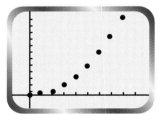

Time in seconds   0   1   2   3   4   5   6   7   8

Distance in feet   0   5.5   22   49.5   88   137.5   198   269.5   352

    5.5   16.5   27.5   38.5   49.5   60.5   71.5   82.5 ◄— First differences

      11   11   11   11   11   11   11 ◄— Second differences

The second differences are constant, so the relationship between time and distance that a car travels as it accelerates at a constant rate is quadratic.

How far do you think the race car will travel at this rate of acceleration after 10 seconds?   550 ft

Consumer Economics

25. Graph the function for the shipping charges.

26. What type of function is this? Step function

Rental A car-rental company charges $30 per day and allows 100 free miles. There is an additional charge of 5 cents per mile after 100 miles.

27. Graph the function that models the situation.

28. What kind of function is this? Piecewise function

A catalog company charges $2 per pound or fraction of a pound (less than 16 ounces) for shipping its merchandise.

Biking Supplies

Item 108: Red Shorts and Jersey

Price: $56.00
Shipping Wt: 4 lb. 6 oz.

Item 109: Helmet and Gloves

Price: $98.00
Shipping Wt: 5 lb. 11 oz.

Investments The Rule of 72 is a method of estimating how long it will take to double your money at various interest rates. To make an estimate, divide 72 by the percent interest rate. For example, if the interest rate is 8%, it will take about $\frac{72}{8}$, or 9, years to double your money. Use the Rule of 72 to predict how long it will take to double an investment if the interest rate is

29. 4%. 18 yr **30.** 6%. 12 yr **31.** 9%. 8 yr

32. Use the Rule of 72 to write an equation for the number of years, t, it will take to double an investment at x% interest. $t = \frac{72}{x}$

33. What kind of function is this? Reciprocal function

Look Back

34. The dimensions of a classroom are 32 feet by 20 feet 8 inches. What is the ratio of the length to the width? **[Lesson 3.6]** $\frac{48}{31}$

35. Astronomy The speed of a radio signal is 186,000 miles per second. How long does it take to send a radio signal to Mars, when it is about 35,000,000 miles away from Earth? **[Lesson 7.1]**
About 188 sec, or about 3 min, 8 sec

36. Cultural Connection: Asia Babylonian mathematicians living 3800 years ago solved problems in compound interest. They asked how many years it would take to double money at 20% interest compounded annually. Today bankers answer that question by looking it up in a table. The Babylonians did the same thing. Use your calculator to make a table that starts with $1000.00. Find how many years it would take to double the amount if it is compounded annually at 20% interest. **[Lesson 10.1]** 4 years; $1200.00, $1440.00, $1728.00, $2073.60, $2488.32, . . .

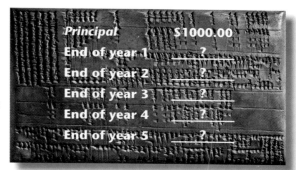

Principal $1000.00
End of year 1   ?
End of year 2   ?
End of year 3   ?
End of year 4   ?
End of year 5   ?

Look Beyond

37. Compare the graphs of the following equations.

a. $y = |x|$ **b.** $y = |x + 2|$ **c.** $y = |x - 1|$ **d.** $y = |x| - 3$ **e.** $y = |x| + 1$

Exploring Transformations

why *In mathematics, transformations are used to model motion.* Motions in geometry can be modeled algebraically. You can use functions to represent these transformations.

Transformations

A **transformation** of a figure is a movement of the figure. A transformation in which you slide a figure to a new position is called a **translation.** Triangle $A'B'C'$ at the left is a translation of triangle ABC.

Triangle $A'B'C'$ is called the **image** of the original triangle ABC. The original triangle ABC is called the **pre-image.**

A transformation in which you flip, or reflect, a figure across a line is called a **reflection.** Triangle $A'B'C'$ at the right is a reflection of triangle ABC across line m. Line m is called a mirror.

A **rotation** is a transformation that turns, or rotates, a figure about a point called a **center.**

Line segment AB has been rotated $180°$ about the turn center D. Line segment $A'B'$ is the image of line segment AB.

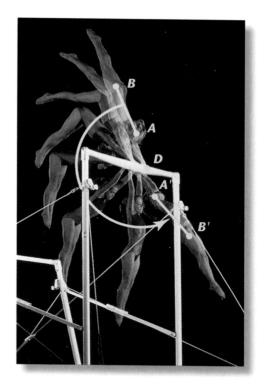

 Does a reflection have the same result as a $180°$ rotation? Explain.

Transformations can be applied to functions in a coordinate plane. The equation of a function can be changed to transform the graph of that function. The following explorations involve translations and reflections in the coordinate plane.

Exploration 1 *Translating Functions*

Graphics Calculator

For this exploration, you can use a graphics calculator, or you can graph the functions by building a table of values and using graph paper.

Part I

 Graph the absolute-value function $y = |x|$.

 On the same coordinate plane, graph the functions $y = |x| + 2$ and $y = |x| - 3$.

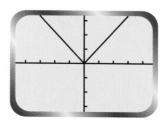

Ⓐ The original function, $y = |x|$, is called the **parent function.** What effects do adding a number to and subtracting a number from the value of y have on the parent function, $y = |x|$?

 What is the equation of each function shown on the coordinate plane at the right?

$y = |x| + 1$ (top)
$y = |x| - 2$ (bottom)

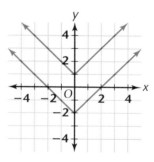

5 Graph the functions $y = |x + 2|$ and $y = |x - 3|$. What effects do adding a number to and subtracting a number from the value of x have on the parent function, $y = |x|$?

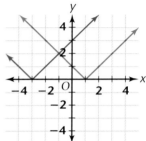

6 What is the equation of each function shown on the coordinate plane at the right? $y = |x + 3|$ (left); $y = |x - 1|$ (right)

7 Explain how to change the equation $y = |x|$ to translate the graph vertically. Explain how to change the equation $y = |x|$ to translate the graph horizontally.

Part II

Graphics Calculator

1 The graph on the calculator screen shows the parent function $y = x^2$. Using a graphics calculator or graph paper, graph the functions $y = x^2 + 1$ and $y = x^2 - 2$.

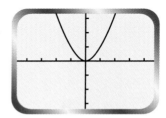

2 Look at your graphs from Step 1. What effects do adding a number to and subtracting a number from the value of y have on the parent function, $y = x^2$?

3 What is the equation of each function shown on the coordinate plane at the right? $y = x^2 + 3$ (top); $y = x^2 - 1$ (bottom)

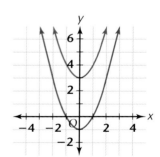

4 Using a graphics calculator or graph paper, graph the functions $y = (x + 2)^2$ and $y = (x - 3)^2$.

5 Look at your graphs from Step 4. What effects do adding a number to and subtracting a number from the value of x have on the graph of the parent function, $y = x^2$?

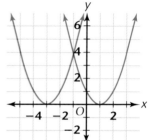

6 What is the equation of each function shown on the coordinate plane at the right? $y = (x + 3)^2$ and $y = (x - 1)^2$

7 Explain how to change the equation $y = x^2$ to translate the graph vertically. Explain how to change the equation $y = x^2$ to translate the graph horizontally. ❖

You can combine both horizontal and vertical translations to transform functions.

Exploration 2 Translations and Reflections

For this exploration, you can use a graphics calculator, or you can graph the functions by building a table of values and using graph paper.

Graphics Calculator

Part I

1 Graph the functions $y = |x|$, $y = |x - 1| + 2$, and $y = |x + 1| - 3$. Compare the three graphs.

2 Write a function to translate the parent function $y = |x|$ two units to the right. $y = |x - 2|$

3 Write a function to translate the parent function $y = |x|$ three units up.
$y = |x| + 3$

4 Write a function to translate the parent function $y = |x|$ both two units to the right *and* three units up. $y = |x - 2| + 3$

5 In the graph at the right, the parent function $y = |x|$ has been translated _?_ units to the left and _?_ units down. What is the equation of the translated function? $2; 1; y = |x + 2| - 1$

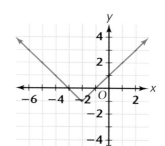

6 Without graphing, describe the graph of $y = (x - 2)^2 + 1$. How has it been translated from the parent function $y = x^2$? Translation of $y = x^2$ by 2 units to the right and 1 unit up.

Part II

Graphics Calculator

1 Graph the functions $y = -|x|$ and $y = -x^2$. What effect does the negative sign have on the parent functions $y = |x|$ and $y = x^2$?

2 Compare the graphs of the function $y = |x| + 1$ and $y = -|x| - 1$. Explain why $y = -|x| - 1$ is a reflection of $y = |x| + 1$ across the x-axis.

3 Compare the graphs of the function $y = |x + 2| + 1$ and $y = -|x + 2| - 1$. Explain why $y = -|x + 2| - 1$ is a reflection of $y = |x + 2| + 1$ across the x-axis.

4 Using a graphics calculator or graph paper, graph the function $y = x^2 - 1$. What is the equation of the function $y = x^2 - 1$ reflected across the x-axis? ❖

Exercises & Problems

Communicate

1. Explain the difference between translation and reflection. Give a geometrical example of each.

2. Give a real-world example of a rotation.

3. Name a function that is a translation of the graph of $y = |x|$ up.

4. Name a function that is a translation of the graph of $y = x^2$ to the left.

5. Name a function that is a translation of the graph of $y = |x|$ to the right and down.

Practice & Apply

Use graph paper to sketch the graph of each function.

6. $y = |x| - 3$

7. $y = |x| + 1$

8. $y = |x| - 4$

9. $y = |x - 1|$

10. $y = (x + 1)^2$

11. $y = |x + 3|$

12. $y = |x - 1| + 2$

13. $y = (x - 1)^2 + 1$

14. $y = |x + 3| - 1$

15. $y = -(x + 1)^2$

16. $y = -|x + 3|$

17. $y = |x + 1| - 2$

18. $y = -|x - 1|$

19. $y = (x - 4)^2 - 3$

20. $y = |x + 2| + 1$

21. $y = -|x| - 3$

22. $y = -|x| + 1$

23. $y = -|x| - 4$

Write the function that translates the parent function $y = |x|$ according to each description below.

24. 3 units up $\quad y = |x| + 3$

25. 4 units down $\quad y = |x| - 4$

26. 6 units to the right $\quad y = |x - 6|$

27. 2 units to the left $\quad y = |x + 2|$

28. 1 unit to the right $\quad y = |x - 1|$

29. 4.5 units up $\quad y = |x| + 4.5$

30. 2 units up and 3 units to the right

31. 1 unit down and 2 units to the left

32. 6 units up and 5 units to the left

33. 100 units down and 17 units to the right

30. $y = |x - 3| + 2$ 31. $y = |x + 2| - 1$ 32. $y = |x + 5| + 6$ 33. $y = |x - 17| - 100$

Write the function that translates the parent function $y = x^2$ according to each description below.

34. 3 units to the right $\quad y = (x - 3)^2$

35. 2 units to the left $\quad y = (x + 2)^2$

36. 1 unit down $\quad y = x^2 - 1$

37. 5 units up $\quad y = x^2 + 5$

38. 40 units down $\quad y = x^2 - 40$

39. 21 units up $\quad y = x^2 + 21$

40. 17 units to the right $\quad y = (x - 17)^2$

41. 19 units to the left $\quad y = (x + 19)^2$

42. 25 units up $\quad y = x^2 + 25$

43. 2 units up and 3 units to the right $\quad y = (x - 3)^2 + 2$

44. 12 units up and 1 unit to the right $\quad y = (x - 1)^2 + 12$

45. 10 units down and 2 units to the left $\quad y = (x + 2)^2 - 10$

46. 6 units up and 9 units to the left $\quad y = (x + 9)^2 + 6$

47. 1 unit down and 7 units to the right $\quad y = (x - 7)^2 - 1$

48. 40 units down and 7 units to the left $\quad y = (x + 7)^2 - 40$

Determine whether each of the following is an example of a translation, a reflection, or a rotation. Explain.

49. a boat gliding across a lake Translation

50. a pendulum in motion Rotation

51. the movement of a clock's second hand Rotation

52. a jet flying across the sky Translation

53. the movement of the Earth around the sun Rotation

54. using a rubber stamp Reflection

Write the transformed function of the parent function $y = |x|$.

55. $y = |x + 2|$

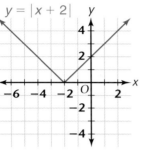

56. $y = |x + 1| - 2$

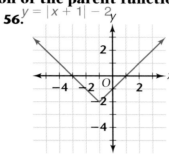

57. $y = |x - 1|$

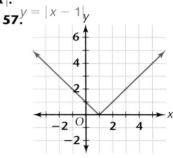

Write the transformed function of the parent function $y = x^2$.

58. $y = -x^2$

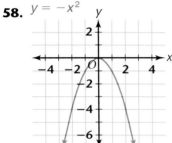

59. $y = x^2 - 3$

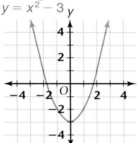

60. $y = (x - 1)^2 - 3$

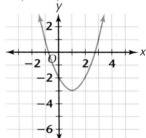

 Look Back

61. Which fraction does not belong in the following set? Why?   **[Lesson 3.2]**

$$\left\{ \frac{2}{4}, \frac{5}{10}, \frac{7}{21}, \frac{11}{22} \right\}$$ $\frac{7}{21}$; because $\frac{7}{21} \neq \frac{1}{2}$

62. Write the equations of two lines that are parallel to the line given by $y = 2x - 1$.   **[Lesson 8.7]** Answers may vary. $y = 2x$ and $y = 2x + 1$

63. Write the equations of two lines perpendicular to the line given by $y = 2x - 1$.   **[Lesson 8.7]** Answers may vary. $y = -\frac{1}{2}x$ and $y = -\frac{1}{2}x - 1$

Look Beyond

64. **Coordinate Geometry** Graph the right triangle with coordinates $(-2, 2)$, $(-2, -4)$, and $(6, -4)$. Multiply each coordinate by 2, and graph. Multiply each original coordinate by $\frac{1}{2}$, and graph. The transformed triangles are the same shape as but different sizes than the original triangle. What are the scale factors?

65. Describe the effect of changes in a on the graph of $y = ax^2$.

Cubes and Pyramids

▼ ◆ ■ ◆ ▼ ◆ ■ ◆ ▼ ◆ ■ ◆ ▼ ◆ ■ ◆ ▼ ◆ ■

*T*he cube below is made from 8 small unit cubes. All sides have been painted bright red, including the top and bottom. When the cubes are separated you can see that each small cube has red paint on 3 sides.

What happens as the length of each side of the red cube increases by 1?

Activity ❶
What is the formula?

Use a set of small unpainted cubes (such as sugar cubes) to build a large cube with 3 units on a side. All six faces of the big cube, including the top and bottom, are painted.

How many small cubes will be needed to make the big cube? How many of the small cubes will have paint on

3 faces? 2 faces?
1 face?   0 faces?

Copy and complete this table for cubes of different lengths. The last row gives the general formula for a cube with sides of length *n*.

Number of small cubes on each bottom edge	Total Number of small cubes	Number of small cube faces painted			
		3	2	1	0
2	8	8	0	0	0
3	27	12			
4					
5					
6					
n					

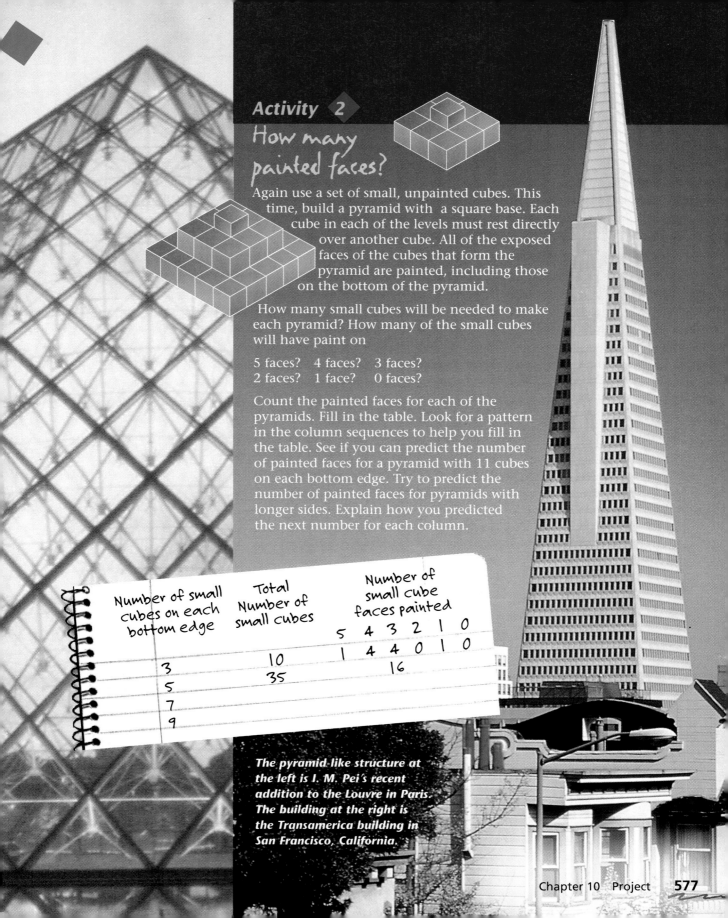

Activity 2

How many painted faces?

Again use a set of small, unpainted cubes. This time, build a pyramid with a square base. Each cube in each of the levels must rest directly over another cube. All of the exposed faces of the cubes that form the pyramid are painted, including those on the bottom of the pyramid.

How many small cubes will be needed to make each pyramid? How many of the small cubes will have paint on

5 faces? 4 faces? 3 faces?
2 faces? 1 face? 0 faces?

Count the painted faces for each of the pyramids. Fill in the table. Look for a pattern in the column sequences to help you fill in the table. See if you can predict the number of painted faces for a pyramid with 11 cubes on each bottom edge. Try to predict the number of painted faces for pyramids with longer sides. Explain how you predicted the next number for each column.

Number of small cubes on each bottom edge	Total Number of small cubes	Number of small cube faces painted					
		5	4	3	2	1	0
	10	1	4	4	0	1	0
3	35			16			
5							
7							
9							

The pyramid-like structure at the left is I. M. Pei's recent addition to the Louvre in Paris. The building at the right is the Transamerica building in San Francisco, California.

Chapter 10 Review

Vocabulary

absolute-value function	559	linear relationship	540	reciprocal function	551
compound interest	540	parabola	546	reflection	570
exponential function	540	parent function	571	rotation	571
exponential growth and decay	540	piecewise function	566	transformation	570
image	570	pre-image	570	translation	570
greatest-integer function	560	quadratic function	545	vertex	546

Key Skills & Exercises

Lesson 10.1

➤ **Key Skills**

Identify exponential sequences.

Examine 4, 8, 16, 32, 64, . . . Each term is 2 times the previous term. To find the next three terms, multiply by 2. The next three terms are $64 \cdot 2 = 128$, $128 \cdot 2 = 256$, and $256 \cdot 2 = 512$. Since the sequence grows at the same rate, the sequence is exponential.

Identify an exponential function as showing growth or decay.

The sequence 50, 25, 12.5, 6.25, 3.125, . . . shows exponential decay. The multiplier is 0.5, and the terms decrease. When the multiplier is greater than 1, a sequence shows exponential growth.

➤ **Exercises**

Find the next three terms of each sequence. Then identify each sequence as linear or exponential.

1. 45, 43, 41, 39, 37, . . .
35, 33, 31
Linear

2. 9000, 900, 90, 9, 0.9, . . .
0.09, 0.009, 0.0009
Exponential

3. 3, 9, 27, 81, 243, . . .
729, 2187, 6561
Exponential

4. If the multiplier for an exponential function is 1.2, does the function show growth or decay? growth

Lesson 10.2

➤ **Key Skills**

Graph basic quadratic functions.

You can use this data to make a graph and identify its vertex.

x	2	3	5	7	8
y	9	4	0	4	9

The vertex is (5, 0).

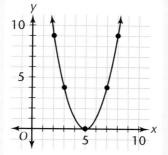

Use the method of differences to identify quadratic relationships.

Examine these sequences.

x	1	2	3	4	5
y	3	5	7	9	11

x	1	2	3	4	5
y	2	5	10	17	26

First differences for the first sequence are constant, so it is linear. Second differences for the second sequence are constant, so it is quadratic.

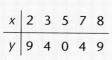

➤ Exercises

5. Which point is the vertex of this parabola? (4, 9)

6. Use the data to graph a parabola.
Which point is the vertex?

x	5	6	7	8	9
y	1	0	1	4	9

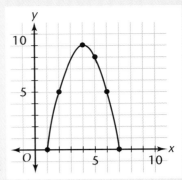

Identify each of the following sequences as quadratic or not quadratic. Explain your answer.

7.

x	1	2	3	4	5
y	2	5	10	17	26

8.

x	1	2	3	4	5
y	6	12	24	48	96

9.

x	1	2	3	4	5
y	3	12	27	48	75

Lesson 10.3

➤ Key Skills

Find the reciprocal of a number.

To find the reciprocal of 4, find a number that, when multiplied by 4, equals 1.

Since $4 \cdot \frac{1}{4} = 1$, the reciprocal of 4 is $\frac{1}{4}$.   Since $\frac{2}{3} \cdot \frac{3}{2} = 1$, the reciprocal of $\frac{2}{3}$ is $\frac{3}{2}$.

➤ Exercises

Find the reciprocal of each number.

10. 6 $\frac{1}{6}$ **11.** $\frac{6}{7}$ $\frac{7}{6}$ **12.** 3.5 $\frac{2}{7}$ **13.** $\frac{1}{15}$ 15

Lesson 10.4

➤ Key Skills

Find the absolute value of a number.

The absolute value of a number is its distance from 0 on a number line.

$$|9.8| = 9.8 \qquad |-9.8| = 9.8$$

9.8 and −9.8 are both 9.8 units from 0.

$$\left|6\frac{1}{4}\right| = 6\frac{1}{4} \qquad \left|-6\frac{1}{4}\right| = 6\frac{1}{4}$$

$6\frac{1}{4}$ and $-6\frac{1}{4}$ are both $6\frac{1}{4}$ units from 0.

Find the greatest integer of a number.

To find INT $\left(3\frac{2}{3}\right)$, round $3\frac{2}{3}$ down to the next integer. So INT $\left(3\frac{2}{3}\right) = 3$.

➤ Exercises

Evaluate each expression. The symbol $|x|$ means the same as ABS(x), the absolute value of x.

14. $|-3|$ 3   **15.** $|6.7|$ 6.7   **16.** $|-1.7|$ 1.7   **17.** ABS(-45) 45

18. INT(45) 45   **19.** INT(3.68) 3   **20.** INT(7.5) 7   **21.** INT(36) 36

Lesson 10.5

➤ **Key Skills**

Identify specific types of functions by their graphs.
Here are the graphs of several types of functions.

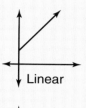
Linear

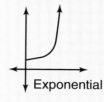
Exponential

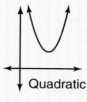
Quadratic

Step

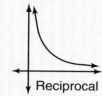
Reciprocal

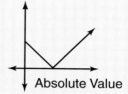
Absolute Value

➤ **Exercises**

What kind of function does each graph model?

22.
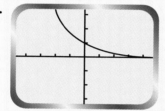
Exponential

23.
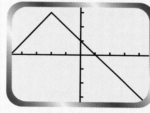
Absolute value

24.
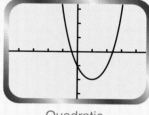
Quadratic

Lesson 10.6

➤ **Key Skills**

Use transformations to graph absolute-value functions and quadratic functions.

Beginning with the parent function $y = |x|$,

the graph of $y = |x - 1|$ is a translation *to the right* 1 unit;

the graph of $y = |x + 1|$ is a translation *to the left* 1 unit;

the graph of $y = |x| + 1$ is a translation *up* 1 unit; and

the graph of $y = |x| - 1$ is a translation *down* 1 unit.

➤ **Exercises**

Sketch the graph of each function.

25. $y = |x| - 2$

26. $y = |x - 2|$

27. $y = -|x + 4|$

28. $y = x^2 - 2$

29. $y = (x - 2)^2$

30. $y = -(x + 4)^2$

Application

31. Investment   Roy invests $1000. His investment earns 7.9% interest per year, which is compounded once each year. How long will it take Roy to double his money?   10 years

Chapter 10 Assessment

Find the next three terms of each sequence. Then identify each sequence as linear or exponential.

1. 10, 20, 40, 80, 160, . . .   **2.** 100, 96, 92, 88, 84, . . .

3. Does the exponential sequence 4000, 400, 40, 4, 0.4, . . . show growth or decay? Explain your answer.

4. Use the following data to graph a parabola. Which point is the vertex?

x	1	3	4	5	7
y	11	3	2	3	11

5. Mica's gymnastics team is raising money to buy equipment. How much does each team member need to raise in order to reach a goal of $500 if there are 10 team members? $50

Find the reciprocal of each number.

6. $6\frac{1}{6}$ **7.** $\frac{1}{12}$ 12   **8.** 5.5 $\frac{2}{11}$

Evaluate each expression.

9. $|-10|$ 10   **10.** ABS(7.9) 7.9   **11.** INT(9.75) 9

Determine whether each relationship is linear, quadratic, or neither.

12.

x	1	2	3	4	5
y	75	70	65	60	55

Linear

13.

x	1	2	3	4	5
y	4	7	12	19	28

Quadratic

Match the graphs with the types of functions.

a. Linear   **b.** Exponential   **c.** Quadratic
d. Reciprocal   **e.** Absolute value   **f.** Step

14. a
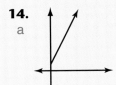

15. f
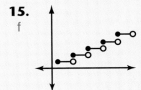

16. c
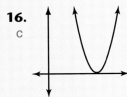

17. d
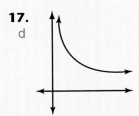

18. e
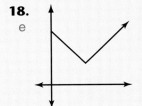

19. b
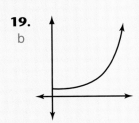

Sketch the graph of each function.

20. $y = |x + 1|$ **21.** $y = (x - 3)^2$ **22.** $y = -|x| + 2$ **23.** $y = x^2 + 4$

Chapters 1 – 10
Cumulative Assessment

College Entrance Exam Practice

Quantitative Comparison For Questions 1–4, write
A if the quantity in Column A is greater than the quantity in Column B;
B if the quantity in Column B is greater than the quantity in Column A;
C if the two quantities are equal; or
D if the relationship cannot be determined from the information given.

	Column A	Column B	Answers
1. A	$(-18)(-4)$	$18(-2)$	Ⓐ Ⓑ Ⓒ Ⓓ **[Lesson 2.5]**
2. C	The solution to $-6 = x + 1$	The solution to $x + 4 = -3$	Ⓐ Ⓑ Ⓒ Ⓓ **[Lesson 5.4]**
3. C	The amount of acid in 15 mL of 20% acid	The amount of acid in 20 mL of 15% acid	Ⓐ Ⓑ Ⓒ Ⓓ **[Lesson 7.3]**
4. A	$|-4.8|$	$|3.2|$	Ⓐ Ⓑ Ⓒ Ⓓ **[Lesson 2.1]**

5. What is the value of $\frac{5^2 + 11}{9 \cdot 3 + 9}$?   **[Lesson 1.7]**

c

   a. 36   **b.** 0   **c.** 1   **d.** $\frac{4}{9}$

6. By which number can you multiply each side of the proportion $\frac{7}{3} = \frac{n}{12}$

b   to solve for n?   **[Lesson 3.6]**

   a. $\frac{3}{7}$ **b.** 12   **c.** 3   **d.** $\frac{1}{7}$

7. Which description classifies the right triangle according

b   to its angles and its sides?   **[Lesson 4.3]**

   a. scalene and obtuse   **b.** scalene and right

   c. isosceles and acute   **d.** scalene and acute

8. Which is a value of x for the absolute-value equation $|x - 6| = 10$?

d   **[Lesson 7.6]**

   a. -16 **b.** 4   **c.** 10   **d.** -4

9. Which sequence is exponential and shows growth?   **[Lesson 10.1]**

d **a.** 2, 4, 6, 8, 10, . . .   **b.** 20, 10, 5, 2.5, 1.25, . . .

   c. 20, 18, 16, 14, 12, . . .   **d.** 2, 4, 8, 16, 32, . . .

10. Which sequence is linear?   **[Lesson 10.1]**

b **a.** 400, 200, 100, 50, 25, . . .   **b.** 400, 350, 300, 250, 200, . . .

   c. 400, 390, 370, 340, 300, . . .   **d.** 400, 800, 1600, 3200, 6400, . . .

11. Change $\frac{7}{8}$ to a percent. **[Lesson 3.7]** 87.5%

12. Simplify $(2x - 4) - (x - 9) - (5x + 2)$.   **[Lesson 5.2]** $-4x + 3$

13. Simplify $\frac{15n^2 - 25n}{-5n}$. **[Lesson 6.1]** $-3n + 5$

14. Graph the list of ordered pairs $(-2, 2)$, $(1, 4)$, and $(-4, 0)$. Tell whether
they lie on a straight line.   **[Lesson 8.1]** Not on a line

15. Write an equation of the line that has slope -2 and that passes through
the point $(-1, 0)$.   **[Lesson 8.4]** $y = -2x - 2$

Solve each system of linear equations.   **[Lessons 9.1, 9.4]**

16. $\begin{cases} 4x - y = 7 \\ -6x + 2y = -14 \end{cases}$ $(0, -7)$ **17.** $\begin{cases} x - 2y = 5 \\ 3x - 2y = 47 \end{cases}$ **18.** $\begin{cases} 9x - 10y = 96 \\ 3x + 8y = -36 \end{cases}$ $(4, -6)$

                                                                    $(21, 8)$

**Identify each graph as linear, exponential, quadratic, reciprocal,
absolute value, or step.**   **[Lesson 10.5]**

19. Absolute value   **20.** Linear

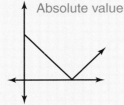   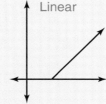

Free-Response Grid   The following questions may be
answered using a free-response grid commonly used by
standardized test services.

21. The lengths of the hypotenuse and one leg of a right
triangle are 78 and 72. Find the length of the other
leg. **[Lesson 4.7]** 30

22. 70% of 45 is _?_. **[Lesson 6.5]** 31.5

23. Solve $8(y - 2) = -6(20 - 4y)$ for y. **[Lesson 7.2]** 6.5

24. What is the slope of a line that is parallel to the line
passing through $(-7, -5)$ and $(1, -3)$? **[Lesson 8.7]** $\frac{1}{4}$

25. How much would each of 16 people need to
contribute toward a gift that costs $136?
[Lesson 10.3] $8.50

CHAPTER 11

Applying Statistics

The need for data collection dates as far back as ancient times, when political and religious leaders gathered information about people and property. The word *statistic* comes from the German word *statistik*, which was commonly used in German universities in the 1700s to describe the comparison of data about different nations.

The field of statistics was important to the United States war effort during World War II. As a result, statistics grew to cover a great variety of areas after the war. Today statisticians can find careers in many fields, including law, education, science, business, sports, and journalism.

shape our lives
Time for a PC?
Home computer usage

4

3
Time (in hours)
2

1

0   **1992**   **1993**   **1994**   **1995**

Shape our lives
The sales race
Top selling video games

Sales (in thousands)
40
30
20
10
0
Dec.   Jan.   Feb.   Mar.

Time (in hours)
2
1
0   1992   1993

USA Facts
How statistics shape our lives
Top CDs purchased
Teens choose alternative over pop

62%
Alternative

12%
Undecided

26%
Pop

USA Facts
How statistics shape our lives
Baseball's a hit!
Stadium attendance
(in thousands)

50

40

Crickets
Midgets
Panthers

Attendance
40
30
20
10
0   1995   1996   1997

PORTFOLIO ACTIVITY

Find an example of statistical information that is presented with tables or graphs in a newspaper or a magazine. Keep a copy of the article and any tables or graphs in your portfolio. Answer the following questions about the data.

1.   Why do you think this data was collected?

2.   Give at least three conclusions that can be drawn from the graphs or tables.

3.   Do you think the presentation of the data leads you to incorrect conclusions? Why or why not?

4.   List other ways that the data could be presented.

You will be asked to complete this activity on page 616. You may wish to save your work for your portfolio.

USA Facts
How statistics shape our lives
Alternative radio on the rise
The number of radio stations are increasing

Number of stations
200
150
100
50
0   **1993**   1994   1995

USA Facts
How statistics shape our lives
Baseball's a hit!
Stadium attendance
(in thousands)

50

Exploring Graphs

Why Graphs are visual representations of data. They are important communication tools in statistics because they allow people to interpret large amounts of data very quickly.

Can you tell which store has the more expensive merchandise?

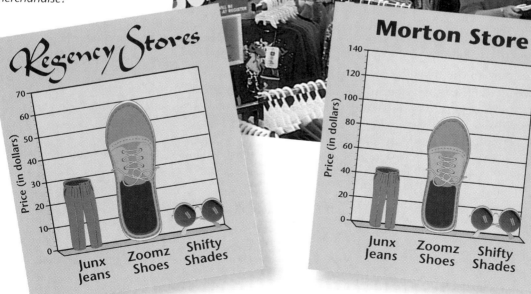

Statistics are often used to persuade, and sometimes graphs are used to intentionally mislead. In the following exploration, you will see that the way data is presented in a graph affects the way it is interpreted.

•Exploration 1 Misleading Graphs

The following bar graphs show the sales (in dollars) of two retail companies for five consecutive years.

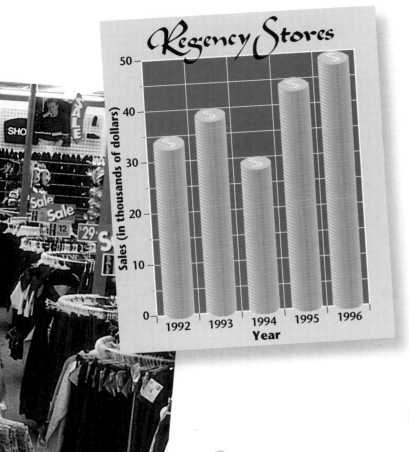

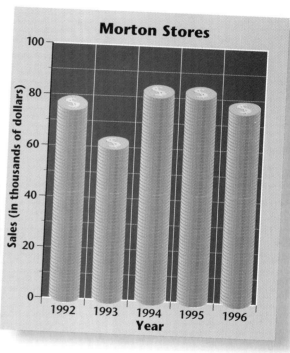

1 What is the approximate amount of sales for the Regency Stores in 1996? What is the approximate amount of sales for the Morton Stores in 1996? Which company's bar graph has a longer bar to represent sales in 1996? Which company actually had greater sales in 1996?

$50,000; $80,000; Regency; Morton

2 What is the approximate amount of sales for the Regency Stores in 1993? What is the approximate amount of sales for the Morton Stores in 1993? Which company's bar graph has a shorter bar to represent sales in 1993? Which company actually had lower sales in 1993?

$40,000; $65,000; Morton; Regency

3 What features of the graphs make it possible to misinterpret the information? How might the information be presented to make a more accurate visual comparison between the two stores?

4 Redraw the graphs to present a more accurate visual comparison between the two stores. ❖

Examine the two graphs on page 586. Explain why one or both of the graphs could be misleading.

Interpreting Line Graphs

A **line graph** is a type of graph that can be used to show changes occurring over time. A line graph can also be used to make predictions based on current trends.

Exploration 2 Line Graphs

Social Studies

For this exploration, use the line graph of the unemployment rate.

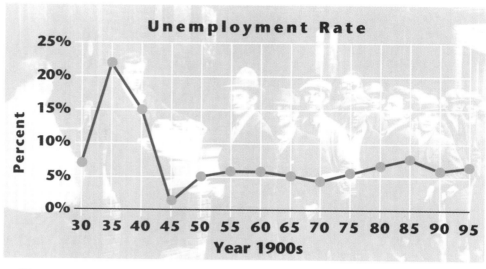

1 Find the point that is above all of the other points. Look at the scale along the bottom of the graph to find the corresponding year. In what year was the unemployment rate the greatest? 1935

2 Find the steepest line segment that goes up. Look at the scale along the bottom of the graph to find the corresponding time period. Between what years did the greatest increase in unemployment occur?
1930 and 1935

3 Find the point that is below all of the other points. Look at the scale along the bottom of the graph to find the corresponding year. In what year was the unemployment rate the lowest? 1945

4 Find the steepest line segment that goes down. Look at the scale along the bottom of the graph to find the corresponding time period. Between what years did the greatest decrease in unemployment occur?
1940 and 1945

5 Look for three points that lie on the same horizontal line. Look at the scale along the bottom of the graph to find the corresponding years. In what three years was the unemployment rate the same?
Any three of these years: 1955, 1960, 1975, 1990

6 Based on the graph, what would you expect the unemployment rate to be in the year 2000? Is this prediction reasonable? What factors must you consider when making predictions based on current data? ❖

CRITICAL Thinking

What historical event might account for the rise in the unemployment rate from 1930 to 1935? What event would account for the decline in the unemployment rate from 1940 to 1945?

Interpreting Circle Graphs

Graphs are used to present data that can be used to analyze current trends. A **circle graph** shows how portions, or percentages, of a *whole quantity* are distributed. The Extension below shows how a circle graph can be used to show the percent of family income spent in different categories.

> ### EXTENSION

Economics   This circle graph shows how the average American family spends its money.

According to the graph, how much might a family with an annual income of $32,000 spend on housing?

From the graph you can see that the average American family spends 31% of its annual income on housing.

Investment 8%
Clothing 6%
Other 22%
Food 15%
Transportation 18%
Housing 31%

Estimation
31% of $32,000 can be estimated as $\frac{1}{3}$ of $30,000.

$\frac{1}{3} \cdot 30,000 = 10,000$

So a family with an annual income of $32,000 might spend about $10,000 on housing. You can use multiplication to find the exact amount.

Multiplication
Multiply 31% times 32,000.

$$0.31 \cdot 32,000 = 9920$$

The exact amount that a family with an annual income of $32,000 might spend on housing is $9920. ❖

Try This   If a family with an annual income of $32,000 spends $7000 on food, are they within the percent given by the graph?   No; $7000 > 15% of $32,000

EXERCISES & PROBLEMS

Communicate

1. Why are graphs important statistical tools?

2. Explain how a graph can be misleading.

3. Describe three different types of graphs that can be used to display data.

4. Describe a situation in which a circle graph is the best way to display a set of data.

5. Describe a situation in which a bar graph is the best way to display a set of data.

6. Explain the difference between a bar graph and a line graph.

7. Nell is doing a science-fair project. She is investigating the change in a person's pulse rate after exercise. Describe a graph that she could make to display this data.

8. Marcus wants to make a graph to show the number of students in the different extracurricular activities in his school. Describe the graph he could make.

Practice & Apply

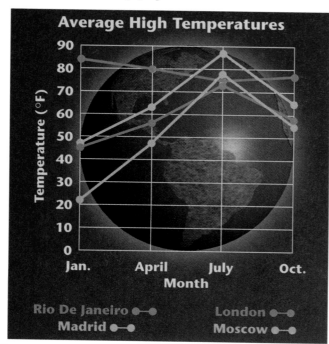

Geography Use the line graph of average high temperatures for Exercises 9–13.

9. Which city had the highest temperature? In which month did this temperature occur? Madrid; July

10. Which city had the lowest temperature? In which month did this temperature occur? Moscow; January

11. Which city had the greatest increase in temperature from January to July? What was the approximate amount of increase? Moscow; approx 55°

12. Which city had the greatest decrease in temperature from July to October? What was the approximate amount of decrease? Moscow; approx 25°

13. Which cities had about the same temperature in July? What was that approximate temperature? Rio de Janeiro, Moscow, and London; approx 75°

Graphing Data

Some professionals spend much of their time researching and collecting data. Graphs are used to display data so that measures of central tendency can be easily viewed, and comparisons can be made.

Gray whales, which live in the northern Pacific Ocean, were recently taken off the endangered species list. Some data, such as the length of full-grown gray whales, have been collected by scientists.

Length of Full-Grown Gray Whales (in meters)

12.7	12.8	12.8	12.8	12.9	13.0	13.1	13.1	13.2	13.3	13.4	13.4
13.5	13.5	13.6	13.6	13.6	13.7	13.7	13.7	13.8	13.8	13.9	13.9
14.0	14.0	14.0	14.0	14.1	14.1	14.1	14.1	14.1	14.1	14.2	14.2
14.2	14.3	14.3	14.3	14.4	14.4	14.4	14.5	14.5	14.5	14.6	15.2

Stem-and-Leaf Plots

A **stem-and-leaf plot** is one way to display data. This stem-and-leaf plot shows the lengths of several full-grown gray whales. The numbers to the left of the vertical line are called **stems**. The numbers to the right of the line are called **leaves**.

Data Set

12.7	13.0	13.6	14.0	14.3	15.2
12.8	13.1	13.6	14.0	14.3	
12.8	13.1	13.7	14.0	14.3	
12.8	13.2	13.7	14.0	14.4	
12.9	13.3	13.7	14.1	14.4	
	13.4	13.8	14.1	14.4	
	13.4	13.8	14.2	14.5	
	13.5	13.8	14.2	14.5	
	13.5	13.9	14.2	14.6	
	13.9				

Stem-and-Leaf Plot

Stems

Leaves

Stem	Leaves
12	7, 8, 8, 8, 9
13	0, 1, 1, 2, 3, 4, 4, 5, 5, 6, 6, 7, 7, 7, 8, 8, 8, 9, 9
14	0, 0, 0, 0, 1, 1, 2, 2, 2, 3, 3, 3, 4, 4, 4, 5, 5, 6
15	2

Data Set		Stem-and-Leaf Plot	
		Stems	Leaves
15	41	1	5, 7, 8, 9
46	18	2	8
52	55	3	3
19	17	4	1, 6
33	28	5	2, 5

To record data in a stem-and-leaf plot, first examine the data set to determine what the stems will be. In the stem-and-leaf plot on page 601, the stems are whole numbers, and the leaves are tenths.

If the data set consists of numbers from 15 to 55, the stems can represent the tens digits, and the leaves can represent the ones digits.

Exploration *Using a Stem-and-Leaf Plot*

1 The data in the stem-and-leaf plot are the length (in meters) of several gray whales. How many measurements are shown by the stem-and-leaf plot? 48

Length of Gray Whales (in meters)

key
12|7 = 12.7

Stems	Leaves
12	7, 8, 8, 8, 9
13	0, 1, 1, 2, 3, 4, 4, 5, 5, 6, 6, 6, 7, 7, 7, 8, 8, 9, 9
14	0, 0, 0, 0, 1, 1, 1, 1, 1, 1, 2, 2, 2, 3, 3, 3, 4, 4, 4, 5, 5, 5, 6
15	2

2 What is the smallest length? What is the greatest length? What is the range of these lengths? 12.7; 15.2; 2.5

3 What is the mode of the lengths? 14.1

4 Estimate the median of the data. What is the actual median? How close was your estimate? Estimate: 14; actual: 13.95; within 0.05 Answers may vary.

5 Estimate the mean. What is the actual mean? How close was your estimate? Estimate: 14; actual: 13.8; within 0.2 Answers may vary.

6 Explain how the arrangement of data in a stem-and-leaf plot provides a convenient way to estimate the measures of central tendency.

7 Suppose that you are using this data to report the average length of gray whales. Which measure of central tendency (mean, median, or mode) would most accurately describe this set of data? ❖

Breathing Intervals (in minutes)

5.6	14.8	6.5	8.4	15.6	7.9	15.6	7.1	16.2
8.6	14.7	15.5	11.0	12.6	9.6	11.4	9.2	14.0
10.3	13.6	8.6	13.5	10.5	10.8	9.2	10.5	12.9
11.3	12.2	13.8	15.8	6.7	11.7	6.8	11.4	11.5
12.3	11.2	10.4	9.8	9.1	12.9	10.5	12.8	9.4
13.7	10.1	12.6	12.2	13.8	14.2	12.7	13.9	10.7
14.8	8.5	11.4	14.7	11.4	16.7	13.9	15.6	7.3

Gray whales breathe every 5 to 15 minutes. They can, however, go as long as 40 minutes without breathing. A scientist studying a group of whales recorded their breathing intervals (in minutes) in a stem-and-leaf plot.

key
$5|6 = 5.6$

To construct the stem-and-leaf plot, first determine the stems. The numbers range from 5.6 to 16.7, so let the stems be whole numbers, from 5 to 16, and let the leaves be tenths. Then record the leaves by the appropriate stems. Arrange the leaves in each row from least to greatest. ❖

Try This Make a stem-and-leaf plot for the following data: 97, 86, 64, 79, 91, 81, 80, 99, 80, 96, 79, 76.

Breathing Intervals (in minutes)

Stems	Leaves
5	6
6	5, 7, 8
7	1, 3, 9
8	4, 5, 6, 6
9	1, 2, 2, 4, 6, 8
10	1, 3, 4, 5, 5, 5, 7, 8
11	0, 2, 3, 4, 4, 4, 4, 5, 7
12	2, 2, 3, 6, 6, 7, 8, 9, 9
13	5, 6, 7, 8, 8, 9, 9
14	0, 2, 7, 7, 8, 8
15	5, 6, 6, 6, 8
16	2, 7

Histograms

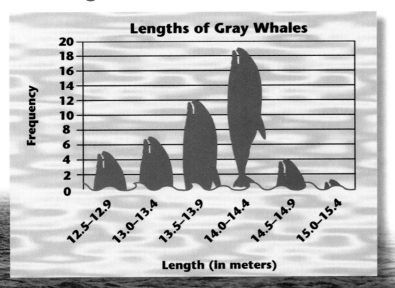

A **histogram** is a bar graph that shows how frequently the numbers in a data set appear. You can construct a histogram from the data in a stem-and-leaf plot. Refer to the stem-and-leaf plot for the length of gray whales on page 602.

The lengths are listed along the horizontal axis of the histogram, and the frequencies of the measurements in the data set are listed along the vertical axis.

Notice that the lengths are listed on the horizontal axis by ranges, such as from 12.5 to 12.9, instead of by each individual measure.

Lesson 11.3 Graphing Data **603**

Is it possible to accurately determine the range, mean, median, or mode from a histogram? Why or why not? Answers may vary. A sample answer is no, because you do not know each individual number or how many times each number is used.

Box-and-Whisker Plots

Box-and-whisker plots show how data is distributed by using the median and the range. The example below shows how to create a box-and-whisker plot.

EXAMPLE

Length of Gray Whales (in meters)

key

| 12|7 = 12.7 |

Stems	Leaves
12	7, 8, 8, 8, 9
13	0, 1, 1, 2, 3, 4, 4, 5, 5, 6, 6, 6, 7, 7, 7, 8, 8, 9, 9
14	0, 0, 0, 0, 1, 1, 1, 1, 1, 1, 2, 2, 2, 3, 3, 3, 4, 4, 4, 5, 5, 5, 6
15	2

Use the stem-and-leaf plot of the length of gray whales (in meters) to construct a box-and-whisker plot.

Solution ➤

First draw a number line. Use points to mark the least value, 12.7, and greatest value, 15.2. Find and mark the median.

Because there are 48 data values, the median is the mean of the 24th and 25th ordered values, 13.95.

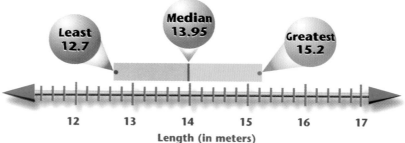

Find the median of all the measurements in the lower half, below 13.95. This measurement, called the **lower quartile**, is 13.45.

Find the median of all the measurements in the upper half, above 13.95. This measurement, called the **upper quartile**, is 14.2.

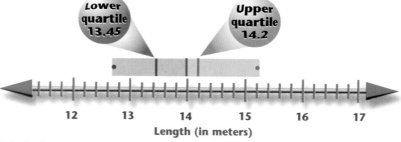

Mark the upper and lower quartiles with vertical marks above the number line.

Draw a rectangular box from the lower quartile to the upper quartile. Complete the box-and-whisker plot by drawing the line segments from each end of the box to the points for the least and greatest values. ❖

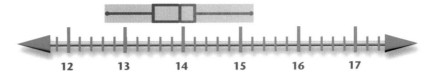

Graphics Calculator

Some graphics calculators create box-and-whisker plots from data that you enter. You can find the lower quartile, the median, and the upper quartile by using the trace feature.

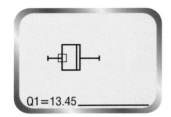

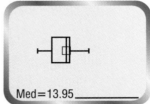

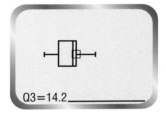

EXERCISES & PROBLEMS

Communicate

1. Explain how to construct a stem-and-leaf plot.

2. Explain how to make a histogram from a stem-and-leaf plot.

3. Explain how to determine the median and the range from a box-and-whisker plot.

4. Explain how to find the upper and lower quartiles from a box-and-whisker plot.

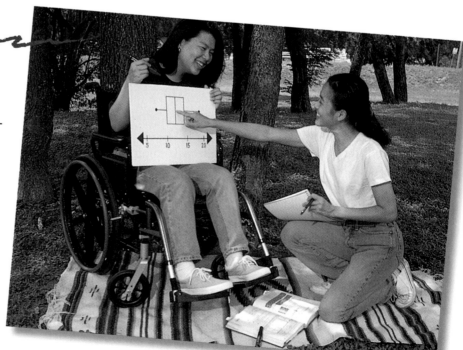

Practice & Apply

Humpback whales communicate with one another by making a variety of sounds called phonations. *These phonations last up to 20 minutes.*

Biology A scientist recorded the length of humpback whale songs. Use the data for Exercises 5–10.

Duration of Humpback Whale Songs (in minutes)							
16.6	19.9	17.2	18.9	19.7	14.1	15.5	19.3
14.5	18.1	15.6	17.8	20.0	17.4	18.2	16.1
15.3	16.3	17.2	16.2	19.3	15.0	14.9	15.2

5. Use the data to make a stem-and-leaf plot.

6. What is the range of the data? 5.9 min

7. What is the median of the data? 16.9 min

8. What is the mean of the data? 17.0 min

9. Is there a mode? If so, what is it? yes; 17.2 and 19.3

10. What is the central tendency of the duration of a humpback whale song, according to this data set? What measure of central tendency do you think best answers this question? Why? Answers may vary. A sample answer is the mean, 17.0 min, since the mean implies average.

Sports The stem-and-leaf plot shows the number of runs credited to the top players in a baseball league. The stems are tens digits, and the leaves are ones digits.

11. Another player had 60 runs. If his total is added to the stem-and-leaf plot, would the range, mean, median, or mode be affected? If so, how?

12. What percent of the players scored more than 80 runs? Round your answer to the nearest tenth of a percent. 17.2%

13. What percent of the players scored fewer than 60 runs? Round your answer to the nearest tenth of a percent. 44.8%

14. Construct a histogram of the data.

15. Make a box-and-whisker plot of the data.

key
3|5 = 35

Baseball Runs

Stems	Leaves
3	5, 5
4	0, 2, 2
5	2, 2, 5, 6, 7, 9, 9, 9
6	0, 0, 1, 2, 3, 5
7	4, 6, 7, 9, 9
8	1, 1, 8, 9
9	
10	4

Academics Use the box-and-whisker plot for Exercises 16–25.

16. Which class had the lowest test score? Class A

17. Which class had the highest test score?
All classes had high score of 100.

18. Which class had the greatest
range of test scores? Class A

19. Which class had the
highest median score? Class A

20. What was the median score
for Class A? 95

21. Is it possible to determine the
mean score for Class B? No

22. A student in Class C scored 83 on
the exam. Where in a box-and-
whisker plot does this score fall?

23. A student in Class B also scored 83
on the exam. Where in a box-and-
whisker plot does this score fall?

24. Is it correct to say that the median score
of Class C is above 80? Why or why not?
No; median score is shown to be less than 80.

25. Can you determine how many students were tested in each class?
Explain. No, the number of items in the data set cannot be
determined from a box-and-whisker plot.

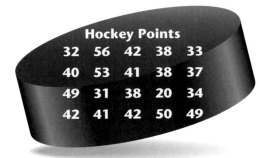

Sports The total number of points scored by hockey teams in one league
are listed. Use this data for Exercises 26–33.

Hockey Points

32	56	42	38	33
40	53	41	38	37
49	31	38	20	34
42	41	42	50	49

26. Make a stem-and-leaf plot of the data.

27. Find the range of the scores. 36

28. What is the median of the data? 40.5

29. Does the data set have a mode? If so, what is it?
yes; 38, 42

30. What is the mean number of points scored? 40.3

31. If the number of points scored by the highest-scoring
team is raised by 5 points, how would the mean be
affected? Would this change the median or mode? If
so, how? mean would change to 40.6; median and
mode would not change.

32. Make a histogram for the hockey data.

33. Construct a box-and-whisker plot for the data.

Health The data represent the average number of dollars
that 40 families spent on dental care for one child in one year.

34. Make a stem-and-leaf plot of the data.

35. What is the median of the data? $163.50

36. What are the lower and upper quartiles of this data? 130.5; 182.5

37. Make a box-and-whisker plot of this data.

Dollars Spent on Dental Care				
130	255	178	141	118
106	245	168	196	184
131	202	146	210	120
141	159	142	119	172
114	124	178	181	208
168	169	125	176	137
106	108	153	152	153
184	194	185	172	170

Biology Dolphins are not usually considered whales, but scientists classify them as toothed whales because they have the same basic body shape as other toothed whales. The best-known dolphin is the bottle-nosed dolphin, which can weigh up to 600 pounds. Use the data below for Exercises 38–43.

Weight of Bottle-nosed Dolphins (in pounds)				
345	574	249	497	588
567	586	383	484	491
275	499	478	343	549
399	512	589	576	497

38. Use the data to make a stem-and-leaf plot.

39. What is the range of the data? 340

40. What is the median of the data? 497

41. What is the mean of the data? 474

42. Is there a mode? If so, what is it? Yes; 497

43. What is the average dolphin weight in this data set?
    474 lb

Look Back

Find each product. [Lesson 3.5]

44. $8 \cdot 1\frac{1}{4}$ 10   **45.** $\frac{2}{3} \cdot 2\frac{3}{5}$ $1\frac{11}{15}$ **46.** $6\frac{1}{5} \cdot 3\frac{3}{10}$ $20\frac{23}{50}$

Write each percent as a fraction in lowest terms. [Lesson 3.7]

47. 80% $\frac{4}{5}$ **48.** 38% $\frac{19}{50}$ **49.** 3.5% $\frac{7}{200}$

Small Business A book dealer held a book fair at a school. She gave the school $150 for allowing her to hold the book fair. She also gave the school 20% of the profits. **[Lesson 3.7]** Let x represent the amount of profit.

50. Write an expression for the school's income. 150 + 0.20x

51. What must the total sales be for the book dealer to have an income of $700? $1062.50

52. **Biology** A wildebeest was observed running a distance of 86.4 meters in 4.5 seconds. Find the speed of the wildebeest in meters per second. **[Lesson 6.2]** 19.2 meters per second

53. Solve $4y - 3 = 6y - 2$ for y. **[Lesson 7.1]** $y = -\frac{1}{2}$

Look Beyond

54. The formula for the circumference of a circle is $C = \pi d$, where d is the diameter. Use this formula to find the sum of the circumferences of the two circles shown.

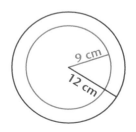

LESSON 11.4 Circle Graphs

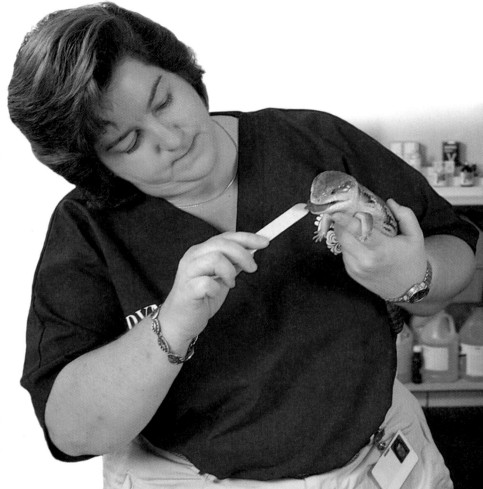

why *Circle graphs, or pie charts, are pictures of how parts of a whole are related. One common use of circle graphs is to compare different groups in a single category.*

Reptiles consist of 6% of the pets seen at the vet clinic.

Constructing a Circle Graph

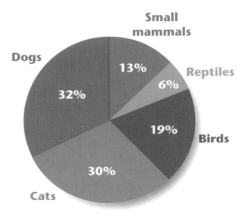

- Dogs 32%
- Small mammals 13%
- Reptiles 6%
- Birds 19%
- Cats 30%

Recall that circle graphs show different parts of a whole. For example, this graph shows the different types of patients seen at one veterinarian's office.

Data in a circle graph are often presented according to percents. The entire graph represents 100%.

To construct a circle graph, it is important to know that a circle consists of 360°. You will use this information when you determine the size of the individual sections of the graph.

EXAMPLE 1

The American Canine Club has many registered breeds. Dogs are registered in seven categories. Make a circle graph showing the percent of dogs registered in each category.

Solution ➤

To construct a circle graph, you will first need to calculate the angle measures for the sections of the graph, or *sectors* of the circle. Look at the data for sporting dogs, which makes up 25% of the registered dogs. This sector will take up 25% of the circle. Multiply 25% by 360°.

$$0.25 \cdot 360 = 90$$

The angle for this sector will be 90°.

Use a compass to draw a circle, marking the center of the circle with a point. Using a protractor, draw a 90° angle in the circle with the vertex at the center point.

Sporting Dogs, 25%
(Yellow labrador)

Terriers, 5%
(West Highland terrier)

Herding Dogs, 15%
(Australian shepherd)

Working Dogs, 5%
(Siberian husky)

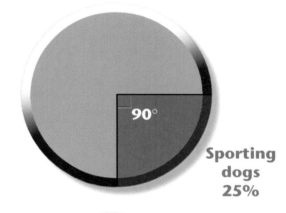

90°

Sporting dogs 25%

Non-sporting Dogs, 25%
(Boston terrier)

Hounds, 10%
(Dachsund)

Toy Dogs, 15%
(Yorkshire terrier)

Calculate the angle measures for the remaining percentages.

Category of breeds	Percent	Angle measure
Sporting dogs	25%	0.25 • 360 = 90°
Hounds	10%	0.1 • 360 = 36°
Working dogs	5%	0.05 • 360 = 18°
Herding dogs	15%	0.15 • 360 = 54°
Terriers	5%	0.05 • 360 = 18°
Toy dogs	15%	0.15 • 360 = 54°
Non-sporting dogs	25%	0.25 • 360 = 90°

Draw the remaining angles in your circle graph, and label each section.

To make a circle graph, you need a compass and a protractor.

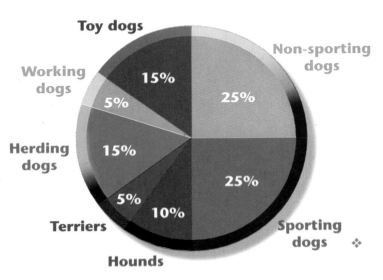

Try This A survey of 500 Americans was conducted to find the position in which they fall asleep. Make a circle graph showing the percent of people in each category.

Sleeping position	Percent
Side	56%
Back	5%
Stomach	11%
Varies	11%
Not sure	17%

When a data set does not contain percents, the percents must be calculated in order to construct a circle graph.

EXAMPLE 2

Make a circle graph for the number of cats in each breed that are in the Feline Fanciers cat show. Use the information in the table.

Breed	Number of cats in show
Persian	99
Abyssinian	30
Balinese	66
Manx	41
Siamese	106
Total	342

The Feline Fanciers Association is having a cat show. Five different cat breeds appear in the show.

Solution ➤

Find the percent for each category.

There are 99 Persian cats in the show. The total number of cats in the show is 342.

$$\frac{99}{342} \approx 0.29, \text{ or approximately } 29\%$$

The Persian breed accounts for 29% of the show.

Find the number of degrees in the circle used to represent that category. A circle contains 360°, and 29% of 360° is

$$0.29 \cdot 360 = 104.4°.$$

Continue for the remaining categories. The original table can be extended to include columns for the percents and the number of degrees.

Breed	Number of cats in show	Percent	Number of degrees
Persian	99	$\frac{99}{342} \approx 0.29 \approx 29\%$	$0.29 \cdot 360 = 104.4°$
Abyssinian	30	$\frac{30}{342} \approx 0.09 \approx 9\%$	$0.09 \cdot 360 = 32.4°$
Balinese	66	$\frac{66}{342} \approx 0.19 \approx 19\%$	$0.19 \cdot 360 = 68.4°$
Manx	41	$\frac{41}{342} \approx 0.12 \approx 12\%$	$0.12 \cdot 360 = 43.2°$
Siamese	106	$\frac{106}{342} \approx 0.31 \approx 31\%$	$0.31 \cdot 360 = 111.6°$
Total	342	100%	360°

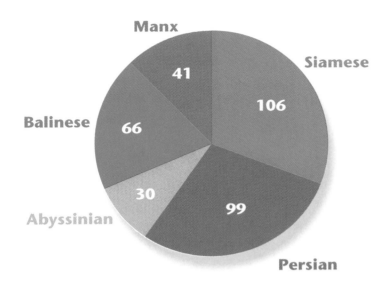

Manx
41
Siamese
106
Balinese
66
Abyssinian
30
Persian
99

Use a compass to construct the circle, and mark the center of the circle with a point. Then use a protractor to draw angles with the indicated measures. It may be necessary to round your angle measures to the nearest degree. ❖

Some computer programs allow you to make many different types of graphs. With some programs, you can make a circle graph for this data by entering the percents. Other programs allow you to enter the original data, and the programs automatically calculate the percents for you.

CRITICAL
Thinking

The circle graph at the right displays the data from Example 2. Explain why the angles in this graph appear to be different from the angles in the graph in Example 2.

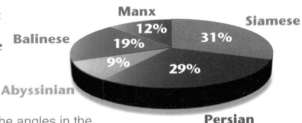

Manx 12%
Balinese 19%
Abyssinian 9%
Siamese 31%
Persian 29%

Answers may vary. A sample answer is that the angles in the second graph appear to be different because of the slant or tilt on the perspective of the graph.

EXERCISES & PROBLEMS

Communicate

1. Describe a situation in which you could use a circle graph to display data.

2. Describe a situation in which you would not use a circle graph to display data.

Hobbies The histogram shows the ages of the stamps in a stamp collection.

3. Describe how you can find the percents that represents each category from the histogram.

4. Explain how to use the percents to make a circle graph.

Stamp Collection

Number of stamps

Age of stamps (in years)
Below 60 | 60–69 | 70–79 | 80–89 | 90–100

Practice & Apply

Economics The Apartment Finders group conducted a survey to determine monthly apartment rents in Mortonville. The circle graph shows the percent of apartments with each indicated monthly rent.

5. How much is the rent for the category with the greatest number of apartments?
$550–$700

6. How much is the rent for the category with the least number of apartments?
less than $400

7. How many categories each account for less than $\frac{1}{4}$ of the apartments? What is the rent for the apartments in each of these categories? more than $1000 and less than $400

8. Which two categories include about the same number of apartments?
$700–$1000 and $400–$550

9. If there are 75,000 apartments in this city, approximately how many would you expect to have rents between $550 and $700? 26,250

10. Kerrville has 80,000 apartments with 15,000 of them renting for $700–$1000 a month. What percent of the apartments in Kerrville rent for $700–$1000 a month? How does this compare with the same category in Mortonville? 18.8%; approx 6% less

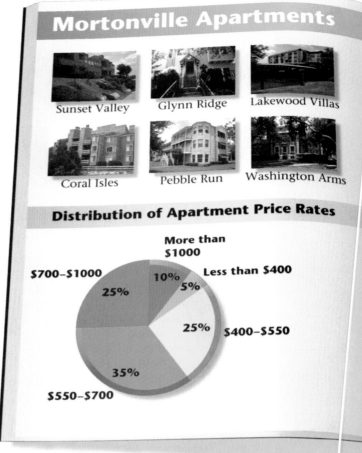

Mortonville Apartments

Sunset Valley   Glynn Ridge   Lakewood Villas

Coral Isles   Pebble Run   Washington Arms

Distribution of Apartment Price Rates

More than $1000 10%
Less than $400 5%
$700–$1000 25%
$400–$550 25%
$550–$700 35%

Communications The following table shows the number of customers who use each of the long distance services available in their community.

	United Long Distance	ABC Service	Advanced Services	Multi Media Services	Fast Phones Long Distance
Number of customers	14,500	8400	6800	4200	3500
Percent	?	?	?	?	?

11. Find the total number of customers using these services. 37,400

12. Copy and complete the table by finding the percent of customers using each service.

13. Use the information from the table to make a circle graph of the percent of customers using each service.

14. Suppose that 400 of the ABC Service customers change to Advanced Services. How would the circle graph you drew in Exercise 13 change?

15. Suppose that 1000 new customers are added and are equally divided among the five services. How would the circle graph you drew in Exercise 13 change? It would not change significantly.

Business The annual report for 1995 for Nichol's Discount Department Store included the circle graph at the right, which shows the percent of business for each department. Suppose that the total sales for the year were $850,000.

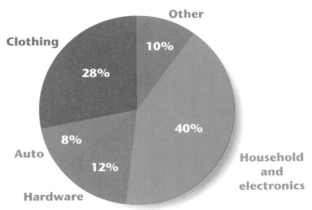

Nichol's Discount Department Store Sales

16. How much were the sales in the hardware department? $102,000

17. How much were the sales in the household and electronics department? $340,000

18. How much were the sales in the auto and clothing departments?
Auto: $68,000; Clothing: $238,000; Total: $306,000

19. Which three departments combined accounted for 50% of the sales?
Hardware, clothing, and other

20. The total sales in 1996 were $980,000. The sales in the household and electronics department were $415,000. Was the percent of sales in this department an increase or decrease in percent from 1995? Justify your answer.

21. The total sales in 1997 were $950,000. The percent of sales in the clothing department was the same in 1997 as in 1995. What were the sales in the clothing department in 1997? $266,000

22. For what situation would a circle graph be the best way to present data?

23. For what situation would a circle graph *not* be the best way to present data? What would be a better way to show the data in this situation?

Sales The table at the right shows the number of cars sold by each salesperson in the first quarter of the year.

24. Complete the table.

25. Construct a circle graph to display the data.

26. What percent of cars were sold by Jarvis? Approx 15%

27. Who sold the greatest percent of the total number of cars? Washington

28. What is the average number of cars sold by the 7 sales people? 21

29. Who sold the median number of cars? Gonzales

30. What is the mode of this data? 17

Car Sales			
	Number of cars	Percent	Degrees
Hart	17	?	?
Morton	26	?	?
Kelley	15	?	?
Washington	30	?	?
Jarvis	22	?	?
Gonzales	19	?	?
Swensen	17	?	?
Total	?	100%	360°

31. Describe a situation in which a bar graph could be used to best display data.

32. Describe a situation in which a bar graph would *not* be the best way to present data. What would be a better way to show the data in this situation?

Wages   The histogram below shows the number of employees in each salary category for a trucking company. Use this information for Exercises 33 and 34.

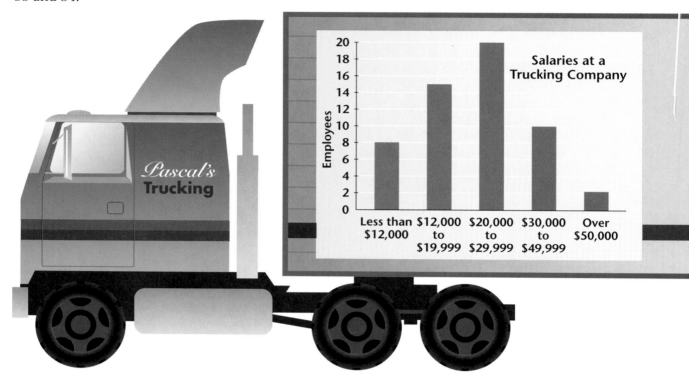

Salaries at a Trucking Company

33. How many employees does the company have?   55

34. Which category has the greatest percent of employees? $20,000 to $29,999; it has the greatest Justify your answer.   number of employees, so it has the greatest percent of employees.

Use the table at the right for Exercises 35–40.

35. Copy and complete the table.

36. Construct a circle graph to display the data.

37. In which category does the median salary lie? $20,000 to $29,999

38. Explain why you cannot determine the category in which the mode or the mean lies.   You do not know exact amounts.

	Amount	Percent	Degrees
Less than $12,000	12	?	?
$12,000 to $19,999	18	?	?
$20,000 to $29,999	32	?	?
$30,000 to $50,000	9	?	?
Over $50,000	1	?	?
Total	?	100%	360°

39. Suppose that three more employees are hired and paid a salary in the $12,000 to $19,999 range. Would this affect your answer to Exercise 37? Explain.   No, the median would still be in the $20,000 to $29,999 category.

40. Describe a situation that would affect the category in which the median salary lies.

41. **Portfolio Activity**   Complete the portfolio activity on page 585.

Look Back

Evaluate each expression if *a* is −2, *b* is 3, and *c* is −6. [Lesson 1.6]

42. abc^2 −216   **43.** $2ab^2c$ 216   **44.** a^2b^3 108

Geometry The lengths of the sides of a triangle are given. Use the converse of the "Pythagorean" Right-Triangle Theorem to determine whether each triangle is a right triangle. **[Lesson 4.7]**

45. 6, 8, 10 yes   **46.** 14, 20, 26 no   **47.** 8, 19, 25 no

Demographics The age distribution of the population of the United States according to the 1990 census is shown by the bar graph at the right. **[Lesson 11.1]**

48. What is the smallest population group in the United States? What is the second smallest group? under 5; 65 and over

49. If the United States population was 248,709,873 according to the 1990 census, explain how you would estimate the number of Americans from 5 to 19 years old. What is your estimate? Multiply 250,000,000 by 20%. Estimate: 50,000,000

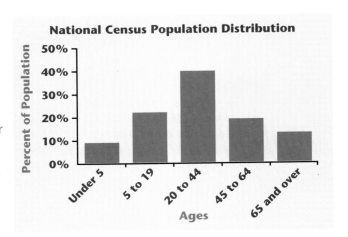

National Census Population Distribution

50. Technology Using a calculator, find the number of Americans from 5 to 19 years old from Exercise 49. Answers may vary. If the percent is 22%, then there are 54,716,172 Americans between 5 and 19 years old.

Look Beyond

Travel Mary drives her car out of a city at a steady rate of 40 miles per hour. The distance that she drives away from the city depends on the number of hours that she drives. Her distance from the city equals the rate times the number of hours.

51. How far away from the city is Mary after 2 hours? 80 mi

52. Write an equation to describe the distance that Mary drives in terms of the number of hours that she drives.
$d = 40h$, where h is the number of hours that Mary drives

53. Copy and complete this table.

54. Graph the points from your table on a coordinate plane. Connect the points. Describe the graph.

Time (in hours)	Distance (in miles)
1	? 40
2	? 80
3	? 120
4	? 160
5	? 200

Lesson 11.4   Circle Graphs   **617**

LESSON 11.5

Exploring

Scatter Plots and Correlation

why *People frequently try to determine relationships between different factors such as math and science aptitude. How can you tell whether these factors are related? Correlations provide a way to measure numerically how well the sets of data are related.*

Scatter Plots

An effective way to see a relationship from data is to display the information as a **scatter plot**. Before you can use a scatter plot, however, you need to

A newspaper article reported that students who do well in math also do well in science.

- have clearly defined variables that you can assign to the axes,
- form ordered pairs from the data you have, and
- locate each data point on the scatter plot in the same way you would graph a point on the coordinate plane.

You can then examine the patterns for clues to see how closely the variables are related.

Exploration Find a Correlation

STATISTICS
Connection

Ten students took a series of aptitude tests. The table shows the scores on two parts of the test. You want to know if students who do well in math also do well in science. Once you create a scatter plot, you can see the pattern for the relationship.

Student	A	B	C	D	E	F	G	H	I	J
Math	61	40	80	21	62	54	20	33	75	51
Science	70	38	92	50	68	41	38	20	73	48

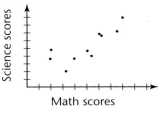

1 Look at the scores of the students who did well in math. How well did they do in science?

2 Look at the scores of the students who did *not* do as well in math. How well did they do in science?

3 Do the data points appear broadly scattered or do they seem to cluster along a line? cluster

4 Does the line of data rise or fall as you go from left to right?
rises from left to right

5 What does the data pattern indicate about the relationship between the variables? ❖ The data pattern shows a correlation between math scores and science scores. Students with high math scores tend to have high science scores, and vice versa.

Correlation

Scatter plots show a picture of how the variables relate to each other by showing how well data points fit a line.

Strong positive correlation As one variable increases, the other also tends to increase. This shows a *positive* correlation. When the points are nearly in a line, there is a *strong* correlation.

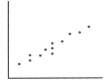

Strong negative correlation As one variable increases, the other tends to decrease. This shows a *negative* correlation. When the points are nearly in a line, there is a *strong* correlation.

Little or no correlation As one variable increases, you cannot tell if the other tends to increase or decrease. There is a weak correlation or none at all.

APPLICATION

Compare the scatter plots with the statements.

1. The variables studied are student's *age* and *distance from home to school.*

2. The variables studied are child's *age* and *time taken to run a fixed distance.*

3. The variables studied are child's *age* and *height.*

Match statements 1, 2, and 3 with the appropriate scatter plots.

Scatter plot A

Scatter plot A shows a positive correlation. A positive correlation occurs when both variables increase. The correlation is strong. The data points are close to the shape of a straight line. This indicates that it is easy to predict the behavior of one variable by the behavior of the other. As a child's age increases, his or her height increases. This scatter plot would fit statement 3.

Scatter plot B

Scatter plot B shows a negative correlation. A negative correlation occurs when one variable increases and the other variable decreases. The data points are close to the shape of a straight line, which is characteristic of a strong correlation. As a child gets older, bigger, and stronger, the time it takes the child to run a fixed distance decreases. This scatter plot would fit statement 2.

Scatter plot C

Scatter plot C shows little or no correlation. The data points are scattered. There is no indication that the correlation is positive or negative. The distance that a student lives from school has no relation to student's age. This scatter plot would fit Statement 1. ❖

Do not confuse a strong positive or negative correlation with a cause-and-effect relationship. Over the past few years there has been a strong positive correlation between annual consumption of diet soda and the number of traffic accidents reported in one Midwestern state. No one would claim that drinking diet soda causes accidents or that the trauma of being in an accident causes people to drink more diet soda. Often a third variable can be responsible for the unexpectedly strong correlation between two otherwise unrelated variables. Can you think of another variable that might be responsible for the correlation in the Midwestern state mentioned? Answers may vary. One possible answer is that an increase in population in the state is responsible for the increase in diet soda consumption and the increase in traffic accidents.

APPLICATION

Agriculture Make a scatter plot for the given data, showing the number of people working on farms for various years. Describe the correlation as strong positive, strong negative, or little to none.

Decline in Number of Farm Workers

Year	Number
1940	8995
1950	6858
1960	4132
1970	2881
1980	2818
1990	2864

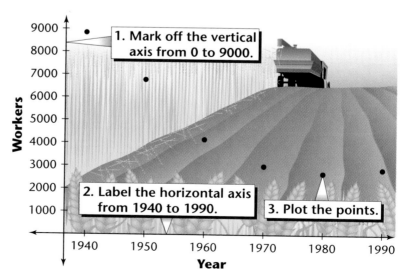

1. Mark off the vertical axis from 0 to 9000.

2. Label the horizontal axis from 1940 to 1990.

3. Plot the points.

As the years increase, the number of workers decreases, so this is a strong negative correlation. However, there has been a leveling off over the last 20 years. ❖

CRITICAL Thinking

Think of two examples of related data. Describe the kind of correlation you would expect from the data. Answers may vary. For example, car-trip mileage and gasoline consumed per trip should show a high positive correlation. A person's height and the distance from the top of his/her head to the ceiling should have a perfect negative correlation.

EXERCISES & PROBLEMS

Communicate

1. Use the scatter plot from the exploration on page 618 to explain how you would decide what type of correlation exists between the math and science scores.

2. How would you label the axes of a scatter plot involving the *cost* of notebooks and the *number of pages* each has? Which axis would you use for the cost? Which axis would you use for the number of pages?

3. Discuss how a correlation provides you with information about the variables.

Describe the correlation for each scatter plot as strong positive, strong negative, or little to none. Explain the reason for your answer.

4.

5.

Consumer Economics The table below shows that the cost of food items that were $1.00 in 1982 has increased over the years.

Year	1982	1985	1988	1991
Cost	1.00	1.06	1.18	1.36

January						
Sun	Mon	Tue	Wed	Thu	Fri	Sat
		1	2	3	4	5
6	7	8	9	10	11	12
13	14	15	16	17	18	19
20	21	22	23	24	25	26
27	28	29	30	31		

6. Which axis you would use for the cost of food, and which you would use for the year? Why?

7. Tell how you would plot the points for the data in the table.

Practice & Apply

Describe each correlation as strong positive, strong negative, or little to none.

8. | Little or none

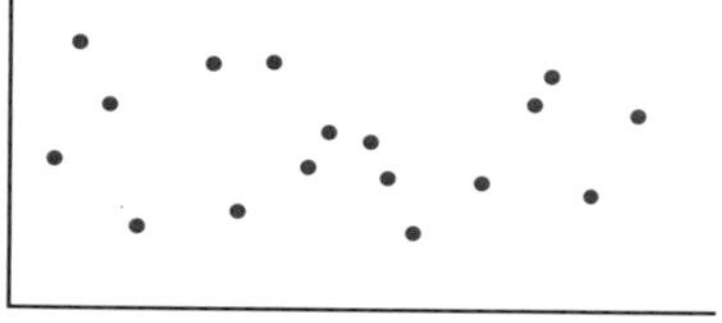

9. | Strong positive

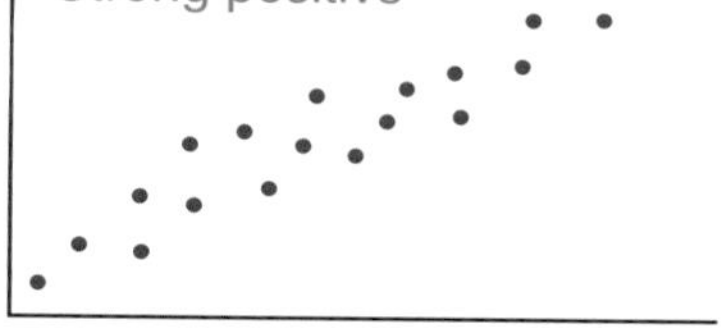

Statistics Use the data in the table for Exercises 10–11.

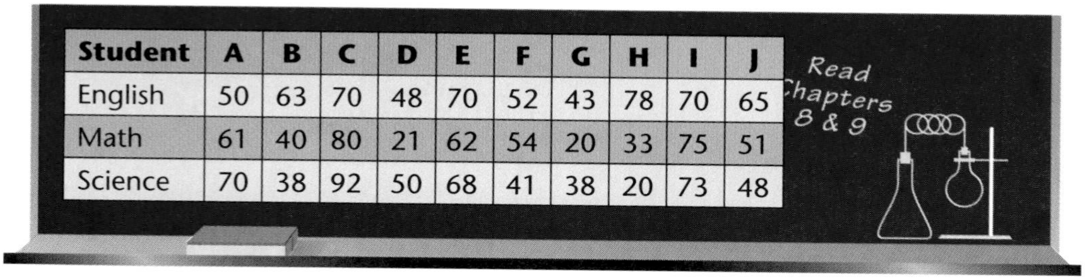

Student	A	B	C	D	E	F	G	H	I	J
English	50	63	70	48	70	52	43	78	70	65
Math	61	40	80	21	62	54	20	33	75	51
Science	70	38	92	50	68	41	38	20	73	48

Read Chapters 8 & 9

10. Graph the scores for math against the scores for English, and explain the correlation.

11. Graph the scores for science against the scores for English, and explain the correlation.

Geography Use the data below to answer Exercises 12–15.

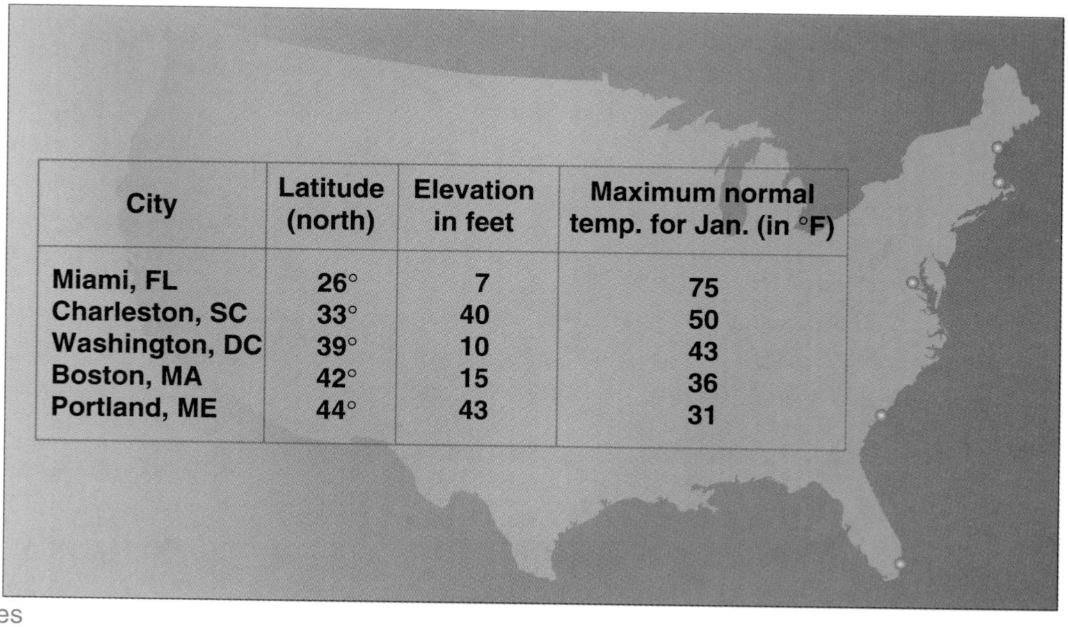

Look down the column for latitude, which is arranged in increasing order. As latitude increases, is there an obvious tendency in what happens to

City	Latitude (north)	Elevation in feet	Maximum normal temp. for Jan. (in °F)
Miami, FL	26°	7	75
Charleston, SC	33°	40	50
Washington, DC	39°	10	43
Boston, MA	42°	15	36
Portland, ME	44°	43	31

12. elevation? No

13. temperature? Yes

14. Describe the correlation between latitude and elevation. Little or no correlation

15. Describe the correlation between latitude and temperature. Strong negative

16. Sports Make a scatter plot of the data below for the winning times in the women's 400-meter run in the Olympics.

Year	1964	1968	1972	1976	1980	1984	1988	1992
Seconds	52.00	52.00	51.08	49.29	48.88	48.83	48.65	48.83

17. Describe the correlation between years and seconds. Strong negative

18. Draw these two coordinate axes on your graph paper. Plot the same points, (1, 2), (2, 3), and (3, 4), on each set of axes. Connect the points with a line.

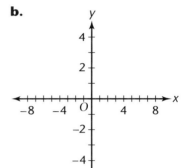

19. Which graph is steeper, **a** or **b**? b

20. What caused one graph to be steeper, even though the points are the same?

21. How does a change in the scale on the axes affect the steepness of the line? *Scale* refers to the distance between units on an axis.

22. Does the scale of the graph affect a correlation? Explain your answer. Not really. If there is a positive correlation, the linear pattern will still go up to the right; the steepness will be different if the scale is changed.

Look Back

Graph each system of inequalities.   [Lesson 9.5]

23. $\begin{cases} 3x + 3y < -3 \\ 4x - 4y \geq 2 \end{cases}$

24. $\begin{cases} 2y > -x - 4 \\ 3y > x + 2 \end{cases}$

25. $\begin{cases} y \leq -x + 3 \\ y \leq 4x - 5 \end{cases}$

26. $\begin{cases} y > \frac{1}{2}x \\ y < 3x \end{cases}$

27. $\begin{cases} y \geq \frac{1}{3}x - 1 \\ y \geq 3x - 2 \end{cases}$

28. $\begin{cases} x \leq 1 \\ y \leq 1 \end{cases}$

Look Beyond

29. Make a scatter plot of the data. Use a straightedge to draw the line that best fits the points on the scatter plot.

x	0	3	6	9	12
y	14	20	26	32	38

30. Use your line to predict the *y*-value for an *x*-value of 15. 44

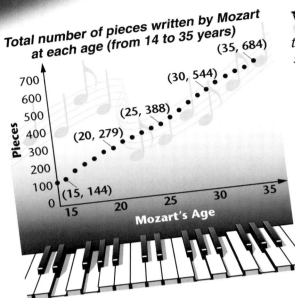

Total number of pieces written by Mozart at each age (from 14 to 35 years)

(35, 684)
(30, 544)
(25, 388)
(20, 279)
(15, 144)

Pieces
Mozart's Age

why *Scatter plots are used to show a trend in the data. This scatter plot presents an overall picture of Mozart's musical output. Because the points lie very nearly on a straight line, it tells you that Mozart wrote at an amazingly steady pace. When a trend is represented by a line that fits the data, you can study the line to see how the data behave in general.*

Music   Wolfgang Amadeus Mozart (1756–1791) composed many musical compositions in his lifetime. The scatter plot shows the total number of compositions that Mozart had written as he reached different ages. For example, the point (35, 684) shows that by age 35 Mozart had written 684 compositions. The graph starts at (14, 121), indicating that by age 14 he had already written 121 musical compositions.

The Line of Best Fit

The **line of best fit** represents an approximation of the data on the scatter plot. The following steps should help you find this line.

Use a straightedge, such as a clear ruler, to model the line.

To fit the line to the points, choose a line that best matches the trend.

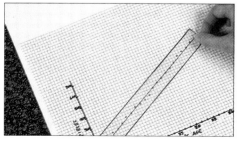

The line does not necessarily have to pass through any of the points.

If you look at the numbers along the Age axis, you see that there is a difference of 21 years between the ages 14 and 35. How many compositions of music did Mozart write during that period of time? Form a ratio that compares the difference in the number of compositions with the difference in the number of years. What does this ratio tell you about Mozart's rate of composition when compared to the given data?

584; 584 to 21; When this ratio is compared to the other ratios formed by the given data, you can see that from the time he was 14 years old until his death at age 35, Mozart composed music at a very steady rate.

Correlation and the Line of Best Fit

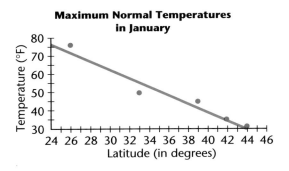

Maximum Normal Temperatures in January

This scatter plot shows the maximum normal temperatures in January for selected cities at different latitudes along the Atlantic coast. It also shows a computer-drawn line that best fits the data. In this scatter plot, the points are close to the line. Use the line of best fit to predict the maximum normal temperature in January for a city at 29° latitude. Approx 65° F

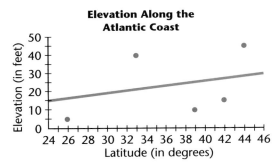

Elevation Along the Atlantic Coast

This scatter plot shows the elevation for selected cities at different latitudes along the Atlantic coast. It shows a computer-drawn line that best fits the data. In this scatter plot, the data points do not fit the line very closely. Use the line of best fit to predict the elevation of a city at 29° latitude. Approx 18 ft

The measure of how closely a set of data points falls along a line is the **correlation coefficient,** r. The correlation coefficient can be computed automatically by a graphics calculator. The possible values for a correlation coefficient range from -1 to 1. The better the data fit the line, the closer the numerical value of the correlation coefficient is to *either* 1 or -1.

In the first scatter plot, the correlation coefficient computed on a calculator is -0.98, which is near -1. This indicates that the data are close to a line that falls from left to right.

In the second scatter plot, the correlation coefficient from a calculator is 0.36, which is not close to 1 or -1. This indicates that the data do not cluster near the best fit line. Because the correlation is positive, the line rises from left to right.

Explain why predictions made from the line of best fit in the Maximum Temperatures scatter plot on page 625 would be more reliable than predictions made from the line of best fit in the Elevations scatter plot.

EXAMPLE

Examine the following scatter plots and the lines. Which line best fits the points, Line 1 or Line 2?

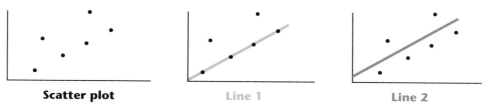

| **Scatter plot** | **Line 1** | **Line 2** |

A line of best fit should take all of the points into consideration.

Solution ➤

Line 1 goes through four of the points but ignores the others.

Line 2 is closer to the four points than it is to the other two, but it takes all of the points into account. **Line 2** is a better fit, even though it doesn't go through any of the points. ❖

EXERCISES & PROBLEMS

Communicate

1. Explain how to use a line of best fit to make a prediction.
2. If the coefficient of correlation is 1 or -1, what does this tell you about the data points and the line of best fit?
3. What does a correlation coefficient near $+1$ tell you?
4. What does a correlation coefficient near -1 tell you?
5. What does a correlation coefficient of 0.23 tell you?

Practice & Apply

Examine the airline timetable at the right.

6. What tends to happen to the number of minutes as the number of miles increases? Increase

7. Make a scatter plot of the data. Is the correlation positive or negative?

City	Miles	Minutes
Jacksonville	270	64
St Louis	484	102
Pittsburgh	526	89
New York	765	120
Minneapolis	906	154
Boston	946	145
Denver	1208	190
Phoenix	1587	236
Los Angeles	1946	286
San Francisco	2139	312

8. The line of best fit rises and fits the points quite well. What does this tell you about the numerical value of the correlation coefficient? Close to 1

Tell whether each correlation coefficient describes a line of best fit that rises or that falls. Also tell whether the line is a good fit.

9. 0.09
Rises, no

10. −0.92
Falls, yes

11. 0.89
Rises, yes

12. −0.45
Falls, no

For Exercises 13 and 14, tell whether the correlation coefficient for the scatter plot is nearest to −1, 0, or 1.

13.
1

14.
0

15. Given the scatter plot, which line best fits the points? Explain. Line 1

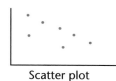
Scatter plot

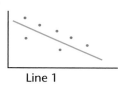
Line 1

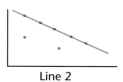
Line 2

For Exercises 16–18, match each scatter plot to a correlation coefficient.   **A.** −0.9   **B.** 0.2   **C.** 0.9

16.
A

17.
C

18.
B

Consumer Economics The table shows that the cost of food items that were $1.00 in 1982 has increased over the years.

Year	1982	1985	1988	1991
Cost	1.00	1.06	1.18	1.36

19. Make a scatter plot, and use a straightedge to estimate the line of best fit.

20. Use your line of best fit to predict the 1998 cost of food items that were $1.00 in 1982. Answers may vary. Approx $1.65

Look Back

What are the next two terms of each sequence?   [Lesson 1.1]

21. 1, 4, 7, 10, 13, . . .
16, 19

22. 2, 5, 10, 17, 26, . . .
37, 50

23. 1, 2, 3, 4, 0, 1, 2, . . .
3, −1

24. **Geometry** The side of a square is 8 centimeters. What is its area?   **[Lesson 4.5]**
64 sq cm

Find the slope of the line through the two points.   [Lesson 8.3]

25. (1, 2), (−4, 3) $-\frac{1}{5}$

26. (0, −2), (−4, 0) $-\frac{1}{2}$

27. (3, 4), (3, −1) Undefined

Look Beyond

28. What value of x in $2x + 4 = 12$ will make the equation true? $x = 4$

29. What value of n in $32 − 2n = 10$ will make the equation true? $n = 11$

Designing a Research Survey

Many questions can be answered by a research survey. In this project you will choose a question and answer it by collecting and analyzing data.

ACTIVITY 1 *Choose a Topic.*

Decide on a topic that you would like to research. State the question that your survey will answer.

When is a polar bear most active?

ACTIVITY 2 *Make a Hypothesis.*

Write a statement to describe the possible outcome of your research. Explain why you think this will be the outcome.

ACTIVITY 3 *Design the Survey.*

Write the questions that you will use in your survey, and explain why each question is necessary. Read through your completed survey, and write a justification of why it is fair.

ACTIVITY 4
Conduct the Survey.
Collect a large sample for your survey. Be sure you collect enough data so that you will be able to accurately test your hypothesis.

Where do teenagers do their homework?

ACTIVITY 5 *Display the Data.*
Summarize the results of your survey in a table. Then use one of the methods you studied in this chapter (bar graph, line graph, histogram, stem-and-leaf plot, box-and-whisker plot, or scatter plot) to display your data.

Does a person's height affect running speed?

ACTIVITY 6
Analyze the Results.
Explain how the data you gathered supports or contradicts the hypothesis you made in Activity 2.

Chapter 11 Review

Vocabulary

box-and-whisker plot	604	histogram	603	mode	595
circle graph	589	line graph	588	quartiles	604
correlation	619	line of best fit	624	range	594
correlation coefficient	625	mean	595	scatter plot	618
frequency table	596	median	595	stem-and-leaf plot	601

Key Skills & Exercises

Lesson 11.1

➤ **Key Skills**

Interpret bar graphs, line graphs, and circle graphs.
Circle graphs show different parts of a whole. Data is presented in percents. The entire graph totals 100%. This circle graph shows the life expectancy of Good Wheel tires.

➤ **Exercises**

Answer the following questions about the circle graph.

1. If Good Wheel tire company makes 1,000 tires, how many of those would be expected to last between 40,000 and 49,000 miles? 200 tires

2. Out of 550 tires, how many would you expect to last at least 30,000 miles?
 About 391 tires

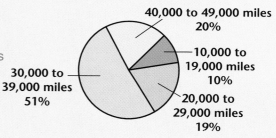

Life Expectancy of Good Wheel Tires (normal wear)

40,000 to 49,000 miles 20%
10,000 to 19,000 miles 10%
20,000 to 29,000 miles 19%
30,000 to 39,000 miles 51%

Lesson 11.2

➤ **Key Skills**

Find the mean, median, mode, and range for a set of data.
Lucy received the following grades on her mathematics tests during a one-month period: 80, 75, 80, 95, and 100. The mean, median, mode and range are shown below.

mean: $\dfrac{80 + 75 + 80 + 95 + 100}{5} = 86$

median: Arrange the grades in order: 75, 80, 80, 95, 100. The middle grade is 80.

mode: The grade that appears most often is 80.

range: The range is $100 - 75$, or 25.

➤ **Exercises**   **3.** 8; 8; 8; 5   **4.** Approx 9.8; 8.5; none; 18   **5.** 97; 90.5; 55; 151
Find the mean, median, mode, and range for each data set.

3. 6, 8, 8, 11, 7   **4.** 5, 7, 10, 13, 21, 3   **5.** 55, 85, 96, 102, 135, 85, 55, 96, 55, 206

Lesson 11.3

➤ Key Skills

Use data to construct stem-and-leaf plots, histograms, and box-and-whisker plots.

Employees in a company are surveyed to find out their ages. A stem-and-leaf plot, histogram, and box-and-whisker plot are each made from the data: 23, 18, 46, 23, 51, 27, 34, 39, 31, 56.

Stems	Leaves
1	8
2	3, 3, 7
3	1, 4, 9
4	6
5	1, 6

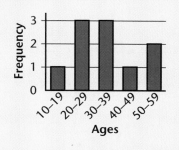

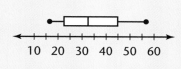

➤ Exercises

For the data set above, identify

6. the lower quartile. 23   **7.** the upper quartile. 46   **8.** the median. 32.5   **9.** the range. 38

The following are 21 test scores.

76, 78, 92, 65, 98, 80, 92, 60, 85, 85, 99,
50, 74, 75, 81, 93, 58, 79, 84, 68, 88.

From the data, construct each of the following.

10. a stem-and-leaf plot   **11.** a histogram   **12.** a box-and-whisker plot

Lesson 11.4

➤ Key Skills

Construct and interpret a circle graph.

A circle graph is constructed from the data in the table below.

Annual Operating Expenses (Dollars)

Gasoline	Maintenance	Repairs	Insurance	Miscellaneous
720	160	300	900	68

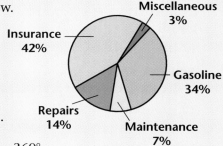

In the circle graph at the right, the cost of gasoline is represented by a sector. Its angle is calculated as follows.

$$\frac{720}{\text{total cost}} \cdot 360° = \frac{720}{720 + 160 + 300 + 900 + 68} \cdot 360°$$

$$= \frac{720}{2148} \cdot 360° \approx 120.7°$$

➤ Exercises

13. Brenda used her car-expense record to budget her expenses for the next year. Make a circle graph of the budget, shown below.

Gasoline	Maintenance	Repairs	Insurance	Miscellaneous
800	200	200	900	100

Lesson 11.5

➤ Key Skills

Identify the correlation in real-world data.
A scatter plot is the graph of a set of ordered pairs based on two matching sets of data. The scatter plot will show whether the correlation between the two data sets is strong, weak, or nonexistent. If there is a strong correlation, then the data points will fall close to a line called the *line of best fit.*

➤ Exercises

Describe each correlation as strong positive, strong negative, or little or none.

14. Strong positive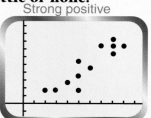

15. Strong negative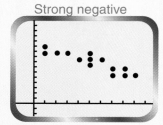

16. Little or none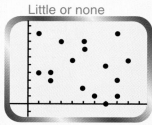

Lesson 11.6

➤ Key Skills

Understand how a correlation coefficient between the values −1 and 1 is related to how closely the data fits a line of best fit.
A correlation coefficient of 0.99 corresponds to a line of best fit that rises. Since 0.99 is close to 1, the fit is good. A correlation coefficient of −0.98 corresponds to a line of best fit that falls. Since −0.98 is close to −1, the fit is good. A correlation coefficient of 0.13 corresponds to a rising line of best fit that does not fit the data very well.

➤ Exercises

Tell whether each correlation coefficient corresponds to a line of best fit that rises or falls. Then tell whether or not the fit is good.

17. 0.02 Line rises, not good

18. −0.95 Line falls, good

19. 0.90 Line rises, good

Tell whether each correlation coefficient in the data in the indicated exercise is nearest to −1, 0, or 1.

20. Exercise 14 1

21. Exercise 15 −1

22. Exercise 16 0

Applications

23. $14,880; $14,500; none; $6800

23. **Education**   The tuition costs at five universities are $11,800; $14,500; $12,500; $17,000; and $18,600. Find the mean, median, mode, and range of the tuition costs.

24. **Consumer Economics**   In late December the McGroty family budgeted the following amounts for its essential expenses during the new year: food, $6500; housing, $10,300; clothing, $2500; transportation, $7000; health and personal care, $1800; miscellaneous, $2000. Draw a circle graph to display the data.

Chapter 11 Assessment

Stocks Don invested $100 each in four companies in 1990. He looks at the change in the value of each stock over several years.

1. $290   **2.** $140   **3.** About $160

1. What was the approximate value of the MRI stock at the end of 1994?
2. What was the approximate value of the MMart stock at the end of 1994?
3. How much did the WMB stock increase in value from 1990 to 1991?
4. How much did the CBC stock *About $60* increase in value from 1990 to 1994?
5. How much did the WMB stock increase in value from 1990 to 1994? *About $50*

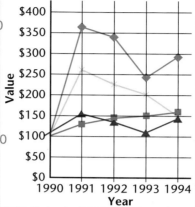

Find the mean, median, mode, and range for each data set.
6. 10, 15, 30, 55, 78
7. 12, 5, 9, 12, 2, 15
8. 83, 23, 29, 31, 41, 43, 53, 83, 23, 83, 6, 9

6. 37.6; 30; none; 68
7. Approx 9.2; 10.5; 12; 13
8. 42.25; 36; 83; 77

The following are the weights in kilograms of 19 students: 90, 77, 52, 81, 58, 84, 66, 69, 52, 62, 81, 84, 77, 73, 74, 82, 65, 59, 60. From the data, construct each of the following.
9. a stem-and-leaf plot
10. a histogram
11. a box-and-whisker plot

A car-expense record is shown below.

Gasoline	Maintenance	Repairs	Insurance	Miscellaneous
$1000	$350	$150	$700	$300

12. Make a circle graph of the data.
13. Which expense is about 30% of the total cost? *Insurance*
14. What percent of the total expenses is due to repairs? *6%*
15. How many degrees is the sector of the graph that represents gasoline? *144°*

16. Make a scatter plot of the data in the table below, which shows the amount of time spent studying and the score on a test the next day.

Time spent studying (in minutes)	15	12	20	30	25	40	60	50	55	45
Test Score	65	68	75	80	79	85	97	80	88	73

17. Describe the correlation of the data above as strong positive, strong negative, or little or none. *Strong positive*
18. Tell whether the correlation coefficient for the data in the scatter plot is nearest to −1, 0, or 1. *1*
19. Use a straightedge to draw a line that you estimate to be the line of best fit for the scatter plot. Use your line of best fit to predict the test score for a student who spends 35 minutes studying.

CHAPTER 12

Applications in Geometry

Many structures, such as igloos, tents, and tepees, are formed like geometric solids. The reason for a particular shape of a structure is often based on the materials available and the surface area or volume desired.

In this chapter, you will learn more about the measures of different figures. You will discover that surface area and volume are useful measures for describing and comparing objects.

These buildings are shaped like **pyramids.**

These tents are shaped very much like a **prism.**

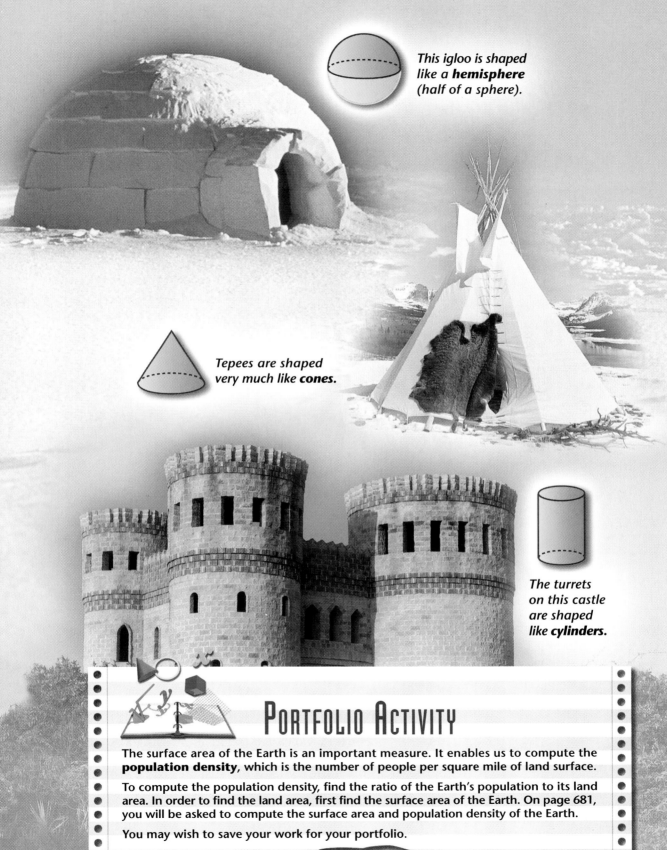

This igloo is shaped like a **hemisphere** (half of a sphere).

Tepees are shaped very much like **cones**.

The turrets on this castle are shaped like **cylinders**.

PORTFOLIO ACTIVITY

The surface area of the Earth is an important measure. It enables us to compute the **population density**, which is the number of people per square mile of land surface.

To compute the population density, find the ratio of the Earth's population to its land area. In order to find the land area, first find the surface area of the Earth. On page 681, you will be asked to compute the surface area and population density of the Earth.

You may wish to save your work for your portfolio.

Exploring Circles

why *Throughout history, the use of circles has improved the quality of living. Early humans discovered that they could use wheels to transport large objects. Today circles are an important part of art, architecture, and technology.*

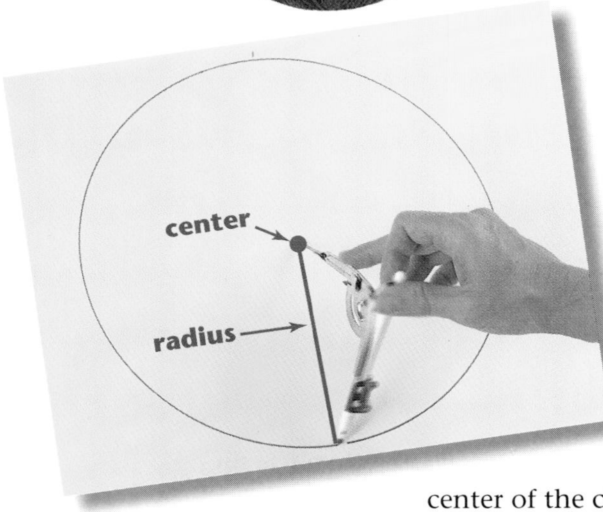

center

radius →

A **circle** is made up of all points in a plane that are a given distance from a point called the **center**. You can draw circles by using a compass. Place the point of the compass at the center of the circle, open the compass to a distance equal to the **radius** of the circle, and then draw the circle.

A **diameter** is a segment through the center of the circle that joins two points on the circle. How can you use the radius to find the diameter? $d = 2r$

The distance around a circle is called the **circumference**. You can use either the diameter or the radius of a circle to find the circumference.

The circumference is shown in red.

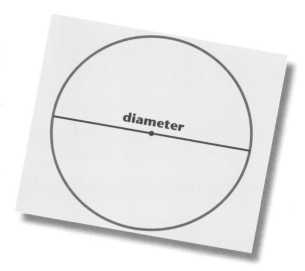

diameter

 •Exploration 1 *Circumference of Circles*

For this exploration, you will need a string, a ruler or tape measure, and various circular objects such as a can, a cup, and a plate.

1–3. Answers may vary.

1 Measure the diameter of each object. Measure the circumference of each object by placing a string around it and measuring the string. Copy and complete the table with your results.

Object	Diameter, *d*	Circumference, *C*
can	?	?
?	?	?
?	?	?

2 Compute the ratio $\frac{C}{d}$ for each circle. Round your answers to the nearest hundredth. How do the ratios compare?

3 Find the average of the three ratios. Compare your average with your classmates' averages. Compute the class average of the ratios.

4 Use the class average to estimate the circumference of a circle with a diameter of 3 yards. Answers may vary. Between 9 and 10 yds

5. Answers may vary. Multiply the diameter by a number approx equal to 3.1.

5 Make a generalization about how to find the circumference of a circle if you know the diameter.

6 Explain how to estimate the diameter of a tree trunk. ❖
Measure the circumference, divide by 3 for π to find approx diameter.

Cultural Connection: Asia, Africa Many civilizations throughout history discovered that the ratio $\frac{\text{circumference}}{\text{diameter}}$ was the same for all circles. The Greek letter π (pi) is used to represent this ratio. The ancient Chinese used the number 3 for this ratio. Ptolemy, a famous astronomer who worked in Egypt in the second century, estimated π at 3.1416.

CRITICAL Thinking

Explain why the formula $C = \pi d$ can also be expressed as $C = 2\pi r$, where C is the circumference, d is the diameter, and r is the radius. Since the diameter of a circle is twice the radius, the variable d can be replaced by $2r$ in the formula for circumference.

Calculator

Most calculators have a ⬚ π ⬚ key. Use the ⬚ π ⬚ key to find the circumference of the circles you measured in Exploration 1.
Answers may vary.

| ⬚ π ⬚ | 3.1415927 |

•Exploration 2 *Area of Circles*

Use the grid to estimate the area of each of the following circles. Then copy and complete the table below.
Answers may vary. Approximate answers are given.

$r = 1$

$r = 2$

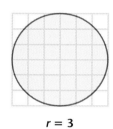
$r = 3$

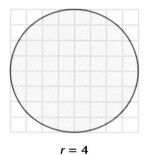
$r = 4$

Radius of circle, r	Estimated area, A	r^2	$\dfrac{A}{r^2}$	
1	?	?	?	3; 1; 3
2	?	?	?	12; 4; 3
3	?	?	?	27 to 28; 9; 3.1
4	?	?	?	48 to 51; 16; 3.2

5 Find the class average of the ratio $\dfrac{A}{r^2}$ when r is 4.
Answers may vary. 3.0 to 3.2

6 Find the circumference, C, when r is 4. Calculate the ratio $\dfrac{C}{d}$ when r is 4. How does the ratio $\dfrac{A}{r^2}$ compare with the ratio $\dfrac{C}{d}$ when r is 4?
$\approx 25.1; \approx 3.1; \dfrac{C}{d} = \dfrac{d\pi}{d} = \pi; \dfrac{A}{r^2}$ is approx the same as $\dfrac{C}{d}$, so $\dfrac{A}{r^2} \approx \pi$.

7 Explain how to estimate the area of a circle with a radius of 5 yards.
Square 5 and multiply by π: $A = (5)^2\pi \approx 78.54$ sq yd

8 Make a generalization about how to find the area of a circle if you know the radius. ❖ Square the radius, and multiply by π.

Draw a circle, and divide the area into 8 equal sections, as shown below. Cut out the sections and put them together to form a shape that is close to a parallelogram. Use the area of a parallelogram to explain why the area of the circle is $A = \pi r^2$. Base is $\frac{1}{2}$ of circumference, or πr. Height is radius, r. Area of parallelogram is base times height, or $\pi r \cdot r = \pi r^2$.

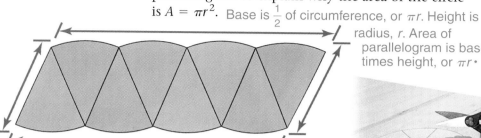

APPLICATION

Alex designed a flower bed with a walkway around it. The radius of the garden is 5 feet and the radius of the outer circle is 7 feet. Find the sum of the circumferences of the inner circle and the outer circle.

5 ft

7 ft

Estimate.
The circumference of a circle is πd, and the diameter of the outer circle is 14. Since the inner circle is smaller, doubling the circumference of the outer circle is an overestimate. Round the value of π down to 3 to compensate for the overestimation.

$$C = \pi \cdot d$$
$$2 \cdot C = 2(\pi \cdot d)$$
$$\approx 2 \cdot 3 \cdot 14$$
$$\approx 84$$

The sum of the circumferences of the inner circle and the outer circle is approximately 84 feet.

Calculator

Calculate.
To find an exact measurement, use a calculator.

Inner circle	**Outer circle**
$C = 2\pi r$	$C = 2\pi r$
$= 2\pi(5)$	$= 2\pi(7)$
$= 10\pi$	$= 14\pi$

$$C_{inner} + C_{outer}$$
$$= 10\pi + 14\pi$$
$$= 24\pi$$

24 ⬚×⬚ ⬚π⬚ ⬚=⬚ | 75.398224

The sum of the circumferences of the inner circle and the outer circle is 24π, which is approximately 75.4 feet. ❖

SUMMARY

Circumference of a Circle
The **circumference**, C, of a circle, where d is the diameter, is

$$C = \pi \cdot d.$$

Area of a Circle
The **area**, A, of a circle, where r is the radius, is

$$A = \pi \cdot r^2.$$

EXERCISES & PROBLEMS

Communicate

1. Explain the relationship between the radius and the diameter of a circle.

2. Explain the difference between the circumference and the area of a circle.

3. Can two circles have the same circumference but different areas? Explain.

4. Can two circles have the same area but different circumferences? Explain.

5. Explain the relationship between the circumference and the diameter of a circle. What is π?

Practice & Apply

Find the circumference and area, to the nearest tenth, of each circle with the given radius or diameter.

6. radius of 2 inches

7. radius of 4.5 centimeters

8. diameter of 6 inches

9. diameter of 4.6 meters

10. diameter of 7 meters

11. radius of $3\frac{1}{4}$ inches

12. Travel Measure the diameter of a bicycle or automobile tire to the nearest inch. Use your calculator to compute the circumference of the tire. How many times does the tire rotate for each mile? (1 mile = 5280 feet)

6. 12.6 in.; 12.6 sq in.
7. 28.3 cm; 63.6 sq cm
8. 18.8 in.; 28.3 sq in.
9. 14.5 m; 16.6 sq m
10. 22.0 m; 38.5 sq m
11. 20.4 in.; 33.2 sq in.

House of Pizza Prices

Medium — 12 in.

Large — 15 in.

Jumbo — 2 ft × 1 ft

Toppings
One
Two
Three

Medium: $5.99 / $6.49 / $6.99

Large: $6.99 / $7.99 / $8.99

Jumbo: $8.99 / $9.49 / $10.99

Consumer Economics The House of Pizza sells three different sizes of pizzas: a medium 12-inch pizza, a large 15-inch pizza, and a jumbo rectangular pizza.

13. Find the area of each pizza in square inches. ≈ 113 sq in.; ≈ 177 sq in.; 288 sq in.

14. Which has a greater area, 3 medium pizzas or 2 large pizzas? How much greater? 2 large; about 14 sq in. more

15. Which has a greater area, 2 medium pizzas or 1 jumbo pizza? How much greater? 1 jumbo; about 62 sq in. more

16. In a special sale the House of Pizza offers 3 medium 1-topping pizzas for the price of 2 medium 1-topping pizzas. Which offers a better price per pizza, the special sale or a jumbo 1-topping pizza? Jumbo 1-topping

Copy and complete the table. Round answers to the nearest tenth.

	Radius	Diameter	Area	Circumference
17.	?4.5 cm	9 centimeters	≈ 63.6? sq cm	≈ 28.3? cm
18.	?2 yd	4 yards	≈ 12.6? sq yd	≈ 12.6? yd
19.	?2.5 in.	5 inches	≈ 19.6? sq in.	≈ 15.7? in.
20.	3 yards	?6 yd	≈ 28.3? sq yd	≈ 18.8? yd
21.	7 inches	?14 in.	≈ 153.9? sq in.	≈ 44.0? in.
22.	16 feet	?32 ft	≈ 804.2? sq ft	≈ 100.5? ft
23.	?1.5 in.	3 inches	≈ 7.1? sq in.	≈ 9.4? in.
24.	15 centimeters	?30 cm	≈ 706.9? sq cm	≈ 94.2? cm
25.	?≈ 2.4 m	?≈ 4.8 m	≈ 17.9? sq m	15 meters
26.	?≈ 2.3 m	?≈ 4.5 m	16 square meters	≈ 14.2? m

Construction A semicircle is $\frac{1}{2}$ of a circle. An outdoor track is designed for the city park using a rectangle and two semicircles.

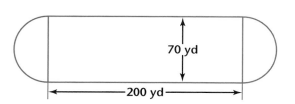

70 yd

200 yd

27. Estimate the shortest distance around the track. Then compute the distance. Actual: ≈ 620 yd

28. The city decides to plant grass inside the track. What is the area that needs to be covered?
Approx 17,848 sq yd

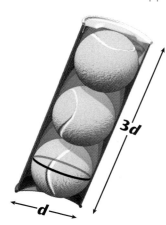

3d

d

29. Sports Three tennis balls fit perfectly in a can that has circular bases. Which is greater, the circumference of a circular base of the can or the height of the can? Explain.
Circumference. Circumference = πd; height 3d; $\pi d > 3d$

Find the area of each shaded region.

30.

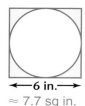

←6 in.→
≈ 7.7 sq in.

31.

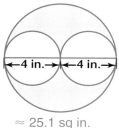

←4 in.→←4 in.→
≈ 25.1 sq in.

32. A semicircle is $\frac{1}{2}$ of a circle. Find the perimeter and area of a semicircle with a diameter of 8 yards. ≈ 20.6 yd; ≈ 25.1 sq yd

Look Back

 Geometry Find the volume of each cube. **[Lesson 1.6]**

33. 2-centimeter edges
8 cu cm

34. 4.5-centimeter edges
≈ 91.1 cu cm

35. 1-yard edges 1 cu yd

36. **Geometry** In parallelogram *ABCD* at the right, find the value of *y*. **[Lesson 7.2]** *y* = 38

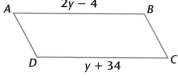

A   2y − 4   B

D   y + 34   C

37. Construction Matt is putting a baseboard around a rectangular gym. The width of the gym is 13 feet less than the length. If he uses 194 feet of baseboard, how long is the gym? **[Lesson 7.4]** 55 ft long

38. Sports Maria and Janel are runners in a relay race. Janel's time is 5 seconds less than Maria's time. Their combined time is 59 seconds. Write and solve a system of equations to find each runner's time.
[Lessons 9.2, 9.3, 9.4] $\begin{cases} m + j = 59 \\ j = m - 5 \end{cases}$ Maria 32 sec, Janel 27 sec

Look Beyond

39. **Probability** What is the theoretical probability that a random dart will hit the shaded region if it is sure to hit the target?
$\frac{3}{16} = 18.75\%$

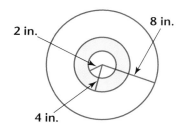

8 in.

2 in.

4 in.

Exploring Surface Area and Volume

why *The shapes and sizes of plants are partly determined by the ratio of surface area to volume that is necessary for their survival. Sometimes a large ratio is advantageous, while other times a small ratio is better.*

Tropical plants have large surface-area-to-volume ratios to receive as much light as possible. Since there is plenty of water in tropical regions, they do not need to conserve water.

Desert plants have small surface-area-to-volume ratios to conserve water. Since light is plentiful in the desert, they do not need a large surface area to receive light.

Cubes

In Lesson 1.6, you studied units of volume called cubic units. Recall that a cube with a 1-centimeter edge has volume of 1 cubic centimeter.

You can use centimeter grid paper to design a *net* for a cubic centimeter. A **net** is a flat figure that can be folded to enclose a particular solid figure.

The number of square units that it takes to cover the solid figure is called the **surface area** of the cube.

·Exploration 1 *Surface Area and Volume of Cubes*

For this exploration, you will need centimeter grid paper and centimeter cubes.

Part I

1 Use centimeter grid paper to create a net for a cube with 1-centimeter edges. What is the surface area of the cube? What is the volume?

6 sq cm; 1 cu cm

2 Use centimeter cubes to build a larger cube with 2-centimeter edges. Create a net for the cube. What is the surface area of the cube? What is the volume of the cube? 24 sq cm; 8 cu cm

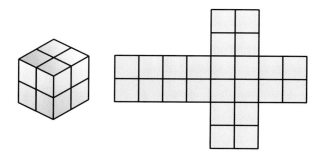

3 Repeat Step 2 for cubes with 3-centimeter and 4-centimeter edges.
3-cm edge: 54 sq cm, 27 cu cm; 4-cm edge: 96 sq cm, 64 cu cm

4 How can you find the surface area of a cube if you know the length of an edge of the cube? Write a formula for the surface area of a cube, using e for edge length.
Square the edge length, and multiply by 6. $S = 6e^2$

5 How can you find the volume of a cube if you know the length of the edge of the cube? Write a formula for the volume of a cube, using e for edge length. Cube the edge length. $V = e^3$

Part II

Use your results from Part I to complete the following table:

Edge length, e	Surface area, S	Volume, V	$\dfrac{S}{V}$
1	6	1	$\dfrac{6}{1}$
2	24 ?	8 ?	? 3
3	54 ?	27 ?	? 2
4	96 ?	64 ?	? 1.5
5	150 ?	125 ?	? 1.2
100	60,000 ?	1,000,000 ?	? 0.06
e	$6e^2$?	e^3 ?	? $6/e$

7 If *e* is 100, what is the ratio of surface area to volume? As the value of *e* gets larger, what happens to the ratio of surface area to volume? 0.06; it approaches zero.

8 Compare the surface-area-to-volume ratios of smaller cubes compared with those of larger cubes. ❖
Larger cubes have smaller ratios.

Warmblooded animals give off heat, which is a byproduct of metabolism, through their skin. A small surface-area-to-volume ratio helps an animal maintain its body temperature with a lower metabolism. Explain why large animals, such as elephants, have low metabolisms, and small animals, such as hummingbirds, have high metabolisms.

Rectangular Solids

A **cube** is a special type of rectangular solid because every edge is the same length. Any solid figure with rectangular sides is called a **rectangular solid**. The sides are called **faces**. Any two parallel faces can be the **bases** of the solid.

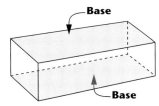

Exploration 2 Surface Area of Rectangular Solids

For this exploration, you will need grid paper and centimeter cubes.

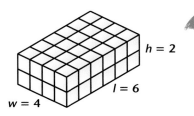

Count the number of squares in the net. This is the surface area of the solid, 88 sq cm.

1 Use cubes to construct a rectangular solid with a 4-centimeter width, a 2-centimeter height, and a 6-centimeter length. Use centimeter grid paper to draw a net for your rectangular solid. How can you use the net to find the surface area?

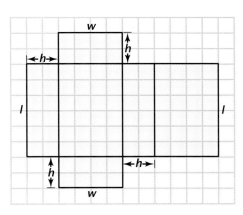

2 What is the surface area of the rectangular solid in Step 1? Describe a strategy that you can use to compute the surface area without using the net. 88 sq cm; $S = 2lw + 2lh + 2wh$

3 Write a formula for the surface area of a rectangular solid.

$$S = 2(lw + lh + wh)$$

4 How many centimeter cubes does it take to build this solid? 48

5 Use the variables l for length, w for width, and h for height to write a formula for the volume of a rectangular solid. $V = lwh$

6 Use the centimeter cubes to find four other rectangular solids with a volume of 48 cubic centimeters, but with different dimensions from those of the solid in Step 1. What is the surface area of each rectangular solid?

7 Make a conjecture about the shape of a rectangular solid with the least surface area for a given volume. ❖

∙Exploration 3 *Surface Area and Constant Volume*

MAXIMUM
MINIMUM
Connection

For this exploration, you will need grid paper and centimeter cubes. Use 12 cubes to build rectangular solids with the dimensions indicated below. Copy and complete the table following the solids.

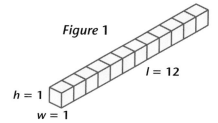

Figure 1 $l = 12$ $h = 1$ $w = 1$

Figure 2 $h = 1$ $l = 6$ $w = 2$

Figure 3 $h = 2$ $l = 6$ $w = 1$

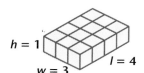

Figure 4 $h = 1$ $w = 3$ $l = 4$

Figure 5 $h = 3$ $l = 4$ $w = 1$

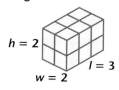

Figure 6 $h = 2$ $l = 3$ $w = 2$

Rectangular solid	Volume	Surface area
Figure 1	12?cu cm	50?sq cm
Figure 2	12?cu cm	40?sq cm
Figure 3	12?cu cm	40?sq cm
Figure 4	12?cu cm	38?sq cm
Figure 5	12?cu cm	38?sq cm
Figure 6	12?cu cm	32?sq cm

7 Are the volumes of each rectangular solid equal or not equal? Explain why.   Equal; each solid is formed with 12 centimeter cubes.

8 Are the surface areas of each rectangular solid equal or not equal? Explain why.   Not equal; the dimensions are different.

9 Compare the surface areas of Figures 2 and 3. Are they equal or not equal? Explain why.   Equal; both solids have the same dimensions in different positions.

10 Compare the surface areas of Figures 4 and 5. Are they equal or not equal? Explain why.   Equal; both solids have the same dimensions in different positions.

11 Explain how to build a rectangular solid with a given volume that has the greatest surface area.

12 Explain how to build a rectangular solid with a given volume that has the least surface area. ❖

CRITICAL
Thinking

On sunny days a snake may lie in the sun to absorb heat. At night, a snake may coil up tightly to retain its body heat. Explain how this strategy involves surface area and volume.

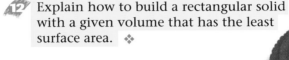

EXERCISES & PROBLEMS

Communicate ～～～

1. Explain the difference between the surface area and volume of a rectangular solid.

2. What is a net? Draw a picture of net on grid paper and use it to show how you can find surface area.

3. Explain why you can use the formula $V = Bh$, where B is the area of the base, to find the volume of a rectangular solid.

4. Explain why you can use the formula $S = 2(lw + lh + wh)$ to find the surface area of a rectangular solid.

Practice & Apply ～～～

Find the surface area and volume of a cube with the given side length.

5. 10 m   **6.** 7 in.   **7.** 15 cm   **8.** 16 ft

9. 2.5 m   **10.** 5.25 m   **11.** 11 in.   **12.** 6.25 cm

Find the surface area and volume of a rectangular solid with the indicated dimensions.

13. 2 m × 3 m × 5 m　　**14.** 3 yd × 4 yd × 2 yd　　**15.** 4 in. × 7 in. × 9 in.

16. 1.5 cm × 8 cm × 3 cm　　**17.** 1 m × 1 m × 2.3 m　　**18.** 3 in. × 2 in. × 1 in.

Find the surface area and volume for each of the rectangular solids described in the table below.

	Length	Width	Height	Surface area	Volume
19.	4 cm	4 cm	4 cm	96 sq?cm	64 cu cm ?
20.	4 in.	3 in.	2 in.	52 sq?in.	24 cu in. ?
21.	3.5 ft	7.2 ft	1.6 ft	84.6?sq ft	40.3 cu ft ?
22.	34 cm	57 cm	88 cm	19,892? sq cm	170,544 cu? cm
23.	$\frac{1}{2}$ in.	$3\frac{3}{4}$ in.	2 in.	$20\frac{3}{4}$ sq?in.	$3\frac{3}{4}$ cu in. ?
24.	102 ft	117 ft	300 ft	155,268?sq ft	3,580,200?cu ft
25.	248 ft	118 ft	12 ft	67,312?sq ft	351,168 cu? ft
26.	20 cm	15 cm	1 cm	670 sq?cm	300 cu cm ?
27.	5 m	5 m	6 m ?	170 sq?m	150 cubic meters
28.	6 m	6 m	?	?	256 cubic meters
29.	4 in.	8 in.	?	589 sq?in.	700 cubic inches
30.	5 cm	8 cm	10 cm?	340 sq?cm	400 cubic centimeters

$7\frac{1}{9}$ m; $242\frac{2}{3}$ sq m

$21\frac{7}{8}$ in.

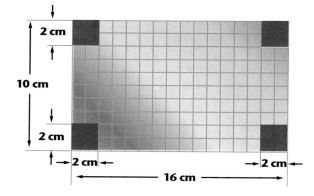

Maximum/Minimum The corners of a 10-centimeter-by-16-centimeter piece of sheet metal are removed to construct an open box. In the diagram on the left, a 2-centimeter-by-2-centimeter square has been removed from each corner to allow the piece of metal to be folded into an open box with a height of 2 centimeters. Use this diagram to help you complete Exercises 31–36.

	Dimensions of removed squares	Height	Length	Width	Volume
31.	1 cm × 1 cm	1 cm	?	?	?
32.	2 cm × 2 cm	2 cm	12 cm	6 cm	?
33.	?	3 cm	?	?	?
34.	?	4 cm	?	?	?

$h = 2$ cm

The height of the box is the same as the side length of each removed square.

35. Which box described in the table above has the greatest volume? Use the guess-and-check method and heights of 1.5, 2.5, 3.5, and 4.5 centimeters to determine whether another box has a greater volume.

36. Explain why you cannot build a box with a height of 5 centimeters from this sheet of metal.

Find the surface area and volume for each rectangular solid.

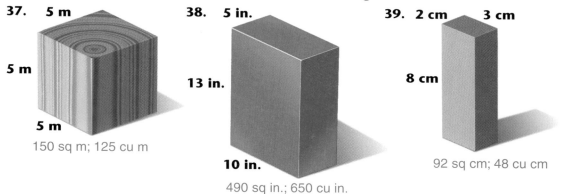

37. 5 m, 5 m, 5 m
150 sq m; 125 cu m

38. 5 in., 13 in., 10 in.
490 sq in.; 650 cu in.

39. 2 cm, 3 cm, 8 cm
92 sq cm; 48 cu cm

40. How many cubic feet are in 1 cubic yard? 27 cu ft

41. How can you change 4 cubic yards to cubic feet? Mult by 27, and change units.

42. How can you change 108 cubic feet to cubic yards? Divide by 27, and change units.

43. Construction   Concrete is sold in cubic yards. How many cubic yards are needed to pour a patio with dimensions of 20 feet by 20 feet by $\frac{1}{2}$ foot? Approx 7.4 cu yd

Look Back

Solve each equation or inequality for x.   [Lessons 5.4, 7.1, 7.5]

44. $2.5x - 2 = 1$ 1.2   **45.** $5x = 2x + 16$ $5\frac{1}{3}$ **46.** $x - 1 \le 2x + 5$ $x \ge -6$

Write an equation for the line containing the given points.
[Lesson 8.4]

47. (0, 4), (4, 0)
$y = -x + 4$

48. (−1, −2), (2, 3)
$y = \frac{5}{3}x - \frac{1}{3}$

49. (4, 3), (5, 3)
$y = 3$

Geometry   Find the circumference and area of each circle.
[Lesson 12.1]

50. radius = 2 cm
≈ 12.6 cm; ≈ 12.6 sq cm

51. diameter = 4.5 cm
≈ 14.1 cm; ≈ 15.9 sq cm

52. diameter = 1 yd
≈ 3.1 yd; ≈ 0.8 sq yd

Look Beyond

Maximum/Minimum   In Exercises 31–36, you designed open boxes from a 10-centimeter-by-16-centimeter rectangle. Refer to the diagrams on page 648 to help you answer Exercises 53–55.

53. Use the diagrams to explain why you can use the equation $V = (16 - 2x)(10 - 2x)x$ to compute the volume.

54. Technology   Use a graphics calculator to create a table to find, to the nearest tenth, the side length of the square, or height, that yields the greatest volume. HINT: Use $Y = (16 - 2X)(10 - 2X)X$ to create the table.

55. Technology   Use a graphics calculator to solve the following related problem: What size square (to the nearest tenth) would you cut out of a 20-centimeter-by-30-centimeter sheet to form an open box with the greatest possible volume?

LESSON 12.3 Prisms

Why *Many familiar objects have the shape of three-dimensional geometric figures known as prisms. Some types of right prisms are storage containers, aquariums, or buildings.*

A **prism** consists of a polygonal region, its translated image, and the connecting line segments. The figure shown at the left suggests a way to think of how a prism is constructed.

- Start with a polygonal region on a plane.
- Translate it onto a parallel plane.
- Connect the image and pre-image with segments.

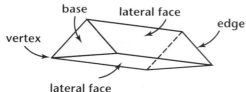

Each flat surface of a prism is called a **face**. Each line segment formed by the intersection of two faces is called an **edge**. Each point of intersection of the edges is called a **vertex**. The polygonal region and its translated image are each called a **base**. Each face of the prism that is *not* a base is called a **lateral face**.

A prism is named according to the shape of its bases. Examples of **right prisms** and **oblique prisms** are shown below.

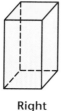

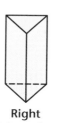

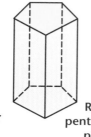

The lateral faces of a prism are always parallelograms. The lateral faces of a right prism are always rectangles.

Why do you think the word *right* is used to describe right prisms? There are right angles at the edges where the lateral faces meet the bases.

Right rectangular prism

Right triangular prism

Right pentagontal prism

Oblique rectangular prism

Oblique triangular prism

CRITICAL *Thinking* — Can cubes and rectangular solids also be classified as right prisms? Explain. Yes, because they have rectangular lateral faces and parallelograms for bases.

In this book all prisms will be right prisms unless stated otherwise.

Surface Area of Right Rectangular Prisms

In Lesson 12.2 you learned the formula for the surface area of a rectangular solid. The same formula is used for the surface area of a right rectangular prism. **Why?** Because adding the areas of the faces gives $2lw + 2lh + 2wh$, which is the surface area of a rectangular solid.

SURFACE AREA OF A RIGHT RECTANGULAR PRISM

The surface area, S, of a right rectangular prism is

$$S = 2(lw + lh + wh),$$

where l is the length, w is the width, and h is the height.

$l = 19$ cm

$h = 27.5$ cm

$w = 6$ cm

A cereal box is shaped like a right rectangular prism.

EXAMPLE 1

Manufacturing The surface area of a box helps the manufacturer determine the amount of material needed to make each box. Find the surface area of the cereal box shown here.

Solution ➤

Substitute 19 for l, 6 for w, and 27.5 for h in the formula for the surface area of a right rectangular prism.

$$\begin{aligned} S &= 2(lw + lh + wh) \\ &= 2(19 \cdot 6 + 19 \cdot 27.5 + 6 \cdot 27.5) \\ &= 2(114 + 522.5 + 165) \\ &= 2(801.5) \\ &= 1603 \end{aligned}$$

The surface area of the cereal box is 1603 square centimeters. ❖

CRITICAL *Thinking* — Why do you think the box manufacturer designed the cereal box with length and height much larger than the width? What advantage does this shape of cereal box have over a cube-shaped cereal box?

Surface Area of Nonrectangular Prisms

In general, the total surface area of *any* prism is the sum of the areas of all the faces. The **lateral surface area** is the sum of the areas of all the lateral faces.

To find a formula for the surface area of a right *nonrectangular* prism, think of flattening the prism to form a net. To find the lateral surface area, *L*, find the area of all the rectangular lateral faces.

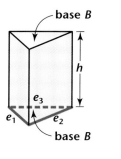
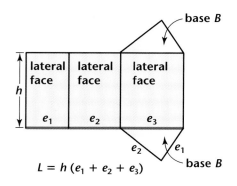

$$L = h(e_1 + e_2 + e_3)$$

Notice that the sum of the edges, $e_1 + e_2 + e_3$, is the perimeter of the base. So you can also write the formula for lateral surface area using *p* for the perimeter of the base.

$$L = h(e_1 + e_2 + e_3) \quad \text{or} \quad L = hp$$

LATERAL SURFACE AREA OF A RIGHT PRISM

The lateral surface area, *L*, of a right prism is

$$L = hp,$$

where *h* is the height and *p* is the perimeter of a base.

To find the total surface area, *S*, add the areas of both bases to the lateral surface area.

SURFACE AREA OF A RIGHT PRISM

The surface area, *S*, of a right prism is

$$S = L + 2B,$$

where *L* is the lateral surface area and *B* is the area of a base.

EXAMPLE 2

This right triangular prism has right triangles for bases. Find the surface area of the right triangular prism.

Notice that this prism is resting on a lateral face. The bases are right triangles.

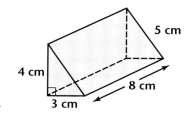

Solution ➤

Find the area of a base, *B*.

The area of a triangle is $\frac{1}{2}bh$.

$$B = \frac{1}{2}bh$$
$$= \frac{1}{2}(3)(4) \qquad \text{Substitute 3 for } b \text{ and 4 for } h.$$
$$= 6$$

The area of a base is 6 square centimeters.

Find the lateral surface area, *L*.

Use the formula for the lateral surface area of a right prism.

$$L = hp$$
$$= 8(3 + 4 + 5) \qquad \text{Substitute 8 for } h \text{ and } 3 + 4 + 5$$
$$\qquad\qquad\qquad\quad \text{for the perimeter, } p.$$
$$= 8 \cdot 12$$
$$= 96$$

The lateral surface area is 96 square centimeters.

Find the total surface area of the prism, *S*.

Use the formula for the surface area of a right prism.

$$S = L + 2B$$
$$= 96 + 2(6) \qquad \text{Substitute 96 for } L \text{ and 6 for } B.$$
$$= 96 + 12$$
$$= 108$$

The tissue paper on Rhoan's Chinese lantern covers the lateral faces of the lantern.

The surface area of the prism is 108 square centimeters. ❖

EXAMPLE 3

Rhoan is making Chinese lantern decorations to hang on the patio when she has a party. The Chinese lantern decorations are made of tissue paper and are shaped like a right prism with regular hexagonal bases. How much tissue paper is needed for each Chinese lantern decoration?

4 in.

10.5 in.

Solution ➤

To find out how much tissue paper is needed, find the lateral surface area of the lantern. Use the formula for the lateral surface area of a right prism, $L = hp$.

The Chinese lantern has 6 congruent lateral faces, each with a 4-inch base edge. So the perimeter, p, is $6 \cdot 4$, or 24 inches. The height, h, is 10.5 inches.

$$L = hp$$
$$= 10.5 \cdot 24 \qquad \text{Substitute 10.5 for } h \text{ and 24 for } p.$$
$$= 252$$

The amount of tissue paper needed for the sides of the Chinese lantern is 252 square inches. ❖

Volume of Rectangular Prisms

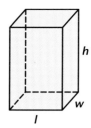

In Lesson 12.2 you learned the formula for the volume of a rectangular solid. The same formula is used for the volume of a right rectangular prism.
Why? Because a rectangular prism is a rectangular solid.

VOLUME OF A RIGHT RECTANGULAR PRISM

The volume, V, of a right rectangular prism is

$$V = lwh,$$

where l is the length, w is the width, and h is the height.

In general, the *volume of a right prism* is the area of the base times the height.

VOLUME OF A RIGHT PRISM

The volume, V, of a right prism is

$$V = Bh,$$

where B is the area of a base.

EXAMPLE 4

The right prism shown here has right triangles for bases. Find the volume of the triangular prism.

Solution ➤

Find the area of a base, B.

The area of a triangle is $\frac{1}{2}bh$.

$$B = \frac{1}{2}bh$$

$$= \frac{1}{2}(6)(8) \qquad \text{Substitute 6 for } b \text{ and 8 for } h.$$

$$= 24$$

The area of a base is 24 square inches.

Find the volume, V, of the prism.

Use the formula for the volume of a right prism.

$$V = Bh$$
$$= 24 \cdot 12 \qquad \text{Substitute 24 for } B \text{ and 12 for } h.$$
$$= 288$$

The volume of the triangular prism is 288 cubic inches. ❖

Try This The triangular prism shown here has right triangles for bases. Find the volume of the triangular prism. 36 cu in.

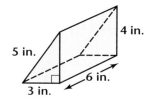

5 in.
4 in.
6 in.
3 in.

Exercises & Problems

Communicate

1. Describe how to construct a prism. Give two examples of prisms.

2. What is the difference between surface area and lateral surface area?

3. Explain the difference between surface area and volume.

4. Can two prisms have the same volume but different surface areas? Give examples to justify your answer.

Practice & Apply

5. Find the volume of a triangular prism with a base area of 4 square meters and a height of 6 meters. 24 cu m

6. Find the lateral surface area of a triangular prism with a base area of 12 square centimeters, a base perimeter of 16 centimeters, and a height of 8 centimeters. 128 sq cm

7. Find the lateral surface area of a triangular prism with a base area of 48 square centimeters, a base perimeter of 32 centimeters, and a height of 4 centimeters. 128 sq cm

8. If a hexagonal prism has a base area of 16.25 square meters and a height of 18.5 meters, what is the volume? 300.6 cu m

9. Find the total surface area of a rectangular prism with a base area of 8 square centimeters, a base perimeter of 17 centimeters, and a height of 7 centimeters. 135 sq cm

10. If a pentagonal prism has a base area of 7 square feet and a height of 8 feet, what is the volume? 56 cu ft

Construction A house plan calls for concrete pillars with regular octagonal bases. Each pillar will be 20 feet tall with a base area of 4.5 square feet.

11. Find the volume of each pillar. 90 cu ft

12. Convert the volume from cubic feet to cubic yards to determine the amount of concrete needed to pour each pillar. $3\frac{1}{3}$ cu yd

13. If concrete costs $65 per cubic yard, how much will it cost to pour four pillars? Approx $867

Find the surface area and volume of each prism. The dimensions for the bases of each prism are given.

14.

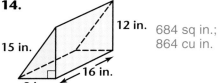

12 in.
15 in.
16 in.
9 in.

684 sq in.; 864 cu in.

15.

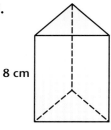

8 cm

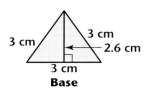

79.8 sq cm; 31.2 cu cm

3 cm   3 cm
2.6 cm
3 cm
Base

16.

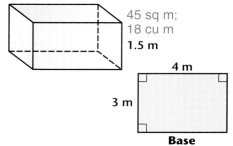

1.5 m

45 sq m; 18 cu m

4 m
3 m
Base

17.

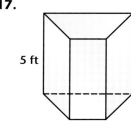

5 ft

6 ft
4 ft   4 ft
5 ft
4 ft
Base

140 sq ft; 125 cu ft

The area of a trapezoid is:
$$A = \frac{(b_1 + b_2)h}{2}$$

18.

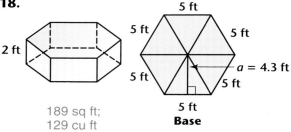

2 ft

189 sq ft; 129 cu ft

5 ft
5 ft   5 ft
5 ft $a = 4.3$ ft
5 ft
5 ft
Base

The area of a regular polygon is:
$$A = \frac{1}{2} ap$$

19.

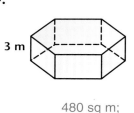

3 m

480 sq m; 504 cu m

8 m
8 m   8 m
8m $a = 7$ m
8 m
8 m
Base

Construction The State Aquarium Society is designing a hexagonal aquarium. The height of the aquarium will be 10 feet and the length of each edge of the hexagonal base will be 10 feet. Each of the six triangles that make up the hexagonal base has a height of 8.6 feet.

20. Find the area of the base. 258 sq ft

21. How much material is needed to build the aquarium if the top is left open? 858 sq ft

22. Find the volume of the aquarium. 2580 cu ft

23. One cubic foot of water is approximately 7.5 gallons. How many gallons would the new aquarium hold? ≈ 19,350 gal

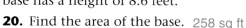

10 ft
10 ft   10 ft
10 ft   10 ft
8.6 ft
10 ft
Base of tank

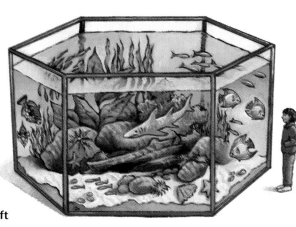

Record the number of vertices, *v*, edges, *e*, and faces, *f*, of each prism described in the table. Then copy and complete the table.

	Shape of base	Vertices, *v*	Faces, *f*	Edges, *e*	*v* + *f*
24.	Triangle	6	5	9	? 11
25.	Rectangle	? 8	? 6	? 12	? 14
26.	Pentagon	? 10	? 7	? 15	? 17
27.	Hexagon	? 12	? 8	? 18	? 20
28.	Heptagon	? 14	? 9	? 21	? 23
29.	Octagon	? 16	? 10	? 24	? 26
30.	*n*-gon	? 2*n*	*n*? + 2	? 3*n*	3*n*? + 2

31. Technology   Use a graphics calculator to create a table showing the number of vertices, faces, edges, and *v* + *f* to verify your results in Exercises 24–30.

32. Find the number of vertices, faces, and edges in a 100-gon. 200; 102; 300

33. Compare *v* + *f* with *e* for the 100-gon. Make a conjecture about the comparison of *v* + *f* with *e* for any *n*-gon. *v* + *f* = 302; *e* = 300; As *v* gets larger, *v* + *f* gets closer to *e*.

Look Back

Sketch the graph of each function using the values −3, −2, −1, 0, 1, 2, and 3 for *x*.   [Lesson 8.1]

34. $y = 1 - 2x$ **35.** $y = -4$ **36.** $y = \frac{x}{2} + 1$

Geometry   Find the circumference and area of each circle described. [Lesson 12.1]

37. radius of 5 centimeters ≈ 31.4 cm; ≈ 78.5 sq cm

38. diameter of 5 centimeters ≈ 15.7 cm; ≈ 19.6 sq cm

39. diameter of 0.5 yard ≈ 1.6 yd; ≈ 0.2 sq yd

Look Beyond

40. Think of putting two triangular prisms together to form a rectangular prism. Show that the volume of one of these triangular prisms is half the volume of the rectangular prism.

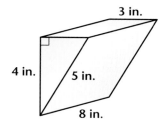
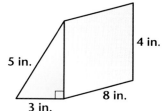

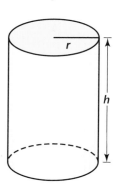

Storage containers, such as industrial drums and cans, are often shaped like cylinders. The cylindrical shape makes the containers rigid and easy to move. Cylindrical containers also hold more than do prisms made of the same amount of material.

Most liquid containers are shaped like cylinders. Drinking cups are often cylinders without one of the bases.

Surface Area of Right Cylinders

A **right cylinder** has two circular bases. The radius, r, is the radius of either base. The height, h, is the perpendicular distance between the bases.

As the number of sides of a regular polygon increases, the figure becomes more and more like a circle. Similarly, as the number of sides of a regular polygonal prism increases, the figure becomes more and more like a cylinder.

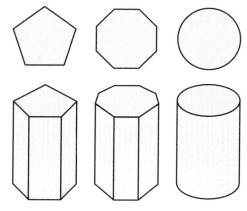

The formulas for the surface areas and volumes of prisms and cylinders are very much alike.

EXAMPLE 1

Manufacturing

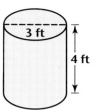

3 ft

4 ft

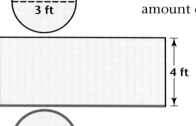

3 ft

4 ft

3 ft

A manufacturer makes cylindrical storage containers and drums. The circular base of a tank has a diameter of 3 feet. The overall height is 4 feet. Find the surface area of the tank.

Solution ➤

To determine the amount of steel needed to make the tank, the manufacturer considers the three pieces shown here.

From this figure, you can see that the surface of a cylinder is made of three parts: two circles and the rectangular lateral surface.

Notice that the length, *l*, of the rectangular surface is the circumference of the circular base, and that the width, *w*, is the height of the cylinder.

Surface area	=	2 times the area of a circular base	+	Lateral surface area

$$S = 2 \cdot \pi r^2 + lw$$
$$S = 2\pi r^2 + 2\pi rh$$

Calculator

Use the formula for the surface area of a cylinder, $S = 2\pi r^2 + 2\pi rh$.

$$S = 2\pi r^2 + 2\pi rh$$
$$= 2\pi(1.5)^2 + 2\pi(1.5)(4)$$ Substitute 1.5 for *r* and 4 for *h*.
$$\approx 51.8$$ Use a calculator.

The surface area of the tank is about 51.8 square feet. ❖

Try This Find the surface area of a cylinder with a height of 6 yards and bases with a diameter of 4.5 yards. ≈ 116.6 sq yd

CRITICAL Thinking Is the expression $2\pi r^2 + 2\pi rh$ the same expression as $2\pi r(r + h)$? Explain why or why not. yes; because of the Distributive Property, $2\pi r(r + h) = 2\pi r^2 + 2\pi rh$

Volume of Right Cylinders

Recall from Lesson 12.1 that you can use a parallelogram to approximate the area of a circle. In the same way, you can use a prism to approximate the volume of a cylinder.

Divide the cylinder into pie-shaped wedges and arrange the wedges to form a prism-like solid.

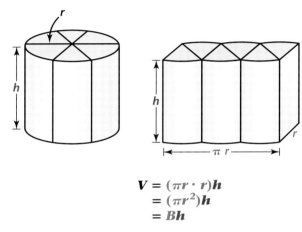

$$V = (\pi r \cdot r)h$$
$$= (\pi r^2)h$$
$$= Bh$$

The formula $V = Bh$ can be used to find the volume of a cylinder.

EXAMPLE 2

Compute the volume, in gallons, of the storage tank in Example 1 on page 659. Use the conversion factor 1 cubic foot $\approx$ 7.5 gallons.

Solution ➤

Find the area of the base, *B*.

The area of the circular base is πr^2.

$$B = \pi r^2$$
$$= \pi(1.5)^2 \qquad \text{Substitute 1.5 for } r.$$
$$\approx 7.1 \qquad \text{Use a calculator.}$$

The area of the base, *B*, is approximately 7.1 square feet.

Find the volume of the cylinder, *V*.

Use the formula $V = Bh$ to find the volume of the tank.

$$V = Bh$$
$$\approx 7.1 \cdot 4 \qquad \text{Substitute 7.1 for } B \text{ and 4 for } h.$$
$$\approx 28.4$$

The volume of the cylindrical tank is approximately 28.4 cubic feet.

Convert the units from cubic feet to gallons.

Write a proportion to convert the 28.4 cubic feet from cubic feet to gallons. Since 1 cubic foot is approximately equal to 7.5 gallons, write and solve the following proportion:

$$\frac{x \text{ gallons}}{28.4 \text{ cubic feet}} \approx \frac{7.5 \text{ gallons}}{1 \text{ cubic foot}}$$

$$\frac{x}{28.4} \approx \frac{7.5}{1}$$

$$x \approx (7.5)(28.4) \qquad \text{Cross multiply.}$$
$$x \approx 213 \qquad\qquad \text{Use a calculator.}$$

The volume of the tank is approximately 213 gallons. ❖

Try This An upright cylindrical water tank has a radius of 5 feet and a height of 20 feet. What is the volume in cubic feet? in gallons? 1570.8 cu ft; 11,781 gal

---

SUMMARY

Lateral Surface Area of a Right Cylinder
The lateral surface area, L, of a right cylinder, where r is the radius and h is the height, is: $L = 2\pi rh$

Surface Area of a Right Cylinder
The surface area, S, of a right cylinder, where L is the lateral surface area and B is the area of a base, is: $S = L + 2B$

Volume of a Right Cylinder
The volume, V, of a cylinder, where B is the area of a base and h is the height, is: $V = Bh$

EXERCISES & PROBLEMS

Communicate

1. What is a cylinder? Give two real-world examples of a cylinder.

2. Describe the lateral surface of a cylinder. Explain how to find the area of the lateral surface.

3. Draw a picture of a cylinder and use it to explain why the formula for its surface area is $S = L + 2B$.

4. Explain why the volume of a cylinder can be calculated with the formula $V = Bh$.

Practice & Apply

Find the surface area and volume of each right cylinder.

5. 2-in. radius and 10-in. height

6. 6-yd radius and 4-yd height

7. 3-cm height and 6-cm diameter

8. 7-m diameter and 8-m height

9. 8.5-ft height and 1.5-ft radius

10. 4-ft radius and 8.2-ft height

Technology   Use a calculator for Exercises 11–13.

11. Convert the volume in Exercise 5 to gallons. ≈ 0.5 gal

12. Convert the volume in Exercise 6 to gallons. ≈ 91,609 gal

13. Convert the volume in Exercise 9 to gallons. ≈ 450.6 gal

Find the surface area and volume of each right cylinder.

14. 15-ft height and 10-ft radius

15. 5-m height and 6-m radius

16. 12-in. height and 8-in. diameter

17. 14-cm height and 11-cm diameter

18. 6.5-yd height and 2.5-yd radius

19. 6.2-m height and 8.2-m diameter

20. 45-ft height and 11-ft radius

21. 5.7-cm height and 6.25-cm radius

22. 22-in. height and 18-in. diameter

23. 1.42-cm height and 3.11-cm diameter

24. 26-m height and 32-m diameter

25. 153-ft height and 100-ft radius

26. 7.5-m height and 4.6-m radius

27. 54-cm height and 110-cm diameter

28. 8-yd height and 7-yd radius

29. 8.9-m height and 5.1-m diameter

30. 6-yd height and 4-yd radius

31. 12-cm height and 1.5-cm diameter

In Exercises 32–38, measure the diameter and height of 7 objects. Then find the surface area and volume of each. In order to calculate surface area, assume all objects have a top and a bottom.

	Object	Radius	Height	Lateral surface area	Surface area	Volume
32.	Soft drink can	?	?	?	?	?
33.	Coffee can	?	?	?	?	?
34.	Vegetable can	?	?	?	?	?
35.	Roll of paper towels	?	?	?	?	?
36.	Drinking cup	?	?	?	?	?
37.	Soda straw	?	?	?	?	?
38.	Drain pipe	?	?	?	?	?

Manufacturing Quench-Aid packages a powdered sports drink in a can with an 8-centimeter diameter and a 10-centimeter height.

8 cm

10 cm

QUENCH AID

DOUBLE

39. Find the volume of the drink can. ≈ 502.7 cu cm

40. Quench-Aid decides to increase the volume by 25%. What is the volume of the new can? ≈ 628.4 cu cm

41. Design three different sized cans, close to the original dimensions, that have approximately 25% more volume than the original can.

42. Find the surface area of each of your cans from Exercise 41. Which can requires the least amount of material to manufacture?

Construction A house plan calls for concrete pillars with circular bases. Each pillar will be 20 feet tall with a radius of 1.5 feet.

43. Find the volume of each pillar. ≈ 141.4 cu ft

44. Convert the volume from cubic feet to cubic yards to determine the amount of concrete needed to pour each pillar. ≈ 5.2 cu yd

45. If concrete costs $65 per cubic yard, how much will it cost to pour four pillars? About $1352

Look Back

46. Use the "Pythagorean" Right-Triangle Theorem to find the hypotenuse of a right triangle with 0.3-meter and 0.4-meter legs.   **[Lesson 4.7]** 0.5 m

47. Write the equations of a line parallel to and a line perpendicular to the line given by $y = 2x + 1$.   **[Lesson 8.7]** Answers may vary.
Slope of line parallel is 2; slope of line perpendicular is $-\frac{1}{2}$.

Statistics In a small community the percent of adults with college degrees has been steadily increasing, as shown in the table below. Use this data for Exercises 48–50.   **[Lessons 11.5, 11.6]**

Year	1950	1955	1960	1965	1970	1975	1980	1990	1995
Percent	12	15	22	25	35	48	52	55	60

48. Make a scatter plot of the data.

49. Draw an estimated line of best fit to predict the percent in the years 2000 and 2010.

50. Is a line the best model for this data? Explain.

Look Beyond

51. Cavalieri's Principle states that if two solids have equal heights and if the cross sections formed by every plane parallel to the bases of both solids have equal areas, then the two solids have the same volume. Use Cavalieri's Principle to demonstrate that the formula for the volume of an oblique cylinder is the same as the formula for a right cylinder. HINT: You may want to use a stack of coins in your demonstration.

LESSON 12.5 Volume of Cones and Pyramids

Why *Cones often appear in nature and pyramids often appear in architecture. Although pyramids and cones appear to be different, they are very much alike.*

Paricutín, located in western Mexico

When a volcano erupts, lava and other debris pile up in layers around the vent, forming a cone-shaped mountain. Paricutín is a type of volcano known as a cinder cone. In 1943, Paricutín began as a crack in the ground that formed in a corn field. One week later the volcano stood 450 feet high. When the eruptions stopped 9 years later, the cone towered 1345 feet above its base.

Cones

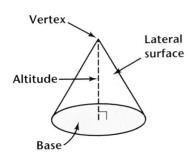

A **cone** is a solid figure that consists of a base that is a circle and a curved **lateral surface** that extends from the base to a single point called the **vertex**. The **altitude** of a cone is the segment from the vertex that is perpendicular to the plane of the base. The height of a cone is the length of its altitude. In this book all cones have circular bases.

Cones can be *right* or *oblique*. In a **right cone** the altitude intersects the base at its center. In an **oblique cone** the altitude intersects the plane of the base at some point other than the center.

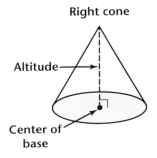

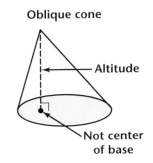

664 CHAPTER 12

Pyramids

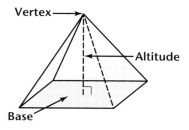

Vertex · Altitude · Base

A **pyramid** is a solid figure consisting of a base that is a polygon and a number of lateral faces that are triangles. The lateral faces meet at a point called the **vertex** of the pyramid. The **altitude** of a pyramid is the segment from the vertex that is perpendicular to the plane of the base. The height of a pyramid is the length of its altitude.

Pyramids can be *right* or *oblique*. In a **right pyramid** the altitude intersects the base at its center. In an **oblique pyramid** the altitude intersects the plane of the base at some point other than the center.

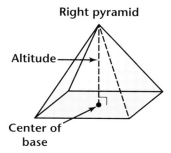

Right pyramid

Altitude · Center of base

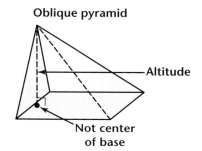

Oblique pyramid

Altitude · Not center of base

In a right, regular pyramid, all the lateral faces are congruent triangles. In an oblique pyramid, the lateral faces are not congruent.

Volume of Cones and Pyramids

Consider a cone and a cylinder with the same base area and the same height.

It takes exactly 3 of these cones to fill this cylinder.

Let V_{cone} represent the volume of this cone. Let V_{cylinder} represent the volume of this cylinder. Then the equation

$$V_{\text{cylinder}} = 3 \cdot V_{\text{cone}} \quad \text{or} \quad \frac{1}{3} \cdot V_{\text{cylinder}} = V_{\text{cone}}$$

describes the relationship between the volume of this cone and the volume of this cylinder.

How can you change the formula for the volume of a cylinder to get the formula for the volume of a cone? Divide each side of the formula by 3.

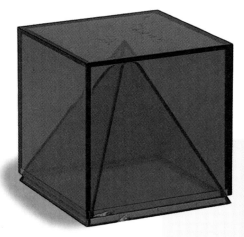

If you have a pyramid and a prism that have the same base area and the same height, you can also show that it takes 3 of the pyramids to fill the prism.

How can you change the formula for the volume of a prism to get the formula for the volume of a pyramid? Divide the product of the base area and height by 3.

VOLUME OF A CONE OR A PYRAMID

The volume, V, of a cone or a pyramid is

$$V = \frac{Bh}{3},$$

where B is the area of the base and h is the height.

EXAMPLE 1

Calculator

Find the volume of the cone.

Solution ➤

Find the area of the base, B.

To find the area of the base, first find the radius of the base. To find the radius, use the formula for the circumference of a circle,

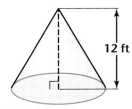

35-ft circumference

$$
\begin{aligned}
C &= 2\pi r \\
35 &= 2\pi r && \text{Substitute 35 for } C. \\
\frac{35}{2\pi} &= \frac{2\pi r}{2\pi} && \text{Division Property of Equality} \\
5.57 &\approx r && \text{Use a calculator.}
\end{aligned}
$$

The radius of the base is approximately 5.57. Use πr^2 to find the area of the base, B.

$$
\begin{aligned}
B &= \pi r^2 \\
&\approx \pi (5.57)^2 && \text{Substitute 5.57 for } r. \\
&\approx 97.47 && \text{Use a calculator.}
\end{aligned}
$$

Find the volume of the cone, V.

Use the formula for the volume of a cone, $V = \frac{Bh}{3}$.

$$
\begin{aligned}
V &= \frac{Bh}{3} \\
&\approx \frac{(97.47)(12)}{3} && \text{Substitute 97.47 for } B \text{ and 12 for } h. \\
&\approx 389.88 && \text{Use a calculator.}
\end{aligned}
$$

The volume of the cone is approximately 390 cubic feet. ❖

Try This   Find the volume of a cone with a base circumference of 27 centimeters and a height of 170 centimeters. ≈ 3291.77 cu cm, or about 3292 cu cm

EXAMPLE 2

A manufacturer designs candles that are made of solid wax and shaped like right pyramids.

The base of a wax candle shaped like a right pyramid is a 6-centimeter-by-6-centimeter square. The height is 4 centimeters. How much wax is needed to make the candle?

Solution ➤

Since the candle is solid, the amount of wax will be the volume of the candle.

Find the area of the base, B.

The base is a square. So use s^2, where s is the length of a side of the square, to find the area of the square base.

$$
\begin{aligned}
B &= s^2 \\
&= 6^2 \qquad \text{Substitute 6 for } s. \\
&= 36
\end{aligned}
$$

The area of the base, B, is 36 square centimeters.

Find the volume of the pyramid, V.

Use the formula for the volume of a pyramid, $V = \frac{Bh}{3}$.

$$
\begin{aligned}
V &= \frac{Bh}{3} \\
&= \frac{36 \cdot 4}{3} \qquad \text{Substitute 36 for } B \text{ and 4 for } h. \\
&= 48
\end{aligned}
$$

The volume of the candle is 48 cubic centimeters. ❖

Try This Find the volume of a pyramid with a 4-centimeter-by-4-centimeter square base and a height of 5 centimeters. ≈ 26.7 cu cm

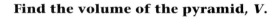

CRITICAL Thinking

How can you use mental math to find the volume of the candle in Example 2?
Divide one side of the base by 3 first.

EXERCISES & PROBLEMS

Communicate

1. What is a cone? How does a cone compare with a cylinder?
2. What is a pyramid? How does a pyramid compare with a prism?
3. Describe how to find the volume of a cone.
4. Describe how to find the volume of a pyramid.

Practice & Apply

Find the volume, to the nearest tenth, of each right cone in Exercises 5–14.

5. radius = 5 in., height = 17 in. ≈ 445.1 cu in.

6. radius = 5 cm, height = 6.5 cm
   ≈ 170.2 cu cm

7. radius = 2.5 m, height = 6 m ≈ 39.3 cu cm

8. diameter = 7.2 m, height = 1.9 m
   ≈ 25.8 cu m

9. radius = $2\frac{1}{2}$ in., height = $3\frac{3}{4}$ in. ≈ 24.5 cu in.

10. radius = $4\frac{1}{5}$ ft, height = $1\frac{2}{5}$ ft
    ≈ 25.9 cu ft

11. diameter = 8.9 ft, height = 7.5 ft ≈ 155.5 cu ft

12. diameter = 16 cm, height = 34 cm
    ≈ 2278.7 cu cm

13. radius = 6 in., height = 7 in. ≈ 263.9 cu in.

14. radius = 12 ft, height = 15 ft
    ≈ 2261.9 cu ft

Find the volume, to the nearest tenth, of each right pyramid in Exercises 15–24.

15. base area = 13 sq in., height = 30 in.
    130 cu in.

16. base area = 12 sq m, height = 9 m
    36 cu m

17. base area = 16 sq yd, height = 15 yd
    80 cu yd

18. base area = 8 sq cm, height = 4 cm
    ≈ 10.7 cu cm

19. base area = 7 sq ft, height = 9 ft
    21 cu ft

20. base area = 14 sq m, height = 10 m
    ≈ 46.7 cu m

21. base area = 25 sq mm, height = 6 mm
    50 cu mm

22. base area = 36 sq ft, height = 6 ft
    72 cu ft

23. base area = 20 sq in., height = 7 in.
    ≈ 46.7 cu in.

24. base area = 26 sq cm, height = 37 cm
    ≈ 320.7 cu cm

25. **Construction** A company specializes in buildings shaped like right pyramids. The pyramid-shaped building shown here is 35 feet high and has a 40-foot-by-40-foot square base. The builder must determine the inside volume to make decisions about the air-conditioning unit necessary to cool the building. What is the volume? ≈ 18,666.7 cu ft

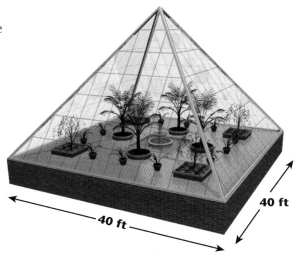

40 ft

40 ft

4 cm

16 cm

Manufacturing Udderly Good Ice Cream Company packages ice-cream cones. For Exercises 28 and 29, each cone has a radius of 4 centimeters and a height of 16 centimeters.

26. How much ice cream does it take to completely fill each cone to the top? ≈ 268.1 cu cm

27. Udderly Good Ice Cream Company decides to reduce the volume of the cones to 56.5 cubic centimeters. Suppose that they decide to keep the same height of the cones. What radius will give the cones a volume of 56.5 cubic centimeters? ≈ 1.8 cu cm

28. Design two different-sized cones that each have a volume of 75π cubic centimeters.

29. Design three different-sized pyramids that each have a volume of 45 cubic meters.

Look Back

Use mental math to simplify each expression. [Lesson 1.8]

30. $(227 + 98) + 273$ 598 **31.** $(25 \cdot 323) \cdot 8$ 64,600 **32.** $(5 \cdot 976) \cdot 20$ 97,600

Estimate the percent of each number. [Lesson 6.5]

33. 26.2% of 3682 ≈ 965 **34.** $6\frac{2}{3}$% of $119.99 ≈ 8 **35.** 39.5% of 200 79

36. 6.5% of 642 ≈ 42 **37.** 137% of $20.00 ≈ 27 **38.** 0.05% of 500 0.25

Solve each equation or inequality.
[Lessons 7.1, 7.2, 7.5]

39. $(2x - 1) + 3x \le -2$ $x \le -\frac{1}{5}$ **40.** $4x - 6 = -x + 1$ $x = \frac{7}{5}$, or $1\frac{2}{5}$

41. $3 - (x - 4) = x$ $x = \frac{7}{2}$, or $3\frac{1}{2}$

Look Beyond

Manufacturing A manufacturer designs an ice-cream cone with a volume of 75 cubic centimeters.

42. Use the volume formula to show why you can use the equation $y = \frac{75 \cdot 3}{\pi x^2}$, where y is the height and x is the radius, to describe this cone.

43. **Technology** Use a graphics calculator to create a table of values for the radius and height of cones with a volume of 75 cubic centimeters. Which dimensions do you think the manufacturer might choose? Explain your reasoning.

Surface Area of Cones and Pyramids

why *The similarities between finding the volumes of pyramids and cones also apply to finding the surface areas of pyramids and cones. The same basic formula can be used for both pyramids and cones.*

Moody Gardens, located in Galveston, Texas

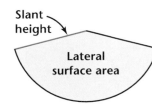

Slant height

Lateral surface area

Base area

Moody Gardens is an indoor botanical garden shaped like a right pyramid. It was necessary to know the surface area of the pyramid-shaped building in order to determine the amount of material necessary to build the structure.

To find the lateral surface area of a right, regular pyramid or cone, you need to know the *slant height, s.*

The **slant height of a right cone** is the length of a segment from the vertex that is perpendicular to the edge of the circular base.

The **slant height of a right regular pyramid** is the altitude of any of the triangular lateral sides.

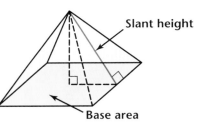
Slant height

Base area

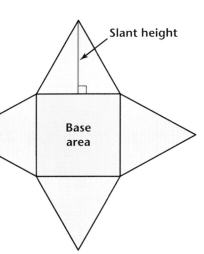
Slant height

Base area

Explain how you can use the "Pythagorean" Right-Triangle Theorem to find the slant height of a right cone if you know its radius and height.

LATERAL SURFACE AREA OF A RIGHT CONE

The lateral surface area, *L*, of a right cone is

$$L = \frac{1}{2}Cs,$$

where *C* is the circumference of the base and *s* is the slant height.

This formula for the lateral surface area of a right cone will also work for right pyramids *if and only if* they have a regular polygon as a base. Instead of using the variable *C* for the circumference of the base, use *p* for the perimeter of the base.

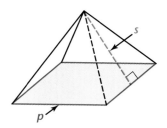

LATERAL SURFACE AREA OF A RIGHT, REGULAR PYRAMID

The lateral surface area, *L*, of a right, regular pyramid is

$$L = \frac{1}{2}ps,$$

where *p* is the perimeter of the base and *s* is the slant height.

The *total surface area* of a cone or a pyramid is the sum of the lateral surface area and the base area.

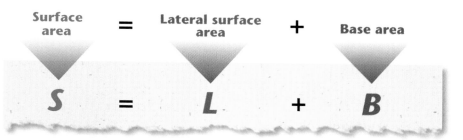

Surface area **=** Lateral surface area **+** Base area

S = L + B

SURFACE AREA OF A CONE OR A RIGHT, REGULAR PYRAMID

The total surface area, *S*, of a cone or a right, regular pyramid is

$$S = L + B,$$

where *L* is the lateral surface area and *B* is the base area.

CRITICAL
Thinking

In terms of the radius of a circle and π, what is the formula for the surface area of a cone? $\pi rs + \pi r^2$, where *r* is the radius and *s* is the slant height

EXAMPLE 1

The renaissance hat is shaped like a right cone. The slant height of the renaissance hat is 23 inches and the radius of the base is 3 inches.

Find the surface area of the renaissance hat shown.

Solution ➤

Find the lateral surface area, *L*, of the cone.

First find the circumference, *C*, of the base.

$$C = 2\pi r$$
$$C = 2\pi(3)$$
$$C \approx 18.8 \text{ inches}$$

Substitute 18.8 for *C* and 23 for *s* in the formula for lateral surface area of a cone.

$$L = \frac{1}{2}Cs$$

$$L \approx \frac{1}{2}(18.8)(23)$$

$$L \approx 216.2 \text{ square inches}$$

Find the area of the base, *B*.

$$B = \pi r^2$$
$$B = \pi \cdot 3^2$$
$$B \approx 28.3 \text{ square inches}$$

Find the total surface areas, *S*, of the cone.

Substitute 216.2 for *L* and 28.3 for *B* in the formula $S = L + B$.

$$S = L + B$$
$$S \approx 216.2 + 28.3$$
$$S \approx 244.5 \text{ square inches}$$

The surface area of the renaissance hat is approximately 244.5 square inches. ❖

EXAMPLE 2

Construction

An architect is designing a tile-covered roof shaped like a right pyramid with a square base. The height of the roof is 10 feet. What area do the tiles need to cover?

26 ft

48 ft

Solution ➤

The tiles need to cover the lateral surface area. Find the lateral surface area of the roof.

The perimeter of the base is 48 · 4, or 192 feet. Use the formula for the lateral surface area, L, of a pyramid.

$$L = \frac{1}{2}ps$$

$$L = \frac{1}{2}(192)(26)$$

$$L = 2496 \text{ square feet}$$

The tiles on the pyramid-shaped roof need to cover 2496 square feet. ❖

CRITICAL *Thinking*

You need to know the length of each edge and the height of each triangular face.

To find the surface area of a right pyramid that does not have a regular polygon for a base, add the areas of the lateral faces to the area of the base. What information do you need to find the area of each face of the pyramids shown?

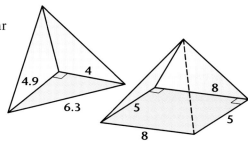

EXERCISES & PROBLEMS

Communicate

1. Describe the slant height of a right, regular pyramid.
2. Explain how to find the lateral surface area of a right, regular pyramid.
3. Explain how to find the surface area of a right cone.
4. Explain how to find the surface area of a right, regular pyramid.
5. Describe how to find the slant height of a right cone using the "Pythagorean" Right-Triangle Theorem.

Practice & Apply

Compute the surface area of each figure described below.

6. A right, regular pyramid with a 4-centimeter-by-4-centimeter base and a height of 5 centimeters. ≈ 59 sq cm
7. A right cone with a radius of 12 centimeters and a height of 8 centimeters. ≈ 996 sq cm

Find the lateral surface area and total surface area of each right cone.

8. a slant height of 8 meters and a diameter of 12 meters *L* ≈ 150.8 sq m, *S* ≈ 263.9 sq m

9. a slant height of 9 inches and a radius of 4 inches *L* ≈ 113.1 sq in., *S* ≈ 163.4 sq in.

10. a slant height of 3.2 centimeters and a radius of 20 centimeters
L ≈ 201.1 sq cm, *S* ≈ 1457.7 sq cm

11. a slant height of 11 meters and a diameter of 10 meters *L* ≈ 172.8 sq m, *S* ≈ 251.3 sq m

12. a slant height of 8 feet and a diameter of 8 feet *L* ≈ 100.5 sq ft, *S* ≈ 150.8 sq ft

13. a slant height of 4 yards and a radius of 2.5 yards *L* ≈ 31.4 sq yd, *S* ≈ 51.1 sq yd

Find the lateral surface area and total surface area of each right regular pyramid.

14. a square base with side lengths of 8 centimeters and a slant height of 6 centimeters *L* = 96 sq cm, *S* = 160 sq cm

15. a square base with side lengths of 16 yards and a slant height of 10 yards
L = 320 sq cm, *S* = 576 sq cm

16. a square base with side lengths of 4 inches and a slant height of 5 inches
L = 40 sq in., *S* = 56 sq in.

17. a square base with side lengths of 7 meters and a slant height of 9 meters
L = 126 sq m, *S* = 175 sq m

Find the total surface area of each right cone.

18. a radius of 6 inches and a slant height of 7 inches ≈ 245 sq in.

19. a radius of 2.5 meters and a slant height of 6 meters ≈ 66.8 sq m

20. a diameter of 8 feet and a slant height of 5 feet ≈ 113.1 sq ft

21. a diameter of 9 centimeters and a slant height of 9 centimeters ≈ 190.9 sq cm

22. a radius of 5 inches and a slant height of 17 inches ≈ 345.6 sq in.

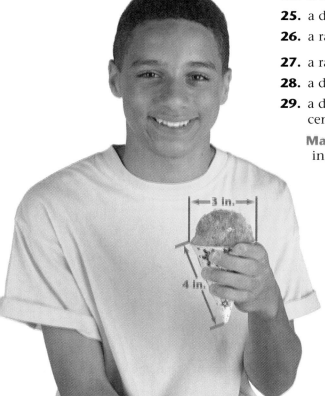

23. a radius of 5 centimeters and a slant height of 6.5 centimeters ≈ 180.6 sq cm

24. a diameter of 4 yards and a slant height of 15 yards ≈ 106.8 sq yd

25. a diameter of 7.2 meters and a slant height of 1.9 meters ≈ 62.2 sq m

26. a radius of $2\frac{1}{2}$ inches and a slant height of $3\frac{3}{4}$ inches ≈ 49.1 sq in.

27. a radius of $4\frac{1}{5}$ feet and a slant height of $1\frac{2}{5}$ feet ≈ 73.9 sq ft

28. a diameter of 8.9 feet and a slant height of 7.5 feet ≈ 167.1 sq ft

29. a diameter of 16 centimeters and a slant height of 34 centimeters ≈ 1055.6 sq cm

Manufacturing Shaved ice at the Frosty Fruity is served in a paper cone that has a slant height of 4 inches.

30. Find the lateral area of one paper cone. ≈ 18.8 sq in.

31. What is the height of the paper cone? ≈ 3.7 in.

32. Suppose that the height of the paper cone is 4 inches. What is the slant height if the diameter is 3 inches? ≈ 4.3 in.

33. Suppose that the paper cones are each made to hold a ball of shaved ice that is only 2 inches in diameter. What is the outer surface area of this type of paper cone if the slant height is 4 inches? ≈ 12.6 in.

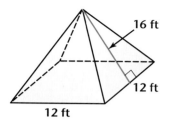

16 ft

12 ft

12 ft

Construction   The roof of a gazebo is a right square pyramid.

34. Find the surface area of the roof. HINT: The base area of the pyramid is *not* part of the roof.   384 sq ft

35. If roofing material costs $4 per square foot, how much will it cost to cover the roof? $1536

A regular tetrahedron is a pyramid that consists of four equilateral triangular faces. Use the regular tetrahedron at the right for Exercises 36–39.

36. Use the "Pythagorean" Right-Triangle Theorem to show that the height of each trianglar face is $\sqrt{48}$ centimeters.

37. What is the area of one trianglar face? What is the total surface area of the tetrahedron?

38. Write a formula for the total surface area of a regular tetrahedron with side length x.

39. Technology   Use your formula from Exercise 38 and a graphics calculator to create a table of total surface areas for side lengths from 1 to 7.

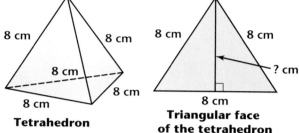

Tetrahedron

Triangular face of the tetrahedron

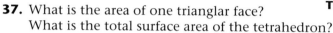

Look Back

Find the circumference and area of each circle.   [Lesson 12.1]

40. radius = 5.78 m

41. radius = $\frac{4}{3}$ in.

42. diameter = $2\frac{3}{8}$ in.

43. diameter = 63 cm

44. radius = 2.01 m

45. diameter = 5.2 ft

Find the surface area and volume for each rectangular solid. [Lesson 12.2]

46. length = 3.4 cm, width = 2.5 cm, height = 1.5 cm $S \approx 34.7$ sq cm, $V \approx 12.8$ cu cm

47. length = 2.4 cm, width = 1.9 cm, height = 3.6 cm $S \approx 40.1$ sq cm, $V \approx 16.4$ cu cm

48. length = 1 cm, width = 1 cm, height = 5 cm $S \approx 22$ sq cm, $V \approx 5$ cu cm

Look Beyond

49. Euler's Formula   Copy and complete the table for pyramids with bases of 3, 4, 5, 6, 7, 8, and n sides. Does the generalization for n satisfy Euler's formula $(V + F - E = 2)$?   yes

Base sides	Vertices, V	Faces, F	Edges, E
3	? 4	? 4	? 6
4	? 5	? 5	? 8
5	? 6	? 6	? 10
6	? 7	? 7	? 12
7	? 8	? 8	? 14
8	? 9	? 9	? 16
n	$n + 1$?	$n +$?1	? $2n$

LESSON 12.7 Spheres

why *A ball is the most common example of a sphere. A fully inflated hot-air balloon is shaped very much like a sphere. You can use the properties of a sphere to get good approximate answers to questions about a hot-air balloon.*

Surface Area of a Sphere

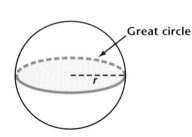

Great circle

A **sphere** is determined by all the points in space that are a certain distance from a given point. Many of the properties of a sphere are like those of a circle.

When a plane slices through the center of a sphere, it intersects the sphere in a circle called a **great circle.** The circumference of the sphere is the circumference of a great circle of the sphere.

Use the definitions of radius and diameter of a circle to describe radius and diameter of a sphere. The radius and diameter of a sphere are the radius and diameter of a great circle of the sphere.

Cultural Connection: Europe The Greek mathematician Archimedes knew that the surface area of a circle is related to the area of a great circle. He found the surface area of a sphere to be 4 times the area of a great circle of the sphere.

SURFACE AREA OF A SPHERE

The surface area, S, of a sphere is

$$S = 4\pi r^2,$$

where r is the radius.

EXAMPLE 1

When fully inflated, the envelope of a hot-air balloon is shaped much like a sphere.

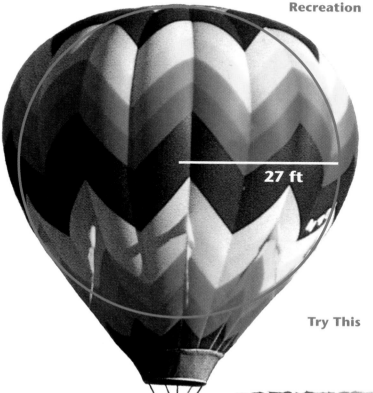

27 ft

Recreation One hot-air balloon has a radius of 27 feet when fully inflated. Approximately how much material was used to make the envelope of this balloon?

Solution ➤

To find an approximate surface area for the hot-air balloon, find the surface area of a sphere with a radius of 27 feet.

Substitute 27 for r in the formula for the surface area of a sphere.

$$S = 4\pi r^2$$
$$= 4\pi(27)^2$$
$$\approx 9160.9 \text{ square feet}$$

Approximately 9161 square feet of material was used to make the envelope of the hot-air balloon. ❖

Try This Find the surface area of a sphere with a diameter of 8 centimeters. $\approx$ 201.1 sq cm

EXAMPLE 2

Find the surface area of the softball below.

Solution ➤

Since the diameter is $3\frac{1}{2}$ inches, the radius is half of $3\frac{1}{2}$.

$$\text{diameter} = 3\frac{1}{2} \div 2$$
$$= \frac{7}{2} \cdot \frac{1}{2}$$
$$= \frac{7}{4}$$

Substitute $\frac{7}{4}$ for r in the formula for the surface area of a sphere.

$$S = 4\pi r^2$$
$$= 4\pi\left(\frac{7}{4}\right)^2$$
$$= 4\pi\left(\frac{49}{16}\right)$$
$$\approx 38.5 \text{ square inches} ❖$$

$3\frac{1}{2}$ in.

Volume of a Sphere

> ## VOLUME OF A SPHERE
>
> The volume, V, of a sphere is
>
> $$V = \frac{4\pi r^3}{3},$$
>
> where r is the radius.

EXAMPLE 3

A spherical water tank has a radius of 20 feet. How many gallons of water does it hold? HINT: 1 cubic foot $\approx$ 7.5 gallons

Solution ➤

Find the volume of the spherical tank. Substitute 20 for r in the formula for the volume of a sphere.

$$V = \frac{4\pi r^3}{3}$$

$$= \frac{4\pi(20)^3}{3}$$

$$\approx 33{,}510.3 \text{ cubic feet}$$

Write and solve a proportion to convert the volume in cubic feet to volume in gallons.

$$\begin{array}{l} \text{gallons} \longrightarrow \\ \text{cubic feet} \longrightarrow \end{array} \quad \frac{x}{33{,}510.3} \approx \frac{7.5}{1}$$

$$x \approx 251{,}327.25$$

The spherical water tank can hold approximately 251,327 gallons of water. ❖

20 ft

Try This The envelope of a hot-air balloon has a radius of 27 feet when fully inflated. Approximately how many cubic feet of gas can it hold?
≈ 82,448 cu ft

Exercises & Problems

Communicate

1. Explain the relationship between a sphere and a circle.

2. What is a great circle? Name some real-world examples of great circles.

3. How is the area of a great circle related to the surface area of a sphere?

4. How can you use the surface area of a sphere to determine the volume?

Practice & Apply

Find the surface area and volume of each sphere.

5. $S \approx 113.1$ sq in.
$V \approx 113.1$ cu in.

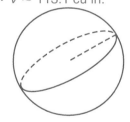

r = 3 in.

6. $S \approx 95$ sq cm
$V \approx 87.1$ cu cm

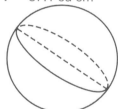

d = 5.5 cm

7. $S \approx 452.4$ sq yd
$V \approx 904.8$ cu yd

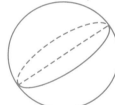

d = 12 yd

8. $S \approx 1017.9$ sq cm
$V \approx 3053.6$ cu cm

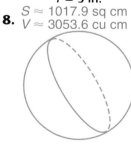

C = 18π cm

9. $S \approx 2123.7$ sq yd
$V \approx 9202.8$ cu yd

C = 26π yd

10. $S \approx 8$ sq ft; $V \approx 2.1$ cu ft

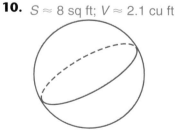

C = 5 ft

11. Half of a sphere is called a **hemisphere.** Write a formula for the surface area of a hemisphere. Use your formula to find the approximate surface area of a bowl shaped like a hemisphere with a radius of 3.5 centimeters.
$S = \dfrac{4\pi r^2}{2}; \approx 77$ sq cm

12. Write a formula for the volume of a hemisphere. Use your formula to find the volume of a hemisphere with a radius of 3.5 centimeters.
$V = \dfrac{4\pi r^3}{6}; \approx 89.8$ cu cm

Find the surface area and volume of each sphere described.

13. radius of 7 in.	**14.** radius of 8 cm	**15.** radius of 10 m
16. radius of 4.25 yd	**17.** radius of 9 in.	**18.** radius of 11.4 cm
19. diameter of 7 ft	**20.** diameter of 4 m	**21.** diameter of 6.4 m
22. radius of 2.5 cm	**23.** radius of 12 in.	**24.** diameter of 3.9 m
25. diameter of 12 ft	**26.** diameter of 6 ft	**27.** diameter of 2.08 m
28. diameter of 8.6 m	**29.** radius of 6 in.	**30.** radius of 6.5 cm
31. radius of 9.35 cm	**32.** radius of 19 in.	**33.** diameter of 14.8 m
34. diameter of 27 ft	**35.** diameter of 7.7 ft	**36.** diameter of 8.46 m
37. radius of 18 cm	**38.** diameter of 34 m	**39.** radius of $1\frac{3}{7}$ ft
40. radius of $5\frac{1}{5}$ ft	**41.** radius of $2\frac{1}{4}$ ft	**42.** radius of $3\frac{3}{4}$ ft

Measure the circumference of three spherical objects (you may use those suggested in the table if you wish). Then find the radius, surface area, and volume of each. Answers may vary.

	Object	Circumference	Radius	Surface area	Volume
43.	tennis ball	9.4? in.	1.5 in.?	28.3? sq in.	14.1? cu in.
44.	basketball	31.4? in.	5 in.?	314.2? sq in.	523.6? cu in.
45.	bowling ball	28.3? in.	4.5 in.?	254.5? sq in.	381.7? cu in.

46. If the radius of a sphere is doubled, what happens to its surface area? 4 times larger

47. If the radius of a sphere is doubled, what happens to its volume? 8 times larger

48. If the radius of a sphere is tripled, what happens to its surface area? 9 times larger

49. If the radius of a sphere is tripled, what happens to its volume? 27 times larger

50. Use your answers to Exercises 46–49 to predict the effect on the surface area and volume of a sphere when the radius is multiplied by n.

Surface area increases by a factor of n^2, and the volume increases by a factor of n^3.

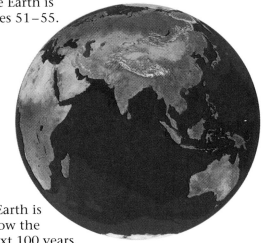

Portfolio Activity The circumference of the Earth is approximately 24,900 miles. Use this information for Exercises 51–55.

51. Use the formula $C = 2\pi r$ to compute the radius of the Earth. ≈ 3963 mi

52. Use the formula $S = 4\pi r^2$ to compute the surface area of the Earth. ≈ 197,359,487.5 sq mi

53. The land area of the Earth is approximately 29% of its surface area. What is the land area of the Earth?
≈ 57,234,251.4 sq mi

54. Use the fact that there are approximately 6 billion (6,000,000,000) people living on Earth and the formula $D = \frac{\text{population}}{\text{land area}}$ to find the population density of Earth.
≈ 104.8 people per sq mi

55. Technology Scientists predict that the population on Earth is increasing at a rate of 1.7% per year. Create a table to show the population and population density on Earth over the next 100 years.

Look Back

Build a table of values using x-values of −3, −2, −1, 0, 1, 2, and 3. [Lesson 2.6]

56. $y = 0.5x + 3$ **57.** $y = x - 1$ **58.** $y = 5 - 2x$

Estimate each square root to the nearest tenth.   [Lesson 4.7]

59. $\sqrt{147}$ ≈ 12.1   **60.** $\sqrt{19}$ ≈ 4.4   **61.** $\sqrt{10}$ ≈ 3.2

Solve each equation.   [Lessons 4.7, 7.1, and 7.2]

62. $x^2 = 15$ ≈ ±3.9   **63.** $x - 6 = -x + 1$ 3.5   **64.** $3 - (x - 4) = x$ 3.5

65. Find the solution to the system $\begin{cases} x - y = 4 \\ 2x - y = -3 \end{cases}$.   **[Lesson 9.4]** (−7, −11)

Look Beyond

Geometry You can solve for the radius of a sphere by using the cube root, which is denoted by $\sqrt[3]{}$.

66. Explain why the formula for the radius of a sphere is $r = \sqrt[3]{\dfrac{3V}{4\pi}}$.

67. Find the radius of a sphere with a volume of 1000 cubic centimeters. ≈ 6.2 cm

68. Find the surface area of a sphere that has a radius of 5 centimeters. Compare it with the surface area of a cube that has an edge length of 10 centimeters. Compare it with the surface area of a cylinder that has a radius of 5 centimeters and a height of 10 centimeters.

69. Technology Use a graphics calculator to create a table of radii (plural for radius) for spheres with volumes of 100 cubic centimeters, 200 cubic centimeters, and so on. What radius would you choose for a tank with a volume between 1200 cubic centimeters and 1300 cubic centimeters?

Investigating SPACE

The solar system contains nine known planets, each a different size and distance from the Sun. You can compare the planets by looking at their surface area, volume, and mass.

ACTIVITY 1 *Comparing Planet Sizes*

The diameters of the planets are listed in the table. Copy and complete the table. Use a spreadsheet or graphics calculator if available.

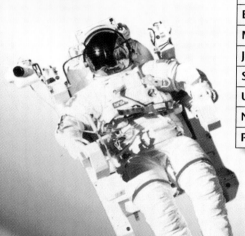

Planet	Diameter (miles)	Radius	Surface area	Volume
Mercury	3000	?	?	?
Venus	7500	?	?	?
Earth	7900	?	?	?
Mars	4200	?	?	?
Jupiter	88,800	?	?	?
Saturn	74,900	?	?	?
Uranus	31,800	?	?	?
Neptune	30,800	?	?	?
Pluto	1400	?	?	?

ACTIVITY 2 *Comparing Rotation Speed*

The "Earth" time it takes each planet to rotate once on its own axis is shown in the table. Compute the circumference of each planet, and use it to find the speed of rotation in miles per hour.

Planet	Diameter (miles)	Circumference	Time to rotate	Rotation Speed (miles per hour)
Mercury	3000	?	59 days	?
Venus	7500	?	243 days	?
Earth	7900	?	24 hours	?
Mars	4200	?	24.5 hours	?
Jupiter	88,800	?	19 hours	?
Saturn	74,900	?	10.6 hours	?
Uranus	31,800	?	17.1 hours	?
Neptune	30,800	?	16.1 hours	?
Pluto	1400	?	6 days	?

ACTIVITY 3 *Modeling Planet Sizes*

Suppose that Earth is compared with a ball that has a 2.5-inch diameter. Using this scale find the relative sizes of any three other planets. What real-world objects can you use to model the sizes of the three planets that you chose?

Chapter 12 Review

Vocabulary

altitude	664	edge	650	pyramid	665
bases	645	faces	645	radius	636
center of a circle	636	great circle	676	rectangular solid	645
circle	636	hemisphere	679	right cylinder	658
circumference	637	lateral face	650	slant height	670
cone	664	lateral surface area	652	sphere	676
cube	645	net	643	surface area	643
diameter	636	prism	650	vertex	650

Key Skills & Exercises

Lesson 12.1

➤ **Key Skills**

Find the circumference and area of a circle.

To find the circumference and area of a circle with a radius of 3 yards, use the formulas below.

Circumference
$C = 2\pi r$
$= 2\pi(3)$
$= 6\pi$
≈ 18.85 yards

Area
$A = \pi r^2$
$= 3^2\pi$
$= 9\pi$
≈ 28.27 square yards

$r = 3$ yd

$C \approx 18.85$ yards
$A \approx 28.27$ square yards

➤ **Exercises**

Find the circumference and area, to the nearest tenth, of each circle with the given radius or diameter.

1. radius = 4 inches

2. diameter = 8.5 feet

3. radius = $5\frac{1}{2}$ centimeters

1. ≈ 25.1 in., ≈ 50.3 sq in.
2. ≈ 26.7 ft, ≈ 56.7 sq ft
3. ≈ 34.6 cm, ≈ 95.0 sq cm

Lesson 12.2

➤ **Key Skills**

Find the surface area and volume of a rectangular solid.

Find the surface area and volume of the rectangular solid shown. Use the formulas below.

Surface area
$S = 2(lw + lh + wh)$
$= 2(12 \cdot 8 + 12 \cdot 2.5 + 8 \cdot 2.5)$
$= 2(96 + 30 + 20)$
$= 292$ square feet

Volume
$V = lwh$
$= 12 \cdot 8 \cdot 2.5$
$= 240$ cubic feet

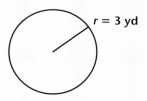

$l = 12$ ft

$h = 2.5$ ft

$w = 8$ ft

➤ **Exercises**

Find the surface area and volume of each rectangular solid with the indicated dimensions.

  4. $l = 5.5$ in., $w = 10$ in., $h = 14$ in. 544 sq in., 770 cu in.

  5. $l = 3$ m, $w = 8.2$ m, $h = 15.6$ m ≈ 398.6 sq m, ≈ 383.8 cu m

  6. $l = 7\frac{1}{2}$ yd, $w = 4\frac{1}{2}$ yd, $h = 6\frac{3}{4}$ yd 229.5 sq yd, ≈ 227.8 cu yd

  7. Find the surface area and volume of a cube with an edge length of 9.4 meters. ≈ 530.2 sq m, ≈ 830.6 cu m

Lesson 12.3

➤ **Key Skills**

Find the surface area and volume of a right prism.

The right triangular prism shown has a base perimeter, p, of 48 meters. Find the surface area and volume of this triangular prism. Use the formulas below.

Surface area

$S = hp + 2B$

   $= \left(9\frac{1}{4}\right)(48) + 2(24)$

   $= 444 + 48$

   $= 492$ square meters

Triangular prism

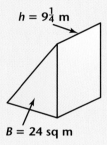

$h = 9\frac{1}{4}$ m

$B = 24$ sq m

Volume

$V = Bh$

   $= (24)\left(9\frac{1}{4}\right)$

   $= 222$ cubic meters

➤ **Exercises**

Find the surface area and volume for each right prism. The dimensions for the bases of each prism are given.

  8. Base: equilateral triangle

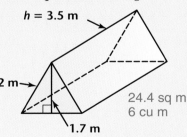

$h = 3.5$ m

2 m

1.7 m

24.4 sq m
6 cu m

  9. Base: parallelogram

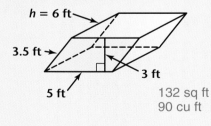

$h = 6$ ft

3.5 ft

3 ft

5 ft

132 sq ft
90 cu ft

  10. Base: trapezoid

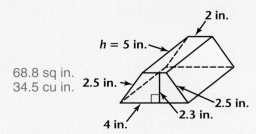

2 in.

$h = 5$ in.

68.8 sq in.
34.5 cu in.

2.5 in.

2.5 in.

2.3 in.

4 in.

Lesson 12.4

➤ Key Skills

Find the surface area and volume of a right cylinder.
Find the surface area and volume of the right cylinder shown.
Use the formulas below.

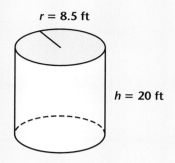

$r = 8.5$ ft

$h = 20$ ft

Surface area	**Volume**
$S = L + 2B$	$V = Bh$

Lateral surface area
$L = 2\pi rh$

The area of the circular base, B, is πr^2.

$S = L + 2B$
$S = 2\pi rh + 2(\pi r^2)$
$\quad = 2\pi(8.5)(20) + 2\pi(8.5)^2$
$\quad = 340\pi + 144.5\pi$
$\quad = 484.5\pi$
$\quad \approx 1522$ square feet

$V = Bh$
$V = (\pi r^2)h$
$\quad = \pi(8.5)^2(20)$
$\quad = 1445\pi$
$\quad \approx 4540$ cubic feet

➤ Exercises

Find the surface area and volume of each right cylinder.

11. height = 6 in., radius = 2.5 in. ≈ 133.5 sq in., ≈ 117.8 cu in.

12. height = 45 cm, diameter = 18 cm ≈ 3053.6 sq cm, $\approx 11,451.1$ cu cm

13. height = 18.5 m, radius = 9.5 m ≈ 1671.3 sq m, ≈ 5245.3 cu m

14. height = 150 yd, diameter = 100 yd $\approx 62,831.9$ sq yd, $\approx 1,178,097.2$ cu yd

Lesson 12.5

➤ Key Skills

Find the volume of a right cone or right pyramid.

Find the volume of
the right cone shown.

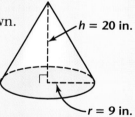

$h = 20$ in.

$r = 9$ in.

Find the volume of this
right pyramid with a
square base.

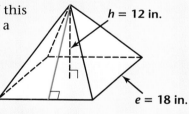

$h = 12$ in.

$e = 18$ in.

Use the formulas below.

Volume of a right cone	**Volume of a right pyramid**
$V = \dfrac{Bh}{3}$	$V = \dfrac{Bh}{3}$

First, find the value of B. Then substitute into the above formulas.

$B = \pi r^2$
$\quad = \pi(9)^2$
$\quad = 81\pi$ square inches

$B = e^2$
$\quad = 18^2$
$\quad = 324$ square inches

$V = \dfrac{81\pi \cdot 20}{3}$
$\quad \approx 1696.5$ cubic inches

$V = \dfrac{324 \cdot 12}{3}$
$\quad = 1296$ cubic inches

➤ Exercises

Find the volume of each right cone.

15. base area = 1200 square meters, height = 25 meters 10,000 cu m

16. base area = $2\frac{1}{4}$ square millimeters, height = $13\frac{1}{3}$ millimeters 10 cu mm

Find the volume of each right pyramid.

17. base area = 15.6 square meters, height = 7 meters 36.4 cu m

18. base area = 30 square inches, height = 13 inches 130 cu in.

Lesson 12.6

➤ Key Skills

Find the lateral surface area and the total surface area of a right cone.

Find the lateral surface area and the total surface area of a right cone shown. Use the formulas below.

$$L = \frac{1}{2}Cs \qquad\qquad S = L + B$$
$$= \frac{1}{2}(2\pi r)s \qquad\quad = L + \pi r^2$$
$$= \pi rs \qquad\qquad = 197.1\pi + \pi(9)^2$$
$$= \pi(9)(21.9) \qquad = 197.1\pi + 81\pi$$
$$= 197.1\pi \qquad\qquad = 278.1\pi$$
$$\approx 619 \text{ square inches} \qquad \approx 873.7 \text{ square inches}$$

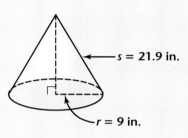

s = 21.9 in.

r = 9 in.

Find the lateral surface area and the total surface area of a right, regular pyramid.

Find the lateral surface area and the total surface area of the right, regular pyramid shown. Use the formulas below.

$$L = \frac{1}{2}ps \qquad\qquad S = L + B$$
$$= \frac{1}{2}(4e)s \qquad\quad = L + e^2$$
$$= 2es \qquad\qquad = 540 + 18^2$$
$$= 2(18)(15) \qquad = 540 + 324$$
$$= 540 \text{ square inches} \qquad = 864 \text{ square inches}$$

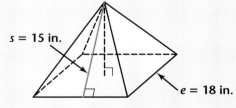

s = 15 in.

e = 18 in.

➤ Exercises

Find the lateral surface area and the total surface area of each right cone described.

19. slant height = 26 yards, radius of base = 10 yards $L \approx$ 816.8 sq yd, $S \approx$ 1131.0 sq yd

20. slant height = 6.4 meters, diameter of base = 8.6 meters $L \approx$ 86.5 sq m, $S \approx$ 144.5 sq m

Find the lateral surface area and the total surface area of each right, regular pyramid described.

21. Square base: edge of base = 6.5 centimeters, $L =$ 130 sq cm, $S \approx$ 172.3 sq cm
slant height = 10 centimeters

22. Regular pentagonal base: edge of base = 20 feet, $L =$ 1500 sq ft, $S =$ 1950 sq ft
area of base = 450 square feet, slant height = 30 feet

Lesson 12.7

➤ Key Skills

Find the surface area and volume of a sphere.

Find the surface area and volume of the sphere shown. Use the formulas below.

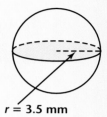

$r = 3.5$ mm

Surface area	**Volume**
$S = 4\pi r^2$	$V = \frac{4}{3}\pi r^3$
$= 4\pi(3.5)^2$	$= \frac{4}{3}\pi(3.5)^3$
≈ 154 square millimeters	≈ 180 cubic millimeters

➤ Exercises

Find the surface area and volume of each sphere.

23. radius = 12 yards $S \approx 1809.6$ sq yd, $V \approx 7238.2$ cu yd

24. diameter = 1.8 inches $S \approx 10.2$ sq in., $V \approx 3.1$ cu in.

25. radius = $5\frac{1}{4}$ miles $S \approx 346.4$ sq mi, $V \approx 606.1$ cu mi

Applications

26. Business   A pizza parlor offers circular pizzas in two sizes, one with a diameter of 10 inches and the other with a diameter of 15 inches. The charge for the smaller pizza is $8 and the percent of profit per square inch is the same for both pizzas. What price should be charged for the second pizza? About $18

27. Manufacturing   The manager of a warehouse that ships miscellaneous office supplies to businesses is selecting flat cardboard sheets that, when folded, form boxes for shipping supplies. He has a choice of two patterns. The first folds into a box with dimensions of 2 inches by 9 inches by 10.5 inches. The second folds into a box with dimensions of 3 inches by 7.5 inches by 8.5 inches. Which box can hold more supplies? second box

Manufacturing   A manufacturer makes cylindrical aquariums. The height of each aquarium is 3 feet and the diameter is 1 foot.

28. Find the base area of each aquarium.   About 0.8 sq ft

29. How much material is needed to build each aquarium if the top is left open?   About 10.2 sq ft

30. Find the volume of each aquarium.   About 2.4 cu ft

31. How many gallons of water will each aquarium hold?
HINT: 1 cubic foot ≈ 7.5 gallons   About 17.7 gal

Chapter 12 Assessment

a. Circle

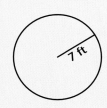

7 ft

b. Rectangular solid

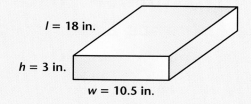

l = 18 in.
h = 3 in.
w = 10.5 in.

c. Right trapezoidal prism

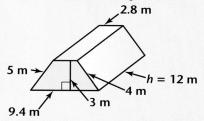

2.8 m
5 m
h = 12 m
4 m
3 m
9.4 m

d. Right cylinder

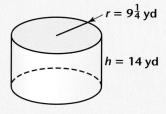

r = 9¼ yd
h = 14 yd

e. Right cone

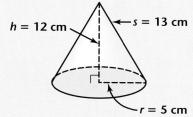

h = 12 cm
s = 13 cm
r = 5 cm

f. Right, regular hexagonal pyramid

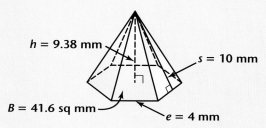

h = 9.38 mm
s = 10 mm
B = 41.6 sq mm
e = 4 mm

Refer to the figures above for Items 1–12.

1. Find the circumference of the circle, to the nearest tenth. ≈ 44.0 ft
2. Find the area of the circle, to the nearest tenth. ≈ 153.9 sq ft
3. Find the surface area of the rectangular solid. 549 sq in.
4. Find the volume of the rectangular solid. 567 cu in.
5. Find the surface area of the right trapezoidal prism. 291 sq m
6. Find the volume of the right trapezoidal prism. 219.6 cu m
7. Find the surface area of the cylinder. ≈ 1351.3 sq yd
8. Find the volume of the cylinder. ≈ 3763.2 cu yd
9. Find the volume of the cone. ≈ 314.2 cu cm
10. Find the surface area of the cone. ≈ 282.7 cu cm
11. Find the volume of the pyramid. 130.1 cu mm
12. Find the surface area of the pyramid.161.6 sq mm

For the sphere shown at the right, find

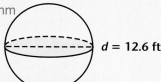

d = 12.6 ft

13. the surface area, to the nearest tenth.
14. the volume, to the nearest tenth. ≈ 498.8 sq ft
≈ 1047.4 cu ft

15. The ice-cream cone shown at the right consists of two parts, a spherical scoop of ice cream and the inverted cone into which the scoop has been inserted. Half of the scoop of ice cream is inside the cone. The radius, *r*, of both the scoop of ice cream and the base of the cone is 2.5 inches. The height, *h*, of the cone is 6 inches, and its slant height, *s*, is 6.5 inches. The entire inner surface of the ice-cream cone is to be coated with a thin layer of chocolate. What will be the approximate area of the chocolate layer? HINT: Find the 51.1 sq in. surface area of the cone, without the base area.

r

Chapters 1 – 12
Cumulative Assessment
College Entrance Exam Practice

Quantitative Comparison For Questions 1–4, write
A if the quantity in Column A is greater than the quantity in Column B;
B if the quantity in Column B is greater than the quantity in Column A;
C if the two quantities are equal; or
D if the relationship cannot be determined from the information given.

	Column A	Column B	Answers
1. C	$-6 \cdot 9$	$6 \cdot (-9)$	Ⓐ Ⓑ Ⓒ Ⓓ **[Lesson 2.5]**
2. B	The number of $15 tickets that you can buy with $90	The number of $12 tickets that you can buy with $84	Ⓐ Ⓑ Ⓒ Ⓓ **[Lesson 6.2]**
3. B	Slope of the graph of $y = -2$	Slope of the graph of $y = x - 2$	Ⓐ Ⓑ Ⓒ Ⓓ **[Lesson 8.2]**
4. A	Correlation coefficient of a line of best fit that rises	Correlation coefficient of a line of best fit that falls	Ⓐ Ⓑ Ⓒ Ⓓ **[Lesson 11.6]**

5. What LCD (least common denominator) would you use to
C   compare the fractions $\frac{4}{15}$ and $\frac{3}{10}$? **[Lesson 3.2]**
   a. 15   **b.** 10   **c.** 30   **d.** 150

6. Which is the best description of an acute angle? **[Lesson 4.1]**
a   **a.** an angle with a measure less than 90°
   b. an angle with a measure less than 180°
   c. an angle with a measure greater than 90°
   d. an angle with a measure of 90° or less

7. What is the solution to $6 - 4t > 18$? **[Lesson 7.5]**
a   **a.** $t < -3$ **b.** $t > -3$
   c. $t > -6$ **d.** $t < -6$

8. What is the slope of a line parallel to the line $2x + 3y = -5$?
b   **[Lesson 8.7]**
   a. 2   **b.** $-\frac{2}{3}$ **c.** $\frac{3}{2}$ **d.** $\frac{2}{3}$

9. Which point is the vertex of the parabola shown at the right?
[**Lesson 10.2**]
d **a.** $(-1, 1)$ **b.** $(0, 0)$ **c.** $(2, 0)$ **d.** $(1, -1)$

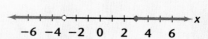

10. Which is the area of a circle that has a diameter of 14 units?
[**Lesson 12.1**]
c **a.** 196π **b.** 14π **c.** 49π **d.** 7π

11. The lengths of the two legs of a right triangle are 12 and 30. The shortest side of a triangle that is similar to this triangle has a length of 18. Find the scale factor and the length of the longer leg of the second triangle. [**Lesson 4.8**] $\frac{3}{2}$; 45

12. Write an inequality that describes the points graphed on the number line at the right. [**Lesson 5.6**] $x < -3$ or $x \geq 3$

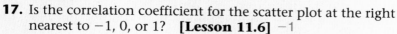

13. Solve $\dfrac{x}{-10} = \dfrac{-3}{5}$. [**Lesson 6.5**] $x = 6$

14. A line has a slope of $\frac{4}{5}$ and passes through the point $(-4, 2)$. $y - 2 = \frac{4}{5}(x + 4)$
Write the point-slope form of the equation of the line. [**Lesson 8.5**]

15. Graph the following system of inequalities: $\begin{cases} y \leq x + 2 \\ x + y > -1 \end{cases}$
[**Lesson 9.5**]

Use the scatter plot for Exercises 16 and 17.
16. Describe the correlation coefficient for the scatter plot at the right as *strong positive, strong negative,* or *little or none.*
[**Lesson 11.5**] Strong negative

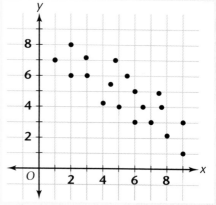

17. Is the correlation coefficient for the scatter plot at the right nearest to -1, 0, or 1? [**Lesson 11.6**] -1

18. Find the volume of a right cone with a height of 15 feet and a base with a radius of 5 feet. [**Lesson 12.5**]
≈ 392.7 cu ft

Free-Response Grid The following questions may be answered using a free-response grid commonly used by standardized test services.

19. Solve the system $\begin{cases} 4x - 3y = 5 \\ 8x + 9y = 3 \end{cases}$ to find the value of x that makes both equations true. [**Lessons 9.2, 9.3, 9.4**] 0.9

20. Find the slope of the line that passes through the points $(-1, -2)$ and $(3, 4)$. [**Lesson 8.2**] $\frac{3}{2}$

21. Find INT(5.6). [**Lesson 10.4**] 5

22. Find the surface area of a right pyramid that has a slant height of 24 inches. The length of an edge of the square base is 12 inches. [**Lesson 12.6**] 720 sq in.

INFO BANK

EXTRA PRACTICE

Chapter 1

Lesson 1.1

1. There are 8 players in a chess tournament. Each player plays every other player. How many games of chess will be played? 28

2. There are 5 people in a room. If each person talks to every other person, how many conversations occur? 10

3. There are 16 people at a meeting. Each person shakes hands with every other person. How many handshakes are there? 120

Find the next three terms in each sequence. Then explain the pattern used to find the terms for each.

4. 6, 9, 12, 15, 18, 21, . . .   **5.** 90, 85, 70, 65, 60, 55, . . .

6. 3, 8, 11, 19, 30, 49, . . .   **7.** 3, 5, 7, 9, 11, 13, . . .

8. 18, 22, 26, 30, 34, 38, . . .   **9.** 100, 93, 86, 79, 72, 65, . . .

10. 34, 38, 42, 46, 50, 54, . . .   **11.** 28, 25, 22, 19, 16, 13, . . .

12. 7, 13, 19, 25, 31, 37, . . .   **13.** 1, 1, 2, 3, 5, 8, . . .

14. 10, 20, 30, 40, 50, 60, . . .   **15.** 8, 15, 22, 29, 36, 43, . . .

Lesson 1.2

Write an algebraic expression for each verbal expression.

1. Each book costs $12. $12x$ **2.** Rebecca makes $50 a day. $50x$

3. Each carton holds 24 oranges. $24x$ **4.** Tim drove 50 miles each hour. $50x$

5. Julie reads 40 pages each day. $40x$ **6.** Each student solved 10 problems. $10x$

Write an equation for each sentence.

7. The total cost of the dinner was $8 per person. $c = 8p$

8. Fran rents a car for $39, plus $0.10 per mile. $r = 39 + 0.10m$

9. Randi paid $2.99 per pound for the cheese. $c = 2.99p$

10. The plumber charges $25, plus $10 an hour. $c = 25 + 10h$

11. Ned makes 10 loaves of bread a day. $n = 10b$

12. T-shirts cost $6.50 each, plus $3 shipping. $c = 6.50t + 3$

Build a table of values by substituting 1, 2, 3, 4, 5, and 10 for x.

13. $y = 3x$ **14.** $y = 5x$ **15.** $y = x + 2$

16. $y = x - 1$ **17.** $y = 2x + 2$ **18.** $y = 4x - 1$

19. $y = 9x + 2$ **20.** $y = 15x$ **21.** $y = 4x + 28$

Lesson 1.3

Make a table of values using 10, 20, 30, and 40 for the variable in each expression. Show a process column.

1. $36x$

2. $2x + 5$

3. $3q + 5$

4. $15a - 2$

5. $6w + 55$

6. $8f - 6$

7. $7w - 3$

8. $10t + 13$

9. $8x + 15$

10. $28r - 1$

11. $500 - 6a$

12. $100 - 2t$

Use a graphics calculator to make a table of values for each equation. Determine the value of x when y has the indicated value.

13. $y = 3x$, $y = 36$ 12

14. $y = 6.8x$, $y = 34$ 5

15. $y = 8x + 5$, $y = 53$ 6

16. $y = 7.8x + 3$, $y = 42$ 5

17. $y = 0.9x - 4$, $y = 5.9$ 11

18. $y = 2.6x - 6$, $y = 17.4$ 9

19. $y = 10.9x + 6$, $y = 115$ 10

20. $y = 15.1x - 6$, $y = 733.9$ 49

Lesson 1.4

Guess-and-check to solve each equation.

1. $4x + 1 = 9$ 2

2. $15 = 5x + 5$ 2

3. $24x - 8 = 112$ 5

4. $10x - 3 = 67$ 7

5. $12 = 2x - 4$ 8

6. $3 + 19x = 98$ 5

7. $7 + 6x = 55$ 8

8. $9 - x = 3$ 6

9. $115 - 3x = 100$ 5

Identify each of the following as an expression or an equation.

10. $5x$ expression

11. $t = 4$ equation

12. $3x - 3$ expression

13. $2 = q$ equation

14. $0.5p$ expression

15. $y + 1$ expression

16. $3w + 4$ expression

17. $3m - 1 = 14$ equation

Pens cost 69 cents each. Write and simplify an expression for the cost of

18. 0 pens. 18. $69(0) = 0$ cents

19. 5 pens. 19. $69(5) = 345$ cents, or $3.45

20. 10 pens. 20. $69(10) = 690$ cents, or $6.90

21. p pens. 21. $69(p) = 69p$ cents

Hats cost \$2.50 each. Write and simplify an expression for the cost of

22. 0 hats. 22. $2.50(0) = \$0$

23. 3 hats. 23. $2.50(3) = \$7.50$

24. 8 hats. 24. $2.50(8) = \$20.00$

25. h hats. 25. $2.50(h) = \$2.50h$

Make a table of values using 1, 2, 3, 4, and 5 for the variable in each expression.

26. $3x$

27. $8y$

28. $2z + 4$

29. $3d - 1$

30. $9q + 3$

31. $15c$

32. $9v - 2$

33. $2 + 8z$

34. If tickets for a concert cost \$12 each, how many tickets can you buy with \$48? 4

35. If videos rent for \$2.99 each, how many videos can you rent with \$11.96? 4

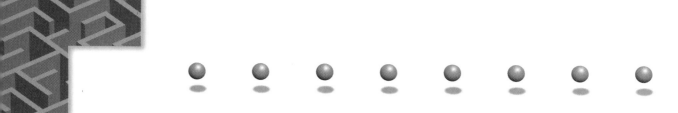

Lesson 1.5

List all of the factors for each number. Circle the prime factors.

1. 15   **2.** 56   **3.** 27   **4.** 49   **5.** 100

6. 12   **7.** 64   **8.** 32   **9.** 54   **10.** 93

11. 35   **12.** 60   **13.** 19   **14.** 125   **15.** 68

If the number is divisible by 2, 5, and 10, write *yes*. If the number is not divisible by all three of these numbers, write *no*.

16. 444 No   **17.** 640 Yes   **18.** 255 No   **19.** 1346 No   **20.** 715 No

21. 3720 Yes   **22.** 15,005 No   **23.** 4788 No   **24.** 8840 Yes   **25.** 25,009 No

If the number is divisible by 2, 3, and 6, write *yes*. If the number is not divisible by all three of these numbers, write *no*.

26. 246 Yes   **27.** 190 No   **28.** 536 No   **29.** 7215 No   **30.** 16,092 Yes

31. 24,558 Yes   **32.** 17,037 No   **33.** 54,000 Yes   **34.** 4698 Yes   **35.** 23,004 Yes

Determine whether each number is prime, composite, or neither.

36. 1 neither   **37.** 21 composite   **38.** 85 composite   **39.** 141 composite   **40.** 39 composite

41. 19 prime   **42.** 73 prime   **43.** 102 composite   **44.** 51 composite   **45.** 88 composite

Lesson 1.6

Use exponents to rewrite each expression.

1. $5 \times 5 \times 5 \times 5$ 5^4 **2.** $3 \times 3 \times 3 \times 3 \times 3 \times 3 \times 3$ 3^7

3. $2 \cdot 2 \cdot 2 \cdot 2 \cdot 7 \cdot 7$ $2^4 \cdot 7^2$ **4.** $3 \cdot 3 \cdot 3 \cdot 5 \cdot 11 \cdot 11$ $3^3 \cdot 5 \cdot 11^2$

5. $x \cdot x \cdot x \cdot x \cdot y \cdot y \cdot y$ $x^4 \cdot y^3$ **6.** $7 \cdot 7 \cdot 7 \cdot r \cdot r \cdot r \cdot r$ $7^3 \cdot r^4$

7. $w \cdot w \cdot w \cdot w + 3$ $w^4 + 3$ **8.** $2 \cdot 2 \cdot a \cdot a \cdot a \cdot a + 4$ $2^2 \cdot a^4 + 4$

Evaluate.

9. 2^3 8   **10.** 4^2 16   **11.** 10^4 10,000   **12.** 3^3 27

13. 10^8 100,000,000   **14.** 7^3 343   **15.** 8^2 64   **16.** 5^4 625

Write the prime factorization, in exponential form, for each number.   **26.** $2^2 \cdot 5 \cdot 11$

17. 16 2^4 **18.** 25 5^2 **19.** 10 $2 \cdot 5$ **20.** 13 already prime

21. 144 $2^4 \cdot 3^2$ **22.** 48 $2^4 \cdot 3$ **23.** 32 2^5 **24.** 125 5^3

25. 90 $2 \cdot 3^2 \cdot 5$ **26.** 220   **27.** 215 $5 \cdot 43$ **28.** 303 $3 \cdot 101$

29. 21 $3 \cdot 7$ **30.** 68 $2^2 \cdot 17$ **31.** 169 13^2 **32.** 93 $3 \cdot 31$

Lesson 1.7

1. $(24 - 6) \cdot (2 + 4) = 108$ **5.** $(15 + 5) \div 5 - 2 = 2$

Place inclusion symbols using the correct order of operations to make each equation true. Use a calculator to check each answer.

1. $24 - 6 \cdot 2 + 4 = 108$

2. $16 \div 2 + 2 - 4 = 0$ $16 \div (2 + 2) - 4 = 0$

3. $6 \cdot 8 + 1 \div 3 = 18$ $6 \cdot (8 + 1) \div 3 = 18$ **4.** $7 + 3 \cdot 8 - 5 = 30$ $(7 + 3) \cdot (8 - 5) = 30$

5. $15 + 5 \div 5 - 2 = 2$

6. $64 + 4 \cdot 2 \div 2 = 68$

7. $39 - 7 \div 4 + 3 = 11$

8. $100 \div 5 \cdot 4 + 15 = 20$ $100 \div (5 \cdot 4) + 15 = 20$

7. $(39 - 7) \div 4 + 3 = 11$ **6.** $(64 + 4) \cdot 2 \div 2 = 68$ or $64 + 4 \cdot 2 \div 2 = 68$

Evaluate each expression using the correct order of operations.

9. $21 - 7 + 2 \cdot 5$ 24

10. $(21 - 7) + 2 \cdot 5$ 24

11. $(21 - 7 + 2) \cdot 5$ 80

12. $32 + 2 \cdot 6 \div 3$ 36

13. $(32 + 2) \cdot 6 \div 3$ 68

14. $32 + 2 \cdot (6 \div 3)$ 36

15. $24 + 4 \div 4$ 25

16. $42 \cdot 20 + 6$ 846

17. $25(30 + 12)$ 1050

18. $15 \div 3 \cdot 3 - 1$ 14

19. $45(2) + 3(16)$ 138

20. $6(4) - 4$ 20

21. $(3 + 7) \div (3 + 2)$ 2

22. $(4 + 5 \cdot 4) \div (2 \cdot 3)$ 4

23. $(7 - 4 \cdot 2 + 2) \div 4$ $\frac{1}{4}$

24. $14 + 4 \div 2 - 5$ 11

25. $6 \div 2 \cdot 5 + 4$ 19

26. $4 + 2^2 - 10 + 3$ 1

Given that a is 4, b is 5, and c is 6, evaluate each expression.

27. $a + b + c$ 15

28. $a \cdot b - c$ 14

29. $b - a \cdot c$ 6

30. $a + b - c$ 3

31. $a \cdot b + c$ 26

32. $a^2 + c^2 - b^2$ 27

33. $(a + b) \cdot c$ 54

34. $a^2 - b + c$ 17

35. $a \cdot b \cdot c^2$ 720

36. $b^2 - a^2$ 9

37. $a^2 + c^2 - b$ 47

38. $a^2 - c - b$ 5

Lesson 1.8

Use mental math to find each sum or product. Show each step, and name each property you use to find the answer.

1. $(56 + 48) + 22$

2. $(25 \cdot 3) \cdot 4$

3. $5 \cdot (6 \cdot 12)$

4. $(137 + 149) + 113$

5. $5 \cdot (138 \cdot 2)$

6. $4 \cdot (76 \cdot 25)$

7. $(157 + 29) + 43$

8. $40 \cdot (36 \cdot 50)$

9. $(167 \cdot 345) \cdot 0$

10. $16 + (24 + 608)$

11. $6(5 + 20)$

12. $25(3 + 4)$

Name the property illustrated.

13. $54 + 15 = 15 + 54$ Comm Prop

14. $6 \cdot (3 \cdot 14) = (6 \cdot 3) \cdot 14$ Assoc Prop

15. $2(2.5 + 5.9) = 2 \cdot 2.5 + 2 \cdot 5.9$ Dist Prop

16. $8 \cdot 40 - 8 \cdot 25 = 8(40 - 25)$ Dist Prop

17. $(6 + 34) + 17 = 6 + (34 + 17)$ Assoc Prop

18. $18 \cdot 3 = 3 \cdot 18$ Comm Prop

19. $7(5x) = (7 \cdot 5)x$ Assoc Prop

20. $7(27.2 - 6.5) = 7 \cdot 27.2 - 7 \cdot 6.5$ Dist Prop

21. $(6 \cdot 9) \cdot 5 = 6 \cdot (9 \cdot 5)$ Assoc Prop

22. $7 + (18 + 26) = (7 + 18) + 26$ Assoc Prop

Chapter

Lesson 2.1

Determine whether each number is an integer.

1. 6 Yes   **2.** −4 Yes   **3.** 3.2 No   **4.** 3 Yes   **5.** $\frac{2}{3}$ No

6. −24 Yes   **7.** −8.2 No   **8.** $-\frac{5}{6}$ No   **9.** 11 Yes   **10.** 173 Yes

Write the opposite of each number.

11. 14 −14   **12.** −7 7   **13.** 66 −66   **14.** −32 32   **15.** −51 51

16. −50 50   **17.** 100 −100   **18.** 18 −18   **19.** −83 83   **20.** 49 −49

Find the absolute value.

21. $|-8|$ 8   **22.** $|5|$ 5   **23.** $|18|$ 18   **24.** $|-26|$ 26   **25.** $|72|$ 72

26. $|-19|$ 19   **27.** $|220|$ 220   **28.** $|-13|$ 13   **29.** $|91|$ 91   **30.** $|-159|$ 159

31. $|6|$ 6   **32.** $|550|$ 550   **33.** $|-75|$ 75   **34.** $|-144|$ 144   **35.** $|97|$ 97

Lesson 2.2

Find each sum.

1. −4 + 2 −2   **2.** 3 + (−9) −6   **3.** −8 + (−5) −13

4. −8 + 10 2   **5.** 8 + (−10) −2   **6.** −8 + (−10) −18

7. −5 + 4 + (−6) −7   **8.** 15 + (−9) + (−12) −6   **9.** −14 + 3 + (−9) −20

10. −15 + 36 21   **11.** 26 + (−40) −14   **12.** −15 + (−34) −49

13. −42 + 16 −26   **14.** −89 + (−23) −112   **15.** −25 + (−25) −50

16. 25 + (−25) 0   **17.** −91 + 78 −13   **18.** −32 + (−40) −72

19. −4 + (−35) + 26 −13   **20.** 16 + (−39) + 21 −2   **21.** −56 + (−56) + (−13)

22. 17 + (−73) + 68 12   **23.** −34 + 43 + (−43) −34   **24.** −8 + (−120) + 30 −98

25. −8 + $|7|$ −1   **26.** $|-8|$ + 7 15   **27.** $|-8|$ + $|7|$ 15

28. $|-9|$ + $|-4|$ 13   **29.** $|-13|$ + $|12|$ 25   **30.** $|-43|$ + $|-18|$ 61

31. −5 + $|-15|$ + $|7|$ 17   **32.** −600 + $|-25|$ + $|-30|$ −545   **33.** 78 + (−43) + $|-5|$ 40

21. −125

Substitute 3 for *m*, −5 for *n*, and 6 for *p*. Evaluate each expression.

34. $m + n + p$ 4   **35.** $m + |n + p|$ 4   **36.** $m + (n + p)$ 4

37. $m + p + |n|$ 14   **38.** $|m + n| + p$ 8   **39.** $|m + p| + n$ 4

40. $|m| + n + p$ 4   **41.** $m + n + |p|$ 4   **42.** $n + (m + p)$ 4

Lesson 2.3

Use guess-and-check to solve each equation.

1. $x + 3 = 4$ 1
2. $f + 5 = 8$ 3
3. $a + 9 = 7$ -2

4. $m + 1 = -1$ -2
5. $x + 6 = -4$ -10
6. $8 + n = -3$ -11

7. $x + 9 = 0$ -9
8. $-7 + y = 4$ 11
9. $6 + z = -2$ -8

10. $-12 + k = 0$ 12
11. $m + 16 = 7$ -9
12. $n + 15 = -10$ -25

13. $-17 + y = -13$ 4
14. $1 - z = 4$ -3
15. $16 - c = -5$ 21

16. $-14 - w = -6$ -8
17. $25 + x = -21$ -46
18. $a + 42 = -24$ -66

19. $3a = 6$ 2
20. $-7z = 21$ -3
21. $5q = -45$ -9

22. $48 \div r = -8$ -6
23. $-36 \div w = 9$ -4
24. $t \div -4 = -7$ 28

25. $4y - 1 = 7$ 2
26. $6x - 18 = 0$ 3
27. $3c - 1 = -10$ -3

Write two inequalities for each pair of integers. Use both the $<$ and $>$ symbols.
  28. $5 > -5$, $-5 < 5$
29. $0 > -8$, $-8 < 0$ **30.** $7 > -14$, $-14 < 7$,   **31.** $-8 < -4$, $-4 > -8$

28. $5, -5$
29. $0, -8$
30. $7, -14$
31. $-8, -4$
32. $16, -15$

33. $8, -9$
34. $2, -20$
35. $-7, 18$
36. $-16, -12$
37. $19, -4$

32. $16 > -15$, $-15 < 16$ **33.** $8 > -9$, $-9 < 8$ **34.** $2 > -20$, $-20 < 2$ **35.** $-7 < 18$, $18 > -7$
36. $-16 < -12$, $-12 > -16$ **37.** $19 > -4$, $-4 < 19$

Lesson 2.4

Evaluate.

1. $7 - 11$ -4
2. $8 - (-7)$ 15
3. $6 - 15$ -9

4. $-5 - 9$ -14
5. $16 - 20$ -4
6. $-7 - (-12)$ 5

7. $30 - (-30)$ 60
8. $-17 - (-18)$ 1
9. $-52 - 19$ -71

10. $-24 - (-24)$ 0
11. $16 - (-90)$ 106
12. $-34 - 43$ -77

13. $-52 - 17$ -69
14. $72 - 75$ -3
15. $-68 - (-10)$ -58

16. $-5 - (-5) + 10$ 10
17. $-3 + 5 - 16$ -14
18. $12 - 18 + 7$ 1

19. $-27 - (-15) + 2$ -10
20. $-100 + 75 - (-75)$ 50
21. $-36 + 21 - 16$ -31

22. $20 - 60 - 32$ -72
23. $42 - (-72) - 6$ 108
24. $-83 + 14 - (-36)$ -33

Substitute 6 for a, -2 for b, and -15 for c. Evaluate each expression.

25. $a + b$ 4
26. $b + c$ -17
27. $a - b$ 8

28. $c - a$ -21
29. $a + b + c$ -11
30. $b - c + a$ 19

31. $c - a - b$ -19
32. $a - (b + c)$ 23
33. $a + c$ -9

34. $a - a - a - a$ -12
35. $c + c + c$ -45
36. $c - b$ -13

37. $b + b + b$ -6
38. $(a + a) + (b + b)$ 8
39. $b - (a - c)$ -23

40. $c + (b - a)$ -23
41. $c - c - c - c$ 30
42. $b - (b - b)$ -2

43. $(a + b) + (c - b)$ -9
44. $(a - b) - (c - b)$ 21
45. $(a - a) - c + c$ 0

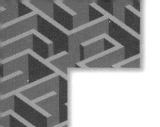

Lesson 2.5

Evaluate.

1. $(3)(-4)$ -12 **2.** $(5)(-6)$ -30 **3.** $(-4)(1)$ -4

4. $(-3)(9)$ -27 **5.** $(-2)(-4)$ 8   **6.** $(7)(-6)$ -42

7. $(16) \div (-4)$ -4 **8.** $(-25) \div (-5)$ 5   **9.** $(-36) \div (9)$ -4

10. $(-100) \div (-10)$ 10   **11.** $(-42) \div (-3)$ 14   **12.** $(36) \div (-3)$ -12

13. $(-56) \div (-2)$ 28   **14.** $(-30) \div (3)$ -10 **15.** $(-5)(14)$ -70

16. $(-12)(-30)$ 360   **17.** $(45)(-5)$ -225 **18.** $(-24) \div (8)$ -3

19. $(-35) \div (-7)$ 5   **20.** $48 \div (-4)$ -12 **21.** $(-2)(-1)(6)$ 12

22. $(-5)(-3)(7)$ 105   **23.** $(-2)[(-3) + (-2)]$ 10   **24.** $(-4)[(-1) - (-4)]$ -12

25. $(-6)[7 + (-5)]$ -12 **26.** $(-36) \div [(-12) - (-3)]$ 4   **27.** $(-12) \div [4 - 6]$ 6

28. $(7)(-2) \div (-7)$ 2   **29.** $(-4)(-18) \div (-9)$ -8 **30.** $(-3)(24)(-10) \div 12$ 60

Lesson 2.6

Make a table of values and ordered pairs for each equation by substituting integer values from −3 to 3 for x.

1. $y = 3 - x$ **2.** $y = 4x + 1$ **3.** $y = 3x - 2$

4. $y = 5x$ **5.** $y = -0.5x$ **6.** $y = -4x$

7. $y = -x + 2$ **8.** $y = 4 - 2x$ **9.** $y = -6x - 1$

Write the ordered pair that represents each point.

10. A (5, 6)   **11.** B (−8, −9)

12. C (−3, 6)   **13.** D (0, −5)

14. E (2, 9)   **15.** F (4, 0)

16. G (−7, −3)   **17.** H (3, −2)

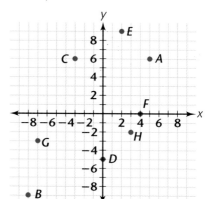

Lesson 2.7

Find the next two terms of each sequence.

1. 54, 50, 46, 42, 38, . . . 34, 30   **2.** 60, 58, 61, 59, 62, 60, . . . 63, 61

3. 13, 18, 16, 21, 19, 24, . . . 22, 27   **4.** 5, 6, 8, 11, 15, 20, . . . 26, 33

5. 1, 2, 5, 10, 17, . . . 26, 37   **6.** 1, 6, 10, 13, 15, . . . 16, 16

7. 1, 2, 3, 5, 8, 13, . . . 21, 34   **8.** 144, 132, 120, 108, 96, . . . 84, 72

9. 60, 50, 40, 30, 20, 10, . . . 0, −10   **10.** 2, 5, 13, 26, 44, . . . 67, 95

Chapter 3

Lesson 3.1

Use fraction bars to determine whether each fraction is closest to $0, \frac{1}{2},$ or 1.

1. $\frac{2}{3}$ $\frac{1}{2}$　　**2.** $\frac{1}{5}$ 0　　**3.** $\frac{7}{12}$ $\frac{1}{2}$　　**4.** $\frac{3}{10}$ $\frac{1}{2}$　　**5.** $\frac{5}{8}$ $\frac{1}{2}$

6. $\frac{5}{6}$ 1　　**7.** $\frac{1}{12}$ 0　　**8.** $\frac{9}{10}$ 1　　**9.** $\frac{4}{5}$ 1　　**10.** $\frac{1}{8}$ 0

Use fraction bars to compare the fractions using the symbol $<, >,$ or =.

11. $\frac{1}{4}$ and $\frac{2}{3}$ $\frac{1}{4} < \frac{2}{3}$　**12.** $\frac{9}{10}$ and $\frac{1}{3}$ $\frac{9}{10} > \frac{1}{3}$　**13.** $\frac{1}{8}$ and $\frac{1}{5}$ $\frac{1}{8} < \frac{1}{5}$　**14.** $\frac{3}{8}$ and $\frac{1}{3}$ $\frac{3}{8} > \frac{1}{3}$

15. $\frac{5}{6}$ and $\frac{3}{4}$ $\frac{5}{6} > \frac{3}{4}$　**16.** $\frac{6}{8}$ and $\frac{2}{3}$ $\frac{6}{8} > \frac{2}{3}$　**17.** $\frac{5}{6}$ and $\frac{3}{10}$ $\frac{5}{6} > \frac{3}{10}$　**18.** $\frac{3}{4}$ and $\frac{6}{8}$ $\frac{3}{4} = \frac{6}{8}$

19. $\frac{5}{12}$ and $\frac{3}{10}$ $\frac{5}{12} > \frac{3}{10}$　**20.** $\frac{5}{6}$ and $\frac{7}{12}$ $\frac{5}{6} > \frac{7}{12}$　**21.** $\frac{3}{5}$ and $\frac{2}{3}$ $\frac{3}{5} < \frac{2}{3}$　**22.** $\frac{3}{5}$ and $\frac{6}{10}$ $\frac{3}{5} = \frac{6}{10}$

Determine whether each fraction is closest to $\frac{1}{4}, \frac{1}{2},$ or $\frac{3}{4}.$

23. $\frac{7}{8}$ $\frac{3}{4}$　　**24.** $\frac{2}{3}$ $\frac{3}{4}$　　**25.** $\frac{3}{10}$ $\frac{1}{4}$　　**26.** $\frac{1}{3}$ $\frac{1}{4}$　　**27.** $\frac{5}{12}$ $\frac{1}{2}$

28. $\frac{7}{10}$ $\frac{3}{4}$　　**29.** $\frac{5}{6}$ $\frac{3}{4}$　　**30.** $\frac{2}{5}$ $\frac{1}{2}$　　**31.** $\frac{6}{11}$ $\frac{1}{2}$　　**32.** $\frac{8}{12}$ $\frac{3}{4}$

Lesson 3.2

Write each fraction in lowest terms.

1. $\frac{3}{6}$ $\frac{1}{2}$　　**2.** $\frac{6}{15}$ $\frac{2}{5}$　　**3.** $\frac{9}{12}$ $\frac{3}{4}$　　**4.** $\frac{8}{20}$ $\frac{2}{5}$　　**5.** $\frac{30}{45}$ $\frac{2}{3}$

6. $\frac{18}{20}$ $\frac{9}{10}$　　**7.** $\frac{14}{21}$ $\frac{2}{3}$　　**8.** $\frac{6}{42}$ $\frac{1}{7}$　　**9.** $\frac{25}{75}$ $\frac{1}{3}$　　**10.** $\frac{4}{18}$ $\frac{2}{9}$

11. $\frac{16}{60}$ $\frac{4}{15}$　　**12.** $\frac{24}{27}$ $\frac{8}{9}$　　**13.** $\frac{12}{54}$ $\frac{2}{9}$　　**14.** $\frac{140}{175}$ $\frac{4}{5}$　　**15.** $\frac{26}{30}$ $\frac{13}{15}$

16. $\frac{35}{50}$ $\frac{7}{10}$　　**17.** $\frac{36}{48}$ $\frac{3}{4}$　　**18.** $\frac{12}{50}$ $\frac{6}{25}$　　**19.** $\frac{90}{102}$ $\frac{15}{17}$　　**20.** $\frac{63}{99}$ $\frac{7}{11}$

Use the LCD to compare each pair of fractions using the symbol $<, >,$ or =.

21. $\frac{1}{4}$ and $\frac{1}{3}$ $\frac{1}{4} < \frac{1}{3}$　**22.** $\frac{9}{10}$ and $\frac{1}{3}$ $\frac{9}{10} > \frac{1}{3}$　**23.** $\frac{1}{8}$ and $\frac{1}{5}$ $\frac{1}{8} < \frac{1}{5}$　**24.** $\frac{3}{8}$ and $\frac{3}{4}$ $\frac{3}{8} < \frac{3}{4}$

25. $\frac{5}{6}$ and $\frac{3}{5}$ $\frac{5}{6} > \frac{3}{5}$　**26.** $\frac{4}{5}$ and $\frac{2}{3}$ $\frac{4}{5} > \frac{2}{3}$　**27.** $\frac{2}{9}$ and $\frac{3}{10}$ $\frac{2}{9} < \frac{3}{10}$　**28.** $\frac{5}{6}$ and $\frac{10}{12}$ $\frac{5}{6} = \frac{10}{12}$

29. $\frac{5}{12}$ and $\frac{3}{8}$ $\frac{5}{12} > \frac{3}{8}$　**30.** $\frac{5}{6}$ and $\frac{9}{10}$ $\frac{5}{6} < \frac{9}{10}$　**31.** $\frac{8}{12}$ and $\frac{2}{3}$ $\frac{8}{12} = \frac{2}{3}$　**32.** $\frac{3}{5}$ and $\frac{5}{11}$ $\frac{3}{5} > \frac{5}{11}$

33. $\frac{4}{5}$ and $\frac{3}{10}$ $\frac{4}{5} > \frac{3}{10}$　**34.** $\frac{3}{6}$ and $\frac{7}{14}$ $\frac{3}{6} = \frac{7}{14}$　**35.** $\frac{8}{9}$ and $\frac{3}{4}$ $\frac{8}{9} > \frac{3}{4}$　**36.** $\frac{7}{15}$ and $\frac{6}{10}$ $\frac{7}{15} < \frac{6}{10}$

Lesson 3.3

5. $10\frac{6}{25}$ **10.** $145\frac{1}{50}$ **15.** $-23\frac{677}{1000}$

Write each decimal as a fraction or mixed number in lowest terms.

1. 0.24 $\frac{6}{25}$ **2.** 0.3 $\frac{3}{10}$ **3.** 0.76 $\frac{19}{25}$ **4.** 4.5 $4\frac{1}{2}$ **5.** 10.24

6. 0.655 $\frac{131}{200}$ **7.** 1.24 $1\frac{6}{25}$ **8.** -82.7 $-82\frac{7}{10}$ **9.** 4.249 $4\frac{249}{1000}$ **10.** 145.02

11. 17.34 $17\frac{17}{50}$ **12.** -0.004 $-\frac{1}{250}$ **13.** 798.84 $798\frac{21}{25}$ **14.** 94.555 $94\frac{111}{200}$ **15.** -23.677

16. -160.327 $-160\frac{327}{1000}$ **17.** 25.682 $25\frac{341}{500}$ **18.** 1.8426 $1\frac{4213}{5000}$ **19.** -36.36 $-36\frac{9}{25}$ **20.** 70.704 $70\frac{88}{125}$

Write each fraction as a terminating or repeating decimal.

21. $\frac{2}{3}$ $0.\overline{6}$ **22.** $\frac{5}{8}$ 0.625 **23.** $-\frac{1}{12}$ $-0.08\overline{3}$ **24.** $1\frac{3}{5}$ 1.6 **25.** $-\frac{7}{9}$ $-0.\overline{7}$

26. $6\frac{1}{2}$ 6.5 **27.** $-\frac{5}{11}$ $-0.\overline{45}$ **28.** $\frac{3}{10}$ 0.3 **29.** $-\frac{3}{7}$ $-0.\overline{428571}$ **30.** $2\frac{4}{9}$ $2.\overline{4}$

31. $1\frac{1}{6}$ $1.1\overline{6}$ **32.** $-3\frac{2}{11}$ $-3.\overline{18}$ **33.** $11\frac{7}{10}$ 11.7 **34.** $-4\frac{7}{8}$ -4.875 **35.** $-\frac{2}{15}$ $-0.13\overline{3}$

Order the numbers from the least to the greatest.

36. $0.5, 0.54, 0.05, 0.04, 0.054$ **37.** $-0.16, 0.06, -0.6, -0.1, 0.66$

38. $\frac{3}{4}, 0.3, 3.4, 1\frac{1}{3}, 0.43, 0.03$ **39.** $-2.8, 2\frac{1}{8}, -2\frac{2}{5}, 0.08, -0.88, -0.28$

40. $-9.25, -9.05, -9.02, -9.2, -9\frac{1}{2}$ **41.** $\frac{13}{10}, 1.9, -0.19, -1.3, -1\frac{1}{3}, 0.3$

42. $\frac{3}{4}, 0.8, 0.075, \frac{2}{3}, \frac{5}{8}$ **43.** $-\frac{2}{5}, -\frac{3}{7}, -0.04, -0.39, -0.41$

Lesson 3.4

Add or subtract. Check each answer using a calculator.

1. $\frac{3}{4} + \frac{1}{3}$ $1\frac{1}{12}$ **2.** $\frac{2}{5} + \frac{1}{2}$ $\frac{9}{10}$ **3.** $\frac{3}{7} - \frac{1}{4}$ $\frac{5}{28}$ **4.** $\frac{7}{12} - \frac{2}{5}$ $\frac{11}{60}$

5. $\frac{7}{10} + \frac{3}{5}$ $1\frac{3}{10}$ **6.** $\frac{1}{11} + \frac{5}{6}$ $\frac{61}{66}$ **7.** $\frac{7}{9} - \frac{2}{3}$ $\frac{1}{9}$ **8.** $\frac{4}{5} - \frac{2}{15}$ $\frac{2}{3}$

9. $\frac{5}{12} + \frac{4}{15}$ $\frac{41}{60}$ **10.** $\frac{7}{8} - \frac{1}{6}$ $\frac{17}{24}$ **11.** $\frac{3}{10} + \frac{5}{12}$ $\frac{43}{60}$ **12.** $\frac{8}{9} - \frac{2}{15}$ $\frac{34}{45}$

13. $1\frac{1}{2} + \frac{2}{3}$ $2\frac{1}{6}$ **14.** $3\frac{2}{5} + 1\frac{1}{3}$ $4\frac{11}{15}$ **15.** $3\frac{9}{10} - 1\frac{2}{3}$ $2\frac{7}{30}$ **16.** $6\frac{7}{8} - \frac{2}{5}$ $6\frac{19}{40}$

17. $3\frac{5}{9} + 2\frac{1}{6}$ $5\frac{13}{18}$ **18.** $2\frac{7}{8} + 1\frac{1}{12}$ $3\frac{23}{24}$ **19.** $4\frac{4}{15} - 1\frac{1}{9}$ $3\frac{7}{45}$ **20.** $7\frac{7}{8} - 2\frac{5}{18}$ $5\frac{43}{72}$

21. $3 - \frac{3}{4}$ $2\frac{1}{4}$ **22.** $-\frac{1}{5} + \frac{3}{4}$ $\frac{11}{20}$ **23.** $-\frac{4}{9} - \frac{2}{3}$ $-1\frac{1}{9}$ **24.** $\frac{1}{2} + \frac{4}{5} - \frac{1}{10}$ $1\frac{1}{5}$

25. $-\frac{5}{6} + 1\frac{2}{3} - \frac{1}{2}$ $\frac{1}{3}$ **26.** $-\frac{4}{9} + 2\frac{4}{5}$ $2\frac{16}{45}$ **27.** $-2\frac{1}{4} - 1\frac{3}{8}$ $-3\frac{5}{8}$ **28.** $-4 - 2\frac{6}{7}$ $-6\frac{6}{7}$

29. $-\frac{3}{11} + 6\frac{4}{5}$ $6\frac{29}{55}$ **30.** $-2\frac{2}{9} - 3\frac{7}{10}$ **31.** $-1\frac{3}{10} + \frac{3}{4} - \frac{2}{5}$ **32.** $-\frac{3}{5} + \left(-2\frac{3}{4}\right) + \left(-\frac{1}{4}\right)$

33. $2\frac{9}{10} - \frac{9}{10}$ **34.** $6\frac{7}{8} - 9\frac{1}{2}$ **35.** $-4\frac{3}{4} - 2\frac{1}{2} + \frac{2}{3}$ **36.** $-1\frac{7}{12} + \left(-2\frac{5}{6}\right) + \left(7\frac{3}{4}\right)$

Lesson 3.5

Estimate each product. Then find the actual product.

1. $4 \cdot \frac{2}{3}$ $2\frac{2}{3}$

2. $\frac{1}{2} \cdot \frac{3}{8}$ $\frac{3}{16}$

3. $2\frac{3}{4} \cdot \frac{8}{9}$ $2\frac{4}{9}$

4. $1\frac{1}{3} \cdot 3\frac{1}{3}$ $4\frac{4}{9}$

5. $4 \cdot 2\frac{3}{4}$ 11

6. $1\frac{5}{6} \cdot 6$ 11

7. $1\frac{4}{5} \cdot 2\frac{1}{4}$ $4\frac{1}{20}$

8. $3\frac{2}{3} \cdot \frac{7}{8}$ $3\frac{5}{24}$

9. $3\frac{1}{8} \cdot 7$ $21\frac{7}{8}$

10. $5\frac{2}{5} \cdot 3\frac{1}{10}$ $16\frac{37}{50}$

11. $8\frac{1}{3} \cdot 1\frac{4}{9}$ $12\frac{1}{27}$

12. $2\frac{3}{8} \cdot 1\frac{4}{5}$ $4\frac{11}{40}$

13. $6\frac{3}{4} \cdot 1\frac{1}{9}$ $7\frac{1}{2}$

14. $6\frac{1}{6} \cdot 5\frac{7}{9}$ $35\frac{17}{27}$

15. $2\frac{4}{7} \cdot 4\frac{2}{3}$ 12

16. $3\frac{6}{11} \cdot 2\frac{4}{9}$ $8\frac{2}{3}$

Find each quotient. Use a calculator to check the quotient.

17. $\frac{1}{3} \div \frac{3}{4}$ $\frac{4}{9}$

18. $\frac{3}{5} \div \frac{7}{10}$ $\frac{6}{7}$

19. $8 \div \frac{4}{5}$ 10

20. $10 \div \frac{5}{6}$ 12

21. $14 \div 1\frac{1}{2}$ $9\frac{1}{3}$

22. $6\frac{1}{4} \div \frac{2}{3}$ $9\frac{3}{8}$

23. $5\frac{3}{4} \div 2$ $2\frac{7}{8}$

24. $\frac{3}{8} \div 1\frac{2}{3}$ $\frac{9}{40}$

25. $2\frac{3}{10} \div \frac{1}{5}$ $11\frac{1}{2}$

26. $1\frac{2}{5} \div 3\frac{1}{10}$ $\frac{14}{31}$

27. $3\frac{2}{3} \div 4\frac{1}{6}$ $\frac{22}{25}$

28. $\frac{8}{9} \div 2\frac{1}{4}$ $\frac{32}{81}$

29. $\frac{4}{7} \div 3\frac{1}{4}$ $\frac{16}{91}$

30. $1\frac{7}{8} \div 3\frac{5}{6}$ $\frac{45}{92}$

31. $\frac{7}{12} \div 6\frac{3}{4}$ $\frac{7}{81}$

32. $2\frac{2}{3} \div 4\frac{3}{5}$ $\frac{40}{69}$

33. $7\frac{1}{5} \div 1\frac{1}{2}$ $4\frac{4}{5}$

34. $\frac{3}{7} \div 5\frac{2}{3}$ $\frac{9}{119}$

35. $2\frac{4}{9} \div 2\frac{1}{3}$ $1\frac{1}{21}$

36. $9\frac{1}{9} \div 1\frac{9}{10}$ $4\frac{136}{171}$

Lesson 3.6

1. $\frac{2}{5}$ **2.** $\frac{9}{10}$ **3.** $\frac{1}{10}$ **4.** $\frac{50}{1}$ **5.** $\frac{7}{1}$ **6.** $\frac{245}{2}$ **7.** $\frac{14}{1}$ **8.** $\frac{3}{5}$

Write a ratio, in lowest terms, to describe each situation.

1. Andrea made 2 out of every 5 shots.　**2.** Shaun answered 9 out of 10 questions correctly.

3. Becki runs 3 miles in 30 minutes.　**4.** Ray drove 150 miles in 3 hours.

5. Aaron works 35 hours in 5 days.　**6.** Wanda earned \$245 in 2 weeks.

7. Matt served 28 customers in 2 hours.　**8.** Sherri returned 24 out of 40 serves.

9. Tim runs 12 miles in 7 days.　**10.** Liam sells 62 of the 100 packages.

9. $\frac{12}{7}$ **10.** $\frac{31}{50}$

Complete each proportion.

11. $\frac{1}{3} = \frac{?}{6}$ 2

12. $\frac{4}{?} = \frac{16}{20}$ 5

13. $\frac{?}{9} = \frac{2}{18}$ 1

14. $\frac{3}{4} = \frac{12}{?}$ 16

15. $\frac{?}{10} = \frac{6}{30}$ 2

16. $\frac{8}{9} = \frac{?}{27}$ 24

17. $\frac{4}{9} = \frac{12}{?}$ 27

18. $\frac{15}{?} = \frac{30}{32}$ 16

19. $\frac{4}{5} = \frac{?}{25}$ 20

20. $\frac{2}{?} = \frac{18}{27}$ 3

21. $\frac{7}{?} = \frac{21}{33}$ 11

22. $\frac{?}{14} = \frac{3}{7}$ 6

23. $\frac{16}{24} = \frac{4}{?}$ 6

24. $\frac{8}{9} = \frac{?}{81}$ 72

25. $\frac{5}{27} = \frac{25}{?}$ 135

26. $\frac{28}{32} = \frac{?}{16}$ 14

27. $\frac{2}{3} = \frac{?}{42}$ 28

28. $\frac{64}{?} = \frac{2}{7}$ 224

29. $\frac{?}{96} = \frac{5}{16}$ 30

30. $\frac{144}{?} = \frac{4}{5}$ 180

31. $\frac{51}{102} = \frac{?}{2}$ 1

32. $\frac{12}{13} = \frac{108}{?}$ 117

33. $\frac{?}{4} = \frac{66}{88}$ 3

34. $\frac{135}{162} = \frac{?}{18}$ 15

35. $\frac{12}{?} = \frac{30}{35}$ 14

36. $\frac{?}{14} = \frac{12}{21}$ 8

37. $\frac{81}{108} = \frac{?}{8}$ 6

38. $\frac{?}{112} = \frac{72}{84}$ 96

Lesson 3.7

Change each decimal to a percent.

1. 0.7 70%   **2.** 0.2 20%   **3.** 0.98 98%   **4.** 0.13 13%   **5.** 0.36 36%

6. 0.29 29%   **7.** 3.6 360%   **8.** 1.7 170%   **9.** 0.235 23.5% **10.** 0.119 11.9%

11. 4.05 405% **12.** 1.98 198% **13.** 6.025 602.5% **14.** 0.786 78.6% **15.** 0.04 4%

16. 0.065 6.5% **17.** 2.113   **18.** 0.009 0.9% **19.** 1.004 100.4% **20.** 0.0033 0.33%
                   17. 211.3%

Change each fraction to a percent.

21. $\frac{1}{2}$ 50%   **22.** $\frac{3}{8}$ 37.5%   **23.** $\frac{2}{5}$ 40%   **24.** $\frac{7}{10}$ 70%   **25.** $\frac{5}{12}$

26. $\frac{150}{250}$ 60%   **27.** $\frac{4}{15}$ **28.** $\frac{17}{20}$ 85%   **29.** $2\frac{2}{3}$ **30.** $1\frac{4}{5}$ 180%

31. $6\frac{1}{10}$ 610%   **32.** $\frac{3}{50}$ 6%   **33.** $\frac{17}{25}$ 68%   **34.** $5\frac{1}{8}$ 512.5%   **35.** $3\frac{5}{12}$

36. $\frac{14}{15}$ **37.** $8\frac{9}{10}$ 890%   **38.** $4\frac{1}{6}$ **39.** $\frac{5}{9}$ **40.** $\frac{3}{2}$ 150%

25. $41.\overline{6}$% or $41\frac{2}{3}$%   **27.** $26.\overline{6}$% or $26\frac{2}{3}$%   **29.** $266.\overline{6}$% or $266\frac{2}{3}$%   **35.** $341.\overline{6}$% or $341\frac{2}{3}$%

36. $93.\overline{3}$% or $93\frac{1}{3}$%   **38.** $416.\overline{6}$% or $416\frac{2}{3}$%   **39.** $55.\overline{5}$% or $55\frac{5}{9}$%

Lesson 3.8

Two number cubes are tossed 20 times with the following results.

Trial	1	2	3	4	5	6	7	8	9	10	11	12	13	14	15	16	17	18	19	20
Cube 1	4	6	1	3	6	1	5	2	2	4	3	6	5	2	1	6	5	4	3	3
Cube 2	3	6	2	1	3	6	1	2	5	4	1	4	5	1	6	5	4	2	2	3

According to the data in this table, find the following experimental probabilities.

1. Both cubes are alike. $\frac{5}{20}$ or $\frac{1}{4}$

2. Both cubes are different. $\frac{15}{20}$ or $\frac{3}{4}$

3. At least one cube is a 3. $\frac{6}{20}$ or $\frac{3}{10}$

4. At least one cube is greater than 3. $\frac{13}{20}$

5. Neither cube is a 4. $\frac{15}{20}$ or $\frac{3}{4}$

6. Both cubes are 4. $\frac{1}{20}$

7. Both cubes are greater than 3. $\frac{6}{20}$ or $\frac{3}{10}$

8. Both cubes are at least 1. $\frac{20}{20}$ or 1

9. At least one cube is less than 3. $\frac{11}{20}$

10. At least one cube is a 1 or a 6. $\frac{11}{20}$

11. Both cubes are 6. $\frac{1}{20}$

12. Both cubes are 7. $\frac{0}{20}$ or 0

13. At least one cube is a 3, 4, or 5. $\frac{14}{20}$ or $\frac{7}{10}$

14. Both cubes are at least 4. $\frac{6}{20}$ or $\frac{3}{10}$

22. 40, 60, 80, 100, 120, 140, 160

Describe the output of each command.

15. A random rational number ranging in value from 0 to 4, not inclusive.

15. RAND*4

16. INT(RAND*4) 0, 1, 2, 3

17. 10*(INT(RAND*2)) 0, 10

18. INT(RAND*2)+1 1, 2

19. 100*(INT(RAND*5))
       0, 100, 200, 300, 400

20. 100*(INT(RAND*2)+2) 200, 300

21. 10*(INT(RAND*3)+4) 40, 50, 60

22. 20*(INT(RAND*7)+2)

Lesson 3.9

A bag contains 15 marbles: 5 red (R), 3 green (G), 4 blue (B), and 3 yellow (Y). One marble is randomly drawn from the bag. Find each probability.

1. $P(G) = $? $\frac{3}{15}$ or $\frac{1}{5}$ **2.** $P(B) = $? $\frac{4}{15}$ **3.** $P(Y) = $? $\frac{3}{15}$ or $\frac{1}{15}$

4. $P(R) = $? $\frac{5}{15}$ or $\frac{1}{3}$ **5.** $P(not\ Y) = $? $\frac{12}{15}$ or $\frac{4}{5}$ **6.** $P(not\ R) = $? $\frac{10}{15}$ or $\frac{2}{3}$

7. $P(not\ B) = $? $\frac{11}{15}$ **8.** $P(not\ G) = $? $\frac{12}{15}$ or $\frac{4}{5}$ **9.** $P(R\ or\ Y) = $? $\frac{8}{15}$

10. $P(R\ or\ G) = $? $\frac{8}{15}$ **11.** $P(R\ or\ B) = $? $\frac{9}{15}$ or $\frac{3}{5}$ **12.** $P(G\ or\ B) = $? $\frac{7}{15}$

		First marble is returned. (independent events)		First marble is not returned. (dependent events)	
13.	P(R, R)	1/9	?	2/21	?
14.	P(R, Y)	1/15	?	1/14	?
15.	P(R, B)	4/45	?	2/21	?
16.	P(G, R)	1/15	?	1/14	?
17.	P(R, G)	1/15	?	1/14	?
18.	P(Y, Y)	1/25	?	1/35	?
19.	P(Y, R)	1/15	?	1/14	?
20.	P(B, B)	16/225	?	2/35	?

Lesson 4.1

Name each figure in all possible ways.

1.

Plane *P*, Plane *XYZ*

2.

Line *m*,
Line *AB*, Line *BA*

3.

Segment *LM*
Segment *ML*

4.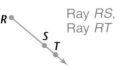

Ray *RS*,
Ray *RT*

5.

Angle 1,
Angle *ABC*,
Angle *CBA*,
Angle *B*

6.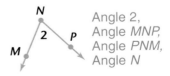

Angle 2,
Angle *MNP*,
Angle *PNM*,
Angle *N*

Use a protractor to draw angles with the following measures. Label each angle as either acute, obtuse, right, or straight. Check students' work.

7. 45° acute   **8.** 68° acute   **9.** 105° obtuse **10.** 83° acute   **11.** 77° acute

12. 120° obtuse **13.** 145° obtuse **14.** 95° obtuse   **15.** 175° obtuse **16.** 24° acute

Lesson 4.2

Use the figure at the right for Exercises 1–12.

1. Find m∠XTY. 53°   **2.** Find m∠ZTW. 143°

3. Find m∠ZTY. 37°   **4.** Find m∠XTW. 127°

5. Find m∠YTV. 127°

6. Name two acute angles. ∠YTX, ∠YTZ

7. Name two obtuse angles. ∠VTY, ∠XTW

8. Name two pairs of vertical angles.
∠YTX, ∠VTW and ∠VTY, ∠XTW

9. Name one pair of complementary angles. ∠ZTY, ∠YTX

10. Name two straight angles. ∠WTY, ∠VTX

11. Name two pairs of supplementary angles.
∠VTY and ∠YTX, ∠WTV and ∠VTY

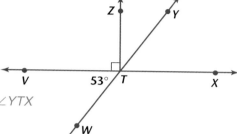

In the figure at the right, m∠1 = 68.5°. Find the measure of each angle.

12. m∠2 90°   **13.** m∠3 21.5°

14. m∠4 68.5°   **15.** m∠5 21.5°

16. m∠6 90°   **17.** m∠7 111.5°

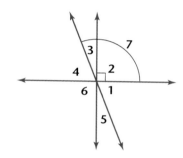

Lesson 4.3

Find the measure of each angle.

1. m∠3 120°   **2.** m∠2 120°   **3.** m∠1 120°   **4.** m∠6 60°

5. m∠7 60°   **6.** m∠8 60°   **7.** m∠4 120°

8. List all pairs of alternate interior angles.

9. List all pairs of alternate exterior angles.

10. List all pairs of consecutive interior angles.

11. Explain why ∠3 and ∠6 are supplementary.

12. Name 10 pairs of supplementary angles.

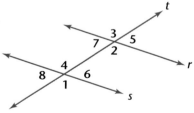

r ∥ s, m∠5 = 60°

Classify each triangle according to its angles and to its sides.

13.

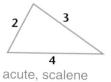

acute, scalene

14.

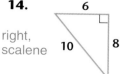

right, scalene

15.

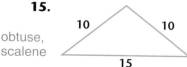

obtuse, scalene

Lesson 4.4

Copy each parallelogram. Then label all of the side lengths and all of the angle measures for each parallelogram.

1.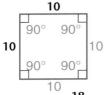

2.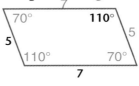

3.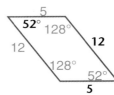

4.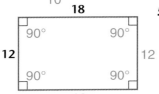

5.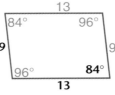

6.

Classify each polygon as concave or convex, and classify the polygon according to the number of sides.

7.

convex
hexagon

8.

concave
septagon

9.

convex
pentagon

10.

concave
hexagon

11.

concave
octagon

Lesson 4.5

Estimate the perimeter and area of each figure.

1.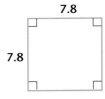

2.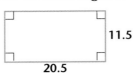

3.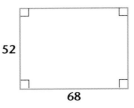

Use formulas for perimeter and area to compute the perimeter and area of each indicated rectangle.

4. $l = 48$ meters, $w = 32$ meters

5. $l = 3.5$ feet, $w = 2.75$ feet

6. $l = 10\frac{5}{8}$ inches, $w = 7\frac{3}{8}$ inches

7. $l = 120$ feet, $w = 120$ feet

8. $l = 1.6$ miles, $w = 0.8$ mile

9. $l = 3\frac{2}{3}$ yards, $w = 1\frac{1}{3}$ yards

10. $l = 2\frac{3}{4}$ inches, $w = \frac{5}{16}$ inch

11. $l = 13.4$ centimeters, $w = 8.8$ centimeters

12. $l = 12\frac{3}{16}$ inches, $w = 12\frac{3}{16}$ inches

13. $l = 8.25$ meters, $w = 3.75$ meters

Lesson 4.6

Find the area of each figure.

1.

$A = 40$
8   10

2.
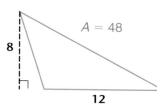
$A = 48$
8
12

3.
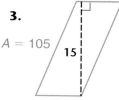
7
$A = 105$
15

4.
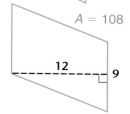
$A = 108$
12
9

5.
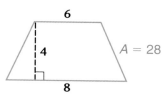
6
4   $A = 28$
8

6.
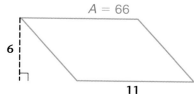
$A = 66$
6
11

7.
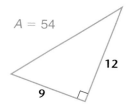
$A = 54$
12
9

8.
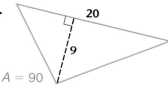
20
9
$A = 90$

9.
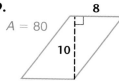
8
$A = 80$
10

Lesson 4.7

Find each square root to the nearest whole number and to the nearest tenth. Then use a calculator to estimate to the nearest hundredth.

1. $\sqrt{7}$ **2.** $\sqrt{13}$ **3.** $\sqrt{24}$ **4.** $\sqrt{40}$ **5.** $\sqrt{75}$

6. $\sqrt{83}$ **7.** $\sqrt{98}$ **8.** $\sqrt{112}$ **9.** $\sqrt{73}$ **10.** $\sqrt{145}$

11. $\sqrt{170}$ **12.** $\sqrt{27}$ **13.** $\sqrt{66}$ **14.** $\sqrt{124}$ **15.** $\sqrt{250}$

The length of the sides of a triangle are given. Use the converse of the "Pythagorean" Right-Triangle Theorem to determine if the triangle is a right triangle.

16. 3, 4, 5 yes   **17.** 6, 10, 14 no   **18.** 12, 35, 37 yes   **19.** 12, 13, 19 no

20. 11, 60, 61 yes   **21.** 21, 28, 35 yes   **22.** 16, 25, 36 no   **23.** 14, 48, 50 yes

For Exercises 24–27, use the "Pythagorean" Right-Triangle Theorem to find the unknown side of each right triangle.

24.
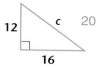
12   c 20
16

25.
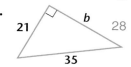
21   b 28
35

26.
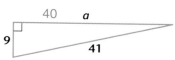
40   a
9   41

Lesson 4.8

Use the scale factor to find the length and width of each rectangle.

	Scale factor	Length	Width	
	original	10 cm	12 cm	
1.	2	?	?	20, 24
2.	4	?	?	40, 48
3.	$\frac{3}{4}$	?	?	7.5, 9
4.	0.9	?	?	9, 10.8
5.	1.2	?	?	12, 14.4
6.	70%	?	?	7, 8.4

	Scale factor	Length	Width	
	original	9 m	16 m	
7.	50%	?	?	4.5, 8
8.	125%	?	?	11.25, 20
9.	0.7	?	?	6.3, 11.2
10.	$\frac{1}{4}$	?	?	2.25, 4
11.	1.5	?	?	13.5, 24
12.	5	?	?	45, 80

The right triangles described in the tables below are similar right triangles. Copy and complete each table.

	Leg	Leg	Scale factor	
	2	6	original	
13.	12	?	?	36, 6
14.	?	36	?	12, 6
15.	1	?	?	3, 0.5
16.	?	42	?	14, 7

	Leg	Leg	Scale factor	
	5	9	original	
17.	10	?	?	18, 2
18.	?	81	?	45, 9
19.	35	?	?	63, 7
20.	?	3	?	$\frac{5}{3}, \frac{1}{3}$

Lesson 5.1

Add.   **12.** $2.75m - 1.6n$ **22.** $10m + 2n + 2$ **23.** $2r - 11t + 13$

1. $(6a + 1) + (2a - 3)$ $8a - 2$ **2.** $(4x - 4) + (3x - 6)$ $7x - 10$ **3.** $(2c - 5) + (3c + 2)$ $5c - 3$

4. $(4y + 2) + (3 - y)$ $3y + 5$ **5.** $(3z - 4) + (7 - 2z)$ $z + 3$ **6.** $(2w + 1) + (4w - 9)$ $6w - 8$

7. $(1 + 4r) + (-5r - 3)$ $-r - 2$ **8.** $(7w + 2) + (-7w + 3)$ 5   **9.** $(4a - 8b) + (7a - 6b)$ $11a - 14b$

10. $(2.2a + 1.1b) + (3a + 0.9b)$ $5.2a + 2b$ **11.** $(3.5x + 4.2y) + (0.6x + y)$ $4.1x + 5.2y$

12. $(2m - 3.2n) + (0.75m + 1.6n)$ **13.** $(0.1v + 6) + (2.3v - 2)$ $2.4v + 4$

14. $(7.5 - 3.4d) + (8 - 1.6d)$ $15.5 - 5d$ **15.** $(-4 - 10.5t) + (6.9 - 5.5t)$ $2.9 - 16t$

16. $\left(\frac{x}{3} + 1\right) + \left(\frac{x}{2} - 2\right)$ $\frac{5x}{6} - 1$ **17.** $\left(\frac{5p}{6} + 5q\right) + \left(\frac{2p}{3} - 4q\right)$ $\frac{3p}{2} + q$

18. $(2x + 3y - 4z) + (x + y + z)$ $3x + 4y - 3z$ **19.** $(7a - 5b + 3c) + (-5a + 2b - 6c)$ $2a - 3b - 3c$

20. $(3x - 5z) + (2x + 3y - 5z)$ $5x + 3y - 10z$ **21.** $(3a + b) + (2b + c) + (a + b + c)$ $4a + 4b + 2c$

22. $(2m - 3) + (4m + 2n) + (4m + 5)$ **23.** $(6r + 2t) + (-4r - 5t + 6) + (7 - 8t)$

Lesson 5.2

6. $-3a - 2b$ 13. $10m - 4n$ 14. $y + 6z$ 16. $-x - y - z$

Find the opposite of each expression.

1. 16 -16
2. -28 28
3. $5z$ $-5z$
4. -7 7

5. $-5r$ $5r$
6. $3a + 2b$
7. $a - 7$ $-a + 7$
8. $-8w + 4$ $8w - 4$

9. $5x - 7y$ $-5x + 7y$
10. $-s - 6$ $s + 6$
11. $8a - c$ $-8a + c$
12. $9 - 3x$ $-9 + 3x$

13. $-10m + 4n$
14. $-y - 6z$
15. $-15 - v$ $15 + v$
16. $x + y + z$

17. $-2c + 4d - 5$
18. $t - w - 5$
19. $8a + 4b - c$
20. $5r - s + 3t$

17. $2c - 4d + 5$ 18. $-t + w + 5$ 19. $-8a - 4b + c$ 20. $-5r + s - 3t$

Perform the indicated operations.

21. $3x - 2x$ x
22. $9r - 4r$ $5r$
23. $6a - 5a$ a
24. $7y - 10y$ $-3y$

25. $8c - (3c + 2)$
26. $9w - (5w - 8)$
27. $6z - (5 - 7z)$ $13z - 5$
28. $8f - (3 - f)$ $9f - 3$

29. $6x - (3x + 1)$
30. $5r - (8 - 4r)$ $9r - 8$
31. $7 - (3y - 4y)$ $y + 7$
32. $16a - (5a - 3)$ $11a + 3$

33. $(4a + 2) + (3a - 6)$ $7a - 4$
34. $(6w - 3) + (3w + 2)$ $9w - 1$

35. $(7 - 3w) + (6 - 8w)$ $-11w + 13$
36. $(9q + 3) - (5 - 5q)$ $14q - 2$

37. $(10q + 8q) - (4m + 2m)$ $18q - 6m$
38. $(3y + 1) - (4y - 1)$ $-y + 2$

39. $(3v - 7t) - (7v + 2t)$ $-4v - 9t$
40. $(2h + 7j) - (-4h - 5j)$ $6h + 12j$

41. $(3d + 1) - (4d + 3) + (5d - 6)$ $4d - 8$
42. $(5m - 4) - (4m + 1) + (8 - m)$ 3

43. $(m + 2n - p) - (3m - 9n + 8p)$
44. $(5r + 7s - t) - (3r - t) + (6s + t)$ $2r + 13s + t$

45. $(3 - 4t + 8r) - (7r + t) - (6 + t)$
46. $(5 + p) - (p + q) - (3p + 4q - 6)$ $-3p - 5q + 11$

25. $5c - 2$ 26. $4w + 8$ 29. $3x - 1$ 43. $-2m + 11n - 9p$
45. $r - 6t - 3$

Lesson 5.3

Solve each equation.

1. $x + 2 = 1$ -1
2. $t + 6 = 2$ -4
3. $w - 4 = -3$ 1
4. $h - 5 = 6$ 11

5. $5 = x + 1$ 4
6. $y + 3 = -4$ -7
7. $5 - w = -5$ 10
8. $x - 9 = -1$ 8

9. $8 = 6 - z$ -2
10. $-7 = 3 - g$ 10
11. $t + 5 = 11$ 6
12. $16 = f - 2$ 18

13. $-8 = 9 - n$ 17
14. $d - 12 = 7$ 19
15. $6 = -9 - c$ -15
16. $(-9) + r = 17$ 26

17. $b + 10 = 2$ -8
18. $8 = h - 3$ 11
19. $x - 3 = 5$ 8
20. $d - 8 = -10$ -2

21. $-6 = 11 - j$ 17
22. $h - 14 = -10$ 4
23. $(-2) + w = -9$ -7
24. $14 = d + (-2)$ 16

25. $x + 9 = -1$ -10
26. $t - 6 = 6$ 12
27. $w - 4 = -2$ 2
28. $h - 5 = 7$ 12

29. $5 = y + 1$ 4
30. $y + 3 = -8$ -11
31. $8 - w = -2$ 10
32. $y - 6 = -1$ 5

33. $3 = 2 - r$ -1
34. $-4 = 3 - h$ 7
35. $t + 6 = 11$ 5
36. $-1 = t - 5$ 4

37. $-6 = 7 - n$ 13
38. $g - 1 = 7$ 8
39. $12 = -9 + y$ 21
40. $(-7) + w = 7$ 14

41. $g + 1 = 1$ 0
42. $8 = x - 6$ 14
43. $y - 7 = 5$ 12
44. $w - 8 = -5$ 3

45. $-4 = 8 - b$ 12
46. $n - 1 = -1$ 0
47. $(-1) + q = -5$ -4
48. $14 = c + (-4)$ 18

Lesson 5.4   Solution methods may vary.

State which property you would use to solve each equation. Then solve.

1. $x - 15 = 20$ 35   **2.** $r + 23 = 10$ -13 **3.** $n - 65 = 59$ 124   **4.** $76 - k = 12$ 64

5. $w - 52 = -45$ 7   **6.** $y + 19 = -10$ -29 **7.** $l + 48 = 50$ 2   **8.** $66 - t = -90$ 156

9. $x + 28 = -20$ **10.** $450 = 100 - f$ **11.** $88 - u = 44$ 44   **12.** $498 - x = 500$ -2

13. $q - 90 = 19$ 109 **14.** $10 - d = 72$ **15.** $s - 25 = 36$ 61   **16.** $54 = 112 - d$ 58

17. $p + 365 = 425$ 60 **18.** $b - 350 = -80$ **19.** $7.8 = h - 9$ 16.8   **20.** $z + 3.2 = 8.8$ 5.6

21. $w - 12.5 = 5.2$ **22.** $30.5 - r = 5$ **23.** $x + \dfrac{1}{4} = \dfrac{3}{4}$ $\dfrac{1}{2}$ **24.** $\dfrac{2}{5} = \dfrac{6}{5} - a$ $\dfrac{4}{5}$

25. $d - \dfrac{5}{6} = \dfrac{1}{3}$ $1\dfrac{1}{6}$ **26.** $\dfrac{5}{8} - g = \dfrac{3}{4}$ $-\dfrac{1}{8}$ **27.** $m - \dfrac{4}{5} = \dfrac{1}{10}$ $\dfrac{9}{10}$ **28.** $3\dfrac{1}{6} + z = 1\dfrac{5}{9}$ $-1\dfrac{11}{18}$

9. -48 10. -350 14. -62 18. 270   21. 17.7   22. 25.5

Lesson 5.5

Simplify each polynomial by combining like terms. Check by substituting two different values for *x*. If the polynomial is already simplified, write *simplified*.

1. $3x - 5x + 3$ $-2x + 3$ **2.** $2 + 4x - 1$ $4x + 1$ **3.** $6x - 4 - 5x$ $x - 4$

4. $3x^2 - 4 - 2x^2$ $x^2 - 4$ **5.** $-4x + 3 - 2x$ $-6x + 3$ **6.** $x^2 - 6x^2 - x$ $-5x^2 - x$

7. $4x^2 - x - 5x$ $4x^2 - 6x$ **8.** $3x^2 - 4x + x$ $3x^2 - 3x$ **9.** $6x - 9 - 9x$ $-3x - 9$

10. $5x + 4 - 2x^2 - x$ **11.** $6x^2 + 1 - 6x$ simplified   **12.** $3x - 2 + 5x^2 + x^2$

13. $6x^2 - (-x^2 - 1)$ $7x^2 + 1$ **14.** $-2x^2 - 2x - (-4x^2 + x)$ **15.** $4x + x^2 - (2x^2 - 1)$

10. $-2x^2 + 4x + 4$ 12. $6x^2 + 3x - 2$ 14. $2x^2 - 3x$ 15. $-x^2 + 4x + 1$

Simplify each binomial addition expression. Check by substituting two different values for *x*.

16. $(4x + 1) + (3x - 5)$ $7x - 4$ **17.** $(5x + 2) + (-4x + 3)$ $x + 5$

18. $(10x - 7) + (7x + 3)$ $17x - 4$ **19.** $(2x - 2) + (-8x - 2)$ $-6x - 4$

20. $(4 - 8x) + (9x - 5)$ $x - 1$ **21.** $(x^2 + 1) + (3x^2 - 2)$ $4x^2 - 1$

22. $(-4x^2 + 3) + (7x^2 - 7)$ $3x^2 - 4$ **23.** $(-2x^2 - 3) + (3x^2 + 5)$ $x^2 + 2$

24. $(3x^2 + x) + (5x^2 - 3x)$ $8x^2 - 2x$ **25.** $(-8x^2 - 6) + (2x^2 - 8)$ $-6x^2 - 14$

26. $(2x^2 + x) + (-7 - x)$ $2x^2 - 7$ **27.** $(-3x - 1) + (-6x^2 - 5x)$ $-6x^2 - 8x - 1$

Simplify each binomial subtraction expression. Check by substituting two different values for *x*.

28. $(4x - 7) - (2x - 5)$ $2x - 2$ **29.** $(-6x - 2) - (-4x + 2)$ $-2x - 4$

30. $(3x - 7) - (7x - 4)$ $-4x - 3$ **31.** $(5x - 2) - (-2x - 6)$ $7x + 4$

32. $(4 - 5x) - (7x - 5)$ $-12x + 9$ **33.** $(-x - 1) - (-3x - 2)$ $2x + 1$

34. $(x^2 - 3) - (3x^2 - 7)$ $-2x^2 + 4$ **35.** $(-5x^2 - 7) - (2x^2 - 5)$ $-7x^2 - 2$

36. $(4x^2 - 2x) + (6x^2 - x)$ $10x^2 - 3x$ **37.** $(-x^2 - 12) - (4 - 2x^2)$ $x^2 - 16$

Lesson 5.6

Graph each inequality on a number line.

1. $x > 2$ **2.** $x \le -4$ **3.** $x \ge 1$ **4.** $x < 6$

5. $x > -1$ **6.** $3 < x$ **7.** $x \le 0$ **8.** $-8 \ge x$

9. $10 \le x$ **10.** $-3 > x$ **11.** $-8 < x$ **12.** $x > -7$

Use any method to solve each inequality.

13. $x - 1 < 2$ $x < 3$ **14.** $y + 5 > 1$ $y > -4$ **15.** $w - 9 \le -1$ $w \ge 8$

16. $p + 6 \ge -10$ $p \ge -16$ **17.** $3 + m > 2$ $m > -1$ **18.** $d - 5 < -6$ $d < -1$

19. $z + 5 \ge -15$ $z \ge -20$ **20.** $k - 3 \le 4$ $k \le 7$ **21.** $h + 6 > 3$ $h > -3$

22. $4 \le 17 + x$ $-13 \le x$ **23.** $q + 0.5 > 9$ $q > 8.5$ **24.** $7 \le r - 4$ $11 \le r$

25. $8 > m - 13$ $21 > m$ **26.** $t + 1.2 < 32.5$ **27.** $10.5 + f \le 1.5$ $f \le -9$

28. $z + 6.2 \le 0.7$ **29.** $r + 11.4 < 0$ **30.** $4.2 - x < -7$ $x > 11.2$

26. $t < 31.3$ **28.** $z \le -5.5$ **29.** $r < -11.4$

Lesson 5.7

State whether each inequality is true or false.

1. $6 > 8 - 4$ true **2.** $-4 \ge 3 - 5$ false **3.** $7 < 12 - 3$ true

4. $-8 \le 2 - 14$ false **5.** $-3 \ge 5 - 8$ true **6.** $2 < 8 - 9$ false

7. $6 \ge -7 + 3$ true **8.** $-8 \le -7 - 1$ true **9.** $-6 \le -19 + 14$ true

Solve each inequality.

10. $x + 9 \le -4$ $x \le -13$ **11.** $x - 7 < 3$ $x < 10$ **12.** $y + \frac{1}{4} \ge 1$ $y \ge \frac{3}{4}$

13. $w - \frac{2}{3} > 2$ $w > 2\frac{2}{3}$ **14.** $x + 0.9 < 3$ $x < 2.1$ **15.** $t - 6.25 > 2.5$ $t > 8.75$

16. $r + 0.05 < 0.01$ **17.** $y + 10 \le 90$ $y \le 80$ **18.** $15 \le h - 1.25$ $16.25 \le h$

19. $\frac{5}{8} \le \frac{1}{4} + y$ $\frac{3}{8} \le y$ **20.** $x - \frac{1}{4} \le \frac{1}{3}$ $x \le \frac{7}{12}$ **21.** $0.85 < -2.65 + n$ $3.5 > n$

22. $-5.7 + x > 3.6$ **23.** $d - 2\frac{1}{3} > -6$ **24.** $-7\frac{1}{2} < 3\frac{1}{4} + x$ $x > -10\frac{3}{4}$

16. $r < -0.04$ **22.** $x > 9.3$ **23.** $d > -3\frac{2}{3}$

Write an inequality that describes the points graphed on each number line.

25. 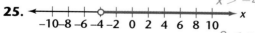 $x > -4$

26. $x \le 5$

27. 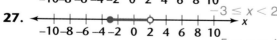 $-3 \le x < 2$

28. $0 \le x \le 6$

29. 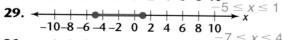 $-5 \le x \le 1$

30. $-2 < x \le 7$

31. $-7 \le x \le 4$

32. 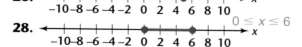 $x < -7$ or $x \ge 6$

33. $x \le -2$ or $x > 0$ **34.** 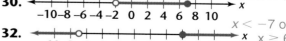 $-3 < x \le 8$

Lesson 6.1

Evaluate 15a for each value of a.

1. -3 -45 **2.** 2.5 37.5   **3.** 8 120   **4.** -4.3 -64.5 **5.** $\frac{1}{3}$ 5

Evaluate 3w + 2 for each value of w.

6. 20 62   **7.** 1.5 6.5   **8.** -7 -19 **9.** -10.6 -29.8 **10.** $\frac{5}{6}$ 4.5

Simplify each expression.

11. $3 \cdot 4x$ $12x$ **12.** $-7x \cdot 5$ $-35x$ **13.** $2x \cdot 5x$ $10x^2$ **14.** $-12x \cdot 4x$ $-48x^2$

15. $-44x \cdot 10x$ **16.** $-4(3x + 5)$ **17.** $7x(3 + 2x)$ $21x + 14x^2$ **18.** $2.5x \cdot 5x$ $12.5x^2$

19. $14x \div 2$ $7x$ **20.** $-64x \div 4$ $-16x$ **21.** $-48x \div (-3)$ $16x$ **22.** $-2(11x - 3)$ $-22x + 6$

23. $5x - (3 - 4x)$ **24.** $4x \cdot 2 + 3x \cdot (-2)$ **25.** $5(x - 2) - (3 - 2x)$ **26.** $(4x - 3) - 9(x + 3)$

27. $(3x + 4y - 3) - 4(y + 2)$ $3x - 11$ **28.** $3(x - 4w) - (5x - 5w)$ $-2x - 7w$

29. $8(2y + 3z) - 5(7y - 7z)$ **30.** $(-4y - 2z) + 7(-y - z)$ $-11y - 9z$

31. $9(2x + 3y) - 6(4x - 2y)$ **32.** $-4(2x - 3w) - 5(3x - 2w)$ $-23x + 22w$

33. $-(y - x) - 4(5y - 2x)$ **34.** $\frac{4x + 20}{4}$ $x + 5$ **35.** $\frac{7 - 21y}{7}$ $1 - 3y$

36. $\frac{-6 + 24w}{-3}$ $2 - 8w$ **37.** $\frac{-18 - 36y}{-9}$ **38.** $\frac{-15x + 10}{5}$ $-3x + 2$

15. $-440x^2$ **16.** $-12x - 20$ **23.** $9x - 3$ **24.** $2x$ **25.** $7x - 13$ **26.** $-5x - 30$
29. $-19y + 59z$ **31.** $-6x + 39y$ **33.** $-21y + 9x$ **37.** $2 + 4y$

Lesson 6.2

Give the property needed to solve the equation. Then solve it.

1. $\frac{x}{2} = -12$ **2.** $\frac{x}{-10} = -4$ **3.** $x - \frac{1}{2} = 2$ **4.** $5x = 4$

5. $-15y = 5$ **6.** $\frac{c}{7} = -70$ **7.** $42x = 6$ **8.** $x + \frac{3}{5} = 3$

9. $-144y = 12$ **10.** $8 = -64z$ **11.** $\frac{x}{-5} = -3.5$ **12.** $7 = 4x$

Solve each equation.   **17.** -45

13. $44x = 55$ 1.25 **14.** $x - 1.5 = 7.5$ 9   **15.** $x + 44 = 55$ 11   **16.** $4t = -36$ -9

17. $d + 0 = -45$ **18.** $7n = 15$ $2\frac{1}{7}$ **19.** $\frac{x}{2} = 17$ 34   **20.** $c - 3 = 19$ 22

21. $q - \frac{1}{5} = 5$ $5\frac{1}{5}$ **22.** $\frac{h}{-1} = -99$ 99   **23.** $9p = -72$ -8 **24.** $t + 3500 = 4000$ 500

25. $0.25 = 0.25f$ 1 **26.** $\frac{x}{0.6} = 7$ 4.2   **27.** $-2y = -35.24$ 17.62 **28.** $\frac{j}{-19} = 0$ 0

29. $0.75v = 75$ 100**30.** $-50 = -2z$ 25   **31.** $\frac{r}{32} = 1$ 32   **32.** $m - 1250 = 2000$ 3250

33. $2.2 = 1.1w$ 2 **34.** $6.5 + y = 7$ 0.5   **35.** $-6v = -582$ 97   **36.** $0.65 = -0.65r$ -1

Lesson 6.3

9. $3y^3 - 2y$ **10.** $10r^2 + 15r$ **11.** $12w^2 - 16w$ **13.** $15r^3 + 5r^2$ **14.** $6r^4 - 4r^2$

Use the Distributive Property to find each product.

1. $5(x + 2)$ $5x + 10$ **2.** $7(3x + 3)$ $21x + 21$ **3.** $6(y + 5)$ $6y + 30$ **4.** $4(w - 8)$ $4w - 32$

5. $y(y + 1)$ $y^2 + y$ **6.** $3x(x - 2)$ $3x^2 - 6x$ **7.** $4f(f - 2)$ $4f^2 - 8f$ **8.** $t^2(t + 5)$ $t^3 + 5t^2$

9. $y(3y^2 - 2)$ **10.** $5r(2r + 3)$ **11.** $4w(3w - 4)$ **12.** $2y^2(y + 1)$ $2y^3 + 2y^2$

13. $5r(3r^2 + r)$ **14.** $r^2(6r^2 - 4)$ **15.** $7w^2(w + 5)$ **16.** $x(12x^2 + 2x)$

17. $-d(5d^3 - 8)$ **18.** $-5r^2(6r + 5)$ **19.** $7q(8q^3 - 3q)$ **20.** $5t^3(2t + 7)$

15. $7w^3 + 35w^2$ **16.** $12x^3 + 2x^2$ **17.** $-5d^4 + 8d$ **18.** $-30r^3 - 25r^2$ **19.** $56q^4 - 21q^2$

Write each polynomial in factored form. **20.** $10t^4 + 35t^3$

21. $6x - 12$ $6(x - 2)$ **22.** $8y + 32$ $8(y + 4)$ **23.** $p^2 - 3p$ $p(p - 3)$ **24.** $2q^2 - 10q$ $2q(q - 5)$

25. $12m^3 + 3m$ **26.** $7a - 7b$ $7(a - b)$ **27.** $n^3 + 6n^2$ $n^2(n + 6)$ **28.** $2x^3 - 8x^2$ $2x^2(x - 4)$

29. $9y^3 + 45y$ **30.** $2r^4 - 14r^2$ **31.** $-4x^3 - 24x$ **32.** $4z^4 - 5z^2$ $z^2(4z^2 - 5)$

33. $7r^3 + 6r$ $r(7r^2 + 6)$ **34.** $10c^2 - 45c$ **35.** $-14y^3 - 21y^2$ **36.** $6a^2 - 14a$ $2a(3a - 7)$

37. $10m^4 + 35m^2$ **38.** $12y^4 - 3y^2$ **39.** $27x^4 + 18x$ **40.** $-5y^3 - 30y$

41. $9x^2 - 18x$ **42.** $8c^2 - 34c$ **43.** $x^4 - 2x$ $x(x^3 - 2)$ **44.** $4x^2 + 7x$ $x(4x + 7)$

45. $15x^3 + 5$ **46.** $6 - 12x$ $6(1 - 2x)$ **47.** $2x - 8y$ $2(x - 4y)$ **48.** $3x + 9$ $3(x + 3)$

25. $3m(4m^2 + 1)$ **29.** $9y(y^2 + 5)$ **30.** $2r^2(r^2 - 7)$ **31.** $-4x(x^2 + 6)$ **34.** $5c(2c - 9)$
35. $-7y^2(2y + 3)$ **37.** $5m^2(2m^2 + 7)$ **38.** $3y^2(4y^2 - 1)$ **39.** $9x(3x^3 + 2)$ **40.** $-5y(y^2 + 6)$
41. $9x(x - 2)$ **42.** $2c(4c - 17)$ **45.** $5(3x^3 + 1)$

Lesson 6.4

Solve each equation.

1. $\frac{2}{3}x = 4$ 6 **2.** $\frac{1}{-4}y = -2$ 8 **3.** $\frac{w}{-8} = 5$ -40 **4.** $\frac{z}{6} = -9$ -54

5. $-\frac{3}{5}t = -12$ 20 **6.** $\frac{7}{8}a = -42$ -48 **7.** $\frac{-w}{5} = 10$ -50 **8.** $\frac{p}{15} = -3$ -45

9. $-\frac{5}{9}r = 25$ -45 **10.** $-\frac{q}{12} = 4$ -48 **11.** $\frac{y}{-4} = -6$ 24 **12.** $\frac{-r}{-3} = -14$ -42

13. $\frac{-y}{5} = -30$ 150 **14.** $\frac{-2}{9}m = -4$ 18 **15.** $-\frac{f}{9} = 11$ -99 **16.** $\frac{w}{-6} = 14$ -84

17. $-\frac{a}{6} = -40$ 240 **18.** $\frac{y}{8} = -21$ -168 **19.** $\frac{r}{-32} = 5$ -160 **20.** $\frac{-z}{7} = -1.8$ 12.6

Solve each equation.

28. -0.5625

21. $\frac{x}{3} = \frac{9}{10}$ 2.7 **22.** $\frac{x}{4} = \frac{-2}{3}$ $-2\frac{2}{3}$ **23.** $\frac{-y}{5} = \frac{3}{7}$ $-2\frac{1}{7}$ **24.** $\frac{w}{-2} = \frac{-3}{11}$ $\frac{6}{11}$

25. $\frac{x}{12} = \frac{-3}{8}$ -4.5 **26.** $\frac{-w}{15} = \frac{-4}{25}$ 2.4 **27.** $\frac{r}{-24} = \frac{5}{16}$ -7.5 **28.** $\frac{-y}{-2} = \frac{-9}{32}$

29. $\frac{z}{-6} = \frac{5}{27}$ $-1\frac{1}{9}$ **30.** $\frac{-k}{12} = \frac{-7}{20}$ 4.2 **31.** $\frac{t}{-14} = \frac{2}{49}$ $-\frac{4}{7}$ **32.** $\frac{q}{21} = \frac{-3}{35}$ -1.8

33. $\frac{a}{1.5} = \frac{3}{-2}$ -2.25 **34.** $\frac{-x}{2.2} = \frac{2}{3}$ $-1\frac{7}{15}$ **35.** $\frac{-h}{7} = \frac{-4.7}{21}$ $1\frac{17}{30}$ **36.** $\frac{j}{-27} = \frac{-4}{7}$ $15\frac{3}{7}$

37. $\frac{m}{-2.1} = \frac{3}{5}$ -1.26 **38.** $\frac{-y}{7} = \frac{-4}{49}$ $\frac{4}{7}$ **39.** $\frac{n}{-10.5} = \frac{4}{-15}$ 2.8 **40.** $\frac{-g}{-11} = \frac{5}{32}$ $1\frac{23}{32}$

41. $\frac{7}{2} = \frac{x}{-3}$ $-10\frac{1}{2}$ **42.** $\frac{x}{9} = \frac{4}{5}$ $7\frac{1}{5}$ **43.** $\frac{x}{10} = \frac{1}{9}$ $1\frac{1}{9}$

44. $\frac{x}{-5} = \frac{3}{4}$ $-3\frac{3}{4}$ **45.** $\frac{4}{x} = \frac{-3}{5}$ $-6\frac{2}{3}$ **46.** $\frac{x}{3} = \frac{4}{7}$ $1\frac{5}{7}$

Lesson 6.5

Solve each proportion.

1. $\frac{16}{18} = \frac{48}{x}$ 54

2. $\frac{28}{15} = \frac{n}{20}$ $37\frac{1}{3}$

3. $\frac{32}{27} = \frac{24}{r}$ 20.25

4. $\frac{m}{42} = \frac{60}{84}$ 30

5. $\frac{20.5}{x} = \frac{82}{46}$ 11.5

6. $\frac{y}{21.4} = \frac{8}{53.5}$ 3.2

7. $\frac{50.8}{124} = \frac{z}{10}$ $4\frac{3}{31}$

8. $\frac{m}{75} = \frac{27}{40}$ 50.625

Determine if each statement is a true proportion.

9. $\frac{36}{7} = \frac{108}{21}$ true

10. $\frac{8}{15} = \frac{15}{20}$ false

11. $\frac{64}{72} = \frac{8}{9}$ true

12. $\frac{12}{7} = \frac{60}{35}$ true

13. $\frac{14}{18} = \frac{9}{20}$ false

14. $\frac{4}{5} = \frac{28}{35}$ true

15. $\frac{13}{26} = \frac{25}{50}$ true

16. $\frac{9}{30} = \frac{10}{40}$ false

Lesson 6.6

Write each percent as a decimal.

1. 65% 0.65 **2.** 5.5% 0.055 **3.** 9% 0.09 **4.** 18% 0.18 **5.** 0.6% 0.006

6. 90% 0.90 **7.** 6.25% 0.0625 **8.** 0.01% 0.0001 **9.** 86.5% 0.865 **10.** 110% 1.10

11. 24% 0.24 **12.** 55.5% 0.555 **13.** 89.9% 0.899 **14.** 125% 1.25 **15.** 16.6% 0.166

Write each decimal as a percent.

16. 0.38 38% **17.** 0.025 2.5% **18.** 7.35 735% **19.** 0.16 16% **20.** 7.00 700%

21. 9.1 910% **22.** 0.188 18.8% **23.** 0.009 0.9% **24.** 0.355 35.5% **25.** 0.0011 0.11%

Find each answer.

26. What number is 30% of 60? 18

27. 130% of what number is 78? 60

28. What percent of 70 is 35? 50%

29. 96 is 80% of what number? 120

30. What percent of 52 is 13? 25%

31. 2 is what percent of 200? 1%

32. What number is 40% of 100? 40

33. What percent of 64 is 16? 25%

34. What number is 75% of 300? 225

35. What number is 4.5% of 130? 5.85

36. 32 is 8% of what number? 400

37. What percent of 45 is 36? 80%

38. What number is 80% of 2? 1.6

39. 5 is what percent of 25? 20%

40. 48 is 15% of what number? 320

41. What percent of 30 is 6? 20%

42. 80 is 100% of what number? 80

43. 92 is 200% of what number? 46

44. What percent of 50 is 10? 20%

45. What percent of 90 is 9? 10%

46. 100 is 40% of what number? 250

47. What percent of 80 is 60? 75%

48. What number is 15% of 45? 6.75

49. 6 is what percent of 90? $6\frac{2}{3}$%

50. What number is 400% of 10? 40

51. What number is 250% of 25? 62.5

52. What number is 25% of 28? 7

53. 69 is what percent of 92? 75%

54. What percent is 58.24 of 52? 112%

55. 30 is what percent of 48? $62\frac{1}{2}$%

56. 21 is 30% of what number? 70

57. 20 is 0.8% of what number? 2500

Lesson 7.1

Solve each equation.

1. $3x + 2 = 23$ 7
2. $8y - 5 = 19$ 3
3. $4x - 2 = 30$ 8

4. $7y - 1 = 27$ 4
5. $3w + 6 = 36$ 10
6. $7 + 7x = 28$ 3

7. $8 - 2z = -4$ 6
8. $6 + 4w = -26$ -8
9. $-10 - 5x = 25$ -7

10. $-2 - 14z = -30$ 2
11. $15 = 2x - 5$ 10
12. $50 = -4y + 2$ -12

13. $-51 = 8x + 5$ -7
14. $-5w + 3 = -72$ 15
15. $14x - 9 = 61$ 5

16. $2(4x + 1) = 74$ 9
17. $3(5x - 2) = 54$ 4
18. $5(4 - 8x) = 100$ -2

19. $4(x - 7) = -36$ -2
20. $40 = 2(2x + 6)$ 7
21. $-30 = 3(4x + 10)$ -5

22. $4x + 2 + 2x = 20$ 3
23. $9x - 5 + x = 45$ 5
24. $3y - 4 - 5y = 4$ -4

25. $4 - 8w - 3 + 2w = -17$ 3
26. $9y + 3y - 5 - 2y = -25$
27. $44 = 3x - 2 + 7x - 4$

28. $6 = 7y - 5 - 8y + 3$ -8
29. $4x + \frac{1}{2} = \frac{1}{4}$ $-\frac{1}{16}$
30. $3y + \frac{2}{5} = \frac{1}{3}$ $-\frac{1}{45}$

31. $\frac{3}{4} + 5y = \frac{1}{8}$ $-\frac{1}{8}$
32. $2w - \frac{1}{9} = \frac{1}{3}$ $\frac{2}{9}$
33. $6x - \frac{7}{10} = -\frac{1}{5}$ $\frac{1}{12}$

34. $5\left(x + \frac{1}{2}\right) = \frac{1}{3}$ $-\frac{13}{30}$
35. $7\left(y - \frac{2}{3}\right) = \frac{1}{4}$ $\frac{59}{84}$
36. $2 = \frac{2}{3} - 5y + 2y$ $-\frac{4}{9}$

37. $\frac{1}{4} + 4x + \frac{2}{5} + x = \frac{3}{10}$ $-\frac{7}{100}$
38. $3w - \frac{1}{7} = \frac{4}{9}$ $\frac{37}{189}$
39. $350(x + 10) = 50$ $-9\frac{6}{7}$

26. -2 **27.** 5

Lesson 7.2

Solve each equation.

1. $7x = 5x + 6$ 3
2. $8w - 14 = 6w$ 7
3. $8y + 1 = 4y + 9$ 2

4. $4x - 5 = 2x - 9$ -2
5. $2 - 4x = 3x - 33$ 5
6. $15 + 2y = 5y + 6$ 3

7. $6x - 5 = 19 - 2x$ 3
8. $24 - 6y = 2y - 40$ 8
9. $3(2x + 6) = 5x + 20$ 2

10. $5(3y + 2) = 8y - 11$ -3
11. $6x - 2 = 4x + 5$ 3.5
12. $8 - 2x = 7x - 4$ $1\frac{1}{3}$

13. $2(4x + 1) = 3(2x + 4)$ 5
14. $7(3x - 5) = 11(2x - 4)$ 9

15. $5x - 2(x - 1) = 2x + 5$ 3
16. $4(y - 5) + 3y + 3 = 3(y + 1)$ 5

17. $6z - 2(z - 5) = 5z + 2$ 8
18. $5(x - 2) + 8x - 14 = 9x - 2$ 5.5

19. $3x + 5 + 2x = 5(2 - x)$ 0.5
20. $8y + 4(2 + y) = 4y - 1$ $-1\frac{1}{8}$

21. $0.75w + 100 = 0.25w - 1$ -202
22. $2.6z - 3.7 = 4.5 + 1.2z$ $5\frac{6}{7}$

23. $4.8(x + 9) = 5.9(x - 2)$ 50
24. $y + \frac{2}{3} + \frac{y}{6} = \frac{1}{3} + 2y$ 0.4

25. $4\left(x + \frac{1}{2}\right) = 8\left(x + \frac{3}{4}\right)$ -1
26. $2y + \frac{4y}{5} - 10 = y + \frac{1}{2}$ $5\frac{5}{6}$

Lesson 7.3

1. Randi is an electrician. She charges $45 per job, plus $25 per hour. If the total labor bill for a job was $95, how many hours did Randi work? 2 hours

2. Jamil earns $150 per week, plus 2.5% of his weekly sales. What must his sales be in order for him to make $175 per week? $1000

3. Karen manages a restaurant. Her base salary is $500 per week, plus 1.1% of the weekly sales at the restaurant. What must the weekly sales be in order for her to make $665 per week? $15,000

4. The basketball booster club spent $1750 to print 2500 basketball programs. They plan to sell each program for $2. How many programs must they sell to make a profit of $650? 1200 programs

5. The community-theater manager determined that the cost of costumes, ticket printing, and theater rental for the winter season will be $3600. If each ticket is sold for $8.50, how many tickets must be sold to make a profit of $3200? 800 tickets

6. A chemist has 80 ounces of a solution that is 5% acid. How much pure acid should he add to make a solution that is 8% acid? ≈2.6 ounces

7. Ariel is mixing 60 milliliters of a solution that is 2% acid with pure water to dilute the solution. How much water should she add to obtain a solution that is 1.5% acid? 20 milliliters

8. Garth plans to mix 4 liters of a solution that is 30% antifreeze with a solution that is 36% antifreeze. How much of the 36% solution should he add to the 30% solution to obtain a new solution that is 33% antifreeze? 4 liters

Lesson 7.4

Find the value of *x*, and the measures of angles 1 and 2.

1. $m\angle 1 = x + 2$; $m\angle 2 = 2x - 2$ 30, 32°, 58°

2. $m\angle 1 = 4x$; $m\angle 2 = 8x + 6$ 7, 28°, 62°

3. $m\angle 1 = 8x + 10$; $m\angle 2 = 18x + 2$ 3, 34°, 56°

4. $m\angle 1 = 4x + 2$; $m\angle 2 = 6x - 2$ 9, 38°, 52°

5. $m\angle 1 = \frac{x}{2} + 6$; $m\angle 2 = x + 9$ 50, 31°, 59°

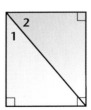

Find the value of *x* and the measures of angles *X*, *Y*, and *Z* in triangle *XYZ*.

6. $m\angle X = 4x$; $m\angle Y = 5x - 6$; $m\angle Z = 10x + 15$ 9, 36°, 39°, 105°

7. $m\angle X = 9x - 9$; $m\angle Y = 12x$; $m\angle Z = 17x - 1$ 5, 36°, 60°, 84°

8. $m\angle X = 3x + 12$; $m\angle Y = 8x + 3$; $m\angle Z = 4x + 15$ 10, 42°, 83°, 55°

9. $m\angle X = \frac{x}{2}$; $m\angle Y = x + 25$; $m\angle Z = \frac{x}{5} + 70$ 50, 25°, 75°, 80°

Lesson 7.5

Write an inequality to represent each statement.

1. W is less than L. $W < L$ **2.** x is greater than 5. $x > 5$

3. V is between 4.5 and 4.6, inclusive.   **4.** w cannot equal 3. $w \ne 3$

5. r is negative. $r < 0$ **6.** d is positive. $d > 0$

7. x is less than or equal to -2. $x \le -2$ **8.** a is greater than or equal to b. $a \ge b$
3. $4.5 \ge V \ge 4.6$

Determine whether each statement is true or false.

9. $5.5 \ge 5.5$ true   **10.** $10.75 < 10.75$ **11.** $4.2 > 4.02$ true   **12.** $6.06 < 6.066$ true

13. $17.17 > 17.107$ **14.** $0 \ge -5$ true   **15.** $\frac{1}{8} \ge \frac{1}{9}$ true   **16.** $\frac{2}{3} \ge \frac{3}{4}$ false

17. $-7 > -4$ false   **18.** $-9 \le 9$ true   **19.** $-6.4 \ge -6.4$ true   **20.** $-7.8 < -8.9$ false

21. $\frac{2}{3} < \frac{5}{9}$ false   **22.** $\frac{16}{5} > \frac{13}{4}$ false   **23.** $-\frac{4}{7} < -\frac{5}{8}$ false   **24.** $\frac{2}{3} \le \frac{-4}{6}$ false
10. false   **13.** true

Solve each inequality.

25. $x + 6 \ge 12$ $x \ge 6$ **26.** $x - 5 < 15$ **27.** $y - 8 \le 7$ $y \le 15$ **28.** $9 + y > 14$ $y > 5$

29. $17 - h > 5$ $h < 12$ **30.** $9 - y \ge -1$ **31.** $-3 - x < -9$ $x > 6$ **32.** $x - 15 \ge -14$ $x \ge 1$

33. $\frac{m}{5} < 2$ $m < 10$ **34.** $\frac{w}{-8} \ge 4$ $w \le -32$ **35.** $\frac{n}{9} > -5$ $n > -45$ **36.** $\frac{-p}{3} \le -7$ $p \ge 21$

37. $6x > 12$ $x > 2$ **38.** $12y \le 16$ $y \le 1\frac{1}{3}$ **39.** $-8x < 20$ $x > -2.5$ **40.** $-9y \ge -90$ $y \le 10$

41. $2z - 2 < 10$ $z < 6$ **42.** $8x + 1 \ge 5$ **43.** $3x + 4 \le 8x - 6$ **44.** $9x + 2 < 4x + 12$

45. $\frac{x}{2} - 1 \le 5$ $x \le 12$ **46.** $18 + \frac{m}{3} > 6$ **47.** $\frac{r}{5} - 10 \ge 2$ $r \ge 60$ **48.** $7 - \frac{c}{6} \le 1$ $c \ge 36$
26. $x < 20$ **30.** $y \le 10$ **42.** $x \ge 0.5$ **43.** $x \ge 2$ **44.** $x < 2$ **46.** $m > -36$

Lesson 7.6   **5.** $-18, -2$ **6.** $-12, -2$ **9.** $-4\frac{2}{3}, 4$ **10.** $-1, 3$ **11.** $-1.5, 2$ **13.** $-0.5, -1$

Find the values of x that solve each absolute-value equation.
Check your answer.   **14.** $-1\frac{1}{7}, 2$ **15.** $-4.4, 2$ **19.** $0.5, 4.5$

1. $|x + 2| = 4$ $-6, 2$ **2.** $|x - 4| = 2$ $2, 6$ **3.** $|x - 1| = 6$ $-5, 7$ **4.** $|x - 5| = 1$ $4, 6$

5. $|x + 10| = 8$ **6.** $|7 + x| = 5$ **7.** $|10 - x| = 2$ $8, 12$ **8.** $|-2 + x| = 2$ $0, 4$

9. $|3x + 1| = 13$ **10.** $|2x - 2| = 4$ **11.** $|4x - 1| = 7$ **12.** $|3x + 6| = 15$ $-7, 3$

13. $|-3 - 4x| = 1$ **14.** $|3 - 7x| = 11$ **15.** $|6 + 5x| = 16$ **16.** $|-5 - 2x| = 13$ $-9, 4$

17. $|4x - 14| = 0$ 3.5 **18.** $|9x - 9| = 9$ $0, 2$ **19.** $|5 - 2x| = 4$ **20.** $|-5 - 6x| = 7$ $-2, \frac{1}{3}$

Find the values of x that satisfy each absolute-value inequality.
Graph the solution on a number line. Check your answer.

21. $|x + 2| > 1$ **22.** $|x - 1| < 2$ **23.** $|x - 7| \ge 4$ **24.** $|x + 5| \le 4$

25. $|x - 3| < 5$ **26.** $|x - 3| \le 9$ **27.** $|x - 2| \le 0$ **28.** $|x + 6| > 2$

29. $|5 - 3x| \ge 9$ **30.** $|x| < 4$ **31.** $|x| \ge 4$ **32.** $|x| > 0$

33. $|x - 5| < 3$ **34.** $|x + 4| < 7$ **35.** $|x - 2| \ge -4$ **36.** $|2x - 8| > 6$

37. $|3x + 3| \le 9$ **38.** $|4x - 8| \ge 20$ **39.** $|7x - 14| < 7$ **40.** $|x + 2| < 5$

Lesson 8.1

Graph each list of ordered pairs. State whether the points lie on a straight line.

1. (1, 2), (2, 3), (3, 4) **2.** (1, 6), (2, 8), (3, 6) **3.** (2, 4), (5, 3), (11, 1)

4. (−1, 2), (3, 2), (−5, 2) **5.** (−4, 3), (−2, 1), (0, 1) **6.** (3, 0), (4, 2), (5, 4)

Make a table of values using 1, 2, 3, 4, and 5 for x.

7. $y = x + 2$ **8.** $y = x + 5$ **9.** $y = x - 6$ **10.** $y = x - 4$

11. $y = 3x$ **12.** $y = -4x$ **13.** $y = 3x + 1$ **14.** $y = 2x - 3$

15. $y = -x + 1$ **16.** $y = -4x - 1$ **17.** $y = 0.5x + 2$ **18.** $y = 2x - 0.5$

Lesson 8.2

Examine the graphs below. Which lines have positive slope? Which have negative slope? Which have neither?

1. positive

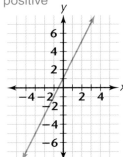

2. neither

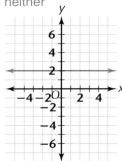

3. negative

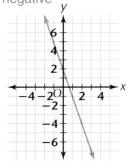

Find the slope for each of the given lines.

4. rise 4, run 2   2 **5.** rise 1, run 8 $\frac{1}{8}$ **6.** rise 4, run 1   4

7. rise −6, run 3   −2 **8.** rise 0, run 10   0 **9.** rise −9, run −5 $\frac{9}{5}$

10. rise 6 − 2, run −4 − 1 $-\frac{4}{5}$ **11.** rise −8 + 3, run 4 + 6 $-\frac{1}{2}$ **12.** rise 5 − 6, run 4 − 7 $\frac{1}{3}$

Each pair of points is on a line. What is the slope of the line?   **24.** $\frac{13}{7}$

13. $M(5, 3)$, $N(3, 4)$ $-\frac{1}{2}$ **14.** $M(1, 7)$, $N(4, 1)$ −2 **15.** $M(-3, 2)$, $N(1, -4)$ $-\frac{3}{2}$

16. $M(-5, 3)$, $N(0, 6)$ $\frac{3}{5}$ **17.** $M(1, -7)$, $N(4, -1)$ 2 **18.** $M(-5, 9)$, $N(3, -3)$ $-\frac{3}{2}$

19. $M(6, 7)$, $N(3, 4)$ 1 **20.** $M(9, -4)$, $N(3, 2)$ −1 **21.** $M(-3, -1)$, $N(6, -4)$ $-\frac{1}{3}$

22. $M(7, 4)$, $N(-5, 8)$ $-\frac{1}{3}$ **23.** $M(2, 5)$, $N(-4, 5)$ 0 **24.** $M(6, 8)$, $N(-1, -5)$

Lesson 8.3

6. $y = \frac{10}{9}x$ 8. $y = -\frac{3}{2}x$ 9. $y = \frac{9}{4}x$ 13. $y = -\frac{5}{3}x$ 14. $y = -\frac{10}{9}x$

Find the equation of the line that passes through the origin and each point.

1. (3, 4) $y = \frac{4}{3}x$ 2. (1, 5) $y = 5x$ 3. (2, 8) $y = 4x$ 4. (7, 4) $y = \frac{4}{7}x$ 5. (6, 1) $y = \frac{1}{6}x$

6. (9, 10)   7. (5, 2) $y = \frac{2}{5}x$ 8. (−2, 3)   9. (−4, −9)   10. (6, −7) $y = -\frac{7}{6}x$

11. (0, 3) $x = 0$ 12. (−4, −4) $y = x$ 13. (3, −5)   14. (9, −10)   15. (−1, −12) $y = 12x$

Describe the graph of each line. Then check your description by graphing.

16. $y = -2x$ 17. $y = 2x$ 18. $y = 3$ 19. $y = -5$

20. $y = x + 2$ 21. $y = -1$ 22. $y = -x + 3$ 23. $y = 2x + 1$

24. $y = x - 1$ 25. $y = -x - 3$ 26. $y = 2x - 1$ 27. $y = 5 - x$

28. $y = -2 + \frac{1}{2}x$ 29. $y = -5 + 0.5x$ 30. $y = \frac{1}{4}x - 1$ 31. $y = 1.5x - 2$

Lesson 8.4

Give the coordinates of the point where the line for each equation crosses the y-axis.

1. $y = 3x + 1$ (0, 1)   2. $y = 4x - 2$ (0, −2)   3. $y = 7x + 5$ (0, 5)   4. $y = -x - 3$ (0, −3)

5. $y = -2x + 4$ (0, 4)   6. $y = -8x - 7$ (0, −7)   7. $y = 0.5x + 10$ (0, 10)   8. $y = \frac{1}{3}x - 9$ (0, −9)

9. $y = -4x + 3$ (0, 3)   10. $y = -\frac{1}{2}x - 7$ (0, −7)   11. $y = 0.2x + 1.8$ (0, 1.8)   12. $y = 4.7x$ (0, 0)

Write an equation for each line described.   14. $y = -3x - 3$ 15. $y = 10x + 12$

13. with slope −1 and y-intercept 3 $y = -x + 3$ 14. through (0, −3) and with slope −3

15. with slope 10 and y-intercept 12   16. with slope −4 and y-intercept 4

17. through (0, 7) and with slope 2 $y = 2x + 7$ 18. through (0, −4) and with slope 1 $y = x - 4$

19. through (0, −1) and with slope −1   20. through (0, 5) and with slope 0 $y = 5$

21. with slope $\frac{3}{4}$ and y-intercept 3 $y = \frac{3}{4}x + 3$ 22. with slope $-\frac{4}{5}$ and y-intercept −2

23. with slope 0.5 and y-intercept −3   24. through (0, 1) and with slope −1.5

16. $y = -4x + 4$ 19. $y = -x - 1$ 22. $y = -\frac{4}{5}x - 2$ 23. $y = 0.5x - 3$ 24. $y = -1.5x + 1$

Write an equation for the line passing through each pair of points.

25. (3, 5), (−3, 1)   26. (1, 0), (2, 1)   27. (−1, 3), (1, −1)

28. (3, 1), (−3, 3)   29. (−2, 0), (2, 4)   30. (0, 1), (1, 3)

31. (2, 1), (−2, −3)   32. (−1, 4), (1, −2)   33. (4, 3), (8, 4)

34. (−5, 1), (5, −3)   35. (−1, −3), (1, −2)   36. (1, 2), (2, 4)

37. (0, 4), (1, 1)   38. (0, −2), (2, 0)   39. (3, 3), (6, 7)

40. (−1, 6), (−3, −4)   41. (5, 4), (−5, −3)   42. (−3, −4), (−1, −4)

43. (0, −3), (5, 0)   44. (2, 0), (−1, 4)   45. (−1, −6), (3, −7)

1. $3x - 6y = 18$ **2.** $4x + 5y = -20$ **3.** $6x + 2y = 36$ **4.** $4x - 8y = 0$
5. $7x + 2y = -14$ **6.** $2x - 3y = 10$ **7.** $x - y = 6$ **8.** $x + 9y = -15$
9. $2x - y = -6$ **10.** $2x - 7y = -20$ **11.** $2x - y = 15$ **12.** $x + 6y = -10$

Lesson 8.5

Write each equation in standard form.

1. $3x = 6y + 18$ **2.** $5y = -4x - 20$ **3.** $-6x = 2y - 36$ **4.** $4x = 8y$

5. $7x + 2y + 14 = 0$ **6.** $2x - 4 = 3y + 6$ **7.** $6 = x - y$ **8.** $3 + 9y = -x - 12$

9. $x = \frac{1}{2}y - 3$ **10.** $\frac{1}{5}x = \frac{7}{10}y - 2$ **11.** $\frac{2}{3}x - 1 = \frac{1}{3}y + 4$ **12.** $0.5x = -3y - 5$

Find the x- and y-intercepts for the graph of each equation. **14.** $(4, 0), (0, -4)$

13. $x + y = 2$ $(2, 0), (0, 2)$ **14.** $x - y = 4$ **15.** $x + 2y = 8$ **16.** $3x + y = 9$ $(3, 0), (0, 9)$

17. $4x - 5y = 20$ **18.** $2x - 7y = 14$ **19.** $9x + y = 18$ **20.** $4x - 6y = 12$

21. $x = 3y$ $(0, 0)$ **22.** $x = -5y + 1$ **23.** $x = \frac{5}{6}y$ $(0, 0)$ **24.** $\frac{x}{2} + y = 4$ $(8, 0), (0, 4)$
15. $(8, 0), (0, 4)$ **17.** $(5, 0), (0, -4)$ **18.** $(7, 0), (0, -2)$ **19.** $(2, 0), (0, 18)$ **20.** $(3, 0), (0, -2)$

Write an equation for each line described.

22. $(1, 0), \left(0, \frac{1}{5}\right)$

25. passing through $(2, 3)$ and with slope -1 $y = -x + 5$

26. crossing the x-axis at $x = 2$ and the y-axis at $y = 2$ $y = -x + 2$

27. passing through $(4, 5)$ and with slope 2 $y = 2x - 3$

28. passing through $(4, 5)$ and with slope 0 $y = 5$

29. crossing the x-axis at $x = -5$ and the y-axis at $y = 0$ $y = 0$

30. passing through $(4, 8)$ and $(-2, -1)$ $y = \frac{3}{2}x + 2$

31. passing through $(1, 4)$ and $(-1, 0)$ $y = 2x + 2$

32. passing through $(-1, -5)$ and with slope 0 $y = -5$

33. crossing the x-axis at $x = 3$ and the y-axis at $y = -2$ $y = \frac{2}{3}x - 2$

34. passing through $(-2, 8)$ and with slope $-\frac{1}{4}$ $y = -\frac{1}{4}x + 7\frac{1}{2}$

Lesson 8.6

Determine whether each line is vertical or horizontal. Then give the slope for each line.

1. $y = 4$ **2.** $x = 3$ **3.** $y = -5$ horizontal, 0 **4.** $x = -7$ **5.** $y = 9$ horizontal, 0
horizontal, 0 vertical, undefined vertical, undefined

Match each equation with the appropriate description.

6. $x + y = 4$ c   **a.** a line through the origin with slope 4

7. $x = 4$ h   **b.** a horizontal line 4 units above the origin

8. $xy = 4$ g   **c.** a line with slope -1 and y-intercept 4

9. $y = 4$ b   **d.** a line through the origin with slope $\frac{1}{4}$

10. $y = 4x$ a   **e.** a line with slope 1 and y-intercept -4

11. $x = 4y$ d   **f.** a line through the origin with slope -4

12. $x - y = 4$ e   **g.** something other than a straight line

13. $4x + y = 0$ f   **h.** a vertical line 4 units to the right of the origin

Lesson 8.7

Write an equation for the line described.

	Contains:	Is parallel to:		Contains:	Is perpendicular to:

1. $(2, -4)$ $3x - y = 5$ $y = 3x - 10$ **2.** $(2, -4)$ $3x - y = 5$ $y = -\frac{1}{3}x - 3\frac{1}{3}$

3. $(-1, 4)$ $y = 2x + 5$ $y = 2x + 6$ **4.** $(-1, 4)$ $y = 2x + 5$ $y = -\frac{1}{2}x + 3\frac{1}{2}$

5. $(0, 5)$ $y = -2x - 1$ $y = -2x + 5$ **6.** $(0, 5)$ $y = -2x - 1$ $y = \frac{1}{2}x + 5$

7. $(-3, -1)$ $2y = x + 2$ $y = \frac{1}{2}x + \frac{1}{2}$ **8.** $(-3, -1)$ $2y = x + 2$ $y = -2x - 7$

9. $(4, -1)$ $y = 5$ $y = -1$ **10.** $(4, -1)$ $y = 5$ $x = 4$

11. $(-2, -2)$ $y = -4x$ $y = -4x - 10$ **12.** $(-2, -2)$ $y = -4x$ $y = \frac{1}{4}x - \frac{3}{2}$

Write the slope of a line that is parallel to each line.

13. $y = 4x + 10$ 4   **14.** $3x + y = 7$ -3 **15.** $10 = -5x + 2y$ $\frac{5}{2}$ **16.** $4x - 3y = 12$ $\frac{4}{3}$

17. $y = \frac{1}{3}x - 3$ $\frac{1}{3}$ **18.** $4x - \frac{1}{4}y = 8$ 16   **19.** $\frac{2}{3}x + 6y = 1$ $-\frac{1}{9}$ **20.** $7x - y = 7$ 7

Write the slope of a line that is perpendicular to each line.

21. $y = 5x + 10$ $-\frac{1}{5}$ **22.** $3x + y = 2$ $\frac{1}{3}$ **23.** $20 = -5x + 2y$ $-\frac{2}{5}$ **24.** $4x - 4y = 12$ -1

25. $y = -\frac{1}{3}x - 3$ 3   **26.** $4x - \frac{1}{4}y = 16$ $-\frac{1}{16}$ **27.** $\frac{2}{3}x + 6y = 6$ 9   **28.** $7x - y = -7$ $-\frac{1}{7}$

Determine whether the two lines are parallel, perpendicular, or neither.

29. $y = 0.5x$
$y = -2x$
perpendicular

30. $y = 3x - 2$
$y = -3x + 4$
neither

27. $x - y = 5$
$x - y = 7$
parallel

28. $x - 3y = 4$
$3x + y = 1$
perpendicular

Lesson 8.8

Graph each linear inequality.

1. $y < 2x + 1$ **2.** $y \leq 3x - 2$ **3.** $y > 5x - 1$

4. $3y \leq 9$ **5.** $x > -4$ **6.** $x - y > 6$

7. $y < \frac{1}{2}x - 2$ **8.** $-6x + 3y - 6 < 0$ **9.** $y \geq -x + 3$

Write a linear inequality for each graph.

10. $y < -2x + 3$ **11.** $y > x - 5$ **12.** $y \geq 2x - 1$

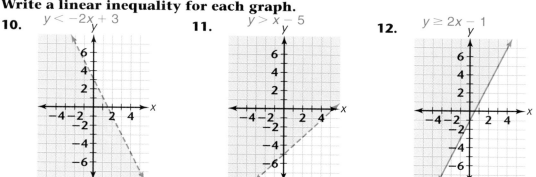

Lesson 9.1

Estimate the solution to each system of linear equations.

1.
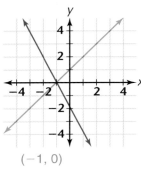
$(-1, 0)$

2.
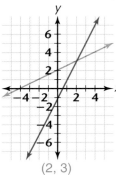
$(2, 3)$

3.
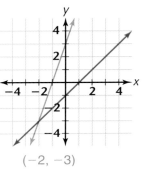
$(-2, -3)$

Solve each system of linear equations by graphing. If necessary, give an approximate solution.

4. $\begin{cases} 2x + y = 7 \\ 4x + y = 13 \end{cases}$ (3, 1)

5. $\begin{cases} x + 3y = 6 \\ 2x + y = -6 \end{cases}$ (−4.8, 3.6)

6. $\begin{cases} 2x - 3y = 5 \\ 3x - 2y = 0 \end{cases}$ (−2, −3)

7. $\begin{cases} 4x + 7y = 13 \\ -5x - 4y = -21 \end{cases}$ (5, −1)

8. $\begin{cases} 4x - 3y = -12 \\ x = -3 \end{cases}$ (−3, 0)

9. $\begin{cases} x = -7 \\ y = 3 \end{cases}$ (−7, 3)

10. $\begin{cases} 2x + 2y = 12 \\ y = -2 \end{cases}$ (8, −2)

11. $\begin{cases} y = 3x - 5 \\ y = -4x + 1 \end{cases}$ (0.9, −2.4)

12. $\begin{cases} y = x - 2 \\ y = 0.5x + 1 \end{cases}$ (6, 4)

13. $\begin{cases} y = \frac{3}{4}x - 3 \\ y = 2x + 1 \end{cases}$ (−3.2, −5.4)

14. $\begin{cases} y = -\frac{1}{5}x - 1 \\ y = 5 \end{cases}$ (−30, 5)

15. $\begin{cases} y = \frac{2}{3}x + 1 \\ y = x - 1 \end{cases}$ (6, 5)

Lesson 9.2

Use the substitution method to solve each system of linear equations. Check your answer.

1. $\begin{cases} x + 4y = 28 \\ y = 7 \end{cases}$ (0, 7)

2. $\begin{cases} m = n \\ 3m + 2n = 15 \end{cases}$ (3, 3)

3. $\begin{cases} t = 3r \\ r + t = 1 \end{cases}$ (0.25, 0.75)

4. $\begin{cases} 2x + 4y = 6 \\ 7x + y = 8 \end{cases}$ (1, 1)

5. $\begin{cases} a - 4b = -14 \\ 2a + b = -1 \end{cases}$ (−2, 3)

6. $\begin{cases} 3x + 5y = 10 \\ 2x + 6y = 4 \end{cases}$ (5, −1)

7. $\begin{cases} x + y = 5 \\ x - 2y = 20 \end{cases}$ (10, −5)

8. $\begin{cases} 2z = 4w \\ z + 4w = -18 \end{cases}$ (−3, −6)

9. $\begin{cases} 2x + 2y = 6 \\ x - y = -9 \end{cases}$ (−3, 6)

10. $\begin{cases} 2p + 5q = 11 \\ 3p - q = -26 \end{cases}$ (−7, 5)

11. $\begin{cases} y = \frac{1}{2}x - 6 \\ x = 4y \end{cases}$ (24, 6)

12. $\begin{cases} x + \frac{1}{5}y = 5 \\ 2x + y = 7 \end{cases}$ (6, −5)

Lesson 9.3

Solve each system of linear equations by using elimination by addition.

1. $\begin{cases} x + y = 2 \\ x - y = 10 \end{cases}$ $(6, -4)$

2. $\begin{cases} 2x + y = 5 \\ x - y = 4 \end{cases}$ $(3, -1)$

3. $\begin{cases} x + 2y = -1 \\ 2x - 2y = 22 \end{cases}$ $(7, -4)$

4. $\begin{cases} 3x + 4y = 5 \\ 2x - 4y = -14 \end{cases}$ $(-1.8, 2.6)$

5. $\begin{cases} x + 2y = 13 \\ 2x - 2y = -4 \end{cases}$ $(3, 5)$

6. $\begin{cases} 2x - y = -5 \\ x - y = -10 \end{cases}$ $(5, 15)$

7. $\begin{cases} 2x + y = 8 \\ 3x + y = 14 \end{cases}$ $(6, -4)$

8. $\begin{cases} 2x + y = 1 \\ 2x - 2y = -26 \end{cases}$ $(-4, 9)$

9. $\begin{cases} 4x - 2y = 12 \\ 5x - 2y = 10 \end{cases}$ $(-2, -10)$

10. $\begin{cases} 2x + 2y = -2 \\ 2x + 3y = 3 \end{cases}$ $(-6, 5)$

11. $\begin{cases} 4x - y = 3 \\ 8x - y = 5 \end{cases}$ $(0.5, -1)$

12. $\begin{cases} 6x + 2y = -4 \\ 3x + 2y = -5 \end{cases}$ $\left(\frac{1}{3}, -3\right)$

13. $\begin{cases} x = 10y + 4 \\ x + 20y = 10 \end{cases}$ $(6, 0.2)$

14. $\begin{cases} 4x + 2 = 4y \\ 6x - y = 7 \end{cases}$ $(1.5, 2)$

15. $\begin{cases} 7x - y = 1 \\ 7x + 2y = -1 \end{cases}$ $\left(\frac{1}{21}, -\frac{2}{3}\right)$

16. $\begin{cases} y - x = \frac{1}{3} \\ 6x + y = 5 \end{cases}$ $\left(\frac{2}{3}, 1\right)$

17. $\begin{cases} 2x + y = 1 \\ 3x - y = \frac{1}{4} \end{cases}$ $(0.25, 0.5)$

18. $\begin{cases} 0.2x + 0.3y = 1 \\ 0.2x + 0.25y = -0.5 \end{cases}$ $(-40, 30)$

19. $\begin{cases} 2x - y = \frac{3}{5} \\ x - y = \frac{1}{5} \end{cases}$ $(0.4, 0.2)$

20. $\begin{cases} 2x + 4y = -1 \\ 2x + y = 0.5 \end{cases}$ $(0.5, -0.5)$

21. $\begin{cases} 3.8x - 2.5y = 0.1 \\ x = 2.5y - 5.5 \end{cases}$ $(2, 3)$

Lesson 9.4

Solve and check each system of linear equations by using elimination.

1. $\begin{cases} 2x + 2y = 10 \\ 3x + y = 7 \end{cases}$ $(1, 4)$

2. $\begin{cases} 2x + 3y = 19 \\ 4x - 2y = -2 \end{cases}$ $(2, 5)$

3. $\begin{cases} 4x - 2y = -14 \\ 2x + 3y = -3 \end{cases}$ $(-3, 1)$

4. $\begin{cases} 4x + 3y = 2 \\ x - y = 4 \end{cases}$ $(2, -2)$

5. $\begin{cases} x + y = 4 \\ 2x - 3y = 13 \end{cases}$ $(5, -1)$

6. $\begin{cases} x = 2y + 2 \\ 3x - y = -9 \end{cases}$ $(-4, -3)$

7. $\begin{cases} y = 8x - 2 \\ 2x - y = -2 \end{cases}$ $\left(\frac{2}{3}, 3\frac{1}{3}\right)$

8. $\begin{cases} 2x + 10y = 3 \\ x = 5y - 9.5 \end{cases}$ $(-4, 1.1)$

9. $\begin{cases} 3x + 2y = 1 \\ 2y = 3x \end{cases}$ $\left(\frac{1}{6}, \frac{1}{4}\right)$

10. $\begin{cases} 4x + y = 3 \\ 2x - 3y = 5.5 \end{cases}$ $\left(1\frac{1}{28}, -1\frac{1}{7}\right)$

11. $\begin{cases} 2x + 5y = 7.5 \\ 3x + 3y = 16.5 \end{cases}$ $\left(6\frac{2}{3}, -1\frac{1}{6}\right)$

12. $\begin{cases} \frac{1}{2}x + \frac{1}{3}y = 5 \\ 4x + y = 14 \end{cases}$ $(-0.4, 15.6)$

13. $\begin{cases} 2.3x + 1.2y = 10.2 \\ 0.9x - 3.5y = 15.9 \end{cases}$ $(6, -3)$

14. $\begin{cases} 3x + 0.5y = 3.75 \\ 0.5x - 0.5y = 3.25 \end{cases}$ $(2, -4.5)$

15. $\begin{cases} \frac{1}{3}x - y = -3 \\ \frac{1}{2}x - 3y = 6 \end{cases}$ $(-30, -7)$

16. $\begin{cases} y = -5x + 2.5 \\ 4x - 0.5y = 2.65 \end{cases}$ $(0.6, -0.5)$

17. $\begin{cases} 3x + 1.2y = 2.1 \\ 1.5x + 0.8y = 0.65 \end{cases}$ $(1.5, -2)$

18. $\begin{cases} 2x + 3y = -3.5 \\ x = y + 2 \end{cases}$ $(0.5, -1.5)$

Lesson 9.5

Determine which of the given points are solutions to the system of linear inequalities.

1. $\begin{cases} 3x - y > 1 \\ x + y < 2 \end{cases}$ a   **a.** $(1, -3)$ **b.** $(2, 1)$ **c.** $(-1, 0)$

2. $\begin{cases} 12x + 3y > 24 \\ 3x + 18y > 24 \end{cases}$ c   **a.** $(-4, 4)$ **b.** $(6, -1)$ **c.** $(1, 12)$

3. $\begin{cases} 32x - 16y < 64 \\ 16x + 32y > 48 \end{cases}$ b   **a.** $(64, 48)$ **b.** $(2, 2)$ **c.** $(-10, 2)$

Write the system of linear inequalities graphed.

4.
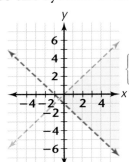
$\begin{cases} y < x \\ y > -x - 1 \end{cases}$

5.
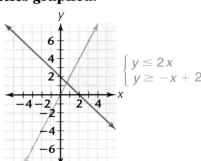
$\begin{cases} y \le 2x \\ y \ge -x + 2 \end{cases}$

Graph each system of linear inequalities.

6. $\begin{cases} y < 3 \\ y < x - 1 \end{cases}$ **7.** $\begin{cases} y > -1 \\ y \ge 5x - 2 \end{cases}$ **8.** $\begin{cases} x \le 4 \\ y \le \frac{1}{2}x - 2 \end{cases}$ **9.** $\begin{cases} y - 3x > 6 \\ 2x + 4y \le 4 \end{cases}$

10. $\begin{cases} 2x - y \le 6 \\ x + y \ge 2 \end{cases}$ **11.** $\begin{cases} x \le -2 \\ y > 4 \end{cases}$ **12.** $\begin{cases} y \le -4 \\ x - y < 1 \\ x + y < 2 \end{cases}$ **13.** $\begin{cases} 3x \ge 9 \\ x - 3y \le 12 \\ 2x + 3y \ge 6 \end{cases}$

Lesson 10.1

3. 31.25, 15.625, 7.8125
7. 729, 2187, 6561
8. 0.015625, 0.0078125, 0.00390625

What are the next three terms of each sequence?

1. 10, 30, 90, 270, 810, . . .   2430, 7290, 21,870   **2.** 20, 40, 60, 80, 100, . . . 120, 140, 160

3. 1000, 500, 250, 125, 62.5, . . .   **4.** 60, 55, 50, 45, 40, . . . 35, 30, 25

5. 30, 28, 26, 24, 22, . . . 20, 18, 16   **6.** 2, 4, 8, 16, 32, . . . 64, 128, 256

7. 3, 9, 27, 81, 243, . . .   **8.** 0.5, 0.25, 0.125, 0.0625, 0.03125, . . .

9. Which of the sequences in Exercises 1–8 show exponential growth? 1, 6, 7, and 8

10. Which of the sequences in Exercises 1–8 show a linear relationship? 2, 4, and 5

Lesson 10.2

Match each table with the correct type of relationship.

a. Linear   **b.** Exponential   **c.** Quadratic   **d.** None of these

1.

x	1	2	3	4	5
y	2	4	7	11	16

c

2.

x	1	2	3	4	5
y	2	8	18	32	50

c

3.

x	1	2	3	4	5
y	10	8	6	4	2

a

4.

x	1	2	3	4	5
y	1	4	7	10	13

a

5.

x	1	2	3	4	5
y	2	4	8	16	32

b

6.

x	1	2	3	4	5
y	4	4	8	12	20

d

Match each relationship with its graph.

a. Linear   **b.** Exponential   **c.** Quadratic   **d.** None of these

7. 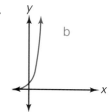 b

8. 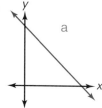 a

9. 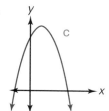 c

10. 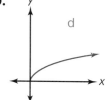 d

Lesson 10.3

How much would each person have to contribute to reach a goal of $500 if there are

1. 5 people? $100   **2.** 10 people? $50   **3.** 50 people? $10

4. 100 people? $5   **5.** 200 people? $2.50   **6.** 1000 people? $0.50

How fast would you have to drive to complete a 300-mile trip in

7. 10 hours? 30 mph   **8.** 6 hours? 50 mph   **9.** 5 hours? 60 mph

10. 3 hours? 100 mph   **11.** 1 hour? 300 mph   **12.** $\frac{1}{2}$ hour? 600 mph

Suppose a string on a guitar is 60 centimeters long and produces a frequency of 300 vibrations per second. What frequency will be produced by a similar guitar string with each length?

13. 50 cm 360 vps   **14.** 40 cm 450 vps   **15.** 80 cm 225 vps   **16.** 100 cm 180 vps

Forty-eight students and teachers from music classes are going on a field trip. How many cars are needed if each car holds the number of passengers shown?

17. 4 passengers 12 cars   **18.** 6 passengers 8 cars   **19.** 7 passengers 7 cars

Lesson 10.4

Evaluate.

1. $|5.6|$ 5.6   **2.** $|-1.9|$ 1.9   **3.** INT(-10.25) -11 **4.** $|12|$ 12   **5.** ABS(-0.65) 0.65

6. INT$(4/5)$ 0   **7.** INT(3.6) 3   **8.** ABS(-7.7) 7.7   **9.** INT$(15/2)$ 7 **10.** ABS$(15/2)$ $\frac{15}{2}$

11. INT$(35/4)$ 8 **12.** INT$(-25/2)$ **13.** ABS$(-25/2)$ **14.** INT(9.99) 9 **15.** ABS(-600) 600

16. $|9|$ 9   **17.** $-|9|$ -9 **18.** $|-9|$ 9   **19.** $-|-9|$ -9 **20.** $-(-|-9|)$ 9

12. -13 **13.** $\frac{25}{2}$

Lesson 10.5

Match each graph with the correct type of function.

a. Linear   **b.** Exponential   **c.** Quadratic

d. Reciprocal   **e.** Absolute value   **f.** Integer

1. 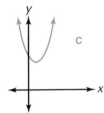 c

2. 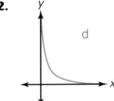 d

3. 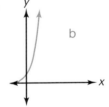 b

4. 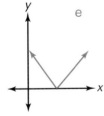 e

5. 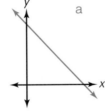 a

6. 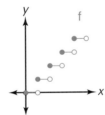 f

Use the method of differences to determine whether the relationship is linear, quadratic, or neither.

7.

x	1	2	3	4	5
y	6	9	12	15	18

linear

8.

x	1	2	3	4	5
y	5	20	45	80	125

quad

9.

x	1	2	3	4	5
y	5	8	14	23	35

quad

10.

x	1	2	3	4	5
y	15	13	11	9	7

linear

11.

x	1	2	3	4	5
y	3	8	15	24	35

quad

12.

x	1	2	3	4	5
y	2	5	10	17	6

neither

13.

x	1	2	3	4	5
y	3	5	7	9	11

linear

14.

x	1	2	3	4	5
y	2	8	18	32	50

quad

15.

x	1	2	3	4	5
y	1	3	8	14	17

neither

16.

x	1	2	3	4	5
y	0	3	8	15	24

quad

17.

x	1	2	3	4	5
y	0	2	6	12	20

quad

18.

x	1	2	3	4	5
y	3	8	13	18	23

linear

Lesson 10.6

Graph each function.

1. $y = |x| + 1$ **2.** $y = |x| - 2$ **3.** $y = |x| - 5$

4. $y = |x + 1|$ **5.** $y = (x - 1)^2$ **6.** $y = (x + 1)^2 - 2$

7. $y = -|x + 2|$ **8.** $y = -|x| - 2$ **9.** $y = -|x - 2|$

Write a function to translate the parent function $y = |x|$ according to each description below.   **11.** $y = |x| - 3$ **13.** $y = |x - 2|$

10. 4 units up $y = |x| + 4$ **11.** 3 units down   **12.** 2 units to the left $y = |x + 2|$

13. 2 units to the right   **14.** 3.5 units up $y = |x| + 3.5$ **15.** 1.5 units down $y = |x| - 1.5$

16. 3 units up and 2 units to the left   **17.** 5 units down and 1 unit to the right

18. 7 units up and 4 units to the right   **19.** 50 units down and 20 units to the left

16. $y = |x + 2| + 3$ **17.** $y = |x - 1| - 5$ **18.** $y = |x - 4| + 7$ **19.** $y = |x + 20| - 50$

Write a function to translate the parent function $y = x^2$ according to each description below.

20. 8 units to the right   **21.** 3 units to the left   **22.** 1 unit up $y = x^2 + 1$

23. 2 units down $y = x^2 - 2$ **24.** 30 units up $y = x^2 + 30$ **25.** 24 units down $y = x^2 - 24$

26. 15 units to the left   **27.** 32 units down   **28.** 16 units to the right

29. 3 units up and 2 units to the right   **30.** 10 units down and 1 unit to the left

31. 9 units down and 5 units to the right   **32.** 8 units up and 6 units to the left

33. 2 units down and 1 unit to the right   **34.** 100 units up and 25 units to the left

20. $y = (x - 8)^2$ **21.** $y = (x + 3)^2$ **26.** $y = (x + 15)^2$ **27.** $y = x^2 - 32$ **28.** $y = (x - 16)^2$

29. $y = (x - 2)^2 + 3$ **30.** $y = (x + 1)^2 - 10$ **31.** $y = (x - 5)^2 - 9$ **32.** $y = (x + 6)^2 + 8$

33. $y = (x - 1)^2 - 2$

Lesson 11.1

The circle graph shows the percent of the population in Carsonville in certain age groups. Suppose the total population in 1995 was about 150,600.

1. How many people were 60 and over? about 28,614

2. How many people were 19 and under? about 52,710

3. How many people were 40–59? about 42,168

4. How many people were 20–59? about 69,276

5. The population is estimated to grow to about 190,000 by the year 2005 and the percent of the population in each age group is expected to remain the same. How many people are expected to be in each age group in the year 2005? See graph at side.

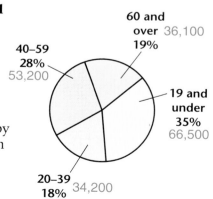

Lesson 11.2

Find the mean, median, mode, and range for each of the following data sets.

1. 12, 12, 18, 14, 8, 9, 10, 20, 19
2. 8, 11, 15, 25, 35, 62, 20, 40 27, 22.5, no mode, 54
3. 4, 2, 5, 1, 1, 9, 8, 4, 6, 3, 2
4. 150, 320, 200, 41, 700, 210, 300 274.4, 210, no mode, 659
5. 130, 135, 132, 120, 145, 136
6. 30, 35, 200, 42, 95, 100, 300, 25 103.375, 68.5, no mode, 275
7. 60, 54, 45, 72, 83, 64, 51, 75, 77, 59, 48, 61 62.42, 60.5, no mode, 38
8. 55, 26, 34, 21, 37, 48, 34, 27, 19, 51, 27, 32 34.25; 33; 27, 34; 36

1. 13.6, 12, 12, 11   3. 4.1; 4; 1, 2, and 4; 8   5. 133, 133.5, no mode, 25

The baseball coaches at Midville Middle School made the list at the right of the runs scored by the members of their team. Use this data for Exercises 9–13.

9. Make a frequency table for the data.
10. Find the median number of runs scored. 4
11. Find the mean of the data. 4.125
12. What is the mode? 2
13. Which measure of central tendency would you use to give the best impression of the players? Explain.
   Answers may vary. The mean gives the best impression because it is the greatest.

Runs Scored

3	5	5	6
7	5	1	0
8	9	2	2
6	1	2	3
0	0	2	4
4	7	8	9

Lesson 11.3

Ms. Smith made the list at the right of the test scores for the last test in her math class.

1. Make a stem-and-leaf plot for the data.
2. What is the median of the data? 78.5
3. What are the lower and upper quartiles for this data? lower quartile = 68.5, upper quartile = 90.5
4. Make a box-and-whisker plot for this data.

Test Scores

80	90	92	68
75	60	68	93
95	69	75	76
80	80	83	91
67	87	77	97
95	58	73	59

Lesson 11.4

The table at the right shows the monthly budget for the Ramirez family.

1. Find the total income for the Ramirez family. $3100
2. Copy and complete the table by finding the percent of the total income that the Ramirez family spends in each category.
3. Use the information from the table to make a circle graph to show the percent of total income that the Ramirez family spends in each category.

Category	Amount	Percent
Housing	$930	?
Food	$775	?
Insurance	$465	?
Transportation	$620	?
Other	$310	?
Total	?	?

Lesson 11.5

Describe the correlation as strong positive, strong negative, or little to none.

1.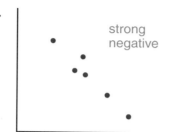
strong negative

2.
little to none

3.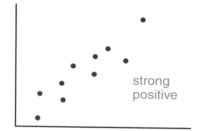
strong positive

4.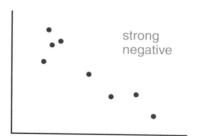
strong negative

5. Make a scatter plot of the following data for the winning times for the women's 200-meter run in the Olympics. Describe the correlation between the years and the winning times.

Year	1964	1968	1972	1976	1980	1984	1988	1992
Seconds	23.00	22.50	22.40	22.37	22.03	21.81	21.34	21.81

Lesson 11.6

Tell whether the given correlation coefficient describes a line of best fit that rises or falls.

1. 0.01 rises   **2.** 0.98 rises   **3.** −0.26 falls   **4.** −0.94 falls   **5.** 0.38 rises

Tell whether the correlation coefficient for the scatter plot is nearest to −1, 1, or 0.

6.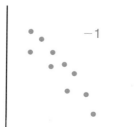
−1

7.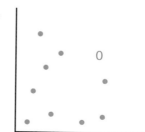
0

8.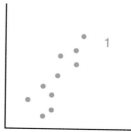
1

Lesson 12.1

Find the circumference and area, to the nearest hundredth, of each circle with the given radius or diameter.

1. radius of 4 inches
2. radius of 5.5 centimeters
3. radius of 15 inches
4. radius of 3 yards
5. diameter of 10 meters
6. diameter of 5.8 centimeters
7. diameter of 9 feet
8. diameter of 19 inches
9. radius of $4\frac{1}{2}$ feet
10. radius of $6\frac{5}{8}$ inches
11. radius of 5.3 centimeters
12. radius of 8.9 meters

Lesson 12.2

Find the surface area and volume of each cube with the given edge length.

1. 5 m
2. 3 in.
3. 12 ft
4. 7 cm
5. 8.5 cm
6. 2.5 ft
7. 9 in.
8. 5.25 cm

Find the surface area and volume of each rectangular solid with the indicated dimensions.

9. 2 m × 4 m × 4 m
10. 3 ft × 5 ft × 6 ft
11. 6 in. × 6 in. × 8 in.
12. 3 cm × 4 cm × 5.5 cm
13. 1 yd × 2 yd × 3 yd
14. 3.2 m × 4.5 m × 6 m
15. 9 m × 10.25 m × 11.5 m
16. 9.25 in. × 10 in. × 13 in.
17. 6 m × 8.3 m × 9.55 m

Lesson 12.3

1. Find the volume of a triangular prism with base area of 8 square meters and a height of 5 meters. 40 cubic meters

2. Find the lateral surface area of a triangular prism with a base area of 6 square centimeters, a base perimeter of 16 centimeters, and a height of 10 centimeters. 160 square centimeters

3. Find the total surface area of a rectangular prism with base area of 10 square inches, a base perimeter of 14 inches, and a height of 2 inches. 48 square inches

4. If a hexagonal prism has a base area of 18.75 square meters and a height of 20.5 meters, what is its volume? 384.375 cubic meters

5. If a pentagonal prism has a base area of 9 square feet and a height of 15 feet, what is its volume? 135 cubic feet

Lesson 12.4

Find the surface area and volume of each right cylinder described.

1. 3-in. radius and 5-in. height
2. 8-yd radius and 6-yd height
3. 5-cm height and 7-cm diameter
4. 9-m diameter and 10-m height
5. 12-ft height and 10-ft diameter
6. 15-cm height and 12-cm diameter
7. 5.5-m height and 3.5-m radius
8. 7.3-m height and 8.4-m diameter
9. 55-ft height and 15-ft radius
10. 56-cm height and 100-cm diameter
11. 35-cm height and 25-cm radius
12. 19-yd height and 5-yd diameter
13. 16-inch height and 11-inch diameter
14. 5-ft height and 5-ft diameter
15. 19.5-cm height and 20.5-cm diameter
16. 55-cm height and 30-cm radius
17. 9.5-m height and 9-m diameter
18. 10-in. height and 2-ft diameter
19. 5-yd height and 14-ft radius
20. 5-ft height and 1-yd diameter

Lesson 12.5

Find the volume of each right cone or right pyramid described.

1. cone with radius of 4 inches and height of 15 inches 251.33 cu in.
2. cone with a radius of 7 centimeters and a height of 5 centimeters 256.56 cu cm
3. cone with a diameter of 8 feet and a height of 10 feet 167.55 cu ft
4. cone with a diameter of 9 yards and a height of 5 yards 106.03 cu yd
5. pyramid with a base area of 25 square centimeters and a height of 15 centimeters 125 cu cm
6. pyramid with a base area of 12 square inches and a height of 8 inches 32 cu in.
7. pyramid with a base area of 9 square yards and a height of 7 yards 21 cu yd
8. pyramid with a base area of 28 square meters and a height of 20 meters 186.7 cu m
9. cone with a diameter of 9.2 feet and a height of 6 feet 132.95 cu ft
10. cone with a radius of 2.25 inches and a height of 3.5 inches 18.56 cu in.
11. cone with a radius of $4\frac{3}{4}$ inches and a height of $5\frac{1}{2}$ inches 129.95 cu in.
12. cone with a diameter of $6\frac{1}{2}$ feet and a height of $3\frac{1}{2}$ feet 38.7 cu ft
13. pyramid with a base area of 35 square feet and a height of 15.5 feet 180.8 cu ft
14. pyramid with a base area of 18 square inches and a height of 9.25 inches 55.5 cu in.
15. pyramid with a base area of 15 square inches and a height of $8\frac{1}{2}$ inches 42.5 cu in.
16. pyramid with a base area of 30 square feet and a height of $19\frac{1}{4}$ feet 192.5 cu ft

Lesson 12.6

Find the lateral surface area and the total surface area of each cone described.

1. a slant height of 4 meters and a diameter of 8 meters
2. a slant height of 7 feet and a diameter of 12 feet
3. a slant height of 5 inches and a radius of 4 inches
4. a slant height of 6 yards and a radius of 7 yards
5. a slant height of 4.8 centimeters and a diameter of 10.4 centimeters
6. a slant height of 7 yards and a radius of 3.5 yards

Find the lateral surface area and the total surface area of each right pyramid described.

7. a square base with side lengths of 6 centimeters and a slant height of 4 centimeters L = 48 sq cm, S = 84 sq cm
8. a square base with sides lengths of 5 inches and a slant height of 4 inches L = 40 sq in., S = 65 sq in.
9. a rectangular base with side lengths of 9 meters and 7 meters and a slant height of 10 meters L = 160 sq m, S = 223 sq m

Find the total surface area of each cone described below.

10. a radius of 8 inches and a slant height of 5 inches 326.73 sq in.
11. a radius of 3.5 meters and a slant height of 4 meters 82.47 sq m
12. a diameter of 12 centimeters and a slant height of 15 centimeters 395.84 sq cm
13. a diameter of 18 feet and a slant height of 10 feet 537.21 sq ft
14. a radius of $5\frac{3}{4}$ inches and a slant height of 7 inches 230.32 sq in.
15. a diameter of 15 meters and a slant height of 11.25 meters 441.79 sq m
16. a radius of 7 meters and a slant height of 15 meters approximately 483.8 sq m

Lesson 12.7

Find the surface area and volume of each sphere described.

1. a radius of 5 m
2. a diameter of 16 in.
3. a diameter of 14 cm
4. a radius of 6 ft
5. a radius of 7.5 m
6. a diameter of 18 ft
7. a diameter of 19 m
8. a radius of 4.25 ft
9. a radius of 7.8 cm
10. a diameter of 15 cm
11. a diameter of 6.5 m
12. a radius of 8.2 cm
13. a diameter of 17 in.
14. a radius of $8\frac{2}{3}$ yd
15. a radius of $15\frac{3}{4}$ in.
16. a diameter of 5.5 m
17. a radius of $7\frac{1}{4}$ ft
18. a diameter of 4.25 m

Functions and Their Graphs

Throughout the text a variety of functions were studied—linear, exponential, absolute value, quadratic, and reciprocal. The simplest form of any function is called the parent function. Each parent function has a distinctive graph. Changes made to the parent function will alter the graph but retain the distinctive features of the parent function. Some changes produce a horizontal or vertical shift, others produce a stretch, and still others a reflection. Pages 734–737 summarize some parent functions and the changes that transform them.

Linear Function

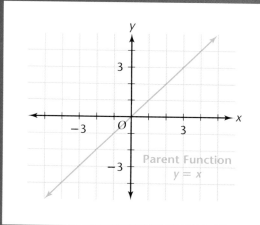

Parent Function
$y = x$

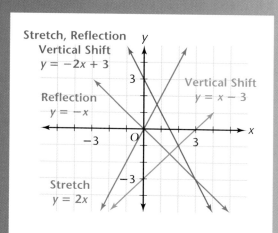

Stretch, Reflection
Vertical Shift
$y = -2x + 3$

Reflection
$y = -x$

Vertical Shift
$y = x - 3$

Stretch
$y = 2x$

Exponential Function

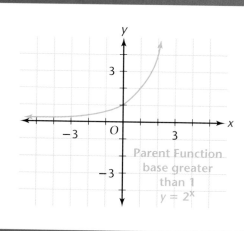

Parent Function
base greater
than 1
$y = 2^x$

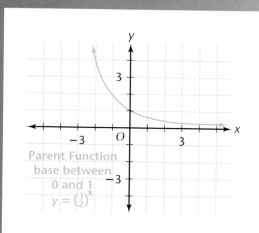

Parent Function
base between
0 and 1
$y = \left(\frac{1}{2}\right)^x$

Absolute-Value Function

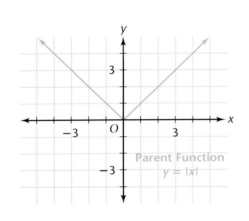

Parent Function
$y = |x|$

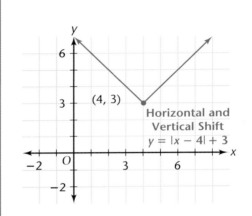

(4, 3)

Horizontal and Vertical Shift
$y = |x - 4| + 3$

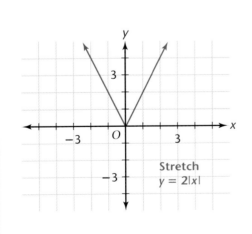

Stretch
$y = 2|x|$

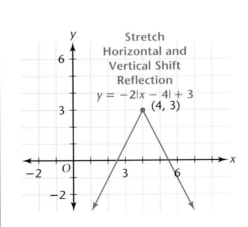

Stretch Horizontal and Vertical Shift Reflection
$y = -2|x - 4| + 3$
(4, 3)

Quadratic Function

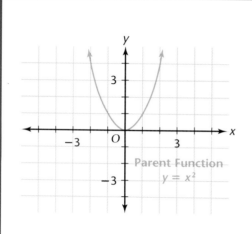

Parent Function
$y = x^2$

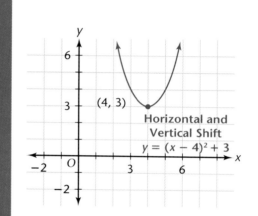

(4, 3)

Horizontal and
Vertical Shift
$y = (x - 4)^2 + 3$

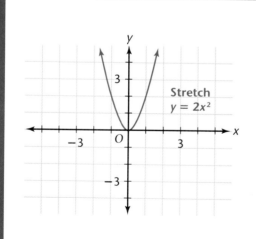

Stretch
$y = 2x^2$

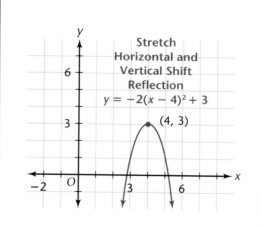

Stretch
Horizontal and
Vertical Shift
Reflection
$y = -2(x - 4)^2 + 3$

(4, 3)

Reciprocal Function

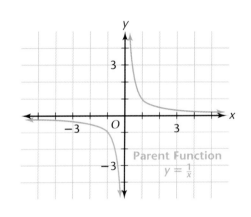

Parent Function
$y = \frac{1}{x}$

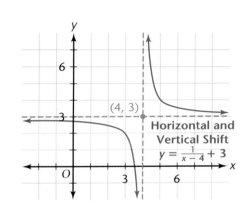

(4, 3)

Horizontal and
Vertical Shift
$y = \frac{1}{x-4} + 3$

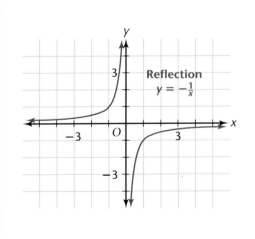

Reflection
$y = -\frac{1}{x}$

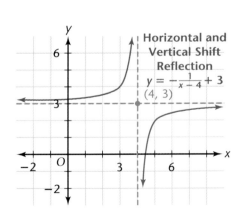

Horizontal and
Vertical Shift
Reflection
$y = -\frac{1}{x-4} + 3$
(4, 3)

Table of Squares, Cubes, and Roots

No.	Squares	Cubes	Square Roots	Cube Roots	No.	Squares	Cubes	Square Roots	Cube Roots
1	1	1	1.000	1.000	51	2,601	132,651	7.141	3.708
2	4	8	1.414	1.260	52	2,704	140,608	7.211	3.733
3	9	27	1.732	1.442	53	2,809	148,877	7.280	3.756
4	16	64	2.000	1.587	54	2,916	157,464	7.348	3.780
5	25	125	2.236	1.710	55	3,025	166,375	7.416	3.803
6	36	216	2.449	1.817	56	3,136	175,616	7.483	3.826
7	49	343	2.646	1.913	57	3,249	185,193	7.550	3.849
8	64	512	2.828	2.000	58	3,364	195,112	7.616	3.871
9	81	729	3.000	2.080	59	3,481	205,379	7.681	3.893
10	100	1,000	3.162	2.154	60	3,600	216,000	7.746	3.915
11	121	1,331	3.317	2.224	61	3,721	226,981	7.810	3.936
12	144	1,728	3.464	2.289	62	3,844	238,328	7.874	3.958
13	169	2,197	3.606	2.351	63	3,969	250,047	7.937	3.979
14	196	2,744	3.742	2.410	64	4,096	262,144	8.000	4.000
15	225	3,375	3.873	2.466	65	4,225	274,625	8.062	4.021
16	256	4,096	4.000	2.520	66	4,356	287,496	8.124	4.041
17	289	4,913	4.123	2.571	67	4,489	300,763	8.185	4.062
18	324	5,832	4.243	2.621	68	4,624	314,432	8.246	4.082
19	361	6,859	4.359	2.668	69	4,761	328,509	8.307	4.102
20	400	8,000	4.472	2.714	70	4,900	343,000	8.367	4.121
21	441	9,261	4.583	2.759	71	5,041	357,911	8.426	4.141
22	484	10,648	4.690	2.802	72	5,184	373,248	8.485	4.160
23	529	12,167	4.796	2.844	73	5,329	389,017	8.544	4.179
24	576	13,824	4.899	2.884	74	5,476	405,224	8.602	4.198
25	625	15,625	5.000	2.924	75	5,625	421,875	8.660	4.217
26	676	17,576	5.099	2.962	76	5,776	438,976	8.718	4.236
27	729	19,683	5.196	3.000	77	5,929	456,533	8.775	4.254
28	784	21,952	5.292	3.037	78	6,084	474,552	8.832	4.273
29	841	24,389	5.385	3.072	79	6,241	493,039	8.888	4.291
30	900	27,000	5.477	3.107	80	6,400	512,000	8.944	4.309
31	961	29,791	5.568	3.141	81	6,561	531,441	9.000	4.327
32	1,024	32,768	5.657	3.175	82	6,724	551,368	9.055	4.344
33	1,089	35,937	5.745	3.208	83	6,889	571,787	9.110	4.362
34	1,156	39,304	5.831	3.240	84	7,056	592,704	9.165	4.380
35	1,225	42,875	5.916	3.271	85	7,225	614,125	9.220	4.397
36	1,296	46,656	6.000	3.302	86	7,396	636,056	9.274	4.414
37	1,369	50,653	6.083	3.332	87	7,569	658,503	9.327	4.431
38	1,444	54,872	6.164	3.362	88	7,744	681,472	9.381	4.448
39	1,521	59,319	6.245	3.391	89	7,921	704,969	9.434	4.465
40	1,600	64,000	6.325	3.420	90	8,100	729,000	9.487	4.481
41	1,681	68,921	6.403	3.448	91	8,281	753,571	9.539	4.498
42	1,764	74,088	6.481	3.476	92	8,464	778,688	9.592	4.514
43	1,849	79,507	6.557	3.503	93	8,649	804,357	9.644	4.531
44	1,936	85,184	6.633	3.530	94	8,836	830,584	9.695	4.547
45	2,025	91,125	6.708	3.557	95	9,025	857,375	9.747	4.563
46	2,116	97,336	6.782	3.583	96	9,216	884,736	9.798	4.579
47	2,209	103,823	6.856	3.609	97	9,409	912,673	9.849	4.595
48	2.304	110,592	6.928	3.634	98	9,604	941,192	9.899	4.610
49	2,401	117,649	7.000	3.659	99	9,801	970,299	9.950	4.626
50	2,500	125,000	7.071	3.684	100	10,000	1,000,000	10.000	4.642

Table of Random Digits

Line\Col	(1)	(2)	(3)	(4)	(5)	(6)	(7)	(8)	(9)	(10)	(11)	(12)	(13)	(14)
1	10480	15011	01536	02011	81647	91646	69179	14194	62590	36207	20969	99570	91291	90700
2	22368	46573	25595	85393	30995	89198	27982	53402	93965	34095	52666	19174	39615	99505
3	24130	48360	22527	97265	76393	64809	15179	24830	49340	32081	30680	19655	63348	58629
4	42167	93093	06243	61680	07856	16376	39440	53537	71341	57004	00849	74917	97758	16379
5	31570	39975	81837	16656	06121	91782	60468	81305	49684	60672	14110	06927	01263	54613
6	77921	06907	11008	42751	27756	53498	18602	70659	90655	15053	21916	81825	44394	42880
7	99562	72905	56420	69994	98872	31016	71194	18738	44013	48840	63213	21069	10634	12952
8	96301	91977	05463	07972	18876	20922	94595	56869	69014	60045	18425	84903	42508	32307
9	89579	14342	63661	10281	17453	18103	57740	84378	25331	12566	58678	44947	05585	56941
10	85475	36857	53342	53988	53060	59533	38867	62300	08158	17983	16439	11458	18593	64952
11	28918	69578	88231	33276	70997	79936	56865	05859	90106	31595	01547	85590	91610	78188
12	63553	40961	48235	03427	49626	69445	18663	72695	52180	20847	12234	90511	33703	90322
13	09429	93969	52636	92737	88974	33488	36320	17617	30015	08272	84115	27156	30613	74952
14	10365	61129	87529	85689	48237	52267	67689	93394	01511	26358	85104	20285	29975	89868
15	07119	97336	71048	08178	77233	13916	47564	81056	97735	85977	29372	74461	28551	90707
16	51085	12765	51821	51259	77452	16308	60756	92144	49442	53900	70960	63990	75601	40719
17	02368	21382	52404	60268	89368	19885	55322	44819	01188	65225	64835	44919	05944	55157
18	01011	54092	33362	94904	31273	04146	18594	29852	71585	85030	51132	01915	92747	64951
19	52162	53916	46369	58586	23216	14513	83149	98736	23495	64350	94738	17752	35156	35749
20	07056	97628	33787	09998	42698	06691	76988	13602	51851	46104	88916	19509	25625	58104
21	48663	91245	85828	14346	09172	30168	90229	04734	59193	22178	30421	61666	99904	32812
22	54164	58492	22421	74103	47070	25306	76468	26384	58151	06646	21524	15227	96909	44592
23	32639	32363	05597	24200	13363	38005	94342	28728	35806	06912	17012	64161	18296	22851
24	29334	27001	87637	87308	58731	00256	45834	15398	46557	41135	10367	07684	36188	18510
25	02488	33062	28834	07351	19731	92420	60952	61280	50001	67658	32586	86679	50720	94953
26	81525	72295	04839	96423	24878	82651	66566	14778	76797	14780	13300	87074	79666	95725
27	29676	20591	68086	26432	46901	20849	89768	81536	86645	12659	92259	57102	80428	25280
28	00742	57392	39064	66432	84673	40027	32832	61362	98947	96067	64760	64584	96096	98253
29	05366	04213	25669	26422	44407	44048	37937	63904	45766	66134	75470	66520	34693	90449
30	91921	26418	64117	94305	26766	25940	39972	22209	71500	64568	91402	42416	07844	69618
31	00582	04711	87917	77341	42206	35126	74087	99547	81817	42607	43808	76655	62028	76630
32	00725	69884	62797	56170	86324	88072	76222	36086	84637	93161	76038	65855	77919	88006
33	69011	65795	95876	55293	18988	27354	26575	08625	40801	59920	29841	80150	12777	48501
34	25976	57948	29888	88604	67917	48708	18912	82271	65424	69774	33611	54262	85963	03547
35	09763	83473	73577	12908	30883	18317	28290	35797	05998	41688	34952	37888	38917	88050
36	91567	42595	27958	30134	04024	86385	29880	99730	55536	84855	29080	09250	79656	73211
37	17955	56349	90999	49127	20044	59931	06115	20542	18059	02008	73708	83517	36103	42791
38	46503	18584	18845	49618	02304	51038	20655	58727	28168	15475	56942	53389	20562	87338
39	92157	89634	94824	78171	84610	82834	09922	25417	44137	48413	25555	21246	35509	20468
40	14577	62765	35605	81263	39667	47358	56873	56307	61607	49518	89656	20103	77490	18062
41	98427	07523	33362	64270	01638	92477	66969	98420	04880	45585	46565	04102	46880	45709
42	34914	63976	88720	82765	34476	17032	87589	40836	32427	70002	70663	88863	77775	69348
43	70060	28277	39475	46473	23219	53416	94970	25832	69975	94884	19661	72828	00102	66794
44	53976	54914	06990	67245	68350	82948	11398	42878	80287	88267	47363	46634	06541	97809
45	76072	29515	40980	07391	58745	25774	22987	80059	39911	96189	41151	14222	60697	59583
46	90725	52210	83974	29992	65831	38857	50490	83765	55657	14361	31720	57375	56228	41546
47	64364	67412	33339	31926	14883	24413	59744	92351	97473	89286	35931	04110	23726	51900
48	08962	00358	31662	25388	61642	34072	81249	35648	56891	69352	48373	45578	78547	81788
49	95012	68379	93526	70765	10592	04542	76463	54328	02349	17247	28865	14777	62730	92277
50	15664	10493	20492	38391	91132	21999	59516	81652	27195	48223	46751	22923	32261	85653

Source: Interstate Commerce Commission

GLOSSARY

absolute error The absolute value of the difference between an actual measurement and a specified measurement. (414)

absolute value For any number x, if x is greater than or equal to 0, $|x| = x$, and if x is less than 0, $|x| = -x$. (75)

absolute-value function A function written in the form $y = |x|$ or $y = \text{ABS}(x)$. (559)

acute angle An angle with a measure of less than 90°. (206)

acute triangle A triangle that has three acute angles. (218)

Addition Property of Equality If equal amounts are added to the expression on each side of an equation, the expressions remain equal. (294)

Addition Property of Inequality If equal amounts are added to the expression on both sides of an inequality, the resulting inequality is still true. (314)

Addition Property of Zero For any number a, $a + 0 = a = 0 + a$. (93)

additive identity The number 0. (93)

adjacent angles Two angles in the same plane with a common vertex and a common side, but with no interior points in common. (213)

algebraic expression Variables combined with numbers and operations. (274)

algebraic logic Refers to the use of the proper order of operations by calculating and computing devices. (51)

alternate exterior angles Two nonadjacent exterior angles which lie on opposite sides of a transversal. (217)

alternate interior angles Two nonadjacent interior angles which lie on opposite sides of a transversal. (217)

altitude of a cone A segment from the vertex perpendicular to the plane of the base. (664)

altitude of a pyramid A segment from the vertex perpendicular to the plane of the base. (665)

angle A figure formed by two rays that intersect at a common point. (205)

approximate solution A reasonable estimate of a point of intersection for a system of equations. (499)

area The number of nonoverlapping unit squares that will cover the interior of a figure. (232)

Associative Property for Addition For all numbers a, b, and c, $(a + b) + c = a + (b + c)$. (60)

average In mathematics, a value that is representative of a set of data. (593)

bar graph A graph used to compare quantities by using horizontal or vertical bars. (587)

base of a percent In a percent problem, the original number from which a percentage is to be determined. (366)

base of a prism The polygon region and its translated image. (650)

base of an exponent In an expression of the form x^a, x is the base. (43)

bases of a parallelogram Either pair of parallel sides of a parallelogram. (239)

bases of a trapezoid The pair of parallel sides of a trapezoid. (243)

binomial A polynomial with two terms. (299)

boundary line A line for a linear inequality that divides the coordinate plane into two half-planes. (478)

box-and-whisker plot A method of showing a distribution of data with the median and the range. (604)

break-even point The point at which total revenue equals total cost. (396)

center The point about which a rotation takes place. (571)

center of a circle The point from which all points on a circle are a given distance. (636)

circle The set of all points in the plane that are equidistant from a given point known as the center of the circle. (636)

circle graph A graph that shows how portions, or percentages, of a whole quantity are distributed. (589)

circumference The distance around a circle. The circumference, C, of a circle, where d is the diameter, is given by the equation $C = \pi \cdot d$. (637)

coefficient The number multiplied by a variable. (274)

Commutative Property for Addition For any numbers a and b, $a + b = b + a$. (276)

complementary angles Two angles with a combined measure equal to 90°. One angle is said to be the complement of the other. Complementary angles may or may not be adjacent. (213)

composite number A natural number that is greater than 2 and that is not prime. (39)

compound interest Money paid or earned on a given amount of money. (540)

concave polygon A polygon that is not convex. (224)

cone A solid figure that consists of one base that is a circle and a curved lateral surface that extends from the base to a single point called the vertex. (664)

congruent segments Segments that are the same length. (218)

conjecture A statement about observations that is believed to be true. (10)

consecutive interior angles Two interior angles which lie on the same side of a transversal. (217)

constant A term in an algebraic expression that represents a fixed amount. (274)

constant function A function of the form $y = b$. (467)

constant of variation In a direct variation of the form $\frac{y}{x} = k$, k is the constant of variation. (454)

converse of the "Pythagorean" Right-Triangle Theorem If the square of the length of one side of a triangle equals the sum of the squares of the lengths of the other two sides, then the triangle is a right triangle. (251)

convex polygon A polygon in which any segment connecting two interior points of the polygon is in the interior of the polygon. (224)

coordinates The numbers in an ordered pair that give the address of a point on a graph. (104)

correlation An indication of the proportion of relationship between two sets of data. (619)

correlation coefficient The measure of how closely a set of data points falls along a line. (625)

corresponding angles Two nonadjacent angles, one interior and one exterior, that lie on the same side of a transversal. (216)

cross product The product of the means and the product of the extremes. In the proportion, $\frac{a}{b} = \frac{c}{d}$, $ad = bc$. (359)

cube A special type of rectangular solid in which every edge is the same length. (645)

cubic unit The basic unit of measurement of volume. (44)

denominator The total number of parts needed to make one whole in a fraction, given by the bottom number of a fraction. (131)

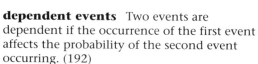

dependent events Two events are dependent if the occurrence of the first event affects the probability of the second event occurring. (192)

dependent variable In a function of two variables, one variable is dependent and the other variable is independent. For example, y is the dependent variable in $y = 2x + 1$. Its value depends on the value of x. (431)

diagonals of a parallelogram Segments that connect the vertices of the opposite angles of a parallelogram. (226)

diameter A segment through the center of a circle that joins two points on the circle. (636)

direct variation If y varies directly as x, then $\frac{y}{x} = k$. The k is the constant of variation. (454)

Distributive Property For all numbers a, b, and c, $a(b + c) = ab + ac$ and $(b + c)a = ba + ca$. (275)

Distributive Property Over Subtraction For all numbers a, b, and c, $a(b - c) = ab - ac$, and $(b - c)a = ba - ca$. (331)

dividing an expression For all numbers a, b, and c, $c \neq 0$, $\frac{a + b}{c} = \frac{a}{c} + \frac{b}{c}$, and $\frac{a - b}{c} = \frac{a}{c} - \frac{b}{c}$. (333)

Division Property of Equality If the expressions on each side of an equation are divided by equal nonzero amounts, the expressions remain equal. (339)

domain The set of first values in the ordered pairs of a function. (431)

edge A segment formed by the intersection of two faces of a geometric solid. (650)

elimination by addition A method used to solve a system of equations in which one variable is eliminated by adding or subtracting opposites. (510)

elimination by multiplication A method used to solve a system of equations in which one variable is eliminated by multiplying one or more of the equations by appropriate values. (517)

equation Two equivalent expressions separated by an equal sign. (31)

equation method A method for solving problems involving percent in which the parts of the problem are translated into an algebraic equation. (366)

equilateral triangle A triangle that has three congruent sides. (218)

equivalent fractions Fractions with the same value. (131)

experimental probability Let t be the number of trials in the experiment. Let f be the number of times a successful event occurs. The experimental probability, P, of the event is given by $P = \frac{f}{t}$. (180)

exponent The number that tells how many times a number is used as a factor. In an expression of the form x^a, a is the exponent. (43)

exponential decay A situation in which a number is repeatedly multiplied by a number between 0 and 1. (540)

exponential function A function represented by an equation, such as $y = 2^x$, in which a base number is raised to a variable exponent. (539)

exponential growth A situation in which a number is repeatedly multiplied by a number greater than 1. (540)

exponential relationship A relationship in which a number grows at a fixed rate. (540)

expression Formed by variables combined with numbers and operations. (31)

exterior angle of a polygon An angle formed between one side of a polygon and the extension of an adjacent side. (216)

extremes In the proportion $\frac{a}{b} = \frac{c}{d}$, a and d are the extremes. (358)

face   Each flat surface of a prism. (650)

factor   Numbers or variables that are multiplied. (35, 274)

first differences   In a sequence of numbers, the value of the differences between consecutive terms. (112)

formula   An equation that describes a numerical relationship. (295)

frequency table   A convenient way to list data. (596)

function   A set of ordered pairs for which no two pairs have the same first coordinate. (431, 536)

great circle   A circle that contains the center of a sphere. (676)

greatest common factor (GCF)   The largest integer that will evenly divide two or more numbers. (137)

height of a parallelogram   The perpendicular length between two bases. (239)

height of a trapezoid   The perpendicular length between the bases. (243)

hemisphere   Half of a sphere that is divided along a great circle. (684)

heptagon   A polygon with seven sides. (228)

hexagon   A polygon with six sides. (224)

histogram   A bar graph that shows how frequently the numbers in a data set appear. (603)

horizontal line   The equation for a horizontal line is written in the form $y = b$. The slope is 0. (468)

hypotenuse   The longest side of a right triangle. (251)

image   The figure that is the result of a transformation. (570)

improper fraction   A fraction whose numerator is greater than its denominator. (150)

independent events   Two events are independent if the occurrence of the first event does not affect the probability of the second event occurring. (188)

independent variable   In a function of two variables, one variable is dependent and the other variable is independent. For example, x is the independent variable in the function $y = 2x + 1$. The value of y depends on the value of x. (431)

inequality   A mathematical statement that contains the symbol $>$, $\geq$, $<$, $\leq$, or $\neq$. (88)

integer function   A function written in the form $y = \text{INT}(x)$. The integer function rounds the number x down to the nearest integer. It is also called the greatest integer function. (560)

integers   The set of all whole numbers and their opposites. (72)

interior angle   An angle on a transversal which lies between the two parallel lines. (216)

intersecting lines   Lines that cross. (205)

isosceles trapezoid   A trapezoid with congruent nonparallel sides. (227)

isosceles triangle   A triangle that has two congruent sides. (218)

lateral face   Each face of a prism that is not a base. (650)

lateral surface area   The sum of the areas of the lateral faces. (652)

least common denominator (LCD)   The smallest number that is a multiple of two or more denominators. (138)

legs   The two sides of a right triangle that form the right angle. (251)

like terms   Terms that contain the same form of a variable. (299)

line   All the points that extend infinitely in two opposite directions. (204)

line graph   A type of graph that can be used to show changes occurring over time or to make predictions based on current trends. (588)

line of best fit   Represents an approximation of the data on a scatter plot. (624)

line of symmetry   The line across which a reflection takes place. (570)

linear equation   An equation whose graph is a line. (105)

linear function   A function whose graph is a straight line. (431)

linear inequality   An inequality with a boundary line that can be expressed in the form $y = mx + b$. The solution to a linear inequality is the set of all ordered pairs that make the inequality true. (478)

linear relationship   A relationship in which a number grows by a fixed amount. (540)

literal equation   An equation that contains a number of different letters representing variables. Many formulas are examples of literal equations. (295, 404)

lowest terms   A fraction is in lowest terms when the only factor common to both the numerator and denominator is 1. (137)

mean   The mean of a set of n numbers is the sum of all the numbers divided by n. The mean is commonly called the average. (595)

means   In the proportion $\frac{a}{b} = \frac{c}{d}$, b and c are the means. (358)

measures of central tendency   The mean, median, and mode of a set of numbers. (593)

median   The middle number when the numbers in a set of data have been arranged in descending or ascending order. When there are an even number of elements, the median is the mean of the two middle numbers. (595)

mixed number   A number written with both a whole number and a fraction. (142)

mode   The number that occurs most often in a set of data. A set of data can have more than one mode. If all the numbers appear the same number of times, the mode does not exist. (595)

monomial   An algebraic expression that is either a constant, a variable, or a product of a constant and one or more variables. (299)

Multiplication and Division Properties of Inequality   If the expressions on each side of an inequality are multiplied or divided by the same positive number, the resulting inequality is still true. If the expressions on each side of an inequality are multiplied or divided by the same negative number and the inequality sign is reversed, the resulting inequality is still true. (410)

Multiplication Property of Equality   If the expressions on each side of an equation are multiplied by equal amounts, the expressions remain equal. (340)

multiplicative inverse   The reciprocal of a number. (353)

Multiplicative Property of -1   For all numbers a, $-1(a) = -a$. (331)

natural numbers The counting numbers such as, 1, 2, 3, . . . (36)

net A flat figure that can be folded to enclose a particular solid figure. (643)

neutral pair A pair of algebra tiles consisting of one positive tile and one negative tile. (78)

numerator The number of parts needed to represent a fraction, given by the top number of a fraction. (131)

oblique cone A cone in which the altitude intersects the plane of the base at some point other than the center. (664)

oblique pyramid A pyramid in which the altitude intersects the plane of the base at some point other than the center. (665)

obtuse angle An angle with a measure greater than 90°. (207)

obtuse triangle A triangle that has one obtuse angle. (218)

octagon A polygon with eight sides. (228)

opposites Two integers that are on opposite sides of zero and that are the same distance from zero. (74)

order of operations The set of rules for computation. (50)

ordered pair The address of a point in the rectangular coordinate system, indicated by two numbers in parentheses, (x, y). (104)

origin The point where the horizontal axis and the vertical axis intersect. (104, 429)

outcome The result of an experiment. (186)

parabola The graph of a quadratic function. (546)

parallel lines Lines that lie in the same plane and do not intersect. (205)

parallelogram A quadrilateral with both pairs of opposite sides parallel. (225)

parent function The original function in a family of functions. (567, 571)

pentagon A polygon with five sides. (224)

percent A ratio that compares a number to 100. (169)

perfect square A square number whose square root is always a rational number. (249)

perimeter The distance around a geometric figure that is contained in a plane. (232)

perpendicular lines Two lines that intersect to form square corners, or right angles. (205)

pi (π) The Greek letter used to represent the ratio $\dfrac{\text{circumference}}{\text{diameter}}$ of a circle. (637)

piecewise function A function that is made up of pieces of other functions. (566)

place value The position, or place, of a digit in a decimal number that determines the value of the digit. (141)

plane All the points on a flat surface that extend infinitely in all directions. (204)

point An exact location. (204)

point of intersection The point where two lines cross. (205)

point-slope form The point-slope form for the equation of a line is $y - y_1 = m(x - x_1)$. The coordinates x_1 and y_1 are taken from a given point, (x_1, y_1), and the slope is m. (461)

polygon A closed figure consisting of segments that intersect only at their endpoints. (224)

polynomial A term or a sum of different terms. (299)

power A term for "exponent." The expression 2^3 is read "two to the third power." (44)

pre-image The original figure in a transformation. (570)

prediction The hypothesized outcome of a probability experiment. (186)

prime factorization The process of writing a natural number as a product of only prime numbers. (46)

prime number A natural number with exactly two factors, itself and 1. (39)

principal square root The positive square root of a number. (249)

prism A solid figure that consists of a polygonal region, its translated image, and the connecting segments. The lateral faces of a right prism are always rectangles. (650)

probability The ratio of the number of successful outcomes to the number of possible outcomes. (178)

Properties of Zero Let a represent any number.
1. The product of any number and zero is zero. $a \cdot 0$ and $0 \cdot a = 0$
2. Zero divided by any nonzero number is zero. $\frac{0}{a} = 0, a \neq 0$
3. A number divided by zero is undefined. That is, it is impossible to divide by zero. (100)

Property of Opposites For any number a, $a + (-a) = 0$. (82)

proportion An equation containing two or more equivalent ratios. (165)

proportion method A method for solving problems involving percent in which the problem is modeled using a proportion. (366)

pyramid A solid figure that consists of one base that is a polygon and a number of lateral faces that are triangles. (665)

"Pythagorean" Right-Triangle Theorem For any right triangle, the square of the length of the hypotenuse is equal to the sum of the squares of the length of the legs. (251)

quadrant One of the four regions in a coordinate plane. A horizontal and vertical number line divide a coordinate plane into four quadrants. (429)

quadratic function A function that involves a square variable, such as $y = x^2$. (545)

quadrilateral A polygon with four sides. (224)

quartile The values used to divide a data set into fourths. (604)

radius A segment which connects the center of a circle with a point on the circle. (636)

range The difference between the greatest number and the least number in a set of data. (594)

range The set of second values in the ordered pairs of a function. (431)

rate of change The amount of increase or decrease of a function. (447)

ratio A comparison of two quantities that uses division. (162)

rational number Any number that can be written as a ratio of two integers in which the denominator is not zero. (162)

ray Part of a line that contains one endpoint and all the points extending in one direction from the endpoint. (204)

reciprocal function A function that involves dividing by a variable, such as $y = \frac{1}{x}$. (551)

Reciprocal Property For any nonzero number r, there is a number $\frac{1}{r}$ such that $r \cdot \frac{1}{r} = 1$. (352)

reciprocals A pair of numbers whose product is 1. (157)

rectangle A parallelogram with four right angles. (226)

rectangular numbers A sequence of numbers that can be represented using dot patterns of progressively larger rectangles. (11)

rectangular solid Any solid figure with rectangular sides. (645)

reflection A transformation that flips, or reflects, a figure across a line called the line of symmetry. (570)

regression line The line of best fit on a graphics calculator. (453)

regular polygon A polygon that has all sides congruent and all angles congruent. (225)

relation Pairing elements from one set with elements of another set. (540)

repeating decimal A nonterminating decimal that repeats in a pattern. (142)

rhombus A parallelogram with four congruent sides. (227)

right angle An angle with a measure of 90°. (206)

right cone A cone in which the altitude intersects the base at its center. (664)

right cylinder A solid figure with two circular bases. (658)

right pyramid A pyramid in which the altitude intersects the base at its center. (665)

right triangle A triangle that has one right angle. (218)

rise The vertical change in a line. (434)

rotation A transformation that turns, or rotates, a figure about a point called a center. (571)

run The horizontal change in a line. (434)

sample space The set of possible outcomes for an experiment. (187)

scale factor A number used to convert measurements in order to enlarge or reduce an object. (256)

scalene triangle A triangle with no congruent sides. (218)

scatter plot A display of data that has been organized into ordered pairs and graphed on the coordinate plane. (618)

second differences In a sequence of numbers, the value of the differences between consecutive first differences. (112)

segment Two endpoints on a line and all the points between the endpoints. (204)

similar figures Figures that have the same shape. (258)

slant height of a right cone The length of the segment from the vertex perpendicular to the circular base of a cone. (670)

slant height of a right regular pyramid The altitude of any of the triangular lateral sides of a pyramid. (670)

slope A measure of the steepness of a line, given by the formula, $m = \frac{y_2 - y_1}{x_2 - x_1}$. (434)

slope formula Given two points with coordinates (x_1, y_1) and (x_2, y_2), the slope, m, is $\frac{\text{difference in } y}{\text{difference in } x} = \frac{y_2 - y_1}{x_2 - x_1}$. (437)

slope-intercept form The slope-intercept form for a line with slope m and y-intercept b is $y = mx + b$. (451)

sphere A solid figure determined by all points that are a certain distance from a given point. (676)

square A rectangle with four congruent sides. (227)

square number A number that has an odd number of factors and that can be represented as a square with an area equal to the square number. (38)

standard form An equation in the form $Ax + By = C$ is in standard form when A, B, and C are integers, A and B are not both zero, and A is not negative. (454)

statistics The science of collecting, analyzing, describing, and interpreting data. (584)

stem-and-leaf plot A display of a set of data in which each piece of data is grouped together on a specific row, and arranged in two columns. The left column is called the stem and the right column is called the leaf. (601)

straight angle An angle consisting of two opposite rays that form a straight line. A straight angle has a measure of 180°. (206)

substitution method A method used to solve a system of equations in which variables are replaced with known values or algebraic expressions. (505)

subtraction For all numbers a and b, $a - b = a + (-b)$. (95)

Subtraction Property of Equality If equal amounts are subtracted from the expressions on each side of an equation, the expressions remain equal. (293)

Subtraction Property of Inequality If equal amounts are subtracted from the expressions on each side of an inequality, the resulting inequality is still true. (314)

supplementary angles Two angles with a combined measure equal to 180°. One angle is said to supplement the other. Supplementary angles may or may not be adjacent. (213)

surface area The number of square units necessary to cover a solid figure. (643)

symbols of inclusion Symbols, such as parentheses, brackets, braces, and fraction marks, that group numbers and variables. (51)

system of equations Two or more equations in two or more variables. (497)

system of linear inequalities Two or more linear inequalities in two or more variables. The solution for a system of linear inequalities is the intersection of the solution sets for each inequality. (524)

terminating decimal A decimal that terminates, or ends. (142)

terms The numbers, variables, or products or quotients of numbers and variables that are added or subtracted in an algebraic expression. (299)

theoretical probability A measure of what you expect to occur. When each outcome listed in a sample space has an equal chance of occurring, the theoretical probability, P, that a particular outcome, or event, E, will occur is given by $P(E) = \dfrac{\text{number of elements in the desired event}}{\text{number of elements in the sample space}}$. (187)

transformation The movement of a figure in a plane. (570)

translation A transformation that moves a geometric figure by sliding. Each of the points of the geometric figure moves the same distance in the same direction. (570)

transversal A line that crosses two other lines in the same plane. (216)

trapezoid A quadrilateral with exactly one pair of parallel sides. (244)

triangle A closed figure consisting of three segments that intersect only at their endpoints. (218)

trinomial A polynomial with three terms. (299)

unlike terms Terms of an algebraic expression, equation, or polynomial that differ by at least one variable. (299)

variable A letter or other symbol that can be replaced by any number or any other expression. (17)

vertex of a cone The single point opposite the base of a cone. (664)

vertex of a parabola The point where a parabola changes direction. (546)

vertex of a prism A point where three or more edges meet. (650)

vertex of a pyramid The point where the lateral triangular faces meet. (665)

vertex of an angle The point where the two rays that form an angle intersect. (205)

vertical angles The opposite angles formed where two lines intersect. All vertical angles have equal measures. (213)

vertical line The equation for a vertical line is written in the form $x = a$. The slope is undefined. (468)

vertices of a polygon The endpoints of the segments that form the polygon. (224)

volume The number of nonoverlapping unit cubes that will fill the interior of a solid figure. (644)

x-axis The horizontal number line in the rectangular coordinate system. (429)

x-coordinate A point on the x-axis in the rectangular coordinate system. (429)

y-axis The vertical number line in the rectangular coordinate system. (429)

y-coordinate A point on the y-axis in the rectangular coordinate system. (429)

y-intercept The point where a line crosses the y-axis. (451)

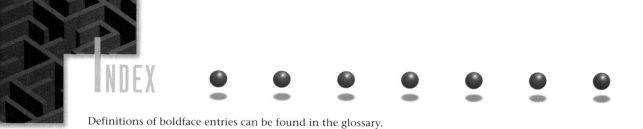

INDEX

Definitions of boldface entries can be found in the glossary.

Absolute error, 414
Absolute value, 74–75
  distance and, 75, 416
  equations, 416
  function, 559–560, 567
  inequalities, 417–418
Absolute-value function, 559–560, 567
Academics, 175, 484, 607
Acute angle, 206
Acute triangle, 218
Addition Property of Equality, 294
Addition Property of Inequality, 309, 314
Addition Property of Zero, 93
Additive identity, 93
Adjacent angles, 213
Advertising, 55, 262
Age, 523
Agriculture, 167, 238, 255, 362, 503, 620
Algebraic expression(s), 17, 274, 328
  adding, 275–277
  dividing, 332–333
  evaluating, 328–330
  multiplying, 330
  subtracting, 281–283
Algebraic logic, 51
Alternate exterior angle, 217
Alternate interior angle, 217
Altitude, 664, 665
Angle(s), 205, 206–207, 211
  acute, 206
  adjacent, 211
  complementary, 211–213
  congruent, 206
  corresponding, 216
  exterior, 216
    alternate, 217
  interior, 216
    alternate, 217
    consecutive, 217
  obtuse, 207
  right, 206
  straight, 206

Angle(s) *(cont.)*
  supplementary, 211–213
  vertex of, 205
  vertical, 211
Applications
  Business and Economics
    Advertising, 55, 262
    Agriculture, 167, 238, 255, 362, 503, 620
    Architecture, 268
    Aviation, 291
    Banking, 102, 124, 310, 311, 312, 351, 543
    Business, 29, 84, 97, 160, 195, 615, 688
    Carpentry, 147, 218, 338, 343, 402
    Construction, 209, 237, 238, 268, 642, 649, 655, 656, 663, 668, 672, 675
    Discounts, 350, 362, 368, 372
    Economics, 589, 614
    Engineering, 231, 441
    Forestry, 240, 394
    Fund-raising, 35, 284, 293, 356, 394, 396, 398, 400, 441, 460, 464, 465, 485, 499, 509, 516, 535, 555
    Income Tax, 371
    Inventory, 62, 285
    Investments, 175, 295, 509, 514, 516, 523, 540, 542, 569, 580
    Manufacturing, 144, 160, 414, 651, 659, 663, 669, 674, 688
    Metalworking, 158
    Rental, 119, 484, 569
    Sales, 27, 61, 175, 324, 503, 522, 529, 535, 615
    Sales Tax, 297, 369, 388
    Savings, 312, 355
    Small Business, 335, 398, 399, 407, 496, 608
    Stocks, 130, 132, 135, 151, 352, 592, 633
    Wages, 61, 62, 69, 140, 175, 334, 394, 395, 398, 485, 616
  Language Arts
    Communicate, 12, 20, 27, 34, 41, 48, 54, 60, 76, 83,

Applications *(cont.)*
    90, 96, 101, 108, 117, 134, 139, 146, 152, 159, 166, 174, 183, 193, 208, 213, 220, 230, 236, 244, 253, 260, 277, 283, 289, 296, 304, 311, 317, 333, 342, 349, 355, 361, 370, 387, 393, 398, 405, 411, 419, 432, 438, 447, 456, 463, 469, 476, 483, 502, 508, 515, 521, 528, 542, 547, 554, 561, 567, 574, 590, 597, 605, 613, 621, 626, 640, 647, 655, 661, 668, 673, 679
    Eyewitness Math, 176, 336, 442, 556
    Language Arts, 470
  Life Skills
    Age, 523
    Auto Maintenance, 145, 147, 363
    Baking, 160
    Consumer Awareness, 22
    Consumer Economics, 35, 145, 147, 312, 367, 356, 362, 384, 388, 413, 457, 491, 548, 562, 569, 621, 627, 632, 641
    Cooking, 167, 297, 364
    Education, 632
    Fire Prevention, 591
    Health, 362, 419, 431, 607
    Home Improvement, 54
    Nutrition, 503
  Other
    Academics, 175, 484, 607
    Cartography, 262
    Communication, 614
    Graphic Arts, 258
    Photography, 110, 153, 262, 278
    Time, 391
  Science
    Astronomy, 569
    Biology, 506, 598, 606, 608
    Chemistry, 397, 399, 424, 448, 485, 503, 509, 518, 522, 523, 530, 555, 592
    Criminology, 175
    Ecology, 32, 448

Face, 650
Factor(s)
    of a natural number, 36
    of an expression, 274
    of polynomials, 345
Fibonacci, Leonardo, 129
Fire Prevention, 591
First differences, 112
Forestry, 240, 394
Formulas, 295
Fractals, 8
Fraction(s), 130
    bar, 131
    comparing, 132
    denominator of, 131
    dividing, 158
    equivalent, 131, 133–136
    improper, 150
    least common denominator of, 138
    lowest terms, 136–137
    multiplying, 154–155
    negative, 151
    numerator of, 131
    ordering, 132
    reciprocals of, 157
    unit, 135, 140
    with like denominators, 149
    with unlike denominators, 150
    written as decimals, 141–143
Frequency table, 596
Function, 431, 536–575
    absolute value, 559–560, 567
    constant, 467
    domain of, 431
    exponential, 538–541, 567
    integer, 560–561, 567
    linear, 430–431, 567
    parent, 571
    piecewise, 566
    quadratic, 544–547, 567
    range of, 431
    rate of change of , 447
    reciprocal, 550–553, 567
Fund-raising, 35, 284, 293, 356, 394, 396, 398, 400, 441, 460, 464, 465, 485, 499, 509, 516, 535, 555

Games, 178, 184
Genetics, 188
Geography, 91, 590, 622
Golden ratio, 129
Golden rectangle, 129
Government, 362, 372, 420, 424
Graphic Arts, 258
Graphics calculator, 11, 22, 26, 28, 44, 51, 157, 229, 238, 279, 303, 383, 396, 399, 403, 431, 461, 464, 470, 473, 474–475, 476, 498, 501, 518, 551, 560, 566, 571, 572, 573, 595, 605
    ABS, 75, 77
    graphing linear functions, 445–446
    graphing linear inequalities, 484, 527–528
    INT, 181
    making tables with, 106, 107, 161, 287, 305, 306, 329, 352, 402, 458, 497, 649, 657, 669, 675, 681, 682
    RAND, 181
    regression line, 453
    solving inequalities with, 310–311
Great circle, 676
Greatest common factor (GCF), 137
Greatest integer function (See *integer function*.)

Health, 362, 419, 431, 607
Height of a parallelogram, 239
Height of a trapezoid, 243
Heptagon, 228
Hexagon, 224
    regular, 225
Histogram, 603
Home Improvement, 54

Horizontal line, 468
Hobbies, 324, 350, 394, 613
Hypotenuse, 251

Image, 570
Improper fraction, 150
Income Tax, 371
Independent events, 188–191
Independent variable, 431
Inequality, 88, 307
    absolute value, 417–418
    graphing, 308, 315–316
    solving, 314
    statements of, 314
    systems of, 524–528
Integer function, 181, 560–561, 567
Integers, 72
    adding, 81–82
    neutral pairs of, 78
    number-line model for, 74
    opposites, 74
    ordering, 88–89
    subtracting, 93–95
    tile models for, 78
Interior angle, 216
Inventory, 62, 285
Investments, 175, 295, 509, 514, 516, 523, 540, 542, 569, 580

Lateral face, 650
Lateral surface area, 652
Least common denominator (LCD), 138
Legs, 251
Life Science, 385, 563
Lifestyles, 166
Like terms, 274
Line(s), 204
    graphs, 588
    intersecting, 205
    of symmetry, 570
    perpendicular, 205

Line(s) *(cont.)*
  parallel, 205, 217, 404
  transversal, 216–217
Line graph, 588
Line of best fit, 624
Linear equation(s), 105
  graphing, 105–106
  point-slope form of, 461, 463
  slope-intercept form of, 451, 463
  standard form of, 459, 463
  systems of, 497–501
  table of values for, 105–106
  using differences to identify, 115–117
  writing, 107
Linear function, 428–485
Linear inequality, 478
  boundary line, 478
  graphing, 478–482
Linear relationship, 540
Literal equation, 295, 404–405
Lowest terms, 137

Manufacturing, 144, 160, 414, 651, 659, 663, 669, 674, 688
Math Connections
  Coordinate Geometry, 470, 560, 575
  Geometry, 11, 13, 14, 15, 28, 41, 55, 84, 118, 133, 140, 141, 147, 155, 159, 160, 167, 185, 277, 278, 279, 290, 295, 297, 298, 306, 325, 334, 339, 342, 344, 347, 349, 350, 361, 363, 372, 376, 388, 394, 400, 401–404, 406, 413, 416, 425, 439, 441, 458, 471, 473, 474, 476, 477, 485, 509, 523, 544, 548, 555, 600, 617, 627, 642, 649, 657, 681
  Maximum Minimum, 233, 234, 238, 415, 646, 648, 649
  Probability, 168, 188, 372, 642
  Statistics, 35, 55, 100, 102, 144, 148, 174, 185, 284, 285,

Math Connections *(cont.)*
  294, 297, 363, 372, 376, 386, 387, 394, 425, 558, 618, 622, 663
Measures of central tendency, 594, 595
  mean, 594, 595
  median, 594, 595
  mode, 594, 595
Mean, 595
Means, 358
Median, 595
Metalworking, 158
Misleading graphs, 587
Mixed number, 142
  dividing, 158
  multiplying, 156–157
Mode, 595
Monomial, 299
Mozart, Wolfgang Amadeus, 624
Multiplication and Division Properties of Inequality, 410
Multiplication Property of Equality, 340
Multiplicative inverse, 353
Multiplicative Property of −1, 331
Music, 554, 624

Natural numbers, 36
Net, 643
Neutral pair, 281
Number line, 74
  distance between points, 95
Numbers
  composite, 38–39
  integers, 72
  natural, 36
  pentagonal, 175
  prime, 38–39
  random, 181–182
  rational, 128, 162, 351
  rectangular, 11
  square, 38–39
Numerator, 131
Nutrition, 503

Oblique cone, 664
Oblique pyramid, 665
Obtuse angle, 207
Obtuse triangle, 218
Octagon, 228
Opposites, 74
  property of, 82
Order of operations, 50, 51
Ordered pair, 104
  graphing, 104
Origin, 104, 429
Outcome, 186

Parabola, 546
  vertex of, 546
Parallel lines, 473
Parallelogram, 225
  area of, 239–240
  rectangle, 226
  rhombus, 227
  square, 227
Parent function, 571
Pascal's Triangle, 15
Patterns
  differences in, 112–115
  fractal, 8
  number, 8
Pentagon, 224
  regular, 225
Percent(s), 169, 365
  estimating with, 173–174
  writing as decimals and fractions, 169–171, 366
  writing fractions and decimals as, 172–173
Perfect square, 249, 232–235
Perimeter, 232–235
  of a rectangle, 232–233, 401–402, 404–405
  of a square, 232
Perpendicular lines, 205, 474–475
Photography, 110, 153, 262, 278

Rectangular numbers, 11
Rectangular solid, 645
   bases of, 645
   faces of, 645
   surface area of, 645–646
   volume of, 646–647
Reflection, 570, 573
Regression line, 453
Regular polygon, 225
Relation, 540
Rental, 119, 484, 569
Repeating decimal, 142
Rhombus, 227
Right angle, 206
Right cone, 664
Right cylinder, 658
Right pyramid, 665
Right triangle, 218
Rise, 434
Rotation, 571
Run, 434

Sales, 27, 61, 175, 324, 503, 522,
   529, 535, 615
Sales Tax, 297, 369, 388
Sample space, 187
Savings, 312, 355
Scale factor, 256
Scalene triangle, 218
Scatter plot, 618–621
   correlation, 618–621
   line of best fit, 624–626
Science, 49
Scuba Diving, 76
Second differences, 112
Segment, 204
   congruent, 218
Sequence
   finding differences, 112–117
   number, 10–11
   terms of, 10
Similar figures, 258, 360
Slant height, 670
Slope, 434–435
   as a difference ratio, 437
   as a rate of change, 447

Slope *(cont.)*
   finding from two points,
     436
   formula, 452
   rise, 434–435
   run, 434–435
   sign of, 435–436
Slope formula, 452
Slope-intercept form, 451
Small Business, 335, 398, 399, 407,
   496, 608
Social Studies, 465, 588
Sphere, 676
   circumference of, 676
   great circle of, 676
   surface area of, 676–677
   volume of, 678
Sports, 83, 84, 92, 124, 149, 167,
     168, 192, 231, 297, 318,
     340, 355, 356, 399, 425,
     452, 485, 523, 534, 568,
     598, 599, 600, 606, 607,
     623, 642
Spreadsheet, 12, 115, 118, 153,
     181–182, 229, 235,
     260, 261, 328, 421, 539,
     541, 558, 563, 600, 626,
     682
Square, 227
Square number, 39
Square root, 248
   principal, 249
Statistics, 584
Standard form, 459
Stem-and-leaf plot, 601–603
Stocks, 130, 132, 135, 151, 352,
   592, 633
Straight angle, 206
Student Government, 362, 394
Substitution method, 504–507
Subtraction, 95
Subtraction Property of
   Equality, 293
Subtraction Property of
   Inequality, 309, 314
Surface area, 643–647
   of cubes, 644–645
   of cylinders, 658–659, 661
   of prisms, 651–653
   of rectangular solids, 645–646
   of spheres, 676–677

Symbols of inclusion, 51
Systems of equations, 496
   solving by elimination,
     510–514, 517–521
   solving by graphing, 496–501
   solving by substitution method,
     504–507
System of linear inequalities,
   524

Temperature, 73, 96, 318, 472
Terms, 274, 299
   like, 274, 299, 300
   unlike, 299
Theater, 398, 516
Time, 391
Transformation, 570–573
   image, 570
   pre-image, 570
   reflection, 570, 573
   rotation, 571
   translation, 570, 571–573
Translation, 570, 571–573
Transversal, 216–217
Trapezoid, 227, 243
   area of, 243
   bases of, 243
   height of, 243
   isosceles, 227
Travel, 29, 297, 329, 343, 439, 490,
   555, 640
Triangle, 218, 403
   acute, 218
   angles of a, 219
   area of, 241–242
   base of, 241–242
   equilateral, 218
   exterior angle of, 223
   height of, 241–242
   isosceles, 218
     base angles of, 219
   obtuse, 218
   remote interior angles of, 223
   right, 218
   scalene, 218
   sides of a, 219–220
Trinomial, 299

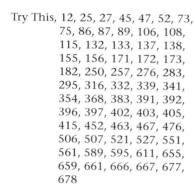

CREDITS

Photos

Abbreviations used: (t) top, (c) center, (b) bottom, (l) left, (r) right, (bckgd) background, (bdr) border.

FRONT COVER: (l), Jake Rajs/Photonica; (c), Ralph Wetmore/Tony Stone Images; (r), Erik Aeder/Pacific Stock.

TABLE OF CONTENTS: Page iv(tr), Dennis Fagan/HRW Photo; iv(bckgd), Art Matrix; iv(br), Courtesy of NASA; v(tr), Michelle Bridwell/HRW Photo; v(br), vi(tr), Michael Young; vi(br), Art Gingert/Comstock; vii(br), Sam Dudgeon/HRW Photo; viii(tr) T. Zimmerman/FPG International; viii(cl), Sam Dudgeon/HRW Photo; viii(br), Michael Young; 1(tr), Dennis Fagan/HRW Photo; 1(cr), Sam Dudgeon/HRW Photo.

CHAPTER ONE: Page 2(cr), (bl) Jerry Jacka; 2-3(bckgd), Steve Vidler/Nawrocki Stock Photo, Inc.; 3(tr),(br), Michelle Bridwell/HRW Photo; 3(c), Jim Newberry/HRW Photo; 4(t), Scott Van Osdol/HRW Photo; 5(tl), Fred Griffin; 5(cr) The Stock Market; 6(br), Scott Van Osdol/ HRW Photo; 7(t), Flip McCririck; 8(c), Art Matrix; 8(cl), Dennis Fagan/HRW Photo; 9(tl),(tr), (cr), Michelle Bridwell/HRW Photo; 12-13(b), Dennis Fagan/HRW Photo; 13(tr), Jonathan Daniel/AllSport; 14(tl), The Bettmann Archive; 15(tr), Alex Bartel/FPG International; 16(tl),18(br), Michelle Bridwell/HRW Photo; 20(br), Steve Ferry/HRW Photo; 21(c), Sam Dudgeon/HRW Photo; 22-23(b), 24(t), Michelle Bridwell/HRW Photo; 26(cl), 28-29(bckgd), Steve Ferry/HRW Photo; 30(c), Sam Dudgeon/HRW Photo; 32(t), Advanced Satellite Productions; 33(t), Sam Dudgeon/HRW Photo; 34(tr), Dennis Fagan/HRW Photo; 36(tl),41(t), Sam Dudgeon/HRW Photo; 43(tr), Courtesy of NASA; 46(t), Steve Ferry/HRW Photo; 48-49(c), Courtesy of NASA; 50(t),(br), Michelle Bridwell/HRW Photo; 51(bl), 53(t), 55(c), Sam Dudgeon/HRW Photo; 56(t), 59(bl), Michelle Bridwell/HRW Photo; 60(br), Steve Ferry/HRW Photo; 62-63(b), Michelle Bridwell/HRW Photo; 64-65(bckgd), The Ninepatch Quilt/The Quilt Complex/Courtesy of the Espirit Quilt Collection. **CHAPTER TWO**: Page 70(l), Stuart Westmorland/Tony Stone Images; 71(r), Alaska Stock Images; 73(tr), Galen Rowell; 76(rc), Tom Campbell/Adventure Photos; 77(tl), (bckgd), SuperStock; 78(t), Fred Bavendam/Peter Arnold, Inc.; 78(bl), (bc), (br), 79(cr), Sam Dudgeon/HRW Photo; 79(r), Michelle Bridwell/HRW Photo; 81(b), Michael Young/HRW Photo; 85(tl), (tr), SuperStock; 87(cr), Michael Young/HRW Photo; 90(r), 92(l), Sam Dudgeon/HRW Photo; 93(t), Sam Dudgeon/HRW Photo; 94(bl), Michelle Bridwell/HRW Photo; 95(b), Sam Dudgeon/HRW Photo; 96(tr), 98(t), Michelle Bridwell/HRW Photo; 100(cl), (cr), David Phillips/HRW Photo; 101(tr), Sam Dudgeon/HRW Photo; 102(tl), Mike Valeri/FPG International; 103(t), (inset), Michelle Bridwell/HRW Photo; 106(bl), Michael Young/HRW Photo; 108(b),110(tl), Michelle Bridwell/HRW Photo; 110(c), John Langford/HRW Photo; 112(tl), Dennis Fagan/HRW Photo; 112-113(tr), Warren Faidley/Weatherstock; 114(c), Sam Dudgeon/HRW Photo; 116(t), Michelle Bridwell/HRW Photo; 118-119(bckgd), Courtesy of NASA; 120(c), Michael Young/HRW Photo. **CHAPTER THREE**: Page 128(c),129(t), Michelle Bridwell/HRW Photo; 128(bckgd), Michael Young/HRW Photo; 128(bl) Sam Dudgeon/HRW Photo; 130(t), (bckgd), Nawrocki Stock Photo, Inc. 130(c), (bckgd), SuperStock; 130(l), Michael Young/ HRW Photo; 133(all photos), Sam Dudgeon/HRW Photo; 134(tl), Michelle Bridwell/HRW Photo; 136(t), Michael Young/HRW Photo; 140(b), Grey Peace/Tony Stone Images; 141(all photos),142-143(bckgd), 144(bl), Sam Dudgeon/HRW Photo; 145(cl), 146(t), Michael Young/HRW Photo; 148(tr),(c), Michelle Bridwell/HRW Photo; 149(tr), Peter Van Steen/HRW Photo; 151(bl), Michael Young/HRW Photo; 152(b), Jon Riley/Tony Stone Images; 153(t), Sam Dudgeon/HRW Photo; 154(bl), Lowell Bridwell/Birdseye Ltd.; 158(bl), Peter Van Steen/HRW Photo; 160(bl), Michelle Bridwell/HRW Photo; 162(tr), Peter Van Steen/HRW Photo; 165(bl), Mark E. Gibson/The Stock Market; 167(br), Philip Bailey/The Stock Market; 168(tr), Ken Levine/AllSport; 168(br), Sam Dudgeon/HRW Photo; 169(cl), Michelle Bridwell/HRW Photo; 171(bl), Michael Young/HRW Photo; 174(tr), David Young Wolf/PhotoEdit; 176(r) Ron Chapple/FPG International; 177(bckgd), J. Zimmerman/FPG International; 178(t), David Phillips/HRW Photo; 178(b), 179(tr), Sam Dudgeon/HRW Photo; 180(tl), Michelle Bridwell/HRW Photo; 180(br), Erich Lessing/Art Resource; 183(r), John Langford/HRW Photo; 184(tr,(bl), 185(br), Sam Dudgeon/HRW Photo; 186(t), Barbara Campbell/Liason International; 187(all photos), 189(b), 190, Michael Young/HRW Photo; 192(cr), Sam Dudgeon/HRW Photo; 193(tr), Michelle Bridwell/HRW Photo; 195(b), Hubbel/HRW Photo; 196(bckgd), Christopher Weil/Photonica; 197(c), (b), Michael Young/HRW Photo. **CHAPTER FOUR**: Page 202(b), Eric Crossan; 202-203(bckgd), Larry Lee/Westlight; 203(t), Michelle Bridwell; 203(b), Sam Dudgeon/HRW Photo 204(t), 205(tl), Michael Young/HRW Photo; 205(b), Sam Dudgeon/HRW Photo: 206(t), 207(b), Stephanie Workman/HRW Photo; 208(t), Steve Ferry/HRW Photo; 211(t), Phillippe Plailly/Science Photo Library/Photo Researchers, Inc.; 211(b), Sam Dudgeon/HRW Photo; 213(br), Michael Young/HRW Photo; 214(bckgd), 215(t), Joe Jaworski; 216(t), Chuck Pefley/Tony Stone Images; 214(tl), 219(tl), Steve Ferry/HRW Photo; 220(tl), Michael Young/HRW Photo; 224(tl), Michelle Bridwell/HRW Photo; 224(tr), Treat Davidson/Photo Researchers, Inc.; 230(tr), Peter Van Steen/HRW Photo; 231(tr), 232(t), 233(bl), 234(b), Michelle Bridwell/HRW Photo; 232(all photos), Sam Dudgeon/HRW Photo; 237(tr), Michelle Bridwell/HRW Photo; 238(tl),(br), Robert J. Wolf; 239(t), Boyd Norton; 240(br), Peter Van

Steen/HRW Photo; 242(bl), 243(br), Michael Young/HRW Photo; 244(tr), Comstock; 248(t), Michelle Bridwell/HRW Photo; 249(br), Sam Dudgeon/HRW Photo; 253(tr), 254(b), Steve Ferry/HRW Photo; 256(t), Dallas & John Heaton/Westlight; 256(b), Tom Bean; 257(b), Sam Dudgeon/HRW Photo; 258(t), Pierre Capretz/HRW Photo; 259(cl),(b), Robert J. Wolf; 261(all photos), Michelle Bridwell/HRW Photo; 262(t),(c), Steve Ferry/HRW Photo, 262(bl), Map Art/Cartesia Software; 264-265(t), Shogoro/Photonica; 265(b), Michelle Bridwell/HRW Photo. **CHAPTER FIVE**: Page 272(bckgd)(l), Martin Rogers/Tony Stone Images; 272(b), SuperStock; 273(bckgd)(l), Alexander Calder Mobil: Red Dragon(1963)/Art Resource/New York; 273(c), Ray Massey/Tony Stone Images; 273(bckgd)(br); SuperStock; 274(t), 280(t), 283(b), Michelle Bridwell/HRW Photo; 286(t), Art Gingert/Comstock; 287(insert), Robert J. Wolf; 287(b), Sam Dudgeon/HRW Photo; 290(b), 291(t),(b), Robert J. Wolf; 292(tr), 293(tr), Michelle Bridwell/HRW Photo; 294(cl),(br), 296(c), Sam Dudgeon/HRW Photo; 302(l), Michelle Bridwell/HRW Photo; 303(r), Sam Dudgeon/HRW Photo; 304(tr), 307(t), Michelle Bridwell/HRW Photo; 311(tr), Peter Van Steen/HRW Photo; 313(t), 314(l), 317(t), Michelle Bridwell/HRW Photo; 320(bckgd), Matthew Neal McVay/Tony Stone Images; 321(tr),(cr), Bob Dammerich; 321(c),(br), Sam Dudgeon/HRW Photo. **CHAPTER SIX**: Page 326(b), 327(c), SuperStock; 326(b), 327(c), Scott Van Osdol/HRW Photo; 328(tr), Michelle Bridwell/HRW Photo; 329(br), 331(tr), Sam Dudgeon/HRW Photo; 333(r), Michelle Bridwell/HRW Photo; 334(bl), 335(tl), Sam Dudgeon/HRW Photo; 335(br), David Philips/HRW Photo; 336(l), Sam Dudgeon/HRW Photo; 336(bckgd), 337(bckgd), Ron Sanford/Tony Stone Images; 338(tr), Michelle Bridwell/HRW Photo; 340(tl), Bob Daemmrich; 342(r), Dennis Fagan/HRW; 343(bckgd), John W. Warden/SuperStock; 343(r), Sam Dudgeon/HRW Photo; 344(bckgd), Philip Habib/Tony Stone Images; 345(r), Dennis Fagan/HRW Photo; 346(b), 347(b), 349(tr), 351(t), 352(tr), 353(c), 355(br), Sam Dudgeon/HRW Photo; 355 (tr), Dennis Fagan/HRW Photo; 356(tl), Sam Dudgeon/HRW Photo; 356(c), Larry Stevens/Nawrocki Stock Photo; 357(tr), Michelle Bridwell/HRW Photo; 360(tl), Michael Young/HRW Photo; 362-63(t), Michelle Bridwell/HRW Photo; 363(b), Robert J. Wolf; 364(tl), Sam Dudgeon/HRW Photo; 365(tl), Colin Pyor/Tony Stone Images; 366(br), Dennis Fagan/HRW Photo; 367(r),369(tl), Michelle Bridwell/HRW Photo; 370(tr), Dennis Fagan/HRW Photo; 373(l), Egyptian Equation Solving/Image #27, Limestone Dynasty XII, c. 1900 B.C./Planet Art Classic Graphics, Ancient Egypt; 373(r), Courtesy of the National Council of Teachers of Mathematics. The Rhind Paprus by Arnold Buffum Chase International. **CHAPTER SEVEN**: Page 380(r), Michael Rosenfeld/Tony Stone Images; 380(b), 381(t), Peter Van Steen/HRW Photo; 382(t), 384(t), Michelle Bridwell/HRW Photo; 385, Robert & Linda Mitchell; 386(tr), 387(tr), Dennis Fagan/HRW Photo; 388(r), SuperStock; 389(r), 390(bl), Michelle Bridwell/HRW Photo; 393(br), Peter Van Steen/HRW Photo; 394(bckgd), Sam Dudgeon/HRW Photo; 395(t), Peter Van Steen/HRW Photo; 397(cl), Sam Dudgeon/HRW Photo; 398(tr), Peter Van Steen/HRW Photo; 399(br), Sam Dudgeon/HRW Photo; 400(tr), 401(t), 405(br), Peter Van Steen/HRW Photo; 407(tr), Steve Ferry/HRW Photo; 408(c), John Terrence Turner/FPG International; 411(tr), Marie Taglient/The Image Bank; 412(bl), The VNR Concise Encyclopedia of Mathematics/Van Nostrand Reinhold Co.; 412(br), Photri; 414(t), Mark Segal/Tony Stone Images; 416(t), Dennis Fagan/HRW Photo; 419(bl), David Philips/HRW Photo; 420(tr), Robert Daemmrich/Tony Stone Images; 421(t), Peter Van Steen/HRW Photo; 421(b), Tony Stone Images. **CHAPTER EIGHT**: Page 426-427(bckgd), Hugh Sitton/Tony Stone Images; 426(c), Robert Harding Associates, London; 426-427(b), John Langford/HRW Photo; 428(t), Sam Dudgeon/HRW Photo; 428-429(t), Printed with the permission of the State of Minnesota; 430(t), Sam Dudgeon/HRW Photo; 431(cl), Jay Thomas/International Stock; 432(tr), Dennis Fagan/HRW Photo; 433(bl), Kathleen Campbell/Allstock; 433(tr), Sam Dudgeon/HRW Photo; 434(all), Michelle Bridwell/HRW Photo; 439(tr), Michelle Bridwell/HRW Photo; 439(br), Lowell Bridwell/Birdseye, Ltd; 440(tl), Tony Stone Images; 442(bckgd), Leo Balteman/FPG International; 444(c), Joe Towers/George Hall/Check Six; 446(bl), Sam Dudgeon/HRW Photo; 448(tr), Dennis Fagan/HRW Photo; 448(bl), SuperStock; 450(tr), Steven E. Sutton/Duomo; 452-453(b), J. Zimmerman/FPG International; 453(t), Richard Dole/Duomo; 454(cl), 455(cr), Michelle Bridwell/HRW Photo; 456(tr), Jim Newberry/HRW Photo; 457 (br), Michelle Bridwell/HRW Photo; 458(tl), Sam Dudgeon/HRW Photo; 459(t), Shooting Star Photo Agency; 460(tl), Michelle Bridwell/Frontera Fotos; 462(cl), Sam Dudgeon/HRW Photo; 463(br), Dennis Fagan/HRW Photo; 464(cr), Michelle Bridwell/Frontera Fotos; 464-465(bckgd), Bob Daemmrich; 466(cl)(tr), David Madison; 469(bl), Dennis Fagan/HRW Photo; 470(br), Sam Dudgeon/HRW Photo; 472(tl), SuperStock; 472(tr), Michael Keller/FPG International; 474(r), Sam Dudgeon/HRW Photo; 478(t), Steve Ferry/HRW Photo; 479(br), Sam Dudgeon/HRW Photo; 483(tr), Steve Ferry/HRW Photo; 485(tr), Sam Dudgeon/HRW Photo; 486(tr), The Bettman Archive; 486-487(bckgd), Carl Yarbrough; 487(tr), Chris Marona. **CHAPTER NINE**: Page 494-495, Robert E. Daemmrich/Tony Stone Images; 494(c), Michelle Bridwell/HRW Photo; 495, Michael Young/HRW Photo; 495(b), C. Aurness/Westlight; 496(tr), Peter Van Steen/HRW Photo; 498-499(c), Michelle Bridwell/HRW Photo; 500-501(b), Robert J. Wolf; 503(cr), Sam Dudgeon/HRW Photo; 504(t), 505(br), Michelle Bridwell/HRW Photo; 506-507(b), Renee Lynn/Davis /Lynn Images; 508(tr), 510(c), Michael Young/HRW Photo; 514(tl), Peter Van Steen/HRW Photo; 516(tl), Sam Dudgeon/HRW Photo; 517(t), David Madison; 521(r), Steve Ferry/HRW Photo; 523(tr), Peter Van

Steen/HRW Photo; 523(tr), Sam Dudgeon/HRW Photo; 524(t), Peter Van Steen/HRW Photo; 525(bl), Mark C. Burnett/Photo Researchers,Inc.; 528(br), Michael Young/HRW Photo; 531(tl), (c1), (br), John Langford/HRW Photo; 531(bl), H. Friedman/HRW Photo; 531(tr), Steve Ferry/HRW Photo; 531(cr) Superstock. **CHAPTER TEN**: Page 536(bl), 536-537(bckgd), Sam Dudgeon/HRW Photo; 537(tr), Michelle Bridwell/HRW Photo; 538(t), 539(bl), Dennis Fagan/HRW Photo; 540(br), Phil Shermiester/Allstock; 542(tr), Dennis Fagan/HRW Photo; 543(tr), Harold E. Edgerton/Palm Press, Inc.; 544(bl), Sam Dudgeon/HRW Photo; 545(cb), Michelle Bridwell/HRW Photo; 547(r), Dennis Fagan/HRW Photo; 548(b), Michelle Bridwell/HRW Photo; 549(r), Laurence Parent; 550(l), Michelle Bridwell/HRW Photo; 552(l), Sam Dudgeon/HRW Photo; 553(tr), Tony Freeman/PhotoEdit; 554(tr),(bl), Sam Dudgeon/HRW Photo; 558(t), 563(r), Michelle Bridwell/HRW Photo; 564(t), Richard T. Bryant/The Stock Source/Atlanta; 565(c), Michelle Bridwell/HRW Photo; 568(br), Robert E. Daemmrich/Tony Stone Images; 569(br), Babylonian Sillabary, 442 B.C., British Museum/Art Resource, N.Y.; 570(t),(br), Jeff Lepore/Photo Researchers, Inc.; 570(cr),(bl), T. Zimmermann/FPG International; 570(tl), 571(tr), Vandystadt/AllSport; 574(tr), Michael Young/HRW Photo; 576(bckgd), Index Stock Photography; 576(r), Nawrocki Stock Photography, Inc. **CHAPTER ELEVEN**: Page 584(l), Michael Young/HRW Photo; 586-587(c), SuperStock; 588(t), Franklin D. Roosevelt Library; 589(r), Bill Hubbell/HRW Photo; 590(tr), Michelle Bridwell/HRW Photo; 591(tl), Telegraph Colour Library/FPG International; 593(c), Michael Young/HRW Photo; 595(cl), Michelle Bridwell/HRW Photo; 597(cr), Michael Young/HRW Photo; 598(tr), Leonard Lee Rue/Animals Animals; 599(cr), William R. Saliaz/Duomo; 601(t), Steve L. Swartz; 602(b), Vince Streano/Tony Stone Images; 605(br), Michael Young/HRW Photo; 606(tr),

Superstock; 607(tr), Michael Young/HRW Photo; 608(tl), Flip Nicklin/Minden Pictures; 609(tr), Michelle Bridwell/HRW Photo; 610(all), Sam Dudgeon/HRW Photo; 611(tl), Michael Young/HRW Photo; 612(bl), Michelle Bridwell/HRW Photo; 614(all), Robert J. Wolf; 618(tl), Sam Dudgeon/HRW Photo; 623(tl), David Madison; 624(tr), Sam Dudgeon/HRW Photo; 626(br), Dennis Fagan/HRW Photo; 628(l), 629(b), Michelle Bridwell/HRW Photo; 629(t) Michael Young/HRW Photo. **CHAPTER TWELVE**: Page 634, Richard Kolar/Earth Scenes; 635(tl), David Rosenberg/Tony Stone Images; 635(r), Steven Gottlieb/FPG International; 635(inset), Alan G. Nelson/Earth Scenes; 635(bl), Schloss Ottis/Earth Scenes; 635(br), John Mead/SPL/Photo Researchers, Inc.; 636(bl), 637(tr), 639(tl), 639(tr), Steve Ferry/HRW Photo; 640(cr), Peter Van Steen/HRW Photo; 643(tr), Willard Clay/Tony Stone Images; 643(cl), David Austen/FPG International; 643(bl), Sam Dudgeon/HRW Photo; 645(tr), Nicholas Parfitt/Tony Stone Images; 645(cl), Sam Dudgeon/HRW Photo; 645(tl), G.C. Kelley/Photo Researchers, Inc.; 647(cr), Paul Freed/Animals Animals; 650(t), Peter Van Steen/HRW Photo; 651(bl), Sam Dudgeon/HRW Photo; 653(bl), 655(tr), 658(c), Peter Van Steen/HRW Photo; 659(tr), R. Ian Lloyd/Westlight; 661(br), 662(b) Peter Van Steen/HRW Photo; 664(t), Robert Adams/SuperStock; 665(all), 666(tl) Steve Ferry/HRW Photo/Cuisenaire Company of America; 667(l),(r), 669(all), Sam Dudgeon/HRW Photo; 670(tl), Robert Mihovil; 672(tl), 674(bl), Peter Van Steen/HRW Photo; 676(tl), Bob Taylor/FPG International; 677(br), Sam Dudgeon/HRW Photo; 678(l), Peter Van Steen/HRW Photo; 679(tr), Michelle Bridwell/HRW Photo; 680(b), Sam Dudgeon/HRW Photo; 681(tr), Tom Van Sant/The Geosphere Project/The Stock Market; 682(all), NASA/Science Source/Photo Researchers, Inc.; 683(b), Michael Young/HRW Photo.

Illustrations

Abbreviated as follows: (t) top; (b) bottom; (l) left; (r) right; (c) center. All art, unless otherwise noted, by Holt, Rinehart and Winston.

Design

Chapter Openers, Nishi Kumar; 2-3, 128-129, 202-203, 272-273, 326-327, 380-381, 426-427, 494-495, 536-537, 584-585, 634-635 Chapter Projects, Paradigm Design; 176, 197, 421, 442-443, 531, 556-557, 628-629, 682-683

Permissions

Grateful acknowledgment is made to the following sources for permission to reprint copyrighted material.

The Associated Press: From "Atlanta Architect Steps Up Quest for Safe Staircases" by Lauren Neergaard from the *Albuquerque Journal*, 1993. Copyright © by the Associated Press.

Alfred A Knopf, Inc.: Text and charts from *Jurassic Park*, by Michael Crichton. Copyright © 1990 by Michael Crichton.

Karol V. Menzie and Randy Johnson: From "Count Your Way to Stair Success" by Karol V. Menzie and Randy Johnson from "Homework" column from *The Baltimore Sun*, 1993. Copyright © by Karol V. Menzie and Randy Johnson.

The National Council of Teachers of Mathematics: From page 84 from *The Rhind Mathematical Papyrus*, translations by Arnold Buffum Chace. Published by The National Council of Teachers of Mathematics, 1979.

Newsweek, Inc.: From "Finding Order in Disorder" by Sharon Begley from Newsweek, December 21, 1987. Copyright © 1987 by Newsweek, Inc. All rights reserved.

The New York Times Company: From "'Hot Hands' Phenomenon: A Myth?" from *The New York Times*, April 19, 1988. Copyright © 1988 by The New York Times Company.

Springer-Verlag New York Inc.: From "What People Mean by 'The Hot Hand' and 'Streak Shooting'" (Retitled: "Researchers' Survey of 100 Basketball Fans") from "The Cold Facts About The 'Hot Hand' in Basketball" by Tversky, Amos, and Gilovich from *Change: New Directions for Statistics, and Computing*, vol. 2, no.1, 1989. Copyright © 1989 by Springer-Verlag New York, Inc.

ADDITIONAL ANSWERS

Lesson 1.0
Exploration, page 7

5. Each group shows a cyclist traveling 50 miles in 5 hours. In the first graph, the cyclist speeds up during the second hour, then stops during the third hour. In the second graph, the cyclist slows down during the second and fourth hours, and speeds up during the third and fifth hours. The third graph shows that the cyclist starts out quickly, rests during the second and fourth hours, and goes slowly during the fifth hour. During the third hour, the cyclist starts at one pace and then accelerates for the second half of the hour.

Lesson 1.1
Ongoing Assessment, page 10

Testing the conjecture gives the following results.

s	11	12	13	14
c	55	66	78	91

Communicate, page 12

1. $A-B; B-C; C-D; A-C; B-D; A-D$

2. The numbers 1 through 20 can be paired into 10 groups which equal 21 ($1 + 20$, $2 + 19$, $3 + 18$, etc.). Since $10 \cdot 21 = 210$, the sum of the numbers 1 through 20 equal 210.

3. Each term is 4 more than the previous term.

4. Each term equals the sum of the two preceding terms.

5. Strategies: Think of a simpler problem; Look for a pattern; Reason logically

from the pattern; Make a picture or diagram of the problem.

Extra Practice, page 694

4. 24, 27, 30; Each term is 3 more than the one before it.

5. 60, 55, 50; Each term is 5 less than the one before it.

6. 79, 128, 207; Each term is the sum of the two terms before it.

7. 15, 17, 19; Each term is 2 more than the one before it.

8. 42, 46, 50; Each term is 4 more than the one before it.

9. 58, 51, 44; Each term is 7 less than the one before it.

10. 58, 62, 66; Each term is 4 more than the one before it.

11. 10, 7, 4; Each term is 3 less than the one before it.

12. 43, 49, 55; Each term is 6 more than the one before it.

13. 13, 21, 34; Each term is the sum of the two terms before it.

14. 70, 80, 90; Each term is 10 more than the one before it.

15. 50, 57, 64; Each term is 7 more than the one before it.

Lesson 1.2
Communicate, page 20

1. A variable is a letter that represents a number. Variables are used to translate sentences into equations.

2. Using h for the number of hours, we may translate as follows: $5 plus $2 per hour, or $5 + 2 \cdot h$, or $5 + 2h$

3. If 6 is substituted for h in $5 + 2h$, we get $5 + 2(6)$, or $17. That means that

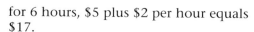

for 6 hours, $5 plus $2 per hour equals $17.

4. The cost of renting a go-cart is $5 plus $2 per hour.

5. A table is useful for finding a pattern in a related series of problems.

6. An expression for the cost of large-sized medals is $25 + 1.25m$. To find the cost of 100 large-sized medals substitute 100 for m: $25 + 1.25(100) = 150$.

Practice and Apply, **pages 21–23**

26.

Hours h	Cost: airport lot
0.5	$0.65
1.0	$1.30
1.5	$1.95
2.0	$2.60
2.5	$3.25
3.0	$3.90
3.5	$4.55
4.0	$5.20
4.5	$5.85
5.0	$6.50

27.

Hours h	Cost: private lot
0.5	$2.30
1.0	$2.60
1.5	$2.90
2.0	$3.20
2.5	$3.50
3.0	$3.80
3.5	$4.10
4.0	$4.40
4.5	$4.70
5.0	$5.00

28. The airport lot is cheaper when parking for 2.5 hours or less. The private lot is cheaper for parking 3 hours or more.

29.

x	$y = 2x$
1	2
2	4
3	6
4	8
5	10
10	20

30.

x	$y = x - 1$
1	0
2	1
3	2
4	3
5	4
10	9

31.

x	$y = x + 3$
1	4
2	5
3	6
4	7
5	8
10	13

32.

x	$y = 2x - 2$
1	0
2	2
3	4
4	6
5	8
10	18

33.

x	$y = 5x$
1	5
2	10
3	15
4	20
5	25
10	50

34.

x	$y = x + 2$
1	3
2	4
3	5
4	6
5	7
10	12

35.

x	$y = 3x - 3$
1	0
2	3
3	6
4	9
5	12
10	27

36.

x	$y = 2x + 1$
1	3
2	5
3	7
4	9
5	11
10	21

37.

x	$y = 4x$
1	4
2	8
3	12
4	16
5	20
10	40

38.

x	$y = 2x - 1$
1	1
2	3
3	5
4	7
5	9
10	19

39.

x	y = 3x + 6
1	9
2	12
3	15
4	18
5	21
10	36

40.

x	y = 4x − 2
1	2
2	6
3	10
4	14
5	18
10	38

41.

x	y = 3x
1	3
2	6
3	9
4	12
5	15
10	30

42.

x	y = 3x + 2
1	5
2	8
3	11
4	14
5	17
10	32

43.

x	y = 4x + 3
1	7
2	11
3	15
4	19
5	23
10	43

44.

x	y = −0.2x − 7
1	−7.2
2	−7.4
3	−7.6
4	−7.8
5	−8
10	−9

45.

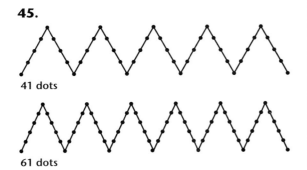

41 dots

61 dots

Extra Practice, **page 694**

13.

x	1	2	3	4	5	10
y	3	6	9	12	15	30

14.

x	1	2	3	4	5	10
y	5	10	15	20	25	50

15.

x	1	2	3	4	5	10
y	3	4	5	6	7	12

16.

x	1	2	3	4	5	10
y	0	1	2	3	4	9

17.

x	1	2	3	4	5	10
y	4	6	8	10	12	22

18.

x	1	2	3	4	5	10
y	3	7	11	15	19	39

19.

x	1	2	3	4	5	10
y	11	20	29	38	47	92

20.

x	1	2	3	4	5	10
y	15	30	45	60	75	150

21.

x	1	2	3	4	5	10
y	32	36	40	44	48	68

Lesson 1.3

Communicate, **page 27**

1. The sale price equals the original price times 0.60. Let *s* be the sale price, and let *p* be the original price. Then *s* = 0.60*p*.

2. Make a table of original prices at $10 increments with corresponding sale prices. Look for a sale price of $72.00.

Original price p	Sale price s = 0.60p
$80	$48.00
$90	$54.00
$100	$60.00
$110	$66.00
$120	$72.00

The original price is $120.

3. List the original prices from the table in Exercise 2 horizontally and sale prices vertically.

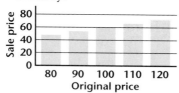

4. Find the original price corresponding to a sale price of $90. The original price should be $150.

5. Answers may vary. Example: A bowling alley rents lanes at $2 per hour in addition to a $4 initial charge. How many hours can a person rent a lane for $32?

Practice and Apply, **pages 28–29**

6.

x	2.7x
10	27
20	54
30	81
40	108

7.

x	3x − 2
10	28
20	58
30	88
40	118

8.

h	4.25h + 10
10	52.5
20	95
30	137.5
40	180

9.

c	12c − 4
10	116
20	236
30	356
40	476

14.

h	5 + 2.25h
1	$7.25
2	$9.50
3	$11.75
4	$14.00
5	$16.25

16.

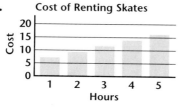

Cost of Renting Skates

19.

x	Process	Amount
0	2(40) + 2(40 + 0)	160
5	2(40) + 2(40 + 5)	170
10	2(40) + 2(40 + 10)	180
15	2(40) + 2(40 + 15)	190
20	2(40) + 2(40 + 20)	200
25	2(40) + 2(40 + 25)	210

22.

Time (hours)	Distance (miles)
1	25
2	50
3	75
4	100
5	125
6	150
7	175
8	200

29.

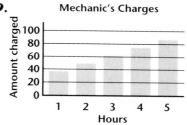

Mechanic's Charges

Extra Practice, **page 695**

1.

x	Process	Value
10	36(10)	360
20	36(20)	720
30	36(30)	1080
40	36(40)	1440

2.

x	Process	Value
10	2(10) + 5	25
20	2(20) + 5	45
30	2(30) + 5	65
40	2(40) + 5	85

3.

q	Process	Value
10	3(10) + 5	35
20	3(20) + 5	65
30	3(30) + 5	95
40	3(40) + 5	125

4.

x	Process	Value
10	15(10) − 2	148
20	15(20) − 2	298
30	15(30) − 2	448
40	15(40) − 2	598

5.

w	Process	Value
10	6(10) + 55	115
20	6(20) + 55	175
30	6(30) + 55	235
40	6(40) + 55	295

6.

x	Process	Value
10	8(10) − 6	74
20	8(20) − 6	154
30	8(30) − 6	234
40	8(40) − 6	314

7.

w	Process	Value
10	7(10) − 3	67
20	7(20) − 3	137
30	7(30) − 3	207
40	7(40) − 3	277

8.

x	Process	Value
10	10(10) + 13	113
20	10(20) + 13	213
30	10(30) + 13	313
40	10(40) + 13	413

9.

x	Process	Value
10	8(10) + 15	95
20	8(20) + 15	175
30	8(30) + 15	255
40	8(40) + 15	335

10.

x	Process	Value
10	28(10) − 1	279
20	28(20) − 1	559
30	28(30) − 1	839
40	28(40) − 1	1119

11.

a	Process	Value
10	500 − 6(10)	440
20	500 − 6(20)	380
30	500 − 6(30)	320
40	500 − 6(40)	260

12.

x	Process	Value
10	100 − 2(10)	80
20	100 − 2(20)	60
30	100 − 2(30)	40
40	100 − 2(40)	20

Lesson 1.4

Ongoing Assessment, **page 30**

$7\frac{1}{2}$ hours

Communicate, **page 34**

1. A *variable* is a letter (or other symbol) that can be replaced by any number (or other expression). An *expression* is a combination of numbers and variables using operations such as addition, subtraction, multiplication, and division. An *equation* is two equivalent expressions connected by an equal sign.

2. Let Phil's apples be p. Since Jeff has 3 more than Phil, let Jeff's apples be $p + 3$. Together they have $p + p + 3$, or $2p + 3$ apples, which is equal to the number 9. Equation: $2p + 3 = 9$.

3. "12 pencils cost $1.92" becomes "12 times the cost of 1 pencil equals $1.92." Represent this with the equation $12 \cdot p = 1.92$ or $12p = 1.92$.

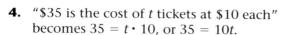

4. "$35 is the cost of t tickets at $10 each" becomes $35 = t \cdot 10$, or $35 = 10t$.

5. Answers may vary. Example: Guess $x = 100$, then $10(100) + 3 = 1003$ too large; Guess $x = 50$, then $10(50) + 3 = 503$ too small; Guess $x = 51$, then $10(51) + 3 = 513$, so the solution is 51.

Practice and Apply, **pages 34–35**

13. Begin with 6, and repeatedly add 2.

14. Begin with 15, and repeatedly add 10.

15. Begin with 100, and repeatedly subtract 10.

16. Begin with 52, and repeatedly subtract 4.

21. Beatrice made a series of guesses and checks, adjusting her answer until arriving at $x = 19$.

26. $a = 3.3$; The exact answer does not solve the problem because you would not buy a fraction of an apple. Therefore, the answer is 3 apples.

Look Back, **page 35**

32. 3 cups pecans or walnuts; $1\frac{1}{2}$ cups dates; $2\frac{1}{4}$ cups sifted all-purpose flour; 9 eggs; $4\frac{1}{2}$ cups brown sugar; $2\frac{1}{4}$ tsp baking powder; $\frac{3}{4}$ tsp salt

Extra Practice, **page 695**

26.

x	3x
1	3
2	6
3	9
4	12
5	15

27.

y	8y
1	8
2	16
3	24
4	32
5	40

28.

z	2z + 4
1	6
2	8
3	10
4	12
5	14

29.

d	3d − 1
1	2
2	5
3	8
4	11
5	14

30.

q	9q + 3
1	12
2	21
3	30
4	39
5	48

31.

c	15c
1	15
2	30
3	45
4	60
5	75

32.

v	9v − 2
1	7
2	16
3	25
4	34
5	43

33.

z	2 + 8z
1	10
2	18
3	26
4	34
5	42

Lesson 1.5

Exploration 2, **page 38**

1.

Natural number	Factors
7	1, 7
8	1, 2, 4, 8
9	1, 3, 9
10	1, 2, 5, 10
11	1, 11
12	1, 2, 3, 4, 6, 12
13	1, 13
14	1, 2, 7, 14
15	1, 3, 5, 15
16	1, 2, 4, 8, 16
17	1, 17
18	1, 2, 3, 6, 9, 18
19	1, 19
20	1, 2, 4, 5, 10, 20

Exploration 3, pages 39–40

3. If the sum of a number's digits is divisible by 3, then so is the number, and vice-versa.

4. If a number is divisible by 6, then it is divisible by 2 and 3 since both 2 and 3 are factors of 6, and 3 and 2 have no common factors. As a result, a number can be tested for divisibility by 6 by testing that number for divisibility by both 2 and 3.

Critical Thinking, page 40

To find all factors of 540, test whether 1, 2, 3, etc. are factors of 540 and then write 540 as a product using each factor found.

$540 = 1 \cdot 540$ $540 = 9 \cdot 60$
$540 = 2 \cdot 270$ $540 = 10 \cdot 54$
$540 = 3 \cdot 180$ $540 = 12 \cdot 45$
$540 = 4 \cdot 135$ $540 = 15 \cdot 36$
$540 = 5 \cdot 108$ $540 = 18 \cdot 30$
$540 = 6 \cdot 90$ $540 = 20 \cdot 27$

The factors of 540 are 1, 2, 3, 4, 5, 6, 9, 10, 12, 15, 18, 20, 27, 30, 36, 45, 54, 60, 90, 108, 135, 180, 270, and 540.

Communicate, page 41

1. Natural numbers are the numbers 1, 2, 3, 4, etc.

2. A number is a *factor* of a second number if it divides into the second number evenly. A *product* is the result of multiplying two numbers or expressions.

3. To find the factors of 72, draw all possible rectangles of area 72 with sides that are natural numbers. The factors of 72 are the lengths and widths of the rectangles.

4. A prime number is a number with exactly two factors. A composite number is a number with more than two factors.

5. List the factors of 1001. If there are more than just 1 and 1001, then 1001 is composite. Otherwise, 1001 is prime.

6. Since 24,570 is even, it is divisible by 2. The sum of the digits of 24,570 is 18, which is divisible by 3, so 24,570 is divisible by 3. Since 24,570 ends in either 0 or 5, it is divisible by 5. Since it ends in 0, it is divisible by 10. 24,570 is divisible by both 2 and 3, so it is divisible by 6. Finally, since 9 divides evenly into the sum of the digits of 24,570, it also divides evenly into 24,570.

Practice and Apply, pages 41–42

10. 1, 2, 3, 4, 6, 8, 12, 16, 24, 48 (circle 2 and 3)

11. 1, 2, 4, 7, 8, 14, 28, 56 (circle 2 and 7)

12. 1, 2, 3, 4, 6, 8, 9, 12, 18, 24, 36, 72 (circle 2 and 3)

13. 1, 2, 3, 4, 6, 7, 12, 14, 21, 28, 42, 84 (circle 2, 3, and 7)

14. 1, 3, 17, 51 (circle 3 and 17)

Look Back, page 42

44.

x	y = 2x + 1
1	3
2	5
3	7
4	9
5	11
6	13
7	15
8	17
9	19
10	21

The *y*-values are all odd numbers.

Extra Practice, page 696

1. 1, 3, 5, 15 (circle 3 and 5)

2. 1, 2, 4, 7, 8, 14, 28, 56 (circle 2 and 7)

3. 1, 3, 9, 27 (*circle 3*)

4. 1, 7, 49 (*circle 7*)

5. 1, 2, 4, 5, 10, 20, 25, 50, 100 (*circle 2 and 5*)

6. 1, 2, 3, 4, 6, 12 (*circle 2 and 3*)

7. 1, 2, 4, 8, 16, 32, 64 (*circle 2*)

8. 1, 2, 4, 8, 16, 32 (*circle 2*)

9. 1, 2, 3, 6, 9, 18, 27, 54 (*circle 2 and 3*)

10. 1, 3, 31, 93 (*circle 3 and 31*)

11. 1, 5, 7, 35 (*circle 5 and 7*)

12. 1, 2, 3, 4, 5, 6, 10, 12, 15, 20, 30, 60 (*circle 2, 3 and 5*)

13. 1, 19 (*circle 19*)

14. 1, 5, 25, 125 (*circle 5*)

15. 1, 2, 4, 17, 34, 68 (*circle 2 and 17*)

Lesson 1.6

Communicate, **page 48**

1. Exponents are numbers indicating how many times a base number is multiplied by itself.

3. The prime factorization of a number can be found by writing the number as a product of two factors and repeating the process on each factor until only prime numbers remain. For example, $48 = 6 \cdot 8 = 2 \cdot 3 \cdot 2 \cdot 4 = 2 \cdot 3 \cdot 2 \cdot 2 \cdot 2$, so $48 = 2^4 \cdot 3$. Another way is to divide a number by 2 repeatedly until it will no longer divide evenly, then by 3, and so on, using only prime numbers as divisors. Eventually the quotient will be a prime number. For example: $48 \div 2 = 24$, $24 \div 2 = 12$, $12 \div 2 = 6$, $6 \div 2 = 3$, so $48 = 2^4 \cdot 3$.

4. The expression "5 squared" is used for 5^2 because 5^2 is the area of a square with sides of length 5.

5. A calculator simplifies the process of dividing large numbers, such as 43,650, by the prime numbers 2, 3, 5, 7, etc. as necessary.

Look Back, **page 49**

47.

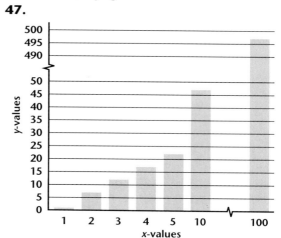

Lesson 1.7

Ongoing Assessment, **page 53**

When 111,111·111,111 is entered in a calculator, the answer is in scientific notation. Your calculator may write the answer as 1.234565432E10. The last digit is not displayed. As the number of ones increases, fewer digits are displayed. Most calculators will not display more then ten digits, if that many.

Communicate, **page 54**

1. $3 + 2 \cdot 4$ equals $3 + 8$, or 11, using the proper order of operations. Some calculators will mistakenly compute $(3 + 2) \cdot 4$ to get 20.

2. $20 \div 2 \cdot 5$ is equal to $10 \cdot 5$, or 50, using algebraic logic. A mistaken evaluation would compute $20 \div (2 \cdot 5)$, or 2.

3. $20 \div 2 \cdot 5$ equals 50.

4. First compute the innermost layer of parentheses. Then work outward.

5. The "order of operation" rules are necessary so that any two people will arrive at the same answer for a given computation.

6. 173 + 223 is approximately 390, and 151 − 21 is 130, so $\frac{173 + 223}{151 - 21}$ is about $\frac{390}{130}$, which equals 3.

Lesson 1.8

Critical Thinking, **page 57**

Grouping does affect subtraction and division. For example, (15 − 6) − 2 = 7 but 15 − (6 − 2) = 11. For division, (16 ÷ 4) ÷ 2 = 2 but 16 ÷ (4 ÷ 2) = 8.

Communicate, **page 60**

1. Answers may vary; a sample answer is given. 4 · 8 · 25 = (4 · 25) · 8 = 100 · 8 = 800 and 26 + 17 + 14 = (26 + 14) + 17 = 40 + 17 = 57

2. **a)** 8(21) is easier to compute as 8(20 + 1), which equals 8 · 20 + 8 · 1, or 168.
b) 11(35) becomes 11(30 + 5), or 11 · 30 + 11 · 5, which equals 385.
c) 14(22) equals (10 + 4) · 22, which is 10 · 22 + 4 · 22, or 220 + 88, which equals 308.

3. Janis and Kayla rented inner tubes for 5 + 3 or 8 hours on Saturday and Sunday. The total rental was 2(5 + 3) = 2 · 8 = 16. By the Distributive Property, 2(5 + 3) = 2 · 5 + 2 · 3 = 10 + 6 = 16; 2 · 5 = 10 is the fee for Saturday, and 2 · 3 = 6 is the fee for Sunday.

4. Suppose the Improvisation Club is selling slices of pizza for a fund raiser with a profit of 50¢ per slice. If the club bought 120 slices and had 17 slices left over unsold, the profit from the slices

they sold would be 50(120 − 17) = 50 · 103 = 5150¢, or $51.50. This could also be computed as 50(120 − 17) = 50 · 120 − 50 · 17 = 6000 − 850 = 5150¢, or $51.50.

Practice and Apply, **pages 61–63**

8. (27 + 98) + 73 = (98 + 27) + 73 = 98 + (27 + 73) = 98 + 100 = 198; Commutative, Associative properties

9. (45 · 32) · 0 = 0; Multiplication Property of Zero

10. (87 · 5) · 2 = 87 · (5 · 2) = 87 · 10 = 870; Associative Property

11. 50 · (118 · 20) = 50 · (20 · 118) = (50 · 20) · 118 = 1000 · 118 = 118,000; Commutative, Associative properties

12. (688 + 915) + 312 = (915 + 688) + 312 = 915 + (688 + 312) = 915 + 1000 = 1915; Commutative, Associative properties

13. (25 · 78) · 4 = (78 · 25) · 4 = 78 · (25 · 4) = 78 · 100 = 7800; Commutative, Associative properties

14. 2 · (129 · 5) = 2 · (5 · 129) = (2 · 5) · 129 = 10 · 129 = 1290; Commutative, Associative properties

15. (133 + 52) + 67 = (52 + 133) + 67 = 52 + (133 + 67) = 52 + 200 = 252; Commutative, Associative properties

22. 9.5(7) + 9.5(8)
= 66.5 + 76
= 142.50
9.5(7 + 8)
= (9.5)(15)
= 142.50

23. 15(4) + 15(6)
= 60 + 90
= 150
15(4 + 6)
= (15)(10)
= 150

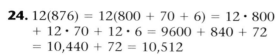

24. $12(876) = 12(800 + 70 + 6) = 12 \cdot 800 + 12 \cdot 70 + 12 \cdot 6 = 9600 + 840 + 72 = 10,440 + 72 = 10,512$

39.

x	1	2	3	4	5	10
y	3	6	9	12	15	30

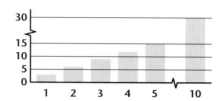

40.

x	1	2	3	4	5	10
y	1	3	5	7	9	19

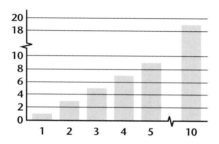

Extra Practice, page 697

1. 126; Associative Property
2. 300; Commutative Property
3. 360; Associative Property
4. 399; Commutative and Associative Properties
5. 1380; Commutative and Associative Properties
6. 7600; Commutative and Associative Properties
7. 229; Commutative and Associative Properties
8. 72,000; Commutative and Associative Properties
9. 0; Mult. Property of Zero
10. 648; Associative Property
11. 150; Distributive Property
12. 175; Distributive Property

Chapter 1 Project

Activity 1, **page 65**

A multiplication table using remainders from division by 5 appears below.

	0	1	2	3	4	5	6	7	8	9	10	11	12
0	0	0	0	0	0	0	0	0	0	0	0	0	0
1	0	1	2	3	4	0	1	2	3	4	0	1	2
2	0	2	4	1	3	0	2	4	1	3	0	2	4
3	0	3	1	4	2	0	3	1	4	2	0	3	1
4	0	4	3	2	1	0	4	3	2	1	0	4	3
5	0	0	0	0	0	0	0	0	0	0	0	0	0
6	0	1	2	3	4	0	1	2	3	4	0	1	2
7	0	2	4	1	3	0	2	4	1	3	0	2	4
8	0	3	1	4	2	0	3	1	4	2	0	3	1
9	0	4	3	2	1	0	4	3	2	1	0	4	3
10	0	0	0	0	0	0	0	0	0	0	0	0	0
11	0	1	2	3	4	0	1	2	3	4	0	1	2
12	0	2	4	1	3	0	2	4	1	3	0	2	4

Activity 2, **page 65**

An addition table using remainders from division by 6 appears below.

	0	1	2	3	4	5	6	7	8	9	10	11	12
0	0	1	2	3	4	5	0	1	2	3	4	5	0
1	1	2	3	4	5	0	1	2	3	4	5	0	1
2	2	3	4	5	0	1	2	3	4	5	0	1	2
3	3	4	5	0	1	2	3	4	5	0	1	2	3
4	4	5	0	1	2	3	4	5	0	1	2	3	4
5	5	0	1	2	3	4	5	0	1	2	3	4	5
6	0	1	2	3	4	5	0	1	2	3	4	5	0
7	1	2	3	4	5	0	1	2	3	4	5	0	1
8	2	3	4	5	0	1	2	3	4	5	0	1	2
9	3	4	5	0	1	2	3	4	5	0	1	2	3
10	4	5	0	1	2	3	4	5	0	1	2	3	4
11	5	0	1	2	3	4	5	0	1	2	3	4	5
12	0	1	2	3	4	5	0	1	2	3	4	5	0

Activity 3, **page 65**

All addition patterns form diagonal lines or stripes. Multiplication patterns form

repeating boxes with zeros on the boundaries. When more numbers are added to either type of pattern, the pattern repeats itself.

Chapter 1 Review, pages 66–68

9.

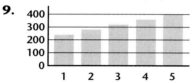

Chapter 1 Assessment, page 69

10.

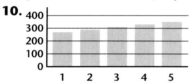

Lesson 2.1

Critical Thinking, **page 75**

$|3| = 3$ and $|-3| = 3$, so both 3 and -3 have the same absolute value. 3 and -3 are opposites. Absolute value finds the distance between a number and 0, and opposite numbers are both the same distance from 0.

Communicate, **page 76**

1. The set of integers is the set of positive counting numbers, their negative values, and zero.

2. Answers may vary. Integers are used to model elevation, temperature, and financial gain or loss.

3. The opposite of a number is a number that is on the opposite side of zero on the number line and the same distance from zero as the original number. For example, the opposite of -5 is 5.

4. The absolute value of a number is the distance from that number to zero on the number line. For example, the absolute value of -6 is 6, since the distance from -6 to 0 is 6 units.

Lesson 2.2

Exploration 1, **page 79**

3. Answers vary. Examples:

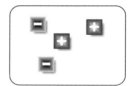 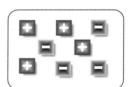

All four models have equal numbers of positive and negative tiles. The total number of tiles is different in each model.

4. 3 and -3 are called opposites because they are on opposite sides of zero on the number line, and they are each the same distance from zero. Also, 3 positive tiles and 3 negative tiles together have a value of zero.

5. The integer represented by a tile arrangement depends only on the number of positive or negative tiles which do not form neutral pairs. For any tile model of an integer, neutral pairs may be added or removed to create a new tile arrangement for that integer.

1. $(-4) + (-2) = -6$

2. $2 + 4 = 6$

3. Both processes in Steps 1 and 2 use sets of 2 and 4 tiles to make a set of 6 tiles. The steps use all positive or all negative tiles, and the tiles are added in different orders.

4. Add the absolute value of 12 and 40 to get 52. Since the addends, 12 and 40, are both positive, the sum 52 should be positive. Thus, $12 + 40 = 52$. Add the absolute values of -25 and -40 to get 65. Since the addends, -25 and -40, are both negative, the sum should be negative. Thus, $(-25) + (-40) = -65$.

6.

−4 + 2 = −2

−4 + 7 = 3

7.

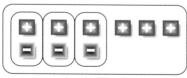

6 + (−3) = 3

6 + (−10) = −4

8. In Steps 6 and 7, positive and negative tiles were placed, then neutral pairs were removed to find the sum. In Step 6, negative tiles were placed first, while in Step 7, positive tiles were placed first.

9. To find $-12 + 40$, subtract $|12|$ from $|40|$ to get 28. The sum $-12 + 40$ is 28. To find $25 + (-40)$, subtract $|25|$ from $|40|$ to get 15. The sum $25 + (-40)$ is -15.

1. Integers may be added by placing the necessary number of positive and negative tiles. All possible neutral pairs are then removed, and the remaining number of tiles give the sum.

2. If two integers have the same sign, add the absolute values and use the sign that they have in common.

3. To find the sum of two integers with unlike signs, first determine their absolute values. Subtract the smaller absolute value from the greater. Then use the sign of the integer with the greater absolute value for the sum.

4. Whenever a negative integer is part of a sum, its absolute value is used to compute the answer. For two negative integers, the sum is found by taking the opposite of the sum of their absolute values. When a negative integer is added to a positive integer, either the absolute value of the negative integer is subtracted from the positive integer or vice-versa.

44. $-2 + (8 + [12 + (-6 + (-1))]) = 11$,
$(8 + 12) - (2 + 6 + 1) = 11$,
$(-2 + 8) + (12 + (-6)) + -1 = 11$

Look Beyond, page 84

49. $7(-2) = (-2) + (-2) + (-2) + (-2) + (-2) + (-2) + (-2) = -14$

Lesson 2.3

Ongoing Assessment, page 85

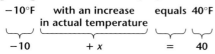

−10°F with an increase equals 40°F
in actual temperature

−10 + x = 40

Ongoing Assessment, page 86

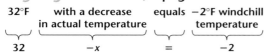

32°F with a decrease equals −2°F windchill
in actual temperature temperature

32 −x = −2

Critical Thinking, page 89

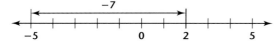

Since $2 + (-7) = -5$, the number line shows that traveling from 2 to −5 is a trip of length 7 units to the left. It follows that a trip from −5 to 2 is of length 7 units to the right; that is, $-5 + 7 = -2$. Since 7 is a positive integer, it follows that $-5 < 2$ and that $2 > -5$.

Communicate, page 90

1. With temperature, the warmer temperature corresponds to the larger number. For example, 12°F, is warmer than −30°F, so $-30 < 12$. For elevation, a higher elevation is represented by a larger number. The top of a tree 35 feet above the ground is higher than one of its roots 4 feet below the ground, so $-4 < 35$.

2.

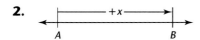

Since A is to the left of B, there is a positive number x which can be added to the number at location A to get the number at location B. According to the rule on page 89, the number at A is less than the number at B.

3. If the solution to $-4 + x = -5$ is positive, then $-4 < -5$ is true and $-4 > -5$ would be false. Because the solution to $-4 + x = -5$ is -1, $-4 > -5$ is true.

4. When guess-and-check shows a guess to be wrong, it helps to know whether the guess was wrong because it was too large or too small. For example, understanding that -3 is larger than -5 is crucial in deciding on another guess for the solution of the equation.

Practice and Apply, pages 90–92

15. $-5 < 5; 5 > -5$

16. $-3 < -1; -1 > -3$

17. $-2 < 0; 0 > -2$

18. $-7 < -3; -3 > -7$

19. $-8 < 4; 4 > -8$

20. $-2 < -1; -1 > -2$

21. $-1 < 4; 4 > -1$

22. $-4 < -2; -2 > -4$

23. $-9 < 7; 7 > -9$

24. $-8 < 10; 10 > -8$

25. $-7 < 8; 8 > -7$

26. $-15 < 15; 15 > -15$

27. $-21 < 8; 8 > -21$

28. $-22 < -16; -16 > -22$

29. $-7 < 14; 14 > -7$

30. $-236 < 101; 101 > -236$

31. $-17 < 16; 16 > -17$

32. $-18 < -9; -9 > -18$

33. $9 < 18; 18 > 9$

34. $-200 < -1; -1 > -200$

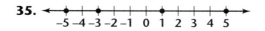

35.
$$-5\ -4\ -3\ -2\ -1\ \ 0\ \ 1\ \ 2\ \ 3\ \ 4\ \ 5$$

36.
$$-10\ -8\ -6\ -4\ -2\ \ 0\ \ 2\ \ 4\ \ 6\ \ 8\ \ 10$$

37.
$$-3\ -2\ -1\ \ 0\ \ 1\ \ 2\ \ 3$$

38.
$$0\ \ 1\ \ 2\ \ 3\ \ 4\ \ 5\ \ 6\ \ 7$$

39.
$$-6\ -5\ -4\ -3\ -2\ -1\ \ 0\ \ 1$$

40.
$$-5\ -4\ -3\ -2\ -1\ \ 0\ \ 1\ \ 2\ \ 3\ \ 4\ \ 5\ \ 6$$

41.
$$10\quad 20\quad 30\quad 40$$

42.
$$-4\ -3\ -2\ -1\ \ 0\ \ 1\ \ 2\ \ 3\ \ 4\ \ 5$$

43.
$$-50\quad -30\quad -10\ \ 0\ \ 10\quad 30\quad 50$$

Lesson 2.4

Ongoing Assessment, **page 95**

The [-] key is used for subtraction. Most calculators have a [+/-] key or [(-)] key to indicate the opposite of a number. $-35 - 27$ would use these keystrokes on most calculators:

[(-)] 35 [-] 27   or   35 [+/-] [-] 27.

Communicate, **page 96**

4. To find $-4 - (-7)$, add 7 positive and 7 negative tiles. That way there will be 7 negative tiles to remove. (However, to ensure that there are 7 negative tiles, it is really only necessary to add 3 positive and 3 negative tiles.)

5. Adding the opposite means that subtracting a number is the same as adding its opposite. For example, subtracting -3 is the same as adding 3.

Practice and Apply, **pages 96–97**

36. Step 1

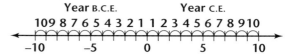

Year B.C.E.   Year C.E.
$$10\ 9\ 8\ 7\ 6\ 5\ 4\ 3\ 2\ 1\ \ 1\ 2\ 3\ 4\ 5\ 6\ 7\ 8\ 9\ 10$$
$$-10\qquad -5\qquad 0\qquad 5\qquad 10$$

Someone born in the year 10 B.C.E. is 1 year old on his or her birthday in 9 B.C.E., 2 years old in 8 B.C.E., etc. until turning 19 in 10 C.E.

Step 2 To find the time from years B.C.E. to years C.E., think of the year B.C.E. as a negative integer and the year C.E. as a positive integer. Subtract the negative integer from the positive, then subtract one from that for the year 0 that was counted but did not occur. For example, from 10 B.C.E. to 10 C.E. we get $10 - (-10)$ or 20, and then $20 - 1$ makes 19 years from 10 B.C.E. to 10 C.E.

Step 3 40 years. $27 - (-14) = 41$, and $41 - 1$ makes 40.

Lesson 2.5

Communicate, **page 101**

5. Starting with a guess of 120 for the average. Then average the differences between 120 and each number as follows:

Number	Difference from 120
95	-25
119	-1
110	-10
130	10
141	11
155	25

The average difference from 120 is $1\frac{2}{3}$.

The actual average is $120 + 1\frac{2}{3}$, or

$121\frac{2}{3}$. So 120 appears to be a reasonable guess.

Lesson 2.6

Try This, **page 106**

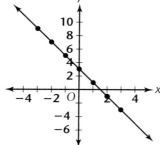

Communicate, **page 108**

2. The first number tells how far to travel left or right from the origin (negative for left, positive for right). The second number tells how far to travel above or below that position (negative for below, positive for above).

3. The entry for 0 gives the initial amount, and the difference between consecutive entries is the amount multiplied by the starting variable. For example, consider these tables and equations.

x	0	1	2	3
y	−4	2	8	14

$y = -4 + 6x$

t	0	1	2	3
d	7	9	11	13

$d = 7 + 2t$

4. Consider the following linear equation and its graph: $c = -1 + 2t$.

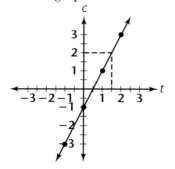

The graph was plotted using $(-1, -3)$, $(0, -1)$, $(1, 1)$, and $(2, 3)$. Using the graph, we can see that if $t = 1\frac{1}{2}$, then c has value 2. A linear graph can be used to find data other than the data used to draw the graph.

Practice and Apply, **pages 109–111**

5.
x	−3	−2	−1	0	1	2	3
y	13	12	11	10	9	8	7

6.
x	−3	−2	−1	0	1	2	3
y	5	6	7	8	9	10	11

7.
x	−3	−2	−1	0	1	2	3
y	2	1	0	−1	−2	−3	−4

8.
x	−3	−2	−1	0	1	2	3
y	8	7	6	5	4	3	2

9.
x	−3	−2	−1	0	1	2	3
y	−2	−1	0	1	2	3	4

10.
x	−3	−2	−1	0	1	2	3
y	−7	−6	−5	−4	−3	−2	−1

11.
x	−3	−2	−1	0	1	2	3
y	−18	−16	−14	−12	−10	−8	−6

12.
x	−3	−2	−1	0	1	2	3
y	21	18	15	12	9	6	3

13.
x	−3	−2	−1	0	1	2	3
y	−45	−40	−35	−30	−25	−20	−15

14.
x	−3	−2	−1	0	1	2	3
y	36	24	12	0	−12	−24	−36

15.
x	−3	−2	−1	0	1	2	3
y	3	2	1	0	−1	−2	−3

16.
x	−3	−2	−1	0	1	2	3
y	1.5	1	0.5	0	−0.5	−1	−1.5

17.

x	−3	−2	−1	0	1	2	3
y	−2	0	2	4	6	8	10

18.

x	−3	−2	−1	0	1	2	3
y	10	7	4	1	−2	−5	−8

19.

x	−3	−2	−1	0	1	2	3
y	−15	−11	−7	−3	1	5	9

20.

x	−3	−2	−1	0	1	2	3
y	−1	1	3	5	7	9	11

25. $y = x - 2$

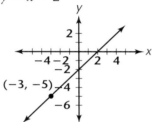

When $x = -3$, $y = -5$.

26. $y = 4 - x$

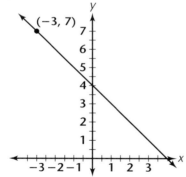

When $x = -3$, $y = 7$.

27. $y = x + 2$

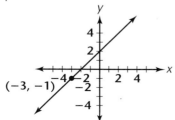

When $x = -3$, $y = -1$.

28. $y = x - 5$

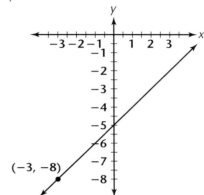

When $x = -3$, $y = -8$.

29. $y = 2x - 3$

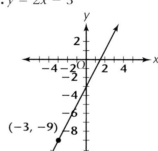

When $x = -3$, $y = -9$.

30. $y = 1 - 2x$

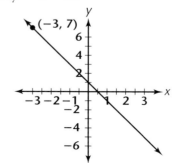

When $x = -3$, $y = 7$.

31. $y = -2x - 1$

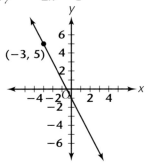

(−3, 5)

When $x = -3$, $y = 5$.

32. $y = 2x - 8$

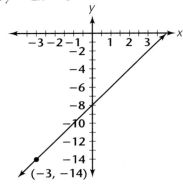

(−3, −14)

When $x = -3$, $y = -14$.

38. $y = 24x$

Look Beyond, **page 111**

57.

x	−3	−2	−1	0	1	2	3
$y = -x + 1$	4	3	2	1	0	−1	−2

x	−3	−2	−1	0	1	2	3
$y = 2x - 8$	−14	−12	−10	−8	−6	−4	−2

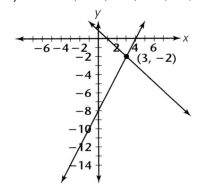

(3, −2)

Extra Practice, **page 700**

1.

x	−3	−2	−1	0	1	2	3
y	6	5	4	3	2	1	0

2.

x	−3	−2	−1	0	1	2	3
y	−11	−7	−3	1	5	9	13

3.

x	−3	−2	−1	0	1	2	3
y	−11	−8	−5	−2	1	4	7

4.

x	−3	−2	−1	0	1	2	3
y	−15	−10	−5	0	5	10	15

5.

x	−3	−2	−1	0	1	2	3
y	1.5	1	0.5	0	−0.5	−1	−1.5

6.

x	−3	−2	−1	0	1	2	3
y	12	8	4	0	−4	−8	−12

7.

x	−3	−2	−1	0	1	2	3
y	5	4	3	2	1	0	−1

8.

x	−3	−2	−1	0	1	2	3
y	10	8	6	4	2	0	−2

9.

x	−3	−2	−1	0	1	2	3
y	17	11	5	−1	−7	−13	−19

Lesson 2.7

Exploration, **page 116**

3. The differences show that each hour adds $7 to the cost. After h hours, $7h$ must be added to the initial $20 cost.

Communicate, **page 117**

4. A second difference can be added to the previous first difference to get the next first difference. This new first difference can be added to the previous term of the original sequence to get the next term of the sequence.

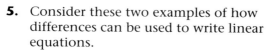

5. Consider these two examples of how differences can be used to write linear equations.

x	0	1	2	3
y	7	4	1	−2

$-3 \quad -3 \quad -3$

So, $y = 7 + (-3)x$.

x	0	1	2	3
y	6	6.5	7	7.5

$0.5 \quad 0.5 \quad 0.5$

So, $y = 6 + 0.5x$.

In each case, to find y for a particular value of x, compute as follows:

$$y = 7 + \underbrace{(-3) + (-3) + \cdots + (-3)}_{x \text{ times}}$$

$$y = 7 + (-3)x$$

$$y = 6 + \underbrace{0.5 + 0.5 + \cdots + 0.5}_{x \text{ times}}$$

$$y = 6 + 0.5x$$

Practice and Apply, **pages 118–119**

6. First differences: 7, 9, 11, 13; Second differences: 2, 2, 2.

Look Back, **page 119**

33.

m	25	50	75	100	125
c	$32.50	$40	$47.50	$55	$62.50

Look Beyond, **page 119**

39. The differences are the same as the sequence.

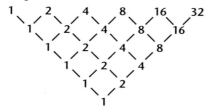

Each term of the sequence is double the one before, so the next 3 terms are 64, 128, 256.

40. The sequence and the differences all follow the pattern of a term being 10 multiplied by the previous term.

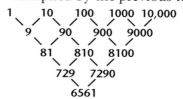

By the pattern shown above, the next three terms are 100,000, 1,000,000, and 10,000,000.

Chapter 2 Project
Ongoing Assessment, **page 120**

The winter temperatures do not vary as much as the summer temperatures in Alberta.

Activity 1, **page 121**

2.

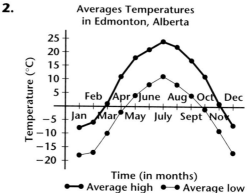

Chapter 2 Review, pages 122–124

12.

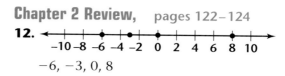

$-6, -3, 0, 8$

13.

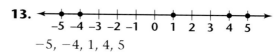

$-5, -4, 1, 4, 5$

38.

x	−2	−1	0	1	2	3
y = x + 3	1	2	3	4	5	6

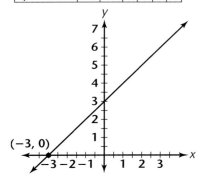

When $x = -3$, $y = 0$.

39.

x	−2	−1	0	1	2	3
y = 3 − x	5	4	3	2	1	0

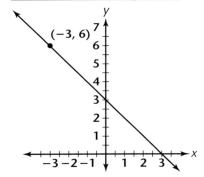

When $x = -3$, $y = 6$.

40.

x	−2	−1	0	1	2	3
y = 2x − 4	−8	−6	−4	−2	0	2

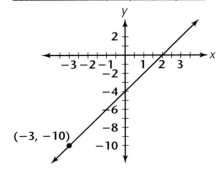

When $x = -3$, $y = -10$.

Chapter 2 Assessment, **page 125**

20.

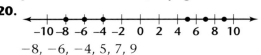

−8, −6, −4, 5, 7, 9

45.

x	−3	−2	−1	0	1	2	3
y	11	9	7	5	3	1	−1

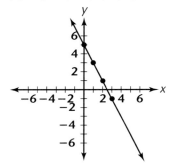

47. The data in the table have a linear relationship because the first differences are constant. $d = 26t$

Chapters 1–2 Cumulative Assessment, pages 126–127

19.

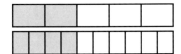

−2, −1, 7, 9

Lesson 3.1

Communicate, **page 134**

1. Equivalent fractions are fractions with equal values. Example: $\frac{2}{5} = \frac{4}{10}$

2. Because $\frac{2}{2} = 1$, you are really multiplying by 1.

3.

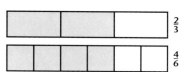

$\frac{2}{3} = \frac{4}{6}$

4. They are equal because $\frac{1}{3}$ and $\frac{2}{6}$ are equivalent fractions.

Lesson 3.3

Exploration 1, **page 143**

2. $\frac{4}{5}, \frac{5}{8}, \frac{3}{4}$

$$\begin{array}{r} .625 \\ 8\overline{)5.000} \\ \underline{48} \\ 20 \\ \underline{16} \\ 40 \\ \underline{40} \\ 0 \end{array}$$

By adding zeros, if the numerator becomes a number that is a multiple of the denominator, the fraction is a terminating decimal.

3. $\frac{4}{11}, \frac{7}{9}, \frac{5}{12}$

$$\begin{array}{r} .3636 \\ 11\overline{)4.000} \\ \underline{33} \\ 70 \\ \underline{66} \\ 40 \\ \underline{33} \\ 70 \\ \underline{66} \\ 4 \end{array} \qquad \begin{array}{r} .4166 \\ 12\overline{)5.000} \\ \underline{48} \\ 20 \\ \underline{12} \\ 80 \\ \underline{72} \\ 80 \\ \underline{72} \\ 8 \end{array}$$

When division produces the same remainder or a series of remainders that repeat, the fraction is a repeating decimal.

Exploration 2, **page 143**

1.

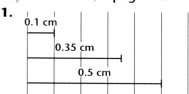

0.1 cm

0.35 cm

0.5 cm

4. $\frac{35}{100} = \frac{350}{1000}$

$\frac{4}{10} = \frac{40}{100} = \frac{400}{1000}$

$\frac{358}{1000}$

Since $\frac{350}{1000} < \frac{358}{1000} < \frac{400}{1000}$, then 0.350 < 0.358 < 0.400. Therefore 0.35 < 0.358 < 0.4.

Communicate, **page 146**

1. Change 0.4 to 0.40. On a 10 × 10 grid, color in 40 squares to represent 0.40 or 0.4. Another 10 × 10 grid, color in 4 squares to represent 0.04. You can see that 0.4 is greater than 0.04.

2. Round 11.6 gallons to 12 gallons. Round $1.199 to $1.20. 12 · 1.20 is $14.40. The gas cost approximately $14.40.

3. one thousand ninety-seven and six hundred eighty-three thousandths

1	0	9	7	.	6	8	3
thousands	hundreds	tens	ones		tenths	hundedths	thousandths

4. $\frac{2}{3}$ is a repeating decimal (0.$\overline{6}$), and $\frac{3}{5}$ is a terminating decimal (0.6).

5. First write the decimals with the same number of decimal places: -5.050, 5.500, -5.200, 5.499. Then arrange the decimals from least to greatest: -5.200, -5.050, 5.499, 5.500, or -5.2, -5.05, 5.499, 5.5.

Practice and Apply, pages 146–148

36. 0.38, 0.4, 0.425, 0.472, 0.49

38. 4.97, 5.037, 5.2, 5.3, 5.32

42. $7\frac{5}{12}$, $7\frac{3}{8}$, $7\frac{1}{3}$, 7.3, 7.045

44. $-5\frac{2}{9}$, -5.55, -5.6, $-5\frac{5}{8}$, $-5\frac{2}{3}$

45. $-\frac{1}{3}$, -0.5, $-\frac{3}{5} = -0.6$, $-\frac{2}{3}$, -0.75

46. 8.72, 8.56, $8\frac{6}{11}$, $8\frac{4}{9}$, $8\frac{1}{9}$

47. $1\frac{1}{3}$, 1.303, $\frac{14}{11}$, $1\frac{1}{5}$, 0.133

Extra Practice, page 702

36. 0.04, 0.05, 0.054, 0.5, 0.54

37. -0.6, -0.16, -0.1, 0.06, 0.66

38. 0.03, 0.3, 0.43, $\frac{3}{4}$, $1\frac{1}{3}$, 3.4

39. -2.8, $-2\frac{2}{5}$, -0.88, -0.28, 0.08, $2\frac{1}{8}$

40. $-9\frac{1}{2}$, -9.25, -9.2, -9.05, -9.02

41. $-1\frac{1}{3}$, -1.3, -0.19, 0.3, $\frac{13}{10}$, 1.9

42. 0.075, $\frac{5}{8}$, $\frac{2}{3}$, $\frac{3}{4}$, 0.8

43. $-\frac{3}{7}$, -0.41, $-\frac{2}{5}$, -0.39, -0.04

Lesson 3.4

Exploration 1, pages 149–150

4.

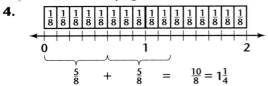

$$\frac{5}{8} + \frac{5}{8} = \frac{10}{8} = 1\frac{1}{4}$$

5.

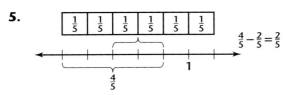

$$\frac{4}{5} - \frac{2}{5} = \frac{2}{5}$$

Exploration 2, pages 150–151

4. $\frac{1}{2} - \frac{1}{3} = \frac{1}{2} \cdot \frac{3}{3} - \frac{1}{3} \cdot \frac{2}{2}$ (Multiply by 1 to

get like denominators.); $= \frac{3}{6} - \frac{2}{6}$ (Subtract.); $= \frac{1}{6}$ (Simplify.)

Communicate, page 152

1. To add $\frac{1}{3}$ and $\frac{1}{4}$, first find the common denominator, 12. Since $\frac{1}{3} = \frac{4}{12}$ and $\frac{1}{4} = \frac{3}{12}$, $\frac{1}{3} + \frac{1}{4} = \frac{7}{12}$ and $\frac{7}{12}$ does not simplify to $\frac{2}{7}$. Therefore $\frac{1}{3} + \frac{1}{4} = \frac{7}{12} \neq \frac{2}{7}$. To get the answer $\frac{2}{7}$, one must add the numerators and denominators, which is incorrect.

2.

| $\frac{1}{3}$ | | $\frac{1}{4}$ | | | | 1 |

$$\frac{1}{3} \quad + \quad \frac{1}{4} \quad = \quad \frac{7}{12}$$

3. Use equivalent fractions to find fractions with the same denominator. Then add the fractions with like denominators. For example, to find $\frac{1}{2} + \frac{2}{3} + \frac{1}{4}$, find fractions with the same denominator: $\frac{1}{2} = \frac{2}{4} = \frac{3}{6} = \frac{4}{8} = \frac{6}{12}$; $\frac{2}{3} = \frac{4}{6} = \frac{8}{12}$; and $\frac{1}{4} = \frac{2}{8} = \frac{3}{12}$. Then add the fractions with like denominators: $\frac{6}{12} + \frac{8}{12} + \frac{3}{12} = \frac{17}{12}$, or $1\frac{5}{12}$.

4. The number of hours an employee works requires an exact answer. Ex.: On Monday Mary worked $3\frac{1}{2}$ hours. On Tuesday she worked $3\frac{1}{4}$ hours. How many hours did she work on those two days?

5. Martin wants to build a picture frame. The lengths are $5\frac{1}{2}$ inches and the widths are $3\frac{3}{4}$ inches. Martin overestimates the perimeter of the

frame to buy the material. Thus, Martin estimates the perimeter, and purchases $6 + 6 + 4 + 4 = 20$ inches of material.

Practice and Apply, **pages 152–153**

23.

Worker	M	T	W	T	F	Weekly
Andrews	36.25	45.31	63.44	32.63	23.56	201.19
Bolla	8.44	23.63	48.94	38.81	30.38	150.20
Garza	56.19	29.06	11.63	19.38	60.06	176.32
Holland	26.53	18.95	32.22	45.48	53.06	176.24
Tate	59.63	57.64	49.69	69.56	19.88	256.40
Wuest	39.90	29.93	43.23	21.61	0	134.67

For Exercises 29-40, answers may vary.

29. Possible answer $4\frac{1}{3} = 2\frac{2}{3} + 1\frac{2}{3}$

30. Possible answer $6\frac{3}{8} = 2\frac{1}{4} + 4\frac{1}{8}$

31. Possible answer $-4\frac{3}{4} + 4\frac{1}{4} = -\frac{1}{2}$

32. Possible answer $11\frac{4}{7} = 4\frac{23}{28} + 6\frac{3}{4}$

33. Possible answer $9\frac{7}{8} = 3\frac{1}{8} + 6\frac{3}{4}$

34. Possible answer $-6\frac{3}{4} + 5\frac{1}{4} = -1\frac{1}{2}$

35. Possible answer $4\frac{1}{3} = 5\frac{2}{3} - 1\frac{1}{3}$

36. Possible answer $6\frac{3}{8} = 8\frac{7}{8} - 2\frac{1}{2}$

37. Possible answer $-\frac{1}{2} = 4\frac{1}{4} - 4\frac{3}{4}$

38. Possible answer $11\frac{4}{7} = 13\frac{6}{7} - 2\frac{2}{7}$

39. Possible answer $9\frac{7}{8} = 12\frac{15}{16} - 3\frac{1}{16}$

40. Possible answer $-1\frac{1}{2} = 5\frac{1}{4} - 6\frac{3}{4}$

Extra Practice, **page 702**

30. $-5\frac{83}{90}$ **34.** $-2\frac{5}{8}$

31. $-\frac{19}{20}$ **35.** $-6\frac{7}{12}$

32. $-3\frac{3}{5}$ **36.** $3\frac{1}{3}$

33. 2

Lesson 3.5

Try This, **page 155**

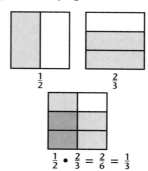

$\frac{1}{2} \cdot \frac{2}{3} = \frac{2}{6} = \frac{1}{3}$

Critical Thinking, **page 155**

Changing the fractions to decimals would result in a less accurate answer because the decimal for $\frac{1}{3}$ is a repeating decimal. On a calculator, such a number would be an approximation of $\frac{1}{3}$, such as 0.333.

Try This, **page 156**

$3\frac{1}{3} \cdot 1\frac{4}{5} = \frac{10}{3} \cdot \frac{9}{5} = \frac{90}{15} = 6$

Check: $3 \cdot 2 = 6$, so the answer is reasonable.

Ongoing Assessment, **page 157**

The keystrokes are the same for both types of calculators.

Critical Thinking, **page 157**

$4\frac{1}{2} \cdot 2\frac{1}{3} = \frac{9}{2} \cdot \frac{7}{3} = \frac{3\cdot3}{2} \cdot \frac{7}{3} = \frac{3\cdot3\cdot7}{2\cdot3} = \frac{3\cdot3\cdot7}{3\cdot2}$
$= \frac{3}{3} \cdot \frac{3\cdot7}{2} = 1 \cdot \frac{21}{2} = \frac{21}{2}$

Communicate, **page 159**

1. You can use an area model to show multiplication with fractions by overlapping a model of one fraction over a model of another fraction, as shown. The area where the fractions

overlap is the solution to the multiplication.

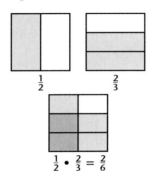

$$\frac{1}{2}$$ $$\frac{2}{3}$$

$$\frac{1}{2} \cdot \frac{2}{3} = \frac{2}{6}$$

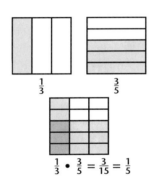

$$\frac{1}{3}$$ $$\frac{3}{5}$$

$$\frac{1}{3} \cdot \frac{3}{5} = \frac{3}{15} = \frac{1}{5}$$

2. Reciprocals are two numbers whose product is 1, such as $\frac{1}{2}$ and 2 or $\frac{3}{4}$ and $\frac{4}{3}$. In division of fractions, multiply by the reciprocal of the divisor. Example: $\frac{3}{4} \div \frac{2}{3} = \frac{3}{4} \cdot \frac{3}{2} = 1\frac{1}{8}$ and $\frac{1}{2} \div \frac{1}{8} = \frac{1}{2} \cdot 8 = 4$

3. Yes, if you divide by a number smaller than one, you are actually multiplying by the reciprocal of that number which will be larger than one. For example $4\frac{1}{2} \div \frac{2}{3} = 4\frac{1}{2} \cdot \frac{3}{2}$. Since $\frac{3}{2}$ is larger than one, the product will be greater than $4\frac{1}{2}$. (The product is $6\frac{3}{4}$.)

4. The reciprocal of -1 is -1 because $\frac{-1}{1} = \frac{1}{-1} = -1$.

Practice & Apply, **pages 159–161**

5. $4 \cdot \frac{1}{3} = \frac{4}{3}$ or $1\frac{1}{3}$

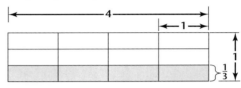

6. $\frac{2}{3} \cdot \frac{1}{2} = \frac{2}{6} = \frac{1}{3}$

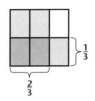

7. $1\frac{1}{2} \cdot 1\frac{1}{3} = \frac{3}{2} \cdot \frac{4}{3}$
$= \frac{12}{6}$
$= 2$

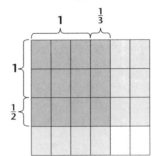

8. $2\frac{1}{2} \cdot 1\frac{1}{2} = \frac{5}{2} \cdot \frac{3}{2}$
$= \frac{15}{4}$
$= 3\frac{3}{4}$

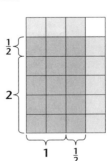

28. $\frac{3}{2}; \frac{2}{3} \cdot \frac{3}{2} = \frac{6}{6} = 1$

29. $\frac{2}{5}; \frac{5}{2} \cdot \frac{2}{5} = \frac{10}{10} = 1$

30. $\frac{3}{5}; \frac{5}{3} \cdot \frac{3}{5} = \frac{15}{15} = 1$

Additional Answers   **783**

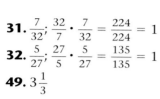

31. $\frac{7}{32}$; $\frac{32}{7} \cdot \frac{7}{32} = \frac{224}{224} = 1$

32. $\frac{5}{27}$; $\frac{27}{5} \cdot \frac{5}{27} = \frac{135}{135} = 1$

49. $3\frac{1}{3}$

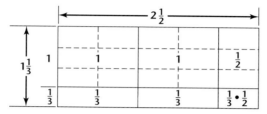

50. $14\frac{7}{8}$

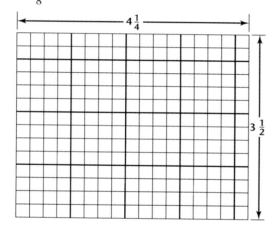

51. $\frac{2}{6}$

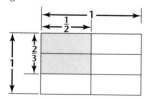

52. $\frac{27}{32}$

53. $\frac{1}{9}$

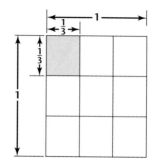

Look Back, **page 161**

76. $y = 2x - 1$

x	y
−3	−7
−2	−5
−1	−3
0	−1
1	1
2	3
3	5

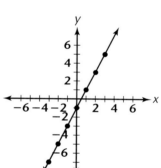

77. $y = 4 - x$

x	y
−3	7
−2	6
−1	5
0	4
1	3
2	2
3	1

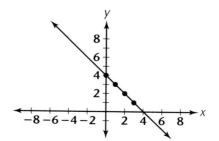

78. $y = -4x + 12$

x	y
-3	24
-2	20
-1	16
0	12
1	8
2	4
3	0

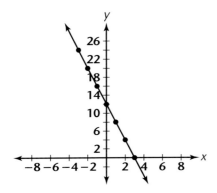

Look Beyond

82. $y = 1/x$ can be used to make a table of reciprocals because if $x = 2$, $y = \frac{1}{2}$; $x = 3$, $y = \frac{1}{3}$; and so forth.

83. There is an error message when $x = 0$ because $y = \frac{1}{0}$ is undefined.

84. 0.5 can be rewritten as $\frac{1}{2}$, and $\frac{1}{2}$ and 2 are reciprocals.

Lesson 3.6

Exploration 1, **page 163**

5. Start with the ratio $\frac{2}{7}$ using 2 yellow tiles and 7 blue tiles. To fill in the second column multiply the number of tiles times 2. For the third column multiply the number of tiles times 3. Continue the pattern.

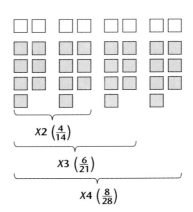

$\times 2 \left(\frac{4}{14} \right)$

$\times 3 \left(\frac{6}{21} \right)$

$\times 4 \left(\frac{8}{28} \right)$

Exploration 2, **page 164**

6. Find the number of attempts on the x-axis and move vertically to the line. The y-coordinate for that point is the number of goals. Find the number of goals on the y-axis and move horizontally to the line. The x-coordinate for that point is the number of attempts.

Ongoing Assessment, **page 165**

Put miles on the x-axis and minutes (or time) on the y-axis. Plot the point (2, 11). Use equivalent fractions to find other ratios equal to $\frac{2}{11}$, such as $\frac{4}{22}$, $\frac{6}{33}$, and $\frac{8}{44}$. Plot these points on your graph and draw a straight line through the points. To find how long it takes for Ty to run 5 miles, look for 5 on the x-axis. Move vertically to the line. The y-coordinate of that point is the answer (27.5 minutes).

Communicate, **page 166**

1. A ratio is the comparison of 2 integers by division, provided that the denominator is not zero.

2. The table could include ratios equivalent to $\frac{7}{2}$. By finding the column for 56 men in the table, you could see

the corresponding number of women (16).

men	7	14	21	28	35	42	49	56
women	2	4	6	8	10	12	14	16

3. In the proportion $\frac{7}{2} = \frac{56}{x}$, the multiplier is 8. Since $7 \times 8 = 56$, $2 \times 8 = 16$. Therefore, $x = 16$.

4. In solving a proportion such as $\frac{7}{2} = \frac{14}{x}$, use equivalent fractions to find x. Since $\frac{7}{2} = \frac{14}{4}$, $x = 4$.

5. If 2 ratios reduce to the same fraction, then the ratios are equivalent. Another way to determine if the ratios are equivalent is by changing both ratios to decimals. If the decimals are equal then the ratios are equivalent.

Practice & Apply, **pages 166–168**

36.

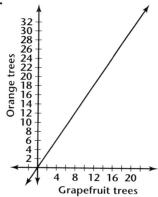

41.

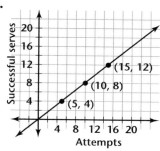

43.

Length	Width	Ratio of dimensions
5	3	$\frac{5}{3} \approx 1.667$
8	5	$\frac{8}{5} = 1.6$
13	8	$\frac{13}{8} = 1.625$

The ratio of length to width for each set of dimensions is very close to 1.618034. Examples of the golden rectangle are shown below.

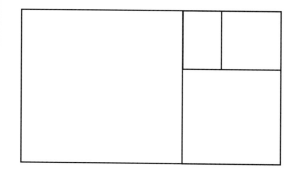

Look Back, **page 168**

47.

x	−2	−1	0	1	2	3	4
y	−1	1	3	5	7	9	11

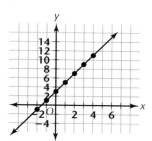

48.

x	−2	−1	0	1	2	3	4
y	−7	−4	−1	2	5	8	11

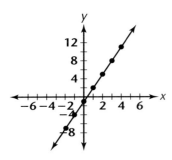

Lesson 3.7

Communicate, **page 174**

1. Show a bar with 10 sections. 80% is 8 of the 10 sections.

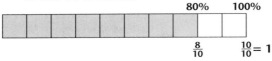

2. To change a percent to a fraction, write the percent over 100 and simplify.

$6\% = \dfrac{6}{100} = \dfrac{3}{50}$

$20.5\% = \dfrac{20.5}{100} = \dfrac{205}{1000} = \dfrac{41}{200}$

3. To change a fraction to a percent, divide the numerator by the denominator to get a decimal. Multiply the decimal by 100 to get a percent.

4. Since 24% is very close to 25% and 397 is very close to 400, $\frac{1}{4}$ of 400 is 100.

5. 40% of 100 is 40. 10% of 400 is 40. The amounts are equal.

Practice & Apply, **pages, 174–175**

6.
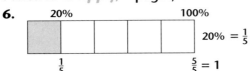

7.
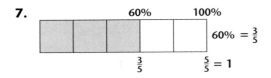

8.

9.
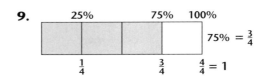

10.
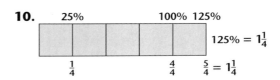

11.
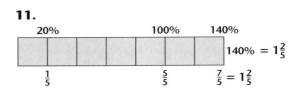

Eyewitness Math, **pages 176–177**

1a. Player A

Table 1: After a hit	
Number of times next shot is made	Number of times next shot is missed
17	19

Player B

Table 1: After a hit	
Number of times next shot is made	Number of times next shot is missed
16	18

1b. Player A hit followed hit: $\dfrac{17}{36} \approx 47.2\%$ or about 47%; hit followed by miss: $\dfrac{19}{36} \approx 52.8\%$ or about 53%. **Player B** hit followed by hit: $\dfrac{16}{34} \approx 47.1\%$ or about 47%; hit followed by miss: $\dfrac{18}{34} \approx 52.9\%$ or about 53%.

1c. About 50% for each because whether the player just hit or missed, the chances of making the next shot would be $\frac{1}{2}$.

2a. Player A

Table 2: After a miss	
Number of times next shot is missed	Number of times next shot is made
15	21

Player B

Table 2: After a miss	
Number of times next shot is missed	Number of times next shot is made
16	18

2b. Player A miss followed by miss: $\frac{15}{36} \approx 41.7\%$ or about 42%; miss followed by hit: $\frac{21}{36} \approx 58.3\%$ or about 58%. **Player B** miss followed by miss: $\frac{16}{34} \approx 47.1\%$ or about 47%; miss followed by hit: $\frac{18}{34} \approx 52.9\%$ or about 53%.

2c. About 50% for each because whether the play just hit or missed, the chances of making the next shot would be $\frac{1}{2}$.

3. Answers may vary. The study showed that the shooting records of players is about what you would get if they hit or missed at random. If a person were coaching, he or she should try to get the ball to the player who has the highest overall shooting percentage.

Lesson 3.8

Communicate, page 183

1. Experimental probability is calculated by performing an experiment and comparing the number of times an event occurs to the number of trials in the experiment.

2. Answers may vary. For example, toss 4 coins 10 times and count the number of times that 3 or 4 heads turn up.

3. Yes. For example, if 2 pairs of players toss 4 coins each, the number of heads showing may be different for both pairs.

4. Yes. If one player tosses 4 coins once and then again, the number of heads showing on each of the 4 coin tosses may be different.

5. integers 1, 2, 3, 4, 5, 6, or 7

6. In the 10 trials, only once were both numbers less than 50. The experimental probability is $\frac{1}{10}$.

Practice & Apply, pages 183–185

7. Toss 4 coins recording the result for each coin. One toss of all coins represents 1 trial. Repeat for 10 trials. Count the number of trials where 4 heads or 4 tails shows up. Divide this number by 10. The quotient represents the experimental probability.

8. Roll 2 number cubes. Record the results. This represents 1 trial. Repeat for 10 trials. Count the number of trials where at least one cube shows a 6. Divide this number by 10. The quotient represents the experimental probability.

9. Yes, both Fred and Ted will have the same result if all trials were successful events or if all trials were unsuccessful events. $\frac{15}{15} = \frac{16}{16}$ or $\frac{0}{15} = \frac{0}{16}$

For Exercises 16–23, answers may vary.

16. Toss a coin. Record the results. Repeat for 10 trials. Divide the number of trials where tails show up by 10.

17. Toss a coin twice, recording the result for each toss. Repeat for 10 trials. Divide the number of trials where 2 tails show up by 10.

18. Roll a number cube; record the result. Repeat for 20 trials. Divide the number of trials where 3, 4, 5, or 6 show up by 20.

19. Roll a number cube; record the result. Repeat for 20 trials. Divide the number of trials where 3 or 6 show up by 20.

20. Roll 2 number cubes, record the result. Repeat for 40 trials. Divide the number of trials where an odd sum shows up by 40.

21. Roll 2 number cubes; record the result. Repeat for 40 trials. Divide the number of trials where the same numbers shows up on both cubes by 40.

22. Toss a coin 5 times, recording the result after each toss. Repeat for 40 trials. Divide the number of trials where at least 3 heads or at least 3 tails appear in a row by 40.

23. Toss a coin 5 times, recording the result after each toss. Repeat for 40 trials. Divide the number of trials where heads and tails alternate by 40.

Lesson 3.9
Critical Thinking, page 191

The probability P(1st number > 5 *and* 2nd number is prime) depends on two independent events happening in a sequence. The probability of selecting a prime number is $\frac{2}{5}$, until it is paired with first selecting a number greater than 5. In that case, the $\frac{2}{5}$ probability is halved to $\frac{1}{5}$. Therefore $\frac{1}{2} \cdot \frac{2}{5}$ gives the correct probability.

Communicate, **page 193**

1. Theoretical probability is the expected outcome in an experiment. Experimental probability is what actually happens in an experiment.

2. Answers may vary. Sample answer. Picking a card from a deck, replacing it, and then picking another card.

3. Answers vary. Sample answer. Picking a card from a deck, not replacing it, and then picking another card.

4. To find the sample space, list all the possibilities for the pair of coins.

H, H	T, H
T, T	H, T

5. The events are independent so there is a $\frac{1}{6}$ chance of getting a 3 on either cube. To get a 3 on both cubes multiply $\frac{1}{6} \cdot \frac{1}{6}$.

Look Back, **page 195**

49. $\frac{1}{7}, \frac{1}{8}, \frac{1}{9}, \frac{1}{10}, \frac{1}{11}, \frac{1}{12}; \frac{1}{2} = 0.500;$
$\frac{1}{3} \approx 0.333; \frac{1}{4} = 0.250; \frac{1}{5} = 0.200;$
$\frac{1}{6} \approx 0.167; \frac{1}{7} \approx 0.143, \frac{1}{8} = 0.125;$
$\frac{1}{9} \approx 0.111; \frac{1}{10} = 0.100; \frac{1}{11} \approx 0.091;$
$\frac{1}{12} \approx 0.083;$ ordered from greatest to
least: $\frac{1}{12}, \frac{1}{11}, \frac{1}{10}, \frac{1}{9}, \frac{1}{8}, \frac{1}{7}, \frac{1}{6}, \frac{1}{5}, \frac{1}{4}, \frac{1}{3}, \frac{1}{2}$

50. $\frac{13}{21}, \frac{21}{34}, \frac{34}{55}, \frac{55}{89}, \frac{89}{144}, \frac{144}{233}; \frac{1}{2} = 0.500;$
$\frac{2}{3} \approx 0.667; \frac{3}{5} = 0.600; \frac{5}{8} = 0.625;$
$\frac{8}{13} \approx 0.615; \frac{13}{21} \approx 0.619; \frac{21}{34} \approx 0.6176;$
$\frac{34}{55} \approx 0.6182; \frac{55}{89} \approx 0.6179; \frac{89}{144} \approx 0.6181;$
$\frac{144}{233} \approx 0.6180;$ ordered from greatest to
least: $\frac{2}{3}, \frac{5}{8}, \frac{13}{21}, \frac{34}{55}, \frac{89}{144}, \frac{144}{233}, \frac{55}{89}, \frac{21}{34}, \frac{8}{13},$
$\frac{3}{5}, \frac{1}{2}$

Chapter 3 Project

Activity 1, **page 197**

Almundy Security	
Opening Value $30\frac{1}{4}$	Net Change
Monday $30\frac{1}{2}$	$+\frac{1}{4}$
Tuesday $29\frac{3}{4}$	$-\frac{3}{4}$
Wednesday $31\frac{1}{4}$	$+1\frac{1}{2}$
Thursday $32\frac{3}{8}$	$+1\frac{1}{8}$
Friday 32	$-\frac{3}{8}$
Net change for the week	$1\frac{3}{4}$
Percent change for the week $\frac{\text{Net change}}{\text{Opening value}}$	$\dfrac{1\frac{3}{4}}{30\frac{1}{4}} = \dfrac{1.75}{30.25} \approx 0.058 \approx 6\%$

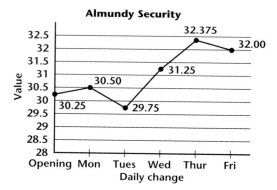

Almundy Security

Lesson 4.1

Critical Thinking, **page 206**
Use the inside numbers when measuring an angle that opens to the right. Use the

outside numbers when measuring an angle that opens to the left.

Communicate, **page 208**

1. Students' descriptions may vary. A *point* shows an exact location. A *line* consists of all points that extend infinitely in two opposite directions. A *ray* is part of a line that contains one endpoint and all the points extending in one direction from the endpoint. A *segment* consists of two endpoints on a line and all the points between the endpoints. A *plane* consists of all the points on a flat surface that extend infinitely in all directions.

2. The hands of a clock, the roof of a house, corners of rooms, and an open door are some examples. Right angles are modeled in the corners of rooms and in the hands of a clock at 3 o'clock. Acute angles are modeled in a door that is open only slightly, and in the hands of a clock at 10 minutes after 3. Obtuse angles are modeled in a wide open door and on roofs of houses.

3. Parallel lines never meet; perpendicular lines meet at right angles.

4. Right angles; they divide two straight angles in half; since a straight angle is 180°, half a straight angle is 90°.

5. Fold a sheet of paper; mark a point on the line that you formed. Fold your paper again on that point so that both parts of your line match up. The angles formed are right angles.

Practice and Apply, **pages 208–210**

32.

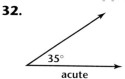

acute

33.

67°
acute

34.

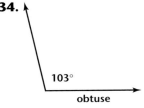

103°
obtuse

35.

170°
obtuse

36.

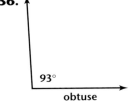

93°
obtuse

37.

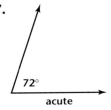

72°
acute

38.

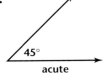

45°
acute

39.

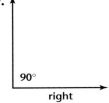

90°
right

40.

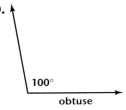

100°
obtuse

41.

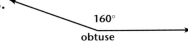

180°
straight

42.

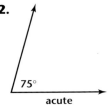

75°
acute

43.

160°
obtuse

44. Roofs with an acute angle retain less snow than roofs which are flat. If a roof accumulates too much snow, the weight of the snow can cause the roof to collapse.

56. No; in any triangle the sum of the measures of the angles equals 180°. $M\angle ABF = 180 - 45 - 50 = 85°$. Since a right angle measures 90°, $\angle ABF$ is not a right angle.

Lesson 4.2
Exploration, **page 211**

5. The sum of the measures of supplementary angles is 180°.

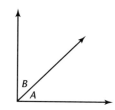

1. Two angles with a combined measure equal to 180°.

2. Two angles with a combined measure equal to 90°.

3. Each vertical angle is supplementary to the adjacent angle that lies between them because they each make a straight angle; therefore the 2 vertical angles must be congruent.

4. Perpendicular lines meet to form all right angles, therefore the adjacent and vertical angles are all 90° and thus, congruent.

5. Answers may vary. Complementary angles may be found in a train trestle and in a flower trellis.

6. Answers may vary. Supplementary angles may be found in the construction of buildings and in architecture.

Practice and Apply, **pages 214–215**

22. ∠CFB and ∠BFA; ∠DFE and ∠CFB

23. ∠EFD and ∠DFB; ∠EFC and ∠CFB; ∠DFC and ∠CFA; ∠DFB and ∠BFA; ∠EFA and ∠AFB; and ∠DFE and ∠EFA

33. The angle measured by the goniometer and the angle between the faces of the crystal will always create a straight angle, so their measures will always be supplementary.

34. Answers may vary. An example is shown.

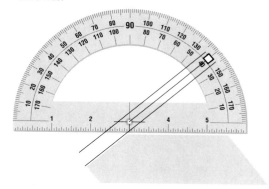

Lesson 4.3
Exploration 1, **pages 216–217**
2. Exterior angles are outside the 2 lines that the transversal crosses, and interior angles are between the 2 lines that the transversal crosses.

Exploration 4, **pages 219–220**
6. A straight line (the third side) is the shortest distance between two points. It is always shorter than any other path (the sum of the other two sides).

Communicate, **page 220**
1. Carpenters use the properties of parallels and transversals to find the measures of other angles in the construction of houses. They use a miter box to cut lumber at certain angles.

2. Possible corresponding congruent angles are ∠4 and ∠6, ∠3 and ∠5, ∠9 and ∠11, and ∠10 and ∠12. Angles that are not corresponding congruent angles are ∠1 and ∠3, ∠1 and ∠5, ∠2 and ∠4, ∠2 and ∠6, ∠7 and ∠9, ∠7 and ∠11, ∠8 and ∠10, or ∠8 and ∠12. Corresponding angles are congruent only when lines are parallel.

3. ∠10 and ∠11 are consecutive interior angles and when the lines are parallel, they are supplementary. Consecutive interior angles of lines that are not parallel are not supplementary.

4. The sum of any two sides must be greater than the length of the remaining side. Since 33 + 33 = 66 and 66 is not greater than 75, the triangle cannot be constructed.

5. In any right triangle, the right angle is the largest angle, so the side opposite the largest angle is the longest.

6. They are opposite congruent sides.

Practice and Apply, **pages 221–223**

16. ∠7 and ∠3; ∠1 and ∠5; ∠2 and ∠8; ∠6 and ∠4

18. The sum of the measures of consecutive interior angles of parallel lines equals 180°.

19. ∠7 and ∠2; ∠2 and ∠6; ∠6 and ∠1; ∠1 and ∠7; ∠3 and ∠8; ∠8 and ∠4; ∠4 and ∠5; ∠5 and ∠3; ∠2 and ∠3; ∠5 and ∠6; ∠1 and ∠4; ∠1 and ∠3; ∠2 and ∠4; ∠5 and ∠7; ∠6 and ∠8; ∠7 and ∠8

35. 1st, 2nd, and 3rd streets are parallel; a transversal (Main Street) of the parallel streets forms right angles at each intersection.

52. corresponding

53. alternate interior

54. alternate exterior

55. none

56. corresponding

57. none

58. corresponding

59. none

60. alternate interior

61. none

62. corresponding

63. alternate interior

64. m∠1 = 45°, m∠2 = 90°, m∠3 = 45°, m∠4 = 135°, m∠1 + m∠2 = m∠4

65. By definition of a straight angle m∠1 + m∠4 + m∠5 = 180°. Since alternate interior angles are congruent, substitute m∠2 for m∠4 and m∠3 for m∠5. By substitution, m∠1 + m∠2 + m∠3 = 180°.

67. The hypotenuse is opposite the right angle (the largest angle in the triangle). That makes the hypotenuse the largest side. Triangles may vary.

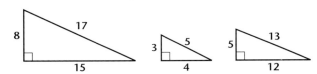

Extra Practice, **page 706**

8. ∠6 and ∠7, ∠2 and ∠4

9. ∠1 and ∠3, ∠5 and ∠8

10. ∠6 and ∠2, ∠7 and ∠4

11. The sum of their measures is 180°.

12. ∠3 and ∠5, ∠2 and ∠7, ∠5 and ∠2, ∠3 and ∠7, ∠6 and ∠1, ∠4 and ∠8, ∠4 and ∠6, ∠8 and ∠1, ∠4 and ∠7, ∠2 and ∠6

Lesson 4.4

Exploration 2, **pages 226–227**

3. A square has right angles and equal diagonals like a rectangle. A square has parallel sides and congruent sides like a rhombus.

Exploration 3, **pages 228–229**

10. The sum is the number of triangles times 180°; $180(n-2)$.

12. A square is a rectangle with four right angles; one angle of a square equals 90°. A regular pentagon has 5 equal angles. Use the formula $(5-2)180$ and divide by the number of angles, 5; a regular pentagon has 108° in one angle.

Communicate, **page 230**

1. Quadrilaterals are classified by their sides and angles.

2. Yes, it has 2 pairs of congruent parallel sides.

3. No, a rectangle is not a square, but a square is a rectangle. A rectangle does not always have all sides equal, so it is not a square. A square always has 4 right angles so it is a rectangle.

4. Count the number of triangles formed from the diagonals from one vertex, subtract two and multiply by 180.

Practice and Apply, **pages 230–231**

32.

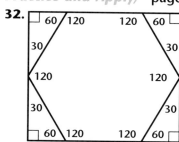

Look Beyond, **page 231**

43. $y = x^2 + 1$

x	−3	−2	−1	0	1	2	3
y	10	5	2	1	2	5	10

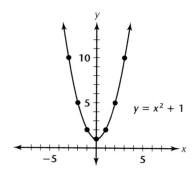

$y = x^2 + 1$

44. $y = x^2 - 2$

x	−3	−2	−1	0	1	2	3
y	7	2	−1	−2	−1	2	7

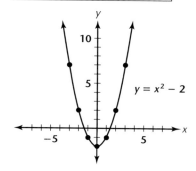

$y = x^2 - 2$

45. $y = x^2 + 2x + 1$

x	−3	−2	−1	0	1	2	3
y	4	1	0	1	4	9	16

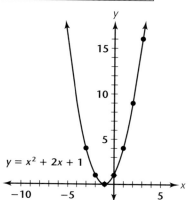

$y = x^2 + 2x + 1$

Lesson 4.5

Exploration 3, **pages 234–235**

2. A rectangle has 2 sides of length *l*, and 2 sides of width *w*, and the formula $P = 2w + 2l$ multiplies 2 times the width and 2 times the length.

11. Square with sides of 20 feet; from the table you find that the shape of a square has the smallest perimeter.

Communicate, **page 236**

1. Answers may vary. The border on a tile floor, a fenced dog kennel

2. Answers may vary. Carpeting a room, tiling a floor

3. Perimeter is the distance around the edge of a figure, while area is the measure of whole surface.

4. Yes; examples may vary.

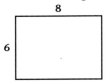
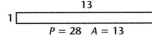

5. If squares have the same perimeter, the measures of the sides are the same and that makes the area the same.

6. Yes; examples may vary.

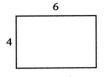
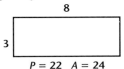

Practice and Apply, **pages 236–238**

14. $P = 286$ meters, $A = 4560$ square meters

15. $P = 16.5$ yards, $A = 14.375$ square yards

16. $P = 60\frac{3}{4}$ inches, $A = 230.66$ square inches

17. $P = 25.8$ centimeters, $A = 37.4$ square centimeters

18. $P = 16.5$ meters, $A = 14.375$ square meters

19. $P = 41$ inches, $A = 45$ square inches

20. $P = 440$ feet, $A = 12{,}100$ square feet

21. $P = 15.4$ miles, $A = 13$ square miles

22. $P = 14\frac{1}{3}$ yards, $A = 8\frac{1}{2}$ square yards

23. $P = 5.25$ inches, $A = 1\frac{17}{32}$ square inches

24. dimensions: 50 feet by 25 feet; perimeter = 150 feet; area = 1250 square feet

25. dimensions: 70 feet by 15 feet; perimeter = 170 feet; area = 1050 square feet

26. dimensions: 70 feet by 425 feet; perimeter = 990 feet; area = 29,750 square feet

27. dimensions: 70 feet by 45 feet; perimeter = 230 feet; area = 3150 square feet

	Fixed Perimeter	Length	Width	Area
34.	250	20	105	2100
35.	250	40	85	3400
36.	250	60	65	3900
37.	250	80	45	3600
38.	250	100	25	2500
39.	250	120	5	600

43. Answers may vary. Dimensions for five rectangular floor plans are 4 ft × 150 ft (with perimeter 308 ft), 8 ft × 75 ft (with perim. 166 ft), 10 ft × 60 ft (with perim. 140 ft), 15 ft × 40 ft (with perim. 110 ft), and 24 × 25 ft (with perim. 98 ft). The 24 ft × 25 ft rectangle has the smallest perimeter because it most closely resembles a

square. The closer the rectangle resembles a square shape, the smaller the perimeter.

Extra Practice, **page 707**

1. $P = (8)(4) = 32$, $A = (8)(8) = 64$

2. $P = (21)(2) + (11)(2) = 64$, $A = (21)(11) = 231$

3. $P = (50)(2) + (70)(2) = 240$, $A = (50)(70) = 3500$

4. $P = 160$ meters, $A = 1536$ square meters

5. $P = 12.5$ feet, $A = 9.625$ square feet

6. $P = 36$ inches, $A = 78\frac{23}{64}$ square inches

7. $P = 480$ feet, $A = 14,400$ square feet

8. $P = 4.8$ miles, $A = 1.28$ square miles

9. $P = 10$ yards, $A = 4\frac{8}{9}$ square yards

10. $P = 6\frac{1}{8}$ inches, $A = \frac{55}{64}$ square inches

11. $P = 44.4$ centimeters, $A = 117.92$ square centimeters

12. $P = 48\frac{3}{4}$ inches, $A = 148\frac{137}{256}$ square inches

13. $P = 24$ meters, $A = 30.9375$ square meters

Lesson 4.6

Exploration 1, **pages 239–240**

5. Answers may vary.

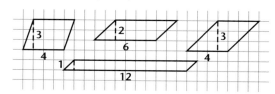

Exploration 2, **pages 241–242**

5. Answers may vary.

6.

A rectangle; an isosceles right triangle; an isosceles triangle.

7. 4 square units; triangles may vary.

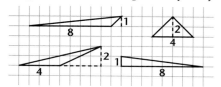

8. Triangles may vary.

right triangles

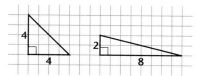

obtuse triangles

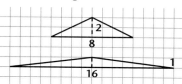

acute triangles

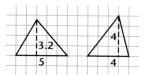

Exploration 3, **page 243**

5. $7\frac{1}{2}$ square units

6. Trapezoids may vary.

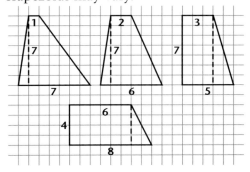

Critical Thinking, **page 244**

The triangles *ABD* and *BCD* form trapezoid *ABCD*.

$$A_{ABD} + A_{BCD} = \text{Area of trapezoid } ABCD$$
$$A_{ABD} + A_{BCD} = \frac{1}{2}(7)(4) + \frac{1}{2}(4)(4)$$
$$= \frac{1}{2}[(7)(4) + (4)(4)]$$
$$= \frac{1}{2}(7 + 4)4$$
$$= \frac{1}{2}(b_1 + b_2)h$$

Communicate, **page 244**

1.

2. A parallelogram can be divided into 2 equal triangles; the area of a parallelogram is $A = bh$, therefore the area of half a parallelogram is $\frac{1}{2}bh$.

3. Draw an upside down copy of a trapezoid next to the original trapezoid to form a parallelogram.

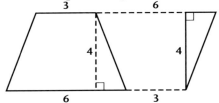

The area of the parallelogram is double the area of the trapezoid. The area of the parallelogram is $A = bh$, therefore the area of the trapezoid can be found by the formula $A = \frac{1}{2}bh$.

4. The method of determining the height for each of these types of triangles is different. To find the height in an acute triangle, measure the inside height. To find the height in an obtuse triangle, measure the height outside the triangle. To find the height of a right triangle, measure the height of the leg of the triangle.

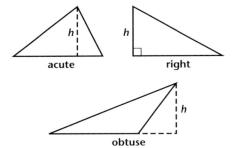

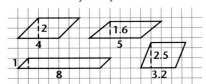

Practice and Apply, **pages 245–247**

14. Answers may vary.

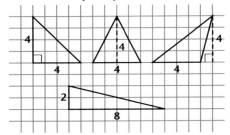

15. Answers may vary.

16. Answers may vary.

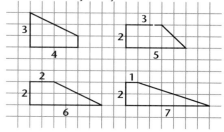

53. $4 \cdot$ Area of one face $= 4\left[\dfrac{(b_1 + b_2)h}{2}\right] = 2(b_1 + b_2)h$

54. 22,500 square feet; since the base of the pyramid is square, find the area of a square with a length of 150 feet on a side.

55. 1600 square feet; since the top of the pyramid is square, find the area of a square with a length of 40 feet on a side.

56. The measure of the base and height stay the same.

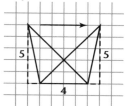

Let triangle *ABC* be the original triangle with $b = 4$ and $h = 5$; triangle *A'BC* is the transformed triangle. The base is still equal to 4 units and the height is still equal to 5 units.

Look Beyond, **page 247**

67. When the base and height of a triangle are both doubled, the area increases by a factor of 4. When the base and height are tripled, the area increases by a factor of nine.

68. When the base and height of a parallelogram are both doubled, the area increases by a factor of four. When the base and height are tripled, the area increases by a factor of nine.

Lesson 4.7

Ongoing Assessment, **page 248**

Yes; since 3.5 yards is between the length of a square 3 yards on a side and a square 4 yards on a side, 3.5 yards is a reasonable estimate.

Ongoing Assessment, **page 250**

3 units, 4 units, 5 units; 9 square units, 16 square units, 25 square units; each area is the square of the length of one side of the triangle; $a^2 + b^2 = c^2$, the sum of the squares of two sides equals the square of the longest side.

Communicate, **page 253**

1. A square root of a given number is greater than 0, such that when you multiply the square root of a number times the square root of the number, the result is the given number; $\sqrt{a} \cdot \sqrt{a} = a$.

2. The sum of the square of the two legs of a right triangle equals the square of the hypotenuse of the right triangle.

3. If you have the lengths of 2 sides of a figure that forms a right triangle, you can find the length of the third side using the "Pythagorean" Right-Triangle Theorem; this method can be used to measure distance.

4. Answers may vary. To find the length of a ladder needed to reach a window; to find the distance from home plate to second base in a baseball game, to find measurements in construction.

Extra Practice, **page 708**

1. 3, 2.6, 2.65

2. 4, 3.6, 3.61

3. 5, 4.9, 4.90

4. 6, 6.3, 6.32

5. 9, 8.7, 8.66

6. 9, 9.1, 9.11

7. 9, 9.9, 9.90

8. 11, 10.6, 10.58

9. 9, 8.5, 8.54

10. 12, 12.0, 12.04

11. 13, 13.0, 13.04

12. 5, 5.2, 5.20

13. 8, 8.1, 8.12

14. 11, 11.1, 11.14

15. 16, 15.8, 15.81

Lesson 4.8

Communicate, **page 260**

1. A scale factor indicates how many times larger or smaller an object is than another; a scale factor is used to convert measurements; examples may vary; constructing models of cars; drawing art objects.

2. Similar figures have the same shape, but do not have to be the same size; all similar figures have proportional sides and congruent angles.

3. Scale factors are used to convert measurement of similar figures by using ratios to enlarge or reduce; the scale factor is multiplied by its corresponding side of an object to enlarge or reduce.

4. Answers may vary.

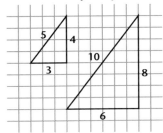

$$\frac{10}{5} = \frac{2}{1}, \frac{8}{4} = \frac{2}{1}, \frac{6}{3} = \frac{2}{1}$$

Practice and Apply, **pages 260–263**

5. 20 cm

6. 24 cm

7. 32 cm; 40 cm

8. 4 cm; 5 cm

9. 6.4 cm; 8 cm

10. 3 cm; 3.75 cm

11. 21 cm; 26.25 cm

12. 7.2 cm; 9 cm

13. 9 cm; 11.25 cm

14. 30 cm; 37.5 cm

15. The ratio of the perimeter of the similar figures to the perimeter of the original rectangle is the same as the scale factor.

16. The ratio of the area of the similar rectangles to the area of the original rectangle is the square of the scale factor.

23. The ratio of the area of the similar triangles to the area of each original triangle is the square of the scale factor.

24. The ratio of the perimeter of the similar triangles to the perimeter of each original triangle is the same value as the scale factor.

29. Scale factors may vary.
Original: 3-4-5;
Scale factor: 2, New Triangle: 6, 8, 10;
Scale factor: 3, New Triangle: 9, 12, 15
Scale factor: $\frac{1}{2}$, New Triangle: $1\frac{1}{2}$, 2, $2\frac{1}{2}$;
Scale factor: $\frac{1}{3}$, New Triangle: 1, $1\frac{1}{3}$, $1\frac{2}{3}$

35. 6 centimeters by 12 centimeters; graphic, perimeter = 24 centimeters, area = 32 square centimeters; enlargement, perimeter = 36 centimeters, area = 72 square centimeters

36. The perimeter is 150% larger and the area is 225% larger.

37.

Scale factor (%)	Scale factor (decimal)	Width	Length	Perimeter	Area
100%	1	4 centimeters	8 centimeters	24 centimeters	32 square centimeters
150%	1.5	6 centimeters	12 centimeters	36 centimeters	72 square centimeters
200%	2	8 centimeters	16 centimeters	48 centimeters	128 square centimeters
300%	3	12 centimeters	24 centimeters	72 centimeters	288 square centimeters
75%	0.75	3 centimeters	6 centimeters	18 centimeters	18 square centimeters
50%	0.5	2 centimeters	4 centimeters	12 centimeters	8 square centimeters

38. Multiply the scale factor times the original perimeter to find the new perimeter; multiply the original area by the square of the scale factor to find the new area.

Look Beyond, page 263

	A	B	C	$\overline{AB}$	$\overline{BC}$
Original	(−2, 1)	(−2, −3)	(1, −3)	4	3

	Scale Factor	A′	B′	C′	$\overline{A'B'}$	$\overline{B'C'}$
50.	2	(−4, 2)	(−4, −6)	(2, −6)	8	6
51.	3	(−6, 3)	(−6, −9)	(3, −9)	12	9
52.	4	(−8, 4)	(−8, −12)	(4, −12)	16	12

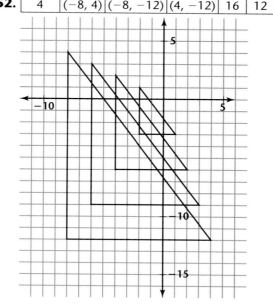

Chapter 4 Project
Activity 1, **page 265**

1. $\sqrt{2}$ miles ≈ 1.41 miles; 8.57 acres

2. 4 miles; 4.82 acres

3. 30 acres

4. 596.61 acres

Activity 2, **page 265**
Answers may vary.

Chapter 4 Review, **pages 266–268**

1.

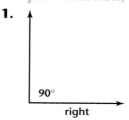

90°

right

2.

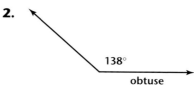

138°

obtuse

3.

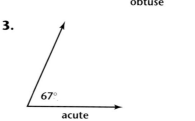

67°

acute

Chapter 4 Assessment, page 269

1.

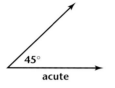

45°
acute

2.

155°
obtuse

3.

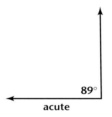

89°
acute

4.

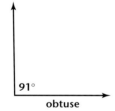

91°
obtuse

12.

13.

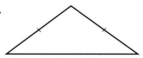

14.

15.

16.

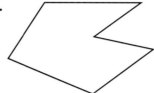

17.

Lesson 5.1

Ongoing Assessment, page 275

$2x - 3$ $4x + 5$

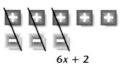

$6x + 2$

The result is $6x + 2$.

Communicate, page 277

1. An algebraic expression is an expression containing numbers and variables. Answers may vary. For example: $4x - 7$ and $3x + 2y$.

2. A coefficient is the factor of, or the number multiplied by, the variable. Answers may vary. For example: The coefficient in the term $4x$ is 4, and the coefficient in the term $-7y$ is -7.

3. For any a and b, $a + b = b + a$. Answers may vary. For example: $3 + x = x + 3$ and $5 + 7 = 7 + 5$.

4. Factor out a common factor of x; then add like terms $3x + 8x = (3 + 8)x = 11x$.

5. $7x$, $2x$, and $3x$ are like terms because they all have only the variable x. $3z$ and $-z$ are like terms because they both have only the variable z. 5 and 23 are like terms because they are both constants. $7y$ and $3y$ are like terms because they both have only the variable y.

6. Use 5 positive x-tiles and 2 positive 1-tiles, and 3 positive x-tiles and 4 negative 1-tiles. Combine like tiles.

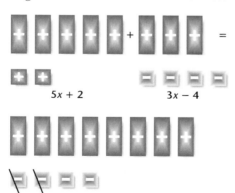

$5x + 2$ $3x - 4$

$8x - 2$

Practice and Apply, pages 277–279

15. 6 positive x-tiles and 2 positive x-tiles combine to form 8 positive x-tiles; 3 positive 1-tiles and 1 positive 1-tile combine to form 4 positive 1-tiles. The result is 8 positive x-tiles and 4 positive 1-tiles. $(6x + 3) + (2x + 1) = 8x + 4$.

28. $(3m + 2r) + (5r - m) + (4r + 2m) + (r - 5m) = 12r - m$

29. $(3a + b + c) + (4b - 2c) + (3a + b + c) + (4b - 2c) = 6a + 10b - 2c$

30. $(3s - 3t) + (5t - 4s + p) + (2p - 5s) = 2t - 6s + 3p$

31. $x(y + z)$: You can distribute the x over the addition of y and z; or $xy + xz$: You can factor out the x to get $x(y + z)$.

Look Beyond, page 279

49.

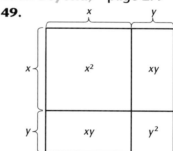

Lesson 5.2

Ongoing Assessment, page 281

Start with 3 positive x-tiles and 5 negative 1-tiles. Give yourself 3 neutral pairs of x-tiles and 2 neutral pairs of 1-tiles. Now you can remove 6 positive x-tiles and 2 positive 1-tiles. This leaves 3 negative x-tiles and 7 negative 1-tiles, or $-3x - 7$.

Critical Thinking, page 282

$-(5x + 2y)$ is the same as the expression $-1(5x + 2y)$, which is simplified by using the Distributive Property and also results in taking the opposite of each term.

Communicate, page 283

1. Use 4 positive *x*-tiles, and take away 3 of them leaving 1 positive *x*-tile.

2. Start with 3 positive *x*-tiles and 2 positive 1-tiles. Add 1 neutral pair of *x*-tiles. Now you can take away 2 positive *x*-tiles leaving 1 positive *x*-tile. Also, take away 1 negative 1-tile leaving 1 positive 1-tile.

3. Start with 5 positive *x*-tiles and 3 positive 1-tiles. Add 1 neutral pair of 1-tiles. Now, take away 2 positive *x*-tiles and 4 positive 1-tiles leaving 3 positive *x*-tiles and 1 negative 1-tile.

4. Let *r* represent the number of reams of paper. The algebraic expression for the information is $(5r + 100) - (2r + 50)$. To simplify: $(5r + 100) + (-2r - 50) = (5r - 2r) + (100 - 50) = (3r + 50)$. Ms. Green had 3 reams of paper and 50 loose sheets.

5. Change the subtraction to addition and then change the signs of the terms being subtracted.

6. $(7y + 9x + 3) + (-3y) - 4x + 1$
            Definition of Subtraction
$(7y - 3y) + (9x - 4x) + (3 + 1)$
                        Rearrange terms.
$4y + 5x + 4$ Simplify.

Practice and Apply, pages 283–285

30. False; the opposite of *z* was not taken.

31. False; the opposite of $-z$ was not taken.

32. False; taking the opposite of $(y - z)$ is incorrect.

33. False; it is not known whether *x* is negative or positive.

50.

51.

Lesson 5.3

Exploration 1, pages 286–287

2.

3.

Ongoing Assessment, page 289

Substitute -1 for *x* in the equation $x - 2 = -3$ to see if it is true. Substitute 1 negative 1-tile for the *x*-tile to see if both sides are equal.

Communicate, page 289

1. Isolate the positive *x*-tile by adding the same number of opposite 1-tiles that are on the *x*-tile side. Answers may vary. For example:

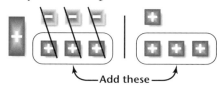

2. Neutral pairs isolate the positive *x*-tile.

3. Set up a positive *x*-tile and 3 negative 1-tiles on the left, and seven positive 1-tiles on the right. Use 3 positive 1-tiles on the left and on the right to make neutral pairs, and cancel the ones on the left. You should be left with a positive *x*-tile on the left and 10 positive 1-tiles on the right. The result is $x = 10$.

4. Start with 1 positive *x*-tile and 4 positive 1-tiles on the left and 2 negative 1-tiles on the right. Add 4 negative 1-tiles to each side to isolate the *x*-tile. Remove neutral pairs. The result is 1 positive *x*-tile on the left and 6 negative 1-tiles on the right. The result is $x = -6$.

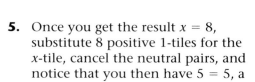

5. Once you get the result $x = 8$, substitute 8 positive 1-tiles for the x-tile, cancel the neutral pairs, and notice that you then have $5 = 5$, a true statement.

Look Beyond, page 291

62. Begin with 4 positive x-tiles and 2 negative 1-tiles on the left side, and 3 positive x-tiles and 1 positive 1-tile on the right side. Remove 3 positive x-tiles from each side. The result is 1 positive x-tile and 2 negative 1-tiles on the left side and 1 positive 1-tile on the right side. To isolate the positive x-tile, add 2 positive 1-tiles to each side, and remove neutral pairs. The result is 1 positive x-tile on the left and 3 positive 1-tiles on the right, or $x = 3$.

Lesson 5.4

Critical Thinking, page 292

Begin with 3 positive x-tiles and 4 positive 1-tiles on the left, and 2 positive x-tiles and 10 positive 1-tiles on the right. Remove 2 x-tiles from each side and 4 1-tiles from each side. The result is 1 positive x-tile on the left and 6 positive 1-tiles on the right, or $x = 6$.

Ongoing Assessment, page 295

Add b to each side, and simplify to get $a + b = x$. Rewrite with x on the left side to get $x = a + b$.

Communicate, page 296

1. Use the Addition Property of Equality with a subtraction equation. Use the Subtraction Property of Equality with an addition equation.

2. Answers may vary. Sample answers are given. a. If you have saved $20 and you need $50, how much more do you

need to save? b. If a friend lends you $20 and has $50 left, how much did the friend have to begin with?

3. a. Begin with 1 positive x-tile and 6 positive 1-tiles on the left side, and 10 positive 1-tiles on the right side. Remove 6 positive x-tiles from each side. The result is 1 positive x-tile on the left side and 4 positive 1-tiles on the right, or $x = 4$. b. Begin with 1 positive x-tile and 6 negative 1-tiles on the left side, and 10 positive 1-tiles on the right side. Add 6 positive 1-tiles to each side, and remove neutral pairs. The result is 1 positive x-tile on the left and 16 positive 1-tiles on the right or $x = 16$.

4. Subtract x from each side. The result is $s = r - x$.

5. Subtract b from each side. The result is $m = n - b$.

Practice and Apply, pages 296–298

30. In Exercise 28 the Subtraction Property of Equality is used and in Exercise 29 the Addition Property of Equality is used.

48. $\begin{cases} s + d = 2c \\ d = 2s + c \end{cases}$ Solve for d in terms of s.

Solutions may vary. A sample is given. First substitute $2s + c$ for d in the first equation.

$s + d = 2c$
$s + (2s + c) = 2c$
$3s = c$

Substitute $3s$ for c in the second equation.

$d = 2s + c$
$d = 2s + 3s$
$d = 5s$

The weight of 5 squirrels equals the weight of 1 dog.

Look Back, page 298

49. Subtract 6, add 12, subtract 6, add 12, and so on. 24, 18, 30

Extra Practice, page 711

1. Add Prop; 35
2. Subt Prop; -13
3. Add Prop; 124
4. Add Prop; Subt Prop; 64
5. Add Prop; 7
6. Subt Prop; -29
7. Subt Prop; 2
8. Add Prop; 156
9. Subt Prop; -48
10. Add Prop; Subt Prop; -350
11. Add Prop; Subt Prop; 44
12. Add Prop; Subt Prop; -2
13. Add Prop; Subt Prop; 109
14. Add Prop; Subt Prop; -62
15. Add Prop; 61
16. Add Prop; Subt Prop; 58
17. Subt Prop; 60
18. Add Prop; 270
19. Add Prop; 16.8
20. Subt Prop; 5.6
21. Add Prop; 17.7
22. Add Prop; Subt Prop; 25.5
23. Subt Prop; $\frac{1}{2}$
24. Add Prop; Subt Prop; $\frac{4}{5}$
25. Add Prop; $1\frac{1}{6}$
26. Add Prop; Subt Prop; $\frac{1}{8}$
27. Add Prop; $\frac{9}{10}$
28. Subt Prop; $-1\frac{11}{18}$

Lesson 5.5

Exploration 1, Part I, page 300

2. $2x$ is represented by 2 positive x-tiles, and $-3x$ is represented by 3 negative x-tiles. Two neutral pairs of x-tiles can be removed, leaving 1 negative x-tile, as shown.

3. $2x - 3x = -x$; 3 = 3 True; $-4 = -4$ True; $-10 = -10$ True

4. Begin with 2 negative x-tiles, 5 positive x-tiles, and 1 negative x-tile. Remove the 3 neutral pairs of x-tiles. This leaves 2 positive x-tiles. So $-2x + 5x - x = 2x$. To check this solution, substitute several different values for x into the equation. For each value of x, the equation should be true.

5.

The left and right sides do not match, so they are not equal.
If $x = -1$, $-1 + 3 \neq -3$; $2 \neq -3$
If $x = 2$, $2 + 3 \neq 6$; $5 \neq 6$
If $x = 4$, $4 + 3 \neq 12$; $7 \neq 12$

6. Let $a = 1$ and $b = 3$.
$1 + 3 \neq 3$
$4 \neq 3$

7. Let $a = -1$ and $b = 2$.
$-2 + 10 \neq -14$
$8 \neq -14$

Exploration 1, Part II, page 301

2. Combine like terms: $-2x^2 + 5x^2 - x^2 = 2x^2$; substitute values of x in the original expression and in the simplified expression to see that both results are equal.

Exploration 2, pages 301–302

2. Answers may vary. Sample answer given.

Let $x = 3$.
$$(2x + 2) + (3x - 3) = 5x - 1$$
$$(2(3) + 2) + (3(3) - 3) = 5(3) - 1$$
$$14 = 14 \text{ True}$$

3. Let $x = -1$.
$$(2x + 2) + (3x - 3) = 5x - 1$$
$$(2(-1) + 2) + (3(-1) - 3) = 5(-1) - 1$$
$$-6 = -6 \text{ True}$$

4. Let $x = 10$.
$$(-3x - 1) + (2x + 5) = -x + 4$$
$$(-3(10) - 1) + (2(10) + 5) = -(10) + 4$$
$$-6 = -6 \text{ True}$$

5. $2x - (3x - 1) = -x + 1$; $4 = 4$ True;
$-3 = -3$ True; $-9 = -9$ True

8. Combine like terms, and simplify:
$-3x^2 - 2x$; use the Definition of
Subtraction (add the opposite), and
combine like terms: $-3x^2 + 2x$; check
by substituting values for x in the
original expression and in the
simplified expression to see that the
results are equal.

Communicate, **page 304**

1. A monomial has 1 term, a binomial has
2 terms, and a trinomial has 3 terms.

2. Answers may vary. $(3x + 2y) - (4x + y)$
$= (3x - 4x) + (2y - y) = -x + y$;
$(5z + 3) + (4z - 1) = (5z + 4z) +$
$(3 - 1) = 9z + 2$

3. Answers may vary. Add the opposite of
each term in the expression. $(4x - 5) -$
$(2x + 6) = (4x - 5) + (-2x - 6) =$
$2x - 11$; $(3y - 4x) - (5y + 6x) =$
$(3y - 4x) + (-5y - 6x) = -2y - 10x$

Practice and Apply, **pages 304–306**

15. $2x^2 - 5x + 3$

17. $-2x^2 + 4x - 2$

21. $-x^2 + 7x$

37.

X	$Y_1 = (X^2 + X) + (-X^2 - 2X)$	$Y_2 = -X$
5	-5	-5

38.

X	$Y_1 = (X^2 + 2X) + (-2X^2 - X)$	$Y_2 = -X^2 + X$
7	-42	-42

Look Back, **page 306**

71. $\frac{3}{1}$, or 3; $\frac{3}{1} \cdot \frac{1}{3} = \frac{3}{3} = 1$

72. $\frac{4}{3}; \frac{4}{3} \cdot \frac{3}{4} = \frac{12}{12} = 1$

73. $\frac{2}{5}; \frac{2}{5} \cdot \frac{5}{2} = \frac{10}{10} = 1$

Look Beyond, **page 306**

74. $-x^2 - 2x + 1$ **75.** $5x^2 - 4x + 9$

76.

x	$-x^2 - 2x + 1$	$(2x^2 - 3x + 5) + (-3x^2 + x - 4)$
7	-62	-62
8	-79	-79
9	-98	-98
10	-119	-119

x	$5x^2 - 4x + 9$	$(2x^2 - 3x + 5) - (-3x^2 + x - 4)$
7	226	226
8	297	297
9	378	378
10	469	469

Lesson 5.6

Ongoing Assessment, **page 308**

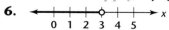

Exploration 3, **page 309**

1. $10 + (-2)$; $10 + (-2) \geq 20$; $8 \geq 20$; no

2. $15 + (-2)$; $15 + (-2) \geq 20$; $13 \geq 20$; no

3. $20 + (-2)$; $20 + (-2) \geq 20$; $18 \geq 20$; no

4. $25 + (-2)$; $25 + (-2) \geq 20$; $23 \geq 20$; yes

5. $30 + (-2)$; $30 + (-2) \geq 20$; $28 \geq 20$; yes

Practice and Apply, **pages 311–312**

6.

7. Divide $5p$ by -5 to get $-p$ and divide -15 by -5 to get 3 so that the result is $-p + 3$.

Look Beyond, **page 335**

72.

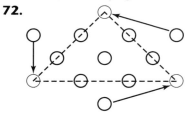

Eyewitness Math, **pages 336–337**

1a. When r is 2 regardless of the values used for P_1, the population will eventually reach 0.5. When r is 4, the population never stabilizes.

1b. There is a pattern when r is 2, but no pattern when r is 4.

1c. All groups should get the same general result, no matter what value between 0 and 1 is chosen for P_1.

1d. When r is 2, you can make a good prediction.

2–3. When r is 4 even minuscule changes in P_1 cause varied results.

Lesson 6.2
Critical Thinking, **page 341**

Both sides of the equation are equal if multiplied or divided by the same numbers. Multiplication and division are opposite operations; to solve a multiplication problem, division is used; to solve a division problem, multiplication is used.

Communicate, **page 342**

1a. Answers may vary. For example, Benjamin makes $5 an hour at his part-time job. If his pay for one week was $100, how many hours did he work?

1b. Answers may vary. For example, when David and Marisa divided their game tokens, each received 10. How many tokens did they have originally?

2. Divide both sides of the equation by 592 for $x \approx 1.4$.

3. Add 246 to both sides of the equation for $x = 774$.

4. Divide both sides of the equation by 5 $\left(\text{or multiply both sides by } \frac{1}{5}\right)$ for $x = \frac{1}{50}$.

6. Divide both sides of the equation by 2π for $r = \frac{C}{2\pi}$.

7. Answers may vary. For example, if Mary worked 4 math problems in 40 minutes, how long did it take her to work 1 problem?

8. Answers may vary. For example, if Mr. Johnson divided his PE class into 4 teams with 20 students on each team, how many students are in his PE class?

9. Answers may vary. For example, in 5 minutes, Adam collected $15 from people coming into the basketball game. If each person paid $2.50, how many people came to the game in that 5 minutes?

10. Answers may vary. For example, Paul organized his magazines by placing them into 4 piles of 20 each. How many magazines did Paul have?

Practice and Apply, **pages 343–344**

11. Multiplication Property of Equality, $y = -39$

12. Multiplication Property of Equality, $x = -702$

13. Addition Property of Equality, $x = 2\frac{1}{3}$

14. Division Property of Equality, $x = 8$

15. Multiplication Property of Equality, $b = -54$

16. Division Property of Equality, $y = -7$

17. Addition Property of Equality, $x = 2\frac{2}{3}$

18. Division Property of Equality, $x = -\frac{8}{7}$

19. Division Property of Equality, $v = 1.25$

20. Division Property of Equality, $-\frac{1}{8} = w$

21. Multiplication Property of Equality, $x = 9.8$

22. Division Property of Equality, $x = \frac{2}{3}$

Extra Practice, **page 713**

1. Multiplication Property of Equality; $x = -24$.

2. Multiplication Property of Equality; $x = 40$.

3. Addition Property of Equality; $x = 2\frac{1}{2}$.

4. Division Property of Equality; $x = \frac{4}{5}$

5. Division Property of Equality; $y = -\frac{1}{3}$

6. Multiplication Property of Equality; $c = -140$

7. Division Property of Equality; $x = \frac{1}{7}$

8. Subtraction Property of Equality; $x = 2\frac{2}{5}$

9. Division Property of Equality; $y = -\frac{1}{12}$

10. Division Property of Equality; $z = -\frac{1}{8}$

11. Multiplication Property of Equality; $x = 17.5$

12. Division Property of Equality; $x = \frac{7}{4}$

Lesson 6.3

Communicate, **page 349**

1. Use two x-tiles as one factor. Use one x-tile and five 1-tiles as the other factor. The product rectangle will be 2 x^2-tiles and 10 x-tiles.

2. Multiply each term in $x + 5$ by $2x$;

3.

4. Write the factors whose product is $x^3 + x$. The factors are x and $x^2 + 1$; $x^3 + x = x(x^2 + 1)$.

Practice and Apply, **pages 349–350**

5.

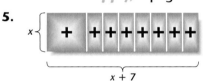

6.

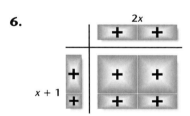

40. The area of the flower bed plus the area of the lawn is equal to the area of the backyard.

Look Beyond, **page 350**

48. $(2x + 1)(x - 5) = 2x^2 - 9x - 5$

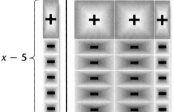

49. $(x + 3)(3x - 4) = 3x^2 + 5x - 12$

Lesson 6.4

Critical Thinking, **page 352**

Zero and -2 are integers by definition. Since both can be expressed as the ratio of two integers, both are also rational numbers. For example, $\frac{0}{4} = 0$ and $\frac{-2}{1} = -2$.

Ongoing Assessment, **page 353**

Answers may vary.

[( ] [(-)] [2] [÷] [5] [ )] [x^{-1}] [ENTER]

Critical Thinking, **page 354**

Both examples are solved by first applying the Multiplication Property of Equality. Method A multiplies each side by the reciprocal of the coefficient of x. Method B multiplies each side by a common denominator, not the reciprocal.

Communicate, **page 355**

1. A rational number is a number that can be expressed as the ratio of 2 integers with 0 excluded from the denominator. Examples include $\frac{2}{3}, \frac{-6}{1}, \frac{94}{3}$.

2. An integer is a rational number because it can be expressed as the ratio of itself and 1.

3. Start at 0 and move right $\frac{1}{2}$ of a unit. Then move left 3 units. The result is $-2\frac{1}{2}$.

4. Start at 0 and move right $\frac{1}{2}$ of a unit. Then move left $\frac{5}{6}$ of a unit. The result is $\frac{-2}{6}$ or $\frac{-1}{3}$.

5. For any number r other than 0, $r \cdot \frac{1}{r} = 1$. The reciprocal of -3 is $\frac{-1}{3}$ because $-3 \cdot \frac{-1}{3} = \frac{3}{3} = 1$.

6. Multiply both sides by 8 (the reciprocal of the coefficient of the y-term) for $y = \frac{-16}{3} \approx -5.33$. Multiply both sides by 24 (the least common denominator) for $3y = -16$. Then divide both sides by 3 for $y = \frac{-16}{3} \approx -5.33$.

Lesson 6.5

Exploration 1, **pages 357–358**

2. For every 2 revolutions of the larger gear, the smaller gear turns 3 revolutions.

3. Set the ratios equal to each other and solve for x: $\frac{2}{3} = \frac{14}{x}$; $x = 21$

4. Set the ratios equal to each other and solve for x: $\frac{2}{3} = \frac{x}{42}$; $x = 28$.

Ongoing Assessment, **page 358**

a. Means: 5, 6; Extremes: 1, 30
b. Means: 12, 1; Extremes: 4, 3
c. Means: 10, 8; Extremes: 20, 4

Critical Thinking, **page 359**

Solve $\frac{6}{0.75} = \frac{8}{x}$.

$$6x = 6.00$$
$$x = 1.00$$

From the proportion, 6 for 75¢ is the same as 8 for $1.00. Therefore, 8 for 99¢ is the better buy.

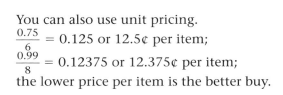

You can also use unit pricing.

$\frac{0.75}{6} = 0.125$ or 12.5¢ per item;

$\frac{0.99}{8} = 0.12375$ or 12.375¢ per item;

the lower price per item is the better buy.

Exploration 3, **page 360**

6. The ratio of the perimeters of similar rectangles is equal to the ratio of the sides of similar rectangles.

Communicate, **page 361**

1. Answers may vary. A proportion is an equation that states that 2 ratios are equal. The cross products of a proportion are equal.

2. Multiply both sides of the equation by 12 to solve for *n*; cross multiply and then solve for *n*.

3. Cross multiply to find if the cross products are equal; reduce both fractions to lowest terms to find if they are equal.

4. Proportions can be used to find the lengths of corresponding sides of similar triangles.

5. Answers may vary. Mary makes 2 free throws out of 3 that she shoots. How many successful shots did she make when she attempted 36 shots?

Practice and Apply, **pages 361–363**

24. $\frac{2}{3} = \frac{6}{9}$; $2 \cdot 9 = 3 \cdot 6$; $18 = 18$; Since $18 = 18$ is a true statement, this is a true proportion; $\frac{3}{2} = \frac{9}{6}$, $\frac{9}{3} = \frac{6}{2}$, and $\frac{3}{9} = \frac{2}{6}$.

29. $\frac{3}{2} = \frac{36}{24}$, $\frac{2}{24} = \frac{3}{36}$ and $\frac{24}{2} = \frac{36}{3}$

30. $\frac{36}{14} = \frac{54}{21}$, $\frac{54}{36} = \frac{21}{14}$ and $\frac{14}{36} = \frac{21}{54}$

31. $\frac{48}{64} = \frac{27}{36}$, $\frac{64}{48} = \frac{36}{27}$ and $\frac{27}{48} = \frac{36}{64}$

32. $\frac{15}{10} = \frac{9}{6}$, $\frac{10}{15} = \frac{6}{9}$ and $\frac{9}{15} = \frac{6}{10}$

33. $\frac{12}{27} = \frac{20}{45}$, $\frac{27}{12} = \frac{45}{20}$ and $\frac{20}{12} = \frac{45}{27}$

34. $\frac{24}{54} = \frac{8}{18}$, $\frac{54}{24} = \frac{18}{8}$ and $\frac{8}{24} = \frac{18}{54}$

35. $\frac{2}{10} = \frac{13}{65}$, $\frac{10}{2} = \frac{65}{13}$ and $\frac{13}{2} = \frac{65}{10}$

36. $\frac{12}{18} = \frac{16}{24}$, $\frac{18}{12} = \frac{24}{16}$ and $\frac{16}{12} = \frac{24}{18}$

41. No; the pricing ratio of the item in the problem is $\frac{48}{60} = \frac{4}{5}$ and the pricing ratio of the item in the graph is $\frac{6}{8} = \frac{3}{4}$.

Lesson 6.6

Critical Thinking, **page 367**

a% of *b* is *c* can be rewritten as $\frac{a}{100} \cdot b = c$. To solve for *b*, multiply each side by $\frac{100}{a}$; $\frac{100c}{a} = b$. To solve for *a*, multiply each side by $\frac{100}{b}$; $\frac{100c}{b} = a$.

Communicate, **page 370**

2. Draw a horizontal bar and label the lower left and right corners 0% and 100%, respectively. Label the top left corner 0. On the bottom of the bar, near the left of the half-way point, write 40% and draw a line through the bar at this point. Write *x* on the top of the bar at this line. At the top right corner of the bar, write 50. Shade the portion of the bar to the left of the 40% line.

3. Draw a horizontal bar and label the lower left and right corners 0% and 200% respectively. Label the top left corner 0. At the half-way point on the bottom of the bar, write 100% and draw a line through the bar at this point. Write 50 on the top of the bar at this line. At the top right corner write *x*. Shade the entire bar.

4. Draw a horizontal bar and label the lower left and right corners 0% and 100%, respectively, and the top left corner 0. At the half-way point on the bottom, write 50% and draw a line

through the bar at this point. Write 30 on the top of the bar at this line. At the top right corner write x. Shade the portion of the bar to the left of the 50% line.

5. Draw a horizontal bar and label the lower left and right corners 0% and 100% respectively. Label the top left corner 0. Label the top right corner 80. Just right of 50% of the way across the bottom of the bar, write x%. Draw a line through the bar at this point, and write 60 at the top of the line. Shade the portion of the bar to the left of the x% line.

6. less; the base (50) is multiplied by a fraction $\left(\frac{40}{100}\right)$ whose value is less than 1.

7. more; the base (50) is multiplied by a fraction $\left(\frac{200}{100}\right)$ whose value is greater than 1.

8. more; 30 is half of 60, which is more than 50.

9. more; 50% of 80 is the same as $\frac{1}{2}$ of 80, which is equal to 40, so 60 must be more than 50%.

Practice and Apply, pages 370–372

21.
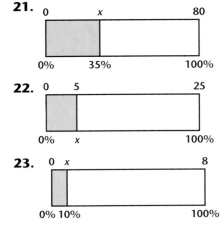

22.

23.

24.

25.

26.

	Taxable income	Amount of tax on income	Amount withheld	Amount or owed	Amount to be refunded
53.	$65,750	$15,904.50	$1,856	$14,048.50	
54.	$55,930	$12,860.30	$25,036		$12,175.70
55.	$43,100	$9,195.00	$5,850	$3,345.00	
56.	$108,426	$29,134.06	$42,865		$13,730.94

Look Back, page 372
68–71.
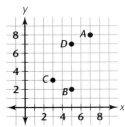

72. $x + 2$ and $x + 4$

Chapter 6 Project, page 373
a.

m	40m
1	40
100	4000
20	800

$m = 100 + 20 = 120$

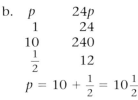

b.

p	$24p$
1	24
10	240
$\frac{1}{2}$	12

$$p = 10 + \frac{1}{2} = 10\frac{1}{2}$$

c.

j	$60j$
1	60
10	600
5	300
$\frac{1}{2}$	30

$$j = 10 + 5 + \frac{1}{2} = 15\frac{1}{2}$$

d.

w	$10w$
1	10
10	100
100	1000
200	2000
40	400
3	30
$\frac{1}{5}$	2

$$w = 200 + 40 + 3 + \frac{1}{5} = 243\frac{1}{5}$$

Answers may vary. For example, $36x = 504$.

x	$36x$
1	36
10	360
4	144

$$x = 10 + 4 = 14$$

Chapter 6 Assessment, page 377

15. Not all rational numbers are integers. For example $\frac{1}{2}$ is a rational number but $\frac{1}{2}$ is not an integer.

16. 0 is the only rational number which does not have a reciprocal. The reason for this is that 0 in the denominator of a fraction is undefined.

Lesson 7.1
Communicate, **page 387**

1. Use 2 x-tiles and 4 positive 1-tiles on the left of the equal sign. Use 6 negative 1-tiles on the right of the equal sign. Add 4 negative 1-tiles to both sides of the equal sign and cancel all neutral pairs on the left to find $2x = -10$. Divide the remaining tiles into 2 equal parts to get $x = -5$.

2. Use the Distributive Property to multiply 7 times $(8 - 2x)$ to find $56 - 14x$. Subtract 56 from both sides and then divide both sides by -14.

3. Multiply 90 by 3 to find the total number needed for the 3 tests, 270. Add the first 2 scores together, subtract the sum of the first 2 scores from 270; $270 - 170 = 100$. Brian needs to score 100 on his third test to have a 90 average.

4. Substitute 7.2 for x in the original equation, multiply 7.2 by 5, and simplify. If you get a *true* statement $x = 7.2$ is the solution to the equation. If the two numbers are not equal, then $x = 7.2$ is not the solution.

5. Enter $6x + 4$ for Y_1 and -2 for Y_2. Find the point of intersection. The x-value of the point of intersection is the solution to the equation.

Lesson 7.2

Exploration, **page 390**

5.
$$4x - 2 = x + 4$$
$$4x - 2 + 2 = x + 4 + 2$$
$$4x = x + 6$$
$$4x - x = x + 6 - x$$
$$3x = 6$$
$$\frac{3}{3}x = \frac{6}{3}$$
$$x = 2$$

6.
$$4x - 2 = x + 4$$
$$4(2) - 2 = (2) + 4$$
$$8 - 2 = 6$$
$$6 = 6 \text{ True}$$

Try This, **page 392**

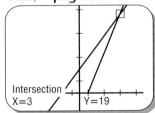

Intersection
X=3   Y=19

Communicate, **page 393**

1. Use 2 positive *x*-tiles and 3 negative 1-tiles on the left side; use 5 positive *x*-tiles and 9 positive 1-tiles on the right side. Form neutral pairs of *x*-tiles on the left by adding 2 negative *x*-tiles to each side; remove neutral pairs on each side. Form neutral pairs of 1-tiles on the right side by adding 9 negative tiles to each side; remove neutral pairs. Divide the remaining 12 negative 1-tiles into 3 equal groups; $x = -4$.

2. Use 4 positive *x*-tiles and 6 negative 1-tiles on the left side; use 1 negative *x*-tile and 4 positive 1-tiles on the right side. Form neutral pairs of *x*-tiles on the right side by adding 1 positive *x*-tile to each side; remove neutral pairs. Form neutral pairs of 1-tiles by adding 6 positive 1-tiles to each side; remove neutral pairs. Divide the 10 remaining 1-tiles, into 5 equal groups; $x = 2$.

3. $5x - 3 = 4x + 7$, Given; $5x - 3 - 4x = 4x + 7 - 4x$, Subtraction Property of Equality; $x - 3 = 7$, Simplify; $x - 3 + 3 = 7 + 3$, Addition Property of Equality; $x = 10$, Simplify.

4. $-2(x - 3) = 3x + 10$, Given: $-2x + 6 = 3x + 10$, Distributive Property; $-2x + 6 + 2x = 3x + 10 + 2x$, Addition Property of Equality; $6 = 5x + 10$, Simplify; $6 - 10 = 5x + 10 - 10$, Subtraction Property of Equality; $-4 = 5x$, Simplify; $-\frac{4}{5} = \frac{5}{5}x$, Division Property of Equality; $-\frac{4}{5} = x$, Simplify.

Lesson 7.3

Communicate, **page 398**

1. Cost is the total amount paid for materials; revenue is the total amount received from sales; profit is the total revenue minus the total cost.

2. Create a chart:

	First Solution	Second Solution	New Solution
Percent Acid	1%	5%	*x*
Amount of Solution	200 milliliters	50 milliliters	250 milliliters
Amount of Acid	200(0.01)	50(0.05)	200(0.01) + 50(0.05) = 250x

Write an equation from the information in the chart, and solve for *x* to find the percent of new solution: $200(0.01) + 50(0.05) = 250x$; $x = 1.8\%$.

3. Solve the equation $16.5h + 90 = 620$; let *h* represent the number of hours worked.
$$16.5h + 90 = 620$$
$$16.5h + 90 - 90 = 620 - 90$$
$$16.5h = 530$$
$$\frac{16.5}{16.5}h = \frac{530}{16.5}$$
$$h \approx 32$$

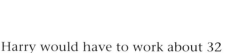

Harry would have to work about 32 hours.

4. Graph $Y_1 = 16.5X + 90$ and $Y_2 = 620$, and find the point of intersection. The point of intersection is approximately (32.12, 620); approximately 32 hours.

Look Back, **page 400**

37. The sum of all the angles in a parallelogram is 360°; opposite angles of a parallelogram are congruent; the sum of adjacent angles is 180°.

Look Beyond, **page 400**

42. The set of all real numbers x, such that $-10 < x < 10$.

43. The set of all real numbers x, such that $-8 < x < 12$.

44. The set of all real numbers x, such that $x \le -3$ or $x \ge 3$.

Lesson 7.4
Communicate, **page 405**

1. Substitute 84 for the area and 12 inches for the base in the formula $A = \frac{1}{2}bh$.

Simplify $\frac{1}{2}(12)(h)$; divide both sides by 6; $h = 14$ inches.

2. Multiply each side by 2 to find $2A = h(b_1 + b_2)$; divide each side by $(b_1 + b_2)$ to find $\frac{2A}{(b_1 - b_2)} = h$.

3. The length is 3 more than twice the width.

4. The width is 6 less than the length.

Practice and Apply, **pages 406–407**

22. $\frac{2A}{h} - b_2 = b_1$

23. $\frac{S}{180} + 2 = n$

24. $h = \frac{t - y - 3x - g}{2}$

25. $y = \frac{r - x}{4}$

26. $x = \frac{4y}{5}$

27. $C = \frac{5}{9}(F - 32)$

30. The heat transfer through a concrete wall is greater since the value of u is higher for concrete than brick.

31. Since the heat transfer is positive, there will be a heat loss in the interior.

Look Beyond, **page 407**

39.

Lesson 7.5
Exploration 1, **page 408**

4. If the expressions on each side is multiplied by 0, both sides are then equal to 0. If the inequality sign was >, <, or ≠, the inequality then becomes false. If the inequality sign was ≤ or ≥, the inequality remains true. Neither side can be divided by 0 because division by 0 is undefined.

Exploration 2, **page 409**

4.

$-3 > -4$

a.

$2 \cdot -3 > 2 \cdot -4; -6 > -8$

b.

$$-2 \cdot -3 > -2 \cdot -4$$
$$6 < 8$$

Exploration 3, page 410

4.

$$-4 < 2$$

a.

$$-4 \div 2 < 2 \div 2$$
$$-2 < 1$$

b.

$$-4 \div (-2) > 2 \div (-2)$$
$$2 > -1$$

Critical Thinking, page 410

$|a| + |b|$ is always greater than or equal to $|a + b|$ because if one or both of a and b is zero, then equality holds. If neither is zero, then $|a| + |b|$ may be the absolute value of the sum of two positive numbers or some combination of positive and negative numbers. The sum of a positive and a negative number is always less than or equal to the sum of the absolute values of the two numbers.

Communicate, page 411

5. Subtract 1 from each side of the inequality to find that $x > 3$.

6. Add 3 to each side of the inequality to find that $x \le 16$.

7. Divide both sides of the inequality by -3 and reverse the inequality sign to $>$: $p > -4$.

8. Subtract $2x$ from each side of the inequality, $2x - 2 \ge 3$. Add 2 to each side to find that $2x \ge 5$. Divide each side by 2 for $x \ge \dfrac{5}{2}$.

Lesson 7.6
Critical Thinking, page 418

When you try to solve the inequality you get a true inequality when $x < 4$ and when $x > 6$. There are no values to make the inequality true. Testing some values verifies this because the left side of the inequality is always positive or 0. There are no positive numbers less than -1.

Communicate, page 419

1. A measurement specification that falls between or equal to a minimum of 44.999 centimeters and a maximum of 45.001 centimeters; it means an error of ± 0.001 is allowed.

2. Choose a variable, for example x, and subtract 45 from x. Write the absolute value of this quantity as less than 0.001, $|x - 45| \le 0.001$.

3. You must consider the possibility that the quantity inside the absolute value symbols could be positive or negative.

4. Write the absolute value of a variable x minus 7. Make this quantity less than or equal to 3 because 3 is the farthest from -7 than any of the desired numbers will be. Therefore the inequality is $|x - (-7)| \le 3$ or $|x + 7| \le 3$.

5. Solve the inequality to determine the boundary numbers. Choose numbers on both sides of these numbers and substitute them into the inequality.

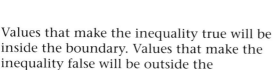

Values that make the inequality true will be inside the boundary. Values that make the inequality false will be outside the boundary.

Practice and Apply, **pages 419–420**

14. $-4 < x < 10$

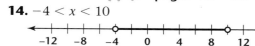

15. $x > 4$ or $x < -12$

16. $4 \le x \le 12$

17. $x \le 3$ or $x \ge 7$

18. $x < -4$ or $x > 8$

19. $-8 \le x \le 12$

20. $-6 < x < 4$

21. $x < 2$ or $x > 6$

22. Acceptable weights are indicated with bold.

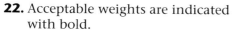

The boundary values are 123 pounds and 113 pounds.

25.

28. $-1 < x < 7$

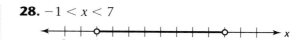

29. The solution for $|x - 3| < 4$ are all numbers between but not including -1 and 7.

32. $x < -1$ or $x > 7$

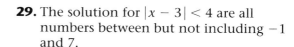

33. The solution is all numbers that are greater than 7 or less than -1.

38. The graph $y = 2x - 6$ is a straight line of slope 2 and a y-intercept of -6. The graph $y = |2x - 6|$ is a V that opens upward and whose lowest point (vertex) is at $(3, 0)$. The portion of the line $y = 2x - 6$ that lies above the x-axis is the right half of the V.

39. The graph $y = -x + 5$ is a straight line of slope -1 and a y-intercept of 5. The graph $y = |-x + 5|$ is a V that opens upward and whose lowest point (vertex) is at $(5, 0)$. The portion of the line $y = -x + 5$ that lies above the x-axis is the left side of the V.

Extra Practice, **page 718**

21. $x < -3$ or $x > -1$

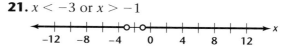

22. $-1 < x < 3$

23. $x \le 3$ or $x \ge 11$

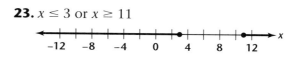

24. $-9 \le x \le -1$

25. $-2 < x < 8$

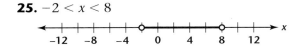

26. $-6 \le x \le 12$

27. $x = 2$

28. $x < -8$ or $x > -4$

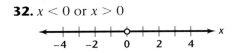

29. $x \le -1\frac{1}{3}$ or $x \ge 4\frac{2}{3}$

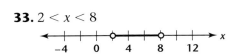

30. $-4 < x < 4$

31. $x \le -4$ or $x \ge 4$

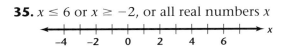

32. $x < 0$ or $x > 0$

33. $2 < x < 8$

34. $-11 < x < 3$

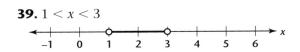

35. $x \le 6$ or $x \ge -2$, or all real numbers x

36. $x < 1$ or $x > 7$

37. $-4 \le x \le 2$

38. $x \le -3$ or $x \ge 7$

39. $1 < x < 3$

40. $-7 < x < 3$

Project, **page 421**
Activity 1, **page 421**

1. The length is five more than the width, or the width is five less than the length.

2. $w = l - 5$

3. $P = 2w + 2l$

4. The length is one more than the previous one.

5.

	A	B	C	D
1	Length	Width	Perimeter	Area
2	15	=A2−5	=2*A2+2*B2	=A2*B2
3	=A2+1	=A3−5	=2*A3+2*B3	=A3*B3
4	=A3+1	=A4−5	=2*A4+2*B4	=A4*B4
5	=A4+1	=A5−5	=2*A5+2*B5	=A5*B5

6. $114 = 2w + 2l$, or $114 = 2(l - 5) + 2l$

Activity 2, **page 421**

Ratio of city park = 500:250 or 2:1.
Answers may vary.

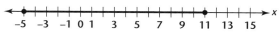

	A	B	C	D
1	Length	Width	Perimeter	Area
2	60 ft	=A2*2	=2*A2+2*B2	=A2*B2
3	=A2+20	=A3*2	=2*A3+2*B3	=A3*B3
4	=A3+20	=A4*2	=2*A4+2*B4	=A4*B4
5	=A4+20	=A5*2	=2*A5+2*B5	=A5*B5

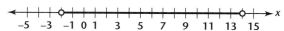

Length	Width	Perimeter	Area
60 ft	120 ft	252 ft	7200 sq ft
80 ft	160 ft	480 ft	12,800 sq ft
100 ft	200 ft	600 ft	20,000 sq ft
120 ft	240 ft	720 ft	28,800 sq ft

Chapter 7 Review, page 422

29. $-5 \leq x \leq 11$

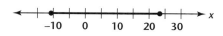

30. $-2 < x < 14$

32. $47.5\% \leq x \leq 54.5\%$; yes; possible for Mr. Green to obtain 47.5%, or less than 50%, of the vote.

Chapter 7 Assessment, page 425

13. $x = 20$, m$\angle A = 100°$; m$\angle B = 71°$; m$\angle C = 9°$

14. $x = 35$, m$\angle A = 60°$; m$\angle B = 75°$; m$\angle C = 45°$

15. $x = 40$, m$\angle A = 30°$; m$\angle B = 100°$; m$\angle C = 50°$

24. $-11 \leq x \leq 23$

25. $x > 2$ or $x < -6$

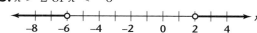

Lesson 8.1

Critical Thinking, **page 429**

Responses may vary. Sample: A point can be located using distance and direction. Let one coordinate represent the distance of the point from the origin, and let the other coordinate represent the angle between the positive *x*-axis and the line through the origin and the point.

Communicate, **page 432**

1. Start at the point in question. Count the number of units that the point is *above* (positive) or *below* (negative) the *x*-axis for the *y*-value. Count the number of units the point is to the *right* (positive) or to the *left* (negative) of the *y*-axis for the *x*-value.

2. No. (6, 7) and (7, 6) are ordered pairs, and thus do not specify the same point in the plane. Both the *x*-coordinate and the *y*-coordinate are different.

3. Begin at the origin. Count 7 units to the right along the *x*-axis, then count 3 units up, parallel to the *y*-axis. This locates the point (7, 3).

4. In any ordered pair, the first value is an element of the domain, and the second value is an element of the range.

5. The vertical axis represents the dependent variable since its value is determined by the value chosen on the horizontal axis.

6. In the left column of the table, list several *x*-values. Evaluate the

expression $2x + 5$ for each listed value.
Enter these y-values for their
corresponding x-values in the right
column.

x	$2x + 5$	y	(x, y)
-2	$2(-2) + 5$	1	$(-2, 1)$
-1	$2(-1) + 5$	3	$(-1, 3)$
0	$2(0) + 5$	5	$(0, 5)$
1	$2(1) + 5$	7	$(1, 7)$

Practice and Apply, **pages 432–433**

7. Yes

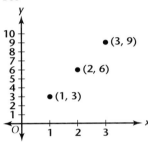

8. No

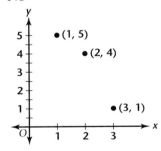

9. No

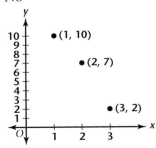

10. Yes

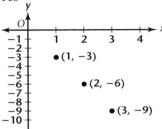

11. Yes

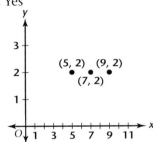

12. Yes

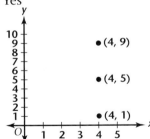

17–18.

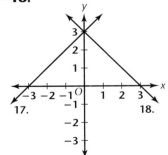

When there is a $+$ sign in front of x the
line goes up from left to right. When
there is a $-$ sign in front of x the line
goes down from left to right.

19–22.

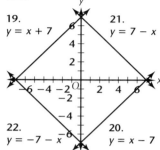

19. $y = x + 7$
21. $y = 7 - x$
22. $y = -7 - x$
20. $y = x - 7$

They are all linear functions. The lines $y = x + 7$ and $y = x - 7$ are parallel to each other. The lines $y = 7 - x$ and $y = -7 - x$ are parallel to each other. The lines $y = x + 7$ and $y = x - 7$ appear perpendicular to the lines $y = 7 - x$ and $y = -7 - x$.

23.

x	x + 3	y
1	1 + 3	4
2	2 + 3	5
3	3 + 3	6
4	4 + 3	7
5	5 + 3	8

24.

x	x + 4	y
1	1 + 4	5
2	2 + 4	6
3	3 + 4	7
4	4 + 4	8
5	5 + 4	9

25.

x	2x	y
1	2 · 1	2
2	2 · 2	4
3	2 · 3	6
4	2 · 4	8
5	2 · 5	10

26.

x	2x + 5	y
1	2 · 1 + 5	7
2	2 · 2 + 5	9
3	2 · 3 + 5	11
4	2 · 4 + 5	13
5	2 · 5 + 5	15

28.

h	3h	d
0	3 · 0	0
1	3 · 1	3
2	3 · 2	6
3	3 · 3	9

$d = 3h$

33.

x	8x + 3	C
1	8(1) + 3	11
2	8(2) + 3	19
5	8(5) + 3	43

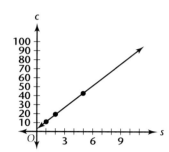

Yes, the points do lie on a straight line.

Look Back, page 433

41.
Sequence   1   1   2   3   5   8
First differences   0   1   1   2   3
Second differences   1   0   1   1

The next two terms of the sequence are 13 and 21. The first differences are the same as the Fibonacci sequence. The pattern is that to obtain the next term you add the previous two terms.

Look Beyond, page 433

44.

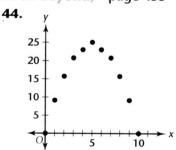

The vertex is (5, 25).

Extra Practice, page 719

1. Yes

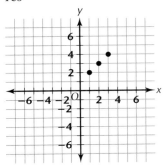

2. No

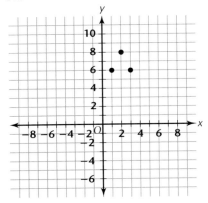

3. Yes

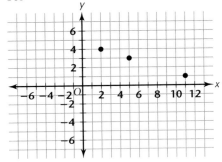

4. Yes

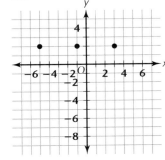

5. No

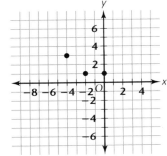

6. Yes

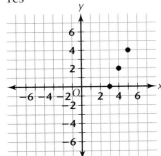

7.

x	1	2	3	4	5
y	3	4	5	6	7

8.

x	1	2	3	4	5
y	6	7	8	9	10

9.

x	1	2	3	4	5
y	−5	−4	−3	−2	−1

10.

x	1	2	3	4	5
y	−3	−2	−1	0	1

11.

x	1	2	3	4	5
y	3	6	9	12	15

12.

x	1	2	3	4	5
y	−4	−8	−12	−16	−20

13.

x	1	2	3	4	5
y	4	7	10	13	16

14.

x	1	2	3	4	5
y	−1	1	3	5	7

15.

x	1	2	3	4	5
y	0	−1	−2	−3	−4

16.

x	1	2	3	4	5
y	−5	−9	−13	−17	−21

17.

x	1	2	3	4	5
y	2.5	3	3.5	4	4.5

18.

x	1	2	3	4	5
y	1.5	3.5	5.5	7.5	9.5

Lesson 8.2

Exploration 1, **page 435**

6. The slope of the line is positive if the rise and run have the same signs. In this case, the line rises from left to right. The slope of the line is negative if the rise and run have opposite signs. In this case, the line falls from left to right.

Exploration 2, **page 436**

1.

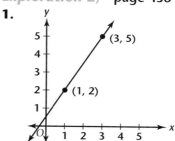

Exploration 3, **page 437**

5–6.

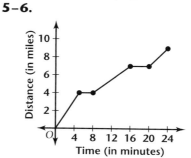

The point (20, 7) shows Jeff's position after a stop of 3 minutes. The point (24, 9) shows Jeff's final position. His rate on the last 4 minutes of his trip is $\frac{1}{2}$ of a mile per minute.

Communicate, **page 438**

1. Start at any point, go up 4 units (vertically), and then move right 3 units (horizontally). Start from this new point. Repeat the process (up 4, right 3) once more, and then draw a line through the three points.

3. Locate two coordinates on line *AC*, then write the differences in *y*-coordinates as the numerator of a fraction, and write the difference in *x*-coordinates as the denominator of the fraction.

4. Points *A*, *B*, and *C* are on the same line.

6. Calculate the difference in the *y*-coordinates, divided by the difference in the *x*-coordinates. Subtract the

x-coordinates in the same order as the corresponding y-coordinates. Slope of line $k = \frac{-3 - (-4)}{5 - 9}$ or $\frac{-4 - (-3)}{9 - 5}$, which simplifies to $-\frac{1}{4}$.

Practice and Apply, **pages 438–441**

10. Answers may vary. The following are sample graphs.

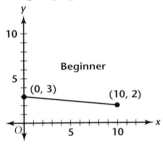

$$\text{slope} = \frac{2 - 3}{10 - 0} = -\frac{1}{10}$$

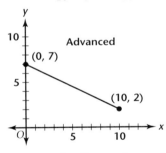

$$\text{slope} = \frac{2 - 7}{10 - 0} = \frac{-5}{10} \text{ or } -\frac{1}{2}$$

17. Answers may vary. A sample answer is shown.

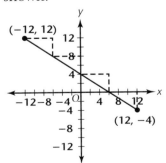

The shape is a straight line; slope $= -\frac{2}{3}$.

18.

Difference in x	Difference in y
$2 - 1 = 1$	$6 - 3 = 3$
$3 - 2 = 1$	$9 - 6 = 3$
$4 - 3 = 1$	$12 - 9 = 3$
$8 - 4 = 4$	$24 - 12 = 12$

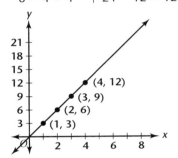

$$\text{slope} = \frac{\text{rise}}{\text{run}} = \frac{6}{2} = 3$$

$$\text{slope} = \frac{\text{difference in } y}{\text{difference in } x} = \frac{6}{2} = \frac{9}{3} = \frac{12}{4} = \frac{24}{8} = 3$$

The slope can be found using any corresponding sets of points on the line.

19. Answers may vary. The slope of the roof for the house in the picture is about 0.5. The angle is about 26°.

52. Answers may vary. The points should fit the line very closely. The slope should be between 3.1 and 3.2. A possible conjecture is that the ratio of circumference to diameter of a circle is about 3.14, or π.

Look Beyond, **page 441**

66.

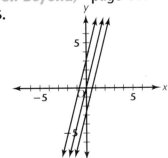

The slope of each line is 4. Since they have the same slope, but they cross the

y-axis at different points, the lines are parallel.

67.

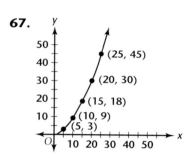

The curve goes up and to the right. It goes up faster than it moves right.

Eyewitness Math, **pages 442–443**

1. No; 72 to 75 is actually the range of a product (tread width · riser height). In the subsequent paragraph, the article refers to this product more appropriately as the "rise and run guidelines."

3. Stairway c; because it has the smallest slope, it will extend farthest into the room.

4. Answers may vary. For example, the stairway in (c) is about 2 feet longer than the stairway in (a). To make up for that lost space, the room would need to be about 2 feet longer. In a house 40 feet wide, that's an additional 80 square feet. Since placing the wall 2 feet further out also affects the second story, that's another 80 additional square feet, for a total extra area of 160 square feet.

Lesson 8.3

Exploration 1, **page 445**

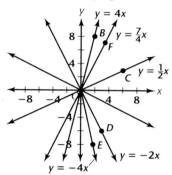

The slope, m, is the y-coordinate divided by the x-coordinate.

Exploration 2, **page 445**

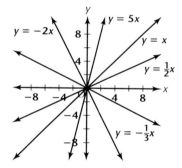

4.

The larger the absolute value of m, the steeper the line. Lines with positive slopes slope upward from left to right, and lines with negative slopes slope downward from left to right.

5. The larger the absolute value of m, the steeper the line. Lines with positive slopes slope upward from left to right, and lines with negative slopes slope downward from left to right.

Exploration 3, page 446

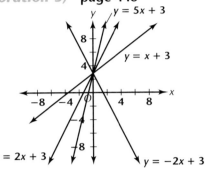

4.

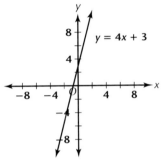

5. Adding 3 raises the graph so that the y-intercept moves from (0, 0) to (0, 3).

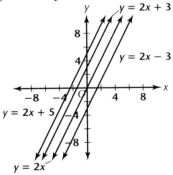

9.

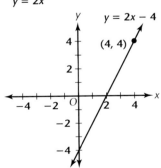

Communicate, page 447

1. The slope of a line through the origin is the value of the y-coordinate divided by the value of the x-coordinate. In the equation of a line $y = mx + b$, m represents the slope: the larger the value of m, the steeper the line. If m is positive, the line slopes upward from left to right, if m is negative, the line slopes downward from left to right. b represents the y-intercept of the line.

6. The two points are (0, 0) and (3, 6). The slope of a line passing through the origin is the value of the y-coordinate divided by the x-coordinate, $\frac{6}{3} = 2$, or $m = \frac{6 - 0}{3 - 0} = 2$. Since the y-intercept is at 0, $b = 0$. The equation for the line is $y = 2x$.

Practice and Apply, pages 448–449

8.

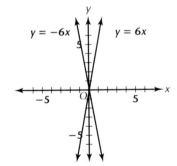

The lines both have y-intercept at (0, 0). They have the same steepness, but one has a positive slope and the other has a negative slope.

9.

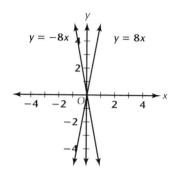

The lines both have y-intercept at $(0, 0)$. They have the same steepness, but one has a positive slope and the other has a negative slope.

10.

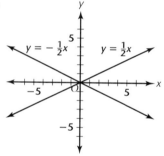

The lines both have y-intercept at $(0, 0)$. They have the same steepness, but one has a positive slope and the other has a negative slope.

11.

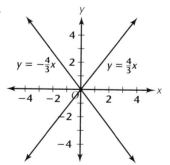

The lines both have y-intercept at $(0, 0)$. They have the same steepness, but one has a positive slope and the other has a negative slope.

12.

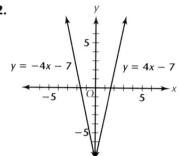

The lines both have y-intercept at $(0, -7)$. They have the same steepness, but one has a positive slope and the other has a negative slope.

13.

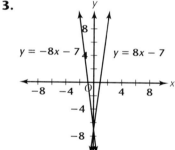

The lines both have y-intercept at $(0, -7)$. They have the same steepness, but one has a positive slope and the other has a negative slope.

15. $M = w + 30$

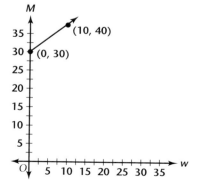

21.

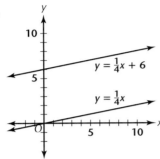

$$y = \tfrac{1}{4}x + 6$$
$$y = \tfrac{1}{4}x$$

The lines are parallel, both have slope $\frac{1}{4}$. They cross the y-axis at different points: (0, 0) and (0, 6).

22.

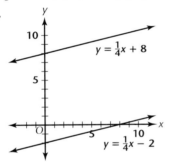

$$y = \tfrac{1}{4}x + 8$$
$$y = \tfrac{1}{4}x - 2$$

The lines are parallel, both have slope $\frac{1}{4}$. They cross the y-axis at different points: (0, 8) and (0, −2).

24. $l = 0.8d + 32$

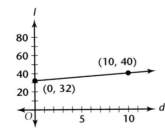

(10, 40)
(0, 32)

25. $l = 34 - 0.5d$; −0.5 feet per day; 16 days

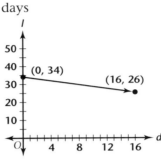

(0, 34)
(16, 26)

26. slope −5, crosses the y-axis at 0

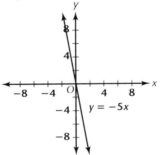

$$y = -5x$$

27. slope −6, crosses the y-axis at 3

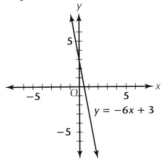

$$y = -6x + 3$$

28. $y = 7$ is a horizontal line at $y = 7$

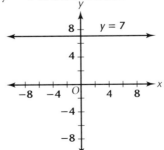

$$y = 7$$

29. horizontal line at $y = -6$

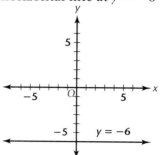

$y = -6$

30. slope -5, crosses y-axis at -1

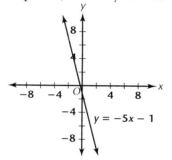

$y = -5x - 1$

31. horizontal line at $y = -2$

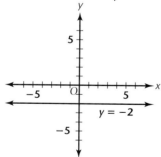

$y = -2$

32. slope -1, crosses y-axis at -3

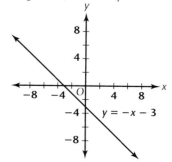

$y = -x - 3$

33. slope 2, crosses y-axis at 3

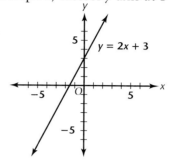

$y = 2x + 3$

34. slope 3, crosses y-axis at 7

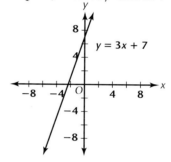

$y = 3x + 7$

35. slope -1, crosses y-axis at 7

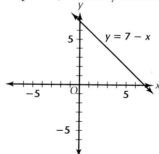

$y = 7 - x$

36. slope $\frac{1}{2}$, crosses y-axis at -3

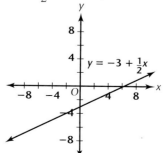

$y = -3 + \frac{1}{2}x$

37. slope 1, crosses *y*-axis at 4

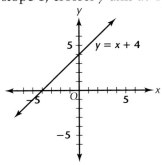

38. (1.) slope $= \dfrac{250 \text{ cubits}}{180 \text{ cubits}} \approx 1.4$

(2.) slope $= \dfrac{429 \text{ feet}}{309 \text{ feet}} \approx 1.4$

(3.) The slopes are approximately the same using either measuring unit.

Extra Practice, **page 720**

16. *y*-intercept (0, 0) and slope −2

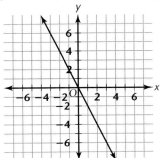

17. *y*-intercept (0, 0) and slope 2

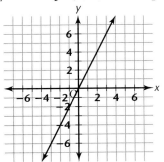

18. *y*-intercept (0, 3) and slope 0

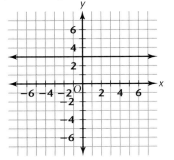

19. *y*-intercept (0, −5) and slope 0

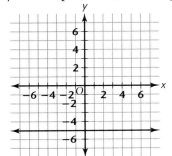

20. *y*-intercept (0, 2) and slope 1

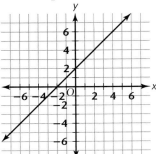

21. *y*-intercept (0, −1) and slope 0

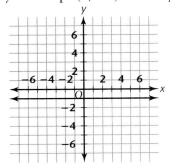

22. y-intercept $(0, 3)$ and slope -1

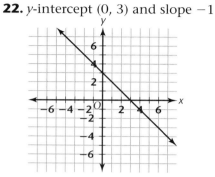

23. y-intercept $(0, 1)$ and slope 2

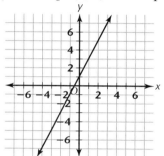

24. y-intercept $(0, -1)$ and slope 1

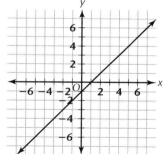

25. y-intercept $(0, -3)$ and slope -1

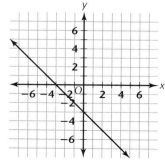

26. y-intercept $(0, -1)$ and slope 2

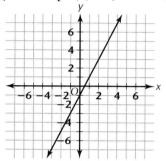

27. y-intercept $(0, 5)$ and slope -1

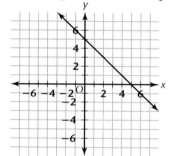

28. y-intercept $(0, -2)$ and slope $\frac{1}{2}$

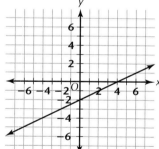

29. y-intercept $(0, -5)$ and slope $\frac{1}{2}$

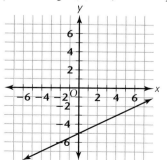

30. y-intercept $(0, -1)$ and slope $\frac{1}{4}$

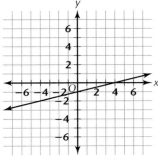

31. y-intercept $(0, -2)$ and slope $\frac{3}{2}$

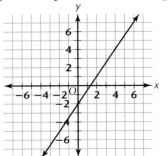

Lesson 8.4

Ongoing Assessment, **page 453**

Using the equation for the regression line, the estimated average speeds in miles per hour are: 161.04, 166.84, 172.64, 178.44, while the actual speeds for those years were 142.86, 152.98, 185.99, 153.62. The variation occurs because the line of best fit gives an estimate of the *average* speed; it does not give an exact answer. Changes in the weather, differing track conditions, and penalties might account for the variation.

Communicate, **page 456**

3. Find the difference in the y-coordinates divided by the difference in the x-coordinates.
$$m = \frac{5 - 2}{3 - 7} = -\frac{3}{4} \text{ or}$$
$$m = \frac{2 - 5}{7 - 3} = -\frac{3}{4}$$

5. The slope is the numerical coefficient of x when the equation is written in the form $y = mx + b$. For the line $y = -x + 10$ the slope is -1.

6. Mark the y-intercept $(0, -3)$. Move 1 unit to the right for the run and 2 units up for the rise. Mark this point. Draw a line through the 2 points.

7. Calculate the slope:
$$m = \frac{4 - 5}{-2 - 1} = \frac{1}{3}.$$
Substitute the slope and the coordinates of one of the points into the equation $y = mx + b$. Solve for b to find the y-intercept:
$$(5) = \left(\frac{1}{3}\right)(1) + b, \text{ or } \frac{14}{3} = b.$$
The equation of the line is $y = \frac{1}{3}x + \frac{14}{3}$.

Practice and Apply, **pages 456–458**

10. Since the slope is 3 and the line crosses the y-axis at $(0, 4)$, the equation is $y = 3x + 4$.

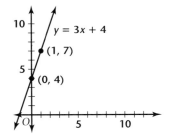

45.

47. Answers may vary. For example:

Weight in Newtons	Length in Millimeters
0	10
1	12
2	15.5
3	21
4	27

Line of best fit: $y = 4.3x + 8.5$

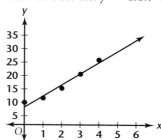

51. Approximately 11.29 years

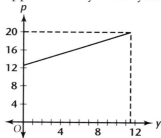

Extra Practice, **page 720**

25. $y = \frac{2}{3}x + 3$

26. $y = x - 1$

27. $y = -2x + 1$

28. $y = -\frac{1}{3}x + 2$

29. $y = x + 2$

30. $y = 2x + 1$

31. $y = x - 1$

32. $y = -3x + 1$

33. $y = \frac{1}{4}x + 2$

34. $y = -\frac{2}{5}x - 1$

35. $y = \frac{1}{2}x - \frac{5}{2}$

36. $y = 2x$

37. $y = -3x + 4$

38. $y = x - 2$

39. $y = \frac{4}{3}x - 1$

40. $y = 5x + 11$

41. $y = \frac{7}{10}x + \frac{1}{2}$

42. $y = -4$

43. $y = \frac{3}{5}x - 3$

44. $y = -\frac{4}{3}x + 2\frac{2}{3}$

45. $y = -\frac{1}{4}x - 6\frac{1}{4}$

Lesson 8.5

Communicate, **page 463**

1. Add $3x$ to both sides, and then add 2 to both sides. The equation is $3x + 5y = 2$.

2. Find the x-intercept by substituting 0 for y and solving for x.
$$3x + 6(0) = 18$$
$$x = 6$$
Find the y-intercept by substituting 0 for x and solving for y.
$$3(0) + 6y = 18$$
$$y = 3$$
The intercepts are $(6, 0)$ and $(0, 3)$.

3. Find the intercepts as shown in Exercise 2. They are $(6, 0)$ and $(0, 4)$. Connect these points to graph the line $2x + 3y = 12$.

4. Subtract x from each side of the equation: $-3y = 9 - x$. Divide both sides of the equation by the coefficient of y.
$$y = -\frac{9}{3} - \frac{x}{-3} \text{ or } y = \frac{1}{3}x - 3$$

5. Use the two points given to calculate the slope, then substitute the slope value and one of the points into the

equation $y - y_1 = m(x - x_1)$, and solve for y.

$$m = \frac{4 - (-8)}{-2 - 4} = -2$$
$$y - 4 = -2(x + 2)$$
$$y = -2x - 4 + 4 \text{ or}$$
$$y = -2x$$

6. Rewrite the equation in the form $y = mx + b$. The value of m is the slope. $y = -\frac{5}{2}x + 20$, $m = -\frac{5}{2}$.

Practice and Apply, **pages 464–465**

19.

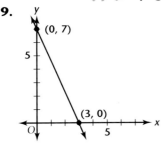

20. Intercepts: $(0, 3)$, $(9, 0)$

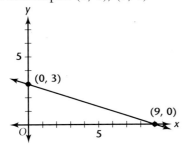

45. October 1, 1908; 10-01-08

46. 1-31-58; 580131

47. July 20, 1969; 690720

Look Beyond, **page 465**

64. Answers may vary. One example: Fill the 3 L bottle. Pour the contents into the 5 L bottle. Fill the bottle again. Pour water from it into the 5 L bottle— it needs 2 L more to be full. Then 1 L will remain in the 3 L bottle.

Lesson 8.6

Try This, **page 467**

The graph of $y = -3$ consists of all points with a y-coordinate of -3. This is a horizontal line. The graph of $x = -3$ consists of all points with an x-coordinate of -3. This is a vertical line.

Communicate, **page 469**

1. Use any two points on the line, for example $(2, 3)$ and $(4, 3)$; $m = 0$. The line is horizontal.

2. Use any two points on the line, for example $(1, 0)$ and $(1, 2)$; m is undefined. The line is vertical.

3. The slope is the numerical coefficient of x, 5. The line crosses the y-axis at $(0, 0)$.

4. Write the equation in the form $y = mx + b$: $y = \frac{1}{5}x$. The slope is $\frac{1}{5}$. The line crosses the y-axis at $(0, 0)$.

5. The equation can be written as $y = 0x + 15$. The slope is zero and the b-value is a constant 15. The value of the y-coordinate is always the same, 15.

6. The run of a vertical line is always 0. The slope, $\frac{\text{rise}}{\text{run}}$, is undefined.

7. The rise of a horizontal line is always 0. The slope, $\frac{\text{rise}}{\text{run}}$, is zero.

Practice and Apply, **pages 470–471**

37. Points; you cannot visit the pool a fraction of times. Another example could be population growth. To indicate a constant population growth during the year for a community you would show the growth with the graph of lines, not points. You cannot indicate a partial person.

Look Back, **page 471**

47. $y = x + 2$

Lesson 8.7

Ongoing Assessment, **page 473**

The slopes are the same.

Ongoing Assessment, **page 475**

2; $-\frac{1}{2}$; the signs are opposite; the absolute values, 2 and $\frac{1}{2}$, are reciprocals. The product of the slopes is $2\left(-\frac{1}{2}\right) = -1$.

Communicate, **page 476**

1. The slope of $y = 4x + 3$ is 4. Any line of the form $y = 4x + b$, where $b \neq 3$, will be parallel to the given line.

2. Write the reciprocal of $\frac{3}{2}$, with the opposite sign, $-\frac{2}{3}$.

3. The slope of the line $y = \frac{1}{3}x + 2$ is $\frac{1}{3}$ so the slope of the line perpendicular to $y = \frac{1}{3}x + 2$ is $-\frac{3}{1}$ or -3.

4. The slope of the line $y = 4x + 3$ is 4. Any line of the form $y = -\frac{1}{4}x + b$ will be perpendicular to the given line.

5. Any line parallel to a line with slope 3 will have the same slope. Since $b = -4$, use the slope-intercept form of the line: $y = 3x - 4$. Now place the x and y terms on the same side of the equation: $3x - y = 4$.

6. First, determine the slope of any line perpendicular to $y = -6x + 12$. This slope is $m = \frac{1}{6}$, so the slope-intercept form of the line with slope $\frac{1}{6}$ and $b = 12$ is $y = \frac{1}{6}x + 12$. In standard form, the equation is $x - 6y = -72$.

7. Find the slope of the line $x - 5y = 15$ by rewriting the equation in the form $y = mx + b$: $y = \frac{1}{5}x - 3$, $m = \frac{1}{5}$. Use the negative reciprocal of this slope and

$b = 0$ to write the new equation: $y = -5x + 0$, or $y = -5x$.

Practice and Apply, **page 477**

8–9. Answers may vary. Examples are shown.

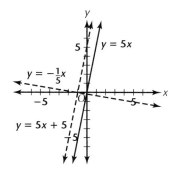

10. Both lines have the same y-intercept, $(0, 2)$. Also, both lines have the same steepness, one uphill and one downhill. The lines are not perpendicular.

15. $5x - 2y = 25$

16. $2x + 5y = -19$

17. $3x - y = -13$

18. $x + 3y = 19$

19. $y = 4$

20. $x = 2$

Lesson 8.8

Exploration 1, **pages 478–479**

2. 4, 5

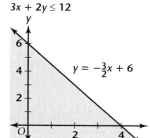

3.

Point	Substitute: $3x + 2y \le 12$	Simplify	Is inequality true?
(1, 1)	$3(1) + 2(1) \le 12$	$5 \le 12$	yes
(2, 1)	$3(2) + 2(1) \le 12$	$8 \le 12$	yes
(2, 2)	$3(2) + 2(2) \le 12$	$10 \le 12$	yes
(3, 1)	$3(3) + 2(1) \le 12$	$11 \le 12$	yes
(3, 2)	$3(3) + 2(2) \le 12$	$13 \le 12$	no
(3, 3)	$3(3) + 2(3) \le 12$	$15 \le 12$	no
(4, 4)	$3(4) + 2(4) \le 12$	$20 \le 12$	no

7. yes; there are a finite number of whole-number points in the real-world shaded region: (0, 0), (0, 1), (0, 2), (0, 3), (0, 4), (0, 5), (0, 6), (1, 0), (1, 1), (1, 2), (1, 3), (1, 4), (2, 0), (2, 1), (2, 2), (2, 3), (3, 0), (3, 1), (4, 0)

Exploration 2, **page 481–482**

2. dashed

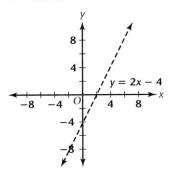

3.

Point	Substitute: $y > 2x - 4$	Simplify	Is inequality true?
(2, 3)	$3 > 2(2) - 4$	$3 > 0$	yes
(0, 5)	$5 > 2(0) - 4$	$5 > -4$	yes
(1, 6)	$6 > 2(1) - 4$	$6 > -2$	yes
(5, 6)	$6 > 2(5) - 4$	$6 > 6$	no
(6, 2)	$2 > 2(6) - 4$	$2 > 8$	no
(3, 2)	$2 > 2(3) - 4$	$2 > 2$	no
(5, 1)	$1 > 2(5) - 4$	$1 > 6$	no

4. above

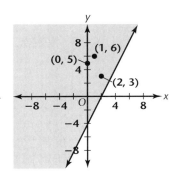

Communicate, **page 483**

1. Graph the solid boundary line $y = -2x + 5$. Test points on both sides of this line, and shade the side where the points satisfy the inequality.

2. Test points on both sides of the boundary line. The side where the points make the inequality true is the side to shade.

3. If the inequality symbol is $\le$ or $\ge$, include the boundary line; if the inequality is $<$ or $>$, do not include the boundary line.

4. A real-world example, such as the amount of time available to study and to work, would not include negative values even though these values satisfy the inequality.

5. When you are looking for an exact quantity, use an equation. When you are looking for a range of possibilities, use an inequality.

Practice and Apply, **pages 483–485**

6. $y < 2x - 1$

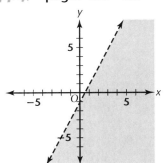

7. $y \leq -\frac{3}{4}x - 3$

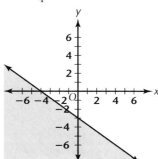

8. $y \geq \frac{1}{3}x - 3$

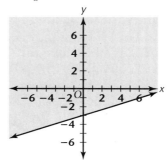

9. $y > -\frac{2}{5}x + 3$

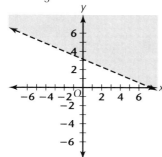

10. $y < -2x + 8$

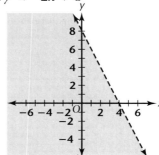

11. $y \geq \frac{5}{3}x - 2$

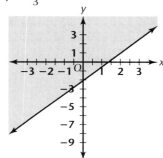

12. $y \geq -x + 5$

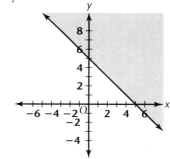

13. $y < -2x - 2$

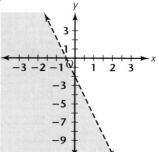

14. $y \leq -\frac{1}{4}x + 1$

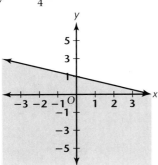

15. $y \leq -\frac{3}{2}x + 1$

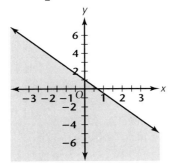

16. $y > -3x - 3$

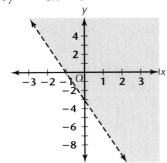

17. $y < -\frac{4}{5}x - 4$

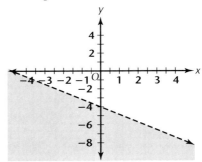

18. $y > 5$

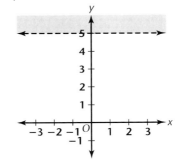

19. $y \leq -3$

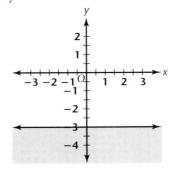

20. $x \geq 3$

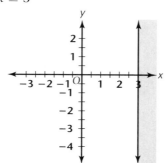

21. $x < -3$

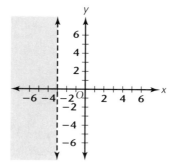

22. $y > 3x + 5$

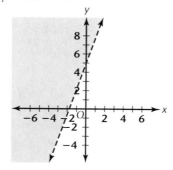

23. $y < -2x - 3$

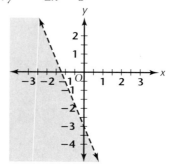

24. $y \leq -4x + 6$

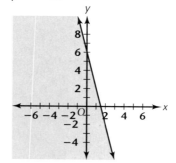

25. $y \geq 5x + 2$

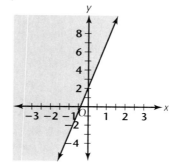

26. $y \geq x + 3$

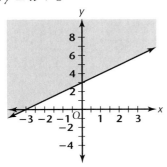

27. $y < 3x - 15$

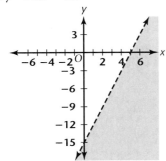

28. $y \geq 5x - 10$

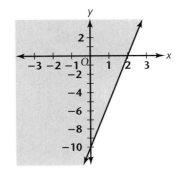

29. $y < \frac{2}{3}x - 2$

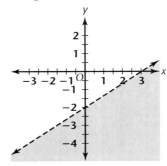

30. $y > \frac{1}{3}x + 8$

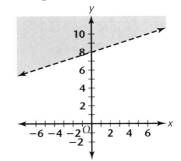

31. $y \geq \frac{5}{2}x - 2$

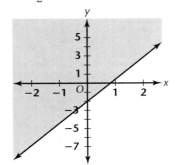

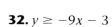

32. $y \geq -9x - 3$

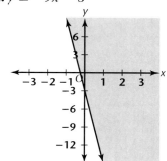

33. $y < 3x - 1$

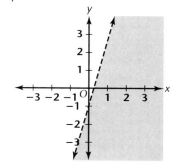

34. $y < -2x + 2$

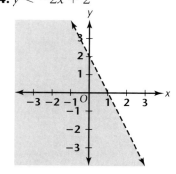

35. $y > -x + 3$

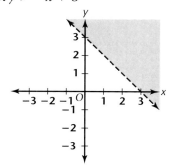

36. $y > 3$

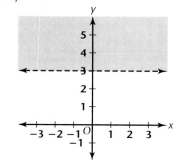

37. $x > 3$

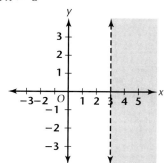

38. $y \geq \frac{1}{2}x - 3$

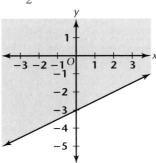

39. $y > \frac{3}{2}x + 1$

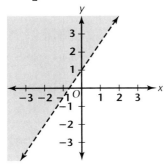

40. $y \leq \frac{1}{3}x + 1$

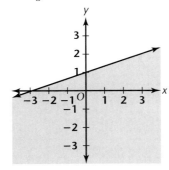

41. $y \leq -x - 3$

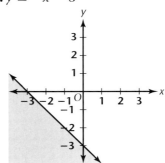

42. $y \geq x$

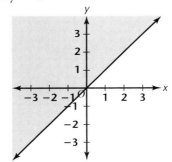

43. $y \leq -\frac{3}{5}x + 8$

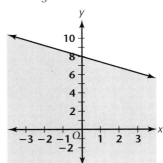

44. $y \geq \frac{1}{2}x - 4$

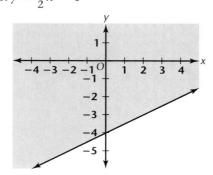

45. $y > \frac{2}{3}x + 5$

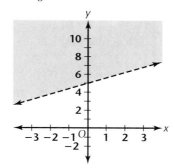

52. Let y represent the number of gold coins and x represent the number of silver coins: $14y + 7x \geq 25{,}000$. Since you cannot have fractional pieces of gold or silver, the reasonable domain is all whole numbers such that $x \geq 0$, and the range is all whole numbers such that $y \geq 4$, $y \geq -\frac{1}{2}x + 1785.7$.

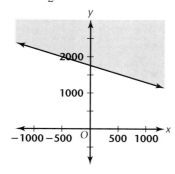

53. Let d represent the number of days the car is to be rented and m represent the number of miles: $25d + 0.30m < 140$. Miles and days are never negative so the reasonable domain is all real numbers such that $m \geq 0$, and the reasonable range is all real numbers such that $0 \leq d < -0.012m + 5.6$.

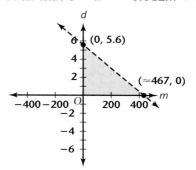

54. Let c represent a correct answer, and let w represent a wrong answer. Then $5c - 2w \geq 80$. The number of correct answers and the number of wrong answers are never negative.

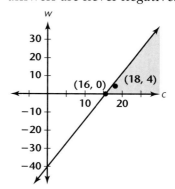

Yes, Alice will score at least an 80 if she gets 18 correct answers and 4 incorrect answers.

55. Let s represent the number of student tickets and a represent the number of adult tickets: $4s + 5a \geq 2000$. The number of adult and student tickets will never be negative. No, they will not meet their goal.

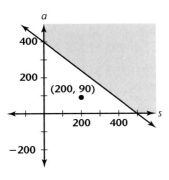

Yes, Jenna can earn at least $60 selling 8 blouses and 7 shirts since the point (8, 7) is within the shaded region.

58. Let x represent the number of 2-point baskets and y represent the number of 3-point baskets: $2x + 3y \geq 24$. The number of baskets scored cannot be negative.

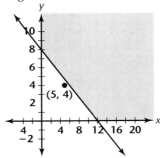

No, Mike could not have scored five 2-point field goals and four 3-point field goals in one game since the point (5, 4) is outside the shaded region.

56. Let p represent the number of pizzas and d represent the number of soft drinks. The number of pizzas and the number of soft drinks will never be negative. Then $8p + 2d \leq 60$.

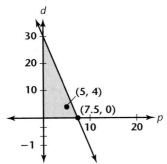

Yes, Will can purchase 5 pizzas and 4 sodas for less than $60.

57. Let b represent the number of blouses and s represent the number of shirts. Then $5b + 4s \geq 60$. The number of shirts and the number of blouses will never be negative.

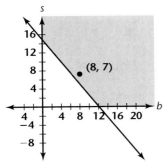

Look Beyond, **page 485**

64.

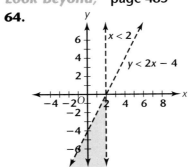

65.

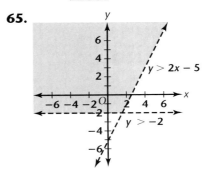

Extra Practice, page 722

1.

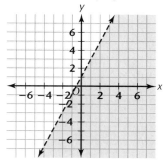

2.

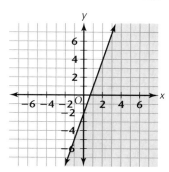

3.

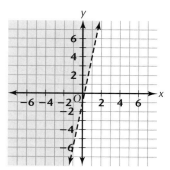

4.

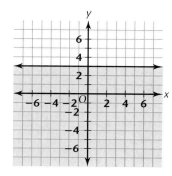

5.

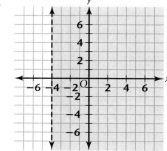

6.

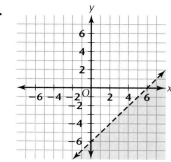

7.

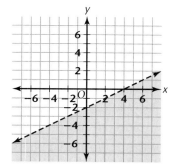

8.

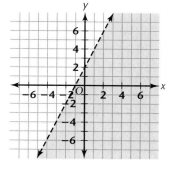

9.

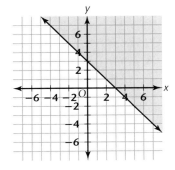

Chapter 8 Project

Activity 1, **page 487**

There are two possibilities: four 2-pound bags plus two 5-pound bags, or nine 2-pound bags and no 5-pound bags. These are the only non-negative integer coordinates that lie on the graph in the first quadrant.

Activity 2, **page 487**

$9a + 8b = 100$; Ms. Smiley should give 4 points for each question in part I and 8 points for each question in part II. This is the only whole-number solution to the equation. $9a + 8b = 200$:

a	8	16
b	16	7

Activity 3, **page 487**

$4x + 6y = 112$ has no whole-number solutions because the left side is always an even number and 125 is not a whole-number multiple of 2.

$5x + 10y = 112$ has no whole-number solutions because the left side is always a multiple of 5 and 112 is not a whole-number multiple of 5.

$6x + 9y = 100$ has no whole-number solutions because the left side is always a multiple of 3 and 100 is not a whole-number multiple of 3.

Chapter 8 Review, **pages 488–490**

1.

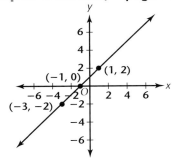

Yes, the points lie on a straight line.

2.

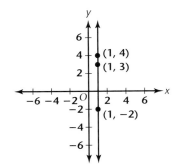

Yes, the points lie on a straight line.

3.

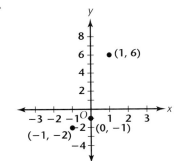

No, the points do not lie on a straight line.

11. $y = 2x + 1$

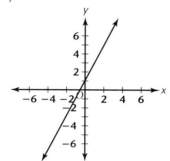

12. $y = -1$

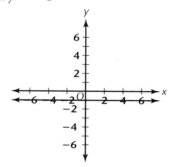

27. $y > x + 1$

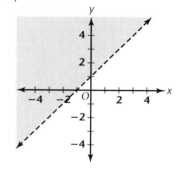

28. $y \geq -\frac{1}{2}x - 1$

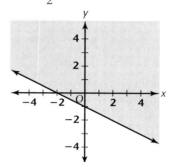

29. $x + y < 4$

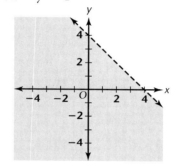

30. $y \leq 3$

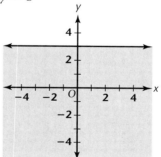

Chapter 8 Assessment, page 491

1.

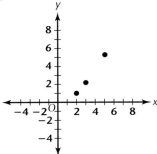

No, they are not on a straight line.

8.

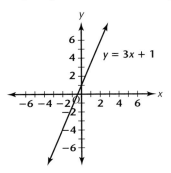

$y = 3x + 1$

12.

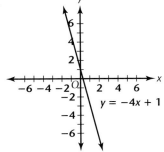

$y = -4x + 1$

18.

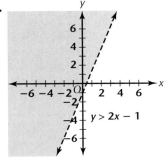

$y > 2x - 1$

19.

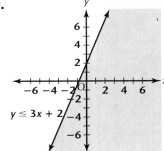

$y \leq 3x + 2$

20.

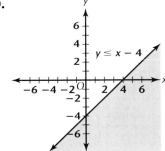

$y \leq x - 4$

21.

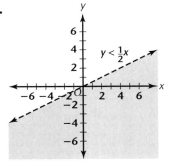

$y < \frac{1}{2}x$

Chapters 1–8 Cumulative Assessment, pages 492–493

20.

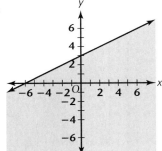

Lesson 9.1

Critical Thinking, **page 501**

The point (93, 107) is the best approximation to the situation since it results in earnings of $701.45 and 93 + 107 = 200. The students *barely* earned $700, and only a whole number of tickets could be sold. The points (92, 108), (91, 109), (90, 110) result in earnings of $703.80, $706.15, and $708.50 respectively. The sum of each of these points is 200, but the earnings move farther away from $700.

Ongoing Assessment, **page 501**

The calculator finds the exact solution to the equations you enter. It does not know what the original equations are. You must enter their equivalents, not their approximate equivalents, in the calculator in order to find the *exact* solution to the original system of equations.

Communicate, **page 502**

1. To solve the system of equations
$$\begin{cases} 2x - y = 5 \\ y = 3x - 7 \end{cases}$$
You must first solve for y in each equation.
$$2x - y = 5$$
$$-y = -2x + 5$$
$$y = 2x - 5$$
Then graph these two equations. The point of intersection is the solution.

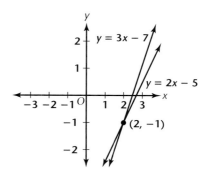

The solution is $(2, -1)$.

2. You check your solution by substituting the x- and y-values of the solution into each equation to verify that the original equations are satisfied.

3. To solve a system of equations on a graphics calculator, first solve for y in each equation. Then enter each equation in the calculator and look at their graphs. Either estimate the solution from the graph, or use the solve feature that automatically finds the point of intersection.

Practice and Apply, **pages 502–503**

22. Let x represent the number of heifer calves and y represent the number of bull calves. Then
$$\begin{cases} y = 2x \\ y + x \approx 50 \end{cases}$$
Answers may vary. One solution is that 17 heifer calves and 34 bull calves were born. The solution indicates that 51 total calves were born.

25. Answers may vary. Let a represent the number of adult tickets sold, and let s represent the number of student tickets sold.
$$\begin{cases} a + s = 1746 \\ 5a + 2s = 5766 \end{cases}$$
Graph the two equations, and approximate the intersection point.

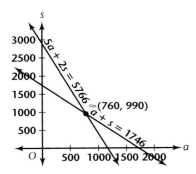

From the graph, it appears that approximately 760 adult tickets and 990 student tickets were sold.

Check: (760, 990):

$760 + 990 \overset{?}{=} 1746$
$\qquad 1750 \neq 1746$

$5(760) + 2(990) \overset{?}{=} 5766$
$\qquad\qquad 5780 \neq 5766$

The approximate solution sells 4 tickets too many and raises \$14 too much money. Try selling 2 fewer adult tickets and 2 fewer student tickets. Check: (758, 988):

$758 + 988 \overset{?}{=} 1746$
$\qquad 1746 = 1746$ True

$5(758) + 2(988) \overset{?}{=} 5766$
$\qquad\qquad 5766 = 5766$ True

The exact solution is 758 adult tickets and 988 student tickets were sold.

Lesson 9.2

Exploration, **page 505**

4. By the exact solution, the Green Team should order $16\frac{2}{3}$ large trees and $33\frac{1}{3}$ small trees. That is impossible, because you cannot plant $\frac{2}{3}$ of a tree or $\frac{1}{3}$ of a tree.

5. Some reasonable approximate solutions are (16, 32), (16, 33), (16, 34), (17, 33)

and (17, 34). None of these satisfy both equations. (16, 32) and (17, 34) satisfy $y = 2x$, and (16, 34) and (17, 33) satisfy $x + y = 50$.

6. It appears that $x = 17$ large trees and $y = 33$ small trees best solves this problem. First of all, these are the rounded versions of the exact solution. Second, (17, 33) satisfies at least one equation. Third, the requirement that there are twice as many smaller trees seems more flexible than the requirement that there are 50 trees total.

Try This, **page 507**

Let w be the width and l be the length. Then

$$\begin{cases} l = 3w \\ 2(w) + 2(l) = 1200 \end{cases}$$

The width is 150 yards, and the length is 450 yards.

Communicate, **page 508**

1. One reason to solve by substitution is to find exact solutions when they can not be found easily by graphing. Another reason is that substitution is often easier, especially when the graphs are difficult to draw (or when it is hard to find the best window on a graphing calculator). Also, substitution makes it easier to use variables other than x and y.

2. $\begin{cases} x + y = 7 \\ 2x - y = 12 \end{cases}$

To solve this system by substitution, first solve for a variable in one of the equations. For example, solve for y in $x + y = 7$:

$x + y = 7$
$\qquad y = 7 - x$

Then substitute $7 - x$ in place of y in

Additional Answers   **853**

the other equation, $2x - y = 12$. Solve for x.

$$2x - (7 - x) = 12$$
$$2x - 7 + x = 12$$
$$3x - 7 = 12$$
$$3x = 19$$
$$x = \frac{19}{3} \text{ or } 6\frac{1}{3}$$

Now substitute $\frac{19}{3}$ for x in $y = 7 - x$.

$$y = 7 - x$$
$$y = 7 - \left(\frac{19}{3}\right)$$
$$y = \frac{21}{3} - \frac{19}{3}$$
$$y = \frac{2}{3}$$

The solution is $\left(\frac{19}{3}, \frac{2}{3}\right)$, or $\left(6\frac{1}{3}, \frac{2}{3}\right)$

Check your solution in both original equations to make sure it is correct.

3. Check your solution by substituting the x- and y- vaues of your solution for x and y in the original equations. If variables other than x or y are being used, substitute as appropiate.

Practice and Apply, **pages 508–509**

4. $x = 4, y = 16$
5. $x = 15, y = 3$
6. $x = 1, y = 4$
7. $m = 12, n = 5$
8. $x = 7, y = 2$
9. $t = -2, z = 13$
10. $w = -5\frac{6}{7}, y = 5\frac{3}{7}$
11. $c = -15, d = -8$
12. $x = -3, y = 17$
13. $w = 4\frac{1}{4}, z = 1\frac{1}{4}$
14. $m = 6, n = -7$
15. $x = -1, y = 6$
16. $x = 8, y = 5$
17. $m = 15, n = -1$

18. $z = 10, h = -4$
19. $z = -18, w = -17$
20. $x = 6, y = 2$
21. $x = 1, y = 1$
22. $x = 1, y = 1$
23. $x = 2, y = 0$
24. $x = 14, y = 10$
25. $x = -2, y = -5$
26. $x = 2, y = 6$
27. $x = 1, y = 1$
28. $x = 3, y = -4$
29. $x = 2, y = 5$
30. $x = 10, y = 6$
31. $x = 5, y = -5$
32. $x = -1, y = 3$
33. $x = 0.5, y = -0.4$
34. $x = -1.2, y = 1.2$
35. $x = 0, y = -2$
36. $x = 2, y = 0$
37. $x = \frac{4}{5}, y = \frac{1}{5}$
38. $x = 718.125, y = -398.125$
39. $x = -\frac{1}{2}, y = 5$
48. $\begin{cases} a + s = 1746 \\ 5a + 2s = 5766 \end{cases}$

$$a + s = 1746$$
$$s = 1746 - a$$
$$5a + 2s = 5766$$
$$5a + 2(1746 - a) = 5766$$
$$5a + 3492 - 2a = 5766$$
$$3a = 5766 - 3492$$
$$3a = 2274$$
$$a = \frac{2274}{3}$$
$$a = 758$$

$$s = 1746 - a$$
$$s = 1746 - 758$$
$$s = 988$$

758 adults and 988 students

Lesson 9.3

Exploration 2, **pages 512–513**

9. To solve a system of equations using the Subtraction Property of Equality, subtract one equation from the other. If one of the variables is eliminated, this method will succeed in solving the system. Solve for one variable using the equation just obtained, then substitute that answer into one of the original equations to solve for the other variable. Check this solution by substituting your answers into both the original equations.

Critical Thinking, **page 513**

No; as long as both equations use both variables, it does not matter which equation you use to find the second variable. This is because the solution found has to satisfy *both* of the original equations.

Communicate, **page 515**

1. To model the solution to $\begin{cases} 3x + y = 6 \\ x - y = 2 \end{cases}$, first use algebra tiles to model each equation. Combine, and remove neutral pairs. There remain 4 positive *x*-tiles and 8 positive 1-tiles, so each *x*-tile represents 2 positive 1-tiles. That is, $x = 2$. In either equation, you can substitute 2 positive 1-tiles for each *x*-tile. When you use the first equation for this, you see 6 positive 1-tiles and 1 positive *y*-tile make 6 positive 1-tiles, so $y = 0$. The solution is (2, 0).

2. The Addition Property of Equality will only help solve a system of equations when coefficients of one of the variables in the two equations are opposites.

3. To determine whether (5, 3) is a solution to $\begin{cases} 2x - y = 7 \\ 2x + y = 10 \end{cases}$, see if $x = 5$ and $y = 3$ satisfy both equations.

$$2x - y = 7$$
$$2(5) - (3) \stackrel{?}{=} 7$$
$$7 = 7$$

$$2x + y = 10$$
$$2(5) + (3) \stackrel{?}{=} 10$$
$$13 \neq 10$$

Since one equation is not satisfied, (5, 3) is not a solution to the system of equations.

4. Graph each line. Since (5, 3) is not the point of intersection, (5, 3) is not a solution.

Look Beyond, **page 516**

43. The top and bottom each have area 55 square inches. Two sides each have area 30 square inches, and the other two sides each have area 66 square inches.

44.

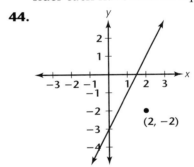

Answers may vary. (2, −2) satisfies the inequality $y < 2x - 3$ because $(-2) < 2(2) - 3$, or $-2 < 1$, is a true statement.

Lesson 9.4

Exploration, **pages 517–518**

3. $x + y = 4$

$$2x + 3y = 11$$

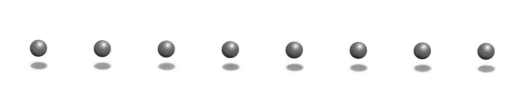

5.

No. After neutral pairs are removed, both x- and y-tiles still remain.

6. Multiply one of the equations by a number to create the same or opposite coefficients; this will eliminate one variable. $-2x - 2y = -8$

7.

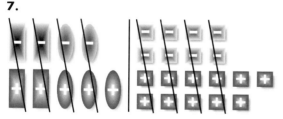

x is eliminated.

9. The first equation was multiplied by -2 to eliminate one of the variables. No; however, if you multiply by 2 instead of -2, subtraction can be used to eliminate the variable x.

Communicate, **page 521**

1. Multiplication is used to solve a system of equations by elimination when there are no opposite coefficients of like terms.

2. If the like terms to be eliminated have the same sign, multiply by a negative number. If the like terms to be eliminated have opposite signs, multiply by a positive number.

3. Model each equation using tiles. Double the number of each type of tile

in the model for $2x - y = 3$ to eliminate the y-variable. Combine the models and remove neutral pairs. Divide the remaining tiles in seven equal groups of 2; $x = 2$. Substitute $x = 2$ into one of the original equations to find $y = 1$.

Methods may vary for Exercises 4–6.

4. Multiply $x + 2y = 8$ by 2 and add the resulting equation to $5x - 4y = 1$; eliminate the variable y and solve for x; substitute the value of x into one of the original equations and solve for y; check.

5. Multiply $3x - y = 10$ by -7 and add the resulting equation to $5x - 7y = 6$; eliminate the y-variable and solve for x; substitute the value of x into one of the original equations and solve for y; check.

6. Multiply $0.8x - 1.2y = 7$ by -2 and add the resulting equation to $1.6x - 2.4y = 2$; eliminate the y-variable and solve for x; substitute the value of x into one of the original equations and solve for y; check.

Practice and Apply, **pages 521–523**

27. $m = -1, n = 2$

28. $a = 3, c = -1$

29. $w = -2, z = 3$

30. $x = 2, y = -2$

31. $w = 9, z = -5$

32. $a = 12, c = 5$

33. $m = 12.5, n = 29.5$

34. $a = 2, c = -1$

35. $x = -2, y = 3$

36. $x = 4\frac{2}{3}, y = 3\frac{1}{3}$

37. $x = -1, y = -2$

38. $x = -\frac{3}{8}, y = -\frac{1}{16}$

39. $x = 3\frac{4}{9}$, $y = -1\frac{1}{9}$

40. $g = 1\frac{1}{7}$, $h = -\frac{2}{7}$

41. $x = \frac{2}{3}$, $y = -4\frac{1}{3}$

42. $w = -4$, $c = 3$

45. Answers may vary. If 2 grams of a 40% solution are mixed with 3 grams of a 30% solution, then there are 1.7 grams of the dissolved substance in 5 grams of solution to make a 34% solution, not 35%.

49. $a = 758$, $s = 988$

Look Beyond, **page 523**

57. Tables may vary.

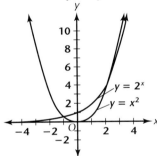

58. Explanations will vary. Sample answer: Both graphs rise and increase quickly in the positive direction; $y = x^2$ rises and increases quickly in the negative direction while $y = 2^x$ decreases in the negative direction and comes closer and closer to the x-axis.

Lesson 9.5

Exploration, **pages 524–525**

4. $\begin{cases} 3x + 2y \leq 15 \\ 2y > x \end{cases}$ $\begin{cases} y \leq -1.5x + 7.5 \\ y > 0.5x \end{cases}$

5. $y = -1.5x + 7.5$ is graphed with a solid line since $y \leq -1.5x + 7.5$ includes the points on the line. $y = 0.5x$ is graphed with a dashed line since $y > 0.5x$ does not include the points on the line.

6.

	$3x + 2y \leq 15$
Point	Is the inequality true?
$A(3, 1)$	$3(3) + 2(1) \leq 15$; Yes
$B(2, 2)$	$3(2) + 2(2) \leq 15$; Yes
$C(4, 4)$	$3(4) + 2(4) \leq 15$; No
$D(6, 1)$	$3(6) + 2(1) \leq 15$; No

	$2y > x$	Are both
Point	Is the inequality true?	inequalities true?
$A(3, 1)$	$2(1) > 3$; No	No
$B(2, 2)$	$2(2) > 2$; Yes	Yes
$C(4, 4)$	$2(4) > 4$; Yes	No
$D(6, 1)$	$2(1) > 6$; No	No

7.

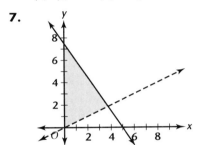

The line $y = -1.5x + 7.5$ contains points that are solutions to the system, but $y = 0.5x$ does not. A dashed line indicates that the points on that line are not included in the solution.

Try This, **page 527**

Let r represent the number of rolls of wrapping paper and c represent the number of cards.

$\begin{cases} 4r + 5c \geq 100 \\ r + c \leq 30 \end{cases}$

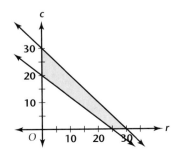

Domain: $0 \leq r \leq 30$ Range: $0 \leq c \leq 30$

$y \leq 20 - \dfrac{4}{5}x$

$y \leq 30 - x$

Communicate, page 528

1. To determine the region that satisfies two inequalities, first graph the boundary line for each inequality. Select a point in each region created by the lines and test whether each point satisfies both inequalities. The region containing the point satisfying both inequalities is the solution region.

2. The $\leq$ sign is appropriate when phrases such as "no more than" or "at the most" occur in a problem.

3. The $\geq$ sign is used when phrases such as "at least" or "no less than" are used in a question.

4. Any problem in which integer answers are required will result in some, but not all, of the points in the solution region that solve the problem. For example, a problem whose solutions are numbers of people requires integer answers.

5. A graphics calculator can be used to graph boundary lines of a system of inequalities as well as shade the solution region.

Practice and Apply, pages 529–530

14. $\begin{cases} 1.25x + 1.75y \geq 7.50 \\ x + y \leq 10 \end{cases}$

Answers may vary; (2, 5) and (4, 4); $0 \leq x \leq 10$, $0 \leq y \leq 10$.

15. Let x represent the number of wreaths; let y represent the number of baskets.

$\begin{cases} 2x + y \leq 40 \\ 3x + 2y \leq 72 \end{cases}$

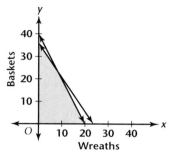

Domain: $0 \leq x \leq 20$
Range: $0 \leq y \leq 36$

16. $\begin{cases} x + y \geq 48 \\ 0.22x + 0.30y \geq 12 \end{cases}$

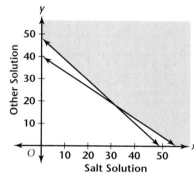

Domain: $x \geq 0$ Range: $y \geq 0$

17.

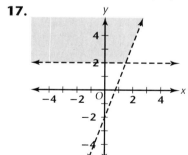

18.

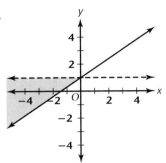

19.

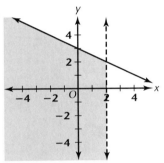

20.

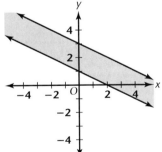

21.

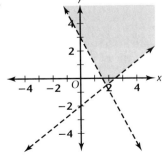

22.

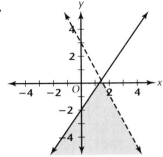

23.

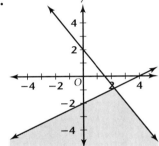

24.

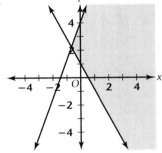

25.

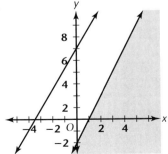

26.

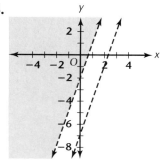

27. no solution

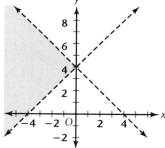

28.

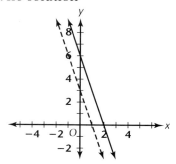

29.

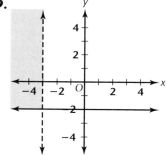

30.

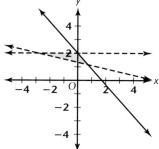

31.

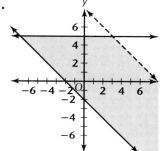

32.

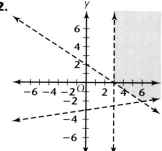

Look Beyond, **page 530**

40.

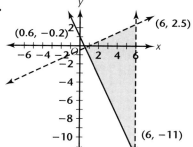

(0.6, −0.2)

(6, 2.5)

(6, −11)

The solution region is shaped like a triangle.

6.
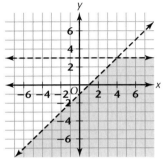

7.
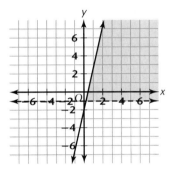

8.
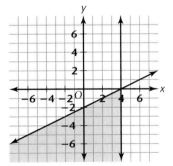

9.
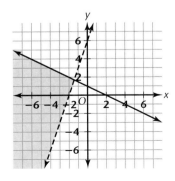

10.
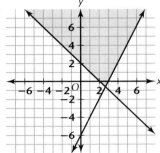

11.
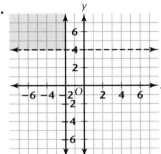

12.
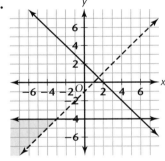

13.
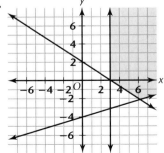

Chapter 9 Project, page 531

x = number of black and white pages
y = number of color pages
Constraints:
$x + y \geq 72$
$\quad x \leq 2y$
$\quad y \leq 40$
Optimization Equation: $200y + 150x =$ Total Cost

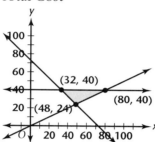

Total Cost at (32, 40) = $12,800
Total Cost at (48, 24) = $12,000
Total Cost at (80, 40) = $20,000
There should be 48 black and white pages and 24 color pages to minimize the cost.

Chapter 9 Review, pages 532–534

17.

18.

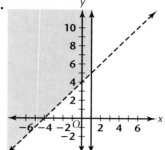

19.

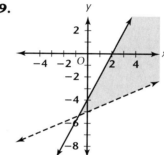

20.

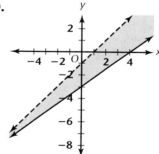

Chapter 9 Assessment, page 535

15.

16.

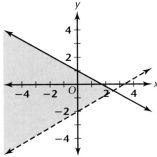

17.

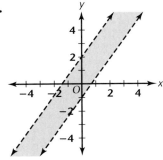

18. $\begin{cases} 6x + 8y \geq 120 \\ x + y \leq 50 \end{cases}$

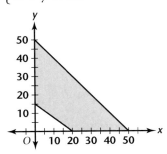

19. Domain: $0 \leq x \leq 50$ Range: $0 \leq y \leq 50$

20. Answers may vary. Possible solutions are 10 boxes of cookies, 30 boxes of brownies; 20 boxes of cookies, 25 boxes of brownies; and 40 boxes of cookies, 5 boxes of brownies.

Lesson 10.1

Exploration 1, **page 538**

3.

Day	Prize A	Prize B
1	$100	$0.01
2	$200	$0.02
3	$300	$0.04
4	$400	$0.08
5	$500	$0.16
6	$600	$0.32
7	$700	$0.64
8	$800	$1.28
9	$900	$2.56
10	$1000	$5.12
11	$1100	$10.24
12	$1200	$20.48
13	$1300	$40.96
14	$1400	$81.92
15	$1500	$163.84
16	$1600	$327.68
17	$1700	$655.36
18	$1800	$1310.72
19	$1900	$2621.44
20	$2000	$5242.88

Prize A: $2000; Prize B: $5242.88

5.

Day	Prize A	Prize B
21	$2100	$10,485.76
22	$2200	$20,971.52
23	$2300	$41,943.04
24	$2400	$83,886.08
25	$2500	$167,772.16
26	$2600	$335,544.32
27	$2700	$671,088.64
28	$2800	$1,342,177.28
29	$2900	$2,684,354.56
30	$3000	$5,368,709.12

A: $3000; B: $5,368,709.12

Exploration 2, page 541

2.

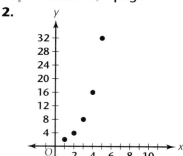

4.

x	1	2	3	4	5
y	$\frac{1}{2}$	$\frac{1}{4}$	$\frac{1}{8}$	$\frac{1}{16}$	$\frac{1}{32}$

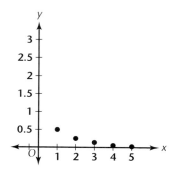

5. In $y = 2^x$ the y-values double each time x increases by 1. In $y = \left(\frac{1}{2}\right)^x$ the y-values are divided by 2 or halved each time x increases by 1.

Communicate, page 542

1. Each term is double the previous term; each term is multiplied by 2.

2. Each term is half the previous term; each term is divided by 2.

3. Each term is 10 more than the previous term; 10 is added to each term.

4. Each term is 20 less than the previous term; 20 is subtracted from each term.

5. Sequences 1 and 2 are exponential because they increase or decrease at a fixed *rate;* each term is a constant multiple of the previous term.

6. Exponential decay is modeled by sequences which decrease and for which each term is a constant multiple of the previous term. Exponential growth is modeled by sequences which increase and for which each term is a constant multiple of the previous term.

7. In exponential decay, the number that is multiplied repeatedly must be greater than 0 and less than 1.

8. Each year Sabrina has to pay 100% of what she already owes plus 10% of that amount, so the principal is multiplied by 1.10. For the interest to be compounded for 4 years the principal will be multiplied by $(1.10)^4$.

Practice and Apply, pages 542–543

26. A fixed rate of change occurs in a sequence when each term is a constant multiple of the previous term.

27. A fixed amount of change occurs in a sequence when a constant amount is added to or subtracted from the previous term.

28. Compound interest is interest that is computed on the current balance after each time period.

30.

	A	B	C
1	Month	Balance	Payment
2	1	1000.00	50.00
3	2	950.00	47.50
4	3	902.50	45.13
5	4	857.37	42.87
6	5	814.50	40.73
7	6	773.77	38.69
8	7	735.08	

After the six months the balance is $735.08.

Look Back, page 543

36–39.

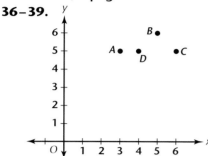

Look Beyond, page 543

40. 100, −200, 400, −800, 1600, −3200, . . . The sequence does not get close to any particular number.

41. 100, −50, 25, −12.5, 6.25, . . . The numbers get closer and closer to 0.

Lesson 10.2

Exploration 1, page 544

1. The surface area of a 2-meter cube is 6 · 4, or 24 square meters. This is 4 times the surface area of a 1-meter cube.

2. The surface area of a 3-meter cube is 6 · 9, or 54 square meters. This is 9 times the surface area of a 1-meter cube.

3. The ratio of the surface area of a 2-meter cube to the surface area of a 3-meter cube is $\frac{24}{54}$ or $\frac{4}{9}$; in simplest form, the elements of the ratio are the squares of the edge lengths.

Exploration 2, pages 546–547

2. 6th term, 36; 7th term, 49; First differences: 9, 11, 13, 15; Second differences: 2, 2, 2

4.

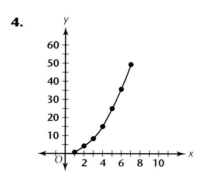

6.

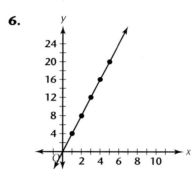

Communicate, pages 547–548

1. The area is 4 times larger.

2. The second differences are constant. The graph is a parabola.

3. The pattern indicates that the function oscillates between −4 and 4.

4. b; it is a curve that grows at a steady rate.

5. a; it is a straight line.

6. d; this is made up of two different linear parts.

7. c; the graph is a parabola.

8. The vertex is the point at which the parabola changes direction. It is the maximum point on the curve shown, at (3, 9). In a table it will be the pair of coordinates with the greatest or least y-value, and the y-values on either side of that point will repeat.

9. If the first differences are constant, the relationship is linear. If the second

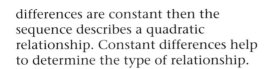

differences are constant then the sequence describes a quadratic relationship. Constant differences help to determine the type of relationship.

Practice and Apply, **pages 548–549**

16. If the length of each side is tripled, the area is 3^2 or 9 times as much.

17. If the length of each side is multiplied by 10, the area is multiplied by 10^2 or 100.

18. If the length of each side is $\frac{1}{3}$ as large, the area is $\left(\frac{1}{3}\right)^2$ or $\frac{1}{9}$ of the original area.

31. Reasonable domain: $t \geq 0$; reasonable range: $h \geq 0$

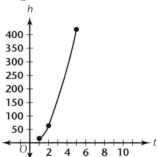

Look Back, **page 549**

41. Points may vary.

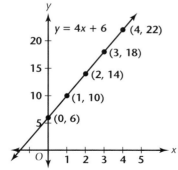

42. Points may vary.

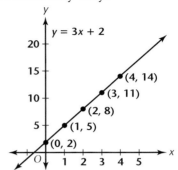

Lesson 10.3

Communicate, **page 554**

1. As one variable increases the other decreases. For example, as your hourly rate of pay increases, the number of hours that you need to work to earn $100 decreases. As your rate of speed decreases, the time it takes to drive a fixed distance increases.

2. Divide 1 by 5; $\frac{1}{5}$

3. Divide 1 by 100; $\frac{1}{100}$

4. Divide 1 by $\frac{1}{4}$; 4

5. Divide 1 by $\frac{1}{6}$; 6

6. Divide $1000 by 4; $250

7. Divide $1000 by 200; $5

8. Let n represent the number of people contributing and a represent the amount per person, then $a = \frac{30}{n}$.

Practice and Apply, **page 554**

9.

x	6	5	3	1	$\frac{1}{3}$	$\frac{1}{5}$	$\frac{1}{6}$
y	$\frac{1}{6}$	$\frac{1}{5}$	$\frac{1}{3}$	1	3	5	6

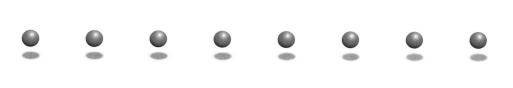

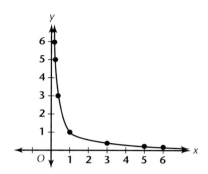

17. There is no number that can be multiplied by 0 to get 1, so 0 has no reciprocal; division by 0 is undefined.

18. One is its own reciprocal because $\frac{1}{1} = 1$.

25. Answers may vary.

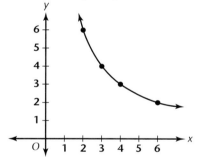

26. Answers may vary.

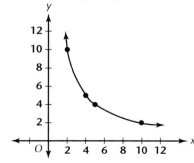

27. Answers may vary.

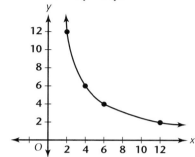

Look Back, **page 555**

47.

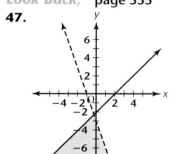

48.

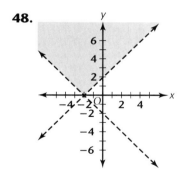

49.

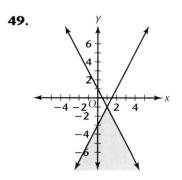

Additional Answers   **867**

Look Beyond, page 555

50.

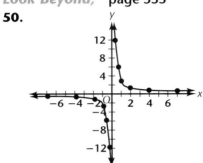

Eyewitness Math, pages 556–557

1. a-d Answers may vary.

BREEDING FREQUENCY DISTRIBUTION

Height	Number of Dinosaurs
0–9	0
10–19	1
20–29	1
30–39	1
40–49	2
50–59	3
60–69	3
70–79	4
80–89	3
90–100	2

BREEDING POPULATION
Dinosaur Height

Wk	A	B	C	D	E	F	G	H	I	J	K	L	M	N	O	P	Q	R	S
1	8	1	0	1	7	4	9	0	2	7	7	9	0	3	1				
2	5	0	9	1	2	0	9	3	9	9	2	3	5	0	1				
3	2	2	6	4	2	6	3	0	8	1	0	8	1	9	1				
4	8	9	4	2	0	6	7	8	0	0	5	5	1	3	7	5			
5	5	1	0	8	5	1	2	7	6	5	5	1	8	2	1	5			
6	1	2	5	9	7	7	4	5	2	1	6	3	0	8	6	0			
7	7	5	6	9	2	1	4	4	3	8	4	4	2	5	3	9	0		
8	0	7	0	4	6	0	6	3	8	9	0	7	5	6	0	1	4		
9	0	7	1	9	0	2	3	6	8	2	1	3	8	2	5	2	4		
10	0	4	6	0	2	6	8	8	9	3	6	8	1	9	3	8	5	5	
11	5	3	2	2	4	4	3	1	9	0	1	1	8	8	6	5	2	5	
12	5	6	4	8	3	5	4	4	9	1	9	0	5	9	4	4	5	5	
13	1	5	8	0	1	0	1	1	5	4	0	9	2	3	3	3	6	2	9
14	0	9	0	4	3	1	2	7	3	0	4	1	4	6	1	8	5	9	4
15	2	9	8	5	2	7	1	5	8	5	8	5	0	3	0	5	1	1	3
16	2	0	1	9	1	5	9	2	7	4	7	6	4	9	5	1	5	2	1
17	2	5	3	9	1	1	4	6	2	6	9	5	8	5	8	6	2	3	2
18	6	1	4	5	1	3	8	3	1	4	5	9	8	7	3	6	2	3	4
	59	76	66	89	49	59	87	73	99	69	79	87	70	97	58	68	41	35	23

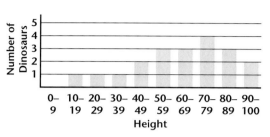

e. the bell-shaped graph on page 556

2. a–c

Dinosaur Height

Wk	A	B	C	D	E	F	G	H	I	J	K	L	M	N	O	P	Q	R	S
1	9	1	5	6	7	4	2	5	9	5									
2	2	7	9	5	8	3	0	1	3	4									
3	0	4	0	2	4	8	6	3	8	5									
4	2	9	8	8	0	9	9	7	3	0									
5	5	5	5	3	6	8	4	8	5	5									
6	2	9	0	8	0	0	8	2	5	0									
7	7	9	6	5	6	7	3	2	1	1									
8	1	7	9	5	5	5	6	3	7	9									
9	9	0	9	9	9	4	9	1	2	7	2	0	0	4	4	5	9	9	3
10	0	6	1	1	5	2	0	5	4	2	1	8	0	5	9	0	2	0	0
11	7	3	7	0	8	8	3	5	1	7	3	6	1	0	3	4	2	7	9
12	4	6	5	0	3	1	8	5	8	4	1	8	8	4	5	4	9	6	1
13	0	2	3	0	4	5	1	0	3	8	2	0	6	5	5	5	8	7	2
14	2	8	1	6	8	1	5	4	7	5	5	6	9	4	2	5	3	3	8
15	2	0	5	6	2	8	7	3	3	8	9	2	1	5	7	8	9	6	3
16	9	4	8	2	4	7	8	1	7	1	8	4	6	1	0	8	2	8	3
	61	80	81	66	79	80	79	55	73	71	31	34	31	28	35	39	44	46	27

CONTROLLED POPULATION FREQUENCY DISTRIBUTION

Height	Number of Dinosaurs
0–9	0
10–19	0
20–29	2
30–39	5
40–49	3
50–59	1
60–69	2
70–79	4
80–89	3
90–100	0

d. Our graph has only two humps, whereas the one in the book has 3. We had only 2 batches of dinosaurs in our simulation.

3. Sample: Dr. Malcolm, because our simulation showed that with a breeding population you would get a graph closer to the single-peak graph on page 556.

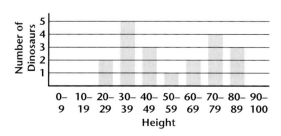

Lesson 10.4

Ongoing Assessment, **page 559**

The signs of the numbers in column D are always positive. The signs of the numbers in column C are either positive or negative, depending on whether the time was overestimated or underestimated. The distance of each number in column C is the same distance from the origin on a number line as the respective number in column D.

Exploration, **page 559**

2.

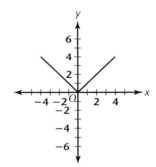

Communicate, **pages 561–562**

1. $|29|$ is 29 since 29 is 29 units from 0 on the number line.

2. $|-34|$ is 34 since -34 is 34 units from 0.

3. $|7.99|$ is 7.99 since 7.99 is 7.99 units from 0.

4. $|-3.44|$ is 3.44 since -3.44 is 3.44 units from 0.

5. INT(7.99) is 7 since 7.99 rounded down makes 7.

6. $\text{INT}\left(\dfrac{-34}{5}\right)$ is -7 since $-\dfrac{34}{5}$, or -6.8, rounded down makes -7.

7. INT(29) is 29 since 29 is the integer 29.

8. INT(-3.44) is -4 since -3.44 rounded down makes -4.

Practice and Apply, **pages 562–563**

32. This is a step function because the graph looks like a flight of stairs. This is not the greatest-integer function because the cost (the *y*-variable or dependent variable) is not always an integer.

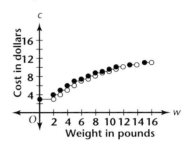

40. Since the person riding always stops when they meet the walking person, and does not pass the walking person, we can assume that they arrive at the same time. Therefore they have walked the same distance and ridden the same distance, that is, halfway. Riding at 5 miles per hour
$$t = \frac{d}{r} = \frac{10}{5} = 2$$
Riding 10 miles takes 2 hours. Walking,
$$t = \frac{d}{r} = \frac{10}{3} = 3\frac{1}{3}$$
Walking 10 miles takes $3\frac{1}{3}$ hours. The trip takes $2 + 3\frac{1}{3}$ hours or 5 hours 20 minutes.

Look Back, **page 563**

53. The vertex is (2, 3).

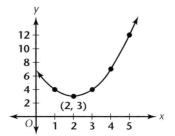

Look Beyond, page 563

54.

x	y
−3	6
−2	5
−1	4
0	3
1	4
2	5
3	6

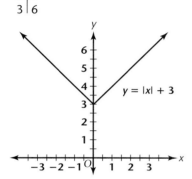

$y = |x| + 3$

55.

x	y
−6	3
−5	2
−4	1
−3	0
−2	1
−1	2
0	3

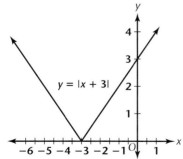

$y = |x + 3|$

Lesson 10.5

Communicate, page 567

1. e. Absolute value; The graph is
V-shaped.

2. b. Exponential; The graph increases
slowly at first, then quickly rises.

3. d. Reciprocal; The graph is a curve for
which y decreases when x increases.

4. f. Step; The graph looks like a staircase.

5. a. Linear; The graph is a line.

6. c. Quadratic; The graph is U-shaped.

7. First differences: 8, 10, 12, 14; Second
differences: 2, 2, 2; Quadratic since the
2nd differences are constant.

8. First differences: 1, 8, 12, 24; Second
differences: 7, 4, 12; Neither linear nor
quadratic. Neither the 1st nor 2nd
differences are constant.

9. First differences: −1.2, −1.2, −1.2;
Linear since the first differences are
constant.

10. First differences: 2, 2, 2, 2; Linear since
the 1st differences are constant.

Practice and Apply, pages 568–569

23. The points fit nicely in a line with one
exception. For the players on the line,
they all score about the same number
of points per minute. Trina does not
score at the same rate as her
teammates.

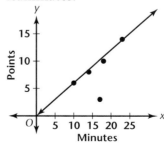

24. When the x-values are positive, $y = x$.
When the x-values are negative,
$y = -x$. The absolute-value function
can be described as two linear functions
with $y \geq 0$, that meet at a vertex.

25.

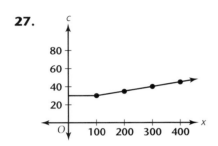

27.

Look Beyond, page 569

37.

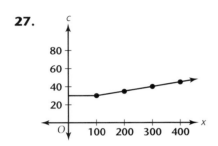

b. The graph of $y = |x + 2|$ is the same as the graph of $y = |x|$ shifted left 2 spaces.

c. The graph of $y = |x - 1|$ is the same as the graph of $y = |x|$ shifted right 1 space.

d. The graph of $y = |x| - 3$ is the same as the graph of $y = |x|$ shifted down 3 spaces.

e. The graph of $y = |x| + 1$ is the same as the graph of $y = |x|$ shifted up 1 space.

Lesson 10.6

Critical Thinking, **page 571**

No; a reflection is a mirror image of the pre-image across the *x*-axis. A rotation of 180° is same as a reflection across the *x*-axis and then a reflection across the *y*-axis.

Exploration 1, **pages 571–572**
Part I

1. and 2.

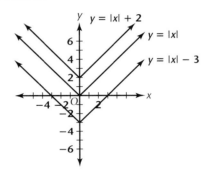

3. Adding or subtracting a number from $|x|$ in the function $y = |x|$ moves the graph up or down by the number of units that were added or subtracted.

5. Adding to or subtracting from *x* in $y = |x|$ translates the graph left or right.

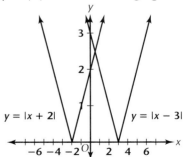

7. Adding a number to or subtracting a number from *x* in $y = |x|$ translates $y = |x|$ that many units left or right. Adding to or subtracting from $|x|$ translates the function vertically that many units up or down, respectively.

Part II

1.

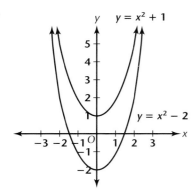

2. Adding a number to or subtracting a number from x^2 in $y = x^2$ translates the graph that many units up or down.

4.

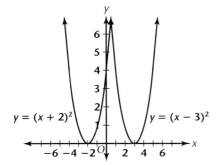

5. Adding to or subtracting from x in $y = x^2$ translates the graph left or right.

7. To translate $y = x^2$ vertically, add to or subtract from x^2. To translate $y = x^2$ horizontally, add to or subtract from x in $y = x^2$.

Exploration 2, page 573
Part I

1.

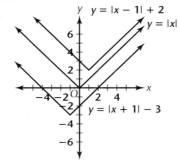

$y = |x - 1| + 2$ is a translation of $y = |x|$ by 1 unit to the right and up 2 units. $y = |x + 1| - 3$ is a translation of $y = |x|$ by 1 unit to the left and 3 units down.

Part II

1.

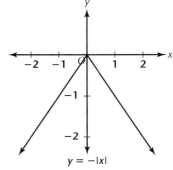

This is a reflection of $y = |x|$ across the x-axis.

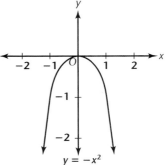

This is a reflection of $y = x^2$ across the x-axis.

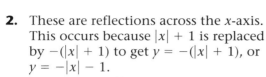

2. These are reflections across the x-axis. This occurs because $|x| + 1$ is replaced by $-(|x| + 1)$ to get $y = -(|x| + 1)$, or $y = -|x| - 1$.

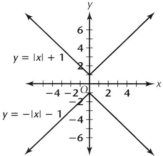

$y = |x| + 1$

$y = -|x| - 1$

3. The reflection occurs because $|x + 2| + 1$ is replaced by $-(|x + 2| + 1)$ to get $y = -(|x + 2| + 1)$, or $y = -|x + 2| - 1$.

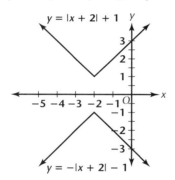

$y = |x + 2| + 1$

$y = -|x + 2| - 1$

4. The equation of $y = x^2 - 1$ reflected across the x-axis is $y = -x^2 + 1$.

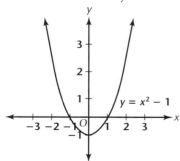

$y = x^2 - 1$

Communicate, **page 574**

1. A translation is a shift in position without changing shape or orientation. A reflection is a mirror-image that does not change the shape of an image but reverses it as if seen from the other side.

translation

reflection

2. Hanging a poster upside-down in the very same spot that it was rightside-up is a real-world example of a rotation.

3. $y = |x| + 4$ is a translation of $y = |x|$ upward.

4. $y = (x + 2)^2$ is a translation of $y = x^2$ to the left.

5. $y = |x - 1| - 2$ is a translation of $y = |x|$ to the right and downward.

Practice and Apply, **pages 574–575**

6. $y = |x| - 3$

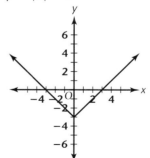

7. $y = |x| + 1$

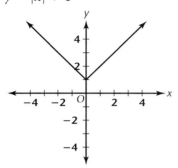

8. $y = |x| - 4$

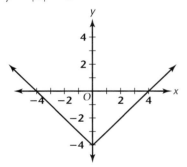

9. $y = |x - 1|$

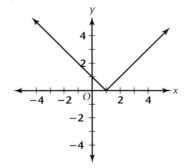

10. $y = (x + 1)^2$

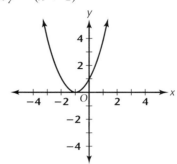

11. $y = |x + 3|$

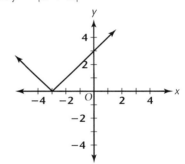

12. $y = |x - 1| + 2$

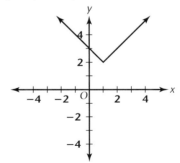

13. $y = (x - 1)^2 + 1$

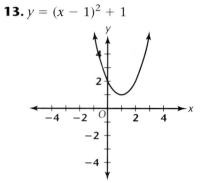

14. $y = |x + 3| - 1$

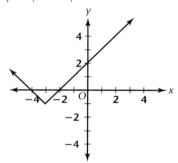

15. $y = -(x + 1)^2$

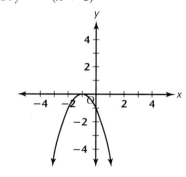

16. $y = -|x + 3|$

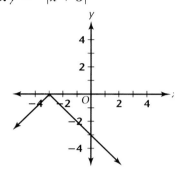

17. $y = |x + 1| - 2$

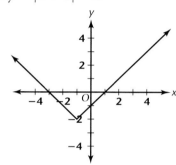

18. $y = -|x - 1|$

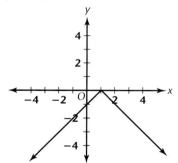

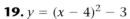

19. $y = (x - 4)^2 - 3$

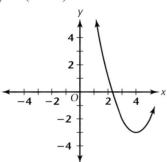

20. $y = |x + 2| + 1$

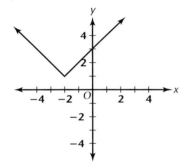

21. $y = -|x| - 3$

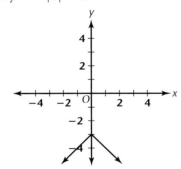

22. $y = -|x| + 1$

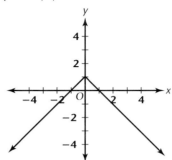

23. $y = -|x| - 4$

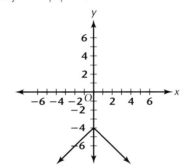

Look Beyond, **page 575**

64.

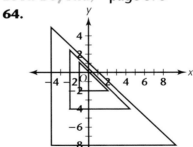

The scale factors are 2 and $\frac{1}{2}$.

65. As $|a|$ gets large, the parabola gets more narrow. As $|a|$ approaches a value of 0, the parabola gets wider and more open.

Additional Answers **877**

1. $y = |x| + 1$

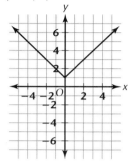

2. $y = |x| - 2$

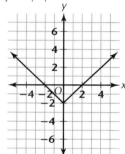

3. $y = |x| - 5$

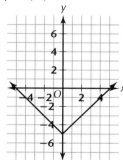

4. $y = |x + 1|$

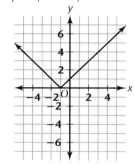

5. $y = (x - 1)^2$

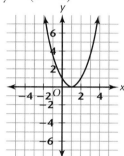

6. $y = (x + 1)^2 - 2$

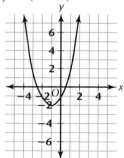

7. $y = -|x + 2|$

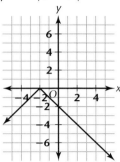

8. $y = -|x| - 2$

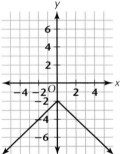

9. $y = -|x - 2|$

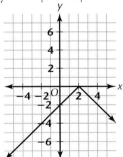

Chapter 10 Project
Activity 1　page 577

Length of sides	Number of cubes	Number of sides painted			
		3	2	1	0
2	8	8	0	0	0
3	27	8	12	6	1
4	64	8	24	24	8
5	125	8	36	54	27
6	216	8	48	96	64
n	n^3	8	$12(n-2)$	$6(n-2)^2$	$(n-2)^3$

Activity 2　page 577

Length of base	Number of cubes	Number of sides painted					
		5	4	3	2	1	0
3	10	1	4	4	0	1	0
5	35	1	4	16	4	9	1
7	84	1	4	28	16	25	10
9	165	1	4	40	36	49	35
11	286	1	4	52	64	81	84

Chapter 10 Review,　pages 578–580

6.

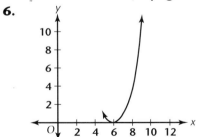

Vertex (6, 0)

7. Quadratic; the second differences are constant.

8. Not quadratic; the second differences are not constant.

9. Quadratic; the second differences are constant.

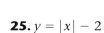

25. $y = |x| - 2$

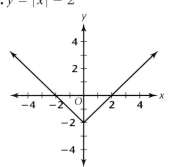

26. $y = |x - 2|$

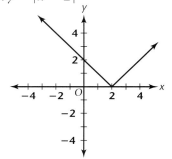

27. $y = -|x + 4|$

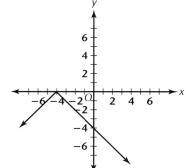

28. $y = x^2 - 2$

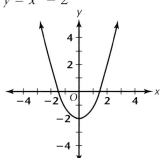

29. $y = (x - 2)^2$

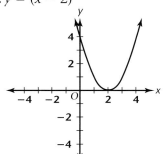

30. $y = -(x + 4)^2$

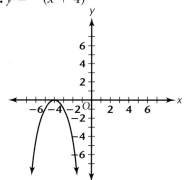

Chapter 10 Assessment, page 581

1. 320, 640, 1280 Exponential

2. 80, 76, 72 Linear

3. Decay. The sequence decreases, getting closer and closer to 0.

4.

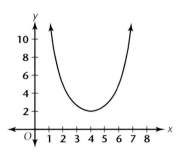

Vertex (4, 2)

20. $y = |x + 1|$

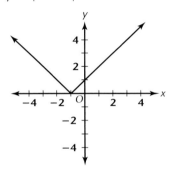

21. $y = (x - 3)^2$

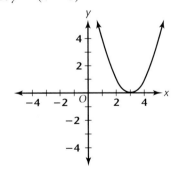

22. $y = -|x| + 2$

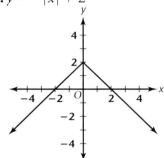

23. $y = x^2 + 4$

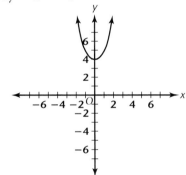

Lesson 11.1

Exploration 1, **page 587**

3. Answers may vary. A sample answer is that the slant of the tops of the bars and the different vertical scales make it possible to misinterpret the data. To make a more accurate visual comparison, the bars should be changed so that they are straight across at the top, and the vertical scales should be changed so that each space represents the same amount of money.

4.

Regency Stores

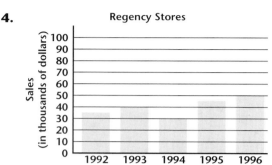

Morton Stores

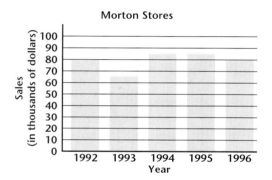

Ongoing Assessment, page 587

Answers may vary. A sample answer is that the vertical scales are different. Also, the shapes of the objects that make up the bars are not even across the top, so it is difficult to determine the value that each bar represents.

Exploration 2, page 588

6. Answers may vary. A sample answer is approximately 8%. This answer is reasonable if you assume that the unemployment rate will rise at the same rate as it has for the past 10 years. However, the data have not been consistent over the time period shown. You must consider the trends over the entire time period. These trends include rises and falls between the years and any cycles that occurred during the time period.

Critical Thinking, page 588

Answers may vary. A sample answer is that the Depression probably accounts for the rise in the unemployment rate from 1930 to 1935. World War II probably accounts for the decline in the unemployment rate from 1940 to 1945.

Communicate, page 590

1. Answers may vary. Graphs are important statistical tools because they allow people to interpret large amounts of data very quickly.

2. Answers may vary. A sample answer is that graphs can be misleading if different scales are used in graphs that are compared.

3. Line graphs connect data values with a line; generally, the data set is displayed over a period of time. Bar graphs use a solid vertical bar to show the data values several different times. Circle graphs show the percentage of times that each data value occurs; the total percentage will always be 100%.

4. Answers may vary. A sample answer is that a circle graph is used if you are trying to show how portions or percentages of a whole quantity are distributed.

5. Answers may vary. A sample answer is that a bar graph is used if you are trying to show how much of something is sold or used over a period of time.

6. Answers may vary. A sample answer is that a bar graph uses a vertical bar to show the quantity of each data value; a line graph uses a line to connect the data values.

7. Answers may vary. A sample answer is that Nell could use a line graph with pulse rates on the vertical axis and

length of time exercising on the horizontal axis.

8. Answers may vary. A sample answer is that Marcus could use a circle graph to show the percentage of students that participate in each type of extracurricular activity.

Practice and Apply, **pages 590–592**

17. Answers may vary. A sample answer is to stop smoking in the house, be careful when cooking and when using open flames (stoves and fireplaces), check the wiring periodically, and watch children carefully.

22. No. Other injuries increased by 100%, spinal cord injuries by 50%, head injuries by 33%, strokes by 29%, and orthopedic injuries by 25%.

Look Beyond, **page 592**

40. Answers may vary. A sample answer is that since the mean of Kevin's grades is 80.2, Kevin can use a grade of 80 to describe his grades.

41. Answers may vary. A sample answer is $\frac{1}{15} + \frac{2}{15} + \frac{1}{5} + \frac{4}{15} + \frac{1}{3} = 1$.

Lesson 11.2
Exploration 1, **page 594**

6. Answers may vary. A sample answer is to remove 6 coins from the 17-stack and place them on the 5-stack. Now, both of these stacks have 11 coins. There is no way to rearrange coins so that every pile will have 11 coins. So, remove 1 coin from the three 11-stacks and 2 coins from the 12-stack. Now, there are 4 stacks with 10 coins each, and 5 coins left over. Place 2 coins on each of the 8-stacks, and 1 coin on the 9-stack. This uses up all the extra coins,

and now there are 10 coins in every stack.

8. All three values are between and do not include the low value of the data set, 5, and the high value, 17. However, all three measures of central tendency are different: the mean is 10, the median is 9, and the mode is 8.

Critical Thinking, **page 595**

Terry cannot raise his average to 98. The highest grade he can make on the remaining three assignments is 100, but 3 grades of 100 would only bring his average up to 84.9.

Exploration 2, **page 596**

5. The mean might be used to see where the students perform in relation to each other; the median might be used as the value the other scores are grouped around; the mode might be used to show that, since so many students scored 90, the test was fair.

6. A frequency table makes it easier to find the mode because of the marks under each number. It is easier to find the median because you can count the marks from each end to find the one in the middle. An estimate can be made of the mean because you can see all the grades and make a reasonable estimate.

Critical Thinking, **page 596**

Answers may vary. A sample answer is that knowing the median and the mode *helps* to estimate the mean because you can see the central score and the most frequent score(s), and then make a reasonable estimate of the answer.

Communicate, **page 597**

1. The mean is the average of the numbers, and the median is the middle number.

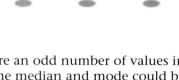

2. If there are an odd number of values in the set, the median and mode could be the same if the value in the middle occurs most often. If there are an even number of values, the median and the mode could be the same if the two middle values are the same and if this value occurs most often.

3. The mean was probably used since it is impossible to have a fraction of a child. The problem with statistics like these is that they do not describe any real family. Any information based on these statistics would therefore not represent information about real families.

4. Since there are an even number of grades, 22, count the marks in the table from the left and from the right. Find the average of the two marks in the middle: $\frac{90 + 90}{2} = 90$.

5. Multiply each value by its frequency, add these products, and divide by the total number of values.

Grade	70	75	80	85	90	95	100	TOTALS
Frequency	I	II	III	IIII	III	ⅦI	IIII	22
Grade × Frequency	70	150	240	340	270	475	400	1945

$\frac{1945}{22}$ is approximately 88.4.
The mean is 88.4.

Practice and Apply, **pages 598–600**

6. mean: 12.4; median: 11; modes: 9, 11, and 12; range: 11

7. mean: 23.6; median: 27; modes: 18, 27, 29; range: 20

8. mean: 4; median: 2; modes: 1 and 2; range: 8

9. mean: 168.4; median: 166; mode: none; range: 213

11.

Points Scored	0	2	3	4	5	6	7	8	9	10	11	12	15	16	18	26	27
Frequency	III	I	II	II	ⅦI	II	II	I		I		I	I	II	IIII	II	I

15. Answers may vary. A sample answer is that the mean would be the best measure of central tendency because it would give the best impression of each player's ability and their ability to work together as a team. The mean is approximately 9.7, which is higher than the median, 7, or the mode, 5.

21.

Female □ Male — bar graph: Number of hours vs. Age Group (in years): 2–5, 6–12, 13–17, 18–24, 25–54, 55+

42.

Frequency vs. Scores: 50, 55, 60, 65, 70, 75, 80, 85, 90, 95, 100

Look Beyond, **page 600**

49. 60 or below: approximately 21%; 61 to 70: approximately 21%; 71 to 80: approximately 27%; 81 to 90: approximately 21%; 91 to 100: approximately 9%

50.

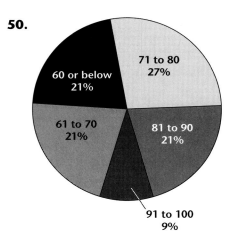

60 or below 21%
71 to 80 27%
61 to 70 21%
81 to 90 21%
91 to 100 9%

Extra Practice, **page 729**

9.

Runs scored	0	1	2	3	4	5	6	7	8	9
Frequency	3	2	4	2	2	3	2	2	2	2

Lesson 11.3

Exploration, **page 602**

6. Answers may vary. A sample is that it is convenient because viewing a stem-and-leaf plot makes it easy to see how many times each number appears.

7. The mean would most accurately describe the average length of gray whales.

Try This, **page 603**

key $6\,|\,4 = 64$

Stems	Leaves
6	4
7	6, 9, 9
8	0, 0, 1, 6
9	1, 6, 7, 9

Communicate, **page 605**

1. Answers may vary. Sample answer: to construct a stem-and-leaf plot, examine the data set to determine what the stems will be. If the data set consists of decimals, the whole numbers will be the stems and the decimals will be the leaves. If the data set consists of numbers between 0 and 100, the tens digits could be the stems and the units digits would then be the leaves. Draw a vertical line. Next to each stem, list the leaves that go with that stem, from smallest to largest, on the right side of the vertical line.

2. Answers may vary. Sample answer: if your data set consists of frequency and length, the frequency could be on the vertical axis and the length along the horizontal axis. List the lengths by a range of numbers that is convenient for the lengths given. Above each range of numbers, draw a bar that has the same width as the range of numbers below it and has the same height as the frequency of all the data values in the given range.

3. The median is the line in the middle of the rectangle. The range is the difference of the value on the far right (the end of the right whisker) to the value on the far left (the end of the left whisker).

4. The upper quartile is the value represented by the right side of the rectangle. The lower quartile is the value represented by the left side of the rectangle.

Practice and Apply, **pages 606–608**

5. key $14\,|\,1 = 14.1$

Duration of Humpback Whale Songs (in minutes)

Stems	Leaves
14	1, 5, 9
15	0, 2, 3, 5, 6
16	1, 2, 3, 6
17	2, 2, 4, 8
18	1, 2, 9
19	3, 3, 7, 9
20	0

11. The range would not be affected. The mean would change from 63.5 to 63.4. The median would still be 60 (the average of 60 and 60). The mode would change from 59 alone to both 59 and 60.

14.
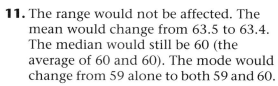

15.
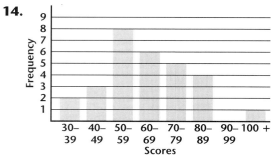

22. Between the median and the upper quartile

23. Between the median and the lower quartile

26. key $2 | 0 = 20$

Hockey Points

Stems	Leaves
2	0
3	1, 2, 3, 4, 7, 8, 8, 8
4	0, 1, 1, 2, 2, 2, 9, 9
5	0, 3, 6

32. Answers will vary depending on the ranges of values used on the horizontal axis.

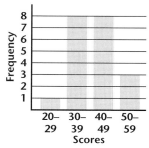

33.
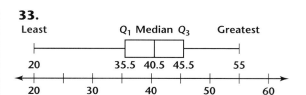

34. key $10 | 6 = 106$

Dollars Spent on Dental Care

Stems	Leaves
10	6, 6, 8
11	4, 8, 9
12	0, 4, 5
13	0, 1, 7
14	1, 1, 2, 6
15	2, 3, 3, 9
16	8, 8, 9
17	0, 2, 2, 6, 8, 8
18	1, 4, 4, 5
19	4, 6
20	2, 8
21	0
22	
23	
24	5
25	5

37.
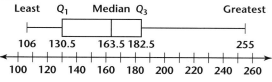

38. key $2 | 49 = 249$

Weight of Bottle-nosed Dolphins (in pounds)

Stems	Leaves
2	49, 75
3	43, 45, 83, 99
4	78, 84, 91, 97, 97, 99
5	12, 49, 67, 74, 76, 86, 88, 89

Look Beyond, page 608

54. smaller circle: 18π centimeters, or approximately 56.5 centimeters; larger circle: 24π, or approximately 75.4 centimeters; sum: $18\pi + 24\pi = 42\pi$, or approx. 131.9 cm.

Extra Practice, page 729

1.

Stems	Leaves
5	8, 9
6	0, 7, 8, 8, 9
7	3, 5, 5, 6, 7
8	0, 0, 0, 3, 7
9	0, 1, 2, 3, 5, 5, 7

4.

Least   Q₁ Median Q₃ Greatest

58   68.5 78.5 90.5 97

50 60 70 80 90 100

Lesson 11.4

Try This, **page 611**

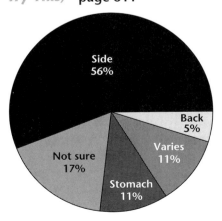

Side 56%

Back 5%

Varies 11%

Not sure 17%

Stomach 11%

Communicate, page 613

Answers may vary. Sample answers are given.

1. A circle graph can be used to compare different groups in a single category.

2. A circle graph would not be used if you are comparing things that change over time.

3. You would first need to find the total number of stamps. Divide the number of stamps in each category by the total number of stamps and change to a percent.

4. You would multiply each percent by 360° to find how many degrees are in each section.

Practice and Apply, pages 614–617

12.

	United Long Distance	ABC Service	Advanced Services	Multi-Media Services	Fast Phones Long Distance
Number of customers	14,500	8400	6800	4200	3500
Percent	≈39%	≈22%	≈18%	≈11%	≈9%

13.

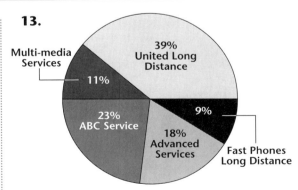

Multi-media Services 11%

39% United Long Distance

9%

23% ABC Service

18% Advanced Services

Fast Phones Long Distance

14. The angle for ABC Service would change to 76° and the angle for Advanced Services would change to 68°. The others would stay the same.

20. Sales in 1996 were 42%. This was an increase of 2% from 1995.

22. Answers may vary. Sample answer: A

circle graph would be the best way to represent data if you are comparing percentages of different categories to the whole.

23. Answers may vary. Sample answer: A circle graph would not be the best way to represent data if you are trying to show how something changes over a period of time. A better way to represent the data would be to use a line or bar graph.

24.

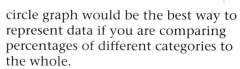

Car Sales

	Number of cars	Percent	Degrees
Hart	17	11.6%	42°
Morton	26	17.8%	64°
Kelley	15	10.3%	37°
Washington	30	20.6%	74°
Jarvis	22	15.1%	54°
Gonzales	19	13.0%	47°
Swensen	17	11.6%	42°
Total	146	100.0%	360°

25.

Car Sales

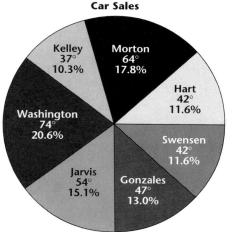

31. Answers may vary. A sample answer is that a bar graph could be used to show the amount of sales for a given time period.

32. Answers may vary. A sample answer is that a bar graph would not be the best way to show how different categories

are related to a whole. A circle graph would be a better way to display the data.

35.

	Amount	Percent	Degrees
Less than $12,000	12	16.$\overline{6}$%	60°
$12,000 to $19,999	18	25%	90°
$20,000 to $29,999	32	44.$\overline{4}$%	160°
$30,000 to $50,000	9	12.5%	45°
Over $50,000	1	1.3$\overline{8}$%	5°
Total:	72	100%	360°

36.

Less than $12,000

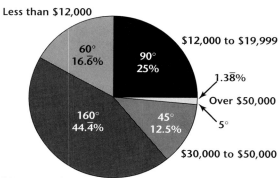

$20,000 to $29,999

40. Answers may vary. A sample answer is that if 22 more employees, each with a salary of $30,000 or more, were added, the median would change.

41. Answers may vary.

Look Beyond, **page 617**

54. The points lie on a straight line. The graph is a ray.

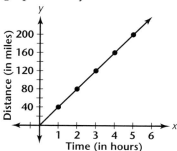

Extra Practice, page 729

2.

Category	Amount	Percent
Housing	$930	30%
Food	$775	25%
Insurance	$465	15%
Transportation	$620	20%
Other	$310	10%
Total	$3100	100%

3.

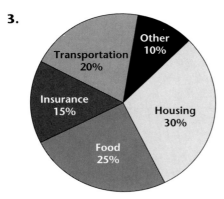

Lesson 11.5

Exploration, pages 618–619

1. Answers may vary. About the same; they also tended to do well in science.

2. Answers may vary. Students with lower math scores tended to also have lower science scores.

Communicate, page 621

1. The data points cluster near a line that rises from left to right. This indicates a strong positive correlation.

2. Since the cost of the notebooks depends on the number of pages, label the horizontal axis to represent the number of pages. The vertical axis then represents cost.

3. A correlation describes the extent to which variables are related, and how closely they are related.

4. Little or no correlation; the data points appear to be scattered randomly.

5. Strong positive; the data points cluster closely to a rising line.

6. Since the cost depends on the year, label the horizontal axis with the year, and the vertical axis with the cost.

7. Plot the ordered pairs (1982, 1.00), (1985, 1.06), (1988, 1.18), (1991, 1.36), choosing an appropriate scale for the cost axis.

Practice and Apply, pages 622–623

10. Little or no correlation

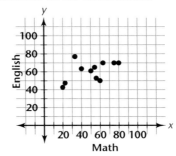

11. Little or no correlation

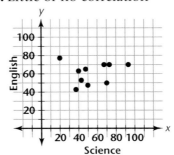

16.

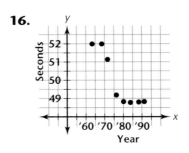

18. a.

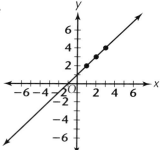

b.

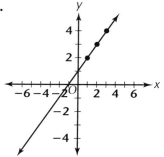

20. The vertical axis has a scale that has been changed. On graph **b**, the vertical axis has two squares for each unit.

21. By increasing the vertical scale, the steepness of the line increases.

23.

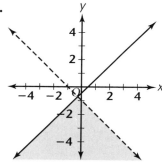

24.

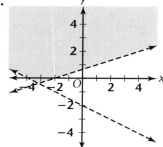

25.

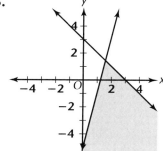

26.

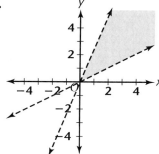

27.

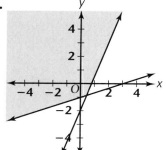

28.

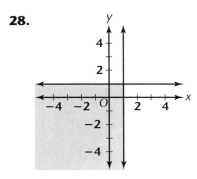

Look Beyond, **page 623**

29.

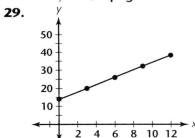

Extra Practice, **page 730**

5. Strong negative correlation

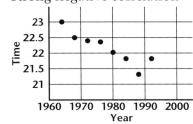

Lesson 11.6

Critical Thinking, **page 626**

Answers may vary. Predictions made from the Temperature scatter plot would be more reliable than the Elevations scatter plot because the points fall closer to the line in the Temperature and the correlation coefficient is close to -1.

Communicate, **page 626**

1. A line of best fit represents an approximation of the data plotted. The points on the line of best fit represent the relationship of the variables and can be used to make a reasonable prediction.

2. A coefficient of correlation of ± 1 means that the data points fit exactly on a line.

3. The data show a strong positive correlation.

4. The data show a strong negative correlation.

5. A correlation coefficient of 0.23 shows little correlation; this means that the data does not cluster close to a line.

Practice and Apply, **pages 626–627**

7. Positive

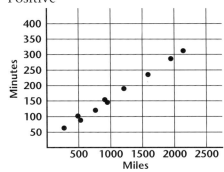

19.

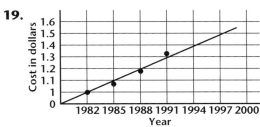

Chapter 11 Project

Answers for this project may vary. Some sample answers are given.

Activity 1, **page 629**

Correlation between a person's height and running speed; Correlation between a person's height and running speed.

Activity 2, **page 629**

Taller people run faster because their stride is longer.

Activity 3, **page 629**

Record the times of 50 randomly selected people in the 50-yard dash at three different times of day. This is fair because the participants are randomly selected, not only students; the distance is short enough for most healthy participants; and the information is collected at three different times of the day to avoid bias.

Activity 4, **page 629**

Survey results.

Activity 5, **page 629**

Organize and summarize data from the survey here, including at least one visual display of the data.

Activity 6, **page 629**

Answers may vary.

Review, **pages 630–632**

10. key $5\,|\,0 = 50$

Stems	Leaves
5	0, 8
6	0, 5, 8
7	4, 5, 6, 8, 9
8	0, 1, 4, 5, 5, 8
9	2, 2, 3, 8, 9

11.

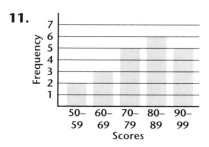

12.

Least   Q₁ Median Q₃ Greatest

50   71 80 90   99

45 50 55 60 65 70 75 80 85 90 95 100

13.

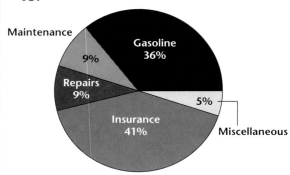

24.

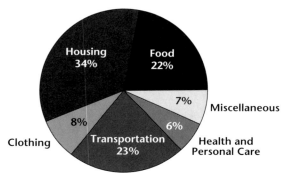

Assessment, **page 633**

9. key $5\,|\,2 = 52$

Stems	Leaves
5	2, 2, 8, 9
6	0, 2, 5, 6, 9
7	3, 4, 7, 7
8	1, 1, 2, 4, 4
9	0

10.

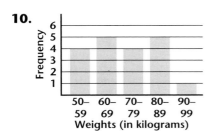

11.

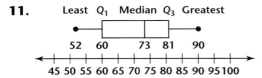

12.

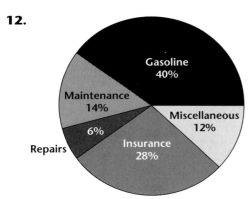

16.

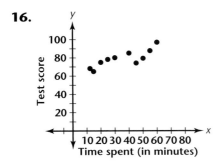

19.

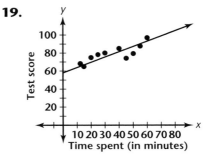

A student who spends 35 minutes studying is predicted to score 78.

Lesson 12.1

Communicate, **page 640**

1. The radius of a circle is one-half of its diameter.

2. The circumference of a circle is the distance around the outside of the circle; the area of a circle is the amount of space a circle covers.

3. No. If two different circles have the same circumference, then $C_1 = C_2$. Since $C = \pi d$, $\pi d_1 = \pi d_2$, where d_1 and d_2 are the diameters. If $d_1 = d_2$, then $r_1 = r_2$, where r is the radius. So, $\pi r_1^2 = \pi r_2^2$ and $A_1 = A_2$. The areas must be the same if the circumferences are the same.

4. No. If two circles have the same area, then $A_1 = A_2$, or $\pi r_1^2 = \pi r_2^2$, where r_1 and r_2 are the radii. If $\pi r_1^2 = \pi r_2^2$, then $r_1^2 = r_2^2$, and $r_1 = r_2$. If r_1 and r_2, then $d_1 = d_2$. So $\pi d_1 = \pi d_2$ and $C_1 = C_2$. The circumference must be the same if the areas are the same.

5. The circumference of a circle is π times the diameter. π is a number very close to 3 that is given by the ratio of the circumference of *any* circle with its diameter; π is a constant in all that is round in nature.

Practice and Apply, pages 640–642

12. Answers may vary; a bicycle tire with a diameter of 27 inches has a circumference of ≈84.82 inches. Assuming the bicycle does not hop or skip, the wheel rotates ≈747 times in one mile.

Extra Practice, page 731

1. $C \approx 25.13$ inches, $A \approx 50.27$ square inches

2. $C \approx 34.56$ centimeters, $A \approx 95.03$ square centimeters

3. $C \approx 94.25$ inches, $A \approx 706.86$ square inches

4. $C \approx 18.85$ yards, $A \approx 28.27$ square yards

5. $C \approx 31.42$ meters, $A \approx 78.54$ square meters

6. $C \approx 18.22$ centimeters, $A \approx 26.42$ square centimeters

7. $C \approx 28.27$ feet, $A \approx 63.62$ square feet

8. $C \approx 59.69$ inches, $A \approx 283.53$ square inches

9. $C \approx 28.27$ feet, $A \approx 63.62$ square feet

10. $C \approx 41.63$ inches, $A \approx 137.89$ square inches

11. $C \approx 33.30$ centimeters, $A \approx 88.25$ square centimeters

12. $C \approx 55.92$ meters, $A \approx 248.85$ square meters

Lesson 12.2

Critical Thinking, page 645

Large animals have a very small surface-area-to-volume ratio, so they lose very little heat through their skin. Therefore, they tend to have low metabolisms. Small animals have a high surface area-to-volume ratio, so they lose more heat through their skin. Therefore, they tend to have high metabolisms in order to maintain body heat.

Exploration 2, pages 645–646

6. Answers may vary. One solid could have $l = 4$, $w = 3$, $h = 4 \Rightarrow$ surface area = 80 square centimeters. Another solid could have $l = 8$, $w = 3$, $h = 2 \Rightarrow$ surface area = 92 square centimeters. Another could have $l = 2$, $w = 2$, $h = 12 \Rightarrow$ surface area = 104 square centimeters. Another could have $l = 1$, $w =$

2, $h = 24 \Rightarrow$ surface area = 148 square centimeters.

7. The length, width, and height should be as close together as possible, so a cube will have the smallest surface area for a given volume.

Exploration 3, pages 646–647

11. Let the height and the width be one unit, and let the length be the same number of units as the volume. Then the surface area will be as large as possible.

12. Let the height = width = length for the smallest surface area for a given volume.

Critical Thinking, page 647

The longer, stretched shape of the snake exposes more of its surface area to the sun's heat, therefore allowing the snake to absorb as much heat through its skin as possible. The compact, coiled shape of the snake exposes less surface area so that less body heat is lost to the surrounding air. For a fixed volume (the snake has fixed volume), the shape that has the least surface area is the most compact shape.

Communicate, page 647

1. Surface area is the amount of material it would take to cover the solid; the volume is the amount of space that the solid takes up.

2. A net is a diagram drawn in 2 dimensions that can be cut out and folded so that it covers the solid. Examples may vary.

An example is   which is a net for the solid shown.

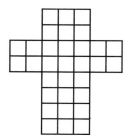 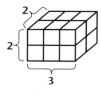

3. Since the area of the base of a solid is length × width, then $V = Bh = l \cdot w \cdot h$.

4. *lw* is the area of one face of a rectangular solid, and there are two faces with these dimensions (the top and bottom). Similarly, *lh* and *wh* are the areas of two other faces of the solid, each of which appears twice on the solid (*lh* is the front and back; *wh* is each of the sides). Therefore, when all these areas are added, the result is the surface area of the rectangular solid.

Practice and Apply, **pages 647–649**

5. $S = 600$ square meters; $V = 1000$ cubic meters

6. $S = 294$ square inches; $V = 343$ cubic inches

7. $S = 1350$ square centimeters; $V = 3375$ cubic centimeters

8. $S = 1536$ square feet; $V = 4096$ cubic feet

9. $S = 37.5$ square meters; $V = 15.6$ cubic meters

10. $S = 165.4$ square meters; $V \approx 144.7$ cubic meters

11. $S = 726$ square inches; $V = 1331$ cubic inches

12. $S = 234.4$ square centimeters; $V \approx 244.1$ cubic centimeters

13. $S = 62$ square meters; $V = 30$ cubic meters

14. $S = 52$ square yards; $V = 24$ cubic yards

15. $S = 254$ square inches; $V = 252$ cubic inches

16. $S = 81$ square centimeters; $V = 36$ cubic centimeters

17. $S = 11.2$ square meters; $V = 2.3$ cubic meters

18. $S = 22$ square inches; $V = 6$ cubic inches

31. 14 centimeters; 8 centimeters; 112 cubic centimeters

32. 144 cubic centimeters

33. 3 centimeters × 3 centimeters; 10 centimeters; 4 centimeters; 120 cubic centimeters

34. 4 centimeters × 4 centimeters; 8 centimeters; 2 centimeters; 64 cubic centimeters

35. The 2 centimeters-high box has the greatest volume; height = 1.5 centimeters, volume = 136.5 cubic centimeters; height = 2.5 centimeters, volume = 137.5 cubic centimeters; height = 3.5 centimeters, volume = 94.5 cubic centimeters; height = 4.5 centimeters, volume = 31.5 cubic centimeters

36. A box with a height of 5 centimeters would have a width = 0. It is impossible to have a box with a width equal to zero.

Look Beyond, **page 649**

53. If *x* represents the length of the square removed, then the height = *x*, length = 16 − 2*x*, and width = 10 − 2*x*. So, volume = $l \cdot w \cdot h = (16 - 2x)(10 - 2x)x$.

54. 2.0 cm

55. 3.9 cm × 3.9 cm

Extra Practice, **page 731**

1. $S = 150$ square meters, $V = 125$ cubic meters

2. $S = 54$ square inches, $V = 27$ cubic inches

3. $S = 864$ square feet, $V = 1728$ cubic feet

4. $S = 294$ square centimeters, $V = 343$ cubic centimeters

5. $S = 433.5$ square centimeters, $V = 614.125$ cubic centimeters

6. $S = 37.5$ square feet, $V = 15.625$ cubic feet

7. $S = 486$ square inches, $V = 729$ cubic inches

8. $S = 165.375$ square centimeters, $V = 144.703125$ cubic centimeters

9. $S = 64$ square meters, $V = 32$ cubic meters

10. $S = 126$ square feet, $V = 90$ cubic feet

11. $S = 264$ square inches, $V = 288$ cubic inches

12. $S = 101$ square centimeters, $V = 66$ cubic centimeters

13. $S = 22$ square yards, $V = 6$ cubic yards

14. $S = 121.2$ square meters, $V = 86.4$ cubic meters

15. $S = 627.25$ square meters, $V = 1060.875$ cubic meters

16. $S = 685.5$ square inches, $V = 1202.5$ cubic inches

17. $S = 372.73$ square meters, $V = 475.59$ cubic meters

Lesson 12.3

Critical Thinking, **page 651**

Answers may vary. Sample answers are given. When the length and height are greater than the width, the surface area of the box is greater than when the width = length = height. A cube has the least surface area. A cubic box would require less material, and thus, lower costs for cereal box manufacturers. However, a cube shaped box is hard to hold; and the existing shape allows most of the surface area to be used for visible attractive design.

Communicate, **page 655**

1. Start with a polygon, and move it a certain distance. Connect the vertices of the original polygon with the vertices of a translated polygon. This makes a prism. Answers may vary. Examples are fish tanks and lampshades.

2. The lateral surface area does not include the surface area of the bases; the surface area includes all faces.

3. Surface area is the amount of material it would take to cover a solid; the volume is the amount of material it would take to fill up a solid.

4. Yes. Examples may vary. The prisms both have volumes that are equal to 30 cubic inches. However, the first prism has a surface area of 72 square inches, and the second prism has a surface area of 62 square inches.

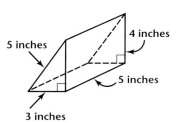

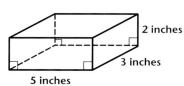

Practice and Apply, **pages 655–657**

31. Make a table where $Y_1 = 2x$, $Y_2 = x + 2$, $Y_3 = 3x$, and $Y_4 = Y_1 + Y_2$; begin the table with an x-value of 3, and increase x-values by 1.

Look Back, **page 657**

34. $y = 1 - 2x$

x	-3	-2	-1	0	1	2	3
y	7	5	3	1	-1	-3	-5

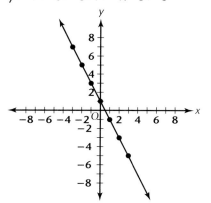

35. $y = -4$

x	-3	-2	-1	0	1	2	3
y	-4	-4	-4	-4	-4	-4	-4

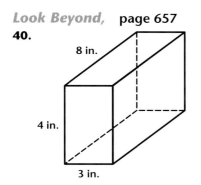

36. $y = \dfrac{x}{2} + 1$

x	-3	-2	-1	0	1	2	3
y	$-\dfrac{1}{2}$	0	$\dfrac{1}{2}$	1	$\dfrac{3}{2}$	2	$\dfrac{5}{2}$

Look Beyond, *page 657*

40.

8 in.

4 in.

3 in.

Volume of a right rectangular
prism = Bh
$= lwh$
$= (8)(3)(4)$
$= 96$ cubic inches
Volume of a right triangular
prism = Bh
$= \left[\dfrac{1}{2}lw\right]h$
$= \left[\dfrac{1}{2}(8)(3)\right]4$
$= 48$ cubic inches

Lesson 12.4

Communicate, **page 661**

1. A cylinder is a figure that has 2 circular
bases. The radius, r, is the radius of either
base. The height, h, is the perpendicular
distance between the bases; thermal mug,
aluminum can.

2. The lateral surface area of a cylinder is the
rectangular surface area of the side of the
cylinder that connects the two circular
bases. To find the area of the lateral surface
area of a cylinder, multiply the length times
the width; the length of the rectangular
surface is the circumference of the circular
base, and the width is the height of the
cylinder.

3. A cylinder is made up of two bases and the
side of the cylinder that connects the two
bases. If the side of the cylinder is rolled out
flat, it forms a rectangular shape. The
surface area is made up then of three parts:
two circles and the rectangular surface area.

4. A cylinder can be divided into pie-shaped
wedges and arranged to form a prism-like
solid, where the area of the base is πr times
r and the volume is then πr^2 times the
height of the cylinder.

5. $S \approx 150.8$ square inches; $V \approx 125.7$ cubic inches

6. $S \approx 377$ square yards; $V \approx 452.4$ cubic yards

7. $S \approx 113.1$ square centimeters; $V \approx 84.8$ cubic centimeters

8. $S \approx 252.9$ square meters; $V \approx 307.9$ cubic meters

9. $S \approx 94.2$ square feet; $V \approx 60.1$ cubic feet

10. $S \approx 306.6$ square feet; $V \approx 412.2$ cubic feet

14. $S \approx 1570.8$ square feet; $V \approx 4712.4$ cubic feet

15. $S \approx 414.7$ square meters; $V \approx 565.5$ cubic meters

16. $S \approx 402.1$ square inches; $V \approx 603.2$ cubic inches

23. $S \approx 29.1$ square centimeters; $V \approx 10.8$ cubic centimeters

24. $S \approx 4222.3$ square meters; $V \approx 20,910.4$ cubic meters

25. $S \approx 158,964.6$ square feet; $V \approx 4,806,636.8$ cubic feet

26. $S \approx 349.7$ square meters; $V \approx 498.6$ cubic meters

27. $S \approx 37,667.7$ square centimeters; $V \approx 513,179.2$ cubic centimeters

28. $S \approx 659.7$ square yards; $V \approx 1231.5$ cubic yards

29. $S \approx 183.5$ square meters; $V \approx 181.8$ cubic meters

30. $S \approx 251.3$ square yards; $V \approx 301.6$ cubic yards

31. $S \approx 60.1$ square centimeters; $V \approx 21.2$ cubic centimeters

	Object	Radius	Height	Lateral surface area	Surface area	Volume
32.	Soft drink can	1.3 in.	5 in.	≈40.8 square in.	≈51.5 square in.	≈26.5 cubic in.
33.	Coffee can	3 in.	10 in.	≈188.5 square in.	≈245 square in.	≈282.7 cubic in.
34.	Vegetable can	1.5 in.	4 in.	≈37.7 square in.	≈51.8 square in.	≈28.3 cubic in.
35.	Roll of paper towels	5 in.	12 in.	≈377 square in.	≈534.1 square in.	≈942.5 cubic in.
36.	Drinking cup	2 in.	6 in.	≈75.4 square in.	≈100.5 square in.	≈75.4 cubic in.
37.	Soda straw	0.2 in.	7 in.	≈8.3 square in.	≈9.0 square in.	≈0.9 cubic in.
38.	Drain pipe	3 in.	120 in.	≈2262 square in.	≈2318.5 square in.	≈3392.9 cubic in.

17. $S \approx 673.9$ square centimeters; $V \approx 1330.5$ cubic centimeters

18. $S \approx 141.4$ square yards; $V \approx 127.6$ cubic yards

19. $S \approx 265.3$ square meters; $V \approx 327.4$ cubic meters

20. $S \approx 3870.4$ square feet; $V \approx 17,106$ cubic feet

21. $S \approx 469.3$ square centimeters; $V \approx 699.5$ cubic centimeters

22. $S \approx 1753$ square inches; $V \approx 5598.3$ cubic inches

41. Answers may vary. Sample answers are given.

New diameter d	New height h	Volume $\pi r^2 h$
Can 1: 8.6 cm	10.8 cm	≈627.4 cubic cm
Can 2: 8.7 cm	10.5 cm	≈624.2 cubic cm
Can 3: 8.8 cm	10.3 cm	≈626.5 cubic cm

42. Answers may vary.
Can 1: $S \approx 408$ square centimeters
Can 2: $S \approx 405.9$ square centimeters
Can 3: $S \approx 406.4$ square centimeters

48.

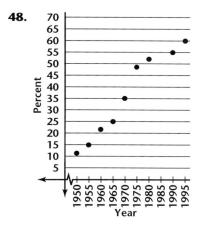

49.

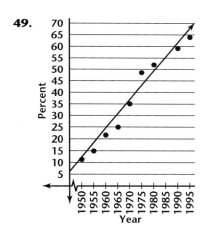

Predictions may vary; Year 2000, about 62%; year 2010, about 67%.

50. No, there are many variables, such as changes in industry or population increase or decrease that can affect the percent of adults with college degrees in a community.

Look Beyond, **page 663**

51. Suppose stack 1 and stack 2 both consist of 8 pennies. Then the volume of each stack will be the same.

If stack 1, which represents a right cylinder, is tilted, then it becomes stack 2, which represents an oblique cylinder. The radius of each stack (cylinder) will be equal because both are made of pennies. The height of

each stack will be equal because both are stacked with 8 pennies.

Therefore, since $h_1 = h_2$ and $r_1 = r_2$, $\pi(r_1)^2 = \pi(r_2)^2 \Rightarrow V_1 = V_2$.

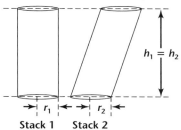

Extra Practice, **page 732**

1. $S = 150.80$ square inches, $V = 141.37$ cubic inches

2. $S = 703.72$ square yards, $V = 1206.37$ cubic yards

3. $S = 186.92$ square centimeters, $V = 192.42$ cubic centimeters

4. $S = 409.98$ square meters, $V = 636.17$ cubic meters

5. $S = 534.07$ square feet, $V = 942.48$ cubic feet

6. $S = 791.68$ square centimeters, $V = 1696.46$ cubic centimeters

7. $S = 197.92$ square meters, $V = 211.66$ cubic meters

8. $S = 303.48$ square meters, $V = 404.55$ cubic meters

9. $S = 6597.34$ square feet, $V = 38,877.21$ cubic feet

10. $S = 33,300.88$ square centimeters, $V = 439,822.97$ cubic centimeters

11. $S = 9424.78$ square centimeters, $V = 68,722.34$ cubic centimeters

12. $S = 337.72$ square yards, $V = 373.06$ cubic yards

13. $S = 742.99$ square inches, $V = 1520.53$ cubic inches

14. $S = 117.81$ square feet, $V = 98.17$ cubic feet

15. $S = 1915.98$ square centimeters, $V = 6436.24$ cubic centimeters

16. $S = 16,022.12$ square centimeters, $V = 155,508.84$ cubic centimeters

17. $S = 395.84$ square meters, $V = 604.36$ cubic meters

18. $S = 1658.76$ square inches, $V = 4523.89$ cubic inches

19. $S = 2550.97$ square feet, $V = 9236.28$ cubic feet

20. $S = 61.26$ square feet, $V = 35.34$ cubic feet

Lesson 12.5

Communicate, page 668

1. Answers may vary. A cone is a solid figure that consists of one circular base and a curved lateral surface that extends from the base to a single point called a vertex. A cylinder has two circular bases the same size and does not come to a point like a cone.

2. Answers may vary. A pyramid is a solid figure consisting of a base that is a polygon and a number of lateral faces that are triangles. The lateral faces meet at a point called the vertex of the pyramid. The lateral faces of a prism are shaped like rectangles, not triangles like a pyramid; the bases of a prism can be triangular, square, or pentagonal.

3. Answers may vary. To find the volume of a cone, multiply the area of the circular base times the height and divide the product by 3. If the circumference is given, find the radius of the base to use to find the area.

4. Answers may vary. To find the volume of a pyramid, multiply the area of the base times the height and divide the product by 3.

Practice and Apply, pages 668–669

28. Answers may vary. Possible answers are given.
Cone 1: $r = 3$ centimeters, $h = 25$ centimeters.
Cone 2: $r = 5$ centimeters, $h = 9$ centimeters

29. Answers may vary. Possible answers are given.
Pyramid 1: $B = 13.5$ square centimeters,

$h = 10$ centimeters, $s \approx 3.7$ centimeters
Pyramid 2: $B = 15$ square centimeters, $h = 9$ centimeters, $s \approx 3.9$ centimeters
Pyramid 3: $B = 18$ square centimeters, $h = 7.5$ centimeters, $s \approx 4.2$ centimeters

Look Beyond, page 669

42. Substitute 75 for V, x for r, and y for h.
$$V = \frac{Bh}{3}$$
$$V = \frac{\pi r^2 h}{3}$$
$$75 = \frac{\pi x^2 y}{3}$$
$$75 \cdot 3 = \left(\frac{\pi x^2 y}{3}\right)3$$
$$\frac{75 \cdot 3}{\pi x^2} = \frac{\pi x^2 y}{\pi x^2}$$
$$\frac{75 \cdot 3}{\pi x^2} = y$$

43. Answers may vary.

X	Y$_1$
1.5	31.831
2	17.905
2.5	11.459
3	7.9577
3.5	5.8465
4	4.4762
4.5	3.5368

Manufacturers might choose cones with a radius between 2.5 centimeters and 3 centimeters for the cone to not be too wide or too skinny.

Lesson 12.6

Ongoing Assessment, page 670

The radius and altitude of the cone make up the legs of a right triangle with the slant height as the hypotenuse. Use the "Pythagorean" Right-Triangle Theorem to find the slant height.

Communicate, page 673

1. The slant height of a right regular pyramid is the altitude or height of any of the triangular lateral sides.

2. To find the lateral surface area of a right, regular pyramid, multiply $\frac{1}{2}$ times the perimeter times the slant height, $L = \frac{1}{2}ps$.

3. To find the surface area of a right cone, add the sum of the lateral surface area to the base area, $S = L + B$.

4. To find the surface area of any right pyramid, add the sum of the lateral surface area to the base area, $S = L + B$.

5. The altitude and the radius of a right cone can be used as measurements of the legs of a right triangle to find the hypotenuse, which is the slant height of a right cone. Use these measurements in the "Pythagorean" Right-Triangle Theorem to find the slant height.

Practice and Apply, **pages 673–675**

36. $a^2 + b^2 = c^2$
$4^2 + b^2 = 8^2$
$b^2 = 8^2 - 4^2$
$b^2 = 48$
$b = \sqrt{48}$

37. approximately 27.7 square centimeters, 110.9 square centimeters

38. $S = x^2\sqrt{3}$

39.

X	Y$_1$
1	1.732
2	6.928
3	15.588
4	27.713
5	43.301
6	62.354
7	84.870

Look Back, **page 675**

40. $C \approx 36.3$ meters, $A \approx 105$ square meters

41. $C \approx 8.4$ inches, $A \approx 5.6$ square inches

42. $C \approx 7.5$ inches, $A \approx 4.4$ square inches

43. $C \approx 197.9$ centimeters, $A \approx 3117.2$ square centimeters

44. $C \approx 12.6$ meters, $A \approx 12.7$ square meters

45. $C \approx 16.3$ feet, $A \approx 21.2$ square feet

Extra Practice, **page 733**

1. $L \approx 50.27$ square meters, $S \approx 100.53$ square meters

2. $L \approx 131.95$ square feet, $S \approx 245.04$ square feet

3. $L \approx 62.83$ square inches, $S \approx 113.10$ square inches

4. $L \approx 131.95$ square yards, $S \approx 285.88$ square yards

5. $L \approx 78.41$ square centimeters, $S \approx 163.36$ square centimeters

6. $L \approx 76.97$ square yards, $S \approx 115.45$ square yards

Lesson 12.7

Communicate, **page 679**

1. The radius and the diameter of a sphere are the radius and diameter of a great circle of a sphere. A sphere is determined by all points in space that are a given distance from a given point. A circle is made up of all points in a plane which are a given distance from a point called a center. The circumference of the sphere is the circumference of a great circle of a sphere.

2. A great circle is the circle formed when a plane intersects the center of a sphere; the equator, a line of longitude.

3. The surface area of a sphere is 4 times the area of a great circle of a sphere.

4. Multiply the surface area of a sphere by the radius of the sphere and divide the product by 3.

Practice and Apply, **pages 679–681**

13. $S \approx 615.8$ square inches, $V \approx 1436.8$ cubic inches

14. $S \approx 804.2$ square centimeters, $V \approx 2144.7$ cubic centimeters

15. $S \approx 1256.6$ square meters, $V \approx 4188.8$ cubic meters

16. $S \approx 227$ square yards, $V \approx 321.6$ cubic yards

17. $S \approx 1017.9$ square inches, $V \approx 3053.6$ cubic inches

18. $S \approx 1633.1$ square centimeters, $V \approx 6205.9$ cubic centimeters

19. $S \approx 153.9$ square feet, $V \approx 179.6$ cubic feet

20. $S \approx 50.3$ square meters, $V \approx 33.5$ cubic meters

21. $S \approx 128.7$ square meters, $V \approx 137.3$ cubic meters

22. $S \approx 78.5$ square centimeters, $V \approx 65.4$ cubic centimeters

23. $S \approx 1809.6$ square inches, $V \approx 7238.2$ cubic inches

24. $S \approx 47.8$ square meters, $V \approx 31.1$ cubic meters

25. $S \approx 452.4$ square feet, $V \approx 904.8$ cubic feet

26. $S \approx 113.1$ square feet, $V \approx 113.1$ cubic feet

27. $S \approx 13.6$ square meters, $V \approx 4.7$ cubic meters

28. $S \approx 232.4$ square meters, $V \approx 333$ cubic meters

29. $S \approx 452.4$ square inches, $V \approx 904.8$ cubic inches

30. $S \approx 530.9$ square centimeters, $V \approx 1150.3$ cubic centimeters

31. $S \approx 1098.6$ square centimeters, $V \approx 3423.9$ cubic centimeters

32. $S \approx 4536.5$ square inches, $V \approx 28{,}730.9$ cubic inches

33. $S \approx 688.1$ square meters, $V \approx 1697.4$ cubic meters

34. $S \approx 2290.2$ square feet, $V \approx 10{,}306$ cubic feet

35. $S \approx 186.3$ square feet, $V \approx 239.0$ cubic feet

36. $S \approx 224.8$ square meters, $V \approx 317$ cubic meters

37. $S \approx 4071.5$ square centimeters, $V \approx 24{,}429$ cubic centimeters

38. $S \approx 3631.7$ square meters, $V \approx 20{,}579.5$ cubic meters

39. $S \approx 25.6$ square feet, $V \approx 12.2$ cubic feet

40. $S \approx 339.8$ square feet, $V \approx 589$ cubic feet

41. $S \approx 63.6$ square feet, $V \approx 47.7$ cubic feet

42. $S \approx 176.7$ square feet, $V \approx 220.9$ cubic feet

55.

Year	Population	Population Density
1	6000000000.0	104.8
2	6102000000.0	106.6
3	6205734000.0	108.4
4	6311231478.0	110.3
5	6418522413.1	112.1
6	6527637294.1	114.1
7	6638607128.1	116.0
8	6751463449.3	118.0
9	6866238328.0	120.0
10	6982964379.5	122.0
99	31304346124.9	547.0
100	31836520009.0	556.2

Look Back, **page 681**

56. $y = 0.5x + 3$

x	−3	−2	−1	0	1	2	3
y	1.5	2	2.5	3	3.5	4	4.5

57. $y = x - 1$

x	−3	−2	−1	0	1	2	3
y	−4	−3	−2	−1	0	1	2

58. $y = 5 - 2x$

x	−3	−2	−1	0	1	2	3
y	11	9	7	5	3	1	−1

Look Beyond, **page 681**

66. If you solve for r in the formula $V = \dfrac{4\pi r^3}{3}$, $r = \sqrt[3]{\dfrac{3V}{4\pi}}$.

68. Surface area of a sphere ≈ 314.2 square centimeters; Surface area of a cube $= 600$ square centimeters; Surface area of a cylinder ≈ 471.2 square centimeters; The surface area of the sphere is the smallest.

69.

X	Y_1
1000	6.2035
1100	6.4038
1200	6.5922
1300	6.7705
1400	6.9398
1500	7.1012
1600	7.2557

A tank with a volume between 1200 and 1300 cubic centimeters would have a radius of about 6.7 centimeters.

Extra Practice, **page 733**

1. $S \approx 314.16$ square meters, $V \approx 523.60$ cubic meters

2. $S \approx 804.25$ square inches, $V \approx 2144.66$ cubic inches

3. $S \approx 615.75$ square centimeters, $V \approx 1436.76$ cubic centimeters

4. $S \approx 452.39$ square feet, $V \approx 904.78$ cubic feet

5. $S \approx 706.86$ square meters, $V \approx 1767.15$ cubic meters

6. $S \approx 1017.88$ square feet, $V \approx 3053.63$ cubic feet

7. $S \approx 1134.11$ square meters, $V \approx 3591.36$ cubic meters

8. $S \approx 226.98$ square feet, $V \approx 321.56$ cubic feet

9. $S \approx 764.54$ square centimeters, $V \approx 1987.80$ cubic centimeters

10. $S \approx 706.86$ square centimeters, $V \approx 1767.15$ cubic centimeters

11. $S \approx 132.73$ square meters, $V \approx 143.79$ cubic meters

12. $S \approx 844.96$ square centimeters, $V \approx 2309.56$ cubic centimeters

13. $S \approx 907.92$ square inches, $V \approx 2572.44$ cubic inches

14. $S \approx 943.87$ square yards, $V \approx 2726.75$ cubic yards

15. $S \approx 3117.25$ square inches, $V \approx 16,365.54$ cubic inches

16. $S \approx 95.03$ square meters, $V \approx 87.11$ cubic meters

17. $S \approx 660.52$ square feet, $V \approx 1596.26$ cubic feet

18. $S \approx 56.75$ square meters, $V \approx 40.19$ cubic meters

Chapter 12 Project

Activity 1, **page 682**

Planet	Diameter (in miles)	Radius	Surface area	Volume
Mercury	3000	1500	28,274,333.88	14,137,166,941.15
Venus	7500	3750	176,714,586.76	220,893,233,455.53
Earth	7900	3950	196,066,797.51	258,154,616,722.21
Mars	4200	2100	55,417,694.41	38,792,386,086.53
Jupiter	88,800	44,400	24,772,840,374.32	366,638,037,539,982.00
Saturn	74,900	37,450	17,624,366,202.57	220,010,838,095,356.00
Uranus	31,800	15,900	3,176,904,155.02	16,837,592,021,585.60
Neptune	30,800	15,400	2,980,240,454.90	15,298,567,668,494.00
Pluto	1400	700	6,157,521.60	1,436,755,040.24

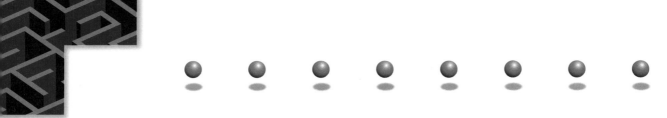

Activity 2, page 683

Planet	Diameter (in miles)	Circumference	Time to Rotate	Rotation Speed (in miles per hour)
Mercury	3000	9424.78	59 days	6.66
Venus	7500	23,561.94	243 days	4.04
Earth	7900	24,818.58	24 days	43.09
Mars	4200	13,194.69	24.5 days	22.44
Jupiter	88,800	278,973.43	19 days	611.78
Saturn	74,900	235,305.29	10.6 days	924.94
Uranus	31,800	99,902.65	17.1 days	243.43
Neptune	30,800	96,761.05	16.1 days	250.42
Pluto	1400	4398.23	6 days	30.54

Activity 3, page 683

Answers may vary. A sample answer is given.
Pluto—a marble; Venus—a tennis ball;
Neptune—a beach ball.

Cumulative Assessment, page 690

15.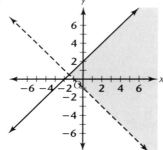